Lecture Notes in Computer Science 9082

Commenced Publication in 1973
Founding and Former Series Editors:
Gerhard Goos, Juris Hartmanis, and Jan van Leeuwen

More information about this series at http://www.springer.com/series/7412

Jón Atli Benediktsson · Jocelyn Chanussot
Laurent Najman · Hugues Talbot (Eds.)

Mathematical Morphology and Its Applications to Signal and Image Processing

12th International Symposium, ISMM 2015
Reykjavik, Iceland, May 27–29, 2015
Proceedings

 Springer

Editors
Jón Atli Benediktsson
University of Iceland
Reykjavik
Iceland

Jocelyn Chanussot
Université de Grenoble Alpes
Grenoble
France

Laurent Najman
ESIEE Paris
Noisy-le-Grand Cedex
France

Hugues Talbot
ESIEE Paris
Noisy-le-Grand Cedex
France

ISSN 0302-9743   ISSN 1611-3349   (electronic)
Lecture Notes in Computer Science
ISBN 978-3-319-18719-8   ISBN 978-3-319-18720-4   (eBook)
DOI 10.1007/978-3-319-18720-4

Library of Congress Control Number: 2015938160

LNCS Sublibrary: SL6 – Image Processing, Computer Vision, Pattern Recognition, and Graphics

Printed on acid-free paper

Springer International Publishing AG Switzerland is part of Springer Science+Business Media
(www.springer.com)

Preface

 UNIVERSITY OF ICELAND

This volume contains the articles accepted for presentation at ISMM 2015, the 12th International Symposium on Mathematical Morphology 2015 held during May 27–29, 2015 in Reykjavik, Iceland. The 11 previous editions of this workshop were all successful and firmly established ISMM as the premier venue for exchanging ideas and presenting results on mathematical morphology.

This time, there were 72 submissions. Each submission was reviewed by between 2 and 4 members of the Program Committee, with an average of 2.4 reviews per paper. The process was timely and efficient. The committee decided to accept a large proportion of these submissions thanks to their high overall quality. In all we accepted 62 papers, spanning all aspects of modern mathematical morphology. Since we had encouraged submitters to showcase their applications, we devoted a significant proportion of the program to the practice of mathematical morphology. The program also included three invited speakers: Coloma Ballester, Yann LeCun, and Sabine Süsstrunk.

We would like to take this opportunity to thank all the people involved in this conference: the invited speakers for devoting some of their precious time to come and talk to us, the authors for producing the necessary, high-quality scientific content, the members of the Program Committee for providing timely evaluations and detailed comments, and the members of the Steering Committee for their necessary guidance and giving us the chance to organize this event. The EasyChair web application was invaluable in handling the review process and putting together this volume.

Last but not least, we would like to thank our sponsors: the University of Iceland and the Faculty of Electrical and Computer Engineering at the University of Iceland.

March 2015

Jón Atli Benediktsson
Jocelyn Chanussot
Laurent Najman
Hugues Talbot

Organization

Organizing Committee

Jean Serra	Université Paris-Est ESIEE, France (Honorary Chair)
Jón Atli Benediktsson	University of Iceland, Iceland (General Chair)
Jocelyn Chanussot	Grenoble Institute of Technology, France (General Co-chair)
Philippe Salembier	Polytechnic University of Catalonia, Barcelona, Spain (General Co-chair)
Laurent Najman	Université Paris-Est ESIEE, France (Program Chair)
Hugues Talbot	Université Paris-Est ESIEE, France (Program Co-chair)

Steering Committee

Jesús Angulo	Mines ParisTech, France
Junior Barrera	University of São Paulo, Brazil
Isabelle Bloch	Télécom ParisTech, France
Gunilla Borgefors	Uppsala University, Sweden
Cris L. Luengo Hendricks	Uppsala University, Sweden
Renato Keshet	Hewlett Packard Laboratories, Israel
Ron Kimmel	Technion–Israel Institute of Technology, Israel
Petros Maragos	National Technical University of Athens, Greece
Christian Ronse	University of Strasbourg, France
Philippe Salembier	Polytechnic University of Catalonia, Spain
Dan Schonfeld	University of Illinois at Chicago, USA
Pierre Soille	European Commission, Joint Research Centre, Italy
Hugues Talbot	Université Paris-Est ESIEE, France
Michael H.F. Wilkinson	University of Groningen, The Netherlands

Program Committee

All reviewers are *de facto* members of the Program Committee.

Angulo, Jesús	Beucher, Serge	Buckley, Michael
Barrera, Junior	Bilodeau, Michel	Burgeth, Bernhard
Benediktsson, Jón Atli	Bloch, Isabelle	Chanussot, Jocelyn
Bertrand, Gilles	Borgefors, Gunilla	Chouzenoux, Emilie

Coeurjolly, David
Cokelaer, François
Couprie, Camille
Couprie, Michel
Cousty, Jean
Crespo, Jose
Curic, Vladimir
Dallamura, Mauro
Debayle, Johan
Decenciere, Etienne
Dokladal, Petr
Duits, Remco
Géraud, Thierry
Iwanowski, Marcin
Jalba, Andrei
Jeulin, Dominique
Keshet, Renato
Kiran, Bangalore
Kiselman, Christer

Koethe, Ullrich
Kurtz, Camille
Lefèvre, Sébastien
Lezoray, Olivier
Lomenie, Nicolas
Luengo Hendriks, Cris L.
Marak, Laszlo
Marcotegui, Beatriz
Masci, Jonathan
Mattioli, Juliette
Matula, Petr
Mazo, Loïc
Merveille, Odyssée
Meyer, Fernand
Monasse, Pascal
Morard, Vincent
Naegel, Benoît
Najman, Laurent
Ouzounis, Georgios

Passat, Nicolas
Perret, Benjamin
Pesquet, Jean-Christophe
Puybareau, Elodie
Roerdink, Jos
Ronse, Christian
Salembier, Philippe
Serra, Jean
Sintorn, Ida-Maria
Talbot, Hugues
Thoonen, Guy
Tochon, Guillaume
Valero, Silvia
van de Gronde, Jasper
Van Droogenbroeck, Marc
Veganzones, Miguel
Westenberg, Michel
Wilkinson, Michael H.F.
Xu, Yongchao

Contents

Evaluations and Applications

Hierarchies

Color, Multivalued and Orientation Fields

Optimization, Differential Calculus and Probabilities

Topology and Discrete Geometry

Algorithms and Implementation

Evaluations and Applications

Hourglass Shapes in Rank Grey-Level Hit-or-miss Transform for Membrane Segmentation in HER2/neu Images

Marek Wdowiak[1], Tomasz Markiewicz[1,2(✉)], Stanislaw Osowski[1,3], Zaneta Swiderska[1], Janusz Patera[2], and Wojciech Kozlowski[2]

[1] Warsaw University of Technology
[2] Military Institute of Medicine
[3] Military University of Technology, Warsaw, Poland
{wdowiakm,markiewt,sto,swidersz}@iem.pw.edu.pl,
{jpatera,wkozlowski}@wim.mil.pl

Abstract. The paper presents an automatic approach to the analysis of images of breast cancer tissue stained with HER2 antibody. It applies the advanced morphological tools to build the system for recognition of the cell nuclei and the membrane localizations. The final results of image processing is the computerized method of estimation of the membrane staining continuity. The important point in this approach is application of the hourglass shapes in rank grey-level hit-or-miss transform of the image. The experimental results performed on 15 cases have shown high accuracy of the nuclei and membrane localizations. The mean absolute error of continuity estimation of the stained membrane between the expert and our system results was 6.1% at standard deviation of 3.2%. These results confirm high efficiency of the proposed solution.

Keywords: Image segmentation · Object recognition · HER2/neu images

1 Introduction

Segmentation of the thin, non-continuous and highly variable objects is a difficult task in mathematical morphology application. To such problems belongs the membrane segmentation of the cells in histopathology Human Epidermal Growth Factor Receptor 2 (HER2/neu) images, related to breast cancer. The histopathological evaluation of the set of immunochemistry stains is the most common task for pathologists.

The HER2/neu biomarker is recognized as a diagnostic, prognostic and predictive factor not only, but especially in the case of breast cancer [1]. It is indicated as aid in an assessment of the breast and gastric cancers for patients for whom trastuzumab treatment is being considered. An over-expression of HER2 protein connected with HER2 gene amplification are diagnosed in approximately 20% of the analyzed breast cancer cases. For such patients, the trastuzumab treatment should be considered. Clinically, trastuzumab binds to the domain IV of the extracellular segment of HER2/neu receptor. Cells treated with this monoclonal antibody undergo arrest in the G1 phase of the cell cycle, so the reduced proliferation is observed. The combination of this

© Springer International Publishing Switzerland 2015
J.A. Benediktsson et al. (Eds.): ISMM 2015, LNCS 9082, pp. 3–14, 2015.
DOI: 10.1007/978-3-319-18720-4_1

antibody with chemotherapy has shown an increased survival and response rate, however, it increases the serious heart problems. It means, that inclusion of a patient to such therapy must be preceded by a reliable histological diagnosis.

The immunohistochemical HER2/neu stain is regarded as a basic step in pathomorphological evaluation of breast cancer. This semi-quantitative examination, performed on the immunostained paraffin section, needs determination of the presence, intensity, and continuity of membrane staining in the tumor cells. Four categories in grade scale are recognized: 0 (no membrane staining is observed or membrane staining is observed in less than 10% of the tumor cells), 1+ (a barely perceptible membrane staining is detected in more than 10% of tumor cells, the cells exhibit incomplete membrane staining), 2+ (a weak to moderate complete membrane staining is observed in more than 10% of tumor cells), and 3+ (a strong complete membrane staining is observed in more than 10% of tumor cells). The case 0 or 1+ indicates no HER2 gene amplification. On the other side grade 3+ indicates immediate HER2 gene amplification. The case 2+ means dubious (undecided) result and directs the case to the fluorescence in situ hybridization (FISH) based quantification of the level of HER2 gene amplification [2]. As shown, the categorization criteria are very subjective and may lead to significant differences in assessment of the particular cases. Especially the distinction between weak and strong, and the continuity of a membrane staining may result in a large variation in their practical interpretation.

The computerized assessment of pathological cases plays an increasingly important role in image analysis, especially when the quantitative result is required for diagnosis. Development of the slide scanners has introduced a lot of software used not only to perform the scanning process, but also for the quantitative analysis of the images. At the same time large data produced by this form of image acquisition has created new challenges for the image analysis algorithms. Whereas a lot of algorithms for the nuclear reactions have been developed [3,4], the HER2/neu membrane reaction is still treated manually or in a very rough way, not taking into account the separate cells. In the last years some new approaches to solve this problem have been proposed. To such methods belong the application of the real-time quantitative polymerase chain reaction (PCR) using LightCycler [5], application of support vector machine [6], or fuzzy decision tree by using Mamdani inference rules [7]. In spite of existing methods new approaches are needed, because of the high variability of a membrane reaction and it's frequent overlapping with a cytoplasm. They create difficult problems justifying for searching the other methods which are more effective to deal with this particular analysis problem.

In this paper novel tools of mathematical morphology for a membrane staining segmentation in HER2/neu images are proposed. Our propositions of the new hourglass shapes expand the traditional and common family of structuring elements, already discussed in mathematical morphology transformations. We used them in the rank grey-level hit-or-miss transform with some modifications, offering the supporting tool for segmentation of immunostained cell membranes. We point some analogies, which connect our propositions with the unconstrained hit-or-miss transform, rank hit-or-miss transform [8] and the grey-level hit-or-miss transform [9], where the inclusion and exclusion of regions in a structural shape are defined. We show the new shapes with fuzzy inclusion and exclusion criteria increase an efficiency of the membrane recognition. Moreover, our ideas can be adapted to other image processing problems with thin and non-continuous distinctive objects.

2   Problem Statement

An automatic evaluation of the HER2/neu membrane staining aims at the recognition of tumor cell nuclei and area of positive membrane reaction. It should specify which parts of recognized membrane come from the specific cells and finally graduate the reaction. Each of these steps requires different algorithms based on various criteria. The cell nuclei detection can be performed as a task of segmentation of the blue rounded and generally non-touching objects. The main problem is a weak staining of the nuclei by the blue hematoxyline. The additional problem is their partial coverage with the brown chromogenic substrate.

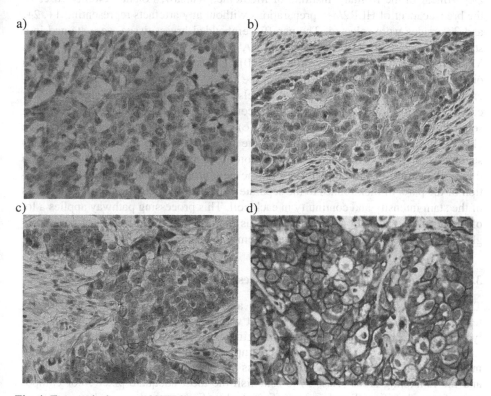

Fig. 1. Four typical cases of HER2/neu images representing the subsequent grades: a) 0, b) 1+, c) 2+, and d) 3+

The important task in this evaluation is recognition of area with a positive membrane reaction, especially when the brown chromogenic substrate is located not only in the cell membranes, but also partially in the cell cytoplasm's. In such cases, there is an identification problem of an appropriate membrane area with staining inside the brown marked regions. An algorithm for assigning parts of recognized membranes to the specific cells must be designed and applied.

A set of typical HER2/neu images with various grades of HER2 status are presented in Fig 1. As we can see the membrane reaction can vary from lower or higher intensity in the location only in a cell membrane (a thin line) to a very high intensity exactly in the membrane location surrounded by slightly colored cytoplasm's of the touching cells. The aim of the presented study is to create method that will be able to identify any membrane positive reactions, irrespective of their intensity, different localization and character of brown chromogenic substrate.

3   Material and Methods

The materials used in experiments come from the archive of the Pathomorphology Department in the Military Institute of Medicine, Warsaw, Poland. Twenty cases of the breast cancer of HER2/neu preparations without any artefacts representing 1+, 2+, and 3+ grades were selected. The paraffin embedded breast tissue were stained in a standard way according to the Ventana PATHWAY anti-HER-2/neu (4B5) Rabbit Monoclonal Primary Antibody protocol [10]. The specimen images were registered on the Olympus BX-61 microscope with the DP-72 color camera under the magnification 400x and resolution 1024x768 pixels. The image processing algorithm was composed in the following steps: a) enhancement of an image contrast, b) creating a set of different color space representations, c) classification of image pixels into three classes: the nuclei, membrane with positive reaction and the remaining regions, d) segmentation of the cell nuclei, e) recognition of the stained cell membranes based on the grey level hit-or-miss transform with the new proposed shape patterns, f) allocation of the parts of membranes to the separate cells and g) calculation of the indicators of the stain intensity and continuity in each cell. This processing pathway applies a lot of methods, however, in this paper we focus mainly on the application of new hourglass shape patterns and fuzzy criteria in membrane localization.

3.1   Image Preprocessing, Colour Spaces and Pixel Classification

The first step of the HER2/neu evaluation algorithm is an image preprocessing intended to enhance the contrast and colors. It was realized by applying an automatically computed contrast stretch and normalizing the color components to fulfill their ranges. In the next step nine sub-image samples (see Fig. 2) were prepared and manually segmented into the blue nuclei (class 1), reactive brown membrane (class 2), and the remaining areas (class 3) to establish the most adequate pixel color components for a classifier. To create an efficient classification system we have to select proper diagnostic features among the possible image representations in the form of pixel intensities. We take into account the following color spaces: RGB, CMYK, HSV, YCbCr, CIE Lab, CIE Lch, CIE uvL, CIE XYZ [11]. The ability of pixel intensity representation to differentiate specific class by the support vector machine (SVM) classifier [12] was evaluated using an area (AUC) under Receiver Operating Characteristics (ROC) curve [13]. In the next step the cross-correlation between the candidate features was studied in order to select not only the best ones, but also of the least correlation between them. The best two features for each of three classes was found and these six features formed the feature vector for a classifier. The classification was

a)   b)

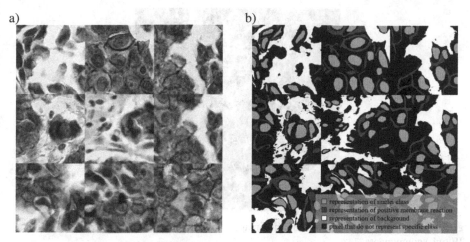

Fig. 2. The representatives of nine samples of image for establishing three classes of data for a classifier: (a) original images, (b) manually marked classes

performed using the SVM with a Gaussian kernel function. After this stage of image processing we obtain three masks representing the nuclei, positive HER2/neu reaction in cell membrane and cytoplasm's, and the remaining parts of image.

3.2   Segmentation of Cell Nuclei

The binary nuclei mask created by the classifier is used for the cell nuclei segmentation. The morphological closing operation was applied to reduce a noise. For separation of the connected nuclei the distance map was build and each extended regional maxima [8] with the selected h value was recognized as the center of cell nucleus. To define area of reaction the series of morphological operations, such as erosion, dilation, closing and hole filling were applied on the binary mask, which represented the positive HER2/neu reaction in cell membrane. Finally, the restriction to a single cell nucleus area was applied in order to eliminate non-cancer cell masks.

3.3   Intensity Map for Membrane Segmentation

To detect the immunoreactive cell membrane the most differentiated intensity map should be defined. It can be done for a single color channel, e.g. luminance, yellow component, or composed from the set of color channels. The selection of them is based on statistical analysis of the training images from Fig. 2. In the following section we will present the results concerning the analysis of ROC curves [13] and correlation between image descriptors to establish the most useful diagnostic features in classification process. The sample intensity map of the image corresponding to Fig. 2 is presented in Fig. 3. It is created as an element-wise product of Y channel from CMYK and u channel form CIE uvL colour space.

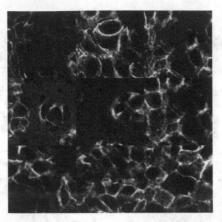

Fig. 3. The sample intensity map of the images from Fig. 2 based on Y (CMYK) and u (CIE uvL) components

3.4   Novel Hourglass Shapes for Discovering the Line Structures

Detection of immunoreactive cell membrane is a complex task due to the significant variability of cell shapes and spatial configuration, stain intensity and continuity, as well as non-specificity of staining. However, one characteristic point can be observed in each case: local domination of the stain intensity in the real cell membrane. This observation stands the basis for the proposition of the novel hourglass and half-hourglass shapes as the composite structuring elements in a detection of the sections of immunoreactive cell membrane (Fig. 4) by using the modified rank grey-level hit-or-miss approach. Thus, we define the horizontal and vertical hourglasses (the first two shapes) and four diagonal half-hourglass shapes as presented in Fig. 4. The central region (marked on red) treated as a foreground (FG) and the hourglass shape marked in dark grey (HG), represent the demanded location of the immunoreactive cell membrane. The other areas in the pattern (marked white in the figure) represent background (BG).

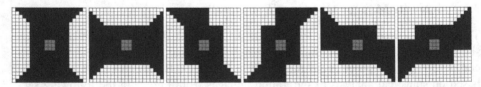

Fig. 4. The hourglass shapes as the proposed structuring elements in immunoreactive membrane recognition

The membrane curve can be treated as a line. The hourglass shapes offer some tolerance for the real line structure. Two cases might happen. In the first one the line structure is mainly included in HG region and only few pixels of the higher intensity values, comparable with values in FG region, are placed in BG area. In the second case a line structure does not match HG region and there are significantly more pixels with a higher intensity value in BG region. The following formula of the rank grey-level hit-or-miss transform based on rank filter ζ [8] has been modified:

$$HMT_{B,k_{BG},h}(X) = \zeta_{B_{BG},k_{BG}}(X < \mu_{FG} - h) \tag{1}$$

where μ_{FG} is the mean value of the image X in FG region, k_{BG} is the rank value, and h is the assumed threshold value. The positive results represent the recognized line structure. The above formula is applied for each pixel of the image and for the sets of masks of Fig. 4 in the line structure detection. Each mask detects the line feature in a given direction. Final result is composed as the logical OR operation of the first two hourglass shape masks or the last four from Fig. 4.

3.5 Allocation of Membranes in Separate Cells

As a result of these steps of algorithm the cell nuclei mask and the immunoreactive membrane regions are recognized in an independent way. The next step is to connect them into one compact system. To realize this task, a watershed algorithm is performed [14]. Its aim is to determine the potential cell membrane location.

a)
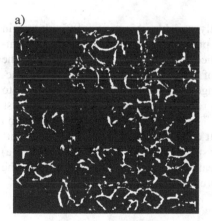

b)
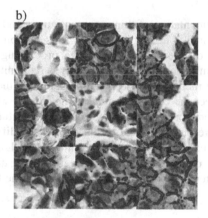

Fig. 5. Final result of application of hourglass shapes for line structure detection of immunoreactive membrane (a) and an input image with the annotated lines of immunoreactive membranes (b)

Series of image filtering operations are performed to reduce the number of false minima. They include (in sequence): the morphological operations of dilation using the cross structuring element of the length equal 3, the gamma correction of $\gamma=0.7$ to reduce the nonlinearity of luminance, the morphological operations of closing using the square 5×5 structuring element to smooth the edges of the objects, and finally h-minima transformation of $h=0.04$ to equalize the values of edge pixels. Watershed segmentation results related to the image of Fig. 3 are presented in Fig. 5. Each watershed region is considered as a separated cell and the region contour as full membrane. Logical AND operation on dilated contour region and detected line structures presented in Fig. 5a were performed to allocate immunoreactive membrane in separated cells.

Final step of image analysis is to calculate the stain continuity in each recognized cell, e.g., the cell with segmented nucleus. The continuity of membrane staining is calculated as a percentage of the recognized stained parts of cell membrane to a cell perimeter length. The exemplary results are presented in the Fig. 6.

a)   b)   c)   d)

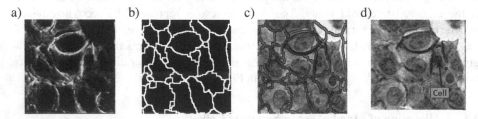

Fig. 6. The immunoreactive membrane representation (a), watershed segmentation results (b), watershed segmentation results presented on the input image (c), allocated membrane for one cell with the calculated stain continuity of 68% (d)

4   Results

Segmentation of thin, non-continuous and highly variable objects is a difficult task in automatic image processing. This section will present the numerical results concerning different stages of image processing, which lead finally to recognition of the membrane. The basic difficulty is recognition of three classes of objects (nuclei, reactive membrane, and remaining areas of the image). To solve this problem we have to select the color mapping of the image, which represents the highest ability of class discrimination (so called diagnostic features). The potential features under selection represent the image pixels intensity in different color representations. Fig. 7 presents the exemplary results concerning the uvL color space in the form ROC curve (Fig. 7a) and correlation family between investigated features (Fig. 7b). The higher AUC the better is the diagnostic significance of the feature.

a)   b)

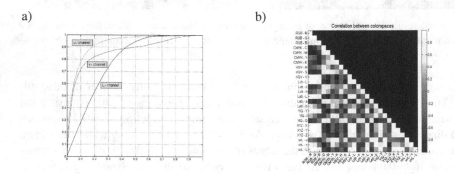

Fig. 7. The ROC curve for uvL colour representation (a) and correlation map between pairs of all 25 features representing different colour spaces (b)

We have applied the significance measure of the feature in the form of difference between the actual AUC and the value 0.5, which corresponds to the random classifier. Fig. 8 presents this significance measure for all features, arranged from the highest to the lowest recognition ability of the class 1 (Fig, 8a), class 2 (Fig, 8b) and class 3 (Fig, 8c).

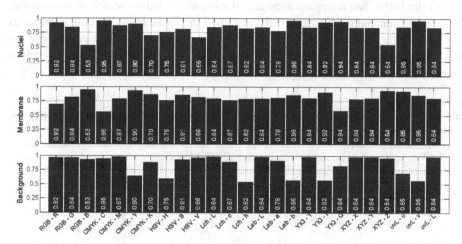

Fig. 8. The diagrams presenting the significance of the succeeding colour features for recognition of a) cell nuclei, b) immunoreactive cell membrane and c) the remaining part of the image (stroma and background)

To limit the number of features used in the class recognition we have chosen 2 highest discriminative ability features for each class. On the basis of the presented results the following features have been selected: R and B components from RGB space, u and L components from uvL representation, b component from CIE Lab and Y component from CMYK. Next, a set of learning data corresponding to nine parts of the image presented in the Fig. 2 was used in learning the SVM classifier of the Gaussian kernel. Finally, based on our experience and the introductory experiments, the threshold value $h=0.2$, used to recognize the stain intensity gradation was selected.

To identify the areas of immunoreactive cell membranes a set of proposed hourglass structuring shapes presented in Fig. 4 was used and evaluated. Based on the preliminary experiments the following parameters for full hourglass shape were adjusted: FG size 3×3 pixels, the distance between FG and BG equal 4 pixels, mask size 21×21 pixels. The second shape, half-hourglass of the structures presented in the Fig. 4, used the same parameters as the full hourglass shape. Moreover, the rank value k_{BG} in (1) was selected in such a way that at least 80% of BG pixels of the image match the relation. The normalized h value has been settled as 0.2 for both shapes.

In the testing phase of the adapted system we have analyzed 15 cases of 2+ HER2/neu status. A set of microscopic images was automatically processed and the results were evaluated in two aspects: a) how many immunoreactive membrane areas were properly recognized and b) what was the mean value of the analyzed immunoreactive cell membrane staining continuity. The detailed numerical results of estimation of the membrane staining continuity related to the application of full hourglass and half-hourglass shapes are presented in Table 1. They represent the testing cases, not taking part in fitting parameters of the image processing. As can be seen, the half-hourglass shapes offer better than hourglass accuracy of a immunopositive membrane recognition. Thus, we recommend them to the HER2/neu image analysis. However both approaches show some little bias, lower than common pathological criteria of between experts results. In the future research larger population of more representative learning images will be used to parameter adjusting and this approach should reduce the bias.

Table 1. The numerical results of membrane staining continuity estimation made by expert and by our system at application of full hourglass and half-hourglass shapes

Case	Continuity [%]				
	Expert's result	Hourglass	Error	Half-hourglass	Error
1	40.4	38.3	-2.0	35.1	-5.3
2	31.6	36.7	5.1	33.5	1.9
3	29.3	42.3	13.1	38.8	9.6
4	35.9	53.4	17.5	47.4	11.6
5	31.6	37.1	5.5	33.2	1.6
6	30.4	44.6	14.2	40.1	9.7
7	32.2	33.0	8	28.9	-3.3
8	35.3	37.8	2.5	33.9	-1.4
9	23.2	30.9	7.7	27.9	4.7
10	32.8	45.0	12.2	40.9	8.0
11	30.8	38.0	7.2	34.3	3.6
12	19.1	29.5	10.4	27.1	8.0
13	25.6	33.9	8.3	31.4	5.8
14	29.8	43.1	13.4	39.6	9.8
15	37.3	49.0	11.6	44.7	7.4
Mean absolute error		8.8%		**6.1%**	
Standard deviation		4.8%		**3.2%**	

The results of image analysis may be also presented in the graphical form. Fig. 9 depicts the exemplary original image under analysis (Fig. 9a) and the results of recognition of the nuclei and membranes made by the expert (Fig. 9b) and by our automatic system (Fig. 8c). The close similarity of the results of image segmentation obtained by our system and expert is confirmed.

It should be noted, that our computerized system is fully automatic. Hence the recognition of the nuclei has been done according to the embedded procedure defined within the algorithm. The user does not interfere in this process. On the other hand the human expert selects the nuclei according to his professional knowledge, blind to the selection results of an automatic system. Therefore, slight differences between the recognized nuclei can be observed in both segmentation results of Fig. 9.

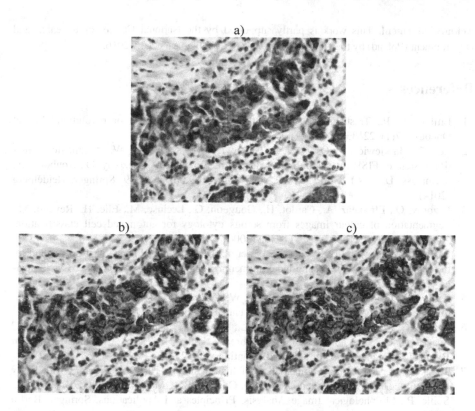

Fig. 9. The original input image (a) and its segmentation made by b) an expert, c) our automatic system with the visible membranes (brown colour lines) and cell nuclei (green points)

5   Conclusions

The paper has presented new approach to the automatic evaluation of the HER2/neu membrane staining in the breast cancer samples, applying the morphological image analysis. The developed algorithm uses few steps of analysis leading to the recognition of the cell nuclei and the membrane localizations. The paper proposed the automatic method of the membrane staining continuity estimation. The important point in this approach is the application of the hourglass shape structuring element in the rank grey-level hit-or-miss transform for the analysis of the image. The experimental results have shown high efficiency of image segmentation with respect to the nuclei and membrane localizations. The mean absolute error of continuity estimation between the expert and our system results, obtained for 15 analyzed cases, was 6.1% at standard deviation of 3.2%. In our opinion this automatic approach, after some additional research, will be able to substitute human expert in this very demanding and tedious task of image processing.

Acknowledgement. This work is partly supported by the National Centre for Research and Development (Poland) by the grant PBS2/A9/21/2013 in the years 2013-2016.

References

1. Littlejohns, P.: Trastuzumab for early breast cancer: evolution or revolution? Lancet Oncology 7(1), 22–33 (2006)
2. Les, T., Markiewicz, T., Osowski, S., Cichowicz, M., Kozlowski, W.: Automatic evaluation system of FISH images in breast cancer. In: Elmoataz, A., Lezoray, O., Nouboud, F., Mammass, D. (eds.) ICISP 2014. LNCS, vol. 8509, pp. 332–339. Springer, Heidelberg (2014)
3. Lezoray, O., Elmoataz, A., Cardot, H., Gougeon, G., Lecluse, M., Elie, H., Revenu, M.: Segmentation of colour images from serous cytology for automated cell classification. Anal. Quant. Cytol. Histol. 22, 311–322 (2000)
4. Grala, B., Markiewicz, T., Kozlowski, W., et al.: New automated image analysis method for the assessment of Ki-67 labeling index in meningiomas. Folia Histochemica et Cytobiologica 47(4), 587–592 (2009)
5. Kim, Y.R., Choi, J.R., Song, K.S., Chong, W.H., Lee, H.D.: Evaluation of HER2/neu status by real-time quantitative PCR in breast cancer. Yonsei Med. J. 43(3), 335–340 (2002)
6. Pezoa, R., Rojas-Moraleda, R., Salinas, L.: Automatic membrane segmentation of IHC images enables the analysis of cancer breast tissues. In: Third Chilean Meeting on Biomedical Engineering, pp. 1–6. JCIB, Santiago (2012)
7. Tabakov, M., Kozak, P.: Segmentation of histopathology HER2/neu images with fuzzy decision tree and Takagi-Sugeno reasoning. Comput. Biol. Med. 49, 19–29 (2014)
8. Soille, P.: Morphological Image Analysis, Principles and Applications. Springer, Berlin (2003)
9. Naegel, B., Passat, N., Ronse, C.: Grey-level hit-or-miss transforms Part I:Unified theory. Pattern Recogn. 40, 635–647 (2007)
10. PATHWAY HER-2/neu (4B5) – user manual. Ventana Medical Systems, Tucson (2009)
11. Gonzalez, R., Woods, R.: Digital image processing. Prentice Hall, New Jersey (2008)
12. Schölkopf, B., Smola, A.: Learning with kernels. MIT Press, Cambridge (2002)
13. Tan, P.N., Steinbach, M., Kumar, V.: Introduction to data mining. Pearson Education Inc., Boston (2006)
14. Matlab Image Processing Toolbox, user's guide. MathWorks, Natick (2012)

Automated Digital Hair Removal by Threshold Decomposition and Morphological Analysis

Joost Koehoorn[1], André C. Sobiecki[1(✉)], Daniel Boda[2], Adriana Diaconeasa[2], Susan Doshi[4], Stephen Paisey[5], Andrei Jalba[3], and Alexandru Telea[1,2]

[1] JBI Institute, University of Groningen, Groningen, The Netherlands
{j.koehoorn,a.c.telea,a.sobiecki}@rug.nl
[2] University of Medicine and Pharmacy 'Carol Davila', Bucharest, Romania
{daniel.boda, adriana.diaconeasa}@umf.ro
[3] Eindhoven University of Technology, Eindhoven, The Netherlands
a.c.jalba@tue.nl
[4] School of Computer Science and Informatics, Cardiff University, Cardiff, UK
doshisk@cardiff.ac.uk
[5] School of Medicine, Cardiff University, Cardiff, UK
paiseysj@cf.ac.uk

Abstract. We propose a method for digital hair removal from dermo-scopic images, based on a threshold-set model. For every threshold, we adapt a recent gap-detection algorithm to find hairs, and merge results in a single mask image. We find hairs in this mask by combining morphological filters and medial descriptors. We derive robust parameter values for our method from over 300 skin images. We detail a GPU implementation of our method and show how it compares favorably with five existing hair removal methods.

Keywords: Hair removal · Threshold sets · Morphology · Skeletonization

1 Introduction

Automatic analysis of pigmented skin lesions [13,7] occluded by hair is a challenging task. Several *digital hair removal* (DHR) methods address this by finding and replacing hairs by plausible colors based on surrounding skin. However, DHR methods are challenged by thin entangled, or low-contrast hairs [18,27,14,11,2,12].

We address the above problems by converting the skin image into a threshold-set and adapting a gap-detection technique to find hairs in each threshold layer. Found gaps are merged into a single hair mask, where we find actual hairs by using 2D medial axes, and finally remove them by image inpainting.

Section 2 reviews related work on digital hair removal. Section 3 details our method. Section 4 presents its implementation. Section 5 compares our results with five DHR methods and also shows an extra application for CBCT image restoration. Section 6 discusses our method. Section 7 concludes the paper.

2 Related Work

In the past decade, many DHR methods have been proposed. DullRazor finds dark hairs on light skin by morphological closing using three structuring elements

© Springer International Publishing Switzerland 2015
J.A. Benediktsson et al. (Eds.): ISMM 2015, LNCS 9082, pp. 15–26, 2015.
DOI: 10.1007/978-3-319-18720-4_2

that model three line orientations [18]. Different morphological operators were used in [22,19]. Hairs are removed by bilinear [18] or PDE-based inpainting [26]. Prewitt edge detection [14] and top-hat filtering [27] help finding low-contrast or thin-and-curled hairs. Huang *et al.* find hairs by multiscale matched filtering and hysteresis thresholding and remove these by PDE-based inpainting [12]. However, this method is quite slow (minutes for a typical dermoscopy image). VirtualShave finds hairs by top-hat filtering, like [27], and uses three density, sphericity, and convex-hull sphericity metrics to separate true positives (hairs) from other high-contrast details (false positives) [11]. Abbas *et al.* find hairs by a derivatives-of-Gaussian (DOG) filter [1,2]. However, this method has many parameters whose setting is complex. Finding other elongated objects such as arterial vessels and fibers is also addressed by path opening methods [8] and grayscale skeletons [10]. The last method also permits filling thin gaps similar to our hairs.

Table 1 captures several aspects of the above DHR methods. As visible, there is little comparison across methods. As method implementations are not publicly available (except [18,12]), comparison is hard. Hence, for our new DHR method outlined next, one main aim is to show how it compares to all reviewed methods.

Table 1. Comparison of existing digital hair removal methods

Method	Hair detector	Inpainting by	Compared with	# test images	Implementation
DullRazor [18]	generalized morphological closing	bilinear interpolation	–	5	available
Huang *et al.* [12]	multiscale matched filters	median filtering	DullRazor	20	available
Fiorese *et al.* [11]	top-hat operator	PDE-based [4]	DullRazor	20	not available
Xie *et al.* [27]	top-hat operator	anisotropic diffusion [20]	DullRazor	40	not available
E-shaver [14]	Prewitt edge detector	color averaging	DullRazor	5	not available
Abbas *et al.* [2]	derivative of Gaussian	coherence transport [5]	DullRazor, Xie *et al.* [27]	100	not available
Our method	gap-detection by multiscale skeletons	fast marching method [24]	DullRazor, Xie *et al.* [27], Huang *et al.* [12], Fiorese *et al.* [11] Abbas *et al.* [2]	over 300	available

3   Proposed Method

Most DHR methods find hairs by local luminance analysis (see Tab. 1, column 2). Such methods often cannot to find hairs that have *variable* color, contrast, thickness, or crispness across an image. Hence, our main idea is to perform a conservative hair detection at all possible luminance values. For this, we propose the following pipeline. First, we convert the input image into a luminance threshold-set representation (Sec. 3.1). For each threshold layer, we find thin hair-like structures using a morphological gap-detection algorithm (Sec. 3.2). Potential hairs found in all layers are merged in a mask image, which we next analyze to remove false-positives (Sec. 3.3). Finally, we remove true-positive hairs by using a classical image inpainting algorithm (Sec. 3.4). These steps are discussed next.

3.1   Threshold-Set Decomposition

We reduce color images first to their luminance component in HSV space. Next, we compute a threshold-set model of the image [28]: Given a luminance image $I : \mathbb{R}^2 \to \mathbb{R}_+$ and a value $v \in \mathbb{R}_+$, the threshold-set $T(v)$ for v is defined as

$$T(v) = \{\mathbf{x} \in \mathbb{R}^2 | I(\mathbf{x}) \geq v\}. \tag{1}$$

For n-bits-per-pixel images, Eqn. 1 yields 2^n layers $T_i = T(i), 0 \leq i < 2^n$. We use $n = 8$ (256 luminances). Note that $T_j \subset T_i, \forall j > i$, i.e. brighter layers are 'nested' in darker ones. If $I(\mathbf{x}) \neq i, \forall \mathbf{x} \in \mathbb{R}^2$, we find that $T_i = T_{i+1}$. In such cases, we simply skip T_i from our threshold-set decomposition, as it does not add any information. Our decomposition $\{T_i\}$ will thus have at most 2^n layers.

3.2   Potential Hair Detection

We find thin-and-long shapes in each layer T_i by adapting a recent gap-detection method [23], as follows.

Original Gap-Detection. Given a binary shape $\Omega \subset \mathbb{R}^2$ with boundary $\partial\Omega$, we compute the *open-close* image $\Omega_{oc} = (\Omega \circ H) \bullet H$ and *close-open* image $\Omega_{co} = (\Omega \bullet H) \circ H$. Here, $\circ$ and $\bullet$ denote morphological opening and respectively closing with a disk of radius H as structuring element. In both Ω_{oc} and Ω_{co}, small gaps get filled; yet, Ω_{co} has more gaps filled than Ω_{oc}, but also fills shallow concavities (dents) along $\partial\Omega$. Next, the skeleton or medial axis $S_{\Omega_{oc}}$ of Ω_{oc} is computed. For this, we first define the distance transform $DT_{\partial\Omega} : \mathbb{R}^2 \to \mathbb{R}_+$ as

$$DT_{\partial\Omega}(\mathbf{x} \in \Omega) = \min_{\mathbf{y} \in \partial\Omega} \|\mathbf{x} - \mathbf{y}\|. \tag{2}$$

The skeleton S_Ω of Ω is next defined as

$$S_\Omega = \{\mathbf{x} \in \Omega | \exists \mathbf{f}_1, \mathbf{f}_2 \in \partial\Omega, \mathbf{f}_1 \neq \mathbf{f}_2, \|\mathbf{x} - \mathbf{f}_1\| = \|\mathbf{x} - \mathbf{f}_2\| = DT_{\partial\Omega}(\mathbf{x})\} \tag{3}$$

where $\mathbf{f}_1$ and $\mathbf{f}_2$ are the contact points with $\partial\Omega$ of the maximally inscribed disc in Ω centered at $\mathbf{x}$. From $S_{\Omega_{oc}}$, the algorithm removes branch fragments that overlap with Ω, yielding a set $F = S_{\Omega_{oc}} \setminus \Omega$ that contains skeleton-fragments located in thin *gaps* that cut *deeply* inside Ω. To find all pixels in the gaps, the proposed method convolve the pixels $\mathbf{x} \in F$ with disk kernels centered at the respective pixels and of radius equal to $DT_{co}(\mathbf{x})$. As shown in [23], this produces an accurate identification of deep indentations, or gaps, in Ω, while ignoring pixels in shallow dents along $\partial\Omega$.

Hair-Detection Modification. We observe that, in a binary image with hairs in foreground, hairs are gaps of surrounding background. We next aim to find robustly hairs in all layers T_i. For this, several changes to [23] are needed. First, we note that [23] uses $DT_{\Omega_{co}}$ as disk-radius values for gap-filling as they argue that Ω_{co} closes more gaps than Ω_{oc}, supported by the observation that $DT_{\Omega_{co}}(\mathbf{x}) \geq DT_{\Omega_{oc}}(\mathbf{x}), \forall \mathbf{x} \in F$. Yet, for our hair-removal context, using $DT_{\partial\Omega_{co}}$ on every layer T_i, and next merging gaps into a single hair-mask, results in too many areas being marked as hair. The resulting mask proves to be too dense – thus, creates too many false-positive hairs for our next filtering step (Sec. 3.3). Using the smaller $DT_{\partial\Omega_{oc}}$ as disk radius prevents this problem, but fails to find many hair fragments – thus, creates too many false-negatives. To overcome these issues, we propose to use a linear combination of $DT_{\partial\Omega_{oc}}$ and $DT_{\partial\Omega_{co}}$. In detail, we define a set of pairs disk-centers $\mathbf{x}$ and corresponding disk-radii ρ as

$$D_\lambda = \{(\mathbf{x}, \rho = (1 - \lambda)DT_{\partial\Omega_{co}}(\mathbf{x}) + \lambda DT_{\partial\Omega_{oc}}(\mathbf{x})) \,|\, \mathbf{x} \in F\} \qquad (4)$$

where $\lambda \in [0, 1]$ gives the effect of $DT_{\partial\Omega_{oc}}$ and $DT_{\partial\Omega_{co}}$ to the disk radius. A value of $\lambda = 0.2$, found empirically (see Sec. 6), avoids finding too many gaps (false-positives), while also preventing too many false-negatives.

Let D be the union of pixels in all disks described by D_λ. We next find the gaps G that potentially describe hairs as the difference

$$G = D \setminus \Omega. \qquad (5)$$

We apply Eqn. 5 to compute a gap G_i from every shape $\Omega_i := T_i$. Next, we merge all resulting gaps G_i together into a single hair-mask image $M = \bigcup_{i=0}^{2^n} G_i$.

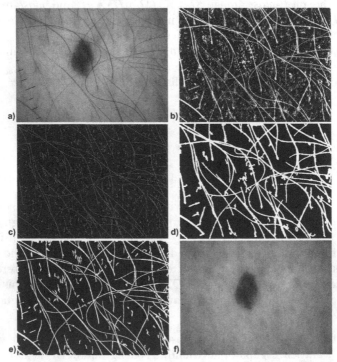

Fig. 1. a) Input image. b) Full hair mask M. c) Simplified mask skeleton S_M^τ. d) Filtered mask M^f. e) Mask created by [12]. f) Inpainted hair using M^f.

Morphological closing finds only hairs darker than skin. To find hairs lighter than skin, we replace closing by morphological opening. Having the dark-hair and light-hair masks M^d and M^l, we can next either combine the two or select one mask to use further. We observed in virtually all our test images that dark and light hairs do not occur together. So, we use next the mask $M \in \{M^d, M^l\}$ that most likely contains hairs, *i.e.*, which maximizes the length of the longest skeleton-branch in $S_{\partial M}$. For example, for the image in Fig. 1 a, which has mainly dark hairs, our method will select to use the mask $M := M^d$ (Fig. 1 b).

3.3 False Positive Elimination

Since we search for gaps on *every* threshold-level, we find more gaps than traditional approaches, *e.g.* [18,27,14,12]. Filtering out 'false positives' (gaps unlikely to be hairs), is thus necessary. We achieve this in four steps, outlined below.

Component Detection. First, we extract from M all 8-connected foreground components $C_i \subset M$. We skip components less than 1% of the size of image M, as these cannot possibly be elongated hairs. Remaining components are analyzed next to see if they are hairs or not.

Hair Skeletons. Hair fragments are long and thin. To measure such properties on our components C_i, we use their skeletons $S_{\partial C_i}$. Yet, components C_i may have jagged borders, due to input-image noise, shadows, or resolution limits (Fig. 1 b), so $S_{\partial C_i}$ have many short spurious branches. We discard these and keep each component 'core' by pruning each $S_{\partial C_i}$ as in [25]: From $S_{\partial \Omega}$, we produce a skeleton $S^\tau_{\partial \Omega}$ which keeps only points in $S_{\partial \Omega}$ caused by details of $\partial \Omega$ longer than τ. By making τ proportional to the component's boundary length $\|\partial C_i\|$, we ensure that longer branches are pruned more than shorter ones. We also impose a minimum τ_{min} to discard tiny spurious fragments, and a maximum τ_{max} to preserve large branches. Hence, the pruning parameter τ for a component C_i is

$$\tau = \max(\tau_{min}, \min(\|\partial C_i\| \cdot \mu, \tau_{max})) \tag{6}$$

where $\mu \in [0, 1]$ is used as a scaling parameter. Figure 1 c shows the simplified skeleton $S^\tau_{\partial M}$ obtained from the mask M in Fig. 1 b.

Hair Detection. In DHR, finding if a component is thin and long is done by *e.g.* (a) fitting lines in a finite number of orientations and checking the length of the longest such line [18]; (b) using principal component analysis to find if the major-to-minor eigenvalue ratio exceeds a threshold [17]; and (c) computing an elongation metric comparing a component's skeleton-length with its area [27]. Xie *et al.* argue that (a) and (b) are limited, as they favor mainly straight hairs and yield false-negatives for curled hairs [27]. They alleviate this by an elongation metric equal to the ratio of the area $\|C_i\|$ to the squared length of the 'central axis' of C_i; but they give no details on how this central-axis (and its length) are computed. In particular, for crossing hairs, *i.e.*, when the skeleton of C_i has multiple similar-length branches, multiple interpretations of the notion of a 'central axis' are possible. We also found that (c) also yields many false-negatives, *i.e.*, marks as hair shapes which do not visually resemble a hair structure at all.

To address such issues, we propose a new metric to find if a thin-and-long shape is likely a hair. Let $J_i = \{\mathbf{x}_i \in S^\tau_{\partial C_i}\}$ be the set of junctions of $S^\tau_{\partial C_i}$, *i.e.*, pixels where at least three $S^\tau_{\partial C_i}$ branches meet. If the maximum distance

$d_{max} = \max_{\mathbf{x} \in J_i, \mathbf{y} \in J_i, \mathbf{x} \neq \mathbf{y}} \|\mathbf{x} - \mathbf{y}\|$ between any two junctions is small, then C_i is too irregular to be a hair. We also consider the average branch-length between junctions $d_{avg} = \|S_{\partial C_i}\|/\|J_i\|$, $i.e.$, the number of skeleton-pixels divided by the junction count. If either $d_{max} < \delta_{max}$ or $d_{avg} < \delta_{avg}$, then C_i has too many branches to be a thin elongated hair (or a few crossing hairs), so we erase $S_{\partial C_i}^\tau$ from the skeleton image. Good values for δ_{max} and δ_{avg} are discussed in Sec. 6.

Mask Construction. We construct the final mask M^f that captures hairs by convolving the filtered skeleton-image (in which false-positives have been removed) with disks centered at each skeleton-pixel $\mathbf{x}$ and of radius equal to $DT_{\partial M}(\mathbf{x})$. Figure 1 d shows the mask M^f corresponding to the skeleton image in Fig. 1 c. Comparing it with the hair-mask produced by [12] (Fig. 1 e), we see that our mask succeeds in capturing the same amount of elongated hairs, but contains fewer small isolated line-fragments (thus, has fewer false-positives).

3.4 Hair Removal

We remove hairs by using classical inpainting [24] on the hair-mask M^f. To overcome penumbras (pixels just outside M^f are slightly darker due to hair shadows), which get smudged by inpainting into M^f, we first dilate M^f isotropically by a 3×3 square structuring element. This tells why hairs in M^f in Fig. 1 d are slightly thicker than those in Fig. 1 b. Figure 1 f shows our final DHR result.

4 Implementation

The most expensive part of our method is computing M, which requires distance transforms and skeletons from up to 256 binary images (Sec. 3.2). As these images can be over 1024^2 pixels for modern dermoscopes, processing a single image must be done within milliseconds to yield an acceptable speed. For this, we use the GPU-based method for exact Euclidean distance transforms in [6]. A simple modification of this method allows us to compute dilations and erosions (by thresholding the distance transform with the radius of the disk structuring element) and simplified skeletons (by implementing the boundary-collapse in [25]). For implementation details, we refer to [28]. We also tested our method on multi-GPU machines by starting k MPI processes for k GPUs. Each process $p \in [0, k)$ does gap-detection on a subset of the threshold-set by launching CUDA threads to parallelize gap-detection at image block level [6]. The k separate masks $M_p, 1 \leq p \leq k$ are merged by process 0 into a single mask M, after which it continues with false-positive removal (Sec. 3.3). Connected component detection, done with union-find [21], and hair inpainting [24], are implemented in C++ on the CPU, as they are done on a single image.

5 Results and Comparison

Material. We have tested our method on over 300 skin images. These cover a wide range of skin lesions; hair thickness, color, length, density; and skin colors, acquired by several types of dermoscopes, by three research groups. Some images contain no hair; they let us see how well can we avoid false positives. This is important, as removing non-hair details may affect subsequent analyses [2,12].

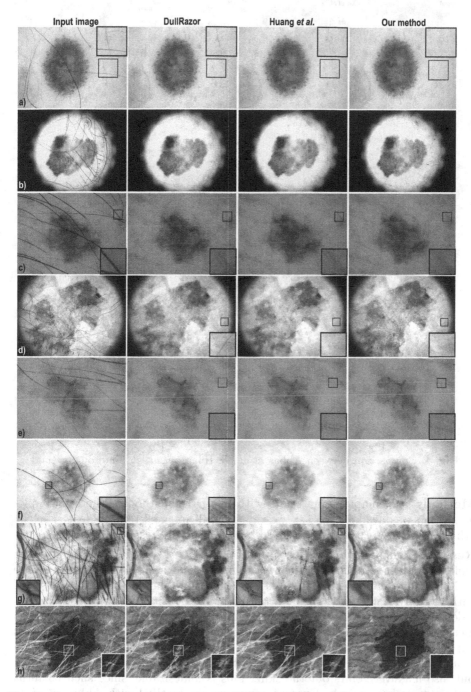

Fig. 2. Comparison of our method with DullRazor [17] and Huang *et al.* [12]. Insets show details.

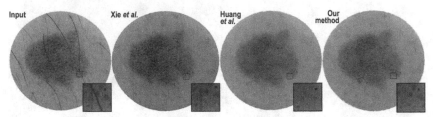

Fig. 3. Comparison between Xie *et al.* [27], Huang *et al.* [12], and our method

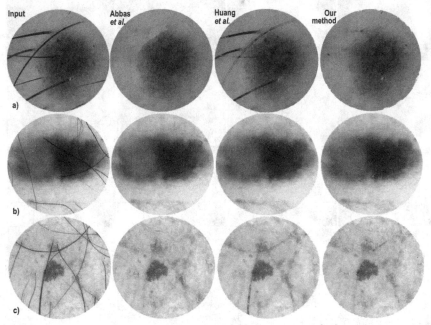

Fig. 4. Comparison between Abbas *et al.* [2], Huang *et al.* [12], and our method

Methods. We compare our results with five DHR methods: Where an implementation was available [18,12], we ran our full image-set through it. For the other methods [27,2,11], we processed images from the respective papers.

Results. Compared to DullRazor and Huang *et al.* [12] (Fig. 2), we see that DullRazor cannot remove low-contrast hairs (a,d); and both methods create 'halos' around removed hairs (c,f;e,f). Images (g,h) show two complex lesions, with hair of variable tints, opacity, thickness, and density. For (g), we create less halos around removed hairs than both DullRazor and Huang *et al.* For (h), our method removes considerably more hair than both methods. Figure 3 compares our results with Xie *et al.* [27] and Huang *et al.* We remove more hairs than Xie *et al.*, but also remove a small fraction of the skin. Huang *et al.* removes all hairs but also massively blurs out the skin. This is undesirable, since such patterns are key to lesion analysis. Figure 4 compares our method with Abbas *et al.* [2] and Huang *et al.* We show comparable results to Abbas *et al.* Huang *et al.* has issues with thick hairs (a) and also creates undesired hair halos (c). Compared

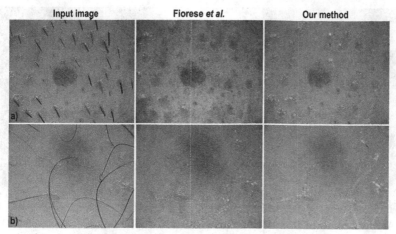

Fig. 5. Comparison between Fiorese *et al.* [11] and our method

to Fiorese *et al.* [11], we show a similar ability in removing both stubble and elongated hairs (Fig. 5). Strikingly, Fiorese *et al.* changes the hue of the input image, which is undesired. Our method correctly preserves the hue of the image.

Validation. We have shown the input images, and obtained DHR results, to two dermatologists having over 11 years of clinical experience. We asked whether the two images would lead them to different interpretations or diagnoses. In all cases, the answer was negative. While a more formal, quantitative, test would bring additional insight, this test tells that our DHR method does not change the images in *undesirable* ways. Separately, hair removal is obviously *desirable*, *e.g.* when using images in automated image-analysis procedures [2,12].

Other Applications. Our method can be used beyond DHR. Figure 6 shows an use-case for cone-beam computed tomography (CBCT) images. Positron emission tomography (PET) is a functional imaging modality used to deduce the spatial distribution of a radio-labelled substance injected into a subject. To put PET data in spatial context, high-resolution CBCT images can be acquired and co-registered with PET data. Two types of sensors are inserted into the subject (a mouse under physiological monitoring): soft plastic tubes (S) and hard metal wires (H). H sensors cause streak artifacts, making the CBCT reconstruction (onto which the PET data is overlaid) unusable. Hence, we want to automatically remove them. Doing this by using the CBCT volume is possible but quite expensive and complex. We remove such artifacts directly from the 2D X-ray images used to create the CBCT volume. Our DHR method is suitable for this, since the H implements appear as thin, elongated, and dark 2D shapes in such projections (see Figs. 6 a-c). Figures 6 e-g show the H implement-removal results. As visible, the H implements present in the input images have been successfully detected and removed. In contrast, the S implements, which have lower contrast and are thicker, are left largely untouched. Figure 6 d shows the 3D reconstruction done from the raw X-ray images (without our artifact removal). In the lower part, the image is massively affected by streak artifacts. Figure 6 h shows the reconstruction done from our DHR-processed images. As visible, most streak

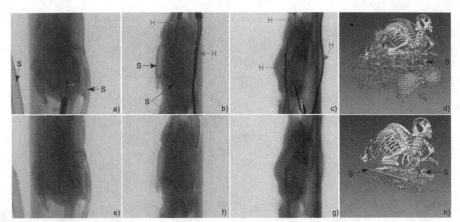

Fig. 6. Artifact removal from CBCT images. (a-c) Input images, with soft (S) and hard metal-wire (H) artifacts. (e-g) Reconstructed images with removed wires. 3D volumetric reconstructions from original images (d) *vs* our images (h).

artifacts have been removed. In contrast, the thick soft (S) tubes have been preserved by our DHR method and the resulting reconstruction.

6 Discussion

Parameters. To obtain full automation, we ran our method on several tens of skin images (at resolution 1024^2), varying all its parameters, and selected those values which visually yielded the best results (most true-positive and least false-positive hairs). Next, we computed final parameters by averaging, and tested that these values give good results on our full image test-set. Table 2 presents the final parameter values, used to produce all images in this paper.

Table 2. Empirically established parameter values

	Description	Definition	Value
H	Structuring element radius	Section 3.2	5.0 pixels
λ	Gap detection parameter	Equation 4	0.2
μ	Skeleton simplification parameter	Equation 6	0.05
$\tau_{\min}$	Minimum skeleton pruning	Equation 6	3.0 pixels
$\tau_{\max}$	Maximum skeleton pruning	Equation 6	40.0 pixels
δ_{max}	Hair detection parameter	Section 3.3	20.0 pixels
δ_{avg}	Hair detection parameter	Section 3.3	10.0 pixels

Robustness. We reliably remove hairs regardless of thickness, curvature, color, or underlying skin pattern. Very thin and low-contrast hairs or stubble may not get (fully) removed, as they are either not found in M^f or do not meet the elongation criteria (Sec. 3.3). Yet, such details do not influence further analysis tasks.

Speed. We compute an open-close, a close-open, a skeletonization, and a skeleton-to-shape reconstruction step for all 256 thresholds. For a 1024^2 pixel image, this takes 28 seconds on a MacBook Pro Core i7 with a GT 750M GPU, and 18 seconds on a comparable desktop PC with a GTX 690. For the same image and desktop PC, DullRazor needs 4 seconds, Fiorese *et al.* 7 seconds, Abbas *et al.* 40 seconds, Xie *et al.* 150 seconds, and Huang *et al.* about 10 minutes.

Implementation. We use [6] to compute distance transforms on the GPU in linear time with the pixel count, and also multiscale skeletons and morphological openings and closings [28]. For inpainting, we use the simple method in [24]. C++ source code of our full method is available at [15].

Limitations. For very dense hairs of varying color on high-contrast skin (*e.g.* Fig. 2 h), we cannot fully remove all hairs. Yet, this image type is extremely atypical. Also, other methods [18,12] remove significantly less hairs in such cases.

7 Conclusions

We have proposed a new approach for digital hair removal (DHR) by detecting gaps in all layers of an image threshold-set decomposition. We find false-positives by using medial descriptors to find thin and elongated shapes. We compared our method against five known DHR methods on a set of over 300 skin images – to our knowledge, is the broadest DHR method comparison published so far.

Machine learning techniques [3,16,9] could improve false-positive filtering. Further false-negative avoidance can be improved by extending our method to use additional input dimensions besides luminance, such as hue and texture.

Acknowledgement. This work was funded by the grants 202535/2011-8 (CNPq, Brazil) and PN-II RU-TE 2011-3-0249 (CNCS, Romania).

References

1. Abbas, Q., Fondon, I., Rashid, M.: Unsupervised skin lesions border detection via two-dimensional image analysis. Comp. Meth. Prog. Biom. 104, 1–15 (2011)
2. Abbas, Q., Celebi, M.E., García, I.F.: Hair removal methods: A comparative study for dermoscopy images. Biomed Signal Proc. Control 6(4), 395–404 (2011)
3. Altman, N.: An introduction to kernel and nearest-neighbor nonparametric regression. The American Statistician 46(3), 175–185 (1992)
4. Bertalmio, M., Sapiro, G., Caselles, V., Ballester, C.: Image inpainting. In: Proc. ACM SIGGRAPH, pp. 417–424 (2000)
5. Bornemann, F., März, T.: Fast image inpainting based on coherence transport. J. Math. Imaging Vis. 28, 259–278 (2007)
6. Cao, T., Tang, K., Mohamed, A., Tan, T.: Parallel banding algorithm to compute exact distance transform with the GPU. In: Proc. ACM I3D, pp. 83–90 (2010)
7. Christensen, J., Soerensen, M., Linghui, Z., Chen, S., Jensen, M.: Pre-diagnostic digital imaging prediction model to discriminate between malignant melanoma and benign pigmented skin lesion. Skin Res. Technol. 16 (2010)
8. Cokelaer, F., Talbot, H., Chanussot, J.: Efficient robust d-dimensional path operators. IEEE J. Selected Topics in Signal Processing 6(7), 830–839 (2012)
9. Cortes, C., Vapnik, V.: Support-vector networks. Mach. Learn. 20(3), 273–297 (1995)

10. Couprie, M., Bezerra, F.N., Bertrand, G.: Topological operators for grayscale image processing. J. Electronic Imag. 10(4), 1003–1015 (2001)
11. Fiorese, M., Peserico, E., Silletti, A.: VirtualShave: automated hair removal from digital dermatoscopic images. In: Proc. IEEE EMBS, pp. 5145–5148 (2011)
12. Huang, A., Kwan, S., Chang, W., Liu, M., Chi, M., Chen, G.: A robust hair segmentation and removal approach for clinical images of skin lesions. In: Proc. EMBS, pp. 3315–3318 (2013)
13. Iyatomi, H., Oka, H., Celebi, G., Hashimoto, M., Hagiwara, M., Tanaka, M., Ogawa, K.: An improved internet-based melanoma screening system with dermatologist-like tumor area extraction algorithm. Comp. Med. Imag. Graph. 32(7), 566–579 (2008)
14. Kiani, K., Sharafat, A.: E-shaver: An improved dullrazor for digitally removing dark and light-colored hairs in dermoscopic images. Comput. Biol. Med. 41(3), 139–145 (2011)
15. Koehoorn, J., Sobiecki, A., Boda, D., Diaconeasa, A., Jalba, A., Telea, A.: Digital hair removal source code (2014), http://www.cs.rug.nl/svcg/Shapes/HairRemoval
16. Kohonen, T.: Learning vector quantization. In: Self-Organizing Maps, pp. 203–217. Springer (1997)
17. Lee, H.Y., Lee, H.-K., Kim, T., Park, W.: Towards knowledge-based extraction of roads from 1m-resolution satellite images. In: Proc. SSIAI, pp. 171–178 (2000)
18. Lee, T., Ng, V., Gallagher, R., Coldman, A., McLean, D.: Dullrazor®: A software approach to hair removal from images. Comput. Biol. Med. 27(6), 533–543 (1997)
19. Nguyen, N., Lee, T., Atkins, M.: Segmentation of light and dark hair in dermoscopic images: a hybrid approach using a universal kernel. In: Proc. SPIE Med. Imaging, pp. 1–8 (2010)
20. Perona, P., Malik, J.: Scale-space and edge detection using anisotropic diffusion. IEEE TPAMI 12(7), 629–639 (1990)
21. Rahimi, A.: Fast connected components on images (2014), http://alumni.media.mit.edu/~rahimi/connected
22. Saugeon, P., Guillod, J., Thiran, J.: Towards a computer-aided diagnosis system for pigmented skin lesions. Comput. Med. Imag. Grap. 27, 65–78 (2003)
23. Sobiecki, A., Jalba, A., Boda, D., Diaconeasa, A., Telea, A.: Gap-sensitive segmentation and restoration of digital images. In: Proc. EG GVC, pp. 136–144 (2014)
24. Telea, A.: An image inpainting technique based on the fast marching method. J. Graphics, GPU, & Game Tools 9(1), 23–34 (2004)
25. Telea, A., van Wijk, J.J.: An augmented fast marching method for computing skeletons and centerlines. In: Proc. VisSym, pp. 251–259 (2002)
26. Wighton, P., Lee, T., Atkins, M.: Dermascopic hair disocclusion using inpainting. In: Proc. SPIE Med. Imaging, pp. 144–151 (2008)
27. Xie, F., Qin, S., Jiang, Z., Meng, R.: PDE-based unsupervised repair of hair-occluded information in dermoscopy images of melanoma. Comp. Med. Imag. Graph. 33(4), 275–282 (2009)
28. Zwan, M.v.d., Meiburg, Y., Telea, A.: A dense medial descriptor for image analysis. In: Proc. VISAPP, pp. 285–293 (2013)

Portal Extraction Based on an Opening Labeling for Ray Tracing

Laurent Noël[✉] and Venceslas Biri

LIGM, Université Paris Est, Creteil, France
Laurent.Noel@u-pem.fr

Abstract. Rendering photo-realistic images from a 3D scene description remains a challenging problem when processing complex geometry exposing many occlusions. In that case, simulating light propagation requires hours to produce a correct image. An opening map can be used to extract information from the geometry of the empty space of a scene, where light travels. It describes local thickness and allows to identify narrow regions that are difficult to traverse with ray tracing. We propose a new method to extract portals in order to improve rendering algorithms based on ray tracing. This method is based on the opening map, which is used to define a labeling of the empty space. Then portals - 2D surfaces embedded in empty space - are extracted from labeled regions. We demonstrate that those portals can be sampled in order to explore the scene efficiently with ray tracing.

Keywords: Ray tracing · Opening map · Labeling · Portals

1 Introduction

Light transport simulation refers to the process of rendering photo-realistic images from a 3D scene description. It remains one of the most challenging problem in computer graphics due to the complexity of scenes (mixture of non-homogeneous materials, complex geometry, non-uniform illumination).

Most state of the art methods rely on Path-Tracing and Monte-Carlo estimation to perform the simulation [1,2].

The color of each pixel can be expressed as an integral over the space of paths [3]. To compute this integral, many random paths traversing the pixel are sampled. A path is computed recursively using ray tracing. A first point x is sampled by tracing a random ray starting at the camera and traversing the pixel. Then the path is extended by sampling a random direction on the hemisphere of its last point. After many reflections, the path eventually reaches a light source and its energy can be computed. All these paths are used as samples for Monte-Carlo integration in order to estimate the integral expressing the color of the pixel. The stochastic aspect of this kind of simulation produces errors in the resulting image which appear as noise. Increasing the number of paths attenuates this noise and the result converges to a noise-free image. To improve the convergence rate of the simulation, one must choose a sampling

© Springer International Publishing Switzerland 2015
J.A. Benediktsson et al. (Eds.): ISMM 2015, LNCS 9082, pp. 27–38, 2015.
DOI: 10.1007/978-3-319-18720-4_3

strategy that samples more frequently paths carrying high energy. Building such a strategy can be challenging when dealing with complex geometry exposing many occluders. In that case, most of the sampled path does not reach a light source and carry no energy.

Digital Geometry and Mathematical Morphology [7] propose many tools to analyze the shape of discretized objects. In particular, for light transport simulation, the shape of the empty space (medium in which the light travels) can be analized to develop efficient rendering algorithms. Surprisingly, just few attempts have been made to use such tools in the rendering community [4,5,6].

In this paper we propose to investigate the use of the opening function [8] to analyze the shape of the empty space of a scene. This function expresses the local thickness of empty space surrounding each point and can be used to identify separations between large and narrow regions of the space. Our goal is to build a labeling on a voxelization of the empty space such that two neighbor regions, having a high difference in their thickness, are separated. This labeling allows us to extract portals, which are surfaces that separate different regions. Those portals can be sampled by a rendering algorithm in order to efficiently explore narrow regions that lead to light sources. We first analyze a simple labeling only based on the opening map of the empty space. Then we point out that it produces many regions that are false positives for our problem. We propose an other labeling built upon the propagation performed by the algorithm that compute the opening map. We show that this labeling defines better regions and enables the extraction of meaningful portals embedded in empty space. We present our results on different scenes and explore the limits of our method on the application to ray tracing.

2   Motivation

We first present a simple scene commonly used in computer graphics for algorithms that deal with complex occlusions and explain why the opening map can be of interest to us.

The Ajar Door scene (figure 1) is composed of two large rooms separated by a door slightly opened. For a configuration where a light source is placed in one room and the camera in the other room, the illumination becomes hard to sample efficiently. Indeed, blindly tracing random paths starting at the camera will end up with a lot of paths that does not reach the room containing the light source.

The opening map of the empty space E tells us for each voxel x the radius of the maximal ball inscribed in E that contains x (figure 2). We observe that the passage highlighted in red is characterized by lower values for this opening map than the two large rooms.

Based on this observation, we present a new labeling method that partitions the empty space in regions based on the opening map. More importantly we extract *portals* which are surfaces that separate a region from its neighbor regions. By sampling those portals we are able to trace rays that leave a given region in order to efficiently explore the empty space.

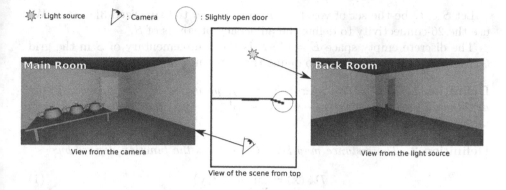

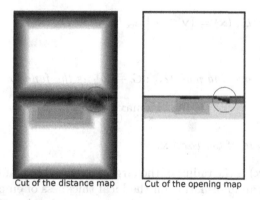

Fig. 1. The Ajar Door scene. The center image is a cut along the vertical axis of a voxelization of the scene. White pixels represent the empty space and black pixels represent the scene. The red circle encloses the small passage that rays must take to go from one room to the other.

Cut of the distance map   Cut of the opening map

Fig. 2. By dilating the distance map (left), we obtain the opening map (right) of the empty space. This map enhances narrow regions which are difficult to explore by tracing random rays.

3   Opening Based Labeling

We first define the opening map of the empty space and the labeling of that map that allows us to identify connected regions of constant opening. We show on an example scene that this method is not efficient enough to get meaningful regions for our problem. In section 3.3 we propose another labeling method based on the same core of ideas. We show in section 4 that it defines better regions for our application to ray tracing.

3.1   Mathematical Background

Let $G = [0, w[\times [0, h[\times [0, d[\subset \mathbb{Z}^3$ a finite 3D grid of voxels of resolution (w, h, d).

Let $S \subset G$ be the set of voxels corresponding to the voxelized 3D scene. We use the 26-connectivity to define the adjacency of voxels of S.

The discrete empty space $E = G \setminus S$ is the complementary of S in the grid G. We use the 6-connectivity to define the adjacency of voxels of E.

Definition 1. *If* $\mathbf{x} = (x_1, x_2, x_3)$, $\mathbf{y} = (y_1, y_2, y_3) \in \mathbb{Z}^3$ *are two voxels, we denote by* $d_\infty(\mathbf{x}, \mathbf{y}) = \max(|x_1 - y_1|, |x_2 - y_2|, |x_3 - y_3|)$ *the 26-distance between the voxels* $\mathbf{x}$ *and* $\mathbf{y}$.

Definition 2. *The distance map* $\mathcal{D}_\infty : G \to \mathbb{Z}^+$ *is the function defined by:*

$$\mathcal{D}_\infty(\mathbf{x}) = \min_{\mathbf{y} \in S} d_\infty(\mathbf{x}, \mathbf{y}) \tag{1}$$

Note that we have $\mathbf{x} \in S \Leftrightarrow \mathcal{D}_\infty(\mathbf{x}) = 0$.

Definition 3. *The maximal ball* $\mathcal{B}_\infty(\mathbf{x})$ *centered in* $\mathbf{x}$ *is the set:*

$$\mathcal{B}_\infty(\mathbf{x}) = \{\mathbf{y} \in G \mid d_\infty(\mathbf{x}, \mathbf{y}) < \mathcal{D}_\infty(\mathbf{x})\} \tag{2}$$

Definition 4. *The opening map* $\Omega_\infty : G \to \mathbb{Z}^+$ *is the function defined by:*

$$\Omega_\infty(\mathbf{x}) = \max_{\mathbf{x} \in \mathcal{B}_\infty(\mathbf{y})} \mathcal{D}_\infty(\mathbf{y}) \tag{3}$$

$\Omega_\infty(\mathbf{x})$ *is the opening of the voxel* $\mathbf{x}$.

For $\mathbf{x} \in E$, $\Omega_\infty(\mathbf{x})$ is the radius of the largest maximal ball of E containing the voxel $\mathbf{x}$. It gives us an information on the local thickness of empty space around $\mathbf{x}$. By looking at the variation of the opening map, we can identify narrow regions that connect large regions.

3.2   The Opening Labeling

Definition 5. *The opening region* $\mathcal{R}(\mathbf{x}) \subset E$ *of the voxel* $\mathbf{x} \in E$ *is the maximal 6-connected set of voxels containing* $\mathbf{x}$ *such that* $\forall \mathbf{y} \in \mathcal{R}(\mathbf{x})$, $\Omega_\infty(\mathbf{x}) = \Omega_\infty(\mathbf{y})$.

The region $\mathcal{R}(\mathbf{x})$ is a connected subset of E of constant opening. We denote by R_Ω the set of all opening regions.

Definition 6. *Let* $L : R_\Omega \to \{1, ..., |R_\Omega|\}$ *be a function that map each opening region to a label. The opening labeling of* E *for* L *is the map* $L_\Omega : E \to \{1, ..., |R_\Omega|\}$ *defined by:*

$$L_\Omega(\mathbf{x}) = L(\mathcal{R}(\mathbf{x})) \tag{4}$$

Limitations. We aim at tracing rays that travel the scene in order to reach regions containing a light source. In particular we want to be able to traverse efficiently narrow regions, such that sampled paths do not remain stuck in a region containing no light. As shown in figure 3, the labeling L_Ω produces false positive regions for our purpose. Regions like A are a problem because they lead directly on a wall. Ideally we want region A to be part of region B. Our experiments show that all voxels of A are contained in maximal balls of region B. We use this information in section 3.3 to define our new labeling.

Fig. 3. The opening labeling of a scene. Region A is a false positive because going from B to A does not grant access to other regions of the scene and leads directly on a wall. Region C is interesting because it represents a passage to enter a corridor. It represents a hole in the 3D scene.

Finding a good merging criterion that meets our expectation would be difficult and would likely depend on parameters that vary from scene to scene. Instead of merging regions, we decided to develop another labeling method based on the maximal balls used to compute the opening map.

3.3   The Opening Forest Labeling

As we noted in section 3.2, maximal balls give us information on the similarity between two neighbor regions. When the union of maximal balls of a region A covers entirely a neighbor region B, it means that it is probably easy to access B from A by tracing random rays.

The computation of the opening map is performed by successive dilations of the distance map on maximal balls of E. This process can be seen as a propagation that explores E by following maximal balls in decreasing order of their radius. Such propagation implicitly defines a forest that respects the properties of definition 7.

Definition 7. *An opening forest for E is a map $\Omega_f : E \to E$ such that if $\Omega_\infty(\mathbf{x}) = \mathcal{D}_\infty(\mathbf{x})$ then $\Omega_f(\mathbf{x}) = \mathbf{x}$. In that case $\mathbf{x}$ is a root of Ω_f. If $\Omega_\infty(\mathbf{x}) \neq \mathcal{D}_\infty(\mathbf{x})$ then $\Omega_f(\mathbf{x})$ is the center of a maximal ball of radius $\Omega_\infty(\mathbf{x})$ containing $\mathbf{x}$. In that case $\Omega_f(\mathbf{x})$ is the parent of $\mathbf{x}$.*

There exist many opening forest for an unique opening map since a voxel $\mathbf{x}$ can be contained in several maximal balls of radius $\Omega_\infty(\mathbf{x})$. Our goal is to build the regions of our new labeling from the trees of an opening forest. Since we want our regions to be 6-connected, we need to use at least an opening forest such that each tree is 6-connected.

Algorithm 1 computes both the opening map and an opening forest Ω_f that meets this criterion. Using a 26-distance map allows to compute the opening map by performing six scans of the 3D grid in each of the six directions (north, south, est, west, top and bottom). Each scan performs an independent dilation of each line of the 26-distance map and can be easily implemented in parallel.

Algorithm 1. Calculate the opening map and the opening forest

---

Input: The distance map $\mathcal{D}_\infty$
Output: The opening map Ω_∞ and the opening forest Ω_f

$\quad \Omega_\infty \leftarrow \mathcal{D}_\infty$
$\quad$ **for** $(i, j, k) \in [\![0, w[\![\times [\![0, h[\![\times [\![0, d[\![$ **do**
$\quad\quad \Omega_f(i, j, k) \leftarrow (i, j, k)$
$\quad$ **end for**
$\quad$ **for** $(j, k) \in [\![0, h[\![\times [\![0, d[\![$ **do**
$\quad\quad$ DILATELINE($\Omega_\infty([0...w[, j, k), \Omega_f([0...w[, j, k))$
$\quad\quad$ DILATELINE($\Omega_\infty(]w...0], j, k), \Omega_f(]w...0], j, k))$
$\quad$ **end for**
$\quad$ **for** $(i, k) \in [\![0, w[\![\times [\![0, d[\![$ **do**
$\quad\quad$ DILATELINE($\Omega_\infty(i, [0...h[, k), \Omega_f(i, [0...h[, k))$
$\quad\quad$ DILATELINE($(\Omega_\infty(i,]h...0], k), \Omega_f(i,]h...0], k))$
$\quad$ **end for**
$\quad$ **for** $(i, j) \in [\![0, w[\![\times [\![0, h[\![$ **do**
$\quad\quad$ DILATELINE($\Omega_\infty(i, j, [0...k[), \Omega_f(i, j, [0...k[))$
$\quad\quad$ DILATELINE($\Omega_\infty(i, j,]k...0]), \Omega_f(i, j,]k...0]))$
$\quad$ **end for** $\qquad\qquad\qquad\qquad \triangleright$ Function DILATELINE is defined in algorithm 2

We use the opening forest to build a new labeling of the empty space E. First we define the notion of *region root*.

Definition 8. *Let* $\mathbf{x} \in E$ *such that* $\Omega_f(\mathbf{x}) = \mathbf{x}$. *The region root* $\mathcal{R}_{root}(\mathbf{x}) \subset E$ *is the maximal 6-connected set of voxels containing* $\mathbf{x}$ *such that* $\forall \mathbf{y} \in \mathcal{R}_{root}(\mathbf{x})$ *we have* $\Omega_f(\mathbf{y}) = \mathbf{y}$ *and* $\Omega_\infty(\mathbf{x}) = \Omega_\infty(\mathbf{y})$.

All voxels of a region root have the same opening and are roots of the opening forest. The figure 4 illustrates different regions roots of a scene.

Let $\mathcal{R}_{root}$ be the set of all region roots of E. The *opening forest labeling* is built from the opening forest and region roots.

Definition 9. *Let* $L : \mathcal{R}_{root} \rightarrow \{1, ..., |\mathcal{R}_{root}|\}$ *be a function that maps each region root to a label. The opening forest label* $L_f(\mathbf{x})$ *of the voxel* $\mathbf{x}$ *is recursively defined by:*

Algorithm 2. Dilate a line of N maximal balls represented by their radius and center

Input: R: a line of N radius value, C: a line of N voxel centers
Output: R and C are dilated according to the radius values of R

```
function DILATELINE(R[N], C[N])
    maxballQueue ← EmptyQueue
    for i ← 0 to N − 1 do
        currentBall ← { index: i, radius: R[i], center: C[i], end: i + R[i] }
        if currentBall.radius = 0 then
            maxballQueue ← EmptyQueue
        else if maxBallQueue not empty and i = maxballQueue.front().end then
            maxballQueue.pop_front()
        end if
        if maxBallQueue is empty then
            maxballQueue.push_back(currentBall)
        else
            if currentBall.radius ≥ maxballQueue.front().radius then
                maxballQueue ← EmptyQueue
                maxballQueue.push_back(currentBall)
            else
                while currentBall.radius ≥maxballQueue.back().radius do
                    maxballQueue.pop_back()
                end while
                if maxballQueue.back().end < currentBall.end then
                    maxballQueue.push_back(currentBall)
                end if
            end if
        end if
        R[i] ← maxballQueue.front().radius
        C[i] ← maxballQueue.front().center
    end for
end function
```

$$\begin{cases} L_f(\mathbf{x}) = L(\mathcal{R}_{root}(\mathbf{x})) \ \textit{if } \Omega_f(\mathbf{x}) = \mathbf{x} \\ L_f(\mathbf{x}) = L_f(\Omega_f(\mathbf{x})) \ \textit{otherwise.} \end{cases} \tag{5}$$

The opening forest labeling is a propagation of the label of a region root to the trees rooted in that region.

The opening forest region $\mathcal{R}_f(\mathbf{x})$ of a voxel $\mathbf{x}$ is the maximal set of voxels such that $\forall \mathbf{y} \in \mathcal{R}_f(\mathbf{x}), L_f(\mathbf{x}) = L_f(\mathbf{y})$. This set is connected.

We denote by R_f the set of all opening forest regions of E.

3.4 Opening Portals

We now define *opening portals*, which are 2D-surfaces that separate opening forest regions. Since opening portals are not composed of voxels, we need to define them as set of 2-faces, which are elements of the *voxel complex framework* [9].

Fig. 4. Some region roots of a scene. Each region is a connected set of centered voxels of the same opening.

Let $X, Y \in R_f$ be two opening forest regions of E. We say that X and Y are neighbor regions if $\exists \mathbf{x} \in X, \exists \mathbf{y} \in Y$ such that $\mathbf{x}$ and $\mathbf{y}$ are 6-neighbors.

Definition 10. *Let* $\mathbf{x} \in \mathbb{Z}^3$ *be a voxel. The 3-face associated to* $\mathbf{x} = (x_1, x_2, x_3)$ *is the set* $F(\mathbf{x}) = \{(x_1, x_2, x_3), (x_1+1, x_2, x_3), (x_1, x_2+1, x_3), (x_1, x_2, x_3+1), (x_1+1, x_2+1, x_3), (x_1, x_2+1, x_3+1), (x_1+1, x_2, x_3+1), (x_1+1, x_2+1, x_3+1)\}$.

The 3-face of a voxel corresponds to its eight corners. This notion allows to define the two dimensional face separating two neighbor voxels.

Definition 11. *Let* $\mathbf{x}$*,* $\mathbf{y}$ *be two 6-adjacent voxels. The set* $F_s(\mathbf{x}, \mathbf{y}) = F(\mathbf{x}) \cap F(\mathbf{y})$ *is the 2-face separating* $\mathbf{x}$ *and* $\mathbf{y}$*. It is composed of four points.*

Definition 12. *Let* $X, Y \in R_f$ *be two neighbor opening forest regions of* E*. The opening portal* $P_{X,Y}$ *separating* X *and* Y *is the set defined by:*

$$f \in P_{X,Y} \Leftrightarrow \exists \mathbf{x} \in X, \exists \mathbf{y} \in Y \text{ such that } f = F_s(\mathbf{x}, \mathbf{y}) \tag{6}$$

We can build rays going from X to Y by sampling points on the 2-faces of $P_{X,Y}$.

4   Results and Discussion

We present the result of our *opening forest labeling* on different scenes and for different resolutions of the voxelization. To illustrate the regions, we display the voxelization of the scene such that the color of a face f of each voxel $\mathbf{x} \in S$ depends on the label of the empty-space voxel $\mathbf{y} \in E$ which is adjacent to $\mathbf{x}$ for the face f ($f = F_s(\mathbf{x}, \mathbf{y})$). We also expose opening portals separating different regions and we compare a local ray sampling strategy (uniform sampling of directions on the hemisphere of the origin point) to the a strategy that sample rays passing through our opening portals.

The application of our sampling strategy to a rendering algorithm is left for a future article. We discuss it in more depth in the conclusion.

Fig. 5. Our opening forest labeling for the Ajar Door scene for a grid resolution of (72, 118, 28). We observe that the back room is composed of one large region (right) separated from the main room by the narrow door entrance, as expected. The main room (left) is split in several regions allowing to travel the scene efficiently by sampling portals (see figure 6).

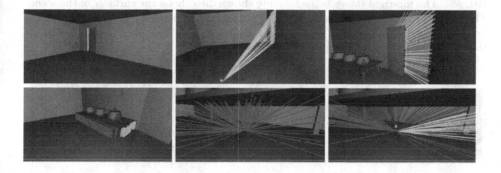

Fig. 6. *Top-left*: the green portal separates the back room from the main room. *Top-center*: 128 rays are sampled through the portal, starting at a point of the back room. *Top-right*: we observe that all rays reach the main room. *Bottom-left*: portals separating the region under the table from neighbor regions. *Bottom-center*: rays sampled with the local strategy. Many of them hit the back of the table. *Bottom-right*: with our strategy, all sampled rays leave the region.

Ajar Door Scene. Figures 5 and 6 demonstrate the application of our method to the Ajar door scene presented in section 2. Using our portals, we observe that we are able to sample rays efficiently to travel from region to region. In particular it makes it easy to pass through the door entrance or to leave the region under the table.

Sibenik Scene. Figure 7 and 8 illustrate our method on the Sibenik scene, which represents a church. This scene is less affected by occlusions than the two others. However, we demonstrate that we can use our sampling strategy to pass through a specific portal in order to reach a particular area of the scene. Being able to achieve this is useful to reach a specific light source by performing less reflections.

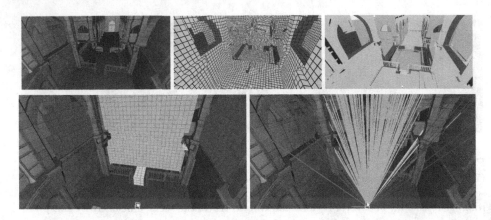

Fig. 7. Top row shows the labeling of the Sibenik scene for a grid resolution of (138, 104, 58). The top-right image is the labeling shown directly on the surfaces of the scene. The bottom-left image exposes three major portals of a region. The bottom-right image demonstrates the sampling of rays leaving that region by selecting portals based on their area (more rays are sampled on large portals).

Fig. 8. Left image shows the local sampling of rays which is not efficient to pass through the left and right portal. Using our method, it is straightforward to sample rays through a specific portal as pointed out in the middle and right image.

Sponza Scene. Figure 9 and 10 show results of our method on the Sponza scene. This scene is composed of several corridors occluded from the main part of the scene by drapes. Starting from a corridor, reaching the main part is hard due to narrow exits. Our sampling strategy enables to do it efficiently. Figure 10 demonstrates the robustness of our method regarding the resolution of the voxel grid. Opening portals remain stable and fit better to the scene as we increase the resolution.

Limitations. Our new labeling still produces neighbor regions with close opening and highly connected by their maximal balls. Such regions are separated because their region roots are not connected. However, these regions exposes better coherency with their opening: a region having a region root with high opening has a high volume and allows to access all of its narrow neighbors using small portals.

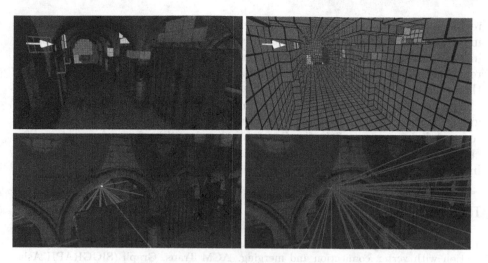

Fig. 9. Top row: portals and labeling of a corridor of the Sponza scene for a resolution of (118, 50, 72). Bottom row: comparison of local sampling (left) and sampling through a chosen portal (right) which allow to leave efficiently the corridor.

Fig. 10. Illustration of portals when we increase the resolution of the voxel grid. We observe that the separation between regions becomes more precise as we increase the resolution of the voxel grid.

A more important issue for our sampling procedure is the presence of highly concave regions. In such a region and without more information, we could sample a portal which is not visible from the origin of the ray. To apply our method to light transport algorithms, it might be unavoidable to use a convex decomposition algorithm or to improve our algorithm to guaranty the convexity of regions.

5   Conclusion and Future Works

We presented a new labeling method to partition the empty space of a 3D scene according to opening and maximal balls. This labeling is used to extract portals between 3D regions that can be efficiently sampled to go from one region to another when tracing rays in the 3D scene.

On a future work, we would want to explore other labeling methods in order to compare them with our on the application to ray tracing. As an example, our region roots can be used as seeds for a watershed propagation on the opening map. More importantly, we aim at enhancing our method to obtain approximately convex regions. As mentioned in section 4, not dealing with convex regions can be a problem because sampling points on portals does not guaranty their visibility from a point of the region.

The purpose of our method is to use our portals to improve sampling strategies of light transport algorithms. The idea is to select portals to sample in order to reach light sources efficiently. Our future work will be focused on this application and the investigation of methods to manage discontinuities that can be introduced by partitioning the scene in discrete regions.

References

1. Georgiev, I., Křivánek, J., Davidovič, T., Slusallek, P.: Light transport simulation with vertex connection and merging. ACM Trans. Graph (SIGGRAPH Asia 2012) 31, 6 (2012)
2. J. Vorba, O. Karlík, M. Šik, T. Ritschel, Jaroslav Křivánek: On-line Learning of Parametric Mixture Models for Light Transport Simulation. ACM Trans. Graph., SIGGRAPH (2014)
3. E. Veach: Robust Monte Carlo Methods for Light Transport Simulation. PhD thesis, Standford Univeristy (1997)
4. Biri, V., Chaussard, J.: Skeleton based importance sampling for path tracing. In: Proceedings of Eurographics (2012)
5. J. Chaussard, L. Noël, V. Biri, M. Couprie: A 3D Curvilinear Skeletonization Algorithm with Application to Path Tracing Discrete Geometry for Computer Imagery, Proceedings (2013)
6. Laurent Noël, Venceslas Biri: Real-Time Global Illumination using Topological Information. Journal on Computing (2014)
7. Azriel Rosenfeld, Reinhard Klette: Digital Geometry Geometric Methods for Digital Picture Analysis. Morgan Kaufmann Series in Computer Graphics and Geometric Modeling (2004)
8. Coeurjolly, D.: Fast and Accurate Approximation of Digital Shape Thickness Distribution in Arbitrary Dimension. Computer Vision and Image Understanding (CVIU) 116(12), 1159–1167 (2012)
9. Bertrand, G., Couprie, M.: Powerful Parallel and Symmetric 3D Thinning Schemes Based on Critical Kernels. Journal of Mathematical Imaging and Vision 48(1), 134–148 (2014)

Evaluation of Morphological Hierarchies
for Supervised Segmentation

Benjamin Perret[1(✉)], Jean Cousty[1], Jean Carlo Rivera Ura[1],
and Silvio Jamil F. Guimarães[1,2]

[1] Université Paris Est, LIGM, ESIEE Paris, France
{benjamin.perret,jean.cousty}@esiee.fr
[2] PUC Minas - ICEI - DCC - VIPLAB, Minas Gerais, Brazil
sjamil@pucminas.br

Abstract. We propose a quantitative evaluation of morphological hierarchies (quasi-flat zones, constraint connectivity, watersheds, observation scale) in a novel framework based on the marked segmentation problem. We created a set of automatically generated markers for the one object image datasets of Grabcut and Weizmann. In order to evaluate the hierarchies, we applied the same segmentation strategy by combining several parameters and markers. Our results, which shows important differences among the considered hierarchies, give clues to understand the behaviour of each method in order to choose the best one for a given application. The code and the marker datasets are available online.

Keywords: Hierarchy · Supervised segmentation · Morphology

1 Introduction

We propose an application driven comparison of several partition hierarchies proposed in the mathematical morphology community: quasi-flat zones hierarchy [1,2], various watershed hierarchies [3,4,5] (by altitude, area, volume, and dynamics), constrained connectivity hierarchy [6], hierarchical observation scale [7]. Compared to the evaluation strategy proposed by Arbelaez et al. [8] which searches for the best segmentation compared to a ground truth segmentation without any prior on the number of regions, we propose to focus on the single object supervised segmentation relying on two marker images: the foreground marker indicates pixels that must be in the segmented object while the background marker gives a set of pixels that is not in the object. Thus, rather than searching if one can find a segmentation that resembles a human segmentation of the whole scene, we evaluate: 1) if a hierarchy contains a set of regions that matches a given object of the scene, and 2) how difficult it is to find this set.

The problem of the automated evaluation of supervised and interactive segmentation algorithms has recently received increasing attention [9,10,11,12]. Following the idea of [12], and in order to perform an objective and quantitative

This work received funding from ANR (ANR-2010-BLAN-0205-03), CAPES/PVE under Grant 064965/2014-01, and CAPES/COFECUB under Grant 592/08.

© Springer International Publishing Switzerland 2015
J.A. Benediktsson et al. (Eds.): ISMM 2015, LNCS 9082, pp. 39–50, 2015.
DOI: 10.1007/978-3-319-18720-4_4

evaluation, we have automatically generated markers from two publicly available one object image datasets: Grabcut [13] and Weizmann [14]. For each image of the databases, we created a set of several foreground and background markers from the ground truth segmentations. Those markers were designed in order to represent different difficulty levels and do not necessarily aim to reproduce possible user interactions. Then, we have selected a simple marked segmentation strategy which consists in searching for the largest regions of the hierarchy which intersect the foreground marker and does not touch the background marker. Our tests also evaluate the importance of several parameters as the adjacency relation and the dissimilarity measure between pixels.

The contributions of this paper are the following. We propose a novel evaluation framework for partition hierarchies relying on the marker based segmentation problem. The test images are taken from public datasets and the generated markers are available online at http://perso.esiee.fr/~perretb/markerdb/. This framework is applied to several morphological hierarchies and allows us to draw some conclusions on relevance of the evaluated hierarchies. A demonstration website where users can segment their own images using the evaluated hierarchies is also available at http://perso.esiee.fr/~perretb/ISeg/.

2   Hierarchies

We give a short description of the hierarchies that will be evaluated and the reader can refer to the cited articles in order to get formal definitions or additional information. A hierarchy of partitions is a sequence of partitions such that each partition is a refinement of the previous partition in the sequence. It is usually represented as a tree or a dendrogram and can be visualized as a saliency map, which is a contour map in which the grey level represents the strength of the contour: *i.e.*, its level of disappearance in the hierarchy.

The links that exist among most of the presented hierarchies and efficient algorithms to construct them are described in [15,16].

The Quasi-Flat Zones (QFZ) hierarchy (Fig. 1(b)) is a classical structure that is constructed by considering the connected components of the level sets of the dissimilarity function [1,2]. More precisely, we say that two adjacent pixels of an image are λ-connected if there dissimilarity is lower than or equal to a value λ. For a given λ in $\mathbb{R}$, the equivalence classes of the relation "is λ-connected" form the λ-partition of the image into its λ-connected components also called λ-flat zones. The set of all λ-partitions for every λ in $\mathbb{R}$ forms the QFZ hierarchy.

Constrained connectivity (CC) hierarchy (Fig. 1(c)) is a filtered version of the QFZ hierarchy [6]. It is constructed by adding additional constraints to the definition of the connectivity. In this work we consider only the global range constraint, which limit the maximal dissimilarity between two pixels of a same connected component. This idea was introduced in order to prevent the chaining effect that may appear in the QFZ hierarchy.

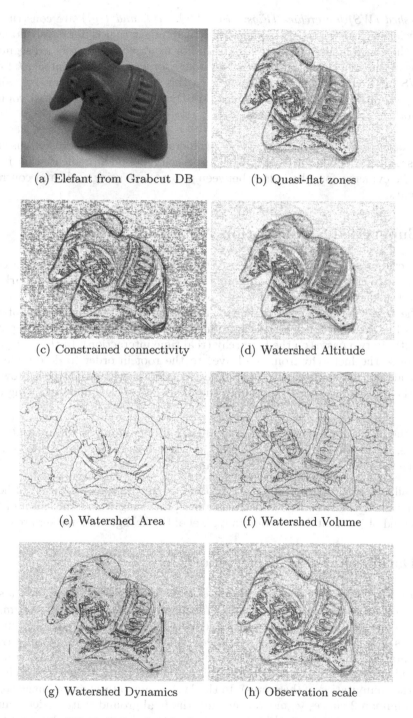

(a) Elefant from Grabcut DB   (b) Quasi-flat zones

(c) Constrained connectivity   (d) Watershed Altitude

(e) Watershed Area   (f) Watershed Volume

(g) Watershed Dynamics   (h) Observation scale

Fig. 1. Examples of saliency maps for each method

Watershed (WS) hierarchies (Figs. 1(d), 1(e), 1(f), and 1(g)) are constructed by considering the watershed segmentation of an image that is iteratively flooded under the control of an attribute [3,4,5,17]. For example, the watershed segmentations of the area closings of size k of an image for every positive integer k form the WS hierarchy by area of the image. In this article, we consider 4 possible attributes: altitude (WSAlt), dynamics (WSDyn), area (WSArea), and volume (WSVol).

Observation scale hierarchy (Fig. 1(h)) is a hierarchical version [7] of Felzenswalb et al. segmentation algorithm [18]. This approach relies on a predicate that measures the evidence for a boundary between two regions using scale and contrast information.

3 Supervised Segmentation Algorithm

In this evaluation we have chosen to use the procedure described in [19] that constructs a two classes segmentation from a hierarchy and two non-empty markers: one for the background and one for the object of interests. Its principle is to identify the object as the largest regions of the hierarchy that intersect the object marker but does not touch the background marker. This result can be computed efficiently in two passes on the hierarchy (real time interaction). In the first pass, we browse the hierarchy from the leaves to the root in order to determine for each node if it intersects each marker. The second pass, where the tree is browsed from the root to the leaves, determines the final class of each node following this rule:

- if the node intersects the background then its final label is *background*;
- else, if the node intersects the object marker then its final label is *object*;
- else the node has the same label as its parent.

Finally, as this procedure tends to produce segmentations with a lot of holes in some hierarchies, we also consider a post-processing where the holes of the segmented object that do not contain a pixel of the background marker are filled.

4 Database and Marker Generation

We consider two publicly available datasets focused on the single object segmentation problem: Grabcut [13] and Weizmann [14]. Grabcut and Weizmann datasets are composed of 50 and of 100 colour images respectively. Each image contains at least one relatively large object which is identified in a ground-truth segmentation. In the Grabcut dataset, the ground truths are stored as tri-maps which identify the object, the background, and mixed pixels (which are excluded from the computation of the scores). In the Weizmann dataset, each image comes with 3 human 2 classes segmentations and the final ground truth is determined with a majority vote (a pixel is classified as object if at least two humans have classified it as object).

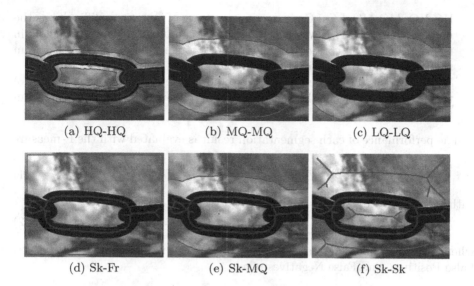

(a) HQ-HQ   (b) MQ-MQ   (c) LQ-LQ

(d) Sk-Fr   (e) Sk-MQ   (f) Sk-Sk

Fig. 2. Different combinations of markers. The combination of markers is indicated in the caption of each sub-figure in the form Background Marker-Object Marker. In each figure the background and foreground marker are respectively depicted in red and blue.

In order to perform a fair and objective evaluation of the different hierarchies we propose several automatic strategies to generate object and background markers from the ground truth. Our main idea here is not to reproduce the interactive segmentation process experienced by a real user but rather to obtain markers representing either various difficulty levels using erosions of various sizes, or markers that resembles to human generated markers using skeletonization. The generated markers are the following (see Fig. 2):

- High Quality (HQ) marker: erosion by a ball of radius 7 pixels;
- Medium Quality (MQ) marker: erosion by a ball of radius 30 pixels;
- Low Quality (LQ) marker: erosion by a ball of radius 45 pixels;
- Skeleton (Sk): morphological skeleton given by [20]; and
- Frame (Fr): frame of the image minus the object ground truth if the object touches the frame (background only). Using the frame as the background marker is nearly equivalent to having no background marker in the sense that it does not depend of the ground truth or the image.

If a connected component is completely deleted by the erosion then a single point located in the ultimate erosion of this connected component is added to the marker.

In the following, the combination of the background marker MB and the foreground marker MF is denoted MB-MF (for example, HQ-MQ stands for the combination of a high quality marker for the background and a medium quality marker for the foreground). Among all the possible combinations of markers, we chose to concentrate on the following ones:

- HQ-HQ, MQ-MQ, LQ-LQ represent a sequence of increasing difficulty as the markers get smaller. Nevertheless, all those combinations are symmetric in the sense that the correct segmentation is roughly at equal distance from the foreground to the background marker;
- Sk-Sk, MQ-Sk, Fr-Sk: here the foreground marker is always the skeleton of the ground-truth, while the background marker gets further and further from the ground-truth.

The performance of each segmentation result is evaluated with the F-measure

$$F = \frac{2.Recall.Precision}{Recall + Precision} \tag{1}$$

with

$$Recall = \frac{TP}{TP + FP} \text{ , and } Precision = \frac{TP}{TP + FN} \tag{2}$$

where TP, FP, and FN stand respectively for the number of True Positives, False Positives, and False Negatives pixels.

5   Results and Discussions

The overall results, combining the scores obtained with the 6 marker combinations on the two datasets (there are thus $6 \times 150 = 900$ measurements for each method), are presented in Fig. 3 using box-and-whisker plots. In these experiments, we considered a 4-adjacency relation with a Lab gradient (that is the Euclidean distance in the L*a*b* colour space) for the dissimilarity measure. We can see that the WSArea and WSVol globally achieve the best performance with very similar results. QFZ and CC are at the bottom of the ranking: while this result is not surprising for QFZ, which is the most basic method, it suggests that the global range constraint in CC, which is supposed to remove spurious regions from the hierarchy, has nearly no effect in this application. The theoretical similarity between QFZ and WSAlt [15] is confirmed by the experiments which show similar results for the two methods. WSDyn provides a small improvement compared to QFZ. Surprisingly, WSVol which can been seen as a combination between WSArea and WSDyn does no seem to take advantage of the information given by the depth measure (similar to the dynamics) compared to WSArea. Finally, OS achieves slightly better results than QFZ, but it does not seem that the segmentation strategy is able to take advantage of the area regularization provided by the method as it remains far from WSArea and WSVol.

The results per marker combination are presented in Fig. 5. The ranking remains the same in the symmetric cases (first row). Nevertheless, we can observe a large gap between the results obtained with low quality and medium quality markers (average increase of 0.1 on the median f-measure). In cases more similar to user interactions implying the skeleton as the foreground marker, we can see that all methods are relatively robust to the quality of the background marker. Nevertheless when the frame is used as the background marker, the combination is strongly asymmetric, and the results of WSVol and especially WSArea become

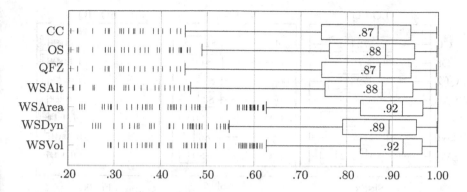

Fig. 3. Results obtained with each hierarchy for all combinations of markers. For each method, we see: 1) the median F-measure (central bar), 2) the first and third quartile (extremities of the box), 3) the lowest datum still within 1.5 inter quartile range (difference between the third and first quartile) of the lower quartile, and the highest datum still within 1.5 inter quartile range of the upper quartile range (left and right extremities), and 4) the outliers (individual points).

less reliable (lower first quartile). This last effect is understandable as the area regularization used in WSArea tends to produce regions of homogeneous sizes which does not always reflect the content of the images. Fig. 4 shows the evolution of the segmentations with respect to the marker combination for the method WSVol on a sample image.

Concerning the influence of the adjacency relation, Fig. 6 presents the result obtained for two hierarchies QFZ and WSArea using a 4- or a 8-adjacency relation (with a Lab gradient and the 6 combinations of markers). We observe

Fig. 4. Segmentation examples using WSVol for the 6 marker combinations. In each figure the background and foreground marker are respectively depicted in red and blue while the contour of the segmentation result is in green.

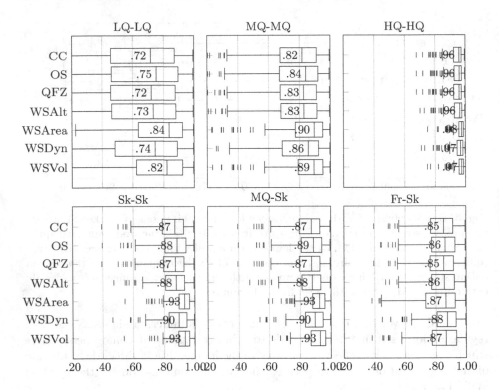

Fig. 5. Results obtained with each hierarchy and each combination of markers. See description of Fig. 3.

that there is nearly always a small gain when using a 8-adjacency instead of a 4-adjacency relation (see Fig. 6). Thus, due to the limited computational overhead, it seems a good idea to favour 8-adjacency over 4-adjacency relation in such segmentation applications.

The importance of the dissimilarity measure is illustrated in Fig. 7. We compare the results obtained with QFZ and WSArea with 3 different dissimilarity measures: 1) the absolute difference of the luminance (grey-scale image computed as the average of the RGB channels), 2) a city-block distance in the RGB space, and 3) an Euclidean distance in the Lab space. While the results are greatly improved by more complex dissimilarity measures for QFZ, the effect is mostly negligible for WSArea.

In Fig. 8, we measure the effect of filling the holes in detected object as a post processing of the segmentation process. As expected, the hierarchies that do not use a form of size regularization criterion are very sensitive to this post-processing, meaning that there is a lot of small contrasted regions (noise, specular, or textures) that lies close to the root in such hierarchies: those regions are thus more probably assigned to the background.

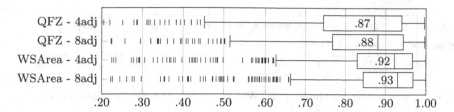

Fig. 6. Comparison between 4- and 8-adjacency relations. See description of Fig. 3.

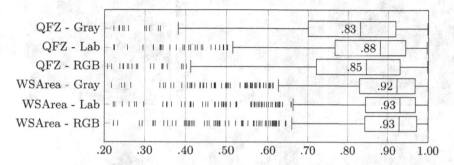

Fig. 7. Comparison between dissimilarity measures. See description of Fig. 3.

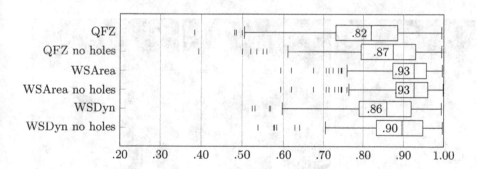

Fig. 8. Effect of filling the holes as a post processing. See description of Fig. 3.

One can also note that there is an important difference of difficulty between Grabut and Weizmann datasets. In Fig. 9, which compiles the overall results (Fig. 3) per dataset (Lab gradient, and 4-adjacency relation), we can observe a significant difference in the median score and even a larger difference in the first quartile. Indeed a visual inspection of the results shows that Weizmann dataset, contrarily to Grabcut dataset, contains images of objects with either very low contrast or with very large scale textures that all the considered methods have difficulties to segment correctly.

Finally, Fig. 10 shows several examples of segmentations.

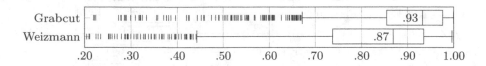

Fig. 9. Comparison between Grabcut and Weizmann datasets. See description of Fig. 3.

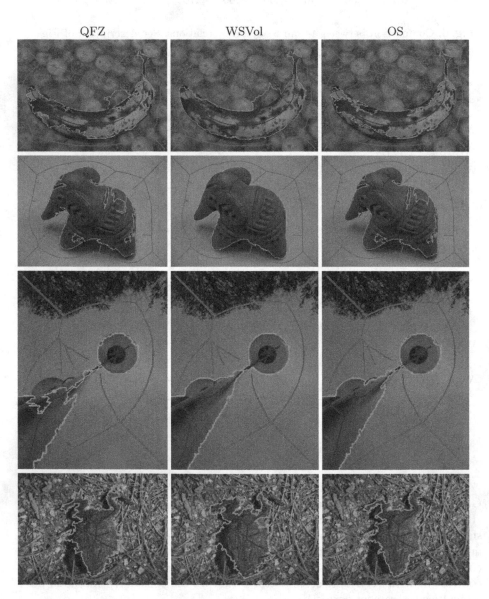

Fig. 10. Examples of segmentation results for QZF, WSVol, and OS. See description of Fig. 4.

6   Conclusion

We performed a systematic and automatic comparison of several morphological hierarchies in the context of the supervised segmentation task. We have designed a framework in order to automatically generates markers from the ground-truth segmentations. The segmentation strategy, which extracts the object from the hierarchy and the markers, is voluntarily simple in order to keep the intelligence in the design of the hierarchy.

Our results suggest that the hierarchical watersheds based on area and volume attributes are generally the best choice among morphological hierarchies for such task. The only limiting case is when the information about the background marker is very weak: in this case, that is nearly the one marker segmentation problem with the adopted segmentation strategy, it is best suited to use a hierarchical watershed by dynamics.

It is of course difficult to say if these results can be generalized to other vision tasks. Nevertheless, we believe that our test framework is a good indicator to see if a given hierarchy is well suited to represent objects in natural scenes and how easy it is to extract such object from it. Moreover, as shown in [15,16], constructing a hierarchical watershed from the QFZ hierarchy is only a linear (with respect to the number of pixels) time post-processing and it may significantly improve results.

The marker datasets are available at http://perso.esiee.fr/~perretb/markerdb/. We have also designed a web site http://perso.esiee.fr/~perretb/ISeg/ that implements the segmentation strategy with an interactive interface and that enables to test the hierarchies on custom images. The site software is implemented in JavaScript and runs entirely in the client web browser: the code to construct the hierarchies and to perform the segmentation is available.

In future works, we plan to test if we obtain similar results in other evaluation frameworks as the one proposed by Arbelaez et al. [8] or more recently by Pont-Tuset et al. [21]. Another question we will have to investigate is also the comparison to non morphological hierarchies and state of the art supervised segmentation methods. Moreover, all those results show that none of the presented methods is perfect and the construction of better hierarchies is an open issue.

References

1. Nagao, M., Matsuyama, T., Ikeda, Y.: Region extraction and shape analysis in aerial photographs. CGIP 10(3), 195–223 (1979)
2. Meyer, F., Maragos, P.: Morphological scale-space representation with levelings. In: Nielsen, M., Johansen, P., Fogh Olsen, O., Weickert, J. (eds.) Scale-Space 1999. LNCS, vol. 1682, pp. 187–198. Springer, Heidelberg (1999)
3. Beucher, S.: Watershed, hierarchical segmentation and waterfall algorithm. In: ISMM, pp. 69–76 (1994)
4. Najman, L., Schmitt, M.: Geodesic saliency of watershed contours and hierarchical segmentation. PAMI 18(12), 1163–1173 (1996)
5. Meyer, F.: The dynamics of minima and contours. In: ISMM, pp. 329–336 (1996)

6. Soille, P.: Constrained connectivity for hierarchical image partitioning and simplification. PAMI 30(7), 1132–1145 (2008)

7. Guimarães, S.J.F., Cousty, J., Kenmochi, Y., Najman, L.: A hierarchical image segmentation algorithm based on an observation scale. In: Gimel'farb, G., Hancock, E., Imiya, A., Kuijper, A., Kudo, M., Omachi, S., Windeatt, T., Yamada, K. (eds.) SSPR & SPR 2012. LNCS, vol. 7626, pp. 116–125. Springer, Heidelberg (2012)

8. Arbelaez, P., Maire, M., Fowlkes, C., Malik, J.: Contour detection and hierarchical image segmentation. PAMI 33(5), 898–916 (2011)

9. Moschidis, E., Graham, J.: A systematic performance evaluation of interactive image segmentation methods based on simulated user interaction. In: IEEE ISBI, pp. 928–931 (2010)

10. McGuinness, K., O'Connor, N.E.: Toward automated evaluation of interactive segmentation. CVIU 115(6), 868–884 (2011)

11. Klava, B., Hirata, N.: A model for simulating user interaction in hierarchical segmentation. In: ICIP (2014)

12. Zhao, Y., Nie, X., Duan, Y., Huang, Y., Luo, S.: A benchmark for interactive image segmentation algorithms. In: IEEE Workshop on POV, pp. 33–38 (2011)

13. Blake, A., Rother, C., Brown, M., Perez, P., Torr, P.: Interactive image segmentation using an adaptive GMMRF model. In: Pajdla, T., Matas, J. (eds.) ECCV 2004. LNCS, vol. 3021, pp. 428–441. Springer, Heidelberg (2004)

14. Alpert, S., Galun, M., Basri, R., Brandt, A.: Image segmentation by probabilistic bottom-up aggregation and cue integration. In: IEEE CVPR (2007)

15. Cousty, J., Najman, L., Perret, B.: Constructive links between some morphological hierarchies on edge-weighted graphs. In: Hendriks, C.L.L., Borgefors, G., Strand, R. (eds.) ISMM 2013. LNCS, vol. 7883, pp. 86–97. Springer, Heidelberg (2013)

16. Najman, L., Cousty, J., Perret, B.: Playing with kruskal: Algorithms for morphological trees in edge-weighted graphs. In: Hendriks, C.L.L., Borgefors, G., Strand, R. (eds.) ISMM 2013. LNCS, vol. 7883, pp. 135–146. Springer, Heidelberg (2013)

17. Cousty, J., Najman, L.: Incremental algorithm for hierarchical minimum spanning forests and saliency of watershed cuts. In: Soille, P., Pesaresi, M., Ouzounis, G.K. (eds.) ISMM 2011. LNCS, vol. 6671, pp. 272–283. Springer, Heidelberg (2011)

18. Felzenszwalb, P.F., Huttenlocher, D.P.: Efficient graph-based image segmentation. IJCV 59(2), 167–181 (2004)

19. Salembier, P., Garrido, L.: Binary partition tree as an efficient representation for image processing, segmentation, and information retrieval. TIP 9(4), 561–576 (2000)

20. Chaussard, J., Couprie, M., Talbot, H.: Robust skeletonization using the discrete lambda-medial axis. PRL 32(9), 1384–1394 (2011)

21. Pont-Tuset, J., Marques, F.: Measures and meta-measures for the supervised evaluation of image segmentation. In: CVPR (2013)

Study of Binary Partition Tree Pruning Techniques for Polarimetric SAR Images

Philippe Salembier[✉]

Technical University of Catalonia, Barcelona, Spain
philippe.salembier@upc.edu

Abstract. This paper investigates several pruning techniques applied on Binary Partition Trees (BPTs) and their usefulness for low-level processing of PolSAR images. BPTs group pixels to form homogeneous regions, which are hierarchically structured by inclusion in a binary tree. They provide multiple resolutions of description and easy access to subsets of regions. Once constructed, BPTs can be used for a large number of applications. Many of these applications consist in populating the tree with a specific feature and in applying a graph-cut called *pruning* to extract a partition of the space. In this paper, different pruning examples involving the optimization of a global criterion are discussed and analyzed in the context of PolSAR images for segmentation. Initial experiments are also reported on the use of Minkowski norms in the definition of the optimization criterion.

Keywords: Binary Partition Tree · PolSAR · Graph-cut · Pruning · Speckle noise · Segmentation

1 Introduction

The application of Binary Partition Trees (BPTs) [18] for remote sensing applications such as Polarimetric SAR (PolSAR) [2] and hyperspectral images [20,21] is currently gaining interest. BPTs are hierarchical region-based representations in which pixels are grouped by similarity. Their construction is often based on an iterative region-merging algorithm: starting from an initial partition, the pair of most similar neighboring regions is iteratively merged until one region representing the entire image support is obtained. The BPT essentially stores the complete merging sequence in a binary tree structure. Once constructed, BPTs can be used for a large number of tasks including image filtering with connected operators, segmentation, object detection or classification [18,3]. Many of these tasks involve the extraction of a partition from the BPT through a graph cut.

In this paper, we focus on low level PolSAR image processing tasks. We study in particular the interest of a specific graph cut called *pruning* in this context.

This work has been developed in the framework of the project BIGGRAPH-TEC2013-43935-R, financed by the Spanish Ministerio de Economía y Competitividad and the European Regional Development Fund (ERDF).

© Springer International Publishing Switzerland 2015
J.A. Benediktsson et al. (Eds.): ISMM 2015, LNCS 9082, pp. 51–62, 2015.
DOI: 10.1007/978-3-319-18720-4_5

We discuss and evaluate various pruning techniques formulated as the search in the BPT of a partition optimizing a certain criterion. The criteria we analyze take into account the specific nature of PolSAR data and the presence of speckle noise resulting from the coherent integration of the electromagnetic waves. The main contributions of this paper compared to [2,3,17] is the proposal of new pruning strategies for PolSAR images as well as the objective evaluation of the resulting partitions thanks to a set of realistic simulated PolSAR images where the underlying ground truth is available [8].

The paper is organized as follows: Section 2 is a short introduction on PolSAR data. Section 3 discusses the BPT creation and its processing with graph cut. Four pruning criteria useful for segmentation of PolSAR images are presented in section 4 and evaluated in section 5. A preliminary study of the interest of Minkowski norms in the definition of the optimization criterion is presented in section 6. Finally, conclusions are reported in section 7.

2   PolSAR Data

Synthetic Aperture Radars (SAR) are active microwave imaging systems. They are becoming increasingly popular for Earth observation because they work independently of the day and night cycle and of weather conditions. A SAR system essentially transmits an electromagnetic wave and records its echo to localize targets. In order to achieve a high spatial resolution, narrow beamwidth or equivalently large antennas are necessary. SAR systems deal with this issue by making use of the relative motion between the sensor and the target. As the radar moves, it repeatedly illuminates the target with electromagnetic pulses. The echoes are coherently recorded and combined in a post-processing that synthesizes a very large array and creates a high resolution image. The speckle noise results from the coherent addition of the scattered electromagnetic waves and is considered as one of the main problems for the exploitation of SAR data.

In the early 90's, multidimensional systems were developed. They provide complex SAR images $[S_1, S_2, \ldots, S_m]$ by introducing some sort of diversity. An important example is Polarimetric SAR (PolSAR) [5,13] where the diversity is based on considering different polarization states for the transmitted and received electromagnetic waves. This makes SAR data sensitive to the target geometry, including vegetation, and to the dielectric properties of the target. For every resolution cell, a PolSAR system measures the scattering matrix:

$$\mathbf{S} = \begin{bmatrix} S_{hh} & S_{hv} \\ S_{vh} & S_{vv} \end{bmatrix} \tag{1}$$

where h and v represent the horizontal and vertical polarization states and S_{pq} for $p, q \in \{h, v\}$ denotes the complex SAR image where the reception (transmission) polarization states is p (q).

Since the dimensions of the resolution cell are normally larger that the wavelength of the electromagnetic wave, the scattered wave results from the coherent combination of many waves. This coherent addition process is known as the speckle. Although the speckle represents a true electromagnetic measurement, its complexity is such that it is considered as a random process. Rewriting the **S** matrix as a vector **k** [6]:

$$\mathbf{k} = [S_{hh}, \sqrt{2}S_{hv}, S_{vv}]^T \tag{2}$$

k is characterized by a three dimensional zero-mean complex Gaussian pdf:

$$p_{\mathbf{k}}(\mathbf{k}) = \frac{1}{\pi^3|\mathbf{C}|} \exp(-\mathbf{k}^H\mathbf{C}^{-1}\mathbf{k}). \tag{3}$$

Therefore, the distribution of **k** is completely described by the Hermitian positive definite covariance matrix:

$$
\begin{aligned}
\mathbf{C} &= E\{\mathbf{k}\mathbf{k}^H\} \\
&= \begin{bmatrix} E\{S_{hh}S_{hh}^H\} & \sqrt{2}E\{S_{hh}S_{hv}^H\} & E\{S_{hh}S_{vv}^H\} \\ \sqrt{2}E\{S_{hv}S_{hh}^H\} & 2E\{S_{hv}S_{hv}^H\} & \sqrt{2}E\{S_{hv}S_{vv}^H\} \\ E\{S_{vv}S_{hh}^H\} & \sqrt{2}E\{S_{vv}S_{hv}^H\} & E\{S_{vv}S_{vv}^H\} \end{bmatrix}
\end{aligned} \tag{4}
$$

where $E\{x\}$ is the statistical expectation of x. The Maximum Likelihood Estimation (MLE) of **C**, i.e., the multilook, under the assumption of statistical ergodicity and homogeneity, is obtained by substituting the statistical expectation by an averaging:

$$\mathbf{Z} = \langle\mathbf{k}\mathbf{k}^H\rangle_n = \frac{1}{n}\sum_{i=1}^{n}\mathbf{k}_i\mathbf{k}_i^H \tag{5}$$

where n indicates the number of independent looks or samples employed to estimate **C** and $\mathbf{k}_i$ is the ith sample vector. The estimated covariance matrix **Z** is statistically characterized by a Wishart distribution [12].

3   BPT Creation and Processing Through Graph Cut

The BPT creation starts by the definition of an initial partition which can be composed of individual pixels as in [2,3]. While this strategy guarantees a high precision as starting point of the merging process, it also implies high computational and memory costs as many regions have to be handled. As an alternative, the initial partition may correspond to an over-segmentation as a super-pixel partition. This initial partition issue was studied in [17] where several alternative strategies were evaluated. The main conclusion of this study is that the use of super-pixel partition as initial partition of the merging process can indeed drastically reduce the computational load of the BPT creation without any significant impact on the quality of the regions and partitions represented by the tree. One of the key point however is to use a denoising filter adapted to PolSAR images such as [14,7] before computing the super-pixel partitions. The best combination found in [17] involves the use of the σ-Lee denoising filter [14]

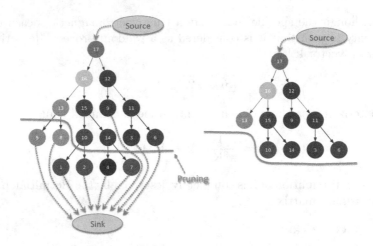

Fig. 1. Illustration of the pruning on a BPT. Left: The root of the BPT is connected to a *source* node and the leaves are connected to a *sink* node. The pruning creates two connect component where siblings belong to the same connected component. Right: the pruned BPT is the connected component that includes the *source*. Its leaves define the extracted partition.

followed by the SLIC algorithm [1] to compute the super-pixel. Only the three diagonal elements of the covariance matrices **Z** (after denoising) have been used to generate the super-pixels.

Once the initial partition is defined, the BPT construction is done by iteratively merging the pair of most similar neighboring regions. In the PolSAR case of interest here, the information carried by pixels (i, j) of an image I corresponds to the covariance matrix $\mathbf{Z}_{ij}^I$ after denoising. To construct the BPT, we need to define a region model and a similarity measure between neighboring regions.

We use the strategy defined in [2] where regions R are modeled by their mean covariance matrix $\mathbf{Z}_R = \frac{1}{|R|} \sum_{i,j \in R} \mathbf{Z}_{ij}^I$, where $|R|$ is the region number of pixels. If the region is homogeneous, this estimation corresponds to the MLE defined by Eq. 5. The distance between neighboring regions, which defines the merging order, relies on the *Symmetric revised Wishart dissimilarity* [2]. The revised Wishart dissimilarity measure [9] is based on a statistical test assuming that the two regions follow a Wishart pdf and that one pdf is known. Thus, it is not symmetric as it depends on which region pdf is assumed to be known. In order to define the merging order, a symmetric version is used:

$$d_{RW}(R_1, R_2) = \left(tr(\mathbf{Z}_{R_1}^{-1}\mathbf{Z}_{R_2}) + tr(\mathbf{Z}_{R_2}^{-1}\mathbf{Z}_{R_1})\right)(|R_1| + |R_2|) \qquad (6)$$

where R_1, R_2 are the two neighboring regions and $tr(\mathbf{A})$ denotes the trace of the **A** matrix and $\mathbf{A}^{-1}$ its inverse.

Once the BPT has been constructed, it can be used for a wide range of applications including filtering, segmentation or classification. In many cases, the application relies on the extraction of a partition from the BPT. This process can be seen as a particular graph cut called *pruning* that can be formally defined as follows: Assume the tree root is connected to a *source* node and that all the tree leaves are connected to a *sink* node. A *pruning* is a graph cut that separates the tree into two connected components, one connected to the source and the other to the sink, in such a way that any pair of siblings falls in the same connected component. The connected component that includes the root node is itself a BPT and its leaves define a partition of the space. This process is illustrated in Fig. 1. In the sequel, we discuss several examples of PolSAR image pruning.

4   Optimum Pruning of BPTs for PolSAR Data

As previously mentioned, the extraction of a partition from the BPT can be defined by a pruning strategy. In this paper, we are interested in pruning techniques that extract partitions optimizing a certain criterion. More precisely, we restrict ourselves to additive criteria, that are criteria defined as:

$$C = \sum_R \phi_R \qquad (7)$$

where R is a set of regions described in the BPT that forms a partition and ϕ_R a measure depending on R.

This type of criterion can be efficiently minimized using an dynamic programing algorithm originally proposed in [18] for global optimization on BPT. The solution consists in propagating local decisions in a bottom-up fashion. The BPT leaves are initially assumed to belong to the optimum partition. Then, one checks if it is better to represent the area covered by two sibling nodes as two independent regions $\{R_1, R_2\}$ or as a single region R (the common parent node of R_1 and R_2). The selection of the best choice is done by comparing the criterion ϕ_R evaluated on R with the sum of the criterion values ϕ_{R_1} and ϕ_{R_2}:

$$\text{If } \phi_R \leq \phi_{R_1} + \phi_{R_2} \begin{cases} \text{then} & \text{select } R \\ \text{else} & \text{select } R_1 \text{ and } R_2 \end{cases} \qquad (8)$$

The best choice (either "R" or "R_1 plus R_2") is stored in the node representing R with the corresponding criterion value (ϕ_R or $\phi_{R_1} + \phi_{R_2}$). The procedure is iterated up to the root and defines the best partition. This algorithm finds the global optimum of the criterion on the tree and the selected regions form a partition of the image.

We discuss now four pruning techniques for low-level processing and grouping of PolSAR data. The main goal of theses pruning techniques is to segment the images so that a precise estimation of the region contours as well as of the polarimetric parameters can be done.

The first and most obvious pruning technique relies on the adaptation of Square Error (SE) to the matrix case, here the covariance matrices. It simply consists in computing the matrix norm of the difference between the covariance matrices $\mathbf{Z}_{ij}^I$ of the pixels belonging to a given region R and the covariance matrix presenting the region model $\mathbf{Z_R}$:

$$\phi_R = \sum_{i,j \in R} \|\mathbf{Z}_{ij}^I - \mathbf{Z}_R\|_F \tag{9}$$

where $\|.\|$ represents the so-called Frobenius norm[1]. This criterion essentially enforces the homogeneity of regions. However, on its own, it is useless because a partition made of the initial leaves of the BPT will be optimum as this is where the deviation of the individual pixels with respect to the region mean will be minimized. Following classical approaches in functional optimization, ϕ_R can be interpreted as a data fidelity term and combined with a data regularization term which encourages the optimization to find partitions with a reduced number of regions. As simple data regularization, we use a constant value λ that penalizes the region presence. Therefore, the final homogeneity-based criterion to be minimized is given by $C = \sum_R \phi_R$ with ϕ_R defined as follows:

$$\phi_R^{SE} = \sum_{i,j \in R} \|\mathbf{Z}_{ij}^I - \mathbf{Z}_R\|_F + \lambda \tag{10}$$

This first pruning criterion may be interesting to extract homogenous regions in terms of the data covariance matrix but it does not take into account the presence of the speckle noise. As discussed above the speckle noise is a random process that is complex to characterize but it is often approximated by a multiplicative noise [6]. Therefore, a second pruning criterion can be derived from the first one normalizing the homogeneity measure by the average norm of the region model. The corresponding ϕ_R can be written as:

$$\phi_R^{SAR\_SE} = \sum_{i,j \in R} \|\mathbf{Z}_{ij}^I - \mathbf{Z}_R\|_F / \|\mathbf{Z}_R\|_F + \lambda \tag{11}$$

The third pruning criterion relies on the region similarity measure used for the construction of the BPT. In section 3, we mentioned that a Wishart-based measure was used to compute the similarity between neighboring regions. Eq. 6 can be adapted to measure the similarity between pixels and the region model. It would lead to an expression such as: $tr((\mathbf{Z}_{i,j}^I)^{-1}\mathbf{Z}_R) + tr((\mathbf{Z}_R)^{-1}\mathbf{Z}_{i,j})$. However, the matrix inversion at the pixel level is computationally demanding and the matrix may even be singular. Therefore, we used a simplified formulation of this measure by taking into account only the diagonal elements of the matrices.

$$\phi_R^{Wishart} = \sum_{i,j \in R} \sqrt{\sum_{k=1,2,3} \left(\frac{\mathbf{Z}_{ij}^I(k,k)^2 + \mathbf{Z}_R(k,k)^2}{\mathbf{Z}_{ij}^I(k,k)\mathbf{Z}_R(k,k)} \right)} + \lambda \tag{12}$$

where $\mathbf{Z}_{ij}^I(k,k)$ and $\mathbf{Z}_R(k,k)$ respectively represent the diagonal elements of covariance matrices $\mathbf{Z}_{ij}^I$ and $\mathbf{Z}_R$.

---

[1] The Frobenius norm of matrix A with elements $[a(k,l)]$ is: $\|A\|_F = \sqrt{\sum_{k,l} a(k,l)^2}$.

Finally, the last pruning criterion relies on a geodesic distance adapted to the cone of positive definite Hermitian matrices [4]. This measure exploits the geometry of the space defined by the covariance matrices and is given by: $\|log\left(\mathbf{Z}_R^{-1/2}\mathbf{Z}_{i,j}^I\mathbf{Z}_R^{-1/2}\right)\|_F$, where $log(.)$ represents the matrix logarithm. As previously, since this measure is quite complex to compute, we use a simplified version taking into account only the diagonal elements of the matrices. The fourth pruning criterion is then given by:

$$\phi_R^{Geodesic} = \sum_{i,j \in R} \sqrt{\sum_{k=1,2,3} ln^2\left(\frac{\mathbf{Z}_{ij}^I(k,k)}{\mathbf{Z}_R(k,k)}\right)} + \lambda \tag{13}$$

where ln represents the natural logarithm.

5   Evaluation

To objectively measure the performances of the pruning discussed in the previous section, we rely on a dataset of PolSAR images on which the ground-truth polarimetric information is available. More precisely, we use the set of simulated PolSAR images [8] where the underlying ground-truth, i.e. the class regions, is modeled by Markov Random Fields. A set of typical polarimetric responses has been extracted from an AIRSAR image (L-band) so that they represent the 8 classes found in the $H/\overline{\alpha}$ plane and randomly assigned to each class. Then, single look complex images have been generated from the polarimetric responses using a Cholesky decomposition [11]. Examples of images and their corresponding ground-truth, denoised images and super-pixel partitions are presented in Fig. 2.

Thanks to this dataset with ground-truth, we can measure the quality of the pruning techniques in the context of segmentation because we know the ideal partition. Fig. 3 shows the evaluation of the segmentation results as classically done in the supervised case through Precision and Recall curves. On the left side, the so-called *Precision and Recall for boundaries* [15] is presented. In this case, each partition is evaluated by considering all pairs of neighboring pixels and by classifying them in either boundary or interior segments. The Precision and the Recall values of this classification are evaluated by comparison with the classification resulting from the ground-truth partition. In addition to this boundary-oriented evaluation, a region-oriented evaluation known as the *Precision and Recall for objects and parts* [16] is presented on the right side of Fig. 3. In this context, regions of the partition are considered as potential candidates to form regions of the ground-truth partition, and are classified as correct or not. In both cases, the curves are formed by modifying the λ value to get coarser or finer partitions. The ideal system has Precision and Recall values equal to one.

As can be seen in Fig. 3, the region-oriented evaluation is more severe than the boundary-oriented evaluation. This is to be expected as the boundary measure simply checks whether boundary elements in the ground-truth partitions match boundary elements of the partitions extracted from the BPT. As the partitions

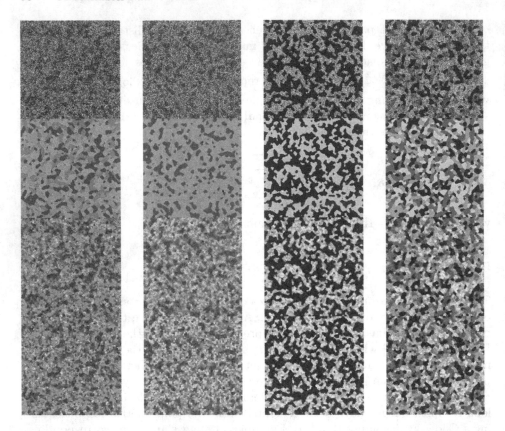

Fig. 2. Examples of original PolSAR images (first row), their corresponding ground-truth (second row), denoised images (third row) and super-pixel partitions (fourth row). RGB-pauli color coding: the polarimetric channels $|HH - VV|$, $|HV|$ and $|HH + VV|$ are assigned to the RGB channels respectively.

are rather dense, it not very difficult to find matching boundary elements. However, the boundary measure does not actually analyze whether the ground-truth regions correspond to regions of the partitions computed from the BPT. This issue is evaluated by the "objects and parts" measure (see [16] for details on this issue). However, the conclusions on both plots are the same: the best pruning technique is the one based on $\phi_R^{SAR\_SE}$ (Eq. 11). The pruning techniques based on the Wishart pdf (Eq. 12) and the geodesic distance (Eq. 13) seems to provide interesting results for very coarse partitions (precision values close to one obtained for high values of λ).

Precision and Recall curves describe the performances for the complete range of pruning parameter values. However, they do not efficiently describe the system sensitivity to the parameter value. To this end, Fig. 4 presents the F value as a function of the pruning parameter. The F value is classically used to summarize the Precision P and Recall R trade-off. It is the harmonic mean of P and R:

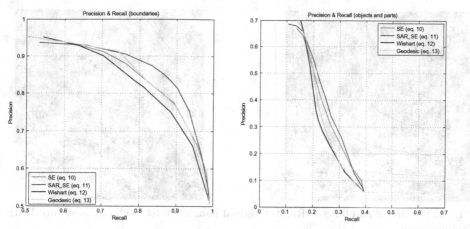

Fig. 3. Precision and Recall (PR) performances of the four pruning techniques (average over the entire dataset). Left: PR for boundaries, Right: PR for objects and parts.

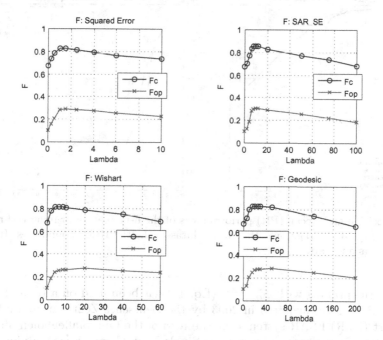

Fig. 4. F value as a function of the pruning parameter. F_b (F_{op}) corresponds to the Precision and Recall for boundaries (object and parts) curves. Top: ϕ_R^{SE} and $\phi_R^{SAR\_SE}$, Bottom: $\phi_R^{Wishart}$ and $\phi_R^{Geodesic}$.

$F = 2PR/(P + R)$. Fig. 4 shows that all pruning techniques provide stable results for a wide range of λ values. If we consider the best pruning approach $\phi_R^{SAR\_SE}$ for example, this means that, in practice, λ values between 5 and 15 will extract similar partitions and there is no need to fine tune the parameter.

Fig. 5. Results on real images. Right: Original images (RGB Pauli composition). Left: Pruning with $\phi_R^{SAR\_SE}$ (Eq. 11).

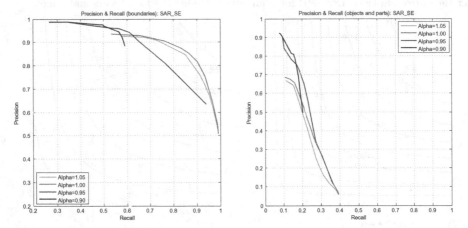

Fig. 6. Precision and Recall (PR) performances of the h-increasing pruning technique defined by Eq. 14 (average over the entire dataset). Left: PR for boundaries, Right: PR for objects and parts.

Finally, the pruning with $\phi_R^{SAR\_SE}$ (Eq. 11) has been used on a L-band fully polarimetric data set acquired in 2003 by the Deutsches Zentrum für Luftund Raumfahrt (DLR) ESAR system over the area of the Oberpfaffenhofen airport near Munich, Germany. The images are Single Look Complex with a pixel size of 1,5x1,5m. Results are shown in Fig. 5 together with the original image. They visually highlight the interest of the BPT to perform low-level processing of PolSAR images while preserving the spatial resolution of the content.

6   Potential on Non-additive Criteria

It has been recently shown that the algorithm identifying optimum partitions as described in section 4 is valid for a larger class of criteria than purely additive criteria such as $C = \sum_R \phi_R$. In particular [19] introduces the notion of h-increasing criterion which is sufficient (and necessary) to guarantee that the search algorithm identifies the optimum partition. Moreover, [10] shows that Minkowski norms: $C = (\sum_R \phi_R^\alpha)^{\frac{1}{\alpha}}$, $\forall \alpha$, are h-increasing as well as their combination by addition or supremum. Therefore the same search algorithm can be used for this larger class of criteria.

As a preliminary experiment to study the potential of h-increasing criteria beyond additive ones ($\alpha = 1$) for PolSAR data, we have studied a simple modification of the best pruning identified in the previous section: $\phi_R^{SAR\_SE}$. We evaluated the performances of the pruning based on the following criterion:

$$C = \sum_R \left(\sum_{i,j \in R} \left(\|\mathbf{Z}_{ij}^I - \mathbf{Z}_R\|_F / \|\mathbf{Z}_R\|_F \right)^\alpha \right)^{\frac{1}{\alpha}} + \lambda \tag{14}$$

The results in terms of precision and recall are shown in Fig. 6 for four values of the α parameter. As can be seen, the classical additive approach ($\alpha = 1$) gives the best results in the sense that it allows to get closer to the ideal point of Precision=1 and Recall=1. However, α values lower than one seem to give better results for coarse partitions (large λ values implying a cut close to the root node). This is an interesting result that suggests that the use of Minkowski norms has to be further investigated at least for PolSAR data.

7 Conclusions

This paper has discussed the interest of Binary Partition Trees (BPTs) for PolSAR images and highlighted the usefulness of a particular type of graph cut called pruning to extract partitions from the BPT. Four specific pruning techniques involving the global optimization of a criterion related the region homogeneity have been evaluated. The best pruning strategy relies on a normalized version of squared error where the normalization takes into account the specific multiplicative nature of the speckle noise. Finally, preliminary results suggest that the use of Minkowski norms has to be further investigated as it proved to provide good results for pruning close to the root of the tree.

References

1. Achanta, R., Shaji, A., Smith, K., Lucchi, A., Fua, P., Süsstrunk, S.: SLIC superpixels compared to state-of-the-art superpixel methods. IEEE Trans. on Pattern Analysis and Machine Intelligence 34(11), 2274–2282 (2012)
2. Alonso-Gonzalez, A., Lopez-Martinez, C., Salembier, P.: Filtering and segmentation of polarimetric SAR data based on binary partition trees. IEEE Trans. on Geoscience and Remote Sensing 50(2), 593–605 (2012)
3. Alonso-Gonzalez, A., Valero, S., Chanussot, J., Lopez-Martinez, C., Salembier, P.: Processing multidimensional SAR and hyperspectral images with binary partition tree. Proceedings of IEEE 101(3), 723–747 (2013)

4. Barbaresco, F.: Interactions between symmetric cone and information geometries: Bruhat-tits and siegel spaces models for high resolution autoregressive doppler imagery. In: Nielsen, F. (ed.) ETVC 2008. LNCS, vol. 5416, pp. 124–163. Springer, Heidelberg (2009)
5. Cloude, S.: Polarisation Applications in Remote Sensing. Oxford Univ. Press (2009)
6. Cloude, S., Pottier, E.: A review of target decomposition theorems in radar polarimetry. IEEE Trans. on Geosc. and Remote Sens. 34(2), 498–518 (1996)
7. Deledalle, C.A., Tupin, F., Denis, L.: Polarimetric SAR estimation based on non-local means. In: IEEE International Geoscience and Remote Sensing Symposium, IGARSS 2010 (2010)
8. Foucher, S., Lopez-Martinez, C.: Analysis, evaluation, and comparison of polarimetric SAR speckle filtering techniques. IEEE Trans. on Image Processing 23(4), 1751–1764 (2014)
9. Kersten, P.R., Lee, J.-S., Ainsworth, T.L.: Unsupervised classification of polarimetric synthetic aperture radar images using fuzzy clustering and EM clustering. IEEE Trans. on Geoscience and Remote Sensing 43(3), 519–527 (2005)
10. Kiran, B.R.: Energetic-Lattice based optimization. PhD thesis, Université Paris-Est (2014)
11. Lee, J.-S., Ainsworth, T.L., Kelly, J.P., López-Martínez, C.: Evaluation and bias removal of multilook effect on entropy/alpha/anisotropy in polarimetric SAR decomposition. IEEE Trans. on Geosc. and Remote Sens. 46(10), 3039–3051 (2008)
12. Lee, J.-S., Hoppel, K., Mango, S., Miller, A.: Intensity and phase statistics of multilook polarimetric and interferometric SAR imagery. IEEE Trans. on Geoscience and Remote Sensing 32(5), 1017–1028 (1994)
13. Lee, J.-S., Pottier, E.: Polarimetric Radar Imaging: From Basics to Applications. CRC Press, Boca Raton (2009)
14. Lee, J.-S., Wen, J.H., Ainsworth, T.L., Chen, K.S., Chen, A.J.: Improved sigma filter for speckle filtering of SAR imagery. IEEE Trans. on Geoscience and Remote Sens. 47(1), 202–213 (2009)
15. Martin, D., Fowlkes, C., Malik, J.: Learning to detect natural image boundaries using local brightness, color, and texture cues. IEEE Trans. on Pattern Analysis and Machine Intelligence 26(5), 530–549 (2004)
16. Pont-Tuset, J., Marques, F.: Measures and meta-measures for the supervised evaluation of image segmentation. In: Computer Vision and Pattern Recognition (CVPR) (2013)
17. Salembier, P., Foucher, S., Lopez-Martinez, C.: Low-level processing of PolSAR images with binary partition trees. In: IEEE International Geoscience and Remote Sensing Symposium, IGARSS 2014, Quebec, Canada (July 2014)
18. Salembier, P., Garrido, L.: Binary partition tree as an efficient representation for image processing, segmentation, and information retrieval. IEEE Trans. on Image Processing 9(4), 561–576 (2000)
19. Serra, J.: Hierarchies and optima. In: Debled-Rennesson, I., Domenjoud, E., Kerautret, B., Even, P. (eds.) DGCI 2011. LNCS, vol. 6607, pp. 35–46. Springer, Heidelberg (2011)
20. Valero, S., Salembier, P., Chanussot, J.: Hyperspectral image representation and processing with binary partition trees. IEEE Trans. on Image Processing 22(4), 1430–1443 (2013)
21. Veganzones, M.A., Tochon, G., Dalla-Mura, M., Plaza, A.J., Chanussot, J.: Hyperspectral image segmentation using a new spectral unmixing-based binary partition tree representation. IEEE Trans. on Image Proc. 23(8), 3574–3589 (2014)

A Comparison Between Extinction Filters and Attribute Filters

Roberto Souza[1(✉)], Letícia Rittner[1], Rubens Machado[2], and Roberto Lotufo[1]

[1] University of Campinas, Campinas, Brazil
{rmsouza,lrittner,lotufo}@dca.fee.unicamp.br
[2] Center for Information Technology Renato Archer, Campinas, Brazil
rubens.campos.machado@gmail.com

Abstract. Attribute filters and extinction filters are connected filters used to simplify greyscale images. The first kind is widely explored in the image processing literature, while the second is not much explored yet. Both kind of filters can be efficiently implemented on the max-tree. In this work, we compare these filters in terms of processing time, simplification of flat zones and reduction of max-tree nodes. We also compare their influence as a pre-processing step before extracting affine regions used in matching and pattern recognition. We perform repeatability tests using extinction filters and attribute filters, set to preserve the same number of extrema, as a pre-processing step before detecting Hessian-Affine and Maximally Stable Extremal Regions (MSER) affine regions. The results indicate that using extinction filters as pre-processing obtain a significantly higher (more than 5% on average) number of correspondences on the repeatability tests than the attribute filters. The results in processing natural images show that preserving 5% of images extrema using extinction filters achieve on average 95% of the number of correspondences compared to applying the affine region detectors directly to the unfiltered images, and the average number of max-tree nodes is reduced by a factor greater than 3. Therefore, we can conclude that extinction filters are better than attribute filters with respect to preserving the number of correspondences found by affine detectors, while simplifying the max-tree structure. The use of extinction filters as a pre-processing step is recommended to accelerate image recognition tasks.

Keywords: Extinction filters · Extinction values · Max-tree · Connected filters · Attribute filters · MSER · Hessian-Affine

1 Introduction

Connected filters act on greyscale images by merging elementary regions called flat zones, which are the largest sets of pixels where the image is constant [1]. Connected filters do not create new contours in the image nor modify their position [2], they can only remove contours. There are two popular techniques used to design connected filters: filters by reconstruction [1], and filters based on a tree representation of the image, such as the component tree [3] and the

© Springer International Publishing Switzerland 2015
J.A. Benediktsson et al. (Eds.): ISMM 2015, LNCS 9082, pp. 63–74, 2015.
DOI: 10.1007/978-3-319-18720-4_6

max-tree [4]. In this work, we focus on the tree based techniques, which have become very popular in the last years. The tree structure we use is the max-tree, which is a compact structure for the component tree representation [5]. The max-tree represents an image through the hierarchical relationship of its connected components. There are algorithms to build it in quasi-linear time [6–9], and filtering the image consists simply in contracting some of the max-tree nodes, which is equivalent to merging sets of flat zones, i.e. it is a connected filter [1]. Max-trees corresponding to natural images usually have many irrelevant nodes, therefore it is interesting to simplify its structure, since features may be extracted from its nodes to be used as input for a classifier.

Attribute filters are the most known kind of connected filters. If the attribute is increasing, they can be implemented through the max-tree pruning of all the nodes that do not attend the threshold used by the filter. The most common attribute filters are contrast filters, *hmax* [10], size filters, *area-open* [11], and a mix of contrast and size filtering, *vmax* [12].

Extinction filters are based on the concept of extinction values, which are a powerful tool to measure the persistence of an increasing attribute, and are useful to discern relevant from irrelevant extrema, usually noise. The concept of extinction values was proposed in [12], and it can be seen as an extension of the concept of dynamics [13]. The most common attributes used to compute extinction values are height, area and volume, and they can be efficiently computed in the max-tree structure [14]. Extinction filters are connected filters that preserve only the extrema with highest extinction values. Unlike attribute filters, they preserve the height of the extrema kept. They have interesting filtering properties, but they are still not much explored by the scientific community. A methodology that allows to compute the extinction values of non-increasing attributes, such as shape attributes, through a second max-tree construction from the original max-tree was proposed in [15]. Therefore, extinction filters can be implemented even for non-increasing attributes.

Contributions: This paper compares the usual attribute filters against extinction filters, both using a single increasing attribute. More specifically the attributes used are height, area and volume. A methodology to set the number of extrema using attribute filters is presented. Extinction and attribute filters, set to preserve the same number of extrema, are analyzed and compared in depth in this work. The robustness of the filters are analyzed through repeatability tests comparing the Maximally Stable Extremal Region (MSER) [16] affine detector, which can be efficiently computed from the max-tree [17], and the MSER detector using attribute and extinction filters as a pre-processing step are shown. This is done also for the Hessian-Affine [18] detector. The results indicate that extinction filters are better than attribute filters with respect to simplification for recognition, since they preserve more the correspondences found by affine detectors.

Paper Organization: Section 2 gives a theoretical background necessary to understand the remaining of this paper. Section 3 explains how to set the number of image extrema using attribute filters. Section 4 illustrates experiments comparing the filters processing times, simplification performances, structural similarities, and their robustness through repeatability tests. Section 5 states our conclusions.

2 Theoretical Background

2.1 Max-Tree

The component tree [3] represents an image through the hierarchical relationship of its connected components. Each node of the component tree stores all the pixels of the connected component it represents. The max-tree is a compact structure for the component tree representation. It is similar to the component tree, but a connected component whose area remained unchanged for a sequence of threshold values is stored in a single node called composite node [5], and each node stores only the pixels of the connected component that are visible in the image. The component tree and the max-tree corresponding to the 1D image $I = [0, 5, 2, 4, 1, 1, 4, 4, 1, 0]$ are illustrated in Figure 1. The max-tree composite nodes are represented by double circles. The leaves of the component tree and the max-tree correspond to regional maxima. In order to process minima with these structures, the duality property must be employed. The height of a max-

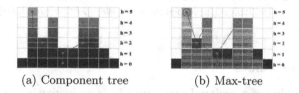

(a) Component tree   (b) Max-tree

Fig. 1. (a) Component tree and (b) max-tree corresponding to the 1D image $I = [0, 5, 2, 4, 1, 1, 4, 4, 1, 0]$

tree node is a contrast attribute, the area is a size attribute, and the volume is a combination of contrast and size attributes. These are all increasing attributes, and they can be computed through the following equations, respectively:

$$\mu_h(C_i) = \max_{\forall k \in descendants(i)} \{h_k - h_i\} , \tag{1}$$

$$\mu_a(C_i) = \sum_{\forall z \in C_i} 1, \tag{2}$$

$$\mu_v(C_i) = \mu_a(C_i) + \sum_{k \in descendants(i)} \mu_a(C_k) \times nlevels(k), \tag{3}$$

where i is a max-tree node, C_i is its corresponding connected component, h_i is the greylevel of node i, $nlevels$ represent the number of sequential threshold values in which the component stayed the same, and $descendants(i)$ is a set containing all the descendants of node i.

2.2   Attribute Filters

Attribute filters (AF) are connected filters that remove the connected components which do not attend a threshold criteria. Attribute filters may either use
a single attribute or a set of attributes to decide which connected components
should be removed. The most usual attribute filters are the *hmax* [10], *vmax* [12],
and the *area-open* [11]. Among these three filters, only the *area-open* is idempotent. The attribute filtering procedure is the following: all the max-tree nodes
for which their attribute is less than the threshold value chosen are contracted.
If the attribute is increasing, this procedure results in a pruning strategy [4]. An
area-open example removing the nodes with area smaller than 3 is depicted in
Figure 2.

(a)   (b)

Fig. 2. (a) Original max-tree with the nodes with area greater than or equal to three
marked in red. The label inside the nodes mean $Node\ ID : h[area]$. (b) The result of
the *area-open* filter.

2.3   Extinction Values

The extinction value of a regional maximum for any increasing attribute is the
maximal size of an attribute filter such that this maximum still exists after the
filtering. The formal definition of extinction value is the following: consider M
a regional maximum of an image I, and $\Psi = (\psi_\lambda)_\lambda$ is a family of decreasing
connected anti-extensive transformations. The extinction value corresponding
to M with respect to Ψ and denoted by $\varepsilon_\Psi(M)$ is the maximal λ value, such
that M still is a regional maxima of $\psi_\lambda(I)$. This definition can be expressed
through the following equation:

$$\varepsilon_\Psi(M) = sup\{\lambda \geq 0 | \forall \mu \leq \lambda, M \subset Max(\psi_\mu(I))\},\tag{4}$$

where $Max(\psi_\mu(I))$ is the set containing all the regional maxima of $\psi_\mu(I)$. Extinction values of regional minima can be defined similarly. An efficient algorithm
for computing extinction values from the max-tree is presented in [14].

2.4   Extinction Filter

The Extinction Filter (EF) is a connected idempotent filter that preserves the
relevant maxima of the image, while reducing the image complexity, i.e. the

number of flat zones. Extinction filters can be thought as an attribute filter
if we consider the node attribute as being the largest extinction value of its
descendant leaves. In this work, we prefer to filter the image based on the number
of relevant extrema to be kept, rather than on the number of nodes with a
particular extinction attribute, which would be the case if we were interpreting
the extinction filter as an attribute filter.

The EF operation is the following: the n leaves with highest extinction values
concerning the increasing attribute being analyzed are chosen. The nodes in
the paths from these leaves to the the root are marked as to be kept. All the
other nodes are pruned. Since the contraction of max-tree nodes is a connected
filter, the EF is a connected filter. Also, it is idempotent, which means that
repeating the filtering procedure does not alter the result. The EF procedure is
illustrated in Figure 3. Suppose that $n = 3$ and nodes 7, 14 and 15 (the yellow
nodes) of Figure 3(a) are the leaves with highest extinction value according to
the attribute being analyzed. The nodes in the paths from these leaves to the
the root are marked in red, Figure 3(b), the remaining nodes are pruned. The
resulting tree is illustrated in Figure 3(c).

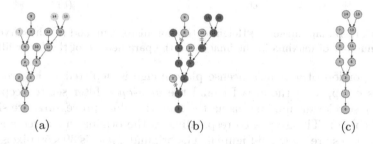

(a)   (b)   (c)

Fig. 3. (a) Original max-tree, the yellow nodes are the three nodes with highest extinction values. (b) Nodes in the path from the three leaves with highest extinction
values to the root are marked in red. (c) The result of the pruning of the nodes not
marked in red.

3   Setting the Number of Extrema with Attribute Filters

Attribute filters, such as the *hmax*, *vmax*, and the *area-open* are usually used
to simplify images in terms of the number of flat zones, but they may also
be used to set the number of extrema (maxima or minima) to be preserved
in an image. In order to do that, first the extinction values of the increasing
attribute being analyzed have to be calculated. Then, the extinction histogram
is computed. The relationship between the number of maxima (minima) in the
image and the parameter of the attribute filter is given by the curve attribute
value versus the number of maxima (minima) minus the cumulative distribution
of the extinction histogram. This curve may have discontinuities, since typically
in a real image there are extrema with the same extinction values, therefore

when using attribute filters often it is not possible to set the exact number of extrema to be preserved. This case is illustrated in Figure 4. The height extinction histogram of the Cameraman image, Figure 4(a), is shown in Figure 4(b). The curve height versus number of maxima is shown in Figure 4(c). The zoom in the plot highlights the discontinuities in this curve. For instance, if the image is filtered with the *hmax* filter with $h = 50$ the number of maxima in the resulting image will be 101, but if it is filtered with $h = 51$ the number of maxima in the resulting image will be 99. It is possible to enhance the method to avoid extinction ties, for simplicity, we prefer not to include this enhancement, as the differences in the filter results would be negligible for the comparative experiments.

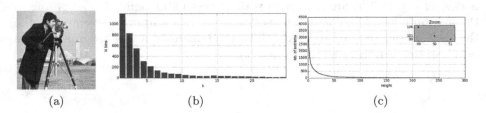

(a)   (b)   (c)

Fig. 4. (a) Cameraman image. (b) Height extinction histogram, and (c) the curve that relates the number of maxima in the image and the parameter h of the *hmax* filter.

A small portion of a vehicle license plate image is depicted in Figure 5(a). The results of applying the area EF and the *area-open* filter set to keep only the three most relevant minima, using the max-tree dual processing, are shown in Figure 5(b)-(c). The graphs corresponding to the original max-tree and the filtered max-trees are shown in Figure 6. The original image is 30×50 pixels. The original max-tree has 322 nodes, of which 54 are leaves. The max-tree filtered with the area EF has 174 nodes, and the max-tree filtered with the *area-open* has 72 nodes. This illustration makes clear the fact that attribute filters erode the heights of the maxima, while the EF preserves it.

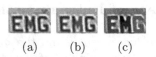

(a)   (b)   (c)

Fig. 5. (a) Original license plate image. (b) License plate filtered with the area EF, and (c) the *area-open* filter.

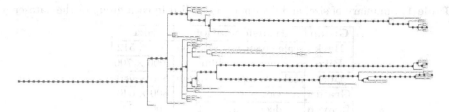

Fig. 6. Graph corresponding to the original max-tree (all nodes), the max-tree after the area EF (green squares and blue circles), and the max-tree after the area-opening filter (green squares)

4   Experiments

In this section, the processing time, structural similarity and simplification performance of AF and EF are analyzed. We also analyze their robustness using them as a pre-processing step before extracting Hessian-Affine and MSER regions and performing repeatability tests. In order to do that, it was used the benchmark dataset proposed in [19]. The dataset is composed of structured and textured scenes divided in eight groups, where in each group a different type of image transformation was applied, such as scale change, blur, and compression. A sample image of each group of the dataset is depicted in Figure 7, and the information concerning their size, and the deformation applied in each group of images are summarized in Table 1.

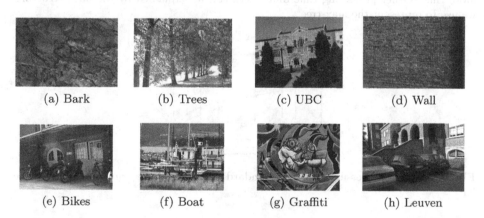

| (a) Bark | (b) Trees | (c) UBC | (d) Wall |

| (e) Bikes | (f) Boat | (g) Graffiti | (h) Leuven |

Fig. 7. Sample images of each group of the dataset

4.1   Processing Time

The processing time of attribute filters and extinction filters is dependent of the image size, the number of max-tree nodes, and the number of tree leaves. In order to compare

Table 1. Summary of size, and deformation applied in each group of the dataset

Group	Transformation	Size
Bark	scale change + rotation	765 × 512
Bike	increasing blur	1000 × 700
Boat	scale change + rotation	850 × 680
Graffiti	viewpoint angle	800 × 640
Leuven	decreasing light	900 × 600
Tree	increasing blur	1000 × 700
UBC	JPEG compression	800 × 640
Wall	viewpoint angle	1000 × 700

their processing times, we chose one sample image of each group in the dataset, and we cropped them so each image would be 512 pixels by 512 pixels. Each image was filtered using the *vmax* and the volume EF using connectivity $C4$ and set to preserve different percentages of extrema. For each percentage and each image of every group the filtering was repeated 20 times. The average processing time and its standard deviation is depicted in Figure 8. The processing time computed encompasses just the filtering of the max-tree, since the max-tree construction, and image restitution times are the same for both filters. The time was measured in a *Intel Core i7* processor with 8GB of memory and a clock frequency of $3.4GHz$. The plot shows that both attribute filters and extinction filters have similar processing times, when set to preserve the same number of extrema, and that the processing time varies little with the number of leaves preserved. The main differences in the EF and the AF algorithms is that the EF needs to sort the leaves extinction values from the highest value to the lowest, and the AF needs only to compute the cumulative extinction histogram, which can be done linearly, while the sorting complexity depends on the algorithm, but usually requires more time. Other processing time differences can be explained by the data structure we use to represent the max-tree[1].

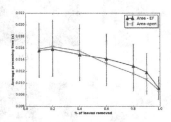

Fig. 8. Average processing time and standard deviation of the *vmax* and the volume EF

4.2   Simplification Performance and Structural Similarity

In order to evaluate the simplification performance, we computed the flat zones simplification rate given by the number of flat zones in the filtered image divided by the

[1] For more details see our max-tree, AF and EF implementation at
https://github.com/rmsouza01/iamxt

number of flat zones in the original image, and the number of max-tree nodes simplification rate, which is the same metric, but using the number of max-tree nodes instead of the number of flat zones. The images similarity was evaluated through the Structural Similarity (SSIM) index [20] between the original image and the filtered image. One sample image of each group in the dataset was used to compute these metrics. The mean values and their standard deviations are illustrated in Figure 9. As expected, AF have a greater simplification power than EF, since they erode the height of the extrema they keep (see Figure 6). On average it reduced almost 5% more the number of flat zones on the image, and twice the number of max-tree nodes when compared to EF. On the other hand, EF obtained a SSIM index 0.7% higher on average than AF.

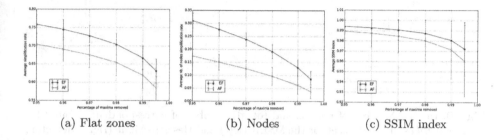

(a) Flat zones   (b) Nodes   (c) SSIM index

Fig. 9. (a) Average flat zones, (b) nodes simplification rates, and (c) SSIM index for different percentages of maxima removed

4.3 Robustness Evaluation through Repeatability Tests

In this subsection, it is evaluated the influence of AF and EF as a pre-processing step before the extraction of MSER and Hessian-Affine regions. We chose specifically these two detectors, in view of the fact that they achieved best results in the survey reported by Mikolajczyk et al. [19], and because of their distinct behaviors, MSER is a region detector, while Hessian-Affine is a point detector.

The protocol used to evaluate the results is the one proposed in [19]. The MSER and Hessian-Affine regions where computed using the same parameters and the binary files provided in [19] . In these paper, the experiments reported used the *area-open* and the area EF as pre-processing filters, but the the same conclusions also hold when using the height and volume attributes[2]. To compute the MSER regions, first the image was filtered using either the EF or the AF. Then, the MSER+ regions were computed. After that, the negative of the original image was filtered and the MSER- regions were computed. For the Hessian-Affine detector, the original image was filtered, then the negative of the filtered image was also filtered before extracting the Hessian-Affine regions.

The percentage of extrema kept was tested as being 10%, 5%, and 3% of the number of extrema of the original images. The number of regions and correspondences found for the group *Bikes* is depicted in Figure 10[3]. The plots shown indicate the overall

---

[2] The results using height and volume attributes are available at
http://adessowiki.fee.unicamp.br/adesso/wiki/code/MainPage/view/
[3] The graphics for all the groups of the dataset are available at
http://adessowiki.fee.unicamp.br/adesso/wiki/code/MainPage/view/

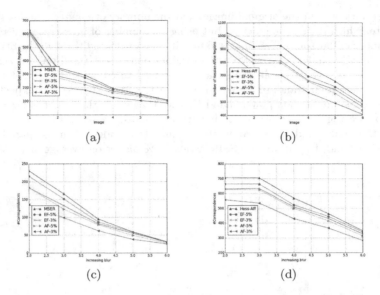

Fig. 10. (a)-(b) Number of regions and (c)-(d) number of correspondences found for the group *Bikes* of the dataset for the MSER (left) and Hessian-Affine (right) detectors

behavior of the filters in the whole dataset. EF preserve a significantly higher number of regions and correspondences, when compared to AF. The number of regions and correspondences found using EF and AF as pre-processing compared to the detectors applied directly to the unfiltered images is summarized in Table 2. Filtering 90% of the extrema using EF preserved on average practically 99% of the correspondences and simplified the number of max-tree nodes by a factor of 2.3, and filtering 95% preserved more than 94% of the correspondences and simplified the max-tree nodes by a factor of 3.2 when compared to the unfiltered results. Also, the EF set to preserve 3% of the extrema obtained on average a higher number of regions detected and matches, compared to the AF set to preserve 5% of the extrema.

Table 2. Summary of the number of regions and correspondences found using EF and AF as pre-processing. The values are percentages computed in relation to the detectors results applied directly to the unfiltered images.

Filter - % extrema kept	MSER		Hessian-Affine	
	$\Delta\#$Regions	$\Delta\#$Corresp.	$\Delta\#$Regions	$\Delta\#$Corresp
EF - 10%	-0.65%	+0.54%	-2.23%	-2.81%
AF - 10%	-3.13%	-1.50%	-8.22%	-8.72%
EF - 5%	-4.57%	-3.61%	-5.45%	-7.17%
AF - 5%	-12.78%	-13.93%	-17.37%	-18.25%
EF - 3%	-12.91%	-15.11%	-10.18%	-12.94%
AF - 3%	-36.64%	-41.56%	-25.83%	-27.45%

5   Conclusions

This paper presented a comparison between attribute filters and extinction filters using the height, area and volume attributes, when they are set to preserve the same or at least the closest value possible of extrema in the image. The methodology to set the number of image extrema using AF was given. It was seen that AF erode the height of the extrema kept. Both filters have similar processing times. AF filters have greater simplification power both in terms of number of flat zones and max-tree nodes, while EF achieve higher structural similarity scores. The repeatability tests performed showed that EF used as a pre-processing preserved a considerably higher number of regions and number of correspondences found by the affine region detectors than using AF as a pre-processing. Although AF has a higher power for simplifying the number of max-tree nodes, it also filters more relevant structures in the image. Therefore, we can conclude there is a trade-off. EF simplify a little less the max-tree structure than AF, but it is better with respect to simplification for recognition, since it preserves better the correspondences found by affine detectors. A max-tree toolbox with the source code to perform both filters is available on GitHub at https://github.com/rmsouza01/iamxt. As a future work, we would like to expand this comparison to non-increasing attributes and filters that use multiple attributes.

Acknowledgments. The authors thank FAPESP, CAPES and CNPQ for the financial support.

References

1. Salembier, P., Serra, J.: Flat zones filtering, connected operators, and filters by reconstruction. IEEE Transactions on Image Processing 4(8), 1153–1160 (1995), doi:10.1109/83.403422.
2. Salembier, P., Wilkinson, M.: Connected operators. IEEE Signal Processing Magazine 26(6), 136–157 (2009), doi:10.1109/MSP.2009.934154
3. Jones, R.: Connected Filtering and Segmentation Using Component Trees. Computer Vision and Image Understanding 75, 215–228 (1999)
4. Salembier, P., Oliveras, A., Garrido, L.: Antiextensive connected operators for image and sequence processing. IEEE Transactions on Image Processing 7(4), 555–570 (1998), doi:10.1109/83.663500
5. Souza, R., Rittner, L., Machado, R., Lotufo, R.: Maximal Max-Tree Simplification. In: 22nd International Conference on Pattern Recognition (ICPR), pp.3132–3137 (August 2014), doi: 10.1109/ICPR.2014.540
6. Berger, Ch., Geraud, T., Levillain, R., Widynski, N., Baillard, A., Bertin, E.:Effective Component Tree Computation with Application to Pattern Recognition in Astronomical Imaging. In: IEEE International Conference on Image Processing, vol. 4, pp. IV-41–IV-44 (2007), doi: 10.1109/ICIP.2007.4379949
7. Najman, L., Couprie, M.: Building the Component Tree in Quasi-Linear Time. IEEE Transactions on Image Processing 15(11), 3531–3539 (2006), doi:10.1109/TIP.2006.877518
8. Wilkinson, M., Gao, H., Hesselink, W., Jonker, J., Meijster, A.: Concurrent Computation of Attribute Filters on Shared Memory Parallel Machines. IEEE Transactions on Pattern Analysis and Machine Intelligence 30(10), 1800–1813 (2008), doi:10.1109/TPAMI.2007.70836

9. Carlinet, E., Geraud, T.: A Comparative Review of Component Tree Computation Algorithms. IEEE Transactions on Image Processing 23(9), 3885–3895 (2014), doi:10.1109/TIP.2014.2336551

10. Salembier, P., Oliveras, A., Maragos, P., Schafer, R., Butt, M.: Practical extensions of connected operators. In: Proc. Int. Symp. Mathematical Morphology, pp. 97–110 (1996)

11. Vincent, L.: Morphological Area Opening and Closings for Greyscale Images. In: Proc. Shape in Picture, 92-NATO Workshop (1992)

12. Vachier, C.: Extinction value: a new measurement of persistence. In: IEEE Workshop on Nonlinear Signal and Image Processing, vol. I, pp. 254–257 (1995)

13. Grimaud, M.: A new measure of contrast: dynamics. In: Proc. SPIE, vol. 1769, pp. 292–305 (1992)

14. Silva, A., Lotufo, R.: New Extinction Values from Efficient Construction and Analysis of Extended Attribute Component Tree. In: XXI Brazilian Symposium on Computer Graphics and Image Processing, October 12-15, pp. 204–211 (2008), doi:10.1109/SIBGRAPI.2008.8

15. Xu, Y., Geraud, T., Najman, L.: Morphological filtering in shape spaces: Applications using tree-based image representations. In: 21st International Conference on Pattern Recognition (ICPR), November 11-15, pp. 485–488 (2012)

16. Matas, J., Chum, O., Urban, M., Pajdla, T.: Robust wide-baseline stereo from maximally stable extremal regions. In: British Machine Vision Conference, pp. 384–393 (2002)

17. Donoser, M., Bischof, H.: Efficient Maximally Stable Extremal Region (MSER) Tracking. In: IEEE Computer Society Conference on Computer Vision and Pattern Recognition, June 17-22, vol. 1, pp. 553–560 (2006), doi: 10.1109/CVPR.2006.107

18. Mikolajczyk, K., Schmid, C.: An affine invariant interest point detector. In: Heyden, A., Sparr, G., Nielsen, M., Johansen, P. (eds.) ECCV 2002, Part I. LNCS, vol. 2350, pp. 128–142. Springer, Heidelberg (2002)

19. Mikolajczyk, K., Tuytelaars, T., Schmid, C., Zisserman, A., Matas, J., Schaffalitzky, F., Kadir, T., Gool, L.: A comparison of affine region detectors. International Journal of Computer Vision 65 (2005)

20. Wang, Z., Bovik, C., Sheikh, H., Simoncelli, E.: Image quality assessment: from error visibility to structural similarity. IEEE Transactions on Image Processing 13(4), 600–612 (2004)

Inner-Cheeger Opening and Applications

Santiago Velasco-Forero[✉]

Centre of Mathematical Morphology, Mines Paris Tech, Paris, France
velasco@cmm.ensmp.fr

Abstract. The aim of this paper is to study an optimal opening in the sense of minimize the relationship perimeter over area. We analyze theoretical properties of this opening by means of classical results from variational calculus. Firstly, we explore the optimal radius as attribute in morphological attribute filtering for grey scale images. Secondly, an application of this optimal opening that yields a decomposition into meaningful parts in the case of binary image is explored. We provide different examples of 2D, 3D images and mesh-points datasets.

Keywords: Attribute filter · 3D Shape · 2D Shape · Cheeger Set

1 Introduction

Many of the methods in mathematical morphology study the relationship between size/shape of objects via max-plus convolution by structuring elements (SE) [1]. One can realize that openings and closings with square, disk or hexagon SEs, are often good enough for some filtering tasks. However, if the structuring elements are able to adapt their shapes and sizes to the image content, some enhancement properties are improved [2,3]. This intuition leads to propose *area openings*[4], and more generally, to introduce *attributes openings*[5] . Recently many other attribute filters have been proposed to more specific problems [6,7,8]. In this paper, we introduce a new attribute named "Inner-Cheeger opening" which is, for a binary shape $\mathbf{S}$, the size t, such that an opening by a disk of t minimize the relationship perimeter over area: t is named the *Cheeger constant* of $\mathbf{S}$ in the theory of Variational Calculus [9]. Additionally, we explore the application of this operator in the context of decomposition for two-dimensional (2D) and three dimensional (3D) shapes into meaningful parts, which is a challenging problem in image processing [10], pattern recognition [11,12], remote sensing [13], and computer vision [14]. Thus, we propose a new shape decomposition method, called *Inner-Cheeger shape decomposition*, denoted by ICSD. This method is characterized by a sequence of Inner-Cheeger openings. Finally, the experimental section includes some examples of 2D, 3D images and mesh-points datasets to illustrate the interest of our method.

2 Inner-Cheeger Opening

The theory of sets of finite perimeter provides a particularly well-suited framework for studying the existence, symmetry, regularity, and structure of minimizers in

© Springer International Publishing Switzerland 2015
J.A. Benediktsson et al. (Eds.): ISMM 2015, LNCS 9082, pp. 75–85, 2015.
DOI: 10.1007/978-3-319-18720-4_7

those geometric variational problems in which surface area is minimized under a volume constraint [9]. In this paper, we consider the follow variational problem and its application to image processing and shape characterization.

2.1 Formulation

Given a domain $\mathbf{S}$ of $\mathbb{R}^d$, with $d \geq 2$, one is asked to find the *Cheeger constant* of $\mathbf{S}$, defined as:

$$h(\mathbf{S}) := \min_{\mathbf{X} \subset \mathbf{S}} \frac{\text{Per}(\mathbf{X})}{\text{Area}(\mathbf{X})}, \tag{1}$$

where $\text{Area}(\mathbf{X})$ is the (d)-dimensional volume of $\mathbf{X}$ and $\text{Per}(\mathbf{X})$ is the perimeter of $\mathbf{X}$ or $(d-1)$-dimensional measure. It is important to note that the minimum in (1) is taken over all nonempty sets of finite perimeter contained in $\mathbf{S}$. Thus, any minimiser of (1) is named *Cheeger set* of $\mathbf{S}$ and it is denoted by $\mathbf{Ch}(\mathbf{S})$. Despite this simple formulation of (1), many no trivial questions arise, for instance, the computation, and estimation of $h(\mathbf{S})$ and the characterization of Cheeger sets of $\mathbf{S}$. They have been the main subject of many recent research papers [15,16]. The following are some of the basic properties which are well-known about the Cheeger problem in (1), the proofs of them can be found for instance in [15,16,17,18].

Proposition 1. *Let $\mathbf{S}_1, \mathbf{S}_2 \subseteq \mathbb{R}^d$ be bounded, open sets. Then the following properties hold.*

1. *(Existence) There exists a (possibly non-unique) Cheeger set $\mathbf{X} \subseteq \mathbf{S}$.*
2. *(Decreasing) If $\mathbf{S}_1 \subset \mathbf{S}_2$ then $h(\mathbf{S}_1) \geq h(\mathbf{S}_2)$*
3. *(Isometry) For any $\lambda > 0$ and any isometry $T : \mathbb{R}^d \to \mathbb{R}^d$, one has $h(T(\mathbf{S})) = \frac{1}{\lambda} h(\mathbf{S})$*
4. *(Intersection with the boundary) The minimum in (1) is attained at a subset $\mathbf{X} \subseteq \mathbf{S}$ such that $\partial \mathbf{X}$ intersect $\partial \mathbf{S}$.*
5. *(Closure with union and intersection) If $\mathbf{X}_1$ and $\mathbf{X}_2$ are Cheeger in $\mathbf{S}$, then $\mathbf{X}_1 \cup \mathbf{X}_2$ and $\mathbf{X}_1 \cap \mathbf{X}_2$ (if it is not empty) are also Cheeger in $\mathbf{S}$.*

Many other properties can be found in [16,18]. However, we have included only the most relevant for our work.

2.2 Convex Case

In the case of $\mathbf{S}$ is convex, we can add to the list of properties in Proposition 1, the uniqueness and convexity of Cheeger sets, which has been proved in [16]. However, finding the Cheeger set of a given $\mathbf{S}$ is a difficult task. This task is simplified if $\mathbf{S}$ is a convex set and $d = 2$. In this situation is follows from [17], Theorem 3.32 that Cheeger $= \mathbf{Ch}(\mathbf{S})$ is convex and uniquely defined as the union of a set of disks of suitable radius [15]. Thus, the Cheeger set in $\mathbf{S}$ is unique and is identified by the following theorem.

Theorem 1. *[15,19] Let* $\mathbf{S} \subseteq \mathbb{R}^2$ *be a nonempty bounded convex set. There exist a unique value* $t > 0$ *such that* $\mathbf{Area}(\mathbf{S}^t) = \pi t^2$. *Then* $h(\mathbf{S}) = \frac{1}{t}$ *and the* $\mathbf{Ch}(\mathbf{S}) = \delta_{B(t)}(\mathbf{S}^t)$, *where* $B(t)$ *denotes a ball of radius* t *and* $\mathbf{S}^t = \{x \in \mathbf{S} : dist(x, \partial \mathbf{S}) > t\}$ *is called the* inner Cheeger set *of* $\mathbf{S}$.

The proof of Theorem 1 is basically based on the *Steiner's formulae* in the case of $d = 2$ [5] and it is included in Appendix 1.

Corollary 1. *For a nonempty bounded convex set* $\mathbf{S} \in \mathbb{R}^2$, *the Cheegger set* $\mathbf{Ch}(\mathbf{S})$ *is an opening with a disk of radius given by the Theorem 1 as structuring element.*

In the sequel, we use the name "Inner-Cheeger opening" to the opening defined in Corollary 1.

2.3 Small Perturbations

Additionally, it is important to note that the uniqueness of Cheeger sets holds up to arbitrary small perturbations of $\mathbf{S}$ [20].

Theorem 2. *[20] Let* $\mathbf{S} \subseteq \mathbb{R}^N$ *be an open set with finite volume. Then for any compact set* $K \in \mathbf{S}$, *there exist a bounded open set* $\mathbf{S}_K \subset \mathbf{S}$ *such that* $K \subset \mathbf{S}_K$ *and* $\mathbf{S}_K$ *has a unique Cheeger set.*

A visual example of this property is illustrated in Fig. 1.

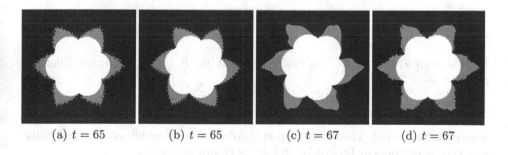

(a) $t = 65$ (b) $t = 65$ (c) $t = 67$ (d) $t = 67$

Fig. 1. Example of invariance of Inner-Cheeger opening to small perturbations of the shape. (b) and (d) are perturbations of (a) and (c) respectively.

2.4 Comparison with Skeleton Transformation

The *skeleton transformation* is a widely used transformation in the field of image processing. The definition of skeleton was introduced by [21], proposing the grass fire analysis: the skeleton consists of the points where different forefronts intersect, or *quench points*. Later, another formal definition of the skeleton was proposed by Calibi in 1965 [22] which relies on the concept of *maximal ball*. The skeleton is defined as the set of the maximal balls. In [22], it was proved that the notion of quench points and centers of maximal balls are equivalent. According to Chapter one of [1], the notion of maximal ball has been known since the 30s [23,24]. A skeleton transformation has to obey the following properties to make its results convenient for the global representation of objects [25]: *Homotopy:* the skeleton transform must preserve the topology of the initial object; *Rotation invariance:* the skeleton transform of a rotated shape should be the rotated of the skeleton; *Reconstruction:* the initial shape should be reconstructed from the skeleton transform. For continuous images, the skeleton $sk(\mathbf{S})$ of a shape $\mathbf{S}$ is given by the *Lantuéjoul's formulate* [26,27] defined as follows:

$$sk(\mathbf{S}) = \bigcup_{\rho>0} \bigcap_{\mu>0} [\varepsilon_{\rho B}(\mathbf{S}) - (\gamma_B(\varepsilon_{\rho B}(\mathbf{S})))] \tag{2}$$

A ball B included in $\mathbf{S}$ is said to be *maximal* if and only if there exist no other ball included in $\mathbf{S}$ and containing B, $\forall B'$ball, $B \subseteq B' \subseteq \mathbf{S} \Rightarrow B' = B$,i.e., a ball is maximal in the shape if it is not included in any single other ball in the shape. We denote $MBT(\mathbf{S}) = \{x \in \mathbf{S}; \exists$ a maximal ball centered at $x\}$, the set of all the centers of the maximal balls included in $\mathbf{S}$, and r_x the corresponding radius of it associated maximal ball. The skeleton $sk(\mathbf{S})$ of a set $\mathbf{S}$ is then defined as the set of the centers of its maximal balls:

$$sk(\mathbf{S}) = \{x \in \mathbf{S}, \exists r_x \geq 0, B(x, r_x) \text{ maximal ball of } \mathbf{S}\}. \tag{3}$$

This collection of maximal balls $MBT(\mathbf{S})$ is equivalent to the shape $\mathbf{S}$ itself since one has the reconstruction formula:

$$\mathbf{S} = \bigcup_{x \in MBT(\mathbf{S})} \delta_{r_x}(x) \tag{4}$$

In this representation, we can easily see that for $\mathbf{S} \subseteq \mathbb{R}^2$ the Inner-Cheeger opening is

$$\mathbf{Ch}(\mathbf{S}) = \bigcup_{x \in MBT(\mathbf{S}), r_x \geq t} \delta_{r_x}(x), \tag{5}$$

where t is defined by Theorem 1. An example of the collection of maximal balls and the correspondent Inner-Cheeger set is shown in Fig. 2.

3   Applications

In this section we present some examples to illustrate our methods. We present two applications: Attribute filtering and shape decomposition in 2D, 3D and mesh shapes.

3.1   Inner-Cheeger Set as an Attribute Filter

Many works about attribute filters have been presented in the literature [28]. Most of them address the problem of filter-out target objects and preserving the

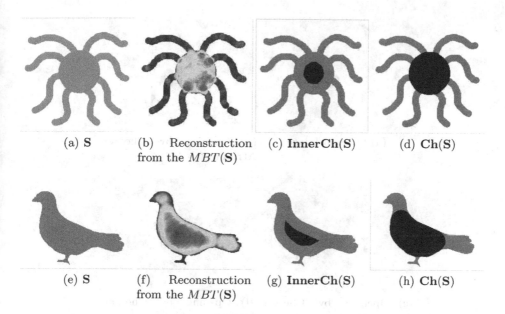

(a) **S**   (b) Reconstruction   (c) **InnerCh(S)**   (d) **Ch(S)**
         from the $MBT(\mathbf{S})$

(e) **S**   (f) Reconstruction   (g) **InnerCh(S)**   (h) **Ch(S)**
         from the $MBT(\mathbf{S})$

Fig. 2. Comparison of Inner-Cheeger opening and Skeleton transformation for two images. Colours in (b) and (f) show of the number of balls in the reconstruction of MBT.

contours of non-interesting objects. In this first application, we start with the description of a grey scale image $\mathbf{I}$ by its upper level set decomposition:

$$\mathbf{I}(x) = \sup(i : \mathbf{I}(x) \geq i \text{ is } \mathtt{true}), \qquad (6)$$

where is discrete images $i = \{0, 1, \dots, t_{max}\}$ and t_{max} stands by the maximum grey scale value of the image. Now, following [5], we can define an attribute filter on $\mathbf{I}$ by the well-known *threshold decomposition principle*:

$$\psi(\mathbf{I})(x) = \sup(i : \Psi(\mathbf{I}(x) \geq i) \text{ is } \mathtt{true}) \qquad (7)$$

where Ψ is a binary attribute filter. The advantage of this representation is the fact that most of the properties that are known to hold for Ψ are also true for its grey-level counterpart ψ. Mainly, if Ψ is an increasing criterion, it was proved in [5] that ψ is a valid opening in the morphological sense. We propose to use the optimal value of t is Theorem 1 as attribute to describe connected components in the threshold decomposition in Eq. 7. In proposition 1, property 2, we have stated that our attribute is increasing, which allows the use a fast implementation (by reducing the search space in Theorem 1), for instance, by means of a max-tree structure [29]. Some example of this attribute filter are illustrated in Fig. 3 for different values of t. This attribute opening preserves contours of object with largest Cheeger set, i.e., objects with large sphericity.

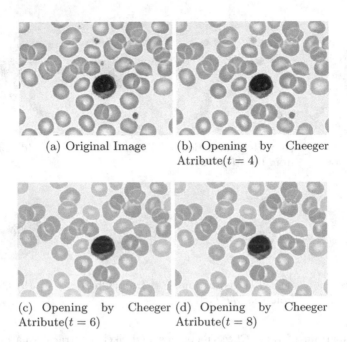

(a) Original Image   (b) Opening by Cheeger Atribute($t = 4$)

(c) Opening by Cheeger Atribute($t = 6$)   (d) Opening by Cheeger Atribute($t = 8$)

Fig. 3. Inner-Cheeger opening for different values of proposed attribute

3.2   Shape Decomposition

Instead of characterizing an object as a whole, *part-based representation* deals with a representation where an object is decompose it into a number of "natural parts" [30]. The aim of this shape decomposition is to simplify a given shape into meaningful parts to make easier its analysis. Many morphological-based operators approaches [31,32,33] have been formulated with the intuition of finding the "best" opening (with a ball as structuring element) by means of an index considering the size "part/shape" ratio and "part/convex hull" ratio. However, these methods are often hard to implement and/or do require the solution of difficult optimization problem.

In this section we propose to use the Inner-Cheeger decomposition to perform shape decomposition. For a given shape **S**, we firstly compute the **Ch(S)** and successively the same transformation is applied to each connected component of the difference between **S** and **Ch(S)**. We note that many of the connected components can have small areas, so a threshold parameter is included to obtain only meaningful parts of the object. We use the term *Inner-Cheeger shape decomposition* (ICSD) to refer this sequential separation, due to the fact that in the case of non-convex shapes, one cannot say that this procedure find a valid Cheeger set. However, as it is illustrated in Fig.4, the decomposition determines relevant parts and as in most of the distance based transformation, the result is robust to articulate transformations. In addition, in our case it is robust to small perturbations of the original shape.

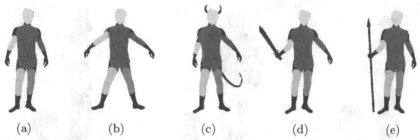

Fig. 4. Example of decomposition by Iterative Cheeger Openings (2 Levels). Color are associated with the order of discovery in the iteration of ICSD. Note that results are robust to modification in the human shape.

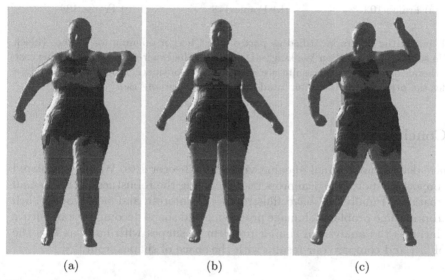

Fig. 5. Three examples of proposed decomposition for 3D shapes. Colour are associated with the order of discovery in the iteration of ICSD. Note that results are robust to changes in the pose of the human shape.

At this point, we can use a similar approach to 3D shapes in Fig. 5 and 3D points clouds in Fig. 6. However, the optimal value of t in Theorem 1 cannot be applied directly for 3D shapes. Thus, assuming that the inner set of a convex 3D shape if composed by the union of balls, we can obtain an equivalent expression for three dimensional objects by using the Steiner's formula as it is shown in Appendix. Accordingly, in 3D shape, we should look for the radius t that $\texttt{Vol}(\mathbf{S}^t) + \texttt{MeanB}(\mathbf{S}^t)t^2 \left[\frac{1}{2\pi} - 1\right] = \frac{8\pi r^3}{3}$, where $\texttt{MeanB}(\mathbf{S})$ stands for mean breadth of the binary shape $\mathbf{S}$. A Fig. 5 illustrates the results of ICSD for three postures for the same 3D human model. Additionally, we explore the application of our approach in mesh of points in three different frames of "Ben Walking Scene"[1].

---

[1] http://4drepository.inrialpes.fr/pages/home

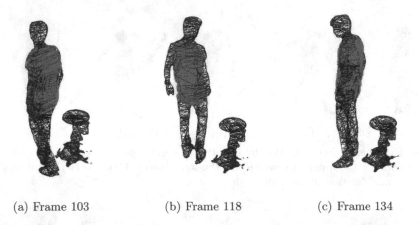

(a) Frame 103   (b) Frame 118   (c) Frame 134

Fig. 6. In red the points identified as part of the Cheeger opening for meshes (black lines) in some frames of "Ben Walking" 4D dataset. The Inner-Cheeger set in this case is the thorax of "Ben". The results are quite robust considering that the input data includes the presence of the "chair" and different postures of "Ben"

4   Conclusion

We have defined an optimal opening via Inner-Cheeger sets. We have explored some important morphological properties including the robustness against small perturbations. Finally, we have illustrated the interest and behavior of such operators in some problems of image processing and shape decomposition. Future work includes the analysis of similar problem in shapes with holes as it is the case of [34] and compare our results with the space of shapes from [35].

A   Appendix

We follow the results from [15,19] for 2D shapes to obtain:

$$h(\mathbf{S}) = \frac{1}{r} = \frac{\mathtt{Per}(\delta_{rB}(\mathbf{S}))}{\mathtt{Area}(\delta_{rB}(\mathbf{S}))} = \frac{\mathtt{Per}(\mathbf{S}) + 2\pi r}{\mathtt{Area}(\mathbf{S}) + \mathtt{Per}(\mathbf{S})r + \pi r^2}$$
$$\mathtt{Area}(\mathbf{S}) + \mathtt{Per}(\mathbf{S})r + \pi r^2 = \mathtt{Per}(\mathbf{S})r + 2\pi r^2$$
$$\mathtt{Area}(\mathbf{S}) = \pi r^2$$

In the case of $d = 3$, the Steiner's formula gives explicitly for a compact convex:

$$\mathtt{Vol}(\delta_{rB}(\mathbf{S})) = \mathtt{Vol}(\mathbf{S}) + \mathtt{Sur}(\mathbf{S})r + \mathtt{MeanCurv}(\mathbf{S})r^2 + \frac{4\pi}{3}r^3 \qquad (8)$$

$$\mathtt{Sur}(\delta_{rB}(\mathbf{S})) = \mathtt{Sur}(\mathbf{S}) + 2\pi\mathtt{MeanCurv}(\mathbf{S})r + 4\pi r^2 \qquad (9)$$

$$\mathtt{MeanCurv}(\delta_{rB}(\mathbf{S})) = \mathtt{MeanCurv}(\mathbf{S}) + 4\pi r \qquad (10)$$

Additionally, it is important to note that the *mean breadth*, denoted by MeanB($\mathbf{I}$), connects the integral of the mean curvature MeanCurv($\mathbf{S}$) for an object $\mathbf{I}$ whose boundary is of class $\mathbb{C}^2$ by the formula, MeanB($\mathbf{I}$) = 2πMeanCurv($\mathbf{S}$).

$$h(\mathbf{S}) = \frac{1}{r} = \frac{\text{Sur}(\delta_{rB}(\mathbf{S}))}{\text{Vol}(\delta_{rB}(\mathbf{S}))} = \frac{\text{Sur}(\mathbf{S}) + 2\pi\text{MeanCurv}(\mathbf{S})r + 4\pi r^2}{\text{Vol}(\mathbf{S}) + \text{Sur}(\mathbf{S})r + \text{MeanCurv}(\mathbf{S})r^2 + \frac{4\pi r^3}{3}}$$

and after simplification, we obtain $\frac{8\pi r^3}{3}$ = Vol($\mathbf{S}$) + MeanB($\mathbf{S}$)$r^2\left[\frac{1}{2\pi} - 1\right]$. We remark that the percolation analysis in three-dimensional objects in [36] uses similar results in random models with spheres and Poisson polyhedra as primary grains.

References

1. Najman, L., Talbot, H.: Mathematical morphology: from theory to applications p–2010. ISTE-Wiley (2010)
2. Angulo, J., Velasco-Forero, S.: Structurally adaptive mathematical morphology based on nonlinear scale-space decompositions. Image Analysis & Stereology 30(2), 111–122 (2011)
3. Ćurić, V., Landström, A., Thurley, M.J., Hendriks, C.L.L.: Adaptive mathematical morphology–a survey of the field. Pattern Recognition Letters 47, 18–28 (2014)
4. Vincent, L.: Grayscale area openings and closings, their efficient implementation and applications. In: Proceedings of the Conference on Mathematical Morphology and its Applications to Signal Processing, pp. 22–27 (May 1993)
5. Breen, E.J., Jones, R.: Attribute openings, thinnings, and granulometries. Computer Vision and Image Understanding 64(3), 377–389 (1996)
6. Talbot, H., Appleton, B.: Efficient complete and incomplete path openings and closings. Image Vision Comput. 25(4), 416–425 (2007)
7. Serna, A., Marcotegui, B.: Attribute controlled reconstruction and adaptive mathematical morphology. In: Hendriks, C.L.L., Borgefors, G., Strand, R. (eds.) ISMM 2013. LNCS, vol. 7883, pp. 207–218. Springer, Heidelberg (2013)
8. Morard, V., Decencière, E., Dokládal, P.: Efficient geodesic attribute thinnings based on the barycentric diameter. Journal of Mathematical Imaging and Vision 46(1), 128–142 (2013)
9. Maggi, F.: Set of Finite Perimeter and Geometric Variational Problems. Cambridge University Press (2012)
10. Mehtre, B.M., Kankanhalli, M.S., Lee, W.F.: Shape measures for content based image retrieval: a comparison. Information Processing & Management 33(3), 319–337 (1997)
11. Cerri, A., Biasotti, S., Abdelrahman, M., Angulo, J., Berger, K., Chevallier, L., El-Melegy, M., Farag, A., Lefebvre, F., Andrea, Giachetti, et al.: Shrec'13 track: retrieval on textured 3d models. In: Proceedings of the Sixth Eurographics Workshop on 3D Object Retrieval, pp. 73–80. Eurographics Association (2013)
12. Velasco-Forero, S., Angulo, J.: Statistical shape modeling using morphological representations. In: 2010 20th International Conference on Pattern Recognition (ICPR), pp. 3537–3540. IEEE (2010)
13. Gueguen, L.: Classifying compound structures in satellite images: A compressed representation for fast queries. Transactions on Geoscience and Remote Sensing, 1–16 (2014)

14. Younes, L.: Spaces and manifolds of shapes in computer vision: An overview. Image and Vision Computing 30(6), 389–397 (2012)
15. Kawohl, B., Lachand-Robert, T.: Characterization of Cheeger sets for convex subsets of the plane. Pac. J. Math. 225(1), 103–118 (2006)
16. Alter, F., Caselles, V.: Uniqueness of the cheeger set of a convex body. Nonlinear Analysis 70(1), 32–44 (2009)
17. Stredulinsky, E., Ziemer, W.: Area minimizing sets subject to a volume constraint in a convex set. The Journal of Geometric Analysis 7(4), 653–677 (1997), http://dx.doi.org/10.1007/BF02921639
18. Leonardi, G.P., Pratelli, A.: On the cheeger sets in strips and non-convex domains. arXiv preprint arXiv:1409.1376 (2014)
19. Caselles, V., Alter, A.C.F.: Evolution of characteristic functions of convex sets in the plane by the minimizing total variation flow. Interfaces and Free Boundaries 7 (2005)
20. Caselles, V., Chambolle, A., Novaga, M.: Some remarks on uniqueness and regularity of cheeger sets. Rend. Semin. Math. Univ. Padova 123, 191–201 (2010)
21. Blum, H.: A transformation for extracting new descriptors of shape. In: Proceedings of a Symposium on Models for the Perception of Speech and Visual Forms. MIT, Boston (November 1967)
22. Calabi, L.: A study of the skeleton of plane figures. Parke Mathematical Laboratories (1965)
23. Durand, G.: Théprie des ensembles. points ordinaires et point singuliers des enveloppes de sphères. Comptes-rendus de l'Acad'emie de Sciences 190, 571–573 (1930)
24. Bouligand, G.: Introduction à la gémétrie infinitésimale directe. Vuibert (1932)
25. Malandain, G., Fernández-Vidal, S.: Euclidean skeletons. Image and Vision Computing 16(5), 317–327 (1998)
26. Lantuejoul, C.: La squelettisation et son application aux mesures topologiques des mosaiques polycristallines. Ph.D. dissertation, Ecole des Mines de Paris (1978)
27. Lantuejoul, C.: Skeletonization in quantitative metallography. Issues of Digital Image Processing 34(107-135), 109 (1980)
28. Salembier, P., Serra, J.: Flat zones filtering, connected operators, and filters by reconstruction. IEEE Transactions on Image Processing 4(8), 1153–1160 (1995)
29. Carlinet, E., Géraud, T.: A comparison of many max-tree computation algorithms. In: Hendriks, C.L.L., Borgefors, G., Strand, R. (eds.) ISMM 2013. LNCS, vol. 7883, pp. 73–85. Springer, Heidelberg (2013)
30. Hoffman, D.D., Richards, W.A.: Parts of recognition. Cognition 18(1), 65–96 (1984)
31. Xu, J.: Morphological decomposition of 2-d binary shapes into convex polygons: A heuristic algorithm. IEEE Transactions on Image Processing 10(1), 61–71 (2001)
32. Yu, L., Wang, R.: Shape representation based on mathematical morphology. Pattern Recognition Letters 26(9), 1354–1362 (2005)
33. Kim, D.H., Yun, I.D., Lee, S.U.: A new shape decomposition scheme for graph-based representation. Pattern Recognition 38(5), 673–689 (2005)
34. Liu, G., Xi, Z., Lien, J.-M.: Dual-space decomposition of 2d complex shapes. In: 27th IEEE Conference on Computer Vision and Pattern Recognition (CVPR), Columbus, OH. IEEE (June 2014)

35. Xu, Y., Géraud, T., Najman, L.: Morphological filtering in shape spaces: Applications using tree-based image representations. In: 2012 21st International Conference on Pattern Recognition (ICPR), pp. 485–488. IEEE (2012)
36. Jeulin, D.: Random structures in physics. In: Space, Structure and Randomness, Contributions in Honor of Georges Matheron in the Fields of Geostatistics, Random Sets, and Mathematical Morphology. Lecture Notes in Statistics, vol. 183, pp. 183–222. Springer, Heidelberg (2005)

Tracking Sub-atomic Particles
Through the Attribute Space

Mohammad Babai[1], Nasser Kalantar-Nayestanaki[1],
Johan G. Messchendorp[1], and Michael H.F. Wilkinson[2]([✉])

[1] KVI-Center for Advanced Radiation Technology, Groningen, The Netherlands
M.Babai@rug.nl
[2] Johann Bernoulli Institute for Mathematics and Computing Science,
University of Groningen, The Netherlands
m.h.f.wilkinson@computer.org

Abstract. In this paper, we present the results of an application of attribute space morphological filters for tracking sub-atomic particles in magnetic fields. For this purpose, we have applied the concept of attribute space and connectivity to the binary images produced by charged particles passing through the tracking detector for the future experiment PANDA. This detector could be considered as an undirected graph with irregular neighbourhood relations. For this project, we rely only on the detector geometry. In addition, we have extended the graph to estimate the z-coordinates of the particle paths. The result is an $O(n^2)$, proof of concept algorithm with a total error of approximately 0.17. The results look promising; however, more work needs to be done to make this algorithm applicable for the real-life case.

Keywords: Attribute space connectivity · Orientation based segmentation · Irregular graph · Sub-atomic particle tracking · Graph morphology

1 Introduction

Many image-analysis problems involve the recognition of thin, line-shaped, oriented structures. When some elongated thin, bright structure needs to be segmented, one possible approach is to remove parts of the image that are neither elongated nor thin. A standard approach is to use of a supremum of openings, where line-shaped structuring elements in different directions could be used [1]. A direct application of this idea will lead to an inefficient implementation of such an operator. One solution to this was proposed by Talbot et al. in [2]. They propose an ordered algorithm for implementing path operators for both complete and incomplete paths with logarithmic complexity as a function of the length of the flexible structuring element used. Most of the proposed solutions to solve the path recognition problems cannot deal with overlapping structures and are not directly capable of grouping pixels into disjoint paths. To solve the problem of overlapping structures, Wilkinson proposed in [3] a new concept of connectivity in higher dimensional spaces, referred to as *attribute spaces*. It is

© Springer International Publishing Switzerland 2015
J.A. Benediktsson et al. (Eds.): ISMM 2015, LNCS 9082, pp. 86–97, 2015.
DOI: 10.1007/978-3-319-18720-4_8

shown that a transformation of an image into a higher dimensional space, for example in an orientation based attribute space, using new computed attributes could be a solution to segment overlapping structures in disjoint segments.

A systematic theory for the construction of morphological operators on graphs was introduced by Toet et al. [4] where structural information could be extracted from graphs using pre-defined probes (structuring graphs). Dilations and erosions were constructed using the graphs neighborhood function and in combination with the probes one could define openings and closings operators on graphs [5].

Here, we present the results of an application of attribute space connected filtering, inspired by graph morphology for recognizing charged particle tracks through the Straw Tube Tracker (STT) for the future experiment PANDA (anti-Proton ANnihilations at DArmstadt). PANDA is one of the experiments at the future Facility for Antiproton and Ion Research (FAIR), which is currently under construction in Darmstadt, Germany. One of the objectives of this experiment is to study the structure of hadronic matter via the annihilation of antiprotons with protons at interaction rates up to 20 MHz. During these collisions, various particles with a large momentum range are produced. The trajectories of charged particles are reconstructed using tracking detectors placed inside a solenoid magnetic field. The curvatures of the reconstructed tracks are used to obtain the momenta of the corresponding particles. The PANDA tracking system consists of central and forward trackers. Here, we consider only one of the central tracker detectors, namely the STT. To cope with the high event rate, the data are processed at runtime and reduced by a factor of about 10^3. To achieve such a reduction factor, intelligent and fast algorithms need to be used that are capable of reconstructing the particle tracks in-situ. There are a number of conventional methods with high precision being used for the recognition of the particle paths through the space [6], but at present none of them is suited for online applications because of their computational complexity. These methods rely on transforming the image points into other spaces (Riemann space, Hough transformation), recognizing the structures, back transformation and fitting curves through the collected space-points where one needs to rely on drift times and corresponding calibrations. In our approach, we use only the geometrical information of the detector. One of the major advantages of our method is that it does not depend on the drift time of the charge through the tubes. Moreover, our algorithm is conceptually easy to implement on embedded architectures such as FPGAs and GPUs with $O(n^2)$ complexity in the number of active pixels.

In this paper, in sec. 2 the geometry of the STT subsystem is discussed, sec. 3 will give a brief summary of the basic concepts of filtering and connectivity, in sec. 3.1 the attribute space and attribute space filters are briefly described. In sec. 4 we present our method and finally we discuss the results and future works in sec. 5.

2 Geometry of the STT

The STT subsystem is the main tracking detector for charged particles and contains 4542 single straw tubes, arranged in a cylindrical volume around the beam-target interaction point. The straws are mounted in the system in two different ways. The axial straws which are parallel to the z-axis (the beam line) and skewed straws which are skewed by a few degrees ($+2.9°$ or $-2.9°$) to the axial direction and are meant to measure the z information of the track. All straws have the same inner diameter of 10 mm and a length of 1500 mm, except for a few outer straws in the border region of each skewed layer, which have a reduced length. The read-out system is designed such that all tubes are read from one side. Figure 1 left shows a schematic view of the STT in the $xy-$plane and the right panel the 3D layout of the STT subsystem [6].

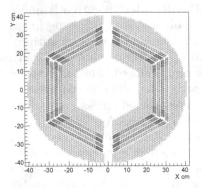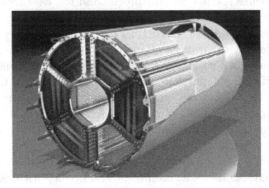

Fig. 1. Left: the front view (along beam direction) of the STT in the xy-plane. The skewed straws are shown in red and blue and green dots represent the axial straws. Right: a 3D view of the STT subsystem.

4A simulation, digitization and reconstruction software package has been developed for this detector. The digitization software provides a realistic model of the hardware readout of the foreseen detector. Here, we use only the data from simulation and digitization parts of this software package. For further details, we refer to the STT technical design report [6].

2.1 STT Graph

The geometry of the STT detector contains the three dimensional coordinates of the center-point of each tube, its half length, the list of its direct neighbors and a direction-vector which describes the slope of the read-out wire; the read-out wire has the same length as the tube itself and passes through the center of the tube. Because of the skewed tubes, the number of neighbors is not equal for all nodes in the graph; it varies between two and 22 neighbours. This way, the detector could be considered as a graph of pixels with varying neighborhood relations. The nodes are described by the tubes and the edges by the neighbouring relations

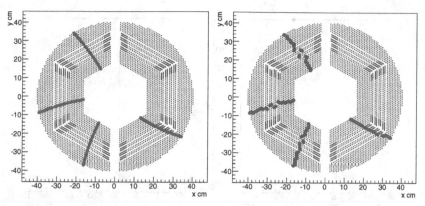

Fig. 2. The position of the STT tubes is shown in blue. Left: the true hit positions (red points) of the particles. Right: red points show the position of the reconstructed hits.

between the nodes. When a charged particle travels through space in a magnetic field, it hits a number of tubes which are turned on while the remaining nodes remain off. This pattern could be considered as a binary image of a number of thin, curved, path-shaped patterns. The curvature, path lengths and multiplicity of the paths depend on the particle type, its momentum and the strength of the magnetic field. An example of such paths is shown in fig. 2. One can see that the axial (z-parallel) straws produce an accurate position in the xy-plane and the skewed nodes have a maximum displacement of ± 3 cm based on their lengths and direction. The latter is due to the fact that the exact hit positions, along the tubes, are not defined. Further, as a consequence of the readout and detector efficiency, the generated paths might have gaps of one or more missing pixels (tubes). These effects are visible in the right panel of fig. 2.

To have a better gradual transition between the layers with a different slope, we have extended the STT-grid by 20398 virtual nodes. These nodes are positioned between the tubes with different slopes at the center of the "virtual" intersection volumes. The virtual nodes have exactly two neighbours and have well defined xyz-coordinates. They are turned on if and only if both parent tubes are hit. Adding these nodes reduces the size of the hit displacements in

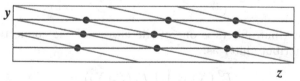

Fig. 3. A schematic view of a number of axial and skewed tubes. Axials are shown in blue, skewed in green. Virtual nodes are represented by red dots. The y and z coordinates of the virtual nodes are determined from the intersection points of the skewed tubes projected onto the plane of the axial tubes. The x coordinate of the virtual nodes is midway between the two planes.

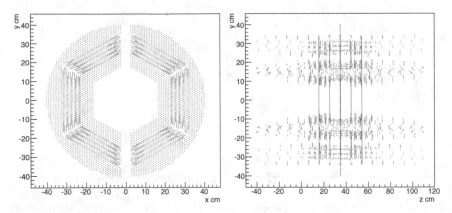

Fig. 4. The structure of the STT graph extended by the virtual nodes. Virtual nodes are shown in red and the center points of the original nodes in blue; the blue dots in the left panel are also the only available information in the z-direction. Left: STT graph in the $xy-$plain. Right: STT graph in the $yz-$plain.

the xy-plane and provides a better estimate of the z-coordinates of the nodes participating in a path. Figure 3 illustrates the positioning of virtual nodes and fig. 4 shows the structure of the grid after the extension.

3   Basic Concepts Filtering and Connections

Consider a universal set E (usually $E = \mathbb{Z}^n$), a binary image X is defined as a subset of E. Define $\mathcal{P}(E)$ to be the power set of E. Connectivity can be defined using the connectivity classes [3,7,8]. Each connectivity class $\mathcal{C} \subseteq \mathcal{P}(E)$ is a set of sets for which holds that the empty set and singletons are connected and for all connected sets with non-empty intersection their union is connected. Using this definition, each image X is a set of connected components C_i with i from some index set I with $\{\forall C_i, \nexists (C \supset C_i) : \ C \subseteq X \wedge C \in \mathcal{C}\}$. A connected component of X is denoted by $\lessdot$. A binary connected opening operator Γ_x on X at the point x could be defined as:

$$\Gamma_x(X) = \begin{cases} C_i : x \in C_i \wedge (C_i \lessdot X) & \text{if } x \in X \\ \emptyset & \text{else.} \end{cases} \tag{1}$$

The group of attribute filters is one of the classes of the connected filters. We can define an attribute filter as:

$$\Gamma^T(X) = \bigcup_{x \in X} \Gamma_T(\Gamma_x(X)). \tag{2}$$

Here, T is a criterion with the property of being increasing. T is usually of the form: $T(C) = Attr(\mathcal{C}) \geq \lambda$. $Attr(\mathcal{C})$ is some attribute of $\mathcal{C}$ and λ is a threshold value [9].

3.1   Attribute Space and Attribute Space Filters

Attribute space connectivity and filters were introduced by Wilkinson as a method to improve attribute filtering when connectivities fail to perceptually group correctly [3]. Here, we summarize only the main concept. In this concept, an image $X \subset E$ is transformed into a higher dimensional space "attribute space" $E \times A$. Here, A is an encoding of the attributes of pixels in X (for example, the local width). The operator $\Omega : \mathcal{P}(E) \to \mathcal{P}(E \times A)$ is defined such that it maps a binary image X into $\mathcal{P}(E \times A)$ space. The increasing inverse operator Ω^{-1} projects $\Omega(X)$ back to X ($\forall X \in \mathcal{P}(E) : \Omega^{-1}(\Omega(X)) = X$). Using this extension, the attribute space connectivity class $\mathcal{A}$ on a universal set E by the transformation pair (Ω, Ω^{-1}) and connectivity class $\mathcal{C}_{E \times A}$ on the transformed image is defined as:

$$\mathcal{A} = \{\mathcal{C} \in \mathcal{P}(E) \mid \Omega(\mathcal{C}) \in \mathcal{C}_{E \times A}\}. \tag{3}$$

Using this and the transformation pair, a connected filter $\Psi^A : \mathcal{P}(E) \to \mathcal{P}(E)$ could be defined as:

$$\Psi^A(X) = \Omega^{-1}(\Psi(\Omega(X))), \tag{4}$$

with $X \in \mathcal{P}(E)$ and $\Psi : \mathcal{P}(E \times A) \to \mathcal{P}(E \times A)$. Using this framework, different operator-pairs based on, for example width, orientation, even a combination of different attributes can be defined to transform X into the attribute space where connected filters could be used for segmentation.

4   Orientation-Based Attribute Space

Using the framework described in sec. 3.1, one can define the operator $\Omega_\alpha : \mathcal{P}(E) \to \mathcal{P}(E \times A)$ that assigns one or more orientations α_i to every active node x in the graph [3]. This yields a function $f(x, \alpha)$ over $E \times A$. Next, we apply the operator Ω_α on the input image X as:

$$\Omega_\alpha(X) = \{x \in X, \alpha \in A \mid f(x, \alpha) > \lambda f_{min} \vee f(x, \alpha) = f_{max}\}. \tag{5}$$

Here, f_{min} is minimum value for $f(x, \alpha)$, f_{max} the maximum value and $\lambda \geq 1$ is a tuneable parameter that determines how strict the selectivity is in assigning a node to an orientation space. Using this concept, the orientation-based attribute space is computed for all active nodes in each event in two different ways, namely dynamically and statically. For the static method, a number of angles are predefined to determine the orientation-based attribute space where in the dynamic method the angles are determined for each instance (binary image) in a pre-sensing step. We have observed that using both methods lead to similar results with negligible differences. This could be explained by the fact that all tubes have a diameter of 1 cm and their relative positions are permanent in the space. Due to this arrangement of the nodes, the number of possible orientations is limited and does not have a broad variation.

To determine the attribute space for the STT-graph, we have applied a slightly modified version of the method proposed in [3] by keeping a record of the participating active nodes for all orientations. In this case, we have used the following method for all active nodes in the STT-graph:

1. Compute an opening transform Ω_X^α using a linear structuring element with orientation α and keep a record of all participating active nodes. The structuring element is chosen such, that its length is maximal for the given orientation α.
2. This yields the gray-level function $f(x, \alpha)$ over the domain $(E \times A)$.
3. For all active nodes in X determine f_{min} and f_{max}.
4. Compute $\Omega_\alpha(X)$ using eq. 5.

4.1 Analysis of Attribute Space Connected Component

Transforming the binary image X (one STT readout instance) to the orientation-based Attribute Space using $\Omega_\alpha : \mathcal{P}(E) \to \mathcal{P}(E \times A)$ produces a binary image with an additional dimension, namely the orientation-based dimension. One can redefine the neighborhood relation between the points in the graph using the computed orientation. This way, we can step through the orientation spaces and collect the participating points in each subspace to construct the connected components. In this case the following method is used:

1. Sort all active nodes based on their layer, decreasing order. Insert all nodes in a priority queue (FIFO).
2. For all nodes in the queue: if it is visited for the first time and $f_{max} >=$ minimum required response, create a component candidate and add all members of the orientation sub-space corresponding with f_{max} to the current candidate. Otherwise the node is added to the list of short response nodes.
3. If the node was visited: Find the candidate to which it was added, then step through the neighboring orientation sub-spaces and add the nodes if Ω_α.

The result of the application of this method is shown in fig. 5. This procedure is shown in alg. 4.1. Note that the singletons are kept in a separate set of node list; this way, those grid points could be accessed if necessary.

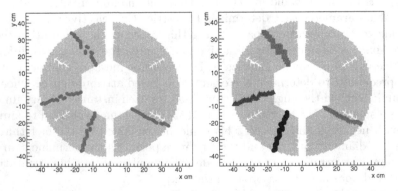

Fig. 5. Left: the digitization output. Right: the composed connected components using alg. 4.1. Each color represents a different path.

Algorithm 4.1 Attribute Space Connected Component

Input: The graph of the tubes, minimum required response.
Output: List of sets "$AllCompList$" containing all connected components.

```
 1: Sort active nodes based on their layers.
 2: Create two FIFOs, ActiveQueue and ShortCompQueue
 3: Insert all active nodes in ActiveQueue
 4: while (¬ empty(ActiveQueue)) do
 5:      Fetch first node "CurentNode"
 6:      if (CurentNode_{fmax} < MinimumResponse) then
 7:          Add node to "ShortCompQueue" and set visited
 8:      else
 9:          if (CurentNode was not visited) then
10:              Create a new empty set S and add CurentNode to it. Set visited
11:              for all nodes active in CurentNode_{fmax} orientation do
12:                  if (¬ visited) then
13:                      Add to S; mark as visited
14:                  end if
15:              end for
16:          else
17:              Create a list of sets CompL where it was added before
18:              for all (cmp ∈ CompL) do
19:                  for all (node ∈ cmp) do
20:                      if (¬ visited ∧ active in current orientation) then
21:                          Add nodes from neighboring orientations to cmp
22:                      end if
23:                  end for
24:              end for
25:          end if
26:      end if
27: end while
28: if (¬ empty(ShortCompQueue)) then
29:      Create an empty set Orph
30:      Fetch element sn
31:      Find cmp ∈ AllCompList the node nn with the shortest distance to en
32:      if (nn and en direct neighbors) then
33:          Add en to cmp
34:      end if
35: else
36:      Add en to the set Orph
37: end if
```

4.2 Determination of z-coordinate

One of the most important features of each track is the evolution of the path in the z direction. The full reconstruction of each path is used to determine the properties of the passing particle. The presence of the skewed tubes helps to estimate the $z-$coordinates of the points along the path. To have a more accurate estimate of the $z-$coordinates of the points, we introduced virtual nodes as described in sec. 2.1. Using the assumption that all the tracks passing through the STT originate from $(0, 0, 0)$, we could estimate the deviations in the z-direction and generated a rough reconstruction of the whole particle path. Starting from the origin to the first virtual layer, we apply linear interpolation by using a constant step size of $\delta z = z_1$, with z_1 being the z-coordinate of the first virtual node in the track. Between the first and the last virtual layer, an adaptive δz is computed, when passing the layers. From the most outer virtual layer to the most outer node we extrapolate linearly and δz remains constant and equal to the latest determined value. Linear interpolation and extrapolation is used

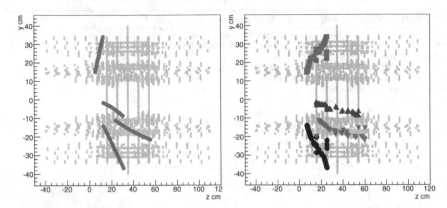

Fig. 6. Left: the true hit positions. Right: the determined paths along the z-axis using linear interpolation, assuming that the paths start at the origin. Because we are not correcting the coordinates of the skewed nodes in the xy-plane, some of the nodes seem to be wrongly positioned.

because there are no forces acting in the z-direction. Figure 6 shows an example of the resulting z reconstruction based on the given description.

5 Results and Discussion

In this section, we present the results of the application of the previously described methods. To obtain the sample test data sets, we have used the PANDA simulation package to produce 10^3 μ^- events with different momenta and a multiplicity of 6 tracks per event[1]. The tracks are generated isotropically around the origin $(0, 0, 0)$. To quantify the segmentation error, we have applied the following method: given an image X with a total area of A, let a reference segmentation divide the image into N regions $\{R_1, \ldots, R_N\}$ with corresponding areas $\{A_1, \ldots, A_N\}$. An example segmentation finds M regions $\{T_1, \ldots, T_M\}$ with areas $\{a_1, \ldots, a_M\}$. For all regions T_j, we find R_k such that $(T_j \bigcap R_k)$ is maximal. One can define the "undermerging" (E_{um}), "overmerging" (E_{om}) and normalized total error (E_{tot}) as:

$$E_{um} = \frac{1}{A} \sum_{j=1}^{M} \frac{\{A_k - \#(T_j \bigcap R_k)\}\#(T_j \bigcap R_k)}{A_k}, \tag{6}$$

$$E_{om} = \frac{1}{A} \sum_{j=1}^{M} \{(a_j - \#(T_j \bigcap R_k)\}, \tag{7}$$

$$E_{tot} = \sqrt{E_{um}^2 + E_{om}^2} \tag{8}$$

---

[1] The software package "pandaroot" is available from: subversion.gsi.de (Revision: 26379). To make this analysis reproducible, we have fixed the starting seed.

in which $\#C$, denotes the cardinality of C. Here, we have slightly modified the measures described in [10]. The total magnitude of error is obviously equal to zero if the two error types are equal to zero or vanishingly small compared to A. Total error does not give much information on the type of the error. Here, the Monte Carlo (MC) truth particle paths are considered as the reference segmentations to determine the values for both error types. As expected the momentum of the passing particle will affect its behavior during travel-time inside the detector due to a stronger interaction with the magnetic field. Particles with lower momenta have a more curved path while the ones with a higher momentum will be straighter. As a consequence of this momentum dependency, we expect that the value of different error measures will be larger for particles with a lower momentum; because high curvature tends to break up paths.

Figure 7 shows the distribution of error values for 10^3 events and table 1 summarises the mean values for a number of arbitrary selected momenta.

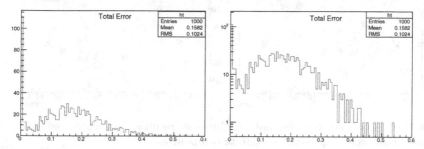

Fig. 7. Distribution of the total error for 10^3, 0.8 GeV events: (left) linear scale; (right) log scale

The under-merging error is mainly caused by the missing nodes or when the paths cross the regions between different sectors where no neighborhood information is available. The over-merging is caused when two or more paths are running through neighboring nodes specially when passing the skewed layers. Because of the fact that there is no direct and accurate information available on the $z-$coordinates when the paths are selected, passing the skewed layers by at least two close neighboring paths will lead to over-merging. Another observation

Table 1. Error estimates for 10^3, μ events. One can observe that E_{tot} shows a small fluctuation and E_{um} drops steadily as a function of momentum

Momentum	E_{um}	E_{om}	E_{tot}
0.4	0.09	0.13	0.18
0.8	0.08	0.12	0.16
1.0	0.08	0.13	0.17
1.5	0.07	0.13	0.16
2.0	0.07	0.12	0.16

is that the value of the total error shows a small fluctuation when moving from low to higher energy particle paths whereas E_{um} drops steadily as momentum increases. The highest values of error are observed for paths of slower particles. These results are preliminary and more statistical tests are needed for a more in depth investigation. Furthermore, we accept a gap-size of zero pixels. Although accepting a larger gap will solve a part of the under-merging problem, it will introduce a more frequent over-merging. Figure 8 shows an event where over- and under-merging effects are visible.

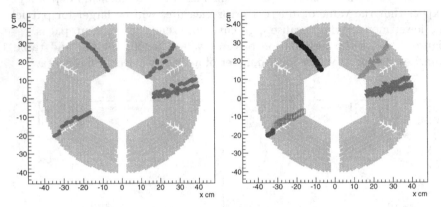

Fig. 8. Left: the detector read-out. Right: the resulting reconstructed paths in xy−plain. Each color represents a separate component.

Considering the structure of the graph and the direct available read-out data based on the detector geometry, the results of the applied method look promising; both in recognition of the tracks in 2D and determining the z−coordinates for the track points. The separation of the crossing paths is still one of the challenges that need to be addressed which will lead into a smaller over-merging error and a better overall performance. The attribute space framework can handle overlaps intrinsically. One can design attribute space filters to treat the image, based on its determined attributes even if there are overlapping structures in the original image domain. However, the presented results show that the direct application of the framework does not provide a complete solution for this problem. Note that it is necessary, in the case of overlap, that a node is a member of at least two separate paths.

One can estimate the computational complexity of Hough transformation by $P^D + P^{D-1}n$, with, P transformation space parameter, D the dimension of P and n is the number of active pixels. We can observe that this method is less sensitive to n. But when using a sensible number of parameters, the complexity will become $O(n^3)$; improvements could be achieved by parallelization. Our method in the current state has a $O(n^2)$ computational complexity and because of its structure we foresee a complexity reduction to $O(n)$ in the future. Because these methods differ intrinsically, we cannot compare them directly; but, we expect that for a sound number of parameters and image size, our method will run with at least the same performance. Future work will consist of a modification of the current method by the inclusion of mechanisms for node splitting and separating crossing paths, a better determination of the z-coordinates of the points in each path and modifying the algorithm to reduce the computational complexity to $O(n)$. The recognition of the paths displaced from the origin is also one of the major challenges for our future work.

Acknowledgment. We thank the PANDA collaboration for their support and special thank to the STT developers group.

References

1. Kurdy, M.B., Jeulin, D.: Directional mathematical morphology operations. In: Proceedings of the Fifth European Congress for Stereology, vol. 8(2), pp. 473–480 (1989)
2. Appleton, B., Talbot, H.: Efficient complete and incomplete path openings and closings. Image and Vision Computing 25(4), 416–425 (2007)
3. Wilkinson, M.H.F.: Attribute-space connectivity and connected filters. Image and Vision Computing 25(4), 426–435 (2007); International Symposium on Mathematical Morphology 2005
4. Toet, A., Heijmans, H.J.A.M., Nacken, P., Vincent, L.: Graph morphology. Journal of Visual Communication and Image Representation 3, 24–38 (1992)
5. Cousty, J., Najman, L., Dias, F., Serra, J.: Morphological filtering on graphs. Computer Vision and Image Understanding 117(4), 370–385 (2013); Special issue on Discrete Geometry for Computer Imagery
6. (PANDA Collaboration) Erni, W., Keshelashvili, I., Krusche, B., et al.: Technical design report for the panda straw tube tracker. The European Physical Journal A 49, 25 (2013); Special Article Tools for Experiment and Theory
7. Ronse, C.: Set-theoretical algebraic approaches to connectivity in continuous or digital spaces. J. Math. Imag. Vis. 8, 41–58 (1998)
8. Serra, J.: Connectivity on complete lattices. J. Math. Imag. Vis. 9(3), 231–251 (1998)
9. Breen, E.J., Jones, R.: Attribute openings, thinnings and granulometries. Comp. Vis. Image Understand. 64(3), 377–389 (1996)
10. Wilkinson, M.H.F., Schut, F.: Digital image analysis of microbes: imaging, morphometry, fluorometry and motility techniques and applications, 1st edn. John Wiley & Sons Ltd., New York (1998)

A Pansharpening Algorithm Based on Morphological Filters

Rocco Restaino[1], Gemine Vivone[2], Mauro Dalla Mura[3](✉),
and Jocelyn Chanussot[3,4]

[1] Dept. of Information Eng., Electrical Eng. and Applied Math. (DIEM),
University of Salerno, Fisciao, Italy
restaino@unisa.it
[2] North Atlantic Treaty Organization (NATO) Science and Technology Organization
(STO) Centre for Maritime Research and Experimentation, La Spceizia, Italy
mauro.dalla-mura@gipsa-lab.grenoble-inp.fr
[3] GIPSA-lab, Grenoble-INP, Saint Martin d'Hères, France
[4] Faculty of Electrical and Computer Engineering, University of Iceland, Reykjavik,
Iceland

Abstract. The fusion of multispectral and panchromatic images acquired by sensors mounted on satellite platforms represents a successful application of data fusion called *Pansharpening*. In this work we propose an algorithm based on morphological pyramid decomposition. This approach implements a Multi Resolution scheme based on morphological gradients for extracting spatial details from the panchromatic image, which are subsequently injected in the multispectral one. Several state-of-the-art methods are considered for comparison. Quantitative and qualitative results confirm the capability of the proposed technique to obtain pansharpened images that outperform state-of-the-art approaches.

Keywords: Mathematical Morphology · Pansharpening · Morphological filtering · Remote sensing · Image enhancement

1 Introduction

Pansharpening refers to a precise problem in data fusion that is the fusion of a MultiSpectral (MS) and a PANchromatic (PAN) image. The MS image is characterized by a high spectral resolution but a low spatial resolution whereas the PAN image has a greater spatial resolution but it acquires an unique spectral band. These images are acquired simultaneously by a PAN and MS sensors mounted on the same platform, as for satellites such as GeoEye and Worldview. The aim of pansharpening is the generation of a synthetic high spatial resolution MS image having the spatial features of the PAN image and the spectral resolution of the MS image. Several applications exploit data provided by pansharpening algorithms such as Google Earth and Microsoft Bing.

The pansharpening literature has been quickly growing in the last years, see [26] for a recent review. Pansharpening algorithms are often grouped into

© Springer International Publishing Switzerland 2015
J.A. Benediktsson et al. (Eds.): ISMM 2015, LNCS 9082, pp. 98–109, 2015.
DOI: 10.1007/978-3-319-18720-4_9

two families: The Component Substitution (CS) and the Multi-Resolution Analysis (MRA) methods. CS methods are based on the transformation of the MS image in another feature space (e.g., with a transformation in the Intensity-Hue-Saturation (IHS) representation [8] or with the Principal Component Analysis (PCA) [8]), where the spatial structure of the data can be more easily decoupled from the spectral information. The fusion process consists in the substitution of the component encompassing the spatial structure by the PAN image and in the subsequent reconstruction of the MS image. The MRA methods instead are based on the generation of a low resolution version of the PAN image for example using a multiresolution decomposition (i.e., a *pyramid*) of the image based on low-pass filters. Spatial details are then extracted from the PAN image as the residuals with respect to the low resolution PAN and injected, opportunely weighted, into the MS bands. The algorithm for constructing the low resolution PAN image and the injection coefficients (i.e., the weights applied to the spatial details) distinguish the different MRA methods.

In this paper we focus on this latter approach and in particular on the way the spatial details are extracted from the PAN. The largest majority of MRA techniques employs linear analysis operators ranging from to the wavelet/contourlet decompositions [20] to pyramids with Gaussian filters [2]. Another distinguishing feature among MRA approaches is the possible presence of decimation at each level of the decomposition that helps for compensating aliasing in the MS image [4]. A possible alternative for obtaining a pyramidal decomposition is offered by non-linear operators such as Morphological Filters (MF) [24]. Indeed some MFs are more effective than linear filters in preserving edges, allowing for superior perfomances in several applications, as, for example, geodesic reconstruction [21] and sharpening [17]. For that reason, morphological pyramids have also been used for fusing images with different characteristics in tasks such as the enhancement of optical imaging systems [6] and medical diagnostics [18]. Nevertheless, morphological pyramids have not received much attention in pansharpening. To the best of our knowledge, only the works of [13,14,1] address the pansharpening problem using approaches based on MFs, but a systematic study on their potentialities in this applicative domain has not been yet carried out. This has motivated us to explore this research direction by analyzing the effect of several different morphological analysis operators and structuring elements.

2   Pansharpening Based on MultiResolution Analysis

Let us denote by **MS** the available MS image, which is a three-dimensional array obtained by stacking N monochromatic images $\mathbf{MS}_k$ (i.e., the single spectral channels), with $k = 1, \ldots, N$, each composed by $n_r^{MS} \times n_c^{MS}$ pixels. Furthermore let us denote by **P** the available PAN image, composed by $n_r^{PAN} \times n_c^{PAN} = rn_r^{MS} \times rn_c^{MS}$ pixels. $r > 1$ is the ratio between the MS and PAN spatial resolution. The objective of the pansharpening process is to produce a synthetic MS image $\widehat{\mathbf{MS}}$ with the same spectral features of **MS** (i.e., N spectral bands) and with the same spatial resolution of **P** (i.e., $n_r^{PAN} \times n_c^{PAN}$).

Classical pansharpening algorithms follow the general scheme based on the injection of spatial details extracted from the PAN image to the MS bands [26]:

$$\widetilde{\mathbf{MS}}_k = \widetilde{\mathbf{MS}}_k + g_k \mathbf{D}_k = \widetilde{\mathbf{MS}}_k + g_k \left(\mathbf{P} - \mathbf{P}_k^{low} \right), \tag{1}$$

in which $k = 1, \ldots, N$ specifies the spectral band, $\widetilde{\mathbf{MS}}_k$ and $\mathbf{D}_k$ are the k-th channels of the MS image upsampled to the PAN scale and of the injected details, respectively, and $\mathbf{g} = [g_1, \ldots, g_k, \ldots, g_N]$ are the *injection gains*. $\mathbf{P}_k^{low}$ denotes a low resolution version of the PAN image that, in general, may depend on the band k and it is obtained as a combination of the MS image spectral bands in CS methods and through a multiresolution decomposition within the MRA class.

In this paper we focus on the latter approach, in which a sequence of images with successively reduced resolution(also called approximations) is constructed through iterations, yielding the required image $\mathbf{P}_k^{low} = \mathbf{P}_k^L$, which is the image of the last level in the decomposition of $\mathbf{P}$ into L levels (i.e., the level of lowest resolution). Starting from the initial image $\mathbf{P}_k^0$ that is typically obtained by equalizing the PAN image $\mathbf{P}$ with respect to $\mathbf{MS}_k$, the decomposition proceeds with the recursion

$$\mathbf{P}_k^l = T_k^l \left[\mathbf{P}_k^{l-1} \right], \qquad l = 1, 2, \ldots L,$$

in which the *decomposition operators* $\left\{ T_k^l \right\}_{l=1,\ldots L}$ can be different, in principle, across the scales and the bands. In general for decimated pyramids, $T^l = R^{\downarrow}\psi_l$, so it comprises a filtering step, which is implemented by the generic *analysis operators* ψ_l, and downsampling with a rate R, denoted by $R^{\downarrow}$ (in general $R = 2$). Usually, the same filter is used across scales, i.e., $\psi_l = \psi$. For decimated decompositions it is necessary to expand the last level of the pyramid to the size of the PAN through an interpolation step $\mathrm{INT}^{\uparrow}$. In undecimated decompositions the spatial support of the decomposed image is not reduced at the different resolution levels, hence no downsampling is performed (i.e., $R = 1$) and no expansion is applied to the last level of the pyramid (i.e., $\mathrm{INT}^{\uparrow} = id$ with id the identity transform). However, the filtering operator ψ_l needs to vary with the levels for obtaining images with progressively reduced resolution. For linear pyramids the analysis operator is obtained through the convolution with a mask h_l, namely $\psi_l[\cdot] = h_l * \cdot$. Typical operators employed in pansharpening are average and Gaussian filters and wavelets [26].

Once the image $\mathbf{D}_k$ is computed, the details need to be injected to the MS bands. The strategies commonly employed rely on a multiplicative or additive injection scheme [26]. The former is referred to *High-Pass Modulation* (HPM) scheme in which the injection gains are given by:

$$g_k = \frac{\widetilde{\mathbf{MS}}_k}{\mathbf{P}_k^{low}}, \quad k = 1, \ldots, N. \tag{2}$$

The latter is called *High-Pass Filtering* (HPF) scheme and is implemented by setting $g_k = 1$.

3 Pansharpening Based on Morphological Filtering

This work is devoted to explore pansharpening algorithms based on morphological pyramidal decompositions. We briefly recall in the following the definition of some MFs and later, how they can be used for performing a decomposition in pansharpening.

3.1 Recall of Morphological Filters

In the following we will introduce the MFs considering a generic scalar image defined in a subset E of $\mathbb{Z}^2$ and with values in a finite subset V of $\mathbb{Z}$, i.e., $\mathbf{I} : E \subseteq \mathbb{Z}^2 \to V \subseteq \mathbb{Z}$. In this work we focus on MFs based on a set called *Structuring Element (SE)* B, that defines the spatial support in which the operator acts on the image [21]. We denote the spatial support of B as $N_B(\mathbf{x})$ that is the neighborhood with respect to the position $\mathbf{x} \in E$ in which B is centered. Flat SEs will be considered, so all elements in the neighborhood have unitary values and the only free parameters for defining B are the origin and N_B. We recall below the definition of the two basic operators of *Erosion* $\varepsilon_B(\mathbf{I})$ and *Dilation* $\delta_B(\mathbf{I})$:

$$\varepsilon_B(\mathbf{I})(\mathbf{x}) = \bigwedge_{y \in N_B(\mathbf{x})} \mathbf{I}(y) ; \quad \delta_B(\mathbf{I})(\mathbf{x}) = \bigvee_{y \in N_B(\mathbf{x})} \mathbf{I}(y) . \qquad (3)$$

Their sequential composition produces *Opening* and *Closing*, respectively defined as:

$$\gamma_B(\mathbf{I})(\mathbf{x}) = \delta_{\check{B}}[\varepsilon_B(\mathbf{I})(\mathbf{x})], \quad \phi_B(\mathbf{I})(\mathbf{x}) = \varepsilon_{\check{B}}[\delta_B(\mathbf{I})(\mathbf{x})], \qquad (4)$$

with $\check{B}$ denoting the SE obtained by reflecting B with respect to its origin. The residual of the application of erosion is referred to as morphological *internal gradient* $\rho_B^-(\mathbf{I}) = \mathbf{I} - \varepsilon_B(\mathbf{I})$. Analogously, we have an *external gradient* considering a dilation: $\rho_B^+(\mathbf{I}) = \delta_B(\mathbf{I}) - \mathbf{I}$. In general, ρ^+ and ρ^- (named also *half-gradients*) are computed with a SE with smallest support. The operators extracting the residuals of opening and closing are referred to as top-hat transforms. Specifically, an opening defines a *White Top Hat*, $\mathrm{WTH}_B(\mathbf{I}) = \mathbf{I} - \gamma_B(\mathbf{I})$ and closing a *Black Top Hat*, $\mathrm{BTH}_B(\mathbf{I}) = \phi_B(\mathbf{I}) - \mathbf{I}$. The composition of erosion and dilation have proven useful for sharpening images through the so called *Toggle Contrast*, TC, mapping [21]:

$$\mathrm{TC}_B(\mathbf{I})(\mathbf{x}) = \begin{cases} \delta_B(\mathbf{I})(\mathbf{x}), & \text{if } \rho_B^+(\mathbf{I})(\mathbf{x}) < \rho_B^-(\mathbf{I})(\mathbf{x}), \\ \varepsilon_B(\mathbf{I})(\mathbf{x}), & \text{otherwise.} \end{cases} \qquad (5)$$

3.2 Morphological Pyramids for Pansharpening

The composition of elementary morphological operators leads to the design of nonlinear pyramidal schemes [10,11,22]. Morphological decomposition approaches date back to Toet *et al.* [23], in which the multiscale representation

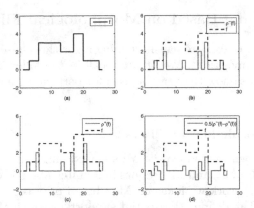

Fig. 1. Examples of morphological gradients, obtained through a flat SE with $N_B = \{-1, 0, 1\}$: (a) function f, (b) internal gradient $\rho^- = f - \epsilon_B(f)$; (c) external gradient $\rho^+ = \delta_B(f) - f$; (d) proposed detail extraction operator i.e., $\overline{\psi}_\rho = 0.5(\rho^- - \rho^+) = 0.5(f - \epsilon_B(f)) + 0.5(f - \delta_B(f))$.

was implemented by the sequential application of opening and closing operators, $\psi_{\text{Toet}} = \phi_B \gamma_B$. The use of ψ_{Toet} in an undecimated approach was later suggested by the same authors for data fusion purposes [24]. In this case the decrease in resolution is obtained by considering SEs with progressively larger support across scales, leading to the operator: $T^l_{\text{Toet}} = \phi_{B_l} \gamma_{B_l}$.

In the works of Laporterie *et al.* [9,15], the decomposition filter was given by the semi-sum of opening and closing $\psi_{\text{TH}} = 0.5(\phi_B + \gamma_B)$. A justification for this choice is clear after some simple algebraic manipulations, showing that this operator is related to top-hats. Indeed the complementary operator $\overline{\psi}_{\text{TH}} = id - \psi_{\text{TH}}$, which has the purpose of extracting the details, can be written as

$$\overline{\psi}_{\text{TH}} = id - 0.5(\phi_B + \gamma_B) = 0.5[(id - \phi_B) + (id - \gamma_B)] \tag{6}$$

$$= 0.5(\text{WTH}_B - \text{BTH}_B) \tag{7}$$

It has also been reported in [21], that the difference between black and white top-hats is commonly used for enhancing the contrast of the images. The pyramidal scheme exploiting this filter was specifically proposed for pansharpening application in [13]. A downsampling is considered at each step for aliasing reduction, leading to the analysis operator $T^l_{\text{Lap}} = R^\downarrow \psi_{\text{TH}} = R^\downarrow 0.5(\phi_B + \gamma_B)$.

An alternative way for obtaining a pyramidal decomposition is to extract high spatial frequency details by means of contrast operators. A notable example is the TC. Spatial details can be derived through the operator

$$\overline{\psi}_{\text{TC}} = \text{DTC}_B - \text{ETC}_B = \max(\text{TC}_B - id, 0) - \max(id - \text{TC}_B, 0), \tag{8}$$

defined in terms of Dilation and Erosion Toggle Contrast operators, DTC and ETC respectively [7]. In the context of data fusion TC has been used for multiscale representation through the application of the scale dependent filter

$\psi_{\mathrm{TC}}^l = id - \overline{\psi}_{\mathrm{TC}}^l$, without downsampling, namely $T_{\mathrm{TC}}^l = \psi_B^l$, where the superscript l indicates the size of the SE [7]. More recently, a decomposition operator combining the effects of both top-hats and toggle contrast operators was proposed by Bai *et al.* [6]: $T_{\mathrm{Bai}}^l = id - (\mathrm{WTH}_{B_l} - \mathrm{BTH}_{B_l}) - (\mathrm{DTC}_{B_l} - \mathrm{ETC}_{B_l})$.

In this work we propose to use *half-gradients* for performing the decomposition. The application of morphological gradients for enhancing the image contrast has been already evidenced [21]. However, to our knowledge, their application for data fusion tasks (e.g., pansharpening) is still unexplored. The effect of the internal and external gradients can be seen in Fig. 1(b,c), where the application to a piecewise-constant mono-dimensional signal f with a flat SE B with neighborhood $N_B = \{-1, 0, 1\}$ is shown. The results of the application of ρ^+ and ρ^- assume positive values in the presence of signal discontinuities. However, while the non zero values of the internal gradient follow the positive discontinuities and preceeds the negative ones, the opposite behavior is shown by the external gradient. Thus, the difference of the two operators reproduces the variations of the function with respect to the local mean, as reported in Fig. 1(d). Accordingly, we define the details extraction operator $\overline{\psi}_\rho$ as the difference of the two gradients

$$\overline{\psi}_\rho = 0.5(\rho^- - \rho^+) = 0.5(id - \epsilon_B) - 0.5(\delta_B - id),$$

in which the factor 0.5 is applied to preserve the dynamic range of the details. This is fundamental in pansharpening since otherwise, spatial artifacts would appear in the pansharpening product. Thus, the corresponding analysis filter is given by

$$\psi_\rho = id - \overline{\psi}_\rho = id - [0.5(id - \epsilon_B) - 0.5(\delta_B - id)] = 0.5(\epsilon_B + \delta_B), \qquad (9)$$

namely, the semi-sum of dilation and erosion. For the decomposition we apply the proposed operator in cascade with a dyadic subsampling, i.e., $T_\rho^l = R^\downarrow 0.5(\epsilon_B + \delta_B)$.

4   Numerical Results

Two data sets have been used to assess the performance of the proposed algorithm. As no High Resolution MS target image is available, the quantitative evaluation of pansharpening algorithm has been carried out with a reduced resolution assessment procedure [26].

4.1   Compared Algorithms

In this section we report the results of the proposed pansharpening method based on morphological multiscale analysis. Some features were fixed *a priori*. The number L of levels in the decomposition is set to $L = \log_2(r)$ with r the resolution ratio between the MS and the PAN image (typically $r = 4$ that implies $L = 2$). We choose the HPM injection scheme and employed a decimated/interpolated decomposition exploiting dyadic subsampling and bilinear interpolation[1] in the analysis and synthesis phase, respectively. Thus the decomposition operator $T = 2^{\downarrow}\psi$ is fully specified by the choices of the analysis operator ψ and of the SE B, on which this work is focused. In the following we refer to the notation: MF-$\{analysis\ operator\}$-$\{SE\}$ for referring to the analyzed algorithms. The four considered analysis operators are:

TH: Top-Hats [9,15]: $\psi_{TH} = id - 0.5[(\text{WTH}_B - \text{BTH}_B)] = 0.5[\gamma_B + \phi_B]$;

TC: Toggle [21]: $\psi_{TC} = id - 0.5[(\text{DTC}_B - \text{ETC}_B)]$;

BAI: Toggle+Top-Hats [6]: $\psi_{Bai} = id - 0.5[(\text{WTH}_B - \text{BTH}_B) - (\text{DTC}_B - \text{ETC}_B)]$;

HGR: Half Gradients [Novel approach]: $\psi_\rho = id - 0.5(\rho^- - \rho^+) = 0.5[\epsilon_B + \delta_B]$.

Note explicitly that, in order to guarantee a fair comparison, a factor 0.5 was added to the detail extraction operator also for the methods proposed in [7] and [6].

Four different SEs were considered:

S: 3×3 squared SE

D: disk-shaped SE with radius 1

H: horizontal linear SE of length 3

V: vertical linear SE of length 3

In addition, test are also performed by considering a SE of size 1 (referred as **MF-1** in the results), which leads to the pyramidal scheme with identity analysis operator $\psi_1 = id$ that can be considered as a baseline for comparison since the decomposition is only based on the decimation step.

The proposed algorithms are also compared to several existing approaches, whose choice has been driven by the performances shown in public contests [5] or in the review [26]. Furthermore, we consider some data fusion strategies based on morphological pyramids. The following acronyms are used for the tested algorithms:

EXP: MS image interpolation (i.e., there is no injection of spatial details), using a polynomial kernel with 23 coefficients [2]

GIHS: *Fast Intensity-Hue-Saturation* (GIHS) image fusion [25]

SFIM: *Smoothing Filter-based Intensity Modulation* [16], based on *High-Pass Modulation* injection scheme and 5×5 box filter (*i.e.*, mean filter) for details extraction

MTF-GLP: *Generalized Laplacian Pyramid* (GLP) [2] with MTF-matched filter [3] with unitary injection model

AWLP: *Additive Wavelet Luminance Proportional* [19]

---

[1] The suitability of bilinear interpolation was already evidenced in [12] and confirmed in this study through the comparison with other interpolation schemes, based on zero-order, cubic, Lanczos-2 and Lanczos-3 kernels. The results are omitted for space constraints.

Table 1. Results obtained on the *Hobart 1* and *Rome data sets*. The comparison among different analysis operators ψ and SEs is reported with dyadic decomposition and bilinear interpolation for the expansion. Values of the Q^{2n}, *SAM*, and *ERGAS* are indicated. For each SE, the best results among operators are marked in bold while the second ones are underlined.

ψ	B	Q4	SAM	ERGAS	Q8	SAM	ERGAS
TH	S	**0.9035**	**4.8304**	3.0838	0.8914	3.9373	3.9785
	D	**0.9023**	4.8321	3.0144	0.8935	3.9691	3.7878
	H	0.8929	4.8416	3.1622	0.8863	4.0137	3.9377
	V	0.8935	4.8424	3.1548	0.8890	3.9956	3.8078
TC	S	0.8729	4.8615	3.3996	0.8731	4.0451	3.8722
	D	0.8731	4.8614	3.4014	0.8723	4.0576	3.9235
	H	0.8711	4.8651	3.4467	0.8705	4.0785	3.9659
	V	0.8707	4.8650	3.4483	0.8706	4.0753	3.9824
BAI	S	0.9020	4.8305	3.0841	0.8906	**3.9248**	3.9381
	D	0.9023	4.8313	2.9974	0.8951	3.9447	3.6673
	H	0.8936	4.8410	3.1449	0.8877	3.9946	3.8570
	V	0.8941	4.8412	3.1373	0.8905	3.9781	3.7352
HGR	S	0.8986	4.8382	**3.0073**	0.8944	3.9364	3.4160
	D	0.9011	4.8321	**2.9444**	0.9014	3.8773	3.2884
	H	**0.8960**	4.8370	3.0491	**0.8960**	3.9367	3.4655
	V	**0.8966**	4.8377	3.0380	**0.8982**	3.8993	3.4144
		Hobart 1 Dataset			*Rome Dataset*		

MF-Toet-HPF, approach proposed in [24]: $T^l_{\text{Toet}} = \phi_{B_l}\gamma_{B_l}$, B_l is a square of size $2l$ pixels, no decimation is done. The HPF injection scheme, or error pyramid, is used.

MF-Toet-HPM, approach proposed in [24]: $T^l_{\text{Toet}} = \phi_{B_l}\gamma_{B_l}$, B_l is a square of size $2l$ pixels, no decimation is done. The HPM injection scheme, or ratio pyramid, is used.

MF-Lap, approach proposed in [13,14,15]: $T_{\text{Lap}} = 2^{\downarrow}\psi_{\text{TH}}$, B is defined as a horizontal line of length two pixels and the expansion is done with a bilinear interpolation. The HPF injection scheme is used.

4.2   Experimental Results

In the reduced scale evaluation protocol [27] the low resolution data sets are simulated by degrading the spatial resolution of the available images by a factor $r = 4$. In greater details, the image scaling has been performed by applying a Gaussian-shaped low pass filter designed for matching the sensor's MTF [3]. In this case the original MS image acts as the *ground truth* (GT) and thus several indices can be used to assess the fused products. We employ here the *Spectral Angle Mapper (SAM)*, to evaluate the spectral quality of the images and two indexes accounting for both spatial and spectral quality: the $Q2^n$-index, and the *Erreur Relative Globale Adimensionnelle de Synthèse (ERGAS)*. Please refer to [26] for details on these indexes. Optimal values are 0 for the SAM and the ERGAS and 1 for the $Q2^n$.

Table 2. Results on the *Hobart 1* and *Rome* data sets. Values of the Q^{2n}, and SAM and $ERGAS$ are reported. For each data set, the best result is marked in bold and the second best are underlined.

	Q4	SAM	ERGAS	Q8	SAM	ERGAS
EXP	0.6744	5.0314	5.1469	0.7248	4.9263	5.4171
GIHS	0.8256	4.9961	3.6727	0.7439	5.1455	4.1691
SFIM	0.8483	4.8961	3.6771	0.8758	4.2457	3.7591
AWLP	0.8910	5.0386	3.1703	0.9011	4.5146	3.3572
MTF-GLP	**0.9030**	**4.8263**	**2.9112**	**0.9016**	4.0957	<u>3.2982</u>
MF-1	0.8693	4.8666	3.4780	0.8684	4.0964	4.0899
MF-Toet-HPF	0.8743	4.9217	3.4821	0.8674	4.5915	4.1589
MF-Toet-HPM	0.8771	4.8739	3.5490	0.8606	4.4041	4.8072
MF-Lap	0.8827	4.8639	3.2440	0.8815	4.3085	3.7357
MF-HGR-S	0.8986	4.8382	3.0073	0.8944	3.9364	3.4160
MF-HGR-D	<u>0.9011</u>	<u>4.8321</u>	<u>2.9444</u>	<u>0.9014</u>	**3.8773**	**3.2884**
MF-HGR-H	0.8960	4.8370	3.0491	0.8960	3.9367	3.4655
MF-HGR-V	0.8966	4.8377	3.0380	0.8982	<u>3.8993</u>	3.4144
	Hobart 1 Dataset			Rome Dataset		

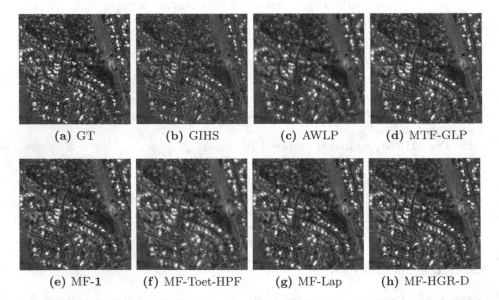

(a) GT   (b) GIHS   (c) AWLP   (d) MTF-GLP

(e) MF-1   (f) MF-Toet-HPF   (g) MF-Lap   (h) MF-HGR-D

Fig. 2. Results for the pansharpening algorithms on the *Hobart 1* data set: (a) GT; (b) GIHS; (c) MTF-GLP; (d) AWLP; (e) MF-1; (f) MF-Toet-HPF; (g) MF-Lap; (h) MF-HGR-D

Two data sets are employed for validation purposes. *Hobart 1* is a sample data set provided from GeoEye and referring to a scene of Hobart, Australia[2]. It is composed by a high resolution panchromatic channel and four MS bands. The

---

[2] Geoeye: Geoeye-1 Geo 8bit 0.5m True Color RGB - Hobart Aust 1, 02/05/2009 (2009).

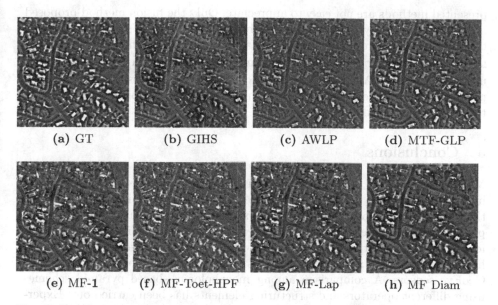

(a) GT   (b) GIHS   (c) AWLP   (d) MTF-GLP

(e) MF-1   (f) MF-Toet-HPF   (g) MF-Lap   (h) MF Diam

Fig. 3. Details injected by pansharpening algorithms on the *Hobart 1* data set: (a) GT; (b) GIHS; (c) MTF-GLP; (d) AWLP; (e) MF-1; (f) MF-Toet-HPF; (g) MF-Lap; (h) MF-HGR-D

images have a spatial resolution of 0.5 m and 2 m for PAN and MS, respectively. The dimension of the PAN image is 512×512 pixels. The *Rome* data set was instead acquired by the WorldView-2 sensor and represents the city of Rome, Italy. It is composed of a high resolution panchromatic channel and eight MS bands. The distributed images are characterized by a spatial resolution of 0.5 m and 2 m for PAN and MS, respectively. The dimension of the PAN image is 300×300 pixels.

The first analysis focuses on the comparison among the analysis operators and the SEs. The results achieved on the Hobart and Rome data sets, using a reduced resolution assessment, are reported in Table 1. A strong dependence of the obtained quality indexes on the ψ is straightforward. The decomposition scheme based on the HGR operator often shows the best performance, thus motivating the novel proposed approach.

A further comparison among the MF algorithms and several classical pansharpening approaches is proposed in Table 2. The first remarkable outcome is that the simplest implementation of the MF pyramidal schemes proposed in [24] and adjusted for the pansharpening applications in [13] yields fused products with low quality. Indeed, on both data sets the scheme proposed in this paper constitutes a significant enhancement with respect to the MF pansharpening literature and it confirms, even for this particular application, the remarkable capability of MFs in preserving the original features of the processed images.

Finally, a visual analysis is performed. We report in Fig. 2 the outcomes of the considered algorithms and the reference image (GT). The differences among the presented methods are not easy to appreciate. Only the fusion method proposed in [24] reveals a very blurred image with an overall quality significantly worse than the other images. To ease the visual inspection only the injected details, i.e. the differences between the fused image $\widehat{MS}$ and the original MS image upsampled at the PAN scale $\widetilde{MS}$, are shown in Fig. 3. In this case the superior performances obtained by the proposed method are much more evident, both in terms of spectral and spatial accuracy.

5   Conclusions

Morphological filters have proven to be useful in several fields of image processing, as, for example, in segmentation and denoising applications. Nevertheless, their application in other research fields such as pansharpening, is rare. In this paper, we have focused our attention on exploring the capabilities of morphological filters in pansharpening. A reduced resolution quality assessment procedure has been performed on two real data sets acquired by the WorldView-2 and the GeoEye sensors. A comparison among morphological-based pyramid schemes using different operators and structuring elements has been carried out. Experiments have shown the validity of the proposed morphological pyramid scheme based on half gradients with respect to several state-of-the-art approaches.

References

1. Addesso, P., Conte, R., Longo, M., Restaino, R., Vivone, G.: A pansharpening algorithm based on genetic optimization of morphological filters. In: Proc. IEEE IGARSS, pp. 5438–5441 (2012)
2. Aiazzi, B., Alparone, L., Baronti, S., Garzelli, A.: Context-driven fusion of high spatial and spectral resolution images based on oversampled multiresolution analysis. IEEE Trans. Geosci. Remote Sens. 40(10), 2300–2312 (2002)
3. Aiazzi, B., Alparone, L., Baronti, S., Garzelli, A., Selva, M.: MTF-tailored multiscale fusion of high-resolution MS and Pan imagery. Photogramm. Eng. Remote Sens. 72(5), 591–596 (2006)
4. Aiazzi, B., Alparone, L., Baronti, S., Garzelli, A., Selva, M.: Advantages of Laplacian pyramids over "à trous" wavelet transforms. In: Bruzzone, L. (ed.) Proc. SPIE Image Signal Process. Remote Sens. XVIII. vol. 8537, pp. 853704-1–853704-10 (2012)
5. Alparone, L., Wald, L., Chanussot, J., Thomas, C., Gamba, P., Bruce, L.M.: Comparison of pansharpening algorithms: Outcome of the 2006 GRS-S data fusion contest. IEEE Trans. Geosci. Remote Sens. 45(10), 3012–3021 (2007)
6. Bai, X.: Morphological image fusion using the extracted image regions and details based on multi-scale top-hat transform and toggle contrast operator. Digit. Signal Process. 23(2), 542–554 (2013)
7. Bai, X., Zhou, F., Xue, B.: Edge preserved image fusion based on multiscale toggle contrast operator. Image and Vision Computing 29(12), 829–839 (2011)
8. Chavez Jr., P.S., Sides, S.C., Anderson, J.A.: Comparison of three different methods to merge multiresolution and multispectral data: Landsat TM and SPOT panchromatic. Photogramm. Eng. Remote Sens. 57(3), 295–303 (1991)

9. Flouzat, G., Amram, O., Laporterie-Déjean, F., Cherchali, S.: Multiresolution analysis and reconstruction by amorphological pyramid in the remote sensing of terrestrial surfaces. Signal Process. 81(10), 2171–2185 (2001)
10. Goutsias, J., Heijmans, H.J.A.M.: Nonlinear multiresolution signal decomposition schemes. I. morphological pyramids. IEEE Trans. Image Process. 9(11), 1862–1876 (2000)
11. Goutsias, J., Heijmans, H.J.A.M.: Nonlinear multiresolution signal decomposition schemes. II. morphological wavelets. IEEE Trans. Image Process. 9(11), 1897–1913 (2000)
12. Laporterie, F.: Représentations hiérarchiques d'images avec des pyramides morphologiques. Application à l'analyse et à la fusion spatio-temporelle de données en observation de la Terre. Ph.D. thesis (2002)
13. Laporterie-Déjean, F., Amram, O., Flouzat, G., Pilicht, E., Gayt, M.: Data fusion thanks to an improved morphological pyramid approach: comparison loop on simulated images and application to SPOT 4 data. In: Proc. IEEE IGARSS, pp. 2117–2119 (2000)
14. Laporterie-Déjean, F., Flouzat, G., Amram, O.: Mathematical morphology multilevel analysis of trees patterns in savannas. In: Proc. IEEE IGARSS, pp. 1496–1498 (2001)
15. Laporterie-Déjean, F., Flouzat, G., Amram, O.: The morphological pyramid and its applications to remote sensing: Multiresolution data analysis and features extraction. Image Anal. Stereol. 21(1), 49–53 (2002)
16. Liu, J.G.: Smoothing filter based intensity modulation: A spectral preserve image fusion technique for improving spatial details. Int. J. Remote Sens. 21(18), 3461–3472 (2000)
17. Maragos, P.: Morphological filtering for image enhancement and feature detection. In: The Image and Video Processing Handbook, 2nd edn., pp. 135–156. Elsevier Academic Press (2005)
18. Mukhopadhyay, S., Chanda, B.: Fusion of 2D grayscale images using multiscale morphology. Pattern Recogn. 34(10), 1939–1949 (2001)
19. Otazu, X., González-Audícana, M., Fors, O., Núñez, J.: Introduction of sensor spectral response into image fusion methods. Application to wavelet-based methods. IEEE Trans. Geosci. Remote Sens. 43(10), 2376–2385 (2005)
20. Shah, V.P., Younan, N.H., King, R.L.: An efficient pan-sharpening method via a combined adaptive-PCA approach and contourlets. IEEE Trans. Geosci. Remote Sens. 46(5), 1323–1335 (2008)
21. Soille, P.: Morphological Image Analysis: Principles and Applications. Springer (2003)
22. Starck, J.-L., Murtagh, F., Fadili, J.M.: Sparse image and signal processing: wavelets, curvelets, morphological diversity. Cambridge University Press (2010)
23. Toet, A.: A morphological pyramidal image decomposition. Pattern Recognition Letters 9(4), 255–261 (1989)
24. Toet, A.: Hierarchical image fusion. Mach. Vision App. 3(1), 1–11 (1990)
25. Tu, T.-M., Su, S.-C., Shyu, H.-C., Huang, P.S.: A new look at IHS-like image fusion methods. Inform. Fusion 2(3), 177–186 (2001)
26. Vivone, G., Alparone, L., Chanussot, J., Dalla Mura, M., Garzelli, A., Licciardi, G., Restaino, R., Wald, L.: A critical comparison among pansharpening algorithms. IEEE Trans. Geosci. Remote Sens. 53(5), 2565–2586 (2015)
27. Wald, L., Ranchin, T., Mangolini, M.: Fusion of satellite images of different spatial resolutions: Assessing the quality of resulting images. Photogramm. Eng. Remote Sens. 63(6), 691–699 (1997)

An Automated Assay for the Evaluation of Mortality in Fish Embryo

Élodie Puybareau[1]([✉]), Marc Léonard[2], and Hugues Talbot[1]

[1] Université Paris-Est / ESIEE,
2 Boulevard Blaise-Pascal, 93162 Noisy-le-Grand Cedex, France
{elodie.puybareau,hugues.talbot}@esiee.fr
[2] L'Oréal Recherche et Développement
Aulnay-sous-Bois, France

Abstract. Fish embryo models are used increasingly for human disease modeling, chemical toxicology screening, drug discovery and environmental toxicology studies. These studies are devoted to the analysis of a wide spectrum of physiological parameters, such as mortality ratio. In this article, we develop an assay to determine Medaka *(Oryzias latipes)* embryo mortality. Based on video sequences, our purpose is to obtain reliable, repeatable results in a fully automated fashion. To reach that challenging goal, we develop an efficient morphological pipeline that analyses image sequences in a multiscale paradigm, from the global scene to the embryo, and then to its heart, finally analysing its putative motion, characterized by intensity variations. Our pipeline, based on robust morphological operators, has a low computational cost, and was experimentally assessed on a dataset consisting of 660 images, providing a success ratio higher than 99%.

Keywords: Toxicology · Medaka · Image stabilisation · Change detection · Connected filtering

1 Introduction

Fish embryo models are used increasingly for human disease modelling, chemical toxicology screening, drug discovery and environmental toxicology studies. Fish embryos (Medaka and Zebrafish) are transparent and their cardio-vascular system is readily visible. They do not require difficult husbandry techniques, they are readily available, belong to the vertebrata subphylum (like humans), and their living milieu makes them a good model to study the consequences of waterways pollution [1]. In addition they are of intermediate size (a few mm long), and are transparent, so they do not require high-power microscopes or sophisticated image acquisition devices. Their circulatory system, in particular, is readily visible. As a result, they have been used in toxicology for a number of

---

This research was partially funded by the French *Agence Nationale de la Recherche* Grant Agreements ANR-10-BLAN-0205.

years [2]. However, the study of such embryos requires a lot of measurements. This is true from the physiological point of view and also from that of the effects of various chemical compounds on their health and behavior. In the past, measurements were conducted manually or semi-automatically, but these approaches do not scale [3]. However, these studies are poised to increase dramatically in both number and scale in the near future.

Recent progress in the automation of image analysis procedures with Medaka [1] or more commonly Zebrafish [4,5] is a natural consequence given the increasing demand in image-based toxicology assays. A beating heart is clearly one of the most basic vital signs. In this article, we propose the design of an image processing pipeline based on mathematical morphology [6,7,8] to assess the presence or absence of a beating heart pattern in fish embryos. We describe, investigate and propose solutions for all the problems and challenges we have encountered in this study. Another objective of this article is to showcase the design of an automated pipeline for processing sequences of fish embryos and more generally to provide some guidelines for the design of robust image analysis pipelines, through the particular example of a real-world problem.

The remainder of the article is as follows. In Section 2, we state our problem and list some of the difficulties we encountered. Section 3 lists our materials and methods. Section 4 provides an overview of our solution. Section 5 shows our results, together with validation and discussion. We conclude in Section 6.

2   Problem Statement and Difficulties

In this article, we study short sequences of Medaka (*Oryzia latipes*) placed in 24-well plates handled by a movable platform under a low-power, fixed macro camera. The plates are illuminated from below. The objective is to find out if every embryo in the plate is alive or not. For this, we rely on the fact that a living embryo should exhibit a visibly beating heart. The uncompressed video sequences are acquired one by one by moving the platform under the fixed camera. The embryos are anesthetized to prevent swimming. Since they are transparent we expect that the heart should be visible under the vast majority of poses.

In many real-world applications, the available hardware was designed for a previous task, and replacement or upgrade is a difficulty. In our case, the platform was initially designed for single-frame acquisition and processing, and not designed to acquire, process or store large streams of video. In particular, the camera is limited to 15 fps, and there is only limited storage. As a result, we faced a real problem of memory management and timely processing due to image size. Hence, we reduced the acquisition to 2 seconds at 15 frames per second and cropped the frames to 800×600, the camera recording only the center of the field of view. The heart of medakas usually beats at around 130 beat per minute (bpm) but in extreme cases it can reach 300 bpm, i.e. 5 Hz. The Nyquist frequency at 15fps is 7.5Hz, so the margin is sufficient but small.

Because of these acquisition and memory restrictions, embryos need to be centered in the field to avoid unusable partial views during acquisition. Compounding this problem, even though we should expect embryos, even live ones, to remain immobile while they are anesthetized, they may still exhibit some residual motion. These can be involuntary reflex swimming (incomplete anesthesia). Artefactual motion can also be due to poorly controlled platform motion inducing sliding, vibrations, shocks due to the general environment of the lab. These may induce false positive on dead embryos. In particular, since the eyes of Medaka are not transparent, this fact can cause significant difficulties. First, while still in egg form, embryos appear tightly wound and eyes can obscure the heart, making the detection of a heartbeat difficult. It is important for the application to know in advance if a well contains an egg or a free swimming eleutheroembryo. On hatched embryos, eyes constitute the darkest part of their body. As a result, their contour has high contrast. In the sequel, we will see that we rely on variance measures to detect the heartbeat. However, in case of even small or insufficiently compensated frame motion, this artificially creates areas of high variance and, in the end, generates false positives results. Unfortunately, the heart is fairly close to the eyes in Medaka, so this problem needs to be handled carefully.

During experimental runs, illumination is manually set by the operator. Hence, it may vary from one experiment to the next, and so the level of electronic noise and image contrast can vary as well. Specifically, strong noise over the embryo body can induce false positives. Simply looking for grey-level variability in the heart region is not robust enough.

As in any real application, sometimes the hardware does not provide expected data. Here incomplete videos can sometimes be acquired, and a whole part of the sequence may appear black or the sequence may be short.

3   Material and Methods

3.1   Material

Organism. Medaka (*Oryzia latipes*) is a small fish member of the genus *Oryzias*. In adulthood, it can reach 4cm in length, and live in both oceans and rivers. It hatches after 7 days at 26. In our study, eggs were purchased from Amagen (UMS 3504 CNRS / UMS 1374 INRA). They were placed in a neutral aqueous medium with methylene blue to detect dead eggs, and studied until 4 days after hatching. Embryos rely on their vitellogenic reserve for their caloric supply, so no nutrients were added to the aqueous medium. For images acquisition, they were placed and manually centered in a new, clear well on a thin gelified bed to prevent sliding shortly before analysis.

Computer and Software. We used the Python 2.7.6 environnement under Windows XP in a HP computer with a Core i7 920 CPU and 4GB of RAM. We used the Numpy, Scipy, Scikit-image[9], Pink [10], and OpenCV for Python. Platform control and image data acquisition was performed using FEI Visilog 7. Inter-process communication was handled through the python COM+ interface via Microsoft Excel.

Hardware. All videos were acquired using a color camera LEICA DFC 300 FX and an acquisition platform purpose-built by FEI .

3.2 Protocol of Acquisition

Fish embryos were anesthetized with tricaïne (0.1g/L) and placed on a support gel in a 24-well plate, one embryo per well. The plate was then placed under the connector board and the acquisition was automatically performed by Visilog. We recorded 2s of uncompressed video sequence at 15 frame per second and a resolution of 800×600 pixels. Platform motion, still images and video sequence acquisition takes 10s per well.

4   Solution Overview

Because of storage constraints, one aim was to achieve the complete sequence analysis online within the same time frame as the acquisition, i.e. in under 10s. We therefore propose a robust pipeline suitable for production usage. It consists of simple operators, which are for the most part readily available and fast. Figure 1 presents the flowchart of our assay. It is split into two phases: a preprocessing step which consists of sequence stabilization and denoising, while the actual processing step consists of detecting significant periodic changes in the embryo (variation of grey level values) assuming they are caused by its beating heart. In this section, we will use standard notations. Let I, J be grey level images defined on a support Ω, taking 8-bit discrete values, i.e. $I, J : \Omega \rightarrow \mathbb{Z} \cap [0, 255]$. $\delta_B(I)$ is the dilation of I by the structuring element B, $\epsilon_B(I)$ is the adjunct erosion, $\gamma_B(I)$ and $\varphi_B(I)$ the corresponding morphological opening and closing [11], $\gamma^r(I, J)$ is the geodesic reconstruction by dilation of I under J (which is an algebraic opening). $\gamma^\lambda(X)$ is the area opening of X with parameter λ [12]. We also introduce the Boolean image $I_{\geq\theta} = \{\forall x \in \Omega, I_{\geq\theta}(x) = (I(x) \geq \theta)\}$, i.e. the thresholded image of I at value θ. To implement the connected operators [13,14], an efficient max-tree/min-tree framework is used [15].

4.1   Segmentation of the Embryo

The first step is an initial segmentation of the embryo. Indeed, this segmentation is crucial for several reasons. In order to reduce the memory usage, we need to crop the area of interest to a small window centered on the embryo. This step allows us to weed out the sequences where an embryo is not visible or intersects the acquisition window. Moreover, we generally need to stabilize the sequence, and this stabilization must be performed on the embryo itself, and not other elements on the field of view like the contours of the well. The background appears white whereas edges of the well and embryo are darker (see Fig. 2(a)), particularly the eyes are very dark. Therefore the embryo is easy to segment as the largest connected component in the min-tree associated with the darkest minimum not connected to the edges of the field of view, simultaneously maximizing

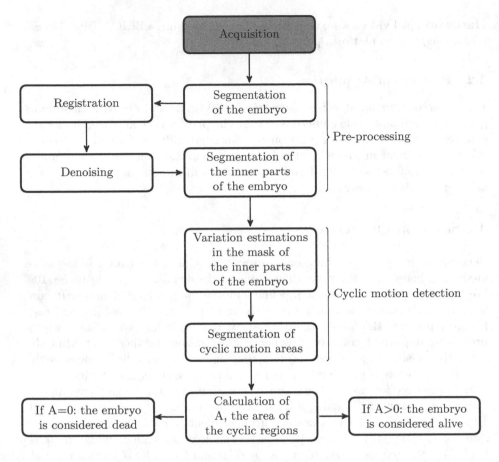

Fig. 1. Flowchart of our embryo mortality image processing assay

the inter-class variance between foreground and background (i.e, following the Otsu criterion [16]). We implemented it as follows. If θ_O is the Otsu threshold for the first input frame S^0, then

$$A^0 = \mathbb{1} - S^0_{\geq(\theta_O+\theta_b)}, \tag{1}$$

where $\mathbb{1}$ is a boolean image consisting only of ones, and θ_b is a baseline value, optimized to 30 experimentally.

Because we are not interested in the thin elements of the embryo (i.e. the tail), a morphological opening is applied, and because a single connected component is desired, an area criterion is also used:

$$A^1 = \varphi_{\mathcal{B}_1}(\varphi_{\lambda_1}(A^0)). \tag{2}$$

Experimentally, we determined the average area of an healthy embryo to 3000 pixel, so we set λ_1 conservatively to 600 pixel. The embryo's body is between

20-60 pixel wide, while artifacts in the well are usually much smaller, so $\mathcal{B}_1$ is set to a discrete Euclidean ball of radius 3. Since the embryo is expected to be near the center of the field of view, we remove all objects connected to the frame of the image, calling the result A^2. For speed, all these operators are implemented on the min-tree structure of the image, except the opening with $\mathcal{B}_1$, which is approximated by a fast, separable dodecagon [17].

This result A^2 is expected to represent the mask of the embryo (see Fig 2(b)). However, if the result is empty, this means that the embryo is not centered in the well and then the sequence cannot be reliably analyzed. If we do find an embryo, we crop the sequence by defining a bounding box around our segmentation, dilated by 10 pixels. The result is a new sequence S^1 centered on the thorax of the embryo (see Fig 2(c)).

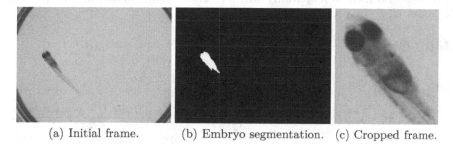

(a) Initial frame.   (b) Embryo segmentation.   (c) Cropped frame.

Fig. 2. Segmentation of the initial frame to locate and the embryo in the well

4.2 Registration

Because of vibrations associated with other equipment around the acquisition platform, the embryo can appear to slide up and down in its well. The sequence needs to be stabilized to avoid false positives. For speed, availability and ease of use, we have chosen the SIFT key-point based method [18]. This method detects remarkable points on each frame and matches sets of pairs of points P_1 and P_2 between two frames. We then solve the rigid transformation equation:

$$P_1 = P_2 \times \mathcal{R} + \mathcal{T} \tag{3}$$

Where $\mathcal{T} = (d_x, d_y)$ is the translation and $\mathcal{R}$ the rotation matrix. We consider only rigid transformations because motion consists only of vibrations and the embryo does not deform. Our model allows to select between both translation-rotations and translation only. If rotation appears negligible (i.e. the sum of square differences between the registration using translation-rotation and translation-only models are not significantly different), then only the translation $\mathcal{T}$ between P_1 and P_2 is calculated while $\mathcal{R}$ is set to the identity. In our experiments, this is the case most of the time.

We then apply the estimated transform between the first cropped frame of the sequence S^1 and all subsequent frames, taking the first as reference. We compute $d_{xmax} = \max |d_x|$ and $d_{ymax} = \max |d_y|$ for all frames, and we crop the bounding box of the whole sequence further by these amounts in x and y respectively. We call S^2 the stabilized, cropped sequence of the embryo.

4.3   Denoising

To eliminate global illumination variation during a sequence, which occurs some-times, we equalize the average intensity of all frames. We obtain a sequence S^3 where the average value of all frames is constant throughout the sequence.

Depending on the average illumination, the sequence can be quite noisy, so we used a 2D+t bilateral filter [19] to remove the noise remaining on S^3.

The bilateral filter is a spatially variant convolution, where for each pixel at coordinate (i, j), in a window W, we have:

$$\mathcal{I}_D(i,j,t) = \frac{1}{\sum_{(k,l,m) \in W} w(i,j,t,k,l,m)} \sum_{(k,l,m) \in W} I(k,l,m) * w(i,j,t,k,l,m), \quad (4)$$

where

$$w(i,j,t,k,l,m) = \exp\left(-\frac{(i-k)^2 + (j-l)^2 + (t-m)^2}{2\sigma_d^2} - \frac{\|I(i,j,t) - I(k,l,m)\|^2}{2\sigma_r^2}\right).$$

$$(5)$$

Here I is the original input image to be filtered; $\mathcal{I}_D$ is the denoised intensity of pixel (i, j). It is crucial not to filter too much, otherwise the heartbeat might be edited out. We have optimized the parameters that removed the noise without removing the heartbeat. These are: window size=$5 \times 5 \times 5$, σ_r=0.75, σ_d=0.9. The result is a denoised sequence S^4. Due to border effects, we remove the first three and the last three frames from this sequence, which is then 24-frames long.

4.4   Segmentation of the Inner Parts of the Embryo

If the embryo is alive, the heart should be detected in its thorax region. To ascertain this, cyclic motion should be detected in this region of the sequence, but not anywhere else where remaining noise or motion might be present. Because eyes are very dark, noise causes detectable variations in them, and remaining motion may be detectable near the contours of the embryo. We developed a mask $\mathcal{M}^1$ of the inner part of the embryo to avoid false detection in these areas.

We define B^1 as the minimum image of the sequence: $B^1 = min(S^4)$. In this image, where the heart and vessels are darker because of the presence of blood, we apply the same procedure as in section 4.1, i.e:

$$B^2 = \varphi_{\mathcal{B}_1}(\varphi_{\lambda_1}(\mathbb{1} - B^1_{\geq(\theta_O + \theta_c)})) \quad (6)$$

In this instance, because of the change of contrast, θ_c is set to 20; θ_O is again obtained using the Otsu criterion; λ_1 and $\mathcal{B}$ are unchanged. The resulting B^2 is a binary mask (with values in $\{0, 1\}$) of the registered body of the embryo. Considering B^2 as a geodesic mask, we now segment the eyes as the one or two more prominent minima from the min-tree. We cannot rely on the eyes being separated. Depending on the pose of the embryo, they might get merged. We write:

$$\mathcal{M}^1 = \epsilon_{\mathcal{B}_3}(\gamma_{\mathcal{B}_2}((B^2.B^1)_{\geq(\theta_O - \theta_c)})) \tag{7}$$

Here the dot in $B^2.B^1$ denotes the pixelwise multiplication. Note that θ_O is re-estimated from the grey-level distribution of the embryo within the geodesic mask. In this formula, $\mathcal{B}_2$ is a ball of radius 1 and $\mathcal{B}_3$ a ball of radius 3. The outline of the resulting mask $\mathcal{M}^1$ are exemplified on Fig. 3.

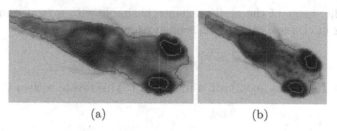

(a)   (b)

Fig. 3. Inner parts segmentation on two embryos. (a) is alive whereas (b) is dead.

4.5   Variation Estimations in the Mask of the Inner Parts of the Embryo

Our challenge is now to detect cyclic motion-induced variations in the sequence assuming it corresponds to a heartbeat. The heartbeat is noticeable by changes of contrast due to blood cell concentration in the heart region. Computing the grey level variance at each pixel of a sequence along the time line show the grey-level variability for that pixel, however this variability may not be due to a cyclic pattern. To account for cyclic variations, we split $\mathcal{S}^4$ into 3 sub-sequences of 8 frames ($\mathcal{S}_1^4$, $\mathcal{S}_2^4$ and $\mathcal{S}_3^4$). An interval of 8 frames is comparable to the expected period of a single heart beat. We compute the pixels variance inside $(\mathcal{M}^1)$ on each sub-sequence, yielding three variance images $\mathcal{V}_1, \mathcal{V}_2, \mathcal{V}_3$. We next compute the median of these three images $\mathcal{V}_i$. In this way, some spurious, potentially even large, residual variation occurring only once in the sequence will produce a large variance only once, and so the median of all the variances in this area will remain small. On the other hand, cyclic variation will exhibit significant variance in all three sub-sequences.

$$\mathcal{S}^4 = \{\mathcal{S}_1^4 \cup \mathcal{S}_2^4 \cup \mathcal{S}_3^4\} \tag{8}$$

$$\forall i \in [1, 3], \ \mathcal{V}_i = \text{variance}(\mathcal{S}_i^4) \tag{9}$$

$$C^1 = \mathcal{M}^1.\,\text{median}(\mathcal{V}_i, i \in [1, 3]) \tag{10}$$

The result of this process is shown in Fig. 4.

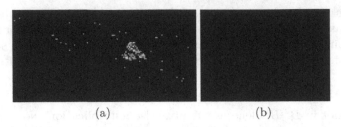

(a)   (b)

Fig. 4. False color rendering of the temporal variance. (a) is for a living embryo, (b) a dead one.

4.6 Segmentation of Cyclic Motion Areas

Binarizing the image of the cyclic variance is simple with area openings and a small closing in classical alternating sequence.

$$D^1 = (\gamma^{\lambda_3}(\varphi_{\mathcal{B}_4}(\gamma^{\lambda_2}(C^1))))_{\geq 1}. \tag{11}$$

Here, $\lambda_2 = 3$, $\mathcal{B}_4$ is a radius-2 ball, and $\lambda_3 = 10$. This image is then thresholded at 1.

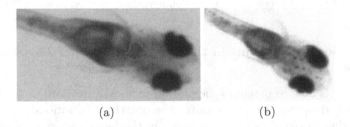

(a)   (b)

Fig. 5. Segmentation of cyclic motion detection on embryos. (a) corresponds to the segmentation of 4 (a) (living embryo), and (b) corresponds to the segmentation of 4 (b) (dead embryo)

If the number of non-zero pixels in D^1 is zero, we consider the embryo is dead, otherwise, it is considered alive.

4.7 Parameter Optimization

In our pipeline, the parameters such as the size of the structuring elements or thresholds need to be optimized. Many of them are dependent on the size of the embryo. For the remaining parameters, they were hand-optimized on a training sample of 100 sequences.

5 Results and Validations

5.1 Sequence Extraction

Videos are read as raw data interpreted as grey-level values. If the maximum intensity of any frame is zero, the video is deemed incomplete and discarded.

5.2 Processing

We tested our program on 651 videos taken over several experimental runs. The protocol was identical each time but conditions such as illumination or marker concentration could change. This number of videos is not insignificant and reflects production usage. We have also tested our protocol on unusual embryos: some with oedemas and with other malformations to ensure the robustness of our protocol (see Fig. 6(a) and (b)). The program processes a sequence in less than 10 seconds, which is in accordance with our initial constraints.

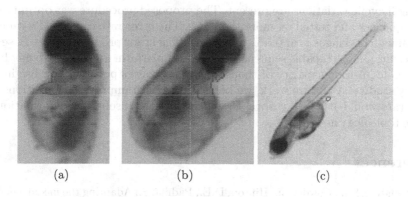

(a)   (b)   (c)

Fig. 6. Heart segmentation in the presence of oedema (a) or malformation (b). Incorrect segmentation in the fin due to fluttering (c).

5.3 Limitations

In some cases, even actually dead embryos can appear to move. This can be due to the gel or the presence of shadows (figure 6(c)). In dark areas, electronic noise is more prevalent, and may appear as motion and in some cases, the embryo may appear to slide on the gel. This happens if we incorrectly do not compensate for rotation in the sequence stabilization phase. The main remaining cause of problem is ambiguity: in some cases, the heart beats so slowly or weakly that we cannot detect it. In most cases, a human operator would also have difficulties in detecting it. We could also argue that embryo with weak heartbeats are not very healthy and likely to die, however this is unconfirmed.

5.4 Validation of the Method

We have run experiments on many sequences: some darker, some lighter, some with varying amounts of tricaine, some with embryo abnormalities, and so on. In total, we processed 651 such sequences, which might be considered sufficient for a first validation step.

From an initial set of 660 videos, 9 were eliminated as incomplete. From the remaining set of 651 sequences, 100 were used for training, i.e. optimizing the parameters of the pipeline. The remaining 551 were used for testing. There was 1 remaining error in the training sample (1% error rate). There were 3 errors in total in the test set, for an error rate of 0.54%. Such an error rate is effectively quite low.

6 Conclusion

This article can be viewed as a solution for the fish embryo version of Schrödinger's problem. We have indeed proposed a simple and effective image analysis pipeline to detect whether a Medaka embryo is alive or not, at a given time, in a particular well, and a specific stage of its development by detecting its heart region and finding if cyclic variations are present. The proposed procedure was optimized on 100 sequences and tested on more than 550. The error rate compared with careful manual checking is near 0.5%, which is sufficient for production use. Based on this result, a more capable sequence acquisition platform is currently being built, with an acquisition capacity of thousands of sequences per week, for use in toxicology studies. As future work, we believe that with improved acquisitions, the heart rate would also be measurable, which can bring useful new information for future toxicology assays.

References

1. Oxendine, S.L., Cowden, J., Hinton, D.E., Padilla, S.: Adapting the medaka embryo assay to a high-throughput approach for developmental toxicity testing. NeuroToxicology 27(5), 840–845 (2006)
2. Dial, N.A.: Methylmercury: Some effects on embryogenesis in the Japanese Medaka,Oryzias latipes. Teratology 17(1), 83–91 (1978)
3. Schindelin, J., Arganda-Carreras, I., Frise, E., Kaynig, V., Longair, M., Pietzsch, T., Preibisch, S., Rueden, C., Saalfeld, S., Schmid, B., Tinevez, J.Y., White, D.J., Hartenstein, V., Eliceiri, K., Tomancak, P., Cardona, A.: Fiji: An open-source platform for biological-image analysis. Nature Methods 9(7), 676–682 (2012)
4. Liu, T., Lu, J., Wang, Y., Campbell, W.A., Huang, L., Zhu, J., Xia, W., Wong, S.T.: Computerized image analysis for quantitative neuronal phenotyping in zebrafish. Journal of Neuroscience Methods 153(2), 190–202 (2006)
5. Xia, S., Zhu, Y., Xu, X., Xia, W.: Computational techniques in zebrafish image processing and analysis. Journal of Neuroscience Methods 213(1), 6–13 (2013)
6. Serra, J.: Image analysis and mathematical morphology. Academic Press (1982)
7. Soille, P.: Morphological Image Analysis, principles and applications. Springer (1999)

8. Najman, L., Talbot, H., (eds.): Mathematical Morphology: From theory to applications. ISTE-Wiley, London, UK (September 2010) ISBN 978-1848212152
9. Van der Walt, S., Schönberger, J.L., Nunez-Iglesias, J., Boulogne, F., Warner, J.D., Yager, N., Gouillart, E., Yu, T.: The scikit-image contributors: Scikit-image: Image processing in Python. Peer J 2(6), e453 (2014)
10. Couprie, M., Marak, L., Talbot, H.: The Pink Image Processing Library. Euroscipy, Paris (2011)
11. Heijmans, H.: Morphological image operators. Advances in Electronics and Electron Physics Series. Academic Press, Boston (1994)
12. Vincent, L.: Grayscale area openings and closings, their efficient implementation and applications. In: Proceedings of the Conference on Mathematical Morphology and its Applications to Signal Processing, Barcelona, Spain, pp. 22–27 (1993)
13. Salembier, P., Oliveras, A., Garrido, L.: Antiextensive connected operators for image and sequence processing. IEEE Transactions on Image Processing 7(4), 555–570 (1998)
14. Jones, R.: Connected filtering and segmentation using component trees. Computer Vision and Image Understanding 75(3), 215–228 (1999)
15. Najman, L., Couprie, M.: Building the component tree in quasi-linear time. IEEE Transactions on Image Processing 15(11), 3531–3539 (2006)
16. Otsu, N.: A threshold selection method from gray-level histograms. Automatica 11(285-296), 23–27 (1975)
17. Jones, R., Soille, P.: Periodic lines: Definition, cascades, and application to granulometries. Pattern Recognition Letters 17(10), 1057–1063 (1996)
18. Lowe, D.G.: Object recognition from local scale-invariant features. In: IEE ICCV, vol. 2, pp. 1150–1157 (1999)
19. Tomasi, C., Manduchi, R.: Bilateral filtering for gray and color images. In: Sixth International Conference on Computer Vision, pp. 839–846. IEEE (1998)

A Top-Down Approach to the Estimation of Depth Maps Driven by Morphological Segmentations

Jean-Charles Bricola[✉], Michel Bilodeau, and Serge Beucher

CMM – Centre de Morphologie Mathématique,
MINES ParisTech – PSL Research University,
35 rue St Honoré, 77300 Fontainebleau, France
{jean-charles.bricola,michel.bilodeau,serge.beucher}@mines-paristech.fr

Abstract. Given a pair of stereo images, the spatial coordinates of a scene point can be derived from its projections onto the two considered image planes. Finding the correspondences between such projections however remains the main difficulty of the depth estimation problem: the matching of points across homogeneous regions is ambiguous and occluded points cannot be matched as their projections do not exist in one of the image planes.

Instead of searching for dense point correspondences, this article proposes an approach to the estimation of depth map which is based on the matching of regions. The matchings are performed at two segmentation levels obtained by morphological criteria which ensure the existence of an hierarchy between the coarse and fine partitions. The hierarchy is then exploited in order to compute fine regional disparity maps which are accurate and free from noisy measurements.

We finally show how this method fits to different sorts of stereo images: those which are highly textured, taken under constant illumination such as Middlebury and those which relevant information resides in the contours only.

Keywords: Watershed · Segmentation hierarchies · Disparity estimation · Joint stereo segmentation · Non-ideal stereo imagery

1 Introduction

The estimation of depth maps from stereoscopic data traditionally comes down to finding pixel correspondences between stereo images. For the sake of simplicity, we assume throughout this paper that stereo images are rectified such that any scene point projects with the same ordinate in the stereo image planes. The difference in abscissa is referred to as the disparity and is inversely proportional to the depth being searched for.

Modern approaches are based on dense pixel correspondences and resort to the framework described in [11]. The first step consists of computing the matching costs between a pair of pixels for each scanline and for each disparity belonging

© Springer International Publishing Switzerland 2015
J.A. Benediktsson et al. (Eds.): ISMM 2015, LNCS 9082, pp. 122–133, 2015.
DOI: 10.1007/978-3-319-18720-4_11

to the search domain. The cost is obtained by means of a dissimilarity measure between the patches centred at the pixels being under study. The reader may find an exhaustive list of such measures in [7]. The second phase of the estimation aims at ensuring a disparity consistency among the scanlines. Local approaches achieve this by diffusing the costs across the scanlines which results in fast algorithms and end with a refinement process. Global approaches generally associate an energy to the disparity map which takes account of both the disparity local costs and the consistency across neighbour pixels. This energy is eventually minimised to yield to final disparity map.

Many of these approaches seek the actual frontiers of objects within their estimation process. For instance, the gradient of the image is used in order to determine depth continuities in [6], whilst a geodesic distance across a relief defined by the image intensity values is used in the context of pixels matching [5] so as to weight the importance of pixels which are in the vicinity of the patch centre with respect to the scene. This leads to a strong interest in using regions for stereo image analysis, because they determine the frontiers of objects as well as the membership of an occluded pixel within a region that is only semi-occluded. Several region-based algorithms have already been made available which either exploit matchings across over-segmentations [15] or fit planes through object-oriented regions given a non-refined disparity map such as [4,14].

In this work, we propose a novel region-based stereo algorithm with the following contributions:

1. The disparity maps are estimated without relying on dense pixel matchings. This aspect is interesting for stereo imagery which is either poorly textured or subject to noise because the assumption that a majority of valid pixel matchings exists with respect to the usual aggregation step may no longer hold. Hence this approach does not assume the existence of a roughly good initial disparity estimate.
2. Disparities are estimated at the region level on a hierarchy of segmentations composed of the coarse level which highlights the objects in the scene and a fine level which is over-segmented. The computations are performed in waterfalls: first at the coarse level where depth planes are highlighted, then at the fine level for which a finer degree of precision is obtained. One can choose at which level to stop the algorithm depending on the speed required for its application.
3. Contours are taken into account within the estimation process after a careful reasoning on occlusion boundaries. Disparities along contours are often misused because they require the knowledge of object membership. This problem has little been raised, with the exception of [13] which exploits boundary junctions to this end.

The mechanism of the depth estimation system is presented in section 2. The method exploits the advantages offered by a marker-driven watershed. Markers are used to produce the coarse and fine segmentations and are transferred across stereo images in order to produce equivalent segmentations which facilitate the establishment of contour point correspondences. These aspects are discussed in section 3.

2 Top-Down Estimation of Disparities

The proposed method relies on the concept of *regional disparity*. We define the regional disparity as a measure attributed to a region of the reference image. This measure represents the average disparity of the pixels that compose the region and is obtained by searching for a displacement which optimally superimposes the region with the target image.

In this approach, regional disparities are initially computed for a coarse partition of the reference image. At this level, region matchings (section 2.1) are quite stable because of their singularity. The resulting disparity map constitutes a gross approximation of the true disparity map, with the most important imprecisions across regions not being fronto-parallel to the camera. At the over-segmented level, matching errors are more frequent, but disparity variation is better captured because regions are smaller. A relaxation process (section 2.2) that relies on the disparity map obtained at the coarse level is applied onto the disparities obtained for the fine partition in order to correct any mistaken measure and ensure some smoothness across the coarse segments. Finally, a regularization of the depth map based on a linear estimation process called kriging (section 2.3) is performed so as to obtain the final disparity maps.

2.1 Region Matchings and Regional Disparities

A region $\mathbf{R}$ is represented as an indicator function R such that $R[x,y] = 1 \Leftrightarrow (x,y) \in \mathbf{R}$. We denote $\mathbf{R}^{(d)}$ the region obtained by shifting $\mathbf{R}$ of d pixels along the horizontal image axis and define its indicator function by $R^{(d)}[x,y] = R[x-d,y]$. A partition is a set that contains all the regions extracted from an image. We describe the coarse partitions of the reference and the second images of the stereo pair as P_{C1} and P_{C2} respectively. We also let P_{F1} to be the fine partition obtained for the reference image. The disparity $d^{\star}$ is assigned to a region of the reference image if its superimposition onto the second image shifted from $d^{\star}$ pixels minimises a dissimilarity cost chosen according to the image acquisition setup.

A popular dissimilarity measure employed on images acquired under the same illumination conditions is the Sum of Absolute Differences (SAD) between the image intensities. It is possible to evaluate the SAD for each region of the reference image and choose the disparity that yields the superimposition with the minimum cost. However, when a region undergoes a semi-occlusion in the second image, the true superimposition is likely to come at a high cost because it partially compares pixels with an occluding object. That results in errors as can be seen in figure 1(b). To circumvent that issue, it is best to compute a mean of absolute differences for each region resulting from the intersections of the reference image partition and a coarse partition of the second image of the stereo pair at every possible disparity and search for the region and the disparity that minimises the mean.

The Jaccard distance [9] measures the overlap between two regions by computing the ratio between their intersection area and union area. The distance between two regions is then expressed as:

$$c_{\text{Jaccard}}(\mathbf{R}_i, \mathbf{R}_j) = 1 - |\mathbf{R}_i \cap \mathbf{R}_j|/|\mathbf{R}_i \cup \mathbf{R}_j| \tag{1}$$

The asymmetric version of this distance which replaces $|\mathbf{R}_i \cup \mathbf{R}_j|$ by $|\mathbf{R}_i|$ in equation 1 can be used to discard the dissimilarity costs obtained across region intersections which cover less than a reasonable threshold of the reference image region. Doing so yields the coarse regional disparity map represented in figure 1(c).

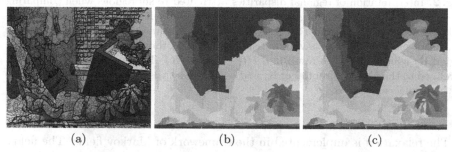

(a)   (b)   (c)

Fig. 1. (a) Coarse and fine segmentations obtained for the reference image of *Teddy*. (b) Regional disparities obtained for the coarse partition without a constraint partition P_{C2} and (c) with a constraint partition P_{C2}. Semi-occluded regions are less affected by measurement errors when using a constraint partition.

When images are poorly textured and acquired under different lighting conditions, the contours appear to remain the most pertinent criterion for matching regions. To this end, the gradients of the stereo images are compared using the SAD taking note of the following observation: a contour that separates two regions always constitutes the physical frontier of one of the regions but might constitute an occlusion border for the other region. For this reason, the regional disparities are computed across subregions of the reference image illustrated in figure 2: every region $\mathbf{R}_i$ of the reference image is split into two subregions denoted by $\mathbf{R}_i-$L and $\mathbf{R}_i-$R along its vertical skeleton. Regions which are likely to be semi-occluded are those for which the disparities of their subregions differ significantly so that the highest disparity equals the disparity of a neighbour region, the one which is occluding. The following rectification is therefore applied on the regional disparities: if $d^\star(\mathbf{R}_i-$R$) \gg d^\star(\mathbf{R}_i-L)$ and $d^\star(\mathbf{R}_i-$R$) \simeq d^\star(\mathbf{R}_j-L)$ such that $\mathbf{R}_i-$R and $\mathbf{R}_j-$L are neighbours, the disparity $d^\star(\mathbf{R}_i-$L$)$ is transferred to $\mathbf{R}_i-$R. And vice-versa.

2.2 Relaxation of Regional Disparities on the Fine Partition

A relaxation process is applied to the disparities initially measured on the fine partition. The process consists of detecting wrong disparity measures, ensuring that disparities evolve smoothly across neighbour regions whilst permitting discontinuities at objects frontiers. For that reason, the relaxation process is applied independently on each cluster of fine regions composing the same coarse region.

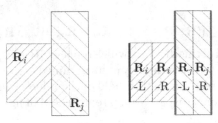

Fig. 2. Interpretation of regional disparities attributed to subregions when comparing the image gradients. Every region $\mathbf{R}_i$ is split along its vertical skeleton into left and right sub-regions $\mathbf{R}_i-\mathrm{L}$ and $\mathbf{R}_i-\mathrm{R}$. A region that is semi-occluded, in this example $\mathbf{R}_i$, has a piece of contour having a disparity equal to the disparity of the occluding region $\mathbf{R}_j$ which is likely to result in $d^\star(\mathbf{R}_i-\mathrm{R}) \simeq d^\star(\mathbf{R}_j-\mathrm{L})$. The disparity $d^\star(\mathbf{R}_i-\mathrm{L})$ is related to the physical frontier of $\mathbf{R}_i$ and constitutes the sole disparity representative of the displacement of $\mathbf{R}_i$.

The relaxation is implemented in the framework of Markov fields. The field's nodes represent the fine regions enclosed in the coarse region under study and the edges model the adjacency relationships between these regions. An objective function is assigned to every Markov field and is expressed as a sum of pairwise terms $P_{i,j}$ which grow quadratically with the difference between disparities assigned to neighbour regions $\mathbf{R}_i$ and $\mathbf{R}_j$, denoted as $d^\circ(\mathbf{R}_i)$ and $d^\circ(\mathbf{R}_j)$ respectively, and a sum of unary terms U_i which penalize the assignment of a disparity $d^\circ(\mathbf{R}_i)$ that strongly contradicts the initial measure $d^\star(\mathbf{R}_i)$. It is of course essential to determine the reliability $\alpha_i \in [0,1]$ of a disparity measure $d^\star(\mathbf{R}_i)$ and modulate the unary terms accordingly. To this end, we use the disparity map obtained for the coarse partition to localise both the fine regions which are likely to be severely occluded in the second image of the stereo pair and those whose disparity is significantly too far from the coarse regional disparity d_C. The measures attributed to these fine regions are assumed to have a low reliability. The pairwise and unary terms are defined by equations 2 and 3 respectively.

$$P_{i,j} \propto (d^\circ(\mathbf{R}_i) - d^\circ(\mathbf{R}_j))^2 \tag{2}$$

$$U_i = \alpha_i|d^\star(\mathbf{R}_i) - d^\circ(\mathbf{R}_i)| + (1 - \alpha_i)|d_C - d^\circ(\mathbf{R}_i)| \tag{3}$$

Finally, the disparity assignments which minimise the objective function are found using the minimum cut algorithm described in [10]. The effect of the relaxation process is illustrated in figures 3(b) and 3(c).

2.3   Regularisation of Disparity Maps

The regularisation of disparity maps is performed by ordinary kriging [2]. Given a set of samples for which the values are known as well as a variogram modelling the variability of the values taken by two points in terms of their distance, the kriging computes an unbiased estimator which minimises the variance of the estimation error.

Since there is no prior information regarding the expected depth map, we assume that the variability between two pixels in terms of disparity is proportional to their euclidean distance provided that these pixels are included in the same coarse segment. The kriging is applied independently on every coarse segment. The seeds originate from points belonging to the watershed of the coarse partition and from pics or holes enclosed in a region (cf. figure 3(e)). A correspondence in the second image is searched for each candidate seed. Only those having a disparity equal to the regional disparity of the fine region in which they are enclosed are eventually retained within the refinement process which yields the result presented in figure 3(f).

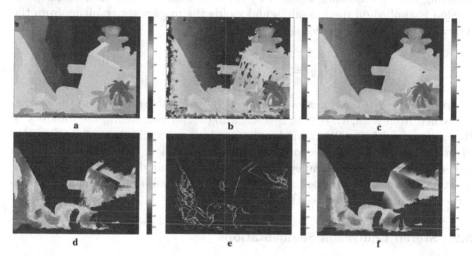

Fig. 3. The top-down approach to depth map computation. (a) Regional disparities of P_{C1} which serve as a basis for detecting wrong measures on the (b) brute regional disparity map of P_{F1}. (c) Regional disparities issued from the relaxation process on P_{F1} and (d) its visualisation with a restricted range of disparities. (e) Contour disparities along the watershed of the coarse segmentation and internal points disparities. (f) Interpolation result obtained by kriging.

3   Joint Stereo Segmentations Using Markers Transfer

The watershed-based segmentation [3] controlled by markers plays an essential role in this approach. This section is devoted to the mechanisms employed for extracting markers which delineate the salient objects in the scene, even if the image gradients suffer from leakages, and for computing equivalent segmentations between the images composing the stereo pair given the regional disparities and matching criteria presented in section 2.

3.1   Markers Extraction

The h-minima of a function g are defined as a binary function $M(g, h)$ which exclusively equals 1 at any point satisfying $\mathbf{R}_g^*(g + h) - g > 0$, where the elevation

h is a positive constant and $R_g^*(g+h)$ stands for the dual geodesic reconstruction of $g+h$ on top of the original function g. Taking the h-minima of a colour gradient yields markers which segment regions based on their frontier contrast. Although h has been fixed in all our experiments, it is worth mentioning that this elevation can be dynamically determined using a method similar to [12].

Markers originating directly from the h-minima have a severe inconvenient because two catchment basins of g merge as soon as a point with the smallest altitude on the watershed of g has been flooded. This favours premature fusions of h-minima as h increases when the gradient is subject to leakages. One way of preventing a flood at such leaking passages is to analyse the shape of the lakes resulting from the flooding induced by the h-minima. To this end, an adaptive erosion is applied on the h-minima and yields the marker set resulting from the indicator function in equation 4:

$$M_\alpha(g,h)[x,y] = \begin{cases} 1 & \text{if } (\mathcal{D} - R_\mathcal{D}(\alpha\mathcal{D}))[x,y] > 0 \\ 0 & \text{otherwise} \end{cases} \tag{4}$$

where $\mathcal{D}$ is the distance function computed by successive erosions on the marker set issued from $M(g,h)$, $\alpha \in [0,1[$ is a scaling factor controlling the intensity of the erosion and $R_\mathcal{D}(\alpha\mathcal{D})$ stands for the geodesic reconstruction of the rescaled distance function under the original distance function. The adaptive erosion splits markers at narrow valleys with respect to their distance function. New markers are contained in the original h-minima and all h-minima can be reconstructed from the new marker set.

3.2 Stereo Equivalent Segmentations

The morphological co-segmentation consists of obtaining equivalent partitions between the images composing a stereo pair. In that context, the watershed segmentation driven by markers remains the tool of choice. The first experiments are presented in [3], where the idea is to propagate the markers obtained for the reference image to the second image. We now propose two mechanisms for obtaining equivalent segmentations of stereo images thanks to the transfer of labels onto image markers. The first one which is asymmetric relies on the regional disparities directly without being concerned about the matching criteria which makes it ideal for non-ideal stereo imagery. The second one which is symmetric revisits the matching criteria discussed in section 2 and in the current form applies only to images taken under the same illumination conditions. However, its mechanism identifies regions which are occluded in the reference image and attributes them a specific label which does not exist in the reference image partition.

Asymmetrical Transfer. In this algorithm, the transfer of markers is guided using the regional disparities presented in section 2. The algorithm first consists of estimating the equivalent partition of the second image, followed by the labelling of the gradient minima from which the watershed is eventually constructed.

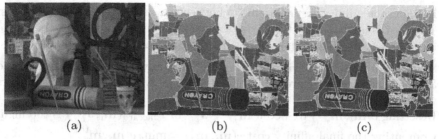

(a)   (b)   (c)

Fig. 4. Asymmetric co-segmentation of (a) *Art* reference image from Middlebury database. (b) Reference image segmentation. (c) Equivalent segmentation of the second image. The labels attributed to the markers are preserved throughout the co-segmentation process and yields a matching between regions.

Given the partitioning of the reference image as a label map $\mathcal{L}_1$, such that $\mathcal{L}_1[x, y] = i \Leftrightarrow R_i[x, y] = 1$, the label map $\mathcal{L}_2$ is estimated according to equation 5.

$$\mathcal{L}_2[x, y] = \arg\max_i \ \{d^*(\mathbf{R}_i) \times R_i[x + d^*(\mathbf{R}_i), y]\} \tag{5}$$

Hence, each region $\mathbf{R}_i$ belonging to the reference image partition is shifted according to its regional disparity. The shifts are of different intensities implying therefore the existence of some overlaps between the shifted regions. In the case of an overlap, only the region which has the smallest depth can be visible to the camera. Hence $\mathcal{L}_2[x, y]$ is set to the label of the region which remains visible at (x, y). The other regions at that point can be marked as being occluded.

The equivalent segmentation is obtained by computing the watershed controlled by the minima of the gradient of the second image. The labels estimated in $\mathcal{L}_2$ are transferred to the markers, i.e. if (x, y) belongs to the minima, then the label $\mathcal{L}_2[x, y]$ is attributed to that point. The preservation of labels through the co-segmentation process yields the matching between the regions of the stereo partitions, as shown in figure 4.

The asymmetrical transfer has one limitation: it is impossible to represent regions in the second image which do not appear in the reference image. Such regions are directly merged in the equivalent segmentation to regions that are visible in the reference image. So another way to tackle the co-segmentation problem is to focus on a symmetrical transfer.

Symmetrical Transfer. The problem now comes down to relabelling the markers obtained independently for each image of the stereo pair. To achieve this, a sequence of back-and-forth label transfers is performed until no label changes after one of the transfer. Let M_{C1} and M_{C2} be the set of markers chosen for computing the coarse segmentations of the reference and the second image respectively. We define as $c(\mathbf{m}_i, \mathbf{m}_j)$ the cost of transferring the label of $\mathbf{m}_i \in M_{C1}$ to $\mathbf{m}_j \in M_{C2}$. A single way transfer consists of:

1. Defining the affinity cost $c(\mathbf{m}_i, \mathbf{m}_j^{(d)})$ between $\mathbf{m}_i$ and $\mathbf{m}_j$ shifted by d pixels. This cost is initialized to the mean of absolute differences between the image intensities across each intersection. If $\mathbf{m}_i \cap \mathbf{m}_j^{(d)} = \emptyset$, the cost is set to $+\infty$.

2. Searching for a set of markers $\mathbf{J}_i^{(d)} = \{\mathbf{m}_k^{(d)}\}$ for any $\mathbf{m}_k \in M_{C2}$ such that the Jaccard distance, as expressed in equation 1, between this union of markers and $\mathbf{m}_i$ is minimised.

3. Resetting $c(\mathbf{m}_i, \mathbf{m}_j^{(d)})$ to $+\infty$ if $\mathbf{m}_j^{(d)} \notin \mathbf{J}_i^{(d)}$ or $c_{\text{Jaccard}}(\mathbf{m}_i, \mathbf{J}_i^{(d)})$ is too high.

4. Computing the final affinity cost $c(\mathbf{m}_i, \mathbf{m}_j) = \min_d c(\mathbf{m}_i, \mathbf{m}_j^{(d)})$

5. Transferring the label of $\mathbf{m}_i$ to $\mathbf{m}_j$ if $c(\mathbf{m}_i, \mathbf{m}_j)$ is reasonably small.

In this procedure, we use the symmetric Jaccard distance, because the markers are chosen with the same flooding criterion and yield segmentations at the same scale of precision. However such markers can split between two stereo images as can be noticed in figure 5. For that reason, step 2 has been introduced in the matching procedure in order to prevent a penalty that would be due to a low Jaccard distance taken between the markers individually. Our procedure hence favours the transfer of a unique label to markers that have split. Step 5 ensures that a transfer can only occur when the regions that are covered by markers are similar in texture and colour. Costs exceeding 0.08 as a mean of the absolute differences of image intensities scaled between 0 and 1 do not generally lead to pertinent matchings.

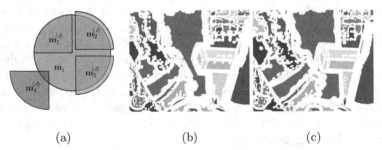

(a)   (b)   (c)

Fig. 5. (a) Illustration of the generalised Jaccard distance. Here, $\mathbf{m}_i$ is a marker of the reference image. The set $\mathbf{J}_i^{(d)}$ that minimises $c_{\text{Jaccard}}(\mathbf{m}_i, \mathbf{J}_i^{(d)})$ is $\{\mathbf{m}_1^{(d)}, \mathbf{m}_2^{(d)}, \mathbf{m}_3^{(d)}\}$. (b)-(c) Using this distance within the symmetric transfer algorithm enables the label transfer from a marker of the source image to its corresponding markers in the target image.

4   Experimental Results and Evaluation

In this section, the results obtained on the Middlebury database are analysed in terms of precision. We also present an application of the proposed method to a particularly challenging stereo pair subject to a considerable amount of noise and homogeneous regions.

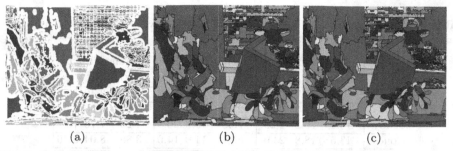

$$(a) \qquad\qquad\qquad (b) \qquad\qquad\qquad (c)$$

Fig. 6. Symmetric co-segmentation of *Teddy* reference image. (a) Reference image markers. (b) Reference image segmentation. (c) Second image segmentation. Regions of the second image which are occluded in the reference image now appear with new labels.

Accuracy. The precision evaluated on the Middlebury database [1] is presented in table 1. Pixels which are badly matched are those having a disparity which differ from a certain threshold with respect to the ground truth [11]. The main source of inaccuracy arises from highly slanted regions, like the ground in Teddy (cf. figure 3(c)). Searching for an optimal superimposition between such regions cannot be reasonably done by performing a simple shift which accounts for the observed instability. One should consider more complex geometrical transformations. Another source of imprecision comes from regions undergoing a severe image border occlusion, like in *Cones* or *Teddy*. Although these regions are detected thanks to the co-segmentation and then merged to neighbour regions according to the image gradient prior to the kriging process (cf. figure 3(f)), the linear variability model doesn't seem to be sufficient to guess the true disparities. Nevertheless, it is interesting to note that methods having the same level of precision usually produce disparity maps which are perceptually less appealing than ours obtained with the present method at the coarse level of the segmentation. The human sensibility to the contrast between the different depth planes is indeed predominant and this is not taken by the accuracy measurement into account.

Micro-stereopsis Imagery. Our method is compared to the semi-global estimation algorithm of [8] on a stereo pair having the following characteristics: low disparity range, acquisition under different illuminations, many homogeneous regions, noise, semi-transparent objects. The establishment of pixel correspondences is therefore particularly difficult and yields to ambiguities across homogeneous regions as shown in figure 7(c). The most pertinent disparities actually arise from the object contours but this information tends to be diffused from either side of the contour in [8]. Our regional disparity map is computed at a coarse segmentation level using the matching criterion based on gradient superimposition and occlusion reasoning from subregions disparities. We obtain the result shown in figure 7(d) which answers to the two aforementioned problems.

Table 1. Percentage of pixels which are badly matched with respect to different tolerances on the Middlebury database

	Tolerance ± 0.5 px.			Tolerance ± 1 px.			Tolerance ± 2 px.		
	NonOcc	*All*	*Disc*	*NonOcc*	*All*	*Disc*	*NonOcc*	*All*	*Disc*
Tsukuba	10.1	10.7	23.7	5.14	5.58	17.3	3.15	3.40	12.3
Venus	9.06	9.50	15.5	2.11	2.46	10.9	0.65	0.80	4.94
Teddy	14.2	20.2	28.6	7.38	15.8	20.8	4.25	7.76	10.7
Cones	13.5	18.8	23.0	6.84	11.9	14.0	3.80	8.04	8.36
Average	16.4			9.50			5.38		

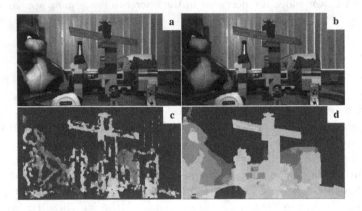

Fig. 7. Micro-stereopsis imagery. (a) reference and (b) second images of the stereo pair, (c) disparity map obtained with [8], (d) our coarse regional disparity map.

5   Conclusion

We have presented an approach to the estimation of depth map controlled by the matching and the superposition of regions. Our depth maps are estimated at two different scales of segmentation: first at a coarse level, then at a fine level. The final disparity maps are obtained by means of an interpolation process, the kriging, which relies on disparity emitter points having a disparity equal to the regional disparity on the fine partition. Points belonging to the watershed are also taken into account as soon as their membership to the appropriate region has been established. The approach strongly relies on the use of the watershed-based segmentation driven by markers in order to obtain the segmentation hierarchies and equivalent segmentations across stereo pairs.

Our method offers good results in terms of precision and perception. It also paves the way to the processing of non-ideal stereo images. Being able to choose the level of precision is of great interest for applications concerned by the process running-time. One could for instance restrain the refinement of disparities to regions having a low average depth only. Furthermore, the availability of efficient watershed implementations based on hierarchical queues, the few nodes involved in each Markov field during the relaxation process and the fact that kriging only comes down to solving systems of linear equations lead to interesting perspectives for computationally efficient implementations of this global depth estimation method. Future work focuses on the exploitation of such mechanisms in the view of processing stereoscopic video sequences.

Acknowledgements. This work has been performed in the project PANORAMA, co-funded by grants from Belgium, Italy, France, the Netherlands, and the United Kingdom, and the ENIAC Joint Undertaking.

References

1. Middlebury stereo database, http://vision.middlebury.edu/stereo/
2. Armstrong, M.: Basic linear geostatistics. Springer (1998)
3. Beucher, S.: Segmentation d'Images et Morphologie Mathématique. Ph.D. thesis, Ecole Nationale Supérieure des Mines de Paris (1990)
4. Bleyer, M., Gelautz, M.: A layered stereo matching algorithm using image segmentation and global visibility constraints. ISPRS Journal of Photogrammetry and Remote Sensing 59(3), 128–150 (2005)
5. De-Maeztu, L., Villanueva, A., Cabeza, R.: Near real-time stereo matching using geodesic diffusion. IEEE Transactions on Pattern Analysis and Machine Intelligence 34(2), 410–416 (2012)
6. Fua, P.: A parallel stereo algorithm that produces dense depth maps and preserves image features. Machine Vision and Applications 6(1), 35–49 (1993)
7. Goshtasby, A.: Similarity and dissimilarity measures. In: Image Registration. Advances in Computer Vision and Pattern Recognition, Springer, London (2012)
8. Hirschmuller, H.: Stereo processing by semiglobal matching and mutual information. IEEE Transactions on Pattern Analysis and Machine Intelligence (2008)
9. Jaccard, P.: Bulletin de la société vaudoise des sciences naturelles. Tech. rep. (1901)
10. Prince, S.: Models for grids. In: Computer Vision: Models, Learning, and Inference. Cambridge University Press (2012)
11. Scharstein, D., Szeliski, R.: A taxonomy and evaluation of dense two-frame stereo correspondence algorithms. International Journal of Computer Vision (2002)
12. Vilaplana, V., Marques, F., Salembier, P.: Binary partition trees for object detection. IEEE Transactions on Image Processing 17(11), 2201–2216 (2008)
13. Yamaguchi, K., Hazan, T., McAllester, D., Urtasun, R.: Continuous markov random fields for robust stereo estimation. In: Fitzgibbon, A., Lazebnik, S., Perona, P., Sato, Y., Schmid, C. (eds.) ECCV 2012, Part V. LNCS, vol. 7576, pp. 45–58. Springer, Heidelberg (2012)
14. Yang, Q., Wang, L., Yang, R., Stewénius, H., Nistér, D.: Stereo matching with color-weighted correlation, hierarchical belief propagation, and occlusion handling. IEEE Transactions on Pattern Analysis and Machine Intelligence 31(3) (2009)
15. Zitnick, C.L., Kang, S.B.: Stereo for image-based rendering using image oversegmentation. International Journal of Computer Vision 75(1), 49–65 (2007)

A Comparison of Some Morphological Filters for Improving OCR Performance

Laurent Mennillo[1], Jean Cousty[2], and Laurent Najman[2(✉)]

[1] UMR6602 - UBP/CNRS/IFMA, Institut Pascal, Aubiére, France
[2] LIGM, Université Paris-Est, Équipe A3SI, Noisy-le-Grand Codex, ESIEE, France
laurent.mennillo@gmail.com, {j.cousty,l.najman}@esiee.fr

Abstract. Studying discrete space representations has recently lead to the development of novel morphological operators. To date, there has been no study evaluating the performances of those novel operators with respect to a specific application. This article compares the capability of several morphological operators, both old and new, to improve OCR performance when used as preprocessing filters. We design an experiment using the Tesseract OCR engine on binary images degraded with a realistic document-dedicated noise model. We assess the performances of some morphological filters acting in complex, graph and vertex spaces, including the area filters. This experiment reveals the good overall performance of complex and graph filters. MSE measures have also been performed to evaluate the denoising capability of these filters, which again confirms the performances of both complex and graph filtering on this aspect.

Keywords: Character recognition · Morphological filtering · Vertex · Graphs · Simplicial complexes

1 Introduction

Mathematical Morphology offers powerful tools that are widely recognized for their utilities for applicative purposes, in particular for filtering out many image defects. The old opening and closing based on structuring elements are still widely used and are described in most image analysis textbooks, although their combination at various scales, namely the granulometries [16], are not as well known. Their main implementation is on the usual 4, 6 or 8 connected grids. However, there exist several recent variations of these operators, depending on the space on which they are defined: we are especially interested in this paper in graphs, first by considering only the vertices (corresponding to the pixels) [18,7] and then, by considering edges (between pixels) and vertices [4,12]. The incentive for using more evolved space representations is to enhance the performance by getting "subpixelic" accuracy. Such an idea has been pushed a step further by considering simplicial complexes [6] (see [5] for a different point of view), a generalization of graphs. Although these new frameworks look promising from a theoretical point of view, to the best of our knowledge, to date, a systematic comparison of these old and novel operators for a dedicated application has not

© Springer International Publishing Switzerland 2015
J.A. Benediktsson et al. (Eds.): ISMM 2015, LNCS 9082, pp. 134–145, 2015.
DOI: 10.1007/978-3-319-18720-4_12

yet been performed. The goal of this paper is to fill that gap, focusing on Optical Character Recognition, or OCR. As it is well known that connected filters, and especially area opening and closing [19] are well adapted to document image analysis, we include them in the present study.

The filtering step is generally just one step in the many ones composing the full application chain. Linear filters can be evaluated by their response to some model of noise. It is more difficult to apply the same evaluation process to the non-linear morphological filters. This is why we choose to assess the performance of an OCR against some model of noise/degradation dedicated to documents. Indeed, OCR is the process of converting a scanned document to machine-encoded text [15]. Such an operation is generally impacted by the quality of the original document and by the introduction of artefacts during the scanning process. Our performance evaluation is hence a measure of the ability of the aforementioned morphological operators to improve OCR performance when used as a preprocessing step on degraded binary document images.

The paper is organized as follows. Section 2 presents the document degradation model used to alter the documents. Section 3 presents the compared morphological filters. Section 4 describes the thorough test protocols of this experiment. Results of the experiments are detailed in Section 5, before, in Section 6, discussing them and concluding the paper.

2   Document Degradation Model

The quality of the document images to be processed is a key point in any document recognition application. Indeed, the accuracy of the results often depend on this quality, and drastic failure can be expected if the quality is too low. For this reason, researchers in the domain have developed models of document image degradation. A state of the art of this research can be found in [3]. Document degradation models are designed to simulate local distortions that are introduced during the processes of document scanning, printing and photocopying. That includes global (perspective and non-linear illumination) and local (speckle, blur, jitter and threshold) effects. Applications of these degradation models are numerous, see [8] for a survey. In this paper, we are using these models to carry out a systematic study of the performance of some morphological filters.

There exist two types of degradation models. As their name implies, physics-based ones [1,2] model the physics of the printing and imaging apparatus, with as much detail as possible. While they lead to accurate models, they might be unnecessary complicated for our purpose. On the other hand, statistics-based models [10] are much simpler, both from an implementation and usage point of views. We thus choose to use this class of models. Relying on statistics of image distributions, they propose a model of real document imaging defects. In the context of this experiment, some of these models have the ability to generate realistic degradations that are appropriate for an OCR performance evaluation. Besides, increasing levels of such degradations can also be produced by adjusting the models parameters, thus allowing for a proper level of comparison.

The binary document degradation model used in this experiment has been presented by Kanungo *et al.* in [9]. This local model, which only applies to binary images, accounts for two types of document degradation, which are *pixel inversion* and *blurring*.

Pixel inversion simulates image noise usually generated by light intensity variations, sensor sensitivity and image thresholding, while *blurring* simulates the point-spread function of the scanner optical system. The *pixel inversion* probability of a background (*resp.* foreground) pixel is modelled following an exponential function of its distance from the nearest foreground (*resp.* background) pixel as:

$$p(0|1, d, \alpha_0, \alpha) = \alpha_0 e^{-\alpha d^2} + \eta, \tag{1}$$

$$p(1|0, d, \beta_0, \beta) = \beta_0 e^{-\beta d^2} + \eta, \tag{2}$$

where parameter d represents the 4-neighbour distance of each background (*resp.* foreground) pixel from its nearest foreground (*resp.* background) pixel, parameter α_0 (*resp.* β_0) is the amount of generated noise on background (*resp.* foreground) pixels, parameter α (*resp.* β) is the decay speed, relatively to distance d, of the background (*resp.* foreground) pixels flipping probability and parameter η represents the lowest flipping probability accounted for all the pixels.

Blurring in document images is due to the point-spread function of the scanner optical system. It is simulated here by a (simple) morphological closing operation using a disk structuring element of diameter k, that accounts for the correlation introduced by the point-spread function.

To summarize, this model with parameters

$$\Theta = (\eta, \alpha_0, \alpha, \beta_0, \beta, k) \tag{3}$$

is used to degrade binary documents by computing the distance map of each pixel, then independently flipping them following their respective probability and finally performing a morphological closing operation.

3 Morphological Filtering

Morphological filters are commonly used to restore or improve the image quality of digitally converted documents. Thus, they can increase OCR performance when used as a preprocessing step. This section roughly presents the four morphological filters that are assessed in our experiment on OCR preprocessing. Due to space restriction, the precise definitions of the operators is not made available in this article but can be found in [11].

3.1 Morphological Operators on Vertices with Structuring Element

Morphological operators defined in vertex spaces act directly on subsets of image pixels through the use of structuring elements. For our experiments, structuring elements corresponding to the 4- and 6-adjacency relations are considered in the 2D grid. In order to actually filter an image, alternate sequential filters are known to be efficient. An alternate sequential filter (ASF) of size λ, denoted by ASF_λ, is a sequence of intermixed morphological openings and closings by balls of increasing size, where the unit ball is the given structuring element and where the balls of larger size are obtained by up to λ dilations of the unit ball with the structuring element. An opening of a set X of white pixels by a ball of radius λ is the union of all balls of radius λ which are included in X. On the other hand, the closing of X is the dual of the opening of X, that is, roughly speaking, an opening of the black-pixels set. Thus, an ASF smooths the object and its complementary in a balanced way while preserving the "most significant balls" of both object and background. In practice, the openings and closings by a ball are obtained by composition of the erosion and the dilation where the given ball is considered as structuring element.

3.2 Morphological Operators on Graph Spaces

Morphological operators on graph spaces have been studied notably in [12,4]. Acting as subpixelic filters with the introduction of edges between each connected vertices, they extend the operators defined in vertex spaces. In order to reach such subpixelic resolution, the basic dilations and erosions involved in the graph based ASFs act from sets of graph vertices to sets of graph edges, and, conversely, from sets of graph edges to sets of graph vertices. More precisely, the filter used in this experiment is the alternate sequential filter denoted by $ASF_{\lambda/2}$ in [4] (Definition 25), which is defined from a combination of these edge/vertex dilations and erosions. Filtering in graph spaces are performed in the graph corresponding the 4- and 6-adjacency relations, as for the operators defined in vertex spaces. It is worth noting that the openings/closings corresponding to the even values of the parameter λ are the same as the openings/closings by balls of size $r = \lambda/2$ involved in the operators on vertex spaces. Therefore, we can roughly say that the subpixelic resolution of graph operators is reached by considering the odd values of λ which in turn implies, two times more iterations of openings/closings for reaching the same filtering size.

3.3 Morphological Operators on Simplicial Complexes

Morphological operators on simplicial complex spaces have been developed in [6] (see also [5]). Simplicial complexes extend graphs to higher dimensions in the sense that a graph is a 1-D simplicial complex made of points and edges considered as 0-D and 1-D elements respectively. For instance, in 2-D, apart from points and edges, a simplicial complex also contains elementary triangles. Considering simplicial complex spaces allows for the design of dimensional morphological operators that can make the distinction between a 0-D element (a point), a 1-D element (an edge) and a 2-D element (a triangle). For application to image processing the well known 6-adjacency relation naturally leads to a 2-D simplicial complex seen as the image domain (see [11]). The filters considered for our experiment with this 2-D simplicial complex space are the ones denoted by $ASF_{\lambda/3}$ in [6] (Definition 9). Note that, in this case the subpixelic resolution is reached to the cost of a multiplication by three of the number of iterations compared to the ASFs in vertex spaces. Cubical complexes (2-D and 3-D basic elements are unit squares and cubes respectively) could be used in order to consider dimensional operators related to the 4-adjacency relation, but operators in this framework are yet to be defined and thus are not yet available for applications.

3.4   Morphological Area Opening and Closing Filters

Morphological area opening γ_λ^a and closing ϕ_λ^a filters for binary and greyscale images have been presented by Vincent in [19]. These operators respectively remove light and dark regions of the image whose area is superior to a parameter $\lambda \in \mathbb{N}$. The 4-adjacency relation was used for these filters in this experiment.

4   Test Protocols

4.1   First Test Protocol

The Tesseract OCR engine, presented in [17], has been used to perform optical character recognition in this experiment. This powerful system has been evaluated by UNLV-ISRI in 1995 (refer to [13]) along with other commercial OCR engines and proved its top-tier performance at the time. Since then, it has been improved extensively by Google. In order to get OCR performance results from this engine on preprocessed documents, the test data and software tools from UNLV-ISRI presented in [14] have been used. More precisely, we have used on a random selection of 100 instances of 300 DPI binary document images. The test procedure is basically the iteration of degradation, filtering, OCR analysis and MSE measure of each document, repeated for each pair (d, λ) of degradation and filtering parameters. Note, however, that the used binary documents are scanned versions of real documents, meaning that they are imperfect and consequently contain noise. Degradation performed on these documents simply allows for a better comparison of the filters efficiency in critical conditions.

Degradation. Degradation levels are specified with parameter $d \in \mathbb{N}$, which acts on the binary document degradation model parameters $\Theta = (\eta, \alpha_0, \alpha, \beta_0, \beta, k)$ as follows:

$$\Theta(d) = (d * 0.02, d * 0.1, 1, d * 0.1, 1, 0). \tag{4}$$

Filtering. The filters used in this experiment are the ASF filters defined in vertices, graphs and simplicial complexes, as well as a combination of area closing and area opening filters. The tests have been conducted on both regular and inverse versions of each document. Furthermore, ASF filters on graph and vertex

Table 1. Experimental set up and results (the degradation level is set up to $d = 4$). For each method, we only display the results obtained with the best filtering level, denoted by λ^*. See text for details.

ID	Filter	Adj.	Inv.	Usc	Thr1	Dsc	Thr2	λ^*	Char acc (%)	Word acc (%)
F00	None								0.02	0.01
F01	Complex	6						5	59.36	61.60
F02	Complex	6	×					3	13.51	32.13
F03	Graph	6						2	14.81	33.77
F04	Graph	6	×					3	50.68	56.89
F05	Graph	6		3/2	×	2/3	×	2	13.82	30.23
F06	Graph	6	×	3/2	×	2/3	×	5	48.54	47.53
F07	Graph	4						2	49.48	46.79
F08	Graph	4	×					2	59.02	57.71
F09	Graph	4		3/2	×	2/3	×	2	38.06	34.84
F10	Graph	4	×	3/2	×	2/3	×	5	45.54	40.66
F11	Vertices	6						2	23.52	22.86
F12	Vertices	6	×					2	25.71	17.57
F13	Vertices	6		3/1		1/3	×	2	63.78	66.18
F14	Vertices	6	×	3/1		1/3	×	3	65.51	69.07
F15	Vertices	4						1	33.98	33.43
F16	Vertices	4	×					2	28.69	24.24
F17	Vertices	4		3/1		1/3	×	2	62.14	60.54
F18	Vertices	4	×	3/1		1/3	×	3	64.93	66.02
F19	area	4						6	47.10	44.82
F20	area	4	×					6	45.66	44.09

spaces have also been evaluated with document resolution scaling of respectively 3/1 and 3/2 (Usc), in order to preserve the same number of iterations between each filter. For instance, the results produced by the ASF_3 on the vertex space whose resolution was upscaled by 3 are comparable (with respect to the size of the removed noise) to the results of $ASF_{3/3}$ where the simplicial complex space is build from the image at the original resolution. Moreover, in both cases, the filters require the same number of iterations (*i.e.*, each one of them needs three opening/closing iterations) for producing the result. In the case of binary document filtering, the corresponding upscaled documents were then binarized with a threshold value of 128 (Thr1), downscaled to their original size after filtering (Dsc) and binarized again with a threshold value of 128 (Thr2) in order to preserve the characters size for OCR processing, since the OCR engine that is used only accepts 300 DPI resolution. Filtering levels were specified for each morphological filter with parameter $\lambda \in \mathbb{N}$. Detailed settings are described in table 1. One can note that binarization after upscaling (Thr1) is not performed in the case of vertex filtering. This is simply due to the fact that these documents are already in binary form after an exact upscaling of 3/1.

OCR Analysis. OCR analysis has been performed by the Tesseract OCR engine in its latest version (3.02). Character and word accuracy obtained from OCR processing of each document, as well as 95% confidence intervals of the obtained accuracy for each set of documents processed with every pair (d, λ) of degradation and filtering levels were then computed with the accuracy, wordacc, accci and wordaccci tools provided in [14].

MSE Measure. Mean squared error has been measured for each processed image I of dimensions $w * h$ with respect to its unprocessed counterpart O considered as the ground truth:

$$MSE = \frac{1}{w * h} \sum_{i=0}^{h-1} \sum_{j=0}^{w-1} [I(i,j) - O(i,j)]^2 \tag{5}$$

4.2   Second Test Protocol

Some observations can be stated about the first test protocol. One can note that MSE measures performed with binary documents that already contain noise as ground truths cannot be considered as a proper evaluation of the filters denoising ability. In addition, OCR analysis and MSE measures are also affected by document scaling, which produce a slight smoothing effect that can impact the results in this situation. This second test protocol has been performed to address these two problems. As the characters size is a crucial factor of OCR analysis, this second test protocol is only focused on MSE performance and has thus been performed on a noise-free binary document that was not downscaled at all. The test procedure is the iteration of degradation, filtering and MSE measure of each document, repeated for each pair (d, λ) of degradation and filtering parameters.

5   Results

5.1   First Test Protocol

In this section, we present the results of the first test protocol in the most critical tested conditions $(d = 4)$, to better compare the efficiency of each filtering setting. In figure 3, only the best performing setting is shown for each filter among regular and inverted document filtering and Table 1 presents the results obtained by each method for the parameter λ^* that maximizes the quality of its results.

As can be observed in figure 1, complex filtering on non inverted documents produces better accuracy results than any other filter at original resolution, with graph filtering closely behind.

One can note that 4-connected and 6-connected vertex filtering at scaled resolution on inverted documents outperforms complex filtering. However, it is at the expense of higher computational time and memory. Indeed, filtering at triple resolution requires to handle a number of pixels multiplied by 9 compared to the original resolution, whereas the size of a simplicial complex space is roughly 6 times the number of pixels. With our implementation, filtering in vertex spaces at triple resolution takes twice the time for filtering in the complex space.

Finally, MSE results of this first test protocol clearly show that complex filtering, graph filtering and area filtering are very close at original resolution. On the other hand, vertex filtering performs best at scaled resolution while graph filtering at scaled resolution is significantly outperformed in this scenario.

A summary of OCR results obtained in this experiment is presented in Table 1 (along with the filtering parameters).

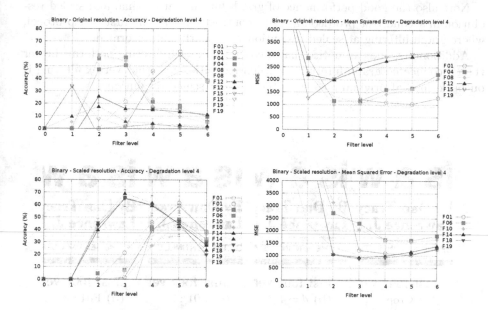

Fig. 1. OCR accuracy and MSE measured in the first test after filtering of 100 binary documents at original and scaled resolution. Dashed lines represent word accuracy while solid lines represent character accuracy.

5.2   Second Test Protocol

In this section we present the results of the second test protocol in the most critical tested conditions, to better compare the efficiency of each filtering setting. As shown in figure 2, where only the best performing setting is shown for each filter among regular and inverted document filtering, complex filtering of the binary image shown in figure 4b produces better MSE results than any other filter tested in these conditions.

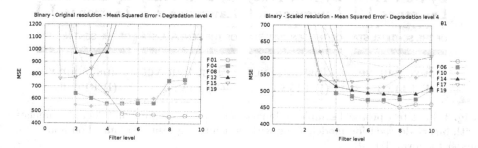

Fig. 2. MSE measured in second test after filtering of the binary image shown in Fig. 4b

Note also the good performance of graph filtering at original and scaled resolution in second test protocol, a result contrasting with the first test protocol, where graph filtering at scaled resolution was clearly outperformed.

Additionally, one can notice that 6-connected graph filtering performs better than 4-connected graph filtering at scaled resolution and that vertex filtering is outperformed at original resolution but close at scaled resolution.

Fig. 3. First test protocol sample on binary documents. Original and degraded images with binary degradation model, along with best filtering results obtained on image 3b, under the form [ID : f].

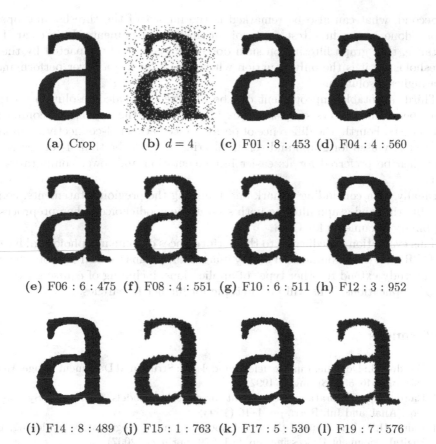

(a) Crop   (b) $d = 4$ (c) F01 : 8 : 453 (d) F04 : 4 : 560

(e) F06 : 6 : 475 (f) F08 : 4 : 551 (g) F10 : 6 : 511 (h) F12 : 3 : 952

(i) F14 : 8 : 489 (j) F15 : 1 : 763 (k) F17 : 5 : 530 (l) F19 : 7 : 576

Fig. 4. Second test protocol sample on a binary document. Original and degraded images with binary degradation model, along with best filtering results obtained on image 4b ($d = 4$), under the form [ID : f : MSE].

6 Discussion and Conclusion

It is clear following this experiment that morphological filtering can greatly improve OCR accuracy when used as a preprocessing step. The different results shown in this experiment are potent indicators of the efficiency of several morphological filters in the context of OCR. Indeed, preprocessing using such filters leads to an increase of respectively up to 65.49% and up to 69.06% in character and word accuracy on binary documents.

However, a few remarks can be stated to further explain the results obtained in this experiment. First, regarding the impact of inverted documents filtering, one can note that a general trend emerges from the results of the first test protocol. Indeed, complex and area filters perform better on non inverted documents, while graph and vertex filters perform better on inverted documents.

Second, what can also be remarked is the impact of the thresholding operations done in the first test protocol on upscaled documents. It is clear, for instance, that graph filtering on such documents is severely impacted by these thresholds, as it is the only situation where this filter has a lower performance at a higher resolution.

Third, a notable improvement can be observed at scaled resolution on the lower performing filters such as vertex filters, but at the expense of computational costs. Fourth, the difference of performance between 4-connectivity and 6-connectivity used for graph and vertex filtering seems to be thin. 4-connectivity may then be preferred for its easier implementation and lower computational cost.

Finally, as a concluding remark and following the previous statements, complex filtering and graph filtering both seem to be good choices when preprocessing binary documents for OCR.

This evaluation is dedicated to the performances of some morphological filters for OCR, and used a data-set of binary images. It remains to see if the conclusions of the study extend to other types of applications, including of course greyscale image applications. This will be the topic for further researches.

References

[1] Baird, H.S.: Document image defect models. In: Structured Document Image Analysis, pp. 546–556. Springer (1992)

[2] Baird, H.S.: Calibration of document image defect models. In: Annual Symp. on Doc. Anal. and Inf. Retr., pp. 1–16 (1993)

[3] Baird, H.S.: The state of the art of document image degradation modelling. In: Digital Document Processing, pp. 261–279. Springer (2007)

[4] Cousty, J., Najman, L., Dias, F., Serra, J.: Morphological filtering on graphs. Computer Vision and Image Understanding 117(4), 370–385 (2013)

[5] Dias, F., Cousty, J., Najman, L.: Dimensional operators for mathematical morphology on simplicial complexes. PRL 47, 111–119 (2014)

[6] Dias, F., Cousty, J., Najman, L.: Some morphological operators on simplicial complex spaces. In: Debled-Rennesson, I., Domenjoud, E., Kerautret, B., Even, P. (eds.) DGCI 2011. LNCS, vol. 6607, pp. 441–452. Springer, Heidelberg (2011)

[7] Heijmans, H.J.A.M., Nacken, P., Toet, A., Vincent, L.: Graph morphology. Journal of Visual Communication and Image Representation 3(1), 24–38 (1992)

[8] Ho, T.K., Baird, H.S.: Evaluation of ocr accuracy using synthetic data. In: Annual Symp. on Doc. Anal. and Inf. Retr. (1995)

[9] Kanungo, T., Haralick, R.M., Baird, H.S., Stuezle, W., Madigan, D.: A statistical, nonparametric methodology for document degradation model validation. PAMI 22(11), 1209–1223 (2000)

[10] Kanungo, T., Haralick, R.M., Phillips, I.: Global and local document degradation models. In: Proceedings of the Second International Conference on Document Analysis and Recognition, pp. 730–734. IEEE (1993)

[11] Mennillo, L., Cousty, J., Najman, L.: Morphological filters for ocr: a performance comparison. Tech. rep. (December 2012),
http://hal.archives-ouvertes.fr/hal-00762631

[12] Meyer, F., Angulo, J.: Micro-viscous morphological operators. In: ISMM 2007, pp. 165–176. INPE (October 2007)
[13] Nartker, T.A., Rice, S.V., Jenkins, F.R.: OCR accuracy: UNLV's fourth annual test. Inform 9(7), 38–46 (1995)
[14] Nartker, T.A., Rice, S.V., Lumos, S.E.: Software tools and test data for research and testing of page-reading ocr systems. In: Document Recognition and Retrieval XII. SPIE, vol. 5676, pp. 37–47 (2005)
[15] Rice, S.V., Nagy, G., Nartker, T.A.: Optical character recognition: An illustrated guide to the frontier. Springer (1999)
[16] Serra, J.: Image analysis and mathematical morphology. Academic Press (1982)
[17] Smith, R.: An overview of the tesseract ocr engine. In: ICDAR 2007, vol. 2, pp. 629–633 (2007)
[18] Vincent, L.: Graphs and mathematical morphology. Signal Processing 16(4), 365–388 (1989)
[19] Vincent, L.: Morphological area openings and closings for greyscale images. In: Shape in Picture. Nato ASI Series, vol. 126, pp. 197–208. Springer, Heidelberg (1992)

Local Blur Estimation Based on Toggle Mapping

Théodore Chabardès and Beatriz Marcotegui[✉]

CMM - Centre for Mathematical Morphology,
MINES ParisTech, PSL Research University,
35 rue Saint Honoré - Fontainebleau, Paris, France
{theodore.chabardes,beatriz.marcotegui}@mines-paristech.fr
http://cmm.mines-paristech.fr

Abstract. A local blur estimation method is proposed, based on the difference between the gradient and the residue of the toggle mapping. This method is able to compare the quality of images with different content and does not require a contour detection step. Qualitative results are shown in the context of the LINX project. Then, quantitative results are given on DIQA database, outperforming the combination of classical blur detection methods reported in the literature.

Keywords: No-reference blur estimation · Local blur estimation · Toggle mapping · DIQA database

1 Introduction

With the proliferation of handheld devices equipped with high resolution cameras and increasing computational power, many mobile applications become possible nowadays. Text is present everywhere in our everyday's life. Visually impaired people have no access to it. In the framework of LINX project we aim at developing a smartphone application making access to textual information possible within everyone's reach, including visually impaired people. This project is funded by the French Interministerial funds (FUI) for competitive clusters.

LINX conditions of use are not under control. The user being visually impaired cannot check the acquired image quality: bad lighting conditions, blur or noise can degrade the acquired image. We focus here on blur, which is a common problem linked to handheld devices and low cost objectives of smartphone cameras.

Blur estimation is an extensively studied topic in the literature. Image quality assessment techniques can be classified in 3 groups:

- Full reference techniques, which compare the image to be evaluated with the undegraded version of it. This comparison can be based on the mean square error, the PSNR, the cross-correlation between both images, the structural similarity index [25] or any other fidelity measure.
- Reduced reference: the comparison is based on some features and not on the whole image.

© Springer International Publishing Switzerland 2015
J.A. Benediktsson et al. (Eds.): ISMM 2015, LNCS 9082, pp. 146–156, 2015.
DOI: 10.1007/978-3-319-18720-4_13

- No-reference: the image quality assessment is based only on the degraded image itself. This is the most difficult case, and corresponds to the LINX context, where only the acquired image is available.

No-reference sharpness metrics have been defined in the spatial and in the frequency domain. Some spatial metrics are based on variance computation [9], total variation, gradient [8], Brenner's gradient [3], multiscale gradient [5], laplacian [16]. Others are based on edge width estimation [15,4] or on histogram measures such as histogram entropy [11,21,17].

Examples of spectral focus measures are the relative high frequency power compared to the low frequency one, the sum of frequency components above a certain threshold [8] or the correlation of wavelet coefficients over scales [6,24]. Other works rely on the phase coherence [10,14,1].

Most of these methods have been designed in an autofocus context. They can rank the quality of an image from an image set as long as their content is similar. Therefore, they are not appropriate for our application. Some attempts have been made in order to drop this restriction. Local approaches have been proposed, computing the metric in a given neighborhood [19], that has to be defined, or after a contour detection step [4,18].

Several papers address the specific problem of document image quality and try to correlate the quality measure with an OCR accuracy [2,18,26]. An interesting database, DIQA, has been introduced in this context [13]. 25 documents have been acquired with a mobile-phone camera. 6 to 8 different images of each document were taken. The camera was focused at varying focal lenghts to generate a series of images with focal blur. The quality of each image is estimated as the OCR accuracy on it.

In this paper we introduce a local blur estimator that does not require a prior contour detection, nor a neighborhood definition. The rest of the paper is organised as follows: section 2 presents the toggle-mapping based blur estimator, section 3 evaluates its performance on DIQA dataset [13]. Finally section 5 concludes the paper and discusses some perspectives of this work.

2   Toggle Mapping Based Blur Estimation

Toggle mapping (TM) operator was introduced in [12]. From an image f, two transformations are computed: the dilation, $\delta_B(f)$ and the erosion $\epsilon_B(f)$. Equation 1 describes the toggle mapping operator:

$$TM_B(f) = \begin{cases} \delta_B(f) & if\ \delta_B(f) - f\ <\ f - \epsilon_B(f) \\ \epsilon_B(f) & otherwise \end{cases} \qquad (1)$$

where $\delta_B(f)$ and $\epsilon_B(f)$ correspond respectively to the dilation and the erosion with structuring element B.

Thus, each pixel is replaced by one of these two transformations, selecting among them the one that is closest to the original pixel value. This process can be iterated until convergence is reached. This principle was generalized in [22], with the use of other transformations involved in the toggle process.

TM was designed as a contrast enhancement operator. It was also used as scene text segmentation tool in [7]: pixels replaced by the dilation are set to 1 and those replaced by the erosion are set to 0. In this paper a local blur estimation is introduced based on TM.

We define the residual image as the absolute difference between the original image and the TM:

$$TM\_Residue_B(f) = \begin{cases} \delta_B(f) - f & if \ \delta_B(f) - f < f - \epsilon_B(f) \\ f - \epsilon_B(f) & otherwise \end{cases} \quad (2)$$

A blur boundary leads to a high residue, and this residue increases with the size of the structuring element involved in the TM. This is true up to a TM size equivalent to the blur edge width. A blur estimation could then be built on the evolution of the TM residues, with a series of structuring elements of increasing size. However a sharp boundary and a homogeneous region would have the same series of low value TM residues. In order to distinguish between these two situations we compare these residues with the gradient value. The morphological gradient is defined as the difference between the maximum and the minimum in the neighborhood of a pixel ($\delta_B(f) - \epsilon_B(f)$), while the TM replaces each pixel with the closest extremum in the neighborhood, that is either the maximum or the minimum. Thus, by definition, the residue of the TM is lower than the gradient. However if the TM uses a larger structuring element (B_M instead of B_N, with $M > N$) its residue can reach the gradient value. In which case, the pixel is classified as blur. Therefore we define the Q image as:

$$Q_{B_N, B_M}(f) = max\,(0, gradient_{B_N}(f) - TM\_Residue_{B_M}(f)) \quad (3)$$

Q image is filtered: first it is thresholded (threshold $= 3$ for all our experiments), then small regions (less than 5 pixels) are removed and finally Q is averaged in each connected component. The result is Q_{filter} image.

Figure 1 shows an example with intermediary images. Figure 1(a) is a crop of an image acquired by a LINX end user with a mobile phone. The image is sharp but we can observe some noise that can be reinforced by the toggle mapping operator. A bilateral filter [23] is applied to get rid of this noise, see figure 1(b). Figure 1(c) shows the toggle mapping of figure 1(b) and figure 1(d) its corresponding residue. This residue is subtracted from the gradient image shown in figure 1(e), leading to pixel-wise quality estimation in figure 1(f). Figure 1(g) illustrates the Q_{filter} image. Figure 2 shows the flow chart of the algorithm. The letters of the diagram (from (a) to (g)) correspond to intermediary images of figure 1. The average of selected pixels for this image is equal to 34, and it can be considered as the quality score q_{score} of the crop, assuming it is homogeneous.

3   Results

In this section we will show qualitative results from LINX examples and quantitative results when applying our algorithm to DIQA database.

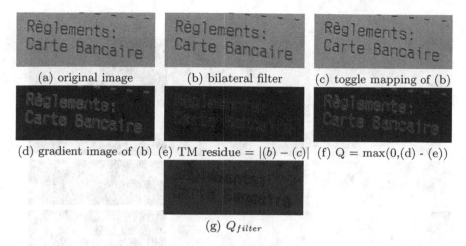

(a) original image   (b) bilateral filter   (c) toggle mapping of (b)

(d) gradient image of (b) (e) TM residue = |(b) − (c)| (f) Q = max(0,(d) - (e))

(g) Q_{filter}

Fig. 1. Intermediary steps of local quality estimation: (b) Bilateral filter $\sigma_{spatial} = 2, \sigma_{gray} = 20$; (c) Toggle mapping of size 2; (d) Gradient of size 1; (e) |(b) − (c)|; (f) Q = max(0,(d)-(e)); (g) Q_{filter} with threshold equal to 3, regions larger than 5 pixels and average Q on each selected connected component

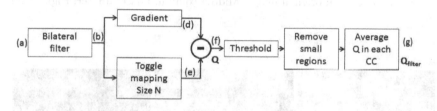

Fig. 2. Flow chart of the algorithm

The parameters used in all our experiments (if not specified otherwise) are:

- Bilateral filter $\sigma_{spatial} = 2, \sigma_{gray} = 20$
- Structuring element: cross (4 neighbors)
- Gradient of size 1 (3x3 pixels)
- Toggle mapping residue of size 2 (5x5 pixels)
- Threshold: 3
- Small regions: 5 pixels

Figure 3 shows some examples with different levels of blur from LINX database. In the first row we can see a good quality image example, with a quality score of 30. In the second row a good quality image with a score of 20 and in the third row a blurred image with a score of 4. We can observe a good correspondence of our score and the observed quality.

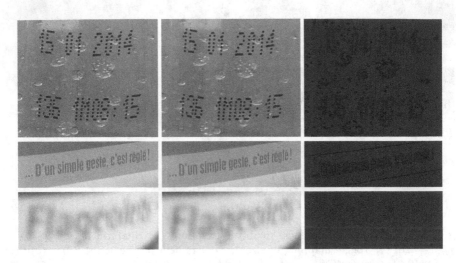

Fig. 3. Example of LINX images with different levels of blur. First row: sharp image (q_{score}=30). Second row: Good quality image (q_{score}=20). Third row: Blurred image (q_{score}=4). Left column: original image, Middle column: bilateral filter, Right column: Q_{filter}.

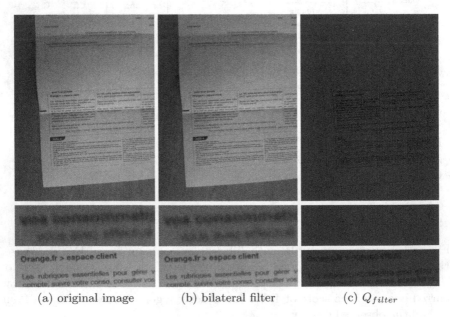

(a) original image   (b) bilateral filter   (c) Q_{filter}

Fig. 4. Example with heterogenous blur. First row, the whole image; second row a crop from the top of the image, in a blurred area (q_{score}=5); third row a crop from the middle of the image, in a good quality area (q_{score}=10).

One of the advantages of our method is that the measure is local. Figure 4 shows an example with heterogeneous blur. The upper part of the image is more out-of-focus than the lower part. We can observe a higher result in the lower part. Two crops from different areas have been extracted, and we confirm a good correlation of our measure with the image quality: the crop from the upper part has a quality score equal to 5 while the crop from the lower part has a quality score equal to 10. Note that the crop has been extracted for visualization purposes. The measure is local and each region has its own score.

In order to have quantitative results we apply our algorithm to the recently published DIQA image database [13]. It is an interesting dataset, composed of 175 document images acquired with a mobile phone. It is noteworthy that images are directly acquired by a camera, so the blur is not simulated. 25 documents have been acquired, 6 to 8 times each, in different conditions: from perfect focus to completely out-of-focus images. The quality of each image is given by the OCR accuracy on the processed image. Three different OCRs have been used: Tesseract, FineReader and Omni. The overall accuracy of these OCRs on the whole database are:

	Tesseract	FineReader	Omni
OCR accuracies:	0.53	0.76	0.72

FineReader leads to the best overall accuracy (0.76) and will be our reference in the rest of the paper, unless stated otherwise. Figure 5 shows 3 crops from this dataset, with an OCR accuracy ranging from 0.97 to 0.24. Our quality score for these images ranges from 5 to 55. Spearman correlation is commonly used to assess the coherence of two variables without taking into account their precise values. It compares their ranking indexes instead of the variables themselves. In our example, the ranking based on the OCR accuracy or on our quality score is the same (1.- (a); 2.-(b); 3.-(c)). Thus, for this simple example with only three images, our quality score is perfectly correlated with the OCR accuracy: the Spearman correlation between them is then equal to 1.

Some works [13,20] have tried to estimate the OCR accuracy based on some features extracted from the image. The score used for comparing different methods is the median of the Spearman correlation applied to each document set. The scores reported are over 0.9. We think that the proposed score overestimates the quality of the OCR accuracy prediction. Indeed, if the Spearman correlation is computed for each document set, the content of images to be compared is similar. The context is then close to an autofocus situation which is much easier than comparing the quality of images with different content. This problem is pointed out in [20]: a failure is reported for an image with a low focus measure whereas the OCR accuracy is over 0.9, provoked by the large white space in the page. Another reported issue concerns an image with a high focus measure with a low OCR accuracy due to a text out-of-focus but a huge headline leading to a focus measure optimistically high. Those are exactly the situations that we address in this paper. Moreover, as stated in [20], by reporting the median value, outlier classes in which the methods might not perform well are disregarded. They propose to compute the Spearman correlation directly for the 175 images

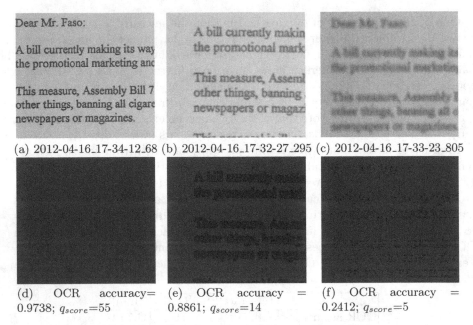

(a) 2012-04-16_17-34-12_68 (b) 2012-04-16_17-32-27_295 (c) 2012-04-16_17-33-23_805

(d)   OCR   accuracy=   (e) OCR accuracy =   (f) OCR accuracy =
0.9738; q_{score}=55   0.8861; q_{score}=14   0.2412; q_{score}=5

Fig. 5. DIQA examples. First row: original images. Second row: Q_{filter} Left (FineReader accuracy = 0.9738; q_{score}=55), Middle (FineReader accuracy =0.8861; q_{score} = 14). Right (FineReader accuracy = 0.2412; q_{score} = 5).

Table 1. Spearman correlation of our quality score against OCR accuracy, with different pre-processing filter

Filter	FineReader	Tesseract	Omni
original	0.745	0.711	0.622
AF 1	0.784	0.748	0.645
bilateral 1	0.816	0.813	0.692
bilateral 2	0.849	0.892	0.725
bilateral 3	0.843	0.902	0.727
median 1	0.787	0.756	0.659
median 2	0.822	0.808	0.690

of the database. We adopt this proposal that we find more reliable. They report in [20] a Spearman correlation of 0.6467 for the whole database as a single set for FineReader OCR.

Table 1 summarizes the results of Spearman correlation between our quality score against the OCR accuracy for 175 images in DIQA dataset. A higher Spearman correlation value indicates the method's ability to rank images according to the expected OCR accuracy. Our score is 0.745 compared to 0.6467 reported in the [20] combining several classical descriptors from [19]. As we observed

Table 2. Spearman correlation on document 1 from DIQA dataset. Spearman correlation(our_score,tesseract) = 0.53. Spearman correlation(our_score,FineReader)=0.89.

	imA	imB	imC	imD	imE	imF	imG
Our score:	16	17	35	38	19	6	37
Our rank:	6	5	3	1	4	7	2
Tesseract accuracy:	0.77	0.78	0.87	0.86	0.70	0.12	0.71
Tesseract rank:	4	3	1	2	6	7	5
Finereader:	0.95	0.97	0.99	0.98	0.97	0.13	0.98
Finereader rank:	6	5	1	2	4	7	3

Fig. 6. Example with large characters in a blurred context, but still readable

a qualitative improvement in LINX database using a prefiltering step, we tried different filters (alternate filter AF, bilateral or median filter) on DIQA database and the best result was 0.849, obtained by a bilateral filter of size 2 (σ_{gray} was fixed to 20, and we have not verified yet the sensitivity of this parameter in our quality measure).

4   Discussion

Spearman correlation estimates the correlation between two ranked variables. For example, if we take the 7 images of the first document of DIQA dataset we get a quality score ranging from 6 to 38. The Spearman correlation between our score and Tesseract accuracy is 0.53. It can be considered as a relative low correlation value. However analyzing the result closer we can observe that imF is classified as the worst image in both cases. imC and imD have quite similar values (0.87 and 0.86 for Tesseract accuracy and 35 and 38 for our score respectively). The ranking is different but the difference in quality between the two images is small. Both measures are able to estimate the different groups of image quality but Spearman correlation does not catch these similarities between images. Somehow it would be interesting to quantize the scores and do not consider an error for ranking differences in the same quantile.

Another issue is related to the scale. Images of figure 6 contain large blurred characters. As they are blurred, no good quality pixel is selected and the image is then rejected. However despite the blur the characters are still readable.

5   Conclusions and Perspectives

In this paper we introduce a local quality estimation approach. It is used as a pre-processing step in a mobile phone application, aiming at avoiding further processing of bad quality images. The method compares the residue of a toggle mapping of size N with the gradient of size M, with $M < N$.

The interest of the approach is shown first through several LINX images. The quality scores obtained are correlated with the perceived quality of the images. Then, in order to evaluate quantitatively the efficiency of our method we rank the quality of the DIQA images. A Spearman correlation of 0.745 is reached, between our score and FineReader accuracy. This performance is to be compared to 0.6467 reported in the literature, combining several classical blur descriptors. Our performance raises to 0.849 if a bilateral filter is applied before estimating the quality. The OCR accuracy is given for the whole image. It would be interesting to verify if the errors appear when the quality is locally lower, for some images with heterogeneous focus.

In the future we will set the quality threshold values for accepting or rejecting a region in the context of the LINX project. Intermediate quality values could be considered: contrast enhancement would be applied in those cases, before further processing. We will also address the problem of large blurred characters, in a multi-scale approach.

Acknowledgments. The work reported in this paper has been performed as part of Cap Digital Business Cluster LINX Project.

References

1. Blanchet, G., Moisan, L.: An explicit sharpness index related to global phase coherence. In: 2012 IEEE International Conference on Acoustics, Speech and Signal Processing (ICASSP), pp. 1065–1068. IEEE (2012)
2. Blando, L.R., Kanai, J., Nartker, T.A.: Prediction of OCR accuracy using simple image features. In: Proceedings of the Third International Conference on Document Analysis and Recognition, vol. 1, pp. 319–322. IEEE (1995)
3. Brenner, J.F., Dew, B.S., Horton, B., King, T., Neurath, P.W., Selles, W.D.: An automated microscope for cytologic research a preliminary evaluation. Journal of Histochemistry & Cytochemistry 24(1), 100–111 (1976)
4. Cao, G., Zhao, Y., Ni, R.: Edge-based blur metric for tamper detection. Journal of Information Hiding and Multimedia Signal Processing 1(1), 20–27 (2010)

5. Chen, M.-J., Bovik, A.C.: No-reference image blur assessment using multiscale gradient. EURASIP Journal on Image and Video Processing 2011(1), 1–11 (2011)
6. Ciancio, A., da Costa, A.L.N.T., da Silva, E.A.B., Said, A., Samadani, R., Obrador, P.: No-reference blur assessment of digital pictures based on multifeature classifiers. IEEE Transactions on Image Processing 20(1), 64–75 (2011)
7. Fabrizio, J., Marcotegui, B., Cord, M.: Text segmentation in natural scenes using toggle-mapping. In: 2009 16th IEEE International Conference on Image Processing (ICIP), pp. 2373–2376. IEEE (2009)
8. Firestone, L., Cook, K., Culp, K., Talsania, N., Preston, K.: Comparison of autofocus methods for automated microscopy. Cytometry 12(3), 195–206 (1991)
9. Groen, F.C.A., Young, I.T., Ligthart, G.: A comparison of different focus functions for use in autofocus algorithms. Cytometry 6(2), 81–91 (1985)
10. Hassen, R., Wang, Z., Salama, M.: No-reference image sharpness assessment based on local phase coherence measurement. In: 2010 IEEE International Conference on Acoustics Speech and Signal Processing (ICASSP), pp. 2434–2437. IEEE (2010)
11. Jarvis, R.A.: Focus optimization criteria for computer image-processing. Microscope 24(2), 163–180 (1976)
12. Kramer, H.P., Bruckner, J.B.: Iterations of a non-linear transformation for enhancement of digital images. Pattern Recognition 7(1), 53–58 (1975)
13. Kumar, J., Ye, P., Doermann, D.: A dataset for quality assessment of camera captured document images. In: Iwamura, M., Shafait, F. (eds.) CBDAR 2013. LNCS, vol. 3857, pp. 113–125. Springer, Heidelberg (2014)
14. Leclaire, A., Moisan, L., et al.: No-reference image quality assessment and blind deblurring with sharpness metrics exploiting fourier phase information (2014)
15. Marziliano, P., Dufaux, F., Winkler, S., Ebrahimi, T.: Perceptual blur and ringing metrics: application to jpeg2000. Signal Processing: Image Communication 19(2), 163–172 (2004)
16. Nayar, S.K., Nakagawa, Y.: Shape from focus. IEEE Transactions on Pattern analysis and machine intelligence 16(8), 824–831 (1994)
17. Chern, N.N.K., Neow, P.A., Ang, V.M.H.: Practical issues in pixel-based autofocusing for machine vision. In: Proceedings of 2001 IEEE International Conference on Robotics and Automation, ICRA, vol. 3, pp. 2791–2796. IEEE (2001)
18. Peng, X., Cao, H., Subramanian, K., Prasad, R., Natarajan, P.: Automated image quality assessment for camera-captured OCR. In: 2011 18th IEEE International Conference on Image Processing (ICIP), pp. 2621–2624. IEEE (2011)
19. Pertuz, S., Puig, D., Garcia, M.A.: Analysis of focus measure operators for shape-from-focus. Pattern Recognition 46(5), 1415–1432 (2013)
20. Rusiñol, M., Chazalon, J., Ogier, J.-M.: Combining focus measure operators to predict OCR accuracy in mobile-captured document images. In: 2014 11th IAPR International Workshop on Document Analysis Systems (DAS), pp. 181–185. IEEE (2014)
21. Schlag, J.F., Sanderson, A.C., Neuman, C.P., Wimberly, F.C.: Implementation of automatic focusing algorithms for a computer vision system with camera control. Carnegie-Mellon Univ., the Robotics Inst. (1983)
22. Serra, J.: Toggle mappings. From pixels to features,61–72 (1988)
23. Tomasi, C., Manduchi, R.: Bilateral filtering for gray and color images. In: Sixth International Conference on Computer Vision, pp. 839–846. IEEE (1998)
24. Wang, Z., Bovik, A.C.: Reduced-and no-reference image quality assessment. IEEE Signal Processing Magazine 28(6), 29–40 (2011)

25. Wang, Z., Bovik, A.C., Sheikh, H.R., Simoncelli, E.P.: Image quality assessment: from error visibility to structural similarity. IEEE Transactions on Image Processing 13(4), 600–612 (2004)
26. Ye, P., Doermann, D.: Learning features for predicting OCR accuracy. In: 2012 21st International Conference on Pattern Recognition (ICPR), pp. 3204–3207. IEEE (2012)

Improved Detection of Faint Extended Astronomical Objects Through Statistical Attribute Filtering

Paul Teeninga[1], Ugo Moschini[1]([✉]), Scott C. Trager[2],
and Michael H.F. Wilkinson[1]

[1] Johann Bernoulli Institute
u.moschini@rug.nl
[2] Kapteyn Astronomical Institute,
University of Groningen, P.O. Box 407, 9700 AK, Groningen, The Netherlands
s.c.trager@rug.nl, m.h.f.wilkinson@computer.org

Abstract. In astronomy, images are produced by sky surveys containing a large number of objects. SExtractor is a widely used program for automated source extraction and cataloguing but struggles with faint extended sources. Using SExtractor as a reference, the paper describes an improvement of a previous method proposed by the authors. It is a Max-Tree-based method for extraction of faint extended sources without stronger image smoothing. Node filtering depends on the noise distribution of a statistic calculated from attributes. Run times are in the same order.

Keywords: Attribute filters · Statistical tests · Astronomical imaging · Object detection

1 Introduction

The processing pipeline of a sky survey includes extraction of objects. With advances in technology more data is available and manually extracting every object is infeasible. A survey example is the Sloan Digital Sky Survey [7] (SDSS) where the DR7 [1] catalogue contains 357 million unique objects. SExtractor [2], a state-of-the-art extraction software, first estimates the image background. With the default settings, to perform a correct segmentation and avoid false positives, objects are identified with the pixels with intensity at a threshold level higher than 1.5 times the standard deviation of the background estimate at that location. We refer here to such mechanism as *fixed threshold*: the threshold value only relies on local background estimates in different sections of the image, ignoring the actual object properties. The downside is that parts of objects below the threshold are discarded (Fig. 1). As improvement to the fixed threshold, we proposed in [8] a method that locally varies the threshold depending on object size by using statistical tests rather than arbitrary thresholds on attributes.

This work was funded by the Netherlands Organisation for Scientific Research (NWO) under project number 612.001.110.

J.A. Benediktsson et al. (Eds.): ISMM 2015, LNCS 9082, pp. 157–168, 2015.
DOI: 10.1007/978-3-319-18720-4_14

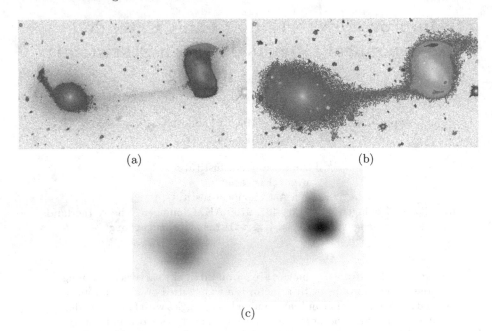

(a)   (b)

(c)

Fig. 1. Result of SExtractor with default settings (a) and the proposed method (b). The filament between the galaxies is not extracted in (a). SExtractor Background estimate (c) shows correlation with objects: a fixed threshold above this already biased estimate would make the segmentation worse.

In this paper, we present an extension of that method modifying the attribute used. In our method, the supporting data structure is a Max-Tree [5] created from the image, where every node corresponds to a connected component for all the threshold levels. The choice is inspired by the simplified component tree used in SExtractor and it was already suggested in [4]. Nodes are marked significant if noise is an unlikely cause, for a given significance level. Nested significant nodes can represent the same object or not and a choice must be made. When deblending objects, other significant branches are considered as new objects. Results are compared with SExtractor. In SExtractor, every connected component above the fixed threshold is re-thresholded at N levels with logarithmic spacing, by default. The range is between the fixed threshold and the maximum value above the component. A tree similar to a Max-Tree is constructed. Branches are considered separate objects if the integrated intensity is above a certain fraction of the total intensity and if another branch with such property exists. An image background, caused by light produced and reflected in earth's atmosphere, is estimated and subtracted before thresholding. SExtractor's estimate shows bias from objects (see Fig. 1), which reduces their intensities. Backgrounds in the SDSS dataset turn out to be nearly flat: constant estimates per image are used. Our background estimate, which also reduces object bias, is briefly described in Section 2.

The data set of 254 monochrome images is a subset of the corrected images in SDSS DR7. All selected images contain merging or overlapping galaxies. These often include faint structures which are difficult to detect with SExtractor. We use r-band images, because they have the best quality [3]. The set has been acquired by CCD. A software bias is subtracted from the images, after which the pixel values are proportional to photo-electron counts [6]. Poissonian noise dominates, but due to the high photon counts at the minimum grey level, the distribution is approximately Gaussian, with a variance which varies linearly with grey level. In the rest of the paper, the extension to our method to identify astronomical objects is described. A comparison with the segmentation performed by SExtractor is presented, followed by conclusions and future directions of research.

2   Background Estimation

The image is assumed to be the sum of a background image B, object image O and Gaussian noise. The noise variance is equal to $g^{-1}(B+O)+R$, with g equivalent to gain in the SDSS catalogue, and R due to other noise sources, such as read noise, dark current and quantisation. First, the image background must be computed. A method giving a constant estimate for the background value that does not correlate with the objects was proposed in [8] and explained more extensively in [9]. With the background removed, the variance of the noise is $g^{-1}O + \sigma_{bg}^2$, where $\sigma_{bg}^2 = g^{-1}B + R$ is approximately equal to our estimate. With $\hat{\mu}_{bg}$ and $\hat{\sigma}_{bg}^2$ the estimates of background and variance, respectively, in case g is not given, it is approximated by $\hat{\mu}_{bg}/\hat{\sigma}_{bg}^2$. It is assumed that R is small compared to $g^{-1}\hat{\mu}_{bg}$. Negative image values are set to 0 and the Max-Tree is constructed. The next step is to identify nodes that are part of objects, referred to as *significant* nodes.

3   Identifying Significant Nodes

To identify the nodes in the tree belonging to objects, let us define P_{anc} as the closest significant ancestor or, if no such node exists, the root, for a node P in the Max-Tree. P is considered significant if it can be shown that $O(x) > f(P_{anc})$ for pixels $x \in P$, given a significance level α. The pixel value associated with P is indicated with $f(P)$. Similarly, let $f(x)$ be the value of pixel x. The power [11] attribute is similar to the definition of object *flux* or the integrated intensity that also SExtractor uses. It is a measure widely used for object identification in astronomy. A definition of the power attribute of P is

$$\text{power}(P) := \sum_{x \in P} (f(x) - f(\text{parent}(P)))^2. \tag{1}$$

An alternative definition that will be used is

$$\text{powerAlt}(P) := \sum_{x \in P} (f(x) - f(P_{anc}))^2. \tag{2}$$

To normalize the power attribute, the values are divided by the noise variance σ^2. Four significance tests that use the attributes above are defined in the following.

Significance Test 1: **power** *given* **area** *of the node.* For node P we use hypothesis

$$H_{\text{power}} := (O(x) \leq f(\text{parent}(P)) \quad \forall x \in P.$$

This test uses the definition of power attribute in Equation 1. Let us assume H_{power} is true and consider the extreme case $O(x) = f(\text{parent}(P))$ for pixels $x \in \text{parent}(P)$. The variance of the noise σ^2 is equal to $g^{-1}f(\text{parent}(P)) + \sigma_{\text{bg}}^2$. For a random pixel x in P, the value $(f(x) - f(\text{parent}(P)))^2/\sigma^2$ has a χ^2 distribution with 1 degree of freedom. As the pixel values in P are independent, $\text{power}(P)/\sigma^2$ has a χ^2 distribution with degrees of freedom equals to the area of P. The χ^2 CDF (or inverse), available in scientific libraries, can be used to test the normalized **power** given significance level α and node area. An example of a rejection boundary of a χ^2 CDF is shown in Fig. 2a. If $\text{power}(P)/\sigma^2 > \text{inverse}\chi^2\text{CDF}(\alpha, \text{area})$, H_{power} is rejected: $O(x) > f(\text{parent}(P)) \geq f(P_{\text{anc}})$, for pixels $x \in \text{parent}(P)$, making P significant.

In all the next three significance tests (all right tailed) the exact distribution of **powerAlt** is not known and it is obtained by simulation. In general, the rejection boundary for a generic **attribute** and significance level α is the result of the inverse CDF, which will be denoted as $\text{inverseAttributeCDF}(\alpha, ...)$. Let r be an integer greater than zero and $n_{\text{samples}}(\alpha, r) := r/\alpha$ with n_{samples} rounded to the closest integer. A number of random independent nodes equals to n_{samples} is generated. On average, for r nodes the **attribute** value is greater than or equal to the rejection boundary. The best estimate of $\text{inverseAttributeCDF}(\alpha, ...)$, without any further information about the distribution, is the average of the two smallest of the $r + 1$ largest **attribute** values.

Significance Test 2: **powerAlt** *given* **area** *and* **distance**. For node P we now use hypothesis

$$H_{\text{powerAlt}} := (O(x) \leq f(P_{\text{anc}})) \quad \forall x \in P.$$

In significance test 1, leaf nodes are less likely to be found significant due to their small area and the low intensity difference with the parent node. The main idea behind significance test 2 is to make the significance level more constant for every node, independently of its height in the tree, by referring to its ancestor rather than the parent node. The definition of power attribute in Equation 2 is used. Let us assume H_{powerAlt} is true and consider the extreme case $O(x) = f(P_{\text{anc}})$. Let us define $\text{distance}(P) := f(P) - f(P_{\text{anc}})$. Let X be a random set of $\text{area}(P)$ - 1 values drawn from a truncated normal distribution with a minimum value of $\text{distance}(P)$. The variance of the normal distribution is set to $\sigma^2 = g^{-1}f(P_{\text{anc}}) + \sigma_{\text{bg}}^2$. Attribute $\text{powerAlt}(P)$ has the same distribution as $\text{distance}^2(P)$ plus the sum of the squared values in X. The rejection boundary for the normalized **powerAlt** attribute is provided by the function $\text{inversePower-AltCDF}(\alpha, \text{area}, \text{distance})$. Hypothesis H_{powerAlt} is rejected if $\text{powerAlt}(P)/\sigma^2 > \text{inversePowerAltCDF}(\alpha, \text{area}, d))$: $O(x)$ at some pixels x in P is higher than $f(P_{\text{anc}})$, making P significant. The minimum area of a significant node is 2 pixels. An estimate is given for $\text{inversePowerAltCDF}$ for

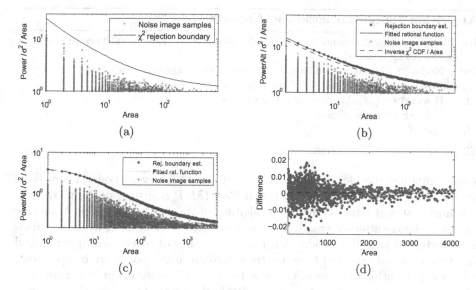

Fig. 2. Rejection boundaries for significance test 1 (a) and the simulated rejection boundaries for test 3 (b) and test 4 (c). Log-log scale. (d) shows the difference between the rational function and the estimate in (c). $\alpha = 10^{-6}$.

constant α, varying `area` and `distance`. For each rejection boundary, varying `distance`, a rational function is fitted to reduce the error and the storage space.

Significance Test 3: `powerAlt` *given* `area`. It is a significance test that has the same goal of significant test 2. It uses the distribution of `powerAlt` given α, `area`. It is independent of the `distance` measure, not used as parameter in the inverse CDF. Using the assumptions from *significance test 2*, `distance`(P) has a truncated normal distribution with a minimum value of 0, the same distribution as a random non-negative pixel value. The rejection boundary is calculated through simulated images. Fig. 2b shows the rejection boundary and the fitted rational function for this significance test.

Significance Test 4: `powerAlt` *given* `area`, *using a smoothing filter.* It is basically equal to significance test 3 with the addition of the smoothing. Smoothing reduces noise. A larger number of objects is detected with this test. The default smoothing filter used in SExtractor is used as in [10]. Filtering is done after background subtraction and before setting negative values to zero. Fig. 2c shows the rejection boundary with its fitted rational function and Fig. 2d shows the difference between the rational function approximation and the estimates for this significance test.

Considerations about the False Positives. Alg. 1 describes the method used for marking nodes as significant. Visiting nodes in non-decreasing order by pixel

---

Algorithm 1. SignificantNodes(M, nodeTest, α, g, σ_{bg}^2)

In: Max-Tree M, significance test t, significance level α, gain g, variance of the noise at the background σ_B^2.
Out: Nodes in M that are unlikely to be noise are marked as significant.
1. **for all** nodes P in M with $f(P) > 0$ in non-decreasing order **do**
2.   **if** nodeTest($M, P, \alpha, g, \sigma_{bg}^2$) is **true then**
3.   Mark P as significant.
4.   **end if**
5. **end for**

---

value simplifies the identification of P_{anc}, if stored for every node. There is no need if the χ^2 test is used. Function nodeTest($M, P, \alpha, g, \sigma_{bg}^2$) performs the significance test and returns true if P is significant, false otherwise. The Max-Tree of a noise image after subtraction of the mean and truncation of negative values is expected to have $0.5n$ nodes, with n the number of pixels. An upper bound on the number of expected false positives is $\alpha 0.5n$ if the nodes are independent, which is not entirely the case. Given a 1489×2048 noise image, the same size of the images in the data set, and $\alpha = 10^{-6}$, the upper bound on the expected number of false positives is approximately 1.52. An estimate of the actual number of false positives is 0.41, 0.72, 0.94 and 0.35 for the four significance tests, respectively, averaged over 1000 simulated noise images.

4 Finding the Objects

Multiple significant nodes could be part of the same object. A significant node with no significant ancestor is marked as an object. Let mainBranch(P) be a significant descendant of P with the largest area. A significant node, with significant ancestor P_{anc}, that differs from mainBranch(P_{anc}) is marked as a new object. The assumption of a new object is not valid in every case: it depends on the used significance test, filter and connectivity, as it will be shown in the comparison section. The method is described in Alg. 2.

Algorithm 2. FindObjects(M)

In: Max-Tree M.
Out: Nodes in M that represent an object are marked.
1. **for all** significant nodes P in M **do**
2.   **if** P has no significant ancestor **then**
3.   Mark P as object.
4.   **else**
5.   **if** mainBranch(P_{anc}) does not equal P **then**
6.   Mark P as object.
7.   **end if**
8.   **end if**
9. **end for**

4.1 Moving up Object Markers

Nodes marked as objects have a number of pixels attached due to noise. The number decreases at a further distance from the noiseless image signal, which can be achieved by moving the object marker up in the tree, λ times the standard deviation of the noise. The obvious choice for an object node P is $\texttt{mainBranch}(P)$, if such a node exists, since it does not conflict with other object markers. Otherwise, it is chosen the descendant of P with the highest p-value found with the related CDF for the $\texttt{power}$ or $\texttt{powerAlt}$ attribute. However, the CDF is not always available or easy to store. Instead, the descendant with the largest $\texttt{power}$ attribute is chosen, if at least one exists. The function that returns the descendant is called $\texttt{mainPowerBranch}(P)$. Alg. 3 describes the method. An alternative to allowing a lower value of $f(P_{\text{final}})$ is to remove those object markers. If the parameter λ is set too low, there are too many noise pixels attached to objects. However, to be able to display faint parts of extended sources a low λ is preferred. After thorough tests on objects simulated with the IRAF software

---

Algorithm 3. $\texttt{MoveUp}(M, \lambda, g, \sigma_{\text{bg}}^2)$

In: Max-Tree M, factor λ, gain g, variance of the noise at the background σ_{bg}^2.
Out: For every object marker that starts in a node P and moves to P_{final}: $f(P_{\text{final}}) \geq f(P_{\text{anc}}) + \lambda$ times the local standard deviation of the noise, when possible. $f(P_{\text{final}})$ might be lower if P_{final} has no descendants.

1. **for all** nodes P in M marked as objects **do**
2.   Remove the object marker from P.
3. $h \leftarrow f(P_{\text{anc}}) + \lambda \sqrt{g^{-1} f(P_{\text{anc}}) + \sigma_{\text{bg}}^2}$
4.   **while** $f(P) < h$ **do**
5.   **if** P has a significant descendant **then**
6. $P \leftarrow \texttt{mainBranch}(P)$.
7.   **else if** P has a descendant **then**
8. $P \leftarrow \texttt{mainPowerBranch}(P)$.
9.   **else**
10.   Break.
11.   **end if**
12.   **end while**
13.   Mark P as object.
14. **end for**

Algorithm 4. $\texttt{MTObjects}(I, \texttt{nodeTest}, \alpha, g, \lambda)$

In: Image I, function $\texttt{nodeTest}$, significance level α, gain g and move factor λ.
Out: Max-Tree M. Nodes in M corresponding to objects are marked.

1. $(\hat{\mu}_{\text{bg}}, \hat{\sigma}_{\text{bg}}^2) \leftarrow \texttt{EstimateBackgroundMeanValueAndVariance}()$
2. $I_{i,j} \leftarrow \max(I_{i,j} - \hat{\mu}_{\text{bg}}, 0)$
3. $M \leftarrow$ create a Max-Tree representation of I.
4. $\texttt{SignificantNodes}(M, \texttt{nodeTest}, \alpha_{\text{nodes}}, g, \hat{\sigma}_{\text{bg}}^2)$.
5. $\texttt{FindObjects}(M)$.
6. $\texttt{MoveUp}(M, \lambda, g, \hat{\sigma}_{\text{bg}}^2)$.

---

with several noise levels, a value of 0.5 was found to be a good compromise. Alg. 4 summarises the whole procedure from background estimation to object identification. The proposed method is called MTObjects.

5   Comparison with Source Extractor

Our method is compared against the segmentation performed by SExtractor 2.19.5. The settings are kept close to their default values:

- Our background and noise root mean square estimates are used. This already improves the segmentation of SExtractor with respect to the original estimate of SExtractor, that correlates too much with objects.
- DETECT_MINAREA = 3. Default in SExtractor 2.19.5.
- FILTER_NAME = default.conv.
- DETECT_THRESH = 1.575σ above the local background. The default threshold of 1.5 is changed to make the expected false positives similar to significance test 4 for noise-only images. Expected false positives per image is approximately 0.38 based on the results of 1000 simulated noise images.

While there is no guarantee that these settings are optimal, our comparison gives an impression of the performance of our method. A more comprehensive comparison test is required.

Object Fragmentation. Another source of false positives is fragmentation of objects due to noise. An example is shown in Fig. 3. Fragmentation appears to happen in relatively flat structures and the chance is increased if different parts of the structure are thinly connected. If one pixel connects two parts, the variation in value due to noise can make a deep cut. In the case of the threshold used by SExtractor, fragmentation is severe if the object values are just below the threshold. The expected number of false positives due to fragmentation for the given data set is unknown. Most images do not clearly show fragmented objects. An image where it does happen is displayed in Fig. 4. While the SExtractor parameter CLEAN_PARAM can be changed to prevent this from happening, it is left to the default as it has a negative effect on the number of objects detected and it causes fragmentation in this image only.

Fig. 3. Fragmented simulated object. Significance test 3 (left), test 4 (middle) and SExtractor (right). The pixels of the object have the value 1.5, close to the SExtractor threshold. Background is 0. Gaussian noise is added with $\sigma = 1$.

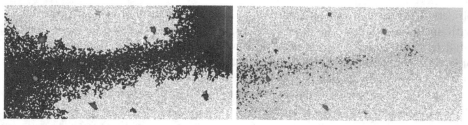

Fig. 4. Fragmentation of an object. Significance test 4 (left) and SExtractor (right). Crop section of `fpC-002078-r1-0157.fits`.

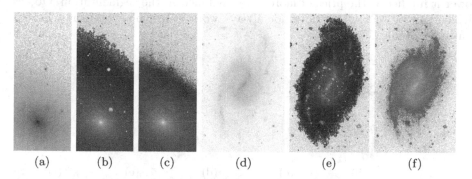

    (a)   (b)   (c)   (d)   (e)   (f)

Fig. 5. MTObjects finds more nested objects. Crop of `fpC-003804-r5-0192.fits`: (a) original image; (b) significance test 4; (c) SExtractor. Crop of `fpC-001332-r4-0066.fits`: (d) original image; (e) significance test 4; (f) SExtractor.

Object Count. All the significance tests were compared against each other and SExtractor. The significance test 4 returns a larger number of objects in about 100% of the images in the dataset w.r.t. significance test 1 and 2 and in about 70% w.r.t. significance test 3 and SExtractor. After inspection of the results, we noticed that MTObjects detects more objects nested in larger objects (galaxies), when the pixel values of the nested objects are above the SExtractor's threshold. Examples are shown in Fig. 5. This is explained by the fact that every node in the Max-Tree is used, while SExtractor uses a fixed number of sub-thresholds. A question is if the better detection of nested objects can explain the performance of significance test 4 compared to SExtractor. To answer that, the data set is limited to smaller objects. This is done by making a sorted area list of the largest connected component of each image at the threshold used by SExtractor. The performance of significance test 4 and SExtractor is now indeed similar: it means that the difference in the total number of all the objects found in the images is explained by the number of nested object detections. We tested then how significance test 4 performs in the case of densely spaced overlapping objects. When two identical objects overlap, one of the nodes marked as object has a lower `power` or `powerAlt` value on average. If overlapping objects are close enough to each other and at SExtractor's threshold they are still detected as separate objects, MTObjects could fail to detect them as separate objects. A grid filled with small stars is generated with the IRAF software. The magnitude is set to −0.2 to make objects barely detectable when noise is added. The diameter of objects is 3 pixels (full width at half maximum). The background equals 1000 at every pixel and the gain is 1. Gaussian noise is added. In this case of densely spaced objects, SExtractors detects a number of stars closer to the actual number than MTObjects with significance test 4.

Faint Structures. The parameter λ used in `MoveUp` is set to 0.5 without adding false positives, which makes detecting fainter structures easier. It is possible to lower it further at the price of more noise included in the segmentation. Object deblending by SExtractor does not always work well if objects do not have a Gaussian profile (Fig. 6).

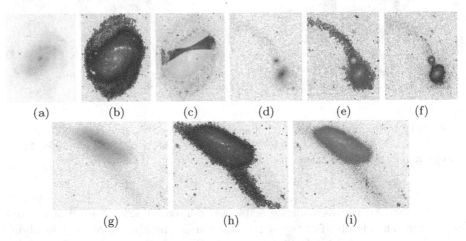

(a)   (b)   (c)   (d)   (e)   (f)

(g)   (h)   (i)

Fig. 6. Comparison of objects with faint extended regions; (a,b,c) crop of `fpC-003903-r2-0154.fits`: (a) original image; (b) significance test 4; (c) SExtractor; (d,e,f) crop of `fpC-004576-r2-0245.fits`: (d) original image; (e) significance test 4; (f) SExtractor; (g,h,i) crop of `fpC-004623-r4-0202.fits`: (g) original image; (h) significance test 4; (i) SExtractor.

Dust Lanes and Artifacts. Dust lanes as in Fig. 7 and artefacts as in Fig. 8 are also a source of false positives. In Fig. 7(f), the galactic core is clearly split due to dust. Fig. 8(a)(b)(c) could represent an artefact or a vertical cut-off. Refraction spikes shown in Fig. 8(d)(e)(f) can also cause false positives as in the wave-like shape in Fig 8(f).

Run Times. The timer is started before background estimation and is stopped after object classification in SExtractor and after executing `MoveUp` in MTObjects. The amount of time spent on classification in SExtractor is unknown. MTObjects does not perform any classification. Tests were done on an AMD Phenom II X4

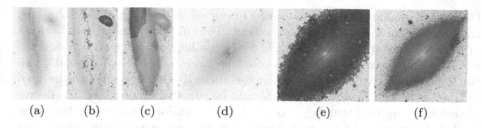

Fig. 7. Dust lanes. Crop of `fpC-004623-r4-0202.fits`: (a) original image; (b) significance test 4; (c) SExtractor. Crop of `fpC-001739-r60308.fits`: (d) original image; (e) significance test 4; (f) SExtractor.

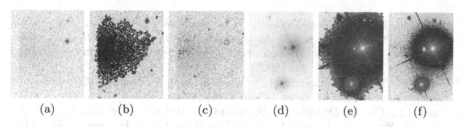

Fig. 8. Artefacts. Crop of `fpC-002326-r4-0174.fits`:(a) original image; (b) significance test 4; (c) SExtractor. Crop of `fpC-001345-r3-0182.fits`: (d) original image; (e) significance test 4; (f) SExtractor.

955. SExtractor is typically 3.6 times faster than MTObjects, but it takes longer time for images that have many pixel values above the fixed threshold. MTObjects is more constant in run time.

6 Conclusions and Future Work

The Max-Tree method (MTObjects) performs better at extracting faint parts of objects compared to state-of-the-art methods like SExtractor. The sensitivity increases with object size. MTObjects appears to be slightly worse in case of densely spaced and overlapping objects, like globular clusters. When an object is defined to have a single maximum pixel value, excluding maxima due to noise, MTObjects is better at finding nested objects. Every possible threshold is tested in MTObjects, whereas SExtractor is bound to a fixed number of thresholds. Deblending objects appears to be better in MTObjects when there is a large difference in size and objects do not have a Gaussian profile. Otherwise, one of the objects will be considered as a smaller branch by MTObjects. Too many pixels are assigned arbitrarily to a single object. The SExtractor method of fitting Gaussian profiles makes more sense in this case and allows for a more even split in pixels. This method could be added as postprocessing step to MTObjects.

The power attribute was initially chosen because in the non-filtered case it has a known scaled χ^2 distribution. Better attribute choices could be investigated.

Deblending similar sized objects can be improved. Nested significant connected components can represent the same object. The current choice, controlled by λ in MoveUp is not ideal. The threshold looks too high for large objects and too low for small objects. Parameter λ could be made variable and dependant on the filter, connectivity and node attributes used. If other noise models are used in other data sets, significance tests should be adjusted accordingly. The degree of smoothing applied that helps to avoid fragmentation could be further investigated. Currently, the rejection boundaries are approximated by simulations which must be recomputed for every filter and significance level. Knowing the exact distributions will speed up this phase.

References

1. Abazajian, K.N., Adelman-McCarthy, J.K., Agüeros, M.A., Allam, S.S., Prieto, C.A., An, D., Anderson, K.S.J., Anderson, S.F., Annis, J., Bahcall, N.A., et al.: The seventh data release of the sloan digital sky survey. The Astrophysical Journal Supplement Series 182(2), 543 (2009)
2. Bertin, E., Arnouts, S.: Sextractor: Software for source extraction. Astronomy and Astrophysics Supplement Series 117, 393–404 (1996)
3. Gunn, J.E., Carr, M., Rockosi, C., Sekiguchi, M., Berry, K., Elms, B., De Haas, E., Ivezić, Ž., Knapp, G., Lupton, R., et al.: The Sloan digital sky survey photometric camera. The Astronomical Journal 116(6), 3040 (1998)
4. Perret, B., Lefevre, S., Collet, C., Slezak, E.: Connected component trees for multivariate image processing and applications in astronomy. In: 2010 20th International Conference on Pattern Recognition (ICPR), pp. 4089–4092 (August 2010)
5. Salembier, P., Oliveras, A., Garrido, L.: Antiextensive connected operators for image and sequence processing. IEEE Transactions on Image Processing 7(4), 555–570 (1998)
6. sdss.org: Photometric flux calibration, http://www.sdss2.org/dr7/algorithms/fluxcal.html
7. Stoughton, C., Lupton, R.H., Bernardi, M., Blanton, M.R., Burles, S., Castander, F.J., Connolly, A.J., Eisenstein, D.J., Frieman, J.A., Hennessy, G.S., et al.: Sloan digital sky survey: early data release. The Astronomical Journal 123(1), 485 (2002)
8. Teeninga, P., Moschini, U., Trager, S.C., Wilkinson, M.H.F.: Bi-variate statistical attribute filtering: A tool for robust detection of faint objects. In: 11th International Conference on Pattern Recognition and Image Analysis: New Information Technologies (PRIA-11-2013), pp. 746–749 (2013)
9. Teeninga, P., Moschini, U., Trager, S.C., Wilkinson, M.H.F.: Improving background estimation for faint astronomical object detection. In: Proc. ICIP 2015 (submitted, 2015)
10. Teeninga, P., Moschini, U., Trager, S.C., Wilkinson, M.H.F.: Bi-variate statistical attribute filtering: a tool for robust detection of faint objects. In: 11th International Conference on Pattern Recognition and Image Analysis: New Information Technologies (PRIA-11-2013), pp. 746–749. IPSI RAS (2013)
11. Young, N., Evans, A.N.: Psychovisually tuned attribute operators for preprocessing digital video. IEE Proceedings-Vision, Image and Signal Processing 150(5), 277–286 (2003)

Automatic Threshold Selection for Profiles of Attribute Filters Based on Granulometric Characteristic Functions

Gabriele Cavallaro[1]([✉]), Nicola Falco[1,2], Mauro Dalla Mura[3],
Lorenzo Bruzzone[2], and Jón Atli Benediktsson[1]

[1] Faculty of Electrical and Computer Engineering, University of Iceland, Reykjavik,
Iceland
gac4@hi.is, nicolafalco@ieee.org
[2] Department of Information Engineering and Computer Science, University of
Trento, Trento, Italy
lorenzo.bruzzone@ing.unitn.it
[3] GIPSA-lab, Grenoble Institute of Technology, Grenoble, France
mauro.dalla-mura@gipsa-lab.grenoble-inp.fr, benedikt@hi.is

Abstract. Morphological attribute filters have been widely exploited for characterizing the spatial structures in remote sensing images. They have proven their effectiveness especially when computed in multi-scale architectures, such as for Attribute Profiles. However, the question how to choose a proper set of filter thresholds in order to build a representative profile remains one of the main issues. In this paper, a novel methodology for the selection of the filters' parameters is presented. A set of thresholds is selected by analysing granulometric characteristic functions, which provide information on the image decomposition according to a given measure. The method exploits a tree (i.e., min-, max- or inclusion-tree) representation of an image, which allows us to avoid the filtering steps usually required prior the threshold selection, making the process computationally effective. The experimental analysis performed on two real remote sensing images shows the effectiveness of the proposed approach in providing representative and non-redundant multi-level image decompositions.

Keywords: Threshold selection · Connected filters · Tree representations · Mathematical morphology

1 Introduction

Developments in remote sensing technology are leading to an increasing availability of images with high spatial resolution. For the analysis of such images, the spatial context information, which is represented by the neighbourhood of each pixel, becomes an important information source that enables precise mapping of complex scenes. In particular, morphological attribute profiles (APs) [1] have been widely used due to their high flexibility in extracting complementary spatial information. APs are defined as the sequential application of morphological Attribute Filters [2]

© Springer International Publishing Switzerland 2015
J.A. Benediktsson et al. (Eds.): ISMM 2015, LNCS 9082, pp. 169–181, 2015.
DOI: 10.1007/978-3-319-18720-4_15

computed considering a set of filtering thresholds, leading to a multi-level decomposition of the image. Attribute filters are connected operators that process an image by removing flat zones according to a given criterion and attribute (i.e. measure computed on a flat zone). The flexibility of such operators relays on the fact that many attributes (e.g., *area*, *moment of inertia* and *standard deviation*) can be used to obtain a different image transformation. Attribute filters work by comparing a component attribute value against a threshold and a decision is made based on a given criterion [3]. Attribute filters are edge-preserving operators since they are able to preserve the geometrical detail of the regions that are not filtered (i.e., connected components can either be removed or fully kept). Moreover, attribute filters can be implemented relying on a tree representation of the image (e.g., the min- and max-tree) [4], where each node represents a region characterized by a certain attribute value. This kind of representation makes the computation of the transformation more computationally effective. Due to the aforementioned properties, APs are very powerful and flexible operators able to richly extract information on the spatial arrangement and characteristics of the objects in a scene. However, the image decomposition relies on the selection of thresholds, which should be tuned in order to provide a profile that is both *representative* (i.e., it contains salient structures in the image) and *non-redundant* (i.e., objects are present only in one or few levels of the profile). The selection of an appropriate range of thresholds remains one of the main open issues. In the literature, few attempts have been done to solve the issue related to the selection of the threshold values. The common approach is based on field-knowledge of the scene, where the values are manually selected by a visual analysis of the scene under consideration [1,5]. In [6] the set of threshold was derived after a preliminary classification and clustering of the input image. In [7] the filter thresholds were chosen based on the analysis of a granulometric curve (i.e., a curve related to the size distribution of the structures in the image [8]). In particular, the thresholds selected are those whose granulometric curve best approximates the one obtained by considering a large set of thresholds. All the aforementioned methods need to compute a large number of filtering steps (potentially with all possible thresholds) in order to be able to perform the selection.

In this paper, a novel strategy for the automatic selection of the thresholds based on the concept of *granulometric characteristic functions* (GCFs) is proposed. The GCF can be seen as an extension of the concept of granulometry. Considering a series of anti-extensive opening operations, a granulometric curve shows the volume (i.e., the sum of the pixel values) of the result of an opening for different thresholds [8]. By duality, a granulometry curve can be derived by a series of extensive closing operations. Here, a GCF is defined as a measure that is computed on the tree representation of the input image showing the evolution of a characteristic measure for increasing values of the thresholds (i.e., increasingly coarser filters). In this framework, different measures (e.g., related to the gray-levels, number of pixels, etc) can be considered, leading to the definition of different GCFs. The proposed method exploits the tree representation of an image; in particular, for gray-scale images, the corresponding tree representation provides useful a-priori information (i.e., prior to the filtering) related to the actual range of the attribute values. Due to this, the computation of the GCF can be performed directly from the tree, without requiring any filtering of the analysed image. In order to show the flexibility of the proposed approach based on the analysis of GCFs, three measures for computing the GCF are considered for testing. Similarly to [7], in the proposed approach, the set of selected thresholds correspond to the one that best approximates the GCF computed on the full set of thresholds. By approximating the GCF curve, we assume that the distribution of a given measure along the profile can be extracted and approximated by using the subset of selected thresholds. In [7], such a strategy was applied on a conventional granulometry curve, requiring to explicitly filter the images accordingly with an initial set of thresholds, which was manually defined prior to the filtering. The advantages of the proposed approach are that the initial range corresponds to the set of the all possible thresholds, thus it does not require any initial selection, and the GCF curve is derived from the tree structure, without involving any filtering step. To identify the thresholds used for building the profiles, the method employs a piecewise linear regression model [9]. The approach is applied on two real remote sensing data, from which different tree representations are extracted and used to derive the GCFs.

This paper is organized as follows: Section 2 provides a briefly introduction to attribute filters and profiles. In Section 3 the proposed method for thresholds selection is described, while the experiment analysis is shown in Section 4. Section 5 discusses the findings of the study and discusses the future research directions.

2   Attribute Profiles

2.1   Attribute Filters

A discrete two-dimensional gray-scale image $f : E \subseteq \mathbb{Z}^2 \to V \subseteq \mathbb{Z}$, with E its spatial domain and V a set of scalar values, can be fully represented as a set of connected components $\mathcal{C}$, defining a partition π_f of E. The way $\mathcal{C}$ is defined leads to different partitions. If we consider a connected operator ψ, by definition it will operate on f only by merging the connected components of the given set $\mathcal{C}$ [3]. Thus, the result of the filtering will be a new partition π_ψ that is coarser (i.e., containing less regions) than the initial one: $\pi_f \sqsubseteq \pi_{\psi(f)}$ meaning that for each pixel $p \in E$, $\pi_f(p) \subseteq \pi_{\psi(f)}(p)$ [10, Ch. 7]. The coarseness of the partition generated by a connected operator is determined by a parameter λ (i.e., a size-related filter parameter). Given two instances of the same connected operator with different filtering parameters, ψ_{λ_i} and ψ_{λ_j}, which we denote for simplicity as ψ_i and ψ_j, respectively, there is an ordering relation between the resulting partitions: $\pi_{\psi_i} \sqsubseteq \pi_{\psi_j}$ given $\lambda_i \leq \lambda_j$. Among the different types of connected operators, attribute filters have largely diffused. AFs filter connected components in $\mathcal{C}$ according to an attribute $\mathcal{A}$ that is computed on each component. In greater details, the value of an attribute $\mathcal{A}$ is evaluated on each connected component in $\mathcal{C}$ and this measure is compared with a reference threshold λ in a binary predicate T_λ (e.g., $T_\lambda := \mathcal{A} \geq \lambda$). In general terms, if the predicate is true the component is maintained otherwise it is removed. According to the attribute considered, different filtering effects can be obtained leading a simplification of the image that are driven by characteristics such as the regions' scale, shape or contrast.

2.2 Tree Representations

Connected operators can be implemented based on representations of an image as a tree, in which the components are the nodes. In fact, connected components in $\mathcal{C}$ can be organized hierarchically in a tree structure since inclusion relations among components can be established. Any two components $A, B \in \mathcal{C}$ are either nested (i.e., $A \subseteq B$ or $B \subseteq A$) or not. The relation between an image and the associated tree is bijective so from a tree it is possible to retrieve the corresponding image.

Attribute filters are among those filters that can be easily implemented on tree representations since they natively work on connected components (conversely to connected filters based on structuring elements). According to the way $\mathcal{C}$ is defined, different tree representations of the same image and hence different filters are obtained. For example, by considering the upper level set $\mathcal{U}(f) = \{X : X \in \mathcal{C}([f \geq v]), v \in V\}$ a *max-tree* results as tree representation. The nodes in the *max-tree* correspond to peak components (connected components of different connectivity classes [11]). By pruning the max-tree, an anti-extensive filter is obtained (i.e., bright regions will be removed), thus, if the operator is also idempotent and increasing, it leads to an opening. Analogously, considering the lower level set $\mathcal{L}(f)$, in which the connected components are defined according to a decreasing ordering relation among the gray-levels (i.e., $\leq$), we obtain a *min-tree* representation and an attribute closing as operator. A different tree representation is given by the *inclusion tree* (or *tree of shapes*) in which the components are defined by a saturation operator that fills holes in components. A hole in a region $X \in \mathcal{C}$ is defined as a component that is completely surrounded by X. The inclusion tree is constructed by progressively saturating the image starting from its regional extrema (i.e., local maxima and minima in the image) until reaching only a single component fully covering E. The inclusion tree can equivalently be obtained by merging the upper and the lower level sets of an image [12]. The sequence of inclusions induced by saturation determines the components in the tree and their links defining the hierarchy. Since the saturation operator is contrast invariant (i.e., bright and dark regions will be treated the same), the filters operating on this tree will be self-dual (quasi self-dual in the case of discrete images).

Furthermore, the implementation of filters based on trees are appealing, since efficient algorithms exist for constructing these representations [13].

2.3 Attribute Profiles

Let us consider a family of L connected operators ψ computed considering a sequence of L either increasing or decreasing values of the filter parameter $\Lambda = \{\lambda_i\}_1^L$ that we call it a *profile* $\mathcal{P}_\psi := \{\psi_i\}_1^L$. Considering the entries of a profile, the absorption property holds on the resulting partitions such that $\psi_j \psi_i$ will lead to π_{ψ_j} for $i \leq j$. So filtered results can be ordered sequentially.

In this work, we will focus on profiles built with attribute filters, so called *attribute profiles* (APs). Profiles considering attribute filters were defined for the analysis of remote sensing images in [1]. By considering a max and a min-tree, attribute opening and closing profiles were defined, respectively as:

$$\mathcal{P}_\gamma = \{\gamma^{T_0}, \gamma^{T_{\lambda_1}}, \ldots, \gamma^{T_{\lambda_L}}\}, \tag{1}$$

$$\mathcal{P}_\phi = \{\phi^{T_0}, \phi^{T_{\lambda_1}}, \ldots, \phi^{T_{\lambda_L}}\}, \tag{2}$$

where γ^T and ϕ^T represent the attribute opening and closing, respectively, $\{T_i\}$ is a criterion evaluated on the set of thresholds Λ and $\phi^{T_0}(f) = \gamma^{T_0}(f) = f$, which is the original image. By denoting with $\mathcal{P}_\phi^-$ the closing profile taken in reverse order (such that each entry is greater or equal than the subsequent one), in [1] its concatenation with an attribute opening profile was named Attribute Profile (AP):

$$\text{AP} = \{\mathcal{P}_\phi^-/\phi^{T_0}, \mathcal{P}_\gamma\}. \tag{3}$$

The AP is composed of $2L + 1$ images (L closings, the original image and L openings).

Analogously, when considering the contrast invariant operator ρ based on the inclusion tree, the profile $\mathcal{P}_\rho$, named *Self-Dual Attribute Profile* (SDAP) [14,15], can be obtained:

$$\mathcal{P}_\rho = \{\rho^{T_0}, \rho^{T_{\lambda_1}}, \ldots, \rho^{T_{\lambda_L}}\}, \tag{4}$$

with $\rho^{T_0}(f) = f$.

3 Threshold Selection Based on Granulometric Characteristic Functions

3.1 Definition of Measure and Granulometric Characteristic Function

In this work, we propose an approach based on descriptive functions computed on a profile. We define a *granulometric characteristic function* (GCF) as a function returning a measure $\mathcal{M}$ which is computed on a profile:

$$\text{GCF}(\mathcal{P}_\psi(f)) = \{\mathcal{M}(\psi_i)\}_{i=1}^L. \tag{5}$$

Thus, if $\mathcal{M} : f \to \mathbb{R}$, $\text{GCF}(\mathcal{P}_\psi(f))$ leads to L scalar values (one for each image in the profile). The definition of the function GCF is inspired to the concept of granulometric curve, but we wanted to not restrain the measure to be a sum of gray-levels (as conventionally done by granulometry) but leave it open to consider other characteristics. As standard granulometric curves show the interaction of the size of the image structures with the filters when the filter parameter varies, so GCFs can provide information on the effect of increasingly coarser filterings with respect to some characteristics of the image.

We propose three definitions of GCFs:

Sum of Gray-level Values. As for the conventional granulometry, this measure provides information related to the effect of the filtering with respect to the changes in terms of gray-levels that are produced in the image.

$$\text{GCF}_{val}(\mathcal{P}_\psi(f)) = \left\{ \sum |f - \psi_i(f)| \right\}_{i=1}^{L}. \tag{6}$$

Number of Changed Pixels. Another possible measure is the number of pixels that change gray-value at different filterings. In this case, the GCF is sensitive to changes in the spatial extent of the regions rather than in gray-levels.

$$\text{GCF}_{pix}(\mathcal{P}_\psi(f)) = \{\text{card}[f(p) \neq \psi_i(f)(p)], \forall p \in E\}_{i=1}^{L}, \tag{7}$$

with card[·] the cardinality of a set.

Number of Changed Regions. This GCF shows the number of connected components that are affected by each filter and it is a topological measure invariant to the spatial extent and gray-level variations induced by the filterings.

$$\text{GCF}_{reg}(\mathcal{P}_\psi(f)) = \{\text{card}[\mathcal{C}(f)] - \text{card}[\mathcal{C}(\psi_i(f))]\}_{i=1}^{L}. \tag{8}$$

According to the definition reported above, the GCF is a monotonic increasing function since the measures that are considered increase for progressively coarser filters. Clearly, other measures can be considered for the definition of different GCFs, if the interest lies in investigating the effects of the filterings with respect to other image characteristics.

3.2 Threshold Selection

The problem we address here is the selection among the set of all possible values of λs $\bar{\Lambda} = \{\lambda_i\}_{i=1}^{L}$ a subset $\hat{\Lambda} = \{\hat{\lambda}_i\}_{i=1}^{\hat{L}}$ with $\hat{L} \ll L$. The full set $\bar{\Lambda}$ is extremely scene dependent and can potentially be very large making the problem of selecting the subset $\hat{\Lambda}$ more complicated to realize since the full set is not readily accessible. A possible strategy for the selection relies on the computation of a profile by considering a relatively large number of λs (considering all of them in real scenarios is impractical) and prune the profile by selecting some of filtered images and related filter parameters so defining $\hat{\Lambda}$. However, such an approach is limited by the need of generating the filtered images in order to perform the selection and by the lack of guarantee that all possible thresholds are considered for selection. Here we propose to consider the GCFs defined in Sec. 3.1 in order to select those values λs that lead to "significant" changes in the effect of the filters (as measured by the considered GCF). This approach is not new since considering the granulometric curve for estimating values of λ that generate salient filtering images has been already proposed in [7]. However, in this work we propose to exploit the tree representation of the image (augmented with the values of the attributes for each node) prior to any filtering. In particular each node, which maps a region of spatially connected pixels in the image, gives information related to the value of attributes, gray-level and number of pixels. This allows us to know all possible values of λ (i.e., to know exactly the full set $\bar{\Lambda}$) and compute the GCFs before any filtering. Similarly to [7], in the proposed approach, the set $\hat{\Lambda}$ of the selected thresholds correspond to the one that best approximates the GCF computed on the $\bar{\Lambda}$ thresholds. By approximating the GCF curve, we assume that the distribution of the measure $\mathcal{M}$ that underlies the GCF can be extracted and approximated by using the selected $\hat{L}$ thresholds. The approximation of the GCF curve is achieved by means of piecewise linear regression [9], in which the independent variable (i.e., the chosen measure) is partitioned into intervals and approximated with separate line segments that fit each interval. The boundaries between the segments are identified by breakpoints, for which projections over the x-axis correspond to the set of selected thresholds $\hat{\Lambda}$.

4   Experimental Results and Discussion

Aiming at showing the flexibility of the proposed approach in providing complementary contextual information, the proposed approach for meaningful threshold selection is evaluated on two real remote sensing images. For each data set, three CDFs (see Sec. 3.1) are derived from the max- and inclusion trees. Due to space issue, the analysis computed on the min-tree is not shown. For a similar reason, in the experimental analysis, only the attribute *area* is considered. However, according to the definition of the CDFs, the method can be computed considering any attribute.

Rome, Italy (Rome): The data set is a panchromatic image acquire by Quick-Bird satellite sensor over the city of Rome, Italy. The data size is 1188×972 pixels with a geometrical resolution of 0.6 m. The acquired scene is a dense heterogeneous urban area, which includes residential buildings, roads, shadows and open areas. The set of thresholds for the attribute *area* is composed of $L = 4513$ and $L = 4995$ unique values for the max-tree and the inclusion tree, respectively.

Pavia, University Area, Italy (Pavia University): In this case, the data set is a hyperspectral image acquired by the optical airborne sensor ROSIS-03 (Reflective Optics Imaging Spectrometer) over the University area of the city of Pavia, Italy. The image is composed of 103 bands with a spectral range between 0.43 and 0.86 μm and a spatial resolution of 1.3 m, showing an area of 610×340 pixels. The set of thresholds for the attribute *area* is composed of $L = 1963$ and $L = 2508$ unique values for the max-tree and the inclusion tree, respectively. The profiles are computed on the first Principal Component of the data.

In the proposed approach, the number of segments used for the approximation are identified by minimizing the reconstruction error (i.e., the mean absolute error, MAE) between the GCF and its approximation. Due to the nature of the employed regression model, the first and the last breakpoints correspond to the first threshold (i.e., equal to 0, resulting in the original input image) and to the last threshold (resulting in all pixels having the same gray-scale value), respectively, which do not

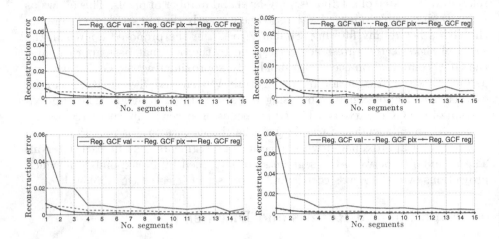

Fig. 1. Reconstruction error computed for the estimation of the GCFs derived by max-tree (left column) and inclusion tree (right column), for Rome (top line) and Pavia (bottom line). The GCF_{val} is denoted in green, the GCF_{pix} is denoted in red, and the GCF_{reg} is denoted in blue.

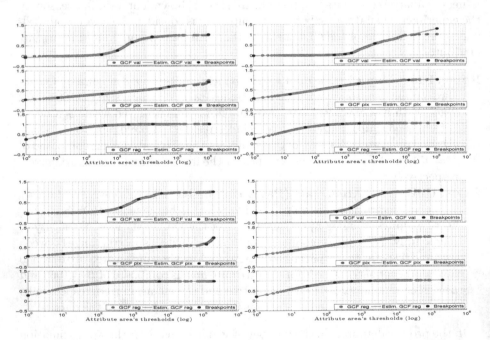

Fig. 2. Comparison between the GCFs derived by max-tree (left column) and inclusion tree (right column), for Rome (top line) and Pavia (bottom line). For each GCF (red circles), the estimated curve (green line) and breakpoints (black circles), which are used to derive the thresholds, are shown.

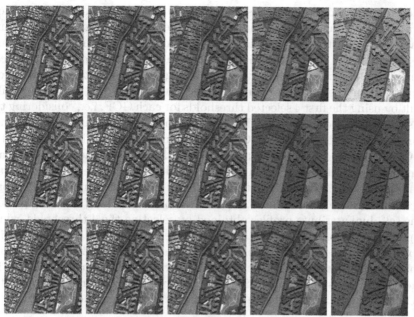

Fig. 3. $\mathcal{P}_\gamma$ computed on Rome based on: (Top line) GCF_{val}; (middle line) GCF_{pix}; (bottom line) GCF_{reg}. Thresholds' value is increasing from left to right, with the first column coinciding with ψ^{T_0} (i.e., the original image).

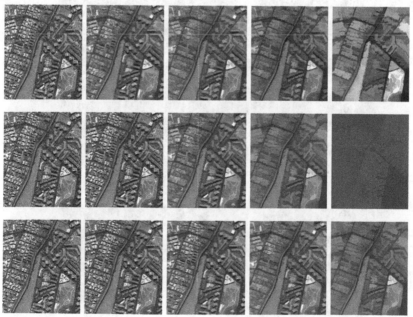

Fig. 4. $\mathcal{P}_\rho$ computed on Rome based on: (Top line) GCF_{val}; (middle line) GCF_{pix}; (bottom line) GCF_{reg}. Thresholds' value is increasing from left to right, with the first column coinciding with ψ^{T_0} (i.e., the original image).

provide useful information. For this reason, these two breakpoints are discarded. Accordingly, the number of thresholds equals to the number of segments minus one. An example of reconstruction error estimation, computed for both the data sets and tree representations, is shown in Fig. 1. There, the reconstruction error has been computed for $\hat{L} = 1, \ldots, 15$ segments. Considering the obtained trend of the errors of reconstruction, the attribute opening profiles and the self-dual attribute profiles are built by using the first 4 selected thresholds for each GCF, i.e., considering the first 5 segments. For each data set, Fig. 2 shows the regression analysis performed to approximate the three derived GCFs. In the figure, the green line represents the real GCFs (computed with all the thresholds in $\bar{\Lambda}$), the blue line denotes the approximation based on 5 segments, and red circles identifies the breakpoints (i.e., the $\hat{\Lambda}$). One can see that the breakpoints are different from one to another GCF, providing different sets $\hat{\Lambda}$ of thresholds. Figs. 3 - 6 show the opening attribute profiles and the self-dual attribute profiles obtained by selecting the 4 thresholds, which corresponds to the breakpoints without the extrema. From the obtained results, it is possible to notice how the considered GCFs are able to model the contextual information according to the chosen measures, while the proposed approach for threshold selection is able to identify those λs values that better characterize the main changes (i.e., changes in slope) in the the distribution of the original GCFs.

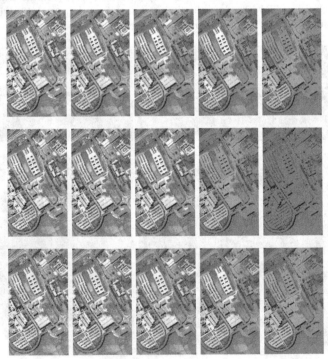

Fig. 5. $\mathcal{P}_\gamma$ computed on Pavia based on: (Top line) GCF_{val}; (middle line) GCF_{pix}; (bottom line) GCF_{reg}. Thresholds' value is increasing from left to right, with the first column coinciding with ψ^{T_0} (i.e., the original image).

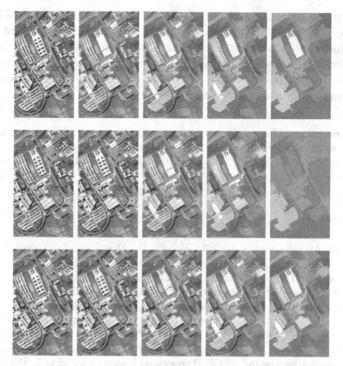

Fig. 6. $\mathcal{P}_\rho$ computed on Pavia based on: (Top line) GCF_{val}; (middle line) GCF_{pix}; (bottom line) GCF_{reg}. Thresholds' value is increasing from left to right, with the first column coinciding with ψ^{T_0} (i.e., the original image).

5   Conclusion

In this paper, a novel methodology for the selection of thresholds based on a new concept of granulometric characteristic functions, is presented. Granulometric characteristic functions are derived from the hierarchical representations of the image, and computed considering a measure of interest. By exploiting the tree (i.e., min-, max-, inclusion- trees) representations, filtering steps prior the selection of the threshold set become unnecessary, making the approach computational efficient. Three GCFs are defined based on different measures, such as, the sum of the gray-level values (i.e., based on the conventional granulometry), the number of changed pixels and the number of changed regions. In addition to the standard granulometry, which is related to the volume (sum of the gray values) of variations, GCF derived from the other measures showed the effects of the decomposition in terms of spatial extent (i.e., how large are the areas that got changed). In order to select the meaningful thresholds, a piecewise linear regression model is employed to approximate the GCF, providing the breakpoints that are used to derive the subset of λ values. The GCFs have been proven useful also for giving hints on the number of threshold to select. The effectiveness of the proposed approach is assessed by a qualitative analysis of the obtained APs and SDAPs built on a panchromatic and hyperspectral images. The obtained image decompositions present effective multi-level characterizations of the original input images according to the chosen attribute (area) and measures. For future research, the presented method will be tested in more applicative contexts, such as feature extraction, supervised classification, and compression of remote sensing images.

Acknowledgements. This research was supported in part by the program J. Verne 2014, project $n°31936TD$ and by EU FP7 Theme Space project North State. The authors are grateful to F. Pacifici for providing the Rome data set, and to Prof. P. Gamba for the Pavia data set.

References

1. Dalla Mura, M., Benediktsson, J.A., Waske, B., Bruzzone, L.: Morphological Attribute Profiles for the Analysis of Very High Resolution Images. IEEE Transaction on Geoscience and Remote Sensing 48, 3747–3762 (2010)
2. Breen, E.J., Jones, R.: Attribute openings, thinnings, and granulometries. Computer Vision and Image Understanding 64(3), 377–389 (1996)
3. Salembier, P., Serra, J.: Flat zones filtering, connected operators, and filters by reconstruction. IEEE Transactions on Image Processing 4, 1153–1160 (1995)
4. Salembier, P., Oliveras, A., Garrido, L.: Antiextensive connected operators for image and sequence processing. IEEE Transacions Image Processing 7, 555–570 (1998)
5. Falco, N., Dalla Mura, M., Bovolo, F., Benediktsson, J.A., Bruzzone, L.: Change Detection in VHR Images Based on Morphological Attribute Profiles. IEEE Geoscience and Remote Sensing Letters 10(3), 636–640 (2013)
6. Mahmood, Z., Thoonen, G., Scheunders, P.: Automatic threshold selection for morphological attribute profiles. In: IEEE International Geoscience and Remote Sensing Symposium (IGARSS), pp. 4946–4949 (July 2012)
7. Franchi, G., Angulo, J.: Comparative study on morphological principal component analysis of hyperspectral images. In: Proceedings of 6th Workshop on Hyperspectral Image and Signal Processing: Evolution in Remote Sensing, WHISPERS (2014)
8. Soille, P.: Morphological image analysis: principles and applications, vol. 49. Springer (2003)
9. McZgee, V.E., Carleton, W.T.: Piecewise regression. Journal of the American Statistical Association 65(331), 1109–1124 (1970)
10. Najman, L., Talbot, H.: Mathematical Morphology: from theory to applications, 520p. ISTE-Wiley (2010) ISBN: 9781848212152
11. Urbach, R., Roerdink, J.B.T.M., Wilkinson, M.H.F.: Connected shape-size pattern spectra for rotation and scale-invariant classification of gray-scale images.. IEEE Trans on PALM, 272–285 (2007), doi:10.1109/TPAMI.2007.28
12. Caselles, V., Monasse, P.: Geometric Description of Images as Topographic Maps, vol. 1984. Springer, Heidelberg (2010)
13. Géraud, T., Carlinet, E., Crozet, S., Najman, L.: A quasi-linear algorithm to compute the tree of shapes of n-D images. In: International Symposium on Mathematical Morphology, pp. 97–108 (2013)

14. Dalla Mura, M., Benediktsson, J.A., Bruzzone, L.: Self-dual attribute profiles for the analysis of remote sensing images. In: Soille, P., Pesaresi, M., Ouzounis, G.K. (eds.) ISMM 2011. LNCS, vol. 6671, pp. 320–330. Springer, Heidelberg (2011)

15. Cavallaro, G., Mura, M.D., Benediktsson, J.A., Bruzzone, L.: A comparison of self-dual attribute profiles based on different filter rules for classification. In: Proceedings of IEEE International Geoscience and Remote Sensing Symposium (IGARSS), pp. 1265–1268 (2014)

Local 2D Pattern Spectra
as Connected Region Descriptors

Petra Bosilj[1]([✉]), Michael H.F. Wilkinson[2], Ewa Kijak[3], and Sébastien Lefèvre[1]

[1] Université de Bretagne-Sud – IRISA, Vannes, France
{petra.bosilj,sebastien.lefevre}@irisa.fr
[2] Johann Bernoulli Institute, University of Groningen, Groningen, The Netherlands
[3] Université de Rennes 1 – IRISA, Rennes, France

Abstract. We validate the usage of augmented 2D shape-size pattern spectra, calculated on arbitrary connected regions. The evaluation is performed on MSER regions and competitive performance with SIFT descriptors achieved in a simple retrieval system, by combining the local pattern spectra with normalized central moments. An additional advantage of the proposed descriptors is their size: being half the size of SIFT, they can handle larger databases in a time-efficient manner. We focus in this paper on presenting the challenges faced when transitioning from global pattern spectra to the local ones. An exhaustive study on the parameters and the properties of the newly constructed descriptor is the main contribution offered. We also consider possible improvements to the quality and computation efficiency of the proposed local descriptors.

Keywords: Shape-size pattern spectra · Granulometries · Max-tree · Region descriptors · CBIR

1 Introduction

Pattern spectra are histogram-like structures originating from mathematical morphology, commonly used for image analysis and classification [12], and contain the information on the distribution of sizes and shapes of image components. They can be efficiently computed using a technique known as granulometry [5] on a max-tree and min-tree hierarchy [9,20].

We study here the application of 2D pattern spectra to Content Based Image Retrieval (CBIR), to retrieve database images describing the same object or scene as the query. Previous success in using the pattern spectra as image descriptors computed at the global [24,25] or pixel scale (known as DMP [3] or DAP [6,19]) convinced us to investigate their behavior as local descriptors.

Standard CBIR systems based on local descriptors consist of region detection, calculation of descriptors and storage in an index. Different indexing schemes are used to perform large scale database search [10,23], but all need powerful local descriptors to achieve good performance [22]. To construct such a descriptor,

The collaboration between the authors was supported by mobility grants from the Université européenne de Bretagne (UEB), French GdR ISIS from CNRS, and an excellence grant EOLE from the Franco-Dutch Network.

© Springer International Publishing Switzerland 2015
J.A. Benediktsson et al. (Eds.): ISMM 2015, LNCS 9082, pp. 182–193, 2015.
DOI: 10.1007/978-3-319-18720-4_16

we want to extend [25] and compute 2D size-shape pattern spectra locally while keeping the good characteristics of the global version (scale, translation and rotation invariance, and computation efficiency). However, to evaluate the quality and properties of our proposed local pattern spectra (LPS) descriptors, we need to reexamine the parameters used with global pattern spectra as well as evaluate the effect of the new parameters introduced by the local descriptor scheme.

We evaluate our descriptors on the MSER regions [13] as they can also be computed on a max-tree [18], using the well-established SIFT descriptors [11] to obtain a baseline CBIR performance. Future work will include comparisons with SIFT extensions which improve performance [1,2]. A competitive precision is achieved when combining the local pattern spectra with normalized central moments, producing a descriptor half the size of SIFT. Two versions of the descriptor are examined - a scale invariant version and a version that is only rotation invariant (deeper interpretation of the results can be found in [4]).

As the goal of this paper is to give an overview of choices and challenges faced when reworking a global pattern spectrum into a local one, we adopt a slightly atypical presentation structure: The background notions are presented in Sec. 2, with the focus on how the max-tree is used in all parts of the CBIR system. The experimental framework used to tune and evaluate the descriptors is explained in Sec. 3. To examine the properties of the proposed LPS descriptor through the influence of parameters used, the main contribution can be found in Sec. 4, where the descriptor performance is also presented. Remarks on possible improvements to the efficiency of LPS computation are given in Sec. 5. Finally, the conclusions are drawn and directions for future work offered in Sec. 6.

2   Background

2.1   Max-tree

The concept of min and max-trees [9,20] is here central for keypoint detection as well as the calculation of feature descriptors. We recall their definition using the *upper* and *lower level sets* of an image, e.g. sets of image pixels p with gray level values $f(p)$ respectively higher and lower than a threshold k.

Given a level k of an image I, each level set is defined as $\mathcal{L}^k = \{p \in I | f(p) \geq k\}$ for the max-tree, or $\mathcal{L}_k = \{p \in I | f(p) \leq k\}$ for the min-tree. Their connected components (also called the *peak components*) $\mathcal{L}^{k,i}$ and $\mathcal{L}_{k,i}$ (i from some index set) are nested and form a hierarchy. The min-tree is usually built as a max-tree of the inverted image $-I$.

2.2   MSER Detection

Peak components of the upper and lower level sets $\{\mathcal{L}^{k,i}\}$ and $\{\mathcal{L}_{k,i}\}$ coincide with the maximal and minimal extremal regions in the context of *Maximally Stable Extremal Regions* (MSER) detector introduced by Matas et al. [13]. The detected regions correspond to bright and dark "blobs" in the image and can be extracted while building the max-tree and the min-tree [18].

Extraction of MSER relies on the stability function $q(\mathcal{L}^{k,i})$, which measures the rate of growth of the region w.r.t. the change of the threshold level k. It is computed for all the elements of nested sequences, and the local minima of this function correspond to the maximally stable regions.

We use here a simplification commonly adopted by many computer vision libraries (e.g. VLFeat [27]) :

$$q(\mathcal{L}^{k,i}) = \frac{A(\mathcal{L}^{k-\Delta,i} \setminus \mathcal{L}^{k,i})}{A(\mathcal{L}^{k,i})}, \tag{1}$$

where the area is denoted by $A(\cdot)$ and Δ is a parameter of the detector. Additional parameters control the allowed region size, limit the appearance of too similar regions and impose a lower limit on the stability score. The implementation parameters were set by comparing the performance to the MSER implementation provided for [15]. For setting these parameters, we used the *viewpoint* dataset and the same measures (repeatability and matching score) provided in [15].

2.3 Attributes and Filtering

Region characteristics can be captured by assigning them *attributes* measuring the interesting aspects of the regions. *Increasing* attributes $K(\cdot)$ give increasing values when calculated on a nested sequence of regions, otherwise they are *nonincreasing*. A value of an increasing attribute on a tree region, $K(\mathcal{L}^{k,i})$, will be greater than the value of that attribute for any of the regions descendants.

Increasing attributes are usually a measure of the *size* of the region. We will simply use the *area* (in pixels) of the region, $A(\mathcal{L}^{k,i})$, as the size attribute. *Strict shape* attributes are the nonincreasing attributes dependent only on the region shape, thus invariant to scaling, rotation and translation [5]. To indicate the shape of a region, we use an elongation measure called *corrected noncompactness*:

$$NC(\mathcal{L}^{k,i}) = 2\pi \left(\frac{I(\mathcal{L}^{k,i})}{A(\mathcal{L}^{k,i})^2} + \frac{1}{6A(\mathcal{L}^{k,i})} \right). \tag{2}$$

$I(\mathcal{L}^{k,i})$ is here the moment of inertia of the region, and the term $\frac{I(\mathcal{L}^{k,i})}{A(\mathcal{L}^{k,i})^2}$ without the correction is equal to the first moment invariant of Hu [8] $I = \mu_{2,0} + \mu_{0,2}$. The correction factor appears when transitioning from the original formula in the continuous space to the discrete image space [28].

We also directly use the normalized central moments $n_{1,1}, n_{2,0}, n_{0,2}, n_{0,4}$ and $n_{4,0}$ of the considered regions. These, and more attributes (e.g. center of mass, covariances, skewness or kurtosis [29]), can be derived from raw region moments.

When the tree is further processed by comparing the region attribute values to a threshold t (or using a more complex criterion), and making a decision to preserve or reject a region based on the attribute value, we are performing an *attribute filtering*. While filtering with an increasing attribute is relatively straightforward, advanced filtering strategies have to be used when performing a filtering with nonincreasing (e. g. shape) attributes [5, 20, 25, 30].

2.4 Granulometries and Global Pattern Spectra

Attribute opening is a specific kind of attribute filtering, in which the attribute used is increasing. Such a transformation is anti-extensive, increasing and idempotent.

A *size granulometry* can be computed from a series of such openings, using increasing values for the threshold t. This series also satisfies the absorption property, since applying an opening with $t' < t$ will have no effect on an image already filtered with an opening using the threshold t. In other words, a size granulometry can be seen as a set of sieves of increasing grades, each letting only details of certain sizes [25] pass through.

Instead of focusing on the details remaining, one can consider the amount of detail removed between consecutive openings. Such an analysis, introduced by Maragos [12] under the name *size pattern spectra*, produces a 1D histogram containing, for each size class or filtering residue, its Lebesgue measure (i. e. the number of pixels in the binary case or the sum of gray levels in the grayscale case). Such histograms can also be computed over different shape classes, leading to the shape-spectra [25]. Finally, shape and size pattern spectra can be combined to build *shape-size pattern spectra* [25], 2D histograms where the amount of image detail for the different shape-size classes is stored in dedicated 2D bins.

Previous work [24, 25] as well as our own experiments suggest that the lower attribute values carry more information. Thus, a logarithmic binning is used for both attributes, producing higher resolution bins for low attribute values. Let v be the attribute value for one of the attributes, N_b the total desired *number of bins* and m the *upper* bound for that attribute (which can be the maximal attribute value in the hierarchy, or a smaller value if we decide to ignore attribute values above a certain threshold). If the minimal value for the attribute is 1 (as with both area and the corrected noncompactness), the base for the logarithmic binning b, and the final bin c, are determined as:

$$b = \sqrt[N_b]{m}, \tag{3}$$

$$c = \lfloor \log_b v \rfloor \tag{4}$$

Enumerating the bins starting from 1, the i-th bin has the range $[b^{i-1}, b^i]$.

Connected pattern spectra are effectively calculated in a single pass over a max-tree [5, 25]. For every region, we calculate both the size attribute $v_1 = A(\mathcal{L}^{k,i})$ and shape attribute $v_2 = NC(\mathcal{L}^{k,i})$, and add the area of the region weighted by its contrast with the parent region δ_h to the spectrum bin $S(c_1, c_2)$. Before using the spectrum as a descriptor, we equalize the sum across all the bins as $S(c_1, c_2) = \sqrt[5]{S(c_1, c_2)}$. More information and discussion about the algorithm used to compute the descriptors is given in Sec. 5.

3 Database and Experimental Setup

To evaluate the retrieval performance of the descriptors without introducing noise in the results with approximate search approaches [10, 23], we chose a relatively small *UCID* database [21], on which we can perform an exact search. The performance of our proposed descriptors is compared to SIFT [11].

The whole *UCID* database contains 1338 images of size 512×384 pixels, divided into 262 unbalanced categories. After region detection and description, a

Table 1. Subsets of the UCID database used in experiments

	# categories / examples	categories selected
ucid5	31 / 5	all UCID categories with ≥ 5 examples
ucid4	44 / 4	all UCID categories ≥ 4
ucid3	77 / 3	all UCID categories ≥ 3
ucid2	137 / 2	all UCID categories ≥ 2
ucid1	262 / 1	all UCID categories

single database entry for every category is constructed, comprising the descriptors from all the images of that category. Therefore, to equalize the database entry sizes as much as possible, different subsets of the *UCID* database were used in the experiments, where the number of examples per category is constant for each database subset (the required number of images is taken from larger categories in order provided by the ground truth). Table 1 summarizes the subsets of the database used for experiments presented herein.

A KD-Tree index [7] is built based on the category descriptors, and stored for querying using the FLANN library [16]. We then perform a query with 1 image for every database category. The index performs a kNN search ($k = 7$) with each descriptor of a region detected on the query image. The final category is given through a voting mechanism where each nearest neighbor d_i of a query descriptor q_j will cast a vote for the category $cat(d_i)$ it belongs to:

$$vote(cat(d_i)) = \frac{1}{(L_1(d_i, q_j) + 0.1) \times |cat(d_i)|^{w_{cat}}}. \tag{5}$$

$L_1(d_i, q_j)$ refers to the distance between these two descriptors and $|cat(d_i)|$ is the number of descriptors in the category of the i-th nearest neighbor. Finally, w_{cat} is a parameter of the experimental setup. A $k > 1$ is chosen to take into account several nearest neighbors if their distance is very similar. As the vote contribution decreases with the distance, very far neighbors will have a negligible contribution even if considered.

The measures we used are mean average precision (MAP) and precision at one (P@1). Performance for different values of w_{cat} are shown in Fig. 1(a) and 2(d), but for all the summarized results, only the performance for the optimal w_{cat} value for each experiment is shown. This choice is made in order to present a fair comparison, and since not all the descriptors reach their peak performance for the same value of w_{cat}. This is additionally justified as this parameter is not present when using an approximate classification scheme.

4   Local Pattern Spectra

Local pattern spectra (LPS) are calculated from the selected MSER regions. As the two trees contain different regions, the descriptor for a maximal MSER will only be based on the max-tree, and similarly for the minimal MSERs.

Table 2. Parameters and their optimal values for the LPS (best alternative parameter choices also given)

symbol	significance	value SI-LPS	value SV-LPS
$m_{\mathcal{A}}$	upper bound for area	region size	
m_{NC}	upper bound for noncompactness	53 (54, 56)	53 (57)
$N_b^{\mathcal{A}}$	number of area bins	10	10 (9)
N_b^{NC}	number of noncompactness bins	6	
M	scale parameter for the size attribute	1000	region size
$w(n_{1,1})$	normalized moment weights	20	
$w(n_{2,0}), w(n_{0,2}),$ $w(n_{4,0}), w(n_{0,4})$		10	

The LPS are calculated like the global ones, except the calculation is done on the corresponding subtree. When calculating the LPS for the MSER region $\mathcal{L}^{k,i}$ in the tree, we only consider the attribute values of the descendants of the node. However, transitioning to the local version of the descriptor will introduce a new parameter influencing the scale invariance property of the descriptors.

To achieve both the desired properties and competitive performance, the proposed descriptor is explained here through examining the experiments used to establish the best parameters. The summary of these parameters, explained individually henceforth, can be found in Tab. 2. Additionally, we consider combining the LPS with normalized central moments and enhancing the performance by adding the global pattern spectra. The influence of the database on the results is also discussed.

4.1 Scale Invariance

When calculating a global pattern spectrum for an entire image, the whole image size is used to determine the base of the logarithmic binning (especially if the database images are the same size [24,25]). If we choose to determine the binning base for each region separately based on the area of that region for the local descriptor scheme, the resulting LPS descriptor is not scale invariant.

Let us consider two version of the same region at different scales, with the area values belonging to the range $[1, m_1]$ and $[1, m_2]$ respectively. The scale invariance property requires that, for a value $v_1 \in [1, m_1]$, the bin c_1 determined in the original scale is the same as the bin c_2 for the value $v_2 = v_1 \frac{m_2}{m_1}$ scaled to the range $[1, m_2]$. However, this is not the case for $m_1 \neq m_2$, as:

$$c_1 = \log_{N\sqrt[N]{m_1}} v_1 \neq c_2 = \log_{N\sqrt[N]{m_2}} v_2. \tag{6}$$

Therefore, to ensure the scale invariance, the area used to determine the binning and the logarithmic base have to be the same for all the regions. This area becomes a parameter of the size attribute in LPS, called the *scale parameter M*.

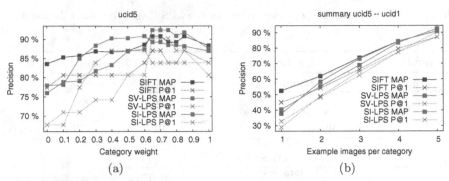

Fig. 1. The results for the final version of the descriptors expressed in terms of mean average precision (MAP) and precision at 1 (P@1) for *ucid5* dataset for varying category weights are shown in 1(a). The results for *ucid5–ucid1* are summarized on 1(b) (performance shown for optimal weight w_{cat} for every dataset).

Using a common scale M can be seen as rescaling all the regions to a reference scale, and has two consequences. First, for a region of size $m > M$, the minimal value v of this region that can contribute to the spectrum when using a common binning is such that $v' = v\frac{M}{m} = 1$, and all the (sub)regions with the area smaller than $\frac{m}{M}$ will be ignored. However, some particular regions with a large enough area can still disappear when rescaling. This is the case for long thin objects with the width (along any dimension) small enough to downscale to under 1 pixel. Such regions should be ignored in the pattern spectrum, even if their attribute values fit with the binning. Because of this, we also determine the maximal possible value of the noncompactness attribute for all of the available area bins.

Second, the minimal area value (1 pixel) of a region of size $m < M$ will be rescaled to the value $v' = \frac{M}{m} > 1$, and the lower area bins at the common scale will be empty. The first area bin $c_{\min}$ that will contain information is then:

$$1 = b^{c_{\min}-1}\frac{m}{M} \rightarrow c_{\min} = \left\lfloor \log_b \frac{M}{m} \right\rfloor + 1. \tag{7}$$

We compare 2 versions of the descriptor: a) the *scale variant version* (SV-LPS), where the area of each region is used as the scale parameter M, and b) the *scale invariant version* (SI-LPS) where M is the same for all regions. Both versions match the SIFT performance in all performed experiments (cf. Fig. 1). The best performance for the SI-LPS was obtained for $M = 1$ (found experimentally) for *UCID* images. However, the UCID database is not very challenging in terms of scale change. We expect the SI-LPS to perform even better when running experiments on a database focusing on scale change, as the performance of the scale invariant SI-LPS should be less affected than that of SV-LPS.

4.2   Binning Parameters

With the area attribute, the upper bound used, m_A, is simply the size of the region: we can plausibly expect regions of all sizes lower than the size of the region itself to be present in its decomposition.

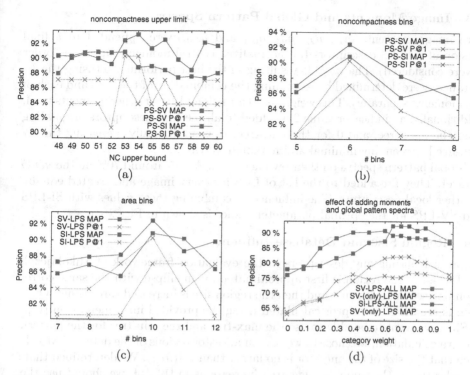

Fig. 2. Parameter tuning on *ucid5* database. The effect of varying the upper bound for noncompactness is shown on 2(a), similar for the amount of noncompactness bins on 2(b), and the area bins on 2(c). The effect of adding the moments and indicator value to the descriptor, with the best parameter settings is shown in 2(d). Note that the global descriptors for the SI-LPS are calculated with the scale value used for the other descriptors, and not using image size.

Examining the values of the noncompactness attribute for several images, we determined that very few regions have high values of this attribute. As such, noncompactness values higher than a certain threshold can be safely ignored. Optimal values m_{NC} for both SV-LPS and SI-LPS were determined by examining the performance of the values close to the ones used in [24, 25]. Similar experiments were done to determine N_b^{NC} and N_b^A. The parameter tuning experiments for the *ucid5* database are shown in Fig. 2.

For both descriptors, we chose $N_b^{NC} = 6$ and $N_b^A = 10$. To choose between several values of m_{NC} performing well on *ucid5*, we compare their performance on *ucid4–ucid1* as well. This was done as the performance for different values of m_{NC} is fairly stable (only about 5% difference for values shown on Fig. 2(a)).

Surprisingly, we also found an alternative set of values for SV-LPS with the lower value of $N_b^A = 9$ but a higher $m_{NC} = 57$. The optimal values as well as the best alternative choices are shown in Tab. 2. As an alternate set of parameters was found producing shorter SV-LPS descriptors, the possibility of shortening the SI-LPS without the loss in performance should also be investigated.

4.3   Image Moments and Global Pattern Spectra

Five image moments, $n_{1,1}, n_{2,0}, n_{0,2}, n_{0,4}$ and $n_{4,0}$, were appended to a final version of all LPS descriptors (all normalized central moments up to the order 5 were considered). The weights resulting in the best performance (using the L_1 distance) were determined by examining the combination of the LPS and each of the moments separately. This weight is 20 for $n_{1,1}$ and 10 for other moments used. Additionally, an indicator value 2 is added to all the LPS descriptors originating from the max-tree, and 0 for the min-tree, thus additionally increasing the L_1 distance between any minimal and maximal MSERs.

Global pattern spectra on their own achieve MAP@5 around 70% on the *ucid5* dataset. They are added to the list of LPS for every image and treated equally to other local descriptors. The influence of combining these values with SI-LPS and SV-LPS for the optimal parameter choice is shown in Fig. 2(d).

4.4   Region Size and Database Influence

Before calculating any descriptors in the evaluation framework of Mikolajczyk et al. [14,15], the region is first approximated by an ellipse with the same corresponding second moments, and then the region size is increased three times. Only then is the SIFT descriptor calculated using the provided implementation [11].

Since we want to be able to use the max-tree and the min-tree for the pattern spectra calculation, we chose to work with ancestor regions of the detected MSER such that the size of the ancestor is no larger than $xA(n_{k,i})$. We determined that, in order to get the same average area increase as in [14,15], we should use the value $x = 7.5$. The reason is that many regions have a much bigger parent region, which is then not considered, and the size increase is often smaller than x times.

Figure 1(b) summarizes the performance on all the subsets from Tab. 1, allowing us to examine the behavior of the descriptors for the increasing database size. The performance expectantly decreases with the increase of database size and decrease of the number of examples provided per category. As the separate influence of these two factors can not be determined just from experiments on these subsets, additional tests were carried out and analyzed in [4].

Besides the performance, it is important to note here that on the largest database subset used, the query speed for LPS is around $4\times$ faster than that for SIFT (when the LPS descriptor of size 66 is used).

5   Remarks on the Algorithm

The system was implemented in C++. The max-tree structure was used for both MSER detection and keypoint description. The non-recursive max-tree algorithm of [18] was used. This allows concurrent computation of the MSER stability function (Eq. (1)), the area attribute and the moment of inertia, and the MSER. The method is as follows:

- Compute the max-tree and min-tree according to [18].
- As the trees are built, compute:

- attribute values for the nodes of the trees,
- local minima of the stability function, forming the sets of MSER regions,
- global pattern spectra [25].
- For each selected MSER region, repeat the computation of the pattern spectra locally in a sub-tree.
- Combine the attribute values, indicator value 0 or 2 and the pattern spectra to form a LPS descriptor for a MSER region.
- Add both global pattern spectra [24] corresponding to the whole image in the collection of descriptors for the image.

Unlike the calculation of global pattern spectra, the local pattern spectra use the constructed hierarchy but can not be computed concurrently because of different upper limits (for area) and binning scaling value.

However, adopting the scale invariant version to concurrent computation can be considered. While it would sacrifice true scale invariance, if the value M is used as a scale parameter, and we are calculating for a region of size m, we can set the largest bin to be $[b^{\lceil log_b m \rceil - 1}, b^{\lceil log_b m \rceil}]$, with the smallest bin having the upper bound $b^{\lceil log_b m \rceil - N_b}$. While not all the values from the whole range of the largest bin will be possible for all the regions, the bin values of the children can be used directly by their parents. When the upper bound of the largest bin changes, the child values can still be used with discarding the values from the smallest bin: the scale of those details is too low to be considered.

6   Discussion and Conclusion

After successfully applying global pattern spectra in CBIR context [24, 26], we now attempt to construct a local region descriptor based on the pattern spectra. On the chosen subsets of the *UCID* database [21], the results obtained were better than when only using global pattern spectra (almost 20% in MAP@5 on *ucid5*), and matched the performance of the SIFT descriptor. The constructed SI-LPS descriptors keep all the invariance properties of the global pattern spectra (translation, rotation and scale invariance).

The proposed descriptors have another advantage. In addition to the description calculation process being slightly faster for the pattern spectra than for the SIFT descriptors, our descriptors length is only half of the length of SIFT. This makes using these descriptors much faster – performing 262 queries on an index of the size 262 (*ucid1* dataset) took 4 times longer using SIFT descriptors. This suggests that (especially in large scale CBIR systems), we can use more example images in order to enhance the precision, while still performing faster than SIFT.

As the performance of the descriptors depends on a lot of parameters, we need to explore a way to determine the optimal parameters automatically. Also, while the LPS descriptors are rotation invariant, enforcing scale invariance introduces an additional parameter. In addition to examining this new parameter closer, both SI-LPS and SV-LPS will be evaluated on a database focused on scale changes to determine the value of true scale invariance in such cases.

Despite the parameters and the descriptor invariance which have to be further studied, matching the SIFT performance on the four subsets of the *ucid* dataset with a descriptor of less than half the length of SIFT is very promising. Additional successful experiments were performed and analyzed in [4]. It also prompts for evaluating the LPS performance with large scale CBIR system. It is probable that the results could be even further improved by combining the current LPS with pattern spectra based on other shape attributes, like in [24].

Lastly, the L_1 distance, designed to compare vectors of scalar values, is not the best choice for comparing histogram-like structures. Using different distances, or even divergences (e.g. [17]) which take into account the nature of the descriptor should also improve the performance.

References

1. Arandjelović, R., Zisserman, A.: Three things everyone should know to improve object retrieval. In: 2012 IEEE Conference on Computer Vision and Pattern Recognition (CVPR), pp. 2911–2918. IEEE (2012)
2. Bay, H., Ess, A., Tuytelaars, T., Van Gool, L.: Speeded-up robust features (SURF). Computer Vision and Image Understanding 110(3), 346–359 (2008)
3. Benediktsson, J.A., Pesaresi, M., Arnason, K.: Classification and Feature Extraction for Remote Sensing Images from Urban Areas based on Morphological Transformations. IEEE Transactions on Geoscience and Remote Sensing 41(9), 1940–1949 (2003)
4. Bosilj, P., Kijak, E., Wilkinson, M.H.F., Lefèvre, S.: Short local descriptors from 2D connected pattern spectra. Submitted to ICIP 2015, https://hal.inria.fr/hal-01134071
5. Breen, E.J., Jones, R.: Attribute openings, thinnings, and granulometries. Computer Vision and Image Understanding 64(3), 377–389 (1996)
6. Dalla Mura, K., Benediktsson, J.A., Waske, B., Bruzzone, L.: Morphological Attribute Profiles for the Analysis of Very High Resolution Images. IEEE Transactions on Geoscience and Remote Sensing 48(10), 3747–3762 (2010)
7. Friedman, J.H., Bentley, J.L., Finkel, R.A.: An algorithm for finding best matches in logarithmic expected time. ACM Transactions on Mathematical Software (TOMS) 3(3), 209–226 (1977)
8. Hu, M.K.: Visual pattern recognition by moment invariants. IRE Transactions on Information Theory 8(2), 179–187 (1962)
9. Jones, R.: Component trees for image filtering and segmentation. In: Coyle, E. (ed.) IEEE Workshop on Nonlinear Signal and Image Processing, Mackinac Island (1997)
10. Lejsek, H., Jónsson, B.T., Amsaleg, L.: NV-Tree: Nearest Neighbors at the Billion Scale. In: Proceedings of the 1st ACM International Conference on Multimedia Retrieval, ICMR 2011, pp. 54:1–54:8 (2011)
11. Lowe, D.G.: Distinctive image features from scale-invariant keypoints. International Journal of Computer Vision 60(2), 91–110 (2004)
12. Maragos, P.: Pattern spectrum and multiscale shape representation. IEEE Transactions on Pattern Analysis and Machine Intelligence 11(7), 701–716 (1989)
13. Matas, J., Chum, O., Urban, M., Pajdla, T.: Robust wide-baseline stereo from maximally stable extremal regions. Image and Vision Computing 22(10), 761–767 (2004)
14. Mikolajczyk, K., Schmid, C.: A performance evaluation of local descriptors. IEEE Transactions on Pattern Analysis and Machine Intelligence 27(10), 1615–1630 (2005)

15. Mikolajczyk, K., Tuytelaars, T., Schmid, C., Zisserman, A., Matas, J., Schaffal-itzky, F., Kadir, T., Van Gool, L.: A comparison of affine region detectors. International Journal of Computer Vision 65(1-2), 43–72 (2005)
16. Muja, M., Lowe, D.G.: Fast Approximate Nearest Neighbors with Automatic Algorithm Configuration. In: International Conference on Computer Vision Theory and Application (VISSAPP 2009), pp. 331–340. INSTICC Press (2009)
17. Mwebaze, E., Schneider, P., Schleif, F.M., Aduwo, J.R., Quinn, J.A., Haase, S., Villmann, T., Biehl, M.: Divergence-based classification in learning vector quantization. Neurocomputing 74(9), 1429–1435 (2011)
18. Nistér, D., Stewénius, H.: Linear time maximally stable extremal regions. In: Forsyth, D., Torr, P., Zisserman, A. (eds.) ECCV 2008, Part II. LNCS, vol. 5303, pp. 183–196. Springer, Heidelberg (2008)
19. Ouzounis, G.K., Pesaresi, M., Soille, P.: Differential Area Profiles: Decomposition Properties and Efficient Computation. IEEE Transactions on Pattern Analysis and Machine Intelligence 34(8), 1533–1548 (2012)
20. Salembier, P., Oliveras, A., Garrido, L.: Antiextensive connected operators for image and sequence processing. IEEE Transactions on Image Processing 7(4), 555–570 (1998)
21. Schaefer, G., Stich, M.: UCID: An Uncompressed Colour Image Database. In: Electronic Imaging 2004, pp. 472–480. International Society for Optics and Photonics (2003)
22. Schmid, C., Mohr, R.: Object recognition using local characterization and semi-local constraints. IEEE Transactions on Pattern Analysis and Machine Intelligence 19(5), 530–534 (1997)
23. Sivic, J., Zisserman, A.: Video google: Efficient visual search of videos. In: Ponce, J., Hebert, M., Schmid, C., Zisserman, A. (eds.) Toward Category-Level Object Recognition. LNCS, vol. 4170, pp. 127–144. Springer, Heidelberg (2006)
24. Tushabe, F., Wilkinson, M.H.F.: Content-based image retrieval using combined 2D attribute pattern spectra. In: Peters, C., Jijkoun, V., Mandl, T., Müller, H., Oard, D.W., Peñas, A., Petras, V., Santos, D. (eds.) CLEF 2007. LNCS, vol. 5152, pp. 554–561. Springer, Heidelberg (2008)
25. Urbach, E.R., Roerdink, J.B.T.M., Wilkinson, M.H.F.: Connected shape-size pattern spectra for rotation and scale-invariant classification of gray-scale images. IEEE Transactions on Pattern Analysis and Machine Intelligence 29(2), 272–285 (2007)
26. Urbach, E.R., Wilkinson, M.H.F.: Shape-only granulometries and grey-scale shape filters. In: Proc. Int. Symp. Math. Morphology (ISMM), vol. 2002, pp. 305–314 (2002)
27. Vedaldi, A., Fulkerson, B.: VLFeat: An Open and Portable Library of Computer Vision Algorithms (2008), http://www.vlfeat.org/
28. Westenberg, M.A., Roerdink, J.B.T.M., Wilkinson, M.H.F.: Volumetric Attribute Filtering and Interactive Visualization using the Max-Tree Representation. IEEE Trans. Image Proc. 16, 2943–2952 (2007)
29. Wilkinson, M.H.F.: Generalized pattern spectra sensitive to spatial information. In: International Conference on Pattern Recognition, vol. 1, pp. 10021–10021. IEEE Computer Society (2002)
30. Xu, Y., Géraud, T., Najman, L.: Morphological filtering in shape spaces: Applications using tree-based image representations. In: 2012 21st International Conference on Pattern Recognition (ICPR), pp. 485–488. IEEE (2012)

Spatial Repulsion Between Markers Improves Watershed Performance

Vaïa Machairas[1]($\boxtimes$), Etienne Decencière[1], and Thomas Walter[2,3,4]

[1] Center for Mathematical Morphology, MINES ParisTech, PSL Research University,
Fontainebleau, France
{vaia.machairas,etienne.decenciere}@mines-paristech.fr
[2] CBIO-Centre for Computational Biology, MINES ParisTech, PSL-Research
University, Fontainebleau, France
[3] Institut Curie, 75248 Paris Cedex, France
[4] INSERM U900, 75248 Paris Cedex, France

Abstract. The Watershed Transformation is a powerful segmentation tool from Mathematical Morphology. Here we focus on the markers selection step. We hypothesize that introducing some kind of repulsion between them leads to improved segmentation results when dealing with natural images. To do so, we compare the usual watershed transformation to waterpixels, i.e. regular superpixels based on the watershed transformation which include a regularity constraint on the spatial distribution of their markers. Both methods are evaluated on the Berkeley segmentation database.

Keywords: Watershed · Waterpixels · Superpixels · Segmentation

1 Introduction

Segmentation aims at partitioning an image such that the resulting regions correspond to meaningful objects in the image. The Watershed Transformation is a powerful tool from Mathematical Morphology for segmentation (Beucher and Lantuéjoul [1], Beucher and Meyer [2]). It can be described as the flooding of a gray level image, considered as a topographic surface, starting from a given set of markers. The number of resulting regions is given by the number of markers used for flooding. When the watershed is applied on the gradient, contours of the resulting partition are likely to fall on object contours. For this reason, and for the low complexity of corresponding algorithms (Meyer [3], Vincent and Soille [4]), this transformation has been widely used for segmentation tasks (Andres et al. [5], Stawiaski and Decencière [6]).

Hence, to obtain a good segmentation with the watershed transformation, we must select carefully two elements: the gradient to be flooded and the markers from which the flooding starts. First, the gradient should convey a pertinent information (e.g. Lab gradient for estimation of visual perception, thick or thin gradients according to the thickness of the objects we would like to detect, oriented gradients,...). Second, the selection of markers is also crucial, as their

© Springer International Publishing Switzerland 2015
J.A. Benediktsson et al. (Eds.): ISMM 2015, LNCS 9082, pp. 194–202, 2015.
DOI: 10.1007/978-3-319-18720-4_17

number will be equal to the number of final regions and their spatial localization will impact the pertinence of the resulting contours. They are often selected among the minima of the gradient, but can also be chosen freely over the image space (marker-controlled watershed, see Meyer and Beucher [7]).

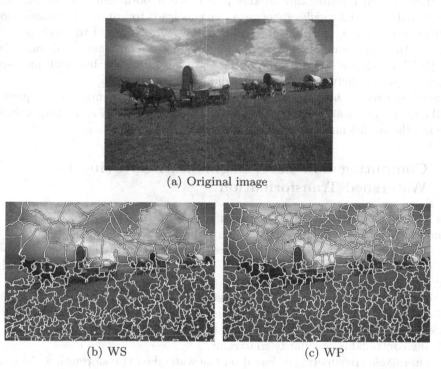

(a) Original image

(b) WS   (c) WP

Fig. 1. Low-level segmentations obtained with watershed (WS) and waterpixels (WP)

We will focus here on the choice of the markers. The success of the watershed transformation critically depends on the choice of markers, which raises the question of how to choose the best set of markers. Either we already have a priori information on the objects we would like to segment. We can then build a measure of pertinence for each marker which takes into account different important and discriminative criteria dependent of the application, and then choose the best according to this measure. This can be done also explicitly thanks to interactive clicking on the image by the user. When no a priori information is available, there are two popular solutions in the literature. One solution is to repeat the process of random selection and combine resulting partitions (m realizations give m partitions) to build a probability density function of contours, as done in the stochastic watershed (Angulo and Jeulin [8]). Another solution is to take general criteria to measure the pertinence: for example extinction values represent the height, area and volume of the corresponding lake before fusion (Vachier and Meyer [9]).

If criteria to evaluate the pertinence of markers have been studied in the literature, the impact of their spatial distribution over the image has not been studied yet. If a given criterion of pertinence has been chosen, say surface extinction values for example, will it change the performance if we add also a criterion on their spatial distribution? In this paper, we hypothesize that introducing some kind of spatial repulsion between markers leads to improved segmentation results when dealing with natural images. We will compare two methods (see Fig. 1): the usual watershed (WS) without regularity constraint, and waterpixels (WP) which are based on the watershed transformation but with imposed spatial repulsion between markers (Machairas et al. [10], [11]).

Section 2 recaps how WS and WP are built. Section 3 compares the quality of the final segmentation for both methods. Eventually, Sec. 4 and 5 respectively discuss the results and conclude on this comparison.

2   Computing Low-Level Segmentations Using the Watershed Transformation

Let f be an image defined on some subset D of $\mathbb{Z}^2$. Let g be the gradient computed on f. We suppose that g is of the form: $g : D \mapsto V$, where $V = \{0, \ldots, 255\}$. Let N be the number of desired regions in the final segmentation.

For each method (WS and WP), N markers must be selected among the minima of the gradient g.

For WS, we will take the N best markers in terms of extinction value (either volume, surface or dynamic). As explained in the next paragraph, the same criterion of pertinence is used for waterpixels computation, but with an additional (ideally controllable) regularity constraint.

Waterpixels are superpixels based on the watershed transformation. As a reminder, superpixels are a special case of low-level segmentation where final regions are rather regular. Since the term was coined by Ren and Malik [12], they have been widely used as primitives for further analysis of the image (detection, segmentation, classification of objects). Waterpixels are a family of fast state-of-the-art methods to compute such superpixels (Machairas et al. [11], [10][1]). In the rest of the paper, we will use m-waterpixels and denote them "waterpixels" for the sake of simplicity. This method is characterized by five steps. First, the gradient g of the image is computed. Second, a grid of regular cells (separated by margins) is created: it will serve as a stake for SP generation. Then follow two steps, inherent to the watershed: the definitions of the N markers (step 3) and of the gradient to be flooded (step 4). Eventually, the watershed is applied on the gradient defined in step 4 with the markers defined in step 3.

Step 3: Selection of the N markers (See Fig.2(b)). A unique marker is selected per grid cell among the minima of the gradient g present in this very cell. If several minima are present, then the one with the highest extinction value is used.

---

[1] An implementation of WP can be found at
http://cmm.ensmp.fr/~machairas/waterpixels.

If no minimum is present, the flat zone with minimum value of the gradient inside this very cell is used as marker. We denote by $\{M_i\}_{1 \leq i \leq N}$, $M_i \in D$ the set of N selected markers. In our experiments [10], surface extinction value consistently led to better results on the Berkeley segmentation database (Martin et al. [13]) than volume and dynamic extinction values, but in principle, any extinction value can be used in this framework.

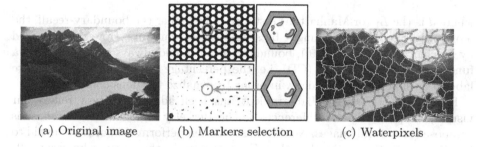

(a) Original image   (b) Markers selection   (c) Waterpixels

Fig. 2. Marker selection of Waterpixels

Step 4: Spatial regularization of the gradient The gradient is spatially regularized by adding the (normalized) distance function to the markers d_M :

$$r_{reg} = g + k \times d_M \tag{1}$$

where g is the gradient of the original image and $k \in \mathbb{R}^+$ is a parameter which enables to weight the importance of the regularization. We have previously shown (Machairas et al. [10]) that waterpixels provide state-of-the-art results on the Berkeley segmentation database.

Experimental comparison between both methods (WS and WP) is presented in the next section. As we want in this paper to study only the impact of the spatial distribution of the markers on adherence to object boundaries (i.e. at step 3 level), we will remove the regularity constraint included in step 4 by setting k to zero.

3 Experiments

3.1 Evaluation Database and Criteria

To measure the impact of the spatial distribution of the markers, both methods were applied on the Berkeley segmentation database (Martin et al. [13]). This database is divided into 3 subsets, "train", "test" and "val", containing respectively 200, 200 and 100 images of sizes 321x481 or 481x321 pixels. Approximatively 6 human-annotated ground-truth segmentations are given for each image. These ground-truth images correspond to manually drawn contours.

To evaluate the adherence to object boundaries, boundary-recall (BR) has been calculated for both methods (WS and WP). This classical measure used in the literature is defined as the percentage of ground-truth contour pixels GT which fall within less than 3 pixels from the segmentation boundaries C (where the segmentation is obtained either by WS or WP):

$$BR = \frac{|\{p \in GT, d(p, C) < 3\}|}{|GT|} \tag{2}$$

where d is the L_1 (or Manhattan) distance. The higher the boundary-recall, the more object contours are caught by segmentation contours. However, as noted by Machairas et al. ([11], [10]), boundary-recall should always be expressed as a function of contour density (CD), i.e. the percentage of image pixels which are labeled as superpixels boundary, in order to penalize tortuous contours.

For both methods, the Lab-gradient has been adopted to best reflect our visual perception of color differences as the Berkeley segmentation database is composed of natural images. No filtering has been performed in order to avoid to favorize one method upon the other. For waterpixels, the margins between cells have been created by reducing cells spread by means of an homothety starting from the center of the cell and of factor ρ, which has been set to $\frac{2}{3}$. Remind that it enables to prevent two markers to be selected if they are too close from each other.

3.2   Results

Results for boundary-recall, expressed as a function of contour density, were averaged over the subset "val" of the database and shown in Fig.3 for each extinction value (volume, surface and dynamic). The red curve corresponds to the watershed without repulsion constraint and the blue curve corresponds to waterpixels (k=0), for different values of N (number of markers, which is also the number of regions in the final partition). The ideal case being the highest boundary-recall for the lowest contour density, we can see that waterpixels out-perform the usual watershed. We have found that the same behavior can be observed when the grid used for waterpixels generation is composed of square cells instead of hexagonal ones (data not shown).

Qualitative results for surface extinction value can be seen in Fig.4. In the first row, we see that the spatial distribution of markers for WP is rather regular thanks to the choice of one marker per grid cell, contrary to the markers of WS. It is interesting to note that most of the markers for WS are trapped in the sea because it is a highly textured region. Thanks to the repulsion constraint between markers for WP, these ones are less numerous in the highly textured regions to the profit of other regions such as the sail: in the left example, we can see a contour which has been caught by WP but missed by WS. By imposing a repulsion between markers, we give a chance to weaker contours to be caught, which would not have been caught when we consider only the N best markers without any spatial constraint. The example on the right, on the other hand,

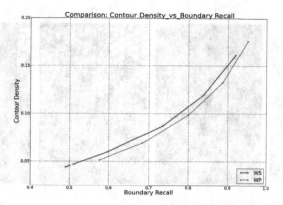

(a) Criterion of pertinence: volume extinction value

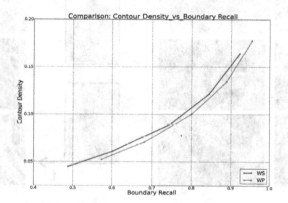

(b) Criterion of pertinence: surface extinction value

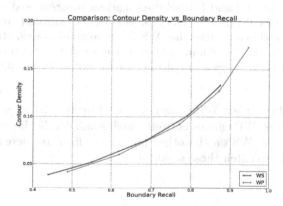

(c) Criterion of pertinence: dynamic extinction value

Fig. 3. Comparison between WS and WP for different extinction values

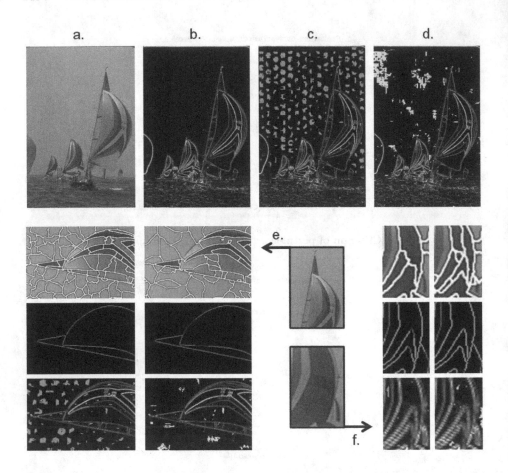

Fig. 4. Selected markers for WS and WP and their impact on the adherence to object boundaries: (a):Original image from the Berkeley segmentation database. (b) Lab-gradient of (a). (c) and (d): selected markers superimposed on the gradient respectively in green for WP and in yellow for WS. **Left example (e):** a region where WP (left column) catches a contour that WS (right column) misses. **Right example (f):** a region where WS (right column) catches a contour that WP (left column) misses. Note that missed contours appear in red whereas reached contours appear in green.

enables to understand when WP is outperformed by WS: by imposing a margin between two markers, WP cannot entirely catch small details, i.e. objects smaller than waterpixels size. WS on the other hand can afford to increase locally the number of markers to catch these small objects.

4   Discussion

With this study, we show that when we select markers among the minima with a criterion of pertinence such as the extinction values, using in addition spatial repulsion for marker selection improves the recovery of pertinent boundaries. As explained, we believe that it is mainly due to the fact that a high percentage of markers are trapped in textured regions when no constraint on their spacing is imposed, which causes misses of weaker but nonetheless meaningful contours.

It is interesting to note that, if we choose randomly the marker of each cell for waterpixels (instead of taking the best in terms of surface extinction value), we will still obtain results as good as the "optimal marker choice" according only to extinction values, i.e. WS.

5   Conclusion

As a conclusion, we have shown that that introducing some regularity into the spatial distribution of the markers used to compute the watershed leads to improved segmentation results when dealing with natural images.

This repulsion between markers can be introduced through waterpixels, but other methods, possibly more general, could be considered. One possible approach would be to find the N most distant markers from the set of all minima, where distance is defined in an augmented space (spatial dimensions, color, etc). In Kulesza [14], determinantal point process are used to model repulsion between pairs of elements : the more they are alike, the more their probability to co-occur decreases. This problem is also known in Operational Research as the p-dispersion problem (find p locations among m existing locations such that their minimum distance from one to another is maximized), as explained in Erkut [15].

It would be also interesting to apply such a repulsion factor to the stochastic watershed in order to improve the quality of the results.

Acknowledgments. Thomas Walter has received funding from the European Community's Seventh Framework Programme (FP7/2007-2013) under grant agreements number 258068 (Systems Microscopy).

References

1. Beucher, S., Lantujoul, C.: Use of watershed in contour detection. In: International Workshop on Image Processing: Real-Time Edge and Motion Estimation (1979)
2. Beucher, S., Meyer, F.: The morphological approach to segmentation: the watershed transformation. In: Dougherty, E. (ed.) Mathematical Morphology in Image Processing, pp. 433–481 (1993)
3. Meyer, F.: Un algorithme optimal pour la ligne de partage des eaux. 8ème congrès de reconnaissance des formes et intelligence artificielle, Lyon, France, vol. 2, pp. 847-857 (November 1991)
4. Vincent, L., Soille, P.: Watersheds in digital spaces: An efficient algorithm based on immersion simulations. IEEE Transactions on Pattern Analysis and Machine Intelligence 13(6), 583–598 (1991)
5. Andres, B., Köthe, U., Helmstaedter, M., Denk, W., Hamprecht, F.: Segmentation of SBFSEM Volume Data of Neural Tissue by Hierarchical Classification. In: Pattern Recognition, vol. D, pp. 142–152 (2008)

6. Stawiaski, J., Decencière, E.: Interactive liver tumor segmentation using watershed and graph cuts. In: Segmentation in the Clinic: A Grand Challenge II (MICCAI 2008 Workshop), New York, USA (2008)
7. Meyer, F., Beucher, S.: Morphological segmentation. JVCIR 1(1), 21–46 (1950)
8. Angulo, J., Jeulin, D.: Stochastic watershed segmentation. In: Banon, G., et al. (eds.) Proc. Int. Symp. Mathematical Morphology, ISSM 2007, pp. 265–276 (2007)
9. Vachier, C., Meyer, F.: Extinction Values: A New Measurement of Persistence. In: IEEE Workshop on Non Linear Signal/Image Processing, pp. 254-257 (1995)
10. Machairas, V., Faessel, M., Cárdenas-Peña, D., Chabardes, T., Walter, T., Decencière, E.: Waterpixels. In: Under revision for IEEE Transaction on Image Processing (2015)
11. Machairas, V., Walter, T., Decencière, E., Waterpixels: Superpixels based on the watershed transformation. In: International Conference on Image Processing (ICIP), pp. 4343–4347 (October 2014)
12. Ren, X., Malik, J.: Learning a classification model for segmentation. In: International Conference on Computer Vision, vol. 1, pp. 10–17 (2003)
13. Martin, D., Fowlkes, C., Tal, D., Malik, J.: A database of human segmented natural images and its application to evaluating segmentation algorithms and measuring ecological statistics. In: Int. Conf. on Computer Vision, vol. 2, pp. 416–423 (July 2001)
14. Kulesza, A., Taskar, B.: Determinantal point processes for machine learning. Foundations and Trends in Machine Learning 5(2-3), 123–286 (2013)
15. Erkut: The discrete p-dispersion problem. European Journal of Operational Research, 6, 48–60 (1990)

Hierarchies

New Characterizations of Minimum Spanning Trees and of Saliency Maps Based on Quasi-flat Zones

Jean Cousty[1](✉), Laurent Najman[1], Yukiko Kenmochi[1], and Silvio Guimarães[1,2]

[1] Université Paris-Est, LIGM, A3SI, ESIEE Paris
{j.cousty,l.najman,y.kenmochi,s.guimaraes}@esiee.fr
[2] PUC Minas - ICEI - DCC - VIPLAB

Abstract. We study three representations of hierarchies of partitions: dendrograms (direct representations), saliency maps, and minimum spanning trees. We provide a new bijection between saliency maps and hierarchies based on quasi-flat zones as used in image processing and characterize saliency maps and minimum spanning trees as solutions to constrained minimization problems where the constraint is quasi-flat zones preservation. In practice, these results form a toolkit for new hierarchical methods where one can choose the most convenient representation. They also invite us to process non-image data with morphological hierarchies.

Keywords: Hierarchy · Saliency map · Minimum spanning tree

1 Introduction

Many image segmentation methods look for a partition of the set of image pixels such that each region of the partition corresponds to an object of interest in the image. Hierarchical segmentation methods, instead of providing a unique partition, produce a sequence of nested partitions at different scales, enabling to describe an object of interest as a grouping of several objects of interest that appear at lower scales (see references in [14]). This article deals with a theory of hierarchical segmentation as used in image processing. More precisely, we investigate different representations of a hierarchy: by a dendrogram (direct set representation), by a saliency map (a characteristic function), and by a minimum spanning tree (a reduced domain of definition). Our contributions are threefold:

1. A new bijection theorem between hierarchies and saliency maps (Th. 1) that relies on the quasi-flat zones hierarchies and that is simpler and more general than previous bijection theorem for saliency maps; and
2. A new characterization of the saliency map of a given hierarchy as the minimum function for which the quasi-flat zones hierarchy is precisely the given hierarchy (Th. 2); and

This work received funding from ANR (contract ANR-2010-BLAN-0205-03), CAPES/PVE (grant 064965/2014-01), and CAPES/COFECUB (grant 592/08).

© Springer International Publishing Switzerland 2015
J.A. Benediktsson et al. (Eds.): ISMM 2015, LNCS 9082, pp. 205–216, 2015.
DOI: 10.1007/978-3-319-18720-4_18

3. A new characterization of the minimum spanning trees of a given edge-weighted graph as the minimum subgraphs (for inclusion) whose quasi-flat zones hierarchies are the same as the one of the given graph (Th. 3).

The links established in this article between the maps that weight the edges of a graph G, the hierarchies on the vertex set $V(G)$ of G, the saliency maps on the edge set $E(G)$ of G, and the minimum spanning trees for the maps that weight the edges of G are summarized in the diagram of Fig. 1.

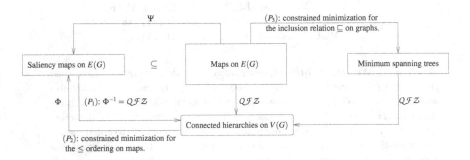

Fig. 1. A diagram that summarizes the results of this article. The solutions to problems (P_1), (P_2), and (P_3) are given by Ths. 1, 2, and 3, respectively. The constraint involved in (P_2) and (P_3) is to leave the induced quasi-flat zones hierarchy unchanged. In the diagram, $\mathcal{QFZ}$ stands for quasi-flat zones (Eq. 3), and the symbols Φ and Ψ stand for the saliency map of a hierarchy (Eq. 5) and of a map respectively (Section 5).

One possible application of these results is the design of new algorithms for computing hierarchies. Indeed, our results allow one to use indifferently any of the three hierarchical representations. This can be useful when a given operation is more efficiently performed with one representation than with the two others. Naturally, one could work directly on the hierarchy (or on its tree-based representation) and finally compute a saliency map for visualization purposes. For instance, in [8,17], the authors efficiently handle directly the tree-based representation of the hierarchy. Conversely, thanks to Th. 1, one can work on a saliency map or, thanks to Th. 3, on the weights of a minimum spanning tree and explicitly computes the hierarchy in the end. In [5,15], a resulting saliency map is computed before a possible extraction of the associated hierarchy of watersheds. In [9], a basic transformation that consists of modifying one weight on a minimum spanning tree according to some criterion is considered. The corresponding operation on the equivalent dendrogram is more difficult to design. When this basic operation is iterated on every edge of the minimum spanning tree, one transforms a given hierarchy into another one. An application of this technique to the observation scale of [7] has been developed in [9] (see Fig. 2).

Another interest of our work is to precise the link between hierarchical classification [16] and hierarchical image segmentation. In particular, it suggests that hierarchical image segmentation methods can be used for classification (the converse being carried out for a long time). Indeed, our work is deeply related to hierarchical classification, more precisely, to ultrametric distances, subdominant ultrametrics and single linkage clusterings. In classification, representation of hierarchies, on which no connectivity hypothesis is made, are studied since the 60's (see references in [16]). The framework presented in this article deals with connected hierarchies and a graph needs to be specified for defining the connectivity of the regions of the partitions in the hierarchies. The connectivity of regions is the main difference between what has been done in classification and in segmentation. Rather than restricting the work done for classification, the framework studied in this article generalizes it. Indeed the usual notions of classification are recovered from the definitions of this article when a complete graph (every two points are linked by an edge) is considered. For instance, when a complete graph is considered, a saliency map becomes an ultrametric distance, which is known to be equivalent to a hierarchy. However, Th. 1 shows that, when the graph is not complete, we do not need a value for each pair of elements in order to characterize a hierarchy (as done with an ultrametric distance) but one value for each edge of the graph is enough (with a saliency map). Furthermore, when a complete graph is considered, the hierarchy of quasi-flat zones becomes the one of single linkage clustering. Hence, Th. 3 allows to recover and to generalize a well-known relation between the minimum spanning trees of the complete graph and single linkage clusterings.

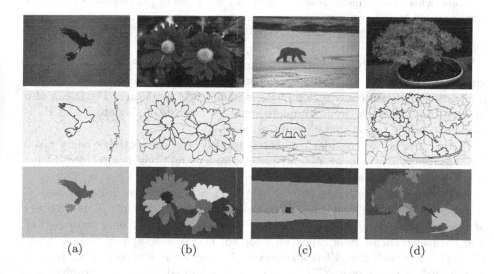

(a)   (b)   (c)   (d)

Fig. 2. Top row: some images from the Berkeley database [1]. Middle row: saliency maps according to [9] developed thanks to the framework of this article. Bottom row: segmentations extracted from the hierarchies with (a) 3, (b) 18, (c) 6 and (d) 16 regions.

2 Connected Hierarchies of Partitions

A *partition* of a finite set V is a set $\mathbf{P}$ of nonempty disjoint subsets of V whose union is V (*i.e.*, $\forall X, Y \in \mathbf{P}$, $X \cap Y = \emptyset$ if $X \neq Y$ and $\cup\{X \in \mathbf{P}\} = V$). Any element of a partition $\mathbf{P}$ of V is called a *region of* $\mathbf{P}$. If x is an element of V, there is a unique region of $\mathbf{P}$ that contains x; this unique region is denoted by $[\mathbf{P}]_x$. Given two partitions $\mathbf{P}$ and $\mathbf{P}'$ of a set V, we say that $\mathbf{P}'$ is a *refinement* of $\mathbf{P}$ if any region of $\mathbf{P}'$ is included in a region of $\mathbf{P}$. A *hierarchy (on V)* is a sequence $\mathcal{H} = (\mathbf{P}_0, \dots, \mathbf{P}_\ell)$ of indexed partitions of V such that $\mathbf{P}_{i-1}$ is a refinement of $\mathbf{P}_i$, for any $i \in \{1, \dots, \ell\}$. If $\mathcal{H} = (\mathbf{P}_0, \dots, \mathbf{P}_\ell)$ is a hierarchy, the integer ℓ is called the *depth of* $\mathcal{H}$. A hierarchy $\mathcal{H} = (\mathbf{P}_0, \dots, \mathbf{P}_\ell)$ is called complete if $\mathbf{P}_\ell = \{V\}$ and if $\mathbf{P}_0$ contains every singleton of V (*i.e.*, $\mathbf{P}_0 = \{\{x\} \mid x \in V\}$). The hierarchies considered in this article are complete.

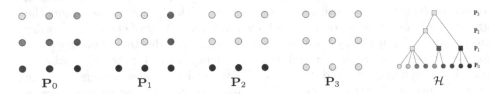

Fig. 3. Illustration of a hierarchy $\mathcal{H} = (\mathbf{P}_0, \mathbf{P}_1, \mathbf{P}_2, \mathbf{P}_3)$. For every partition, each region is represented by a gray level: two dots with the same gray level belong to the same region. The last subfigure represents the hierarchy $\mathcal{H}$ as a tree, often called a dendrogram, where the inclusion relation between the regions of the successive partitions is represented by line segments.

Figure 3 graphically represents a hierarchy $\mathcal{H} = (\mathbf{P}_0, \mathbf{P}_1, \mathbf{P}_2, \mathbf{P}_3)$ on a rectangular subset V of $\mathbb{Z}^2$ made of 9 dots. For instance, it can be seen that $\mathbf{P}_1$ is a refinement of $\mathbf{P}_2$ since any region of $\mathbf{P}_1$ is included in a region of $\mathbf{P}_2$. It can also be seen that the hierarchy is complete since $\mathbf{P}_0$ is made of singletons and $\mathbf{P}_3$ is made of a single region that contains all elements.

In this article, we consider connected regions, the connectivity being given by a graph. Therefore, we remind basic graph definitions before introducing connected partitions and hierarchies.

A *(undirected) graph* is a pair $G = (V, E)$, where V is a finite set and E is composed of unordered pairs of distinct elements in V, *i.e.*, E is a subset of $\{\{x, y\} \subseteq V \mid x \neq y\}$. Each element of V is called a *vertex or a point (of G)*, and each element of E is called an *edge* (of G). A *subgraph of G* is a graph $G' = (V', E')$ such that V' is a subset of V, and E' is a subset of E. If G' is a subgraph of G, we write $G' \sqsubseteq G$. The vertex and edge sets of a graph X are denoted by $V(X)$ and $E(X)$ respectively.

Let G be a graph and let $(x_0, \dots, x_k)$ be a sequence of vertices of G. The sequence $(x_0, \dots, x_k)$ is a *path (in G) from x_0 to x_k* if, for any i in $\{1, \dots, k\}$, $\{x_{i-1}, x_i\}$ is an edge of G. The graph G *is connected* if, for any two vertices x and y of G, there exists a path from x to y. Let X be a subset of $V(G)$. The *graph induced by X (in G)* is the graph whose vertex set is X and whose edge set contains any edge of G which is made of two elements in X. If the graph induced by X is connected, we also say, for simplicity, that X *is connected (for G)* . The subset X of $V(G)$ is a *connected component of G* if it is connected for G and maximal for this property, *i.e.*, for any subset Y of $V(G)$, if Y is a connected superset of X, then we have $Y = X$. In the following, we denote by $\mathbf{C}(G)$ the set of all connected components of G. It is well-known that this set $\mathbf{C}(G)$ of all connected components of G is a partition of $V(G)$. This partition is called the *(connected components) partition induced by G*. Thus, the set $[\mathbf{C}(G)]_x$ is the unique connected component of G that contains x.

Given a graph $G = (V, E)$, a *partition of V is connected (for G)* if any of its regions is connected and a *hierarchy on V is connected (for G)* if any of its partitions is connected.

For instance, the partitions presented in Fig. 3 are connected for the graph given in Fig. 4(a). Therefore, the hierarchy $\mathcal{H}$ made of these partitions, which is depicted as a dendrogram in Fig. 3 (rightmost subfigure), is also connected for the graph of Fig. 4(a).

For image analysis applications, the graph G can be obtained as a pixel or a region adjacency graph: the vertex set of G is either the domain of the image to be processed or the set of regions of an initial partition of the image domain. In the latter case, the regions can in particular be "superpixels". In both cases, two typical settings for the edge set of G can be considered: (1) the edges of G are obtained from an adjacency relation between the image pixels, such as the well known 4- or 8-adjacency relations; and (2) the edges of G are obtained by considering, for each vertex x of G, the nearest neighbors of x for a distance in a (continuous) features space onto which the vertices of G are mapped.

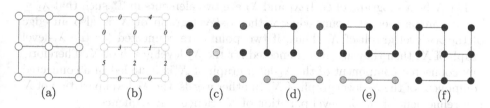

(a)   (b)   (c)   (d)   (e)   (f)

Fig. 4. Illustration of quasi-flat zones hierarchy. (a) A graph G; (b) a map w (numbers in black) that weights the edges of G (in gray); (c, d, e, f) the λ-subgraphs of G, with $\lambda = 0, 1, 2, 3$. The associated connected component partitions that form the hierarchy of quasi-flat zones of G for w is depicted in Fig. 3.

3   Hierarchies of Quasi-flat Zones

As established in the next sections, a connected hierarchy can be equivalently treated by means of an edge-weighted graph. We first recall in this section that the level-sets of any edge-weighted graph induce a hierarchy of quasi-flat zones. This hierarchy is widely used in image processing [12].

Let G be a graph, if w is a map from the edge set of G to the set $\mathbb{R}^+$ of positive real numbers, then the pair (G, w) is called an *(edge-)weighted graph*. If (G, w) is an edge-weighted graph, for any edge u of G, the value $w(u)$ is called the *weight of u (for w)*.

Important Notation. In the sequel of this paper, we consider a weighted graph (G, w). To shorten the notations, the vertex and edge sets of G are denoted by V and E respectively instead of $V(G)$ and $E(G)$. Furthermore, we assume that the vertex set of G is connected. Without loss of generality, we also assume that the range of w is the set $\mathbb{E}$ of all integers from 0 to $|E| - 1$ (otherwise, one could always consider an increasing one-to-one mapping from the set $\{w(u) \mid u \in E\}$ into $\mathbb{E}$). We also denote by $\mathbb{E}^\bullet$ the set $\mathbb{E} \cup \{|E|\}$.

Let X be a subgraph of G and let λ be an integer in $\mathbb{E}^\bullet$. The *λ-level set of X (for w)* is the set $w_\lambda(X)$ of all edges of X whose weight is less than λ:

$$w_\lambda(X) = \{u \in E(X) \mid w(u) < \lambda\}. \tag{1}$$

The *λ-level graph of X* is the subgraph $w_\lambda^V(X)$ of X whose edge set is the λ-level set of X and whose vertex set is the one of X:

$$w_\lambda^V(X) = (V(X), w_\lambda(X)). \tag{2}$$

The connected component partition $\mathbf{C}(w_\lambda^V(X))$ induced by the λ-level graph of X is called the *λ-level partition of X (for w)*.

For instance, let us consider the graph G depicted in Fig. 4(a) and the map w shown in Fig. 4(b). The 0-, 1-, 2- and 3-level sets of G contain the edges depicted in Figs. 4(c), (d), (e), and (f), respectively. The graphs depicted in these figures are the associated 0-, 1- 2- and 3-level graphs of G and the associated 0-, 1-, 2- and 3-level partitions are shown in Fig. 3.

Let X be a subgraph of G. If λ_1 and λ_2 are two elements in $\mathbb{E}^\bullet$ such that $\lambda_1 \leq \lambda_2$, it can be seen that any edge of the λ_1-level graph of X is also an edge of the λ_2-level graph of X. Thus, if two points are connected for the λ_1-level graph of X, then they are also connected for the λ_2-level graph of X. Therefore, any connected component of the λ_1-level graph of X is included in a connected component of the λ_2-level graph of X. In other words, the λ_1-level partition of X is a refinement of the λ_2-level partition of X. Hence, the sequence

$$\mathcal{QFZ}(X, w) = (\mathbf{C}(w_\lambda^V(X)) \mid \lambda \in \mathbb{E}^\bullet) \tag{3}$$

of all λ-level partitions of X is a hierarchy. This hierarchy $\mathcal{QFZ}(X, w)$ is called the *quasi-flat zones hierarchy of X (for w)*. It can be seen that this hierarchy is complete whenever X is connected.

For instance, the quasi-flat zones hierarchy of the graph G (Fig. 4(a)) for the map w (Fig. 4(b)) is the hierarchy of Fig. 3.

For image analysis applications, we often consider that the weight of an edge $u = \{x, y\}$ represents the dissimilarity of x and y. For instance, in the case where the vertices of G are the pixels of a grayscale image, the weight $w(u)$ can be the absolute difference of intensity between x and y. The setting of the graph (G, w) depends on the application context.

4 Correspondence Between Hierarchies and Saliency Maps

In the previous section, we have seen that any edge-weighted graph induces a connected hierarchy of partitions (called the quasi-flat zones hierarchy). In this section, we tackle the inverse problem:

(P_1) given a connected hierarchy $\mathcal{H}$, find a map w from E to $\mathbb{E}$ such that the quasi-flat zones hierarchy for w is precisely $\mathcal{H}$.

We start this section by defining the saliency map of $\mathcal{H}$. Then, we provide a one-to-one correspondence (also known as a bijection) between saliency maps and hierarchies. This correspondence is given by the hierarchy of quasi flat-zones. Finally, we deduce that the saliency map of $\mathcal{H}$ is a solution to problem (P_1).

Until now, we handled the regions of a partition. Let us now study their "dual" that represents "borders" between regions and that are called graph-cuts or simply cuts. The notion of a cut will then be used to define the saliency maps.

Let $\mathbf{P}$ be a partition of V, the *cut of* $\mathbf{P}$ *(for G)*, denoted by $\phi(\mathbf{P})$, is the set of edges of G whose two vertices belong to different regions of $\mathbf{P}$:

$$\phi(\mathbf{P}) = \{\{x,y\} \in E \mid [\mathbf{P}]_x \neq [\mathbf{P}]_y\}. \tag{4}$$

Let $\mathcal{H} = (\mathbf{P}_0, \ldots, \mathbf{P}_\ell)$ be a hierarchy on V. The *saliency map of* $\mathcal{H}$ is the map $\Phi(\mathcal{H})$ from E to $\{0, \ldots, \ell\}$ such that the weight of u for $\Phi(\mathcal{H})$ is the maximum value λ such that u belongs to the cut of $\mathbf{P}_\lambda$:

$$\Phi(\mathcal{H})(u) = \max \{\lambda \in \{0, \ldots, \ell\} \mid u \in \phi(\mathbf{P}_\lambda)\}. \tag{5}$$

In fact, the weight of the edge $u = \{x,y\}$ for $\Phi(\mathcal{H})$ is directly related to the lowest index of a partition in the hierarchy $\mathcal{H}$ for which x and y belong to the same region:

$$\Phi(\mathcal{H})(u) = \min\left\{\lambda \in \{1, \ldots, \ell\} \mid [\mathbf{P}_\lambda]_x = [\mathbf{P}_\lambda]_y\right\} - 1. \tag{6}$$

For instance, if we consider the graph G represented by the gray dots and line segments in Fig. 5(a), the saliency map of the hierarchy $\mathcal{H}$ shown in Fig. 3 is the map shown with black numbers in Fig. 5(a). When the 4-adjacency relation is used, a saliency map can be displayed as an image (Figs. 5(e, f) and Fig. 2) which is useful for visualizing the associated hierarchy at a glance.

We say that a map w from E to $\mathbb{E}$ is a *saliency map* if there exists a hierarchy $\mathcal{H}$ such that w is the saliency map of $\mathcal{H}$ (*i.e.*, $w = \Phi(\mathcal{H})$).

If φ is a map from a set S_1 to a set S_2 and if φ^{-1} is a map from S_2 to S_1 such that the composition of φ^{-1} with φ is the identity, then we say that φ^{-1} is the inverse of φ.

The next theorem identifies the inverse of the map Φ and asserts that there is a bijection between the saliency maps and the connected hierarchies on V.

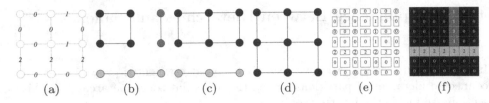

Fig. 5. Illustration of a saliency map. The map (depicted by black numbers) is the saliency map $s = \Phi(\mathcal{H})$ of the hierarchy $\mathcal{H}$ shown in Fig. 3 when we consider the graph G depicted in gray. (b, c, d) the 1-, 2-, and 3-level graphs of G for s. The vertices are colored according to the associated 1-, 2-, and 3-level partitions of G: in each subfigure, two vertices belonging to a same connected components have the same grey level. Subfigures (e) and (f) show possible image representations of a saliency map when one considers the 4-adjacency graph.

Theorem 1. *The map Φ is a one-to-one correspondence between the connected hierarchies on V of depth $|E|$ and the saliency maps (of range $\mathbb{E}$). The inverse Φ^{-1} of Φ associates to any saliency map w its quasi-flat zones hierarchy:* $\Phi^{-1}(w) = \mathcal{QFZ}(G, w).$

Hence, as a consequence of this theorem, we have $\mathcal{QFZ}(G, \Phi(\mathcal{H})) = \mathcal{H}$, which means that $\mathcal{H}$ is precisely the hierarchy of quasi-flat zones of G for its saliency map $\Phi(\mathcal{H})$. In other words, the saliency map of $\mathcal{H}$ is a solution to problem (P_1). For instance, if we consider the hierarchy $\mathcal{H}$ shown in Fig. 3, it can be observed that the quasi-flat zones hierarchy for $\Phi(\mathcal{H})$ (see Fig. 5) is indeed $\mathcal{H}$. We also deduce that, for any saliency map w, the relation $\Phi(\mathcal{QFZ}(G, w)) = w$ holds true. In other words, a given saliency map w is precisely the saliency map of its quasi-flat zones hierarchy.

From this last relation, we can deduce that there are some maps that weight the edges of G and that are not saliency maps. Indeed, in general, a map is not equal to the saliency map of its quasi-flat zones hierarchy. For instance, the map w in Fig. 4 is not equal to the saliency map of its quasi-flat zones hierarchy which is depicted in Fig. 5. Thus, the map w is not a saliency map. The next section studies a characterization of saliency maps.

5   Characterization of Saliency Maps

Following the conclusion of the previous section, we now consider the problem:

(P_2) given a hierarchy $\mathcal{H}$, find the minimal map w such that the quasi-flat zones hierarchy for w is precisely $\mathcal{H}$.

The next theorem establishes that the saliency map of $\mathcal{H}$ is the unique solution to problem (P_2).

Before stating Th. 2, let us recall that, given two maps w and w' from E to $\mathbb{E}$, the map w' is less than w if we have $w'(u) \leq w(u)$ for any $u \in E$.

Theorem 2. *Let $\mathcal{H}$ be a hierarchy and let w be a map from E to $\mathbb{E}$. The map w is the saliency map of $\mathcal{H}$ if and only if the two following statements hold true:*

1. *the quasi-flat zones hierarchies for w is $\mathcal{H}$; and*
2. *the map w is minimal for statement 1, i.e., for any map w' such that $w' \leq w$, if the quasi-flat zones hierarchy for w' is $\mathcal{H}$, then we have $w = w'$.*

Given a weighted graph (G, w), it is sometimes interesting to consider the saliency map of its quasi-flat zones hierarchy. This saliency map is simply called the *saliency map of w* and is denoted by $\Psi(w)$. From Th. 2, the operator Ψ which associates to any map w the saliency map $\Phi(\mathcal{QFZ}(G, w))$ of its quasi-flat zones hierarchy, is idempotent (*i.e.* $\Psi(\Psi(w)) = \Psi(w)$). Furthermore, it is easy to see that Ψ is also anti-extensive (we have $\Psi(w) \leq w$) and increasing (for any two maps w and w', if $w \geq w'$, then we have $\Psi(w) \geq \Psi(w')$). Thus, Ψ is a morphological opening. This operator is studied, in different frameworks, under several names (see [11,13,10,18,16]). When the considered graph G is complete, it is known in classification (see, e.g., [16]) that this operator is linked to the minimum spanning tree of (G, w). The next section proposes a generalization of this link.

6 Minimum Spanning Trees

Two distinct maps that weight the edges of the same graph (see, e.g., the maps of Figs. 4(b) and 5(a)) can induce the same hierarchy of quasi-flat zones. Therefore, in this case, one can guess that some of the edge weights do not convey any useful information with respect to the associated quasi-flat zones hierarchy. More generally, in order to represent a hierarchy by a simple (*i.e.*, easy to handle) edge-weighted graph with a low level of redundancy, it is interesting to consider the following problem:

(P_3) given an edge-weighted graph (G, w), find a minimal subgraph $X \sqsubseteq G$ such that the quasi-flat zones hierarchies of G and of X are the same.

The main result of this section, namely Th. 3, provides the set of all solutions to problem (P_3): the minimum spanning trees of (G, w). The minimum spanning tree problem is one of the most typical and well-known problems of combinatorial optimization (see [3]) and Th. 3 provides, as far as we know, a new characterization of minimum spanning trees based on the quasi-flat zones hierarchies as used in image processing.

Let X be a subgraph of G. The weight of X with respect to w, denoted by $w(X)$, is the sum of the weights of all the edges in $E(X)$: $w(X) = \sum_{u \in E(X)} w(u)$. The subgraph X is a *minimum spanning tree (MST)* of (G, w) if:

1. X is connected; and
2. $V(X) = V$; and
3. the weight of X is less than or equal to the weight of any graph Y satisfying (1) and (2) (*i.e.*, Y is a connected subgraph of G whose vertex set is V).

For instance, a MST of the graph shown in Fig. 4(b) is presented in Fig. 6(a).

Theorem 3. *A subgraph X of G is a MST of (G, w) if and only if the two following statements hold true:*

1. *the quasi-flat zones hierarchies of X and of G are the same; and*
2. *the graph X is minimal for statement 1, i.e., for any subgraph Y of X, if the quasi-flat zones hierarchy of Y for w is the one of G for w, then we have $Y = X$.*

Theorem 3 (statement 1) indicates that the quasi-flat zones hierarchy of a graph and of its MSTs are identical. Note that statement 1 appeared in [6] but Th. 3 completes the result of [6]. Indeed, Th. 3 indicates that there is no proper subgraph of a MST that induces the same quasi-flat zones hierarchy as the initial weighted graph. Thus, a MST of the initial graph is a solution to problem (P_3), providing a minimal graph representation of the quasi-flat zones hierarchy of (G, w), or more generally by Th. 1 of any connected hierarchy. More remarkably, the converse is also true: a minimal representation of a hierarchy in the sense of (P_3) is necessarily a MST of the original graph. To the best of our knowledge, this result has not been stated before.

For instance, the level sets, level graphs and level partitions of the MST X (Fig. 6(a)) of the weighted graph (G, w) (Fig. 4) are depicted in Figs. 6(b), (c), (d). It can be observed that the level partitions of X are indeed the same as those of G. Thus the quasi-flat zones hierarchies of X and G are the same.

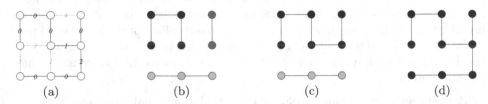

(a)   (b)   (c)   (d)

Fig. 6. Illustration of a minimum spanning tree and of its quasi-flat zones hierarchy. (a) A minimum spanning tree X (black edges and black circled vertices) of the weighted graph of Fig. 4(b); (b, c, d) the 1-, 2-, and 3-level graphs of X. The vertices are colored according to the associated 1-, 2-, and 3-level partitions of X: in each subfigure, two vertices belonging to the same connected components have the same color.

7   Saliency Map Algorithms

When a hierarchy $\mathcal{H}$ is stored as a tree data structure (such as *e.g.*, the dendrogram of Fig. 3), the weight of any edge for the saliency map of this hierarchy can be computed in constant time, provided a linear time preprocessing. Indeed, the weight of an edge linking x and y is associated (see Eq. 6) to the lowest index of a partition for which x and y belongs to the same region. This index can be obtained by finding the index of the least common ancestor of $\{x\}$ and $\{y\}$ in the tree. The algorithm proposed in [2] performs this task in constant time, provided a linear time preprocessing of the tree. Therefore, computing the saliency map of $\mathcal{H}$ can be done in linear $O(|E| + |V|)$ time complexity.

Thus, the complete process that computes the saliency map $\Psi(w)$ of a given map w proceeds in two steps:

i) build the quasi-flat zones hierarchy $\mathcal{H} = \mathcal{QFZ}(G, w)$ of G for w; and
ii) compute the saliency map $\Psi(w) = \Phi(\mathcal{H})$.

On the basis of [6], step i) can be performed with the quasi-linear time algorithm shown in [15] and step ii) can be performed in linear-time as proposed in the previous paragraph. Thus, the overall time complexity of this algorithm is quasi-linear with respect to the size $|E| + |V|$ of the graph G.

The algorithm sketched in [13], based on [4], for computing the saliency map of a given map w has the same complexity as the algorithm proposed above. However, the algorithm of [13] is more complicated since it requires to compute the topological watershed of the map. This involves a component tree (a data structure which is more complicated than the quasi-flat zones hierarchy in the sense of [6]), a structure for computing least common ancestors, and a hierarchical queue [4], which is not needed by the above algorithm. Hence, as far as we know, the algorithm presented in this section is the simplest algorithm for computing a saliency map. It is also the most efficient both from memory and execution-time points of view. An implementation in C of this algorithm is available at www.esiee.fr/~info/sm.

8   Conclusions

In this article we study three representations for a hierarchy of partitions. We show a new bijection between hierarchies and saliency maps and we characterize the saliency map of a hierarchy and the minimum spanning trees of a graph as minimal elements preserving quasi-flat zones. In practice, these results allow us to indifferently handle a hierarchy by a dendrogram (the direct tree structure given by the hierarchy), by a saliency map, or by an edge-weighted tree. These representations form a toolkit for the design of hierarchical (segmentation) methods where one can choose the most convenient representation or the one that leads to the most efficient implementation for a given particular operation. The results of this paper were used in [9] to provide a framework for hierarchicalizing a certain class of non-hierarchical methods. We study in particular a hierarchicalization of [7]. The first results are encouraging and a short term perspective is the precise practical evaluation of the gain of the hierarchical method with respect to its non-hierarchical counterpart.

Another important aspect of the present work is to underline and to precise the close link that exists between classification and hierarchical image segmentation. Whereas classification methods were used as image segmentation tools for a long time, our results incite us to look if some hierarchies initially designed for image segmentation can improve the processing of non-image data such as data coming from geography, social network, etc.. This topic will be a subject of future research.

References

1. Arbelaez, P., Maire, M., Fowlkes, C., Malik, J.: Contour detection and hierarchical image segmentation. PAMI 33(5), 898–916 (2011)
2. Bender, M., Farach-Colton, M.: The LCA problem revisited. In: Latin American Theoretical INformatics, pp. 88–94 (2000)
3. Cormen, T.H., Leiserson, C.E., Rivest, R.L., Stein, C., et al.: Introduction to algorithms, vol. 2. MIT Press, Cambridge (2001)
4. Couprie, M., Najman, L., Bertrand, G.: Quasi-linear algorithms for the topological watershed. JMIV 22(2-3), 231–249 (2005)
5. Cousty, J., Najman, L.: Incremental algorithm for hierarchical minimum spanning forests and saliency of watershed cuts. In: Soille, P., Pesaresi, M., Ouzounis, G.K. (eds.) ISMM 2011. LNCS, vol. 6671, pp. 272–283. Springer, Heidelberg (2011)
6. Cousty, J., Najman, L., Perret, B.: Constructive links between some morphological hierarchies on edge-weighted graphs. In: Hendriks, C.L.L., Borgefors, G., Strand, R. (eds.) ISMM 2013. LNCS, vol. 7883, pp. 86–97. Springer, Heidelberg (2013)
7. Felzenszwalb, P., Huttenlocher, D.: Efficient graph-based image segmentation. IJCV 59, 167–181 (2004)
8. Guigues, L., Cocquerez, J.P., Men, H.L.: Scale-sets image analysis. IJCV 68(3), 289–317 (2006)
9. Guimarães, S.J.F., Cousty, J., Kenmochi, Y., Najman, L.: A hierarchical image segmentation algorithm based on an observation scale. In: Gimel'farb, G., et al. (eds.) SSPR & SPR 2012. LNCS, vol. 7626, pp. 116–125. Springer, Heidelberg (2012)
10. Kiran, B.R., Serra, J.: Scale space operators on hierarchies of segmentations. In: Kuijper, A., Bredies, K., Pock, T., Bischof, H. (eds.) SSVM 2013. LNCS, vol. 7893, pp. 331–342. Springer, Heidelberg (2013)
11. Leclerc, B.: Description combinatoire des ultramétriques. Mathématiques et Sciences Humaines 73, 5–37 (1981)
12. Meyer, F., Maragos, P.: Morphological scale-space representation with levelings. In: Nielsen, M., Johansen, P., Fogh Olsen, O., Weickert, J. (eds.) Scale-Space 1999. LNCS, vol. 1682, pp. 187–198. Springer, Heidelberg (1999)
13. Najman, L.: On the equivalence between hierarchical segmentations and ultrametric watersheds. JMIV 40(3), 231–247 (2011)
14. Najman, L., Cousty, J.: A graph-based mathematical morphology reader. PRL 47(1), 3–17 (2014)
15. Najman, L., Cousty, J., Perret, B.: Playing with kruskal: Algorithms for morphological trees in edge-weighted graphs. In: Hendriks, C.L.L., Borgefors, G., Strand, R. (eds.) ISMM 2013. LNCS, vol. 7883, pp. 135–146. Springer, Heidelberg (2013)
16. Nakache, J.P., Confais, J.: Approche pragmatique de la classification: arbres hiérarchiques, partitionnements. Editions Technip (2004)
17. Kiran, B.R., Serra, J.: Global–local optimizations by hierarchical cuts and climbing energies. PR 47(1), 12–24 (2014)
18. Ronse, C.: Ordering partial partitions for image segmentation and filtering: Merging, creating and inflating blocks. JMIV 49(1), 202–233 (2014)

Braids of Partitions

Bangalore Ravi Kiran[1(✉)] and Jean Serra[2]

[1] Centre de robotique, MINES ParisTech, PSL-Research University, Paris, France
ravi.kiran@mines-paritech.fr
[2] Université Paris-Est, A3SI-ESIEE LIGM, Paris, France
jean.serra@esiee.fr

Abstract. In obtaining a tractable solution to the problem of extracting a minimal partition from hierarchy or tree by dynamic programming, we introduce the braids of partition and h-increasing energies, the former extending the solution space from a hierarchy to a larger set, the latter describing the family of energies, for which one can obtain the solution by a dynamic programming. We also provide the singularity condition for the existence of unique solution, leading to the definition of the energetic lattice. The paper also identifies various possible braids in literature and how this structure relaxes the segmentation problem.

Keywords: Hierarchies · Dynamic programming · Optimization · Lattice

1 Introduction

Hierarchical segmentation methods have been an important tool in providing a simplification of the image domain, following which various operations of filtering, segmentation and labeling become simpler structured problems on hierarchies of partitions (HOP), in comparison to the whole space. These problems are often formulated as optimization problems, where the space of solutions are partitions from a hierarchy. Breiman et al. [4] performed decision tree pruning to obtain a tree-classifier with least complexity to avoid overfitting, which corresponds to a pruning with minimal energy from a tree. This was first extended for the image segmentation problem, by Salembier-Garrido [13], where they calculated an optimal pruning of a binary partition tree by performing a gradient search over the Lagrange multiplier. Further on Guigues [8] introduced the scale-set descriptor, which operates on an input hierarchy of segmentations on an input image, and a parametrized energy. The scale sets are a hierarchy of minimal cuts corresponding to a given Lagrange multiplier. These methods use the Breiman Dynamic programming approach to perform pruning or extract the optimal cut. Further on Guigues provides conditions of sub-additivity of constraint function and super-additivity of objective function as conditions for finding a globally unique optimal cut, which was generalized in [9] to h-increasing energies. The λ-cut or scale-set [8] produces a descriptor based on any input hierarchy and a parametrized energy like Mumford-Shah. The attribute watersheds [6] work on the attributes of volume, area, dynamic of the component-tree hierarchy.[1]

---

[1] This work was partly funded by ANR-2010-BLAN-0205-03 program KIDICO.

© Springer International Publishing Switzerland 2015
J.A. Benediktsson et al. (Eds.): ISMM 2015, LNCS 9082, pp. 217–228, 2015.
DOI: 10.1007/978-3-319-18720-4_19

In this paper firstly introduce a new family of partitions larger than the hierarchy over which the dynamic program is still valid, namely the braids of partitions (BOP). Further on we extend the property of h-increasingness to the braids, and as well prove the energetic lattice and ordering relation over braids.

2   Braids of Partitions (BOP)

We now consider the problem of construction of other families, which no longer form chains, while they share hierarchical properties, and expand the search space for the optimization problem. We propose the braid, which on one hand provides a richer hierarchical model enabling multiple segmentations of a given region of the image domain, while on the other remains in conformance with the dynamic program substructure.

2.1   Definitions

A partition π of space E is a set of subsets of E that are pairwise disjoint and whose union reconstitutes E. A partial partition [12] of support S denoted as $\pi(S)$, is a partition of the subset $S \subset E$. The family of all partitions are denoted by $\Pi(E)$ and that of partial partitions as $\mathcal{D}(E)$. A hierarchy of partitions (HOP) is a chain of partitions $H = \{\pi_i, i \in [0, n]\}$, where $\pi_i \leq \pi_j, i < j$, where $\leq$ denotes refinement ordering. The minimal element π_0 of H is the called leaves partition which contains a finite number of elements. A cut is a partition composed of classes from a hierarchy (or more generally any family of partitions). The cuts of H are denoted by $\Pi(E, H)$.

An energy is a non-negative function on the family of partial partitions, $\omega : \mathcal{D} \rightarrow \mathbb{R}^+$. The energy of a partition or partial partition is usually obtained by the composition product, $\text{comp}(\cdot)$ of energies, by addition, supremum or other laws, over its constituent classes, e.g. $\omega(\pi(S)) = \sum_{a \sqsubset \pi(S)} \omega(a)$. Now the optimal cut in [4], [13], [8], is calculated by aggregating local optima. The local optimum at class S either choses the parent $\{S\}$, or the disjoint union of the optimums over the its children as shown in equation 1.

$$\pi^*(S) = \begin{cases} \{S\}, & \text{if } \omega(S) \leq \text{comp}(\omega(\pi^*(a))), a \in \pi(S) \\ \bigsqcup_{a \in \pi(S)} \pi^*(a), & \text{otherwise} \end{cases} \quad (1)$$

2.2   Braids

A braid is a family of partitions B, where the pairwise refinement supremum of any two elements is a cut of in some hierarchy $\Pi(E, H)$. This leads to the more formal definition:

Definition 1. *Let $\Pi(E)$ be the complete lattice of all partitions of set E; let H be a hierarchy in $\Pi(E)$. A braid B of monitor H is a family in $\Pi(E)$ where the refinement supremum of any pair of distinct partitions $\pi_1, \pi_2 \in B$ is a cut of H, other than $\{E\}$, that is in, $\Pi(E, H) \setminus \{E\}$:*

$$\forall \pi_1, \pi_2 \in B \quad \Rightarrow \pi_1 \vee \pi_2 \in \Pi(E, H) \setminus \{E\} \tag{2}$$

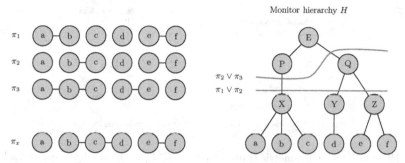

Fig. 1. Space E is partitioned into leaves $\{a, b, c, d, e, f\}$. The family $B_1 = \{\pi_1, \pi_2, \pi_3\}$ forms a braid, whose pairwise supremum is indicated on the dendrogram. Note that $\pi_1(X), \pi_2(X)$ have a common parent X, but $\pi_2(Q), \pi_3(Q)$ a common grand parent Q. However the family $\pi_x \cup B_1$ is not a braid since $\pi_3 \vee \pi_x$ gives the whole space E.

Given three partitions π_1, π_2, π_3 then the classes of suprema partitions $\pi_1 \vee \pi_2, \pi_1 \vee \pi_3$ are nested or disjoint. A braid can posses multiple monitoring hierarchies. One thus still has a scale selection to perform in the context of choosing a monitor hierarchy for a given application.

In Figure 1 we demonstrate a simple example of a braid family with the dendrogram corresponding to its monitor hierarchy. As we can see the classes of partitions π_1, π_2 are neither nested nor disjoint, and basically correspond to different segmentation hypotheses that exist in the stack of segmentations. The set of all cuts of a braid B is denoted by $\Pi(E, B)$. A braid may also contain its monitor H, though this is not necessary. On the other hand, any hierarchy is a braid with itself as monitor. A braid cannot be represented by a single saliency function, except when it reduces to a hierarchy whose classes are connected sets.

The partition with one class $\{E\}$ is not considered in Definition 2, since this would imply that any family of arbitrary partitions would form a braid, with $\{E\}$ as supremum, thus losing any useful structure. In case of a hierarchy the cone or family of classes containing a point $x \in E$ can only be nested or disjoint. While the cone of classes in the BOP, that contain a single point, are not necessarily nested, though their suprema are. This provides the local-global substructure for the dynamic program.

2.3   Underlying Questions

In the process of trying to create a structure where the dynamic program substructure holds, we are in fact posing the following sequence of questions. Given a general

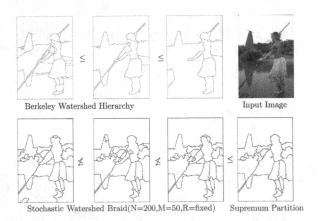

Berkeley Watershed Hierarchy   Input Image

Stochastic Watershed Braid(N=200,M=50,R=fixed)   Supremum Partition

Fig. 2. HOP vs BOP: Ultrametric contour map (UCM [3]), hierarchy (top) and a braid of partitions (bottom). Braids of partitions were produced from multiple instances of random marker based stochastic watershed, with same number of regions. The supremum or monitoring partition, corresponding to these unordered family of partitions is shown. Braids help reorganize partial refinement between partitions.

set of partitions $B = \{\pi_i\}, i \in \{1, 2, 3, ...n\}$: Firstly, how the partial optimum between any two partitions with a non-trivial supremum is calculated ? Over what support are partial partitions compared ? Secondly and more profoundly, given that there are cuts extractable other than these n-partitions, how does one index these different cuts. What are the types of ordering relations observable between any two partitions with a non-trivial supremum ? Furthermore it would also be useful understand the combinatorial nature by calculating the number of optimal cuts can one extract.

When B is a hierarchy, any two partitions are ordered by refinement, i.e. $\pi_i \leq \pi_j$ or $\pi_j \leq \pi_i, \forall i, j, \in \{1, 2, 3, ...n\}$. We now observe the possible ordering relations possible between pairs of partial partitions over a supports from their supremum $S \in \pi_1 \vee \pi_2$, we have: either parent-child/child parent $\pi_i \sqcap S \leq \pi_j \sqcap S$ or a braid structure $\pi_i \sqcap S \neq \pi_j \sqcap S$, though here one must maintain a nested or disjoint supremum S to ensure a local ordering to follow. Given two partitions, we can observe various local ordering between classes. This is discussed and demonstrated in an illustrative example in Figure 3.

There are two problems that are related but that very different in algorithmic complexity when dealing with braids: 1. Generating general braid of partitions and 2. Validating that a given general family of partitions is a braid. In both questions the underlying problem to evaluate is the order of refinement between the partitions. To generate braids one needs to fix some how this choice of partial order, while in case of validation one needs to verify this property of ordering of supremum as evoked in the braids definition in equation (2). We also can easily note that question (2) is a combinatorial problem since partial order across pairs of partitions need to be validated. While question (1) is simpler. We shall use

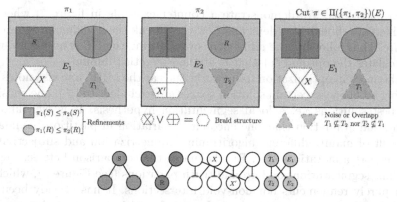

Fig. 3. We show two partitions π_1, π_2 demonstrating four ordering relations: parent→child and child→parent relation (red, blue), a p.p.→p.p. braid structure (hexagon), and finally overlapping classes that aren't inclusions (orange). One can note that once we have a refinement relation between two partitions locally, as in case of classes R, S, this implies that the remaining pairs of classes are either equal, ordered themselves or are partial partitions forming a partial braid structure since they share a common supremum. We also show the intersection graph produced by connecting regions with non-void overlaps to visualize the different ordering relations. The classes corresponding to the components of the intersection graph, gives the supremum of the two partitions. We also show a cut extracted from π_1, π_2.

the stochastic watershed model [2] here to demonstrate how one can control the partial order in generating a braid. Though the generation of braids can be done using a variety of methods. Another simple way to generate a braid would be to fragment/regroup differently an already existing hierarchy of partitions. The disadvantage is that here one fixes the monitoring hierarchy.

2.4 Motivation and Finding Braids in Literature

The need for such models arises in several situations. Firstly we observe that many super-pixel segmentation algorithms, and also multivariate segmentation algorithms [17], [18], operate on agglomerative clustering and region merging. In the former case we obtain a quick super-pixel segmentation by using the clustering tree, while in the latter case we compose partitions of the image domain based on different components of a vectorial image. In a paper close to our work, [5] models the image segmentation problem as the extraction of maximally weighted independent set (MWIS) on the intersection graph. This graph is built over the regions of segmentations produced using various super-pixel low level segmentations. They further associated an energy with each region or node. The algorithm of MWIS consists in calculating the MWIS by dynamic programming. There are two differences between this paper and [5]: Firstly, the segmentations used in [5] do not ensure a stable pairwise supremum, resulting in holes or overlaps. Secondly the intersection graph is blind to the the partial ordering relation between partitions. We demonstrate a counter example in figure 7, where we show different refinement orders, and how they break the dynamic program substructure. Thus following the refinement order during the DP is necessary, when one calculates the optimal cut that is at the energetic infimum.

Furthermore in optimization frameworks such as the MRF, one also notes that in forcing uniqueness, certain solution spaces are excluded. In [15], one considers the K-best solutions i.e. a local segmentation hypothesis. It is well know in segmentation evaluation that one encounters variation in partition boundaries, as a result of mainly different algorithmic parametrization and subjectivity in human expert annotations [16]. Braids enable the comparison between regions of machine segmentations and ground truth partitions (see Figure 3), which are neither purely refinements nor non-void intersections. It has already been well studied that the "segmentation soup" (family of partitions generated from across different algorithms and parameterizations) provided a better support for object detection [11].

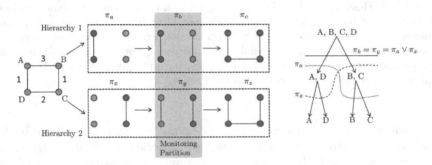

Fig. 4. There can be multiple minimum spanning trees (MSTs) for a given edge-weighted graph [15]. Figure shows a planar weighted graph with two different possible choices in selecting the lowest weighted edge in prim's algorithm. This leads to two different partitions of the nodes set as extracted by the components of the graph. This gives two different hierarchies that can be extracted. The supremum or monitoring partition is created when the second edge is added. Here partition $\{A, B, C, D\}$ monitors over partitions $\{\{A, D\}, \{B\}, \{C\}\}$ and $\{\{B, C\}, \{A\}, \{D\}\}$. We have demonstrated here how the the distinct MST enumeration can used to generated a braid.

We describe shortly the braids found in literature. Angulo et al. [2] accumulate watersheds of stochastically sampled markers chosen from the image domain. This produces an estimate the density function of gradient of the image. The set of partitions produced during the iterations of the stochastic watershed algorithm form a braid structure. This essentially corresponds to a random marker based watershed extracted from the minimum spanning tree. An example is demonstrated in Figure 2. K-Smallest Spanning Tree Segmentations [15] propose multiple distinct segmentations of the image by considering the K-smallest distinct minimum spanning trees. It can be shown easily that the degenerate set of weighted edges with equal weights when permuted over in Prim's algorithm produce different segmentations, which by definition have a common supremum, defined by the heaviest weighted edge governing the degenerate lower weighted edges. This is demonstrated in Figure 4. In a similar line, one can also demonstrate that the attribute watersheds [6] based on area, volume and dynamic, together produce a braid structure with volume hierarchies usually monitoring the other two [10]. Particular versions of braids have appeared in classification problems, for example Diday [7], demonstrates pyramids, where a child may have two parents.

3 Dynamic Programming and h-increasingness

h-increasingness is a property of energies, which preserves the optimal substructure in extracting the minimal cut so that one can use a dynamic program to solve it. It states that the ordering of energies is preserved under concatenation of partial partitions (Figure 5).

Definition 2. *(h-increasingness) Let $\pi_1(S)$, $\pi_2(S)$ be two different p.p. of the same support $S \in E$, be a family of disjoint supports over E. Let π_0 be any partial partition in $\mathcal{D}$ other than $\pi_1(S), \pi_2(S)$. A finite singular energy ω on the partial partitions $\mathcal{D}(E)$ is h-increasing when for every triplet $\{\pi_1(S), \pi_2(S), \pi_0\}$ one has:*

$$\omega(\pi_1(S)) \leq \omega(\pi_2(S)) \Rightarrow \omega(\pi_1(S) \sqcup \pi_0) \leq \omega(\pi_2(S) \sqcup \pi_0) \qquad (3)$$

In implication (3) when the inequality is made strict, we have what we call strict h-increasingness. h-increasingness was first introduced in [9], which generalized the condition of *separable energies* of Guigues [8]. Separability in equation (1), is obtained by replacing comp$(\cdot)$ by a sum of the energies of the constituent classes of a partial partition, to calculate the energy of the partial partition. We can also perform a composition by supremum [14], [17].

Both laws are indeed particular cases of the classical Minkowski expression

$$\omega(\pi(S)) = \left[\sum_{u=1}^{q} \omega(T_u)^\alpha \right]^{\frac{1}{\alpha}} \qquad (4)$$

which is a norm in $\mathbb{R}^n$ for $\alpha \geq 1$. Even though over partial partitions $\mathcal{D}(E)$, it is no longer a norm, it yields strictly h-increasing energies for all $\alpha \in\ ]-\infty, +\infty[$:

Proposition 1. *Let $E \in \mathcal{P}(E)$, let $\omega : \mathcal{P}(E) \to \mathbb{R}$ be a positive or negative energy defined on $\mathcal{P}(E)$. Then the extension of ω to the partial partitions $\mathcal{D}(E)$ by means of Relation (4) is strictly h-increasing.*

Proof. Let $\pi(S)$ $\pi'(S)$ be two p.p. of support S, with q, q' elements each, respectively. When $0 \leq \alpha < \infty$, the mapping $y = x^\alpha$ on $\mathbb{R}^+$ is strictly increasing and, according to Relation (6), the inequality $\omega(\pi(S)) < \omega(\pi'(S))$ implies

$$\sum_{1}^{q} [\omega(T_u)]^\alpha < \sum_{1}^{q'} [\omega(T'_u)]^\alpha \implies \sum_{1}^{q} [\omega(T_u)]^\alpha + [\omega(\pi_0)]^\alpha < \sum_{1}^{q'} [\omega(T'_u)]^\alpha + [\omega(\pi_0)]^\alpha$$

$$(5)$$

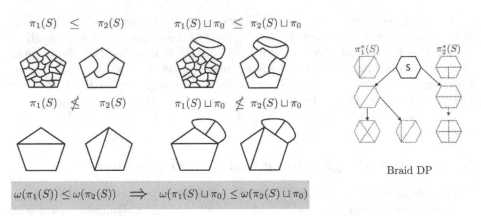

Fig. 5. Left: h-increasingness for HOP (top) versus BOP (bottom). Right: An elementary step of the dynamic program on a partial braid over a support S. The optimal partial partition is the minimum across the partial optima and S, i.e. $\omega(S), \omega(\pi_1^*(S)), \omega(\pi_2^*(S))$.

hence $\omega(\pi_1 \sqcup \pi_0) < \omega(\pi_2 \sqcup \pi_0)$. When $\alpha \leq 0$, the sense of the inequality changes on both sides of implication in (5) but changes again when applying the $(\cdot)^\alpha$. This again leads to $\omega(\pi_1 \sqcup \pi_0) < \omega(\pi_2 \sqcup \pi_0)$, and achieves the proof.   □

One can easily check that the proposition remains true when $\omega : P(E) \to \mathbb{R}^-$ is a negative energy. For $\alpha = +\infty$ (resp.$-\infty$), Minkowski expression yields the supremum (resp. the infimum), which is h-increasing but not strictly. A number of other laws are compatible with h-increasingness, such as weighted sum, alternating compositions varying with level in the hierarchy [10].

Table 1. Table composition laws for different α's in equation (4)

α	Composition laws	Applications
$-\infty$	Infimum	Ground truth energies [9]
0	Number of Classes	CART classifier complexity [4]
+1	Addition	Salembier-Garrido, Guigues [13], [8]
$+\infty$	Supremum	Valero[17], Veganzones[18], Soille[14]

Many other α's that are left open to be explored. The parameter α in fact alike λ-cuts [8] provides a way to control the refinement of the optimal cut [10].

As demonstrated in Figure 5, the dynamic program substructure would now consist in making a choice between the parent supremum (if it is a class of the braid), and the partial partitions that it monitors. We consider in the figure a braid composed of two hierarchies (this is to be able to index the partial partitions.). Equation (6) gives the DP step for BOP shown for HOPs in equation (1). Equation (6) demonstrates a DP sub-structure very similar to the hierarchies except now they are compared over the monitoring supremum class S. When $\omega(\pi_1(S)) = \omega(\pi_2(S))$, and $\omega(\pi_1(S)) < \omega(\{S\}))$, we can pick randomly, as long as we pick one of the partial partitions, so that in a strict sense to keep the energies remain singular.

$$\pi^*(S) = \arg\min\left\{\omega(\{S\}), \omega(\pi_1^*(S)), \omega(\pi_2^*(S))\right\} \tag{6}$$

4   Energetic Ordering and Energetic Lattices

Given the problem of finding an optimal cut, we review separately the requirement of obtaining a unique solution. On the HOP, this has been enforced by many authors [4], [13], [8], [17], [1] as a partition which is either the largest or the smallest, amongst optimal cuts with the same energy. The classical energy based minimization associates an energy with every cut, and takes the cut which has the smallest energy[2]. A hierarchy can have multiple cuts with the same minimal energy, and to ensure a unique solution we introduce the following axiom of singularity:

Definition 3. *Let ω be an energy on the partial partitions $\mathcal{D}(E)$, and B be a braid B with a monitor hierarchy H. Energy ω is singular when*

1. the energies $\omega(\pi(S))$ of all p.p. $\pi(S)$ of H are either strictly smaller, or strictly greater, than the energies of their supports S:

$$\forall\ \pi(S) \in \Pi(S), \omega(\{S\}) < \omega(\pi(S))\}\ or\ \omega(\{S\}) > \omega(\pi(S))\}, \tag{7}$$

2. if $\forall \pi_1, \pi_2 \in B$ and $\forall S \in \pi_1 \vee \pi_2$, we have $\omega(\pi_1 \sqcap S) \neq \omega(\pi_2 \sqcap S)$.

Consider now two partial partitions $\pi(S), \pi'(S)$ over support S, which is also their refinement supremum $S = \pi(S) \vee \pi'(S)$ (see Figure 6). Intuitively, one may assess that partition π_1 is less energetic than π_2 for an energy ω when $\omega(\pi_1 \sqcap \{S\}) \leq \omega(\pi_2 \sqcap \{S\})$ in each class of $\pi_1 \vee \pi_2$.

Theorem 1. *Given $\pi_1, \pi_2 \in \Pi(E)$ two partitions of space E, and an energy ω, the partition π_1 is said to be less energetic than π_2, i.e. $\pi_1 \preceq_\omega \pi_2$ when in each class of supremum $\pi_1 \vee \pi_2$ the energy of the partial partition of π_1 is smaller or equal to that of π_2*

$$\pi_1 \preceq_\omega \pi_2 \Leftrightarrow \{S \in \pi_1 \vee \pi_2 \Rightarrow \omega(\pi_1 \sqcap \{S\}) \leq \omega(\pi_2 \sqcap \{S\})\} \tag{8}$$

The relation $\preceq_\omega$ called energetic ordering, is an ordering relation for all singular energies ω, if and only if the family Π is the set $\Pi(\omega, E, B)$ of all cuts of a braid B. Proof given in thesis [10].

---

[2] A finite set E of only 25 leaves can be partitioned in 0.5×10^{18} different manners, following the Bell's number.

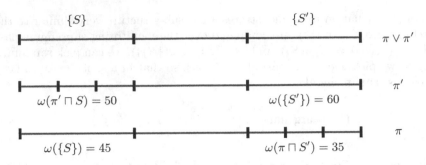

Fig. 6. An example of energetic ordering: We have $\pi \preceq_\omega \pi'$ since in each class of $\pi \vee \pi'$, the energy ω of π is less than or equal to that of π'

To prove that the energetic order yields a complete lattice, we must remark two properties. Firstly, consider a hierarchy H reduced to the two partitions π_0 and π_1, with $\pi_0 \leq \pi_1$. Then the unique smallest partition of $\Pi(E, H)$ is obviously obtained by replacing each class S of π_1 by the corresponding p.p. of π_0 when the latter has an energy smaller than that of S. Denote the resulting minimal partition by $\pi_0 \curlywedge_\omega \pi_1$. In case of a braid, π_0 and π_1 are no longer ordered by refinement, and the energetic comparisons have to be performed in each class of $\pi_0 \vee \pi_1$. Secondly, consider now a standard hierarchy H, (i.e. with $n+1$ levels), and k cuts $\{\pi_j, 1 \leq j \leq k\}$ of H. The sequence 9 generates a new hierarchy H' where each two classes are ordered or disjoint, hence are classes of H.

$$\pi'_1 = \wedge_1^k \pi_j; \ \pi'_2 = \wedge_2^k \pi_j; \ ...; \ \pi'_k = \pi_k \tag{9}$$

Theorem 2. *Let B be a braid of monitor $H = \{\pi_i, 0 \leq i \leq n\}$, and ω a singular energy. The family $\Pi(\omega, H)$ of all cuts of H has a unique minimal element*

$$\pi^* = (((\pi_0 \curlywedge_\omega \pi_1) \curlywedge_\omega \pi_2)...) \curlywedge_\omega \pi_n \tag{10}$$

and a unique maximal element $\pi^{**} = ((\pi_0 \curlyvee_\omega \pi_1)...) \curlyvee_\omega \pi_n$. This property extends to braid B.

Proof. As $\pi_0 \curlywedge_\omega \pi_1$ is the less energtic cut made of classes of π_0 and π_1, the same can be stated with $(\pi_0 \curlywedge_\omega \pi_1) \curlywedge_\omega \pi_2$ for the classes of π_0, π_1, π_2. Thus, by induction, the cut (2) is the unique smallest cut of $\Pi(\omega, H)$. The dual approach leads to the largest energetic cut. Finally, if H is replaced by braid B, then each class of S may have to be compared with several sets of children partial partitions a_1, a_2, etc. but again every minimal (resp. maximal) choice is unique by singularity, which achieves the proof. □

Corollary 1. *When in addition to 2 the energy ω is h-increasing, then $\Pi(\omega, H)$, and further $\Pi(\omega, B)$ turn out to be complete lattices. The infimum and supremum of family $\{\pi_j, 1 \leq j \leq k\}$ are denoted by $\curlywedge_\omega \pi_j$ and $\curlyvee_\omega \pi_j$.*

Finally, we must remark here that given a singular energy on braid, one ensures unique optimal cut, but one which cannot be obtainable by a dynamic program. While a singular and h-increasing energy yields itself to a DP producing an optimal cut, though there can exist other cuts with the same minimal energy. Finally a singular and strictly h-increasing energy is one which yields a unique optimal cut with the DP.

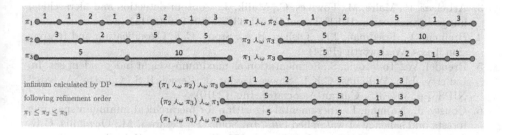

Fig. 7. A counter example showing the breakdown of DP when not following refinement ordering between partitions. Three partitions π_1, π_2, π_3 with their energies over each class. We demonstrate the different infima achievable for different orders of refinement followed across the partitions, while applying the dynamic program. We see for two orders we don't achieve the global infimum. We thus always need an algorithm that works in the order of refinement to keep the DP substructure. This is also why one uses a bottom up pruning [13] or climbing [8]. The actual infimum of the energetic lattice is obtained by following the order of refinement.

5 Conclusion

The paper introduced the new hierarchical structure of the braids of partitions, which expanded the space of hierarchies for the problem of extracting optimal cuts. Furthermore it showed that the braids are the largest family of partitions over which the energetic lattice can be defined. The DP to extract a unique minimal cut consists inherently of an ordering based optimization problem, which is expressed by the energetic lattice structure. A generalized h-increasingness condition for energies operable on braids was also demonstrated. This gives the DP that aggregates local optima to obtain the global optimum. Finally, the paper also provided a short review of braids available in literature, and provides a perspective on how the braid model can be used to become algorithm independent while organizing image domain or space into a family of partitions which preserves the dynamic program substructure. We foresee applications in the domain of multivariate optimization, machine learning and super-pixel segmentation based optimization.

References

1. Akcay, H.G., Aksoy, S.: Automatic detection of geospatial objects using multiple hierarchical segmentations. IEEE T. Geoscience & Remote Sensing 46(7), 2097–2111 (2008)
2. Angulo, J., Jeulin, D.: Stochastic watershed segmentation. In: Banon, G.J.F., Barrera, J., de Mendonça Braga-Neto, U., Hirata, N.S.T. (eds.) Proceedings (ISMM), Rio de Janeiro, vol. 1, pp. 265–276. INPE (October 2007)
3. Arbelaez, P., Maire, M., Fowlkes, C., Malik, J.: Contour detection and hierarchical image segmentation. IEEE Trans. PAMI 33(5), 898–916 (2011)
4. Breiman, L., Friedman, J.H., Olshen, R.A., Stone, C.J.: Classification and Regression Trees. Wadsworth (1984)
5. Brendel, W., Todorovic, S.: Segmentation as maximum-weight independent set. In: Lafferty, J.D., Williams, C.K.I., Shawe-Taylor, J., Zemel, R.S., Culotta, A. (eds.) NIPS, pp. 307–315. Curran Associates, Inc. (2010)
6. Cousty, J., Najman, L.: Incremental algorithm for hierarchical minimum spanning forests and saliency of watershed cuts. In: Soille, P., Pesaresi, M., Ouzounis, G.K. (eds.) ISMM 2011. LNCS, vol. 6671, pp. 272–283. Springer, Heidelberg (2011)
7. Diday, E.: Spatial pyramidal clustering based on a tessellation. In: Banks, D., McMorris, F., Arabie, P., Gaul, W. (eds.) Classification, Clustering, and Data Mining Applications, pp. 105–120. Studies in Classification, Data Analysis, and Knowledge Organisation, Springer, Heidelberg (2004)
8. Guigues, L., Cocquerez, J.P., Men, H.L.: Scale-sets image analysis. International Journal of Computer Vision 68(3), 289–317 (2006)
9. Kiran, B.R., Serra, J.: Global-local optimizations by hierarchical cuts and climbing energies. Pattern Recognition 47(1), 12–24 (2014)
10. Kiran, B.R.: Energetic Lattice based optimization. Ph.D. thesis, Université Paris-Est, LIGM-A3SI, ESIEE (2014)
11. Malisiewicz, T., Efros, A.A.: Improving spatial support for objects via multiple segmentations. In: British Machine Vision Conference (BMVC) (September 2007)
12. Ronse, C.: Partial partitions, partial connections and connective segmentation. J. Math. Imaging Vis. 32(2), 97–125 (2008)
13. Salembier, P., Garrido, L.: Binary partition tree as an efficient representation for image processing, segmentation, and information retrieval. IEEE Trans. on Image Processing 9(4), 561–576 (2000)
14. Soille, P.: Constrained connectivity for hierarchical image partitioning and simplification. IEEE Transactions on Pattern Analysis and Machine Intelligence 30(7), 1132–1145 (2008)
15. Straehle, C., Peter, S., Köthe, U., Hamprecht, F.: K-smallest spanning tree segmentations. In: Weickert, J., Hein, M., Schiele, B. (eds.) GCPR 2013. LNCS, vol. 8142, pp. 375–384. Springer, Heidelberg (2013)
16. Unnikrishnan, R., Pantofaru, C., Hebert, M.: Toward objective evaluation of image segmentation algorithms. IEEE Transactions on Pattern Analysis and Machine Intelligence 29(6), 929–944 (2007)
17. Valero, S.: Hyperspectral image representation and Processing with Binary Partition Trees. Ph.d., Universitat Politècnica de Catalunya, UPC (2011)
18. Veganzones, M., Tochon, G., Dalla-Mura, M., Plaza, A., Chanussot, J.: Hyperspectral image segmentation using a new spectral unmixing-based binary partition tree representation. ITIP 23(8), 3574–3589 (2014)

Constrained Optimization on Hierarchies and Braids of Partitions

Jean Serra[1] and Bangalore Ravi Kiran[2(✉)]

[1] Université Paris-Est, A3SI-ESIEE LIGM, Lneteis, France
`jean.serra@esiee.fr`
[2] Centre de robotique, MINES ParisTech, PSL-Research University, Paris, France
`ravi.kiran@mines-paritech.fr`

Abstract. This theoretical paper provides a basis for the optimality of scale-sets by Guigues [6] and the optimal pruning of binary partition trees by Salembier-Garrido [11]. They extract constrained-optimal cuts from a hierarchy of partitions. Firstly, this paper extends their results to a larger family of partitions, namely the braid [9]. Secondly, the paper shows the dependence of valid constraint function values and multiplier values in a Lagrangian optimization framework. Lastly, but most importantly, it also proposes the energetic order and energetic lattice based solutions for the constraint optimization problem. This approach operates on a partition based constraint thus ensuring the existence of a valid multiplier and constraint value.

Keywords: Hierarchies · Lagrange · Optimization · Lattice

1 Introduction

In this theoretical paper[1] we aim to first demonstrate that the optimal cuts on hierarchies or λ-cuts in the sense of Guigues [6] and Salembier et al.[11], only provide an upper bound on the minimum energy to the original constrained optimization problem on hierarchies. We show that the choice of a suitable Lagrange multiplier λ in fact provides a solution to the perturbed problem first stated in Everett's theorem [5]. Further in a fundamental contribution we demonstrate how the constrained optimization problem in the Lagrange sense can be solved using the energetic lattice [7], by replacing the numerical constraints by partition based constraint. We start with a quick review of notation and definitions.

1.1 Definitions

We denote a partition of space E by π and a partial partition(p.p.) [10] of subset $S \subseteq E$ by $\pi(S)$. The family of all partitions of E is denoted by $\Pi(E)$, while that

---

[1] Please refer to accompanying paper [9] for notions of braids, energetic ordering, energetic lattice, singularity and h-increasingness. This work was partly funded by the French ANR-2010-BLAN-0205-03 program KIDICO.

J.A. Benediktsson et al. (Eds.): ISMM 2015, LNCS 9082, pp. 229–240, 2015.
DOI: 10.1007/978-3-319-18720-4_20

of partial partitions by $\mathcal{D}(E)$). A hierarchy of partitions (HOP) is a finite chain of partitions $H = \{\pi_i, \ i \in [0, n]\}$, with $\pi_i \leq \pi_j, i < j$, where $\leq$ stands for the refinement ordering. The minimal element π_0 of H is called the leaves partition, while the maximal element is the one class partition $\{E\}$, called the root. A cut of hierarchy H is a partition of E whose elements are composed of classes in H. The set of all cuts of H is $\Pi(E, H)$. A braid B is a family of partitions B where the pairwise refinement supremum of any two partial partitions is a cut of some hierarchy H; i.e. belongs to $\Pi(E, H)$ [9].

An energy $\omega : \mathcal{D} \to \mathbb{R}^+$ is a non-negative function that is defined on the family of partial partitions. The energy of a partition or partial partition is obtained by the composition of energies either by addition, supremum or other laws [9] of its constituent classes, $\omega(\pi(S)) = \sum_{a \sqsubset \pi(S)} \omega(a)$, though these might not be the only way. The energy ω is said to be singular when for any p.p. $\pi(S)$ we have $\omega(\{S\}) \neq \omega(\pi(S)), S \subseteq E$. It is said to be h-increasing when

$$\omega(\pi(S)) \leq \omega(\pi'(S)) \Rightarrow \omega(\pi(S) \sqcup \pi_0) \leq \omega(\pi'(S) \sqcup \pi_0), \quad \forall S \subseteq E \qquad (1)$$

where $\sqcup$ indicates the concatenation of any p.p π_0 with support that is disjoint with S. A h-increasing energy becomes strict when the inequality $\leq$ becomes $<$.

Now the optimal cut in [4], [11], [6], is calculated by aggregating local optima. The local optimum at class S either choses the parent $\{S\}$, or the disjoint union of the optimums over the its children as shown in equation 2.

$$\pi^*(S) = \begin{cases} \{S\}, & \text{if } \omega(S) \leq \text{comp}(\omega(\pi^*(a))), a \in \pi(S) \\ \bigsqcup_{a \in \pi(S)} \pi^*(a), & \text{otherwise} \end{cases} \qquad (2)$$

The solution to the dynamic program in equation (2), when aggregated for all $S \in H$, following a lexicographic order gives the optimal cut π^*. One should also note that the composition $\text{comp}(\cdot)$ is performed by addition, or supremum, or many other laws that preserves h-increasingness. It is shown in [7] that this optimal cut is the minimal element of an energetic lattice. We reproduce the theorem in [7]:

Theorem 1. *Let Π be a family of partitions of E, and let $\pi_1, \pi_2 \in \Pi$. Given an energy ω, the partition π_1 is said to be less energetic than π_2, and one writes $\pi_1 \preceq_\omega \pi_2$ when in each class of $\pi_1 \vee \pi_2$ the energy of the partial partition of π_1 is smaller or equal to that of π_2: :*

$$\pi_1 \preceq_\omega \pi_2 \ \Leftrightarrow \ \{S \in \pi_1 \vee \pi_2 \Rightarrow \omega(\pi_1 \sqcap \{S\}) \leq \omega(\pi_2 \sqcap \{S\})\} \qquad (3)$$

The relation $\preceq_\omega$ is an ordering relation for all singular energies ω, called energetic ordering, if and only if the family Π is the set $\Pi(\omega, E, B)$ of all cuts of a braid B.

The set $\Pi(\omega, E, B)$ forms a complete lattice for the energetic ordering $\preceq_\omega$.

This theorem described an energetic ordering and thus an energetic lattice [9], which models the dynamic program to obtain the optimal cut. We will use this lattice structure for the constrained optimization problem on HOP and braids further on in this paper.

1.2   Constrained Optimization on Hierarchies

A lattice structure has been developed in [9], [7] for the dynamic program based minimization of any general non parametrized energy. We now concentrate on the constrained optimization problem, which is achieved by the unconstrained minimization of a Lagrangian function, as is the case with Guigues and Salembier. Consider the constrained optimization problem:

$$\underset{\pi \in \Pi(E,H)}{\text{minimize}} \sum_{S \in \pi} \omega_\varphi(S) \quad \text{subject to} \quad \sum_{S \in \pi} \omega_\partial(S) \leq C \tag{4}$$

The corresponding Lagrangian can now be written as:

$$\omega(\pi, \lambda) = \sum_{S \in \pi} \omega_\varphi(S) + \lambda \cdot (\omega_\partial(S) - C) \tag{5}$$

where $\pi \in \Pi(E, H)$ is a cut from HOP H. The objective function being minimized is denoted by ω_φ, the constraint function by ω_∂, while C is an imposed constraint function value. These energies hold on all partial partitions $\omega_\varphi, \omega_\partial : \mathcal{D} \to \mathbb{R}$ of the working space E. In (5) the multiplier λ is scalar for the sake of simpler notation and pedagogy. This can always extend to vector multiplier and constraint functions.

2   Guigue's λ-cuts are Upper Bounds

We illustrate in Figure 1 a tree with its classes, objective function being minimized, constraint function, as well as the λ or scale function values as defined by Guigues [6]. The objective ω_φ is chosen to be super-additive while constraint ω_∂ is sub-additive, as in [6]. Now the family of Lagranians is $\{\omega(\lambda) = \omega_\varphi + \lambda \omega_\partial, \lambda \geq 0\}$. Energies are composed additively here, i.e. $\omega(\pi(S)) = \sum_{a \in \pi(S)} \omega(a)$. Further when we have equal parent and child energies, we pick the parent, like in [6].

The λ-cut denoted by $\pi^*(\lambda)$ are cuts with least ω_φ, given multiplier λ. There are three such cuts, for three different values of the multiplier calculated by the scale function defined by Guigues [6] $\lambda = -\Delta\omega_\varphi / \Delta\omega_\partial$.

Counter Example: Following Guigues and Salembier (as well as Casselles et al. [2]) we search for the cut with the smallest λ, or λ-cut, which satisfies an input constraint value here set to $C = 7.5$. Here the λ-cut with $\omega_\varphi(\pi^*(\lambda = 3.5)) = 15$ and constraint $\omega_\partial(\pi^*(\lambda = 3.5)) = 6$ is the optimal λ-cut satisfying constraint. We now consider other cuts, which are not λ-cuts: say $\pi = (g, c, d, k)$ with energy $\omega(\pi, \lambda) = 11.5 + 7 \lambda$. The cut π obviously provides a better minimum than the minimal λ-cut (g, h, k) since $\omega_\partial(\pi) = 7$, which is below the constraint $C = 7.5$, for an objective value $\omega_\varphi(\pi) = 11.5$, which also is smaller than the objective function $\omega_\varphi(\pi^*(\lambda = 3.5)) = 15)$. What is worse is that there are two such different cuts $\pi = (g, c, d, k)$ and $\pi' = (a, b, h, k)$, that have the same constraint and objective values, $\omega_\partial(\pi') = 7$ and $\omega_\varphi(\pi') = 11.5$. Thus there are several constrained minimal cuts for the energy ω_φ, and none of them are

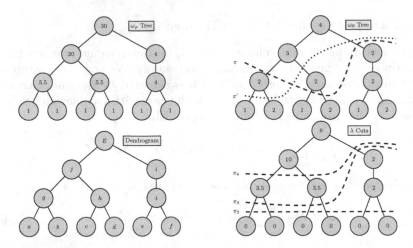

Fig. 1. Bottom Left, a hierarchy H with classes. The two trees in the top row, indicate the two energies $(\omega_\varphi, \omega_\partial)$ associated with the corresponding classes. π and π' are two cuts of H. Bottom right, in the nodes, we depict the λ values by equating parent and child energies, whose level sets give the minimal cuts w.r.t. the ω_λ. They are depicted in the λ-tree for $\lambda = 2, 3.5, 4$ as π_2, π_3, π_4. The λ values for the leaves are assumed to be 0, though in case of Breiman et al. [4] λ for the leaf classes are set to ∞ to avoid over-fitting.

obtained from the sequence of λ-cuts $\pi^*(\lambda)$! And we cannot take their infimum $\pi \wedge \pi' = (a, b, c, d, k)$ because $\omega_\partial(\pi \wedge \pi') = 8$, which is above the constraint $C = 7.5$. The plot of the different energies of the λ-cuts with the constraint are shown in Figure 2.

Observations: For an imposed cost $\omega_\partial(\pi) \leq C$ one is not assured of the existence of a corresponding multiplier value λ. The family of λ-cuts $\{\pi^*(\lambda), \lambda \in \overline{\mathbb{R}}\}$ is not complete to describe all possible constraint values. A cut that minimizes ω_φ may not belong to the λ-cuts $\{\pi^*(\lambda)\}$. One may also remark that the dual problem still remains a combinatorial problem. As we know from convex optimization [3], given a multiplier, the dual Lagrangian serves as an upper bound on the optimum corresponding to the primal Lagrangian. In our words, $\pi^*(\lambda^*)$ is only the upper-bound of the constrained minimal cuts. Furthermore the values of λ can be discrete, real, or rational, while still lacking a $C \to \lambda$ constraint-multiplier map. Finally, one can note that uniqueness is lost, even when ω_φ is strictly h-increasing.

3   Everett's Theorem

Everett's seminal paper [5] studies resource allocation problem by a choosing an *optimal* Lagrangian parameter. In literature one of its earliest uses appears in source-coding by its usage by Shoham-Gersho [12] to study variable rate set

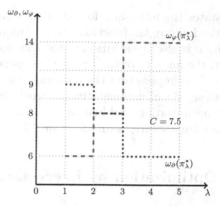

Fig. 2. For $2 < \lambda < 3$ the minimal cut is (a, b, c, d, i) and $\omega_\partial = 8$, for $\lambda \geq 3$ the minimal cut is (g, h, i) and $\omega_\partial = 6$, i.e. ω_∂ is never equal to the cost $C = 7.5$ at any time

quantizers. The function f is an objective being minimized and g, h are inequality and equality constraints, which are three real valued functions defined over an arbitrary abstract set X. X need not be topological. Neither do we require continuity, derivation, or convexity of the functions f, g, h.

Theorem 2. *Given the Lagrangian function, with multipliers λ, μ*

$$\min_{x \in X}\{f(x) + \mu g(x) + \lambda h(x)\}$$

the solution $\bar{x}(\lambda)$ to this unconstrained minimization problem, is also an optimal solution to perturbed primal problem, namely

$$\underset{x \in X}{\text{minimize}} \quad f(x)$$
$$\text{subject to} \quad g(x) \leq g(\bar{x}(\lambda)); \ h(x) = h(\bar{x}(\lambda));$$

For Guigues and Salembier's problem, this can be restated as:

Theorem 3. *Given the multiplier $\lambda \in \mathbb{R}$ and the Lagrangian function,*

$$\min_{\pi \in \Pi(E,H)} \left\{ \sum_{S \in \pi} \omega_\varphi(S) + \lambda \sum_{S \in \pi} \omega_\partial(S) \right\}$$

the solution $\bar{\pi}(\lambda)$ to this unconstrained minimization is also an optimal solution to perturbed primal problem:

$$\underset{\pi \in \Pi(E,H)}{\text{minimize}} \quad \sum_\pi \omega_\varphi(S) \ \text{subject to} \ \sum_{S \in \pi} \omega_\partial(S) \leq \sum_{S \in \bar{\pi}(\lambda)} \omega_\partial(S)$$

Everett's theorem states the following: for any non-negative λ, if an unconstrained minimum of the Lagrangian function can be found, with solution $\bar{x}(\lambda)$ or $\bar{\pi}(\lambda)$, then this solution is also the solution to the constrained problem whose constraints are, in fact, the amount of each resource expended in achieving the unconstrained solution. This implies that the constraints are set by choosing the λ parameter. Any arbitrary set of non-negative λ's works here, notably causing the original constraint optimization problem to be unknown, and is only to be defined once the Lagrangian's solutions are determined.

4   Constrained Optimization by Energetic Lattices

We now present the energetic lattice framework for constrained optimization. We shall first reformulate the minimization of the Lagrangian optimization using the energetic lattice. Further on we develop a constrained optimization model, where the constraint values a are based on refinement of partitions, as against the purely numerical order. We also introduce inf-modularity, which generalizes the sub-additivity of constraint function in Guigues [6], resulting in the important Theorem 4.

4.1   Refinement and Energetic Lattices

There are two types of lattices we refer to in this paper, refinement lattices and energetic lattices. The refinement lattices over a family of partitions, namely $\Pi(E)$ all partitions, $\Pi(E, H)$, partitions from a hierarchy H and finally $\Pi(E, B)$ partitions from a braid B. Here $\Pi(E, H), \Pi(E, B)$ are sub-lattices of $\Pi(E)$. When we say two partitions are ordered, $\pi_i \leq \pi_j$, we refer to the refinement ordering.

Given this family of partitions, and any singular energy ω we can now consider the corresponding energetic order and lattice. Here, $\Pi_{\omega(\lambda)}$ is a based on the energetic ordering $\preceq_{\omega(\lambda)}$ w.r.t the Lagrangian $\omega(\pi, \lambda)$, of order $\preceq_{\omega(\lambda)}$. The minimal cut for this energetic lattice is $\pi^*(\lambda) = \wedge_{\omega(\lambda)}\{\pi, \pi \in \Pi\}$. The value of $\omega(\pi, \lambda)$ for a cut $\pi \in \Pi(E, B)$ is denoted by $\omega(\pi, \lambda)$, and that for the minimal cut by $\omega(\pi^*(\lambda))$. Similarly we have the lattices Π_{ω_φ} and Π_{ω_∂} for $(\preceq_{\omega_\varphi}, \omega_\varphi)$ and $(\preceq_{\omega_\partial}, \omega_\partial)$ energetic order-energy pairs, respectively. In these lattices the family of partitions under study are assumed to be cuts, either from the hierarchy or the braid under study.

4.2   Inf-Modularity

Definition 1. *An energy* $\omega_\partial : \mathcal{D}(E) \to \mathbb{R}^+$ *is said inf-modular when for each p.p.* π *of support* $S \in \mathcal{P}(E)$ *we have*

$$\omega_\partial(\{S\}) \leq \bigwedge\{\omega_\partial(a), a \sqsubseteq \{S\}\}. \tag{6}$$

Inf-modularity provides a non-linear version of sub-additivity [1], where the former acts on partial partitions, while latter on general subsets of the space. An energy ω is sub-additive when, for any p.p. $\pi(S)$ of support S, the energy $\omega(S) \leq \sum_{T_i \in \pi(S)} \omega(T_i)$ for the sake of comparison. Since the energy ω is defined on partial partitions of E, we need to introduce the energy ω'_{∂} on sets such that $\omega'_{\partial}(S) = \omega_{\partial}(\{S\})$. Then any extension ω_{∂} of ω'_{∂} to the p.p. of E which satisfies the following inequality is inf-modular.

$$\omega_{\partial(S)}(\pi) \leq \sum_{j=1}^{j=p} \omega_{\partial}(\{T_j\}) = \sum_{j=1}^{j=p} \omega'_{\partial}(T_j), \qquad (7)$$

Conversely, the restriction ω'_{∂} to sets of an inf-modular energy ω_{∂} is sub-additive [8]. For example, in a partition of $\mathbb{R}^2$ the perimeters ω'_{∂} of the classes generate an inf-modular energy ω_{∂} on the partial partitions.

4.3 Lagrange Families

Definition 2. *A scalar Lagrange family of energies* $\{\omega(\lambda) = \omega_{\varphi} + \lambda\omega_{\partial}, \lambda \in \overline{\mathbb{R}}\}$ *is one where* $\omega(\lambda)$, ω_{φ}, *and* ω_{∂} *are singular and h-increasing, and further* ω_{∂} *is inf-modular.*[2]

Theorem 4. *Let* $\{\omega(\lambda) = \omega_{\varphi} + \lambda\ \omega_{\partial}\}$, *be a scalar Lagrange family of energies on the partial partitions of a space* E, *and suppose* $\lambda > 0$. *Given a braid* B *on space* E, *let* $\Pi_{\omega(\lambda)}, \Pi_{\omega_{\varphi}}$ *and* $\Pi_{\omega_{\partial}}$ *be energetic lattices over the cuts* $\pi \in \Pi(E, B)$ *w.r.t. the Lagrangian* $\omega(\pi, \lambda)$, *the objective* ω_{φ}, *and constraint* ω_{∂} *respectively. The minimal element of* $\Pi_{\omega(\lambda)}$ *is denoted by* $\pi^*(\lambda)$.

$$0 \leq \lambda \leq \mu \ \Rightarrow \ \pi^*(\lambda) \succeq_{\omega_{\partial}} \pi^*(\mu) \ \ and \ \ \pi^*(\lambda) \preceq_{\omega_{\varphi}} \pi^*(\mu) \qquad (8)$$

i.e. as λ *increases, the sequence* $\{\pi^*(\lambda), \lambda > 0\}$ *of the* λ-*cuts w.r.t. the* $\Pi_{\omega(\lambda)}$ *decreases in the energetic lattice* $\Pi_{\omega_{\partial}}$ *and increases in the energetic lattice* $\Pi_{\omega_{\varphi}}$. *Concerning the energies* ω_{∂} *(resp.* ω_{φ}) *we have:*

$$\lambda \leq \mu \ \Rightarrow \ \omega_{\partial}(\pi^*(\lambda)) \geq \omega_{\partial}(\pi^*(\mu)) \ \ (resp. \ \omega_{\varphi}(\pi^*(\lambda)) \leq \omega_{\varphi}(\pi^*(\mu))). \qquad (9)$$

Proof for this theorem is given in [8]. The two energies ω_{φ} and ω_{∂} vary in opposite senses on the minimal cuts. The relation in (9) generalizes the result of Salembier and Guigues over hierarchies and for linear energies ω_{φ} and ω_{∂}, to braids and Lagrange families. But the stronger implications (8) require the energetic lattices $\Pi_{\omega_{\varphi}}$ and $\Pi_{\omega_{\partial}}$. The role of inf-modularity of the constraint function ω_{∂} is demonstrated in Theorem 4.

5 Lagrange Minimization by Energy (LME)

In Everett's theorem one considers a set X and objective f and constraints g, h defined at any point $x \in X$. Our situation slightly differs: the set X is replaced

---

[2] One can write a vectorial version!

by the set of all cuts of a braid B, and this set is equipped with three different lattice structures Π_{ω_φ}, Π_{ω_∂} and $\Pi_{\omega(\lambda)}$, governed by the three energies ω_φ, ω_∂, and $\omega(\lambda)$. The primal and dual problems must be re-stated in the new framework of energetic lattice:

Problem 1. (LME Primal problem): Given a braid B, a constraint value C, and objective and constraint functions ω_φ and ω_∂, find the cut(s) $\pi \in B$ that minimize $\omega_\varphi(\pi)$, subject to the constraint $\omega_\partial(\pi) \le C$.

The domain of the feasible cuts is $\Pi' \subseteq \Pi(E, B)$

$$\Pi' = \{\pi,\ \pi \in \Pi(E, B),\ \omega_\partial(\pi) \le C\} \tag{10}$$

which by its definition is a braid of partitions itself [9]. In the Lagrangian lattice $\Pi_{\omega(\lambda)}(E, B)$ of energy $\omega(\lambda) = \omega_\varphi + \lambda\omega_\partial$, the λ-cut $\pi^*(\lambda)$ is a cut with least energy given the multiplier λ for the Lagrangian energy $\omega(\pi^*(\lambda))$. The energy $\omega(\pi^*(\lambda))$ is a function of λ, ω_φ and ω_∂, but not of the cuts $\pi \in \Pi$. The dual problem is relative to this energy $\omega(\pi^*(\lambda))$:

Problem 2. (Multiplier Problem) Given a braid B on E and two energies ω_φ and ω_∂, find the parameter λ which maximizes $\omega(\pi^*(\lambda))$, subject to the constraint $\lambda > 0$.

The following theorem answers both primal and dual problems (proof in [8])

Theorem 5. *Given a braid B, let $\{\omega(\lambda) = \omega_\varphi + \lambda\omega_\partial, \lambda \in \overline{\mathbb{R}}\}$ be a scalar Lagrange family of energies. Let $\lambda^* = \inf\{\lambda \mid \omega_\partial(\pi^*(\lambda)) \le C\}$. If*

1. *the feasible set Π' is not empty,*
2. *Multiplier-Constraint map: $\omega_\partial(\pi^*(\lambda^*)) - C = 0$,*

then $\pi^*(\lambda^*)$ and λ^* are solutions of the problems 1 and 2 respectively . When ω_φ is strictly h-increasing, then the solution $\pi^*(\lambda^*)$ is unique.

Multiplier-Constraint Mapping: Theorem 5 demonstrates the multiplier dependence of the constraint function as already discussed by Everett's theorem 3. This Condition 2 in theorem (5) requires that for any constraint function value $\omega_\partial(\pi) = C$ there exists a corresponding optimal multiplier λ^* such that $\omega_\partial(\pi(\lambda^*))$. This is a highly unrealistic constraint. Furthermore finding an optimal multiplier in the dual domain returns us back to a combinatorial problem. Even for very simple constraint values, in the counter example 1 there are cases where no multiplier exists for an imposed constraint function value C. Instead these methods, including Guigues, Salembier and other, rely on the capability to approximate the imposed constraint function value by searching for a "good" value of multiplier λ.

6   Lagrange Minimization by Cut-Constraints (LMCC)

In the LME model, the constraint function ω_ϑ and the cost C are numerical, while the minimization itself is expressed in the energetic lattices Π_{ω_φ} and $\Pi_{\omega(\lambda)}$. Can we, alternatively, reformulate the constraint conditions directly with the cuts? We now examine this question, by looking for the cuts smaller than or equal to a given cut $\pi_C \in \Pi_{\omega_\vartheta}$:

Problem 3.   Find minimal cut $\pi_\varphi \in \Pi_{\omega_\varphi}$ subject to the constraint $\pi_\varphi \preceq_{\omega_\vartheta} \pi_C$.

$$\Pi_C = \{\pi \mid \pi \in \Pi,\ \pi \preceq_{\omega_\vartheta} \pi_C\}$$

Here π_C is the set of feasible solutions and we have $\pi_\varphi = \wedge_{\omega_\varphi}\{\pi \mid \pi \in \Pi_C\}$. As before, the Lagrangian $\omega(\lambda) = \omega_\varphi + \lambda\omega_\vartheta, \lambda \in \mathbb{R}$, is introduced. It induces the energetic lattice $\Pi_{\omega(\lambda)}$ of minimal cut $\pi^*(\lambda)$. A new minimal λ is also introduced by

$$\lambda^* = \sup\{\lambda \mid \pi^*(\lambda) \preceq_{\omega_\vartheta} \pi_C\} \tag{11}$$

Problem 4.   LMCC multiplier problem: find the value of the parameter λ which optimizes $\pi^*(\lambda)$ in $\Pi(\omega_\vartheta)$ subject to the constraint $\lambda \geq 0$.

For solving jointly both problems 3 and 4, the following conditions are needed:

Theorem 6. *Given a braid B, let $\{\omega(\lambda) = \omega_\varphi + \lambda\omega_\vartheta, \lambda \in \mathbb{R}\}$ be a scalar Lagrange family of energies. Let $\lambda^* = \sup\{\lambda \mid \pi^*(\lambda) \preceq_{\omega_\vartheta} \pi_C\}$. If*

1. *Constraint satisfaction: the set $\Pi_C = \{\pi \mid \pi \in \Pi, \pi \preceq_{\omega_\vartheta} \pi_C\}$ is not empty,*
2. *Positive multiplier: $\lambda \geq 0$,*
3. *Energetic-Lattice constraint assumption: $\pi_{\omega_\varphi} \succeq_{\omega_\varphi} \pi^*(\lambda^*)$.*

are fulfilled, then the set of feasibility in λ is $\lambda \geq \lambda^*$, and $\pi^*(\lambda^*)$ and λ^* are the unique solutions to the problems 3 and 4 respectively.

Proof. We first prove that the set of feasible λ are $\lambda \geq \lambda^*$. According to relation 11 when $\lambda > \lambda^*$, then $\pi^*(\lambda)$ doest not belong to the feasible set Π_C. Therefore, if λ is such that $\pi^*(\lambda) \preceq_{\omega_\vartheta} \pi_C$, then $\lambda \leq \lambda^*$; if in addition $\lambda \geq 0$, then theorem 4 applies and $\pi^*(\lambda^*) \leq \pi^*(\lambda)$, hence $\pi^*(\lambda^*) \preceq_{\omega_\vartheta} \pi_C$, i.e. $\pi^*(\lambda^*) \in \Pi_C$. Consequently $\pi^*(\lambda^*) \succeq_{\omega_\varphi} \pi(\omega_\varphi)$, and by assumption 3, $\pi_\varphi = \pi^*(\lambda^*)$. The minimal cut (of Lattice $\Pi_{\omega(\lambda^*)}$ is a solution of problem 3, and even the unique one, since π_φ is the minimal element of a lattice. Concerning the multiplier problem we can apply Theorem 4 since $\lambda \geq 0$, which gives:

$$\Upsilon_{\omega_\vartheta}\pi^*(\lambda) \preceq_{\omega_\vartheta} \pi^*(\lambda^*),\ \forall \lambda \geq \lambda^* \geq 0.$$

As $\pi^*(\lambda^*)$ is also an element of left hand side of the above inequality, thus we obtain equality which solves the the multiplier problem.   □

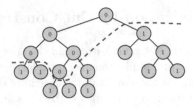

Fig. 3. Constraint cut π_0 shown in dotted line, which takes all classes below it. The 1-classes have $\omega_\partial > C$, the 0-classes have $\omega_\partial \leq C$. First index the classes of H by a lexicographic ordering from the root E to the leaves, and go top-down beginning at the root. When a class S has all its sons T such that $\omega_\partial(T) \leq C$, then replace S by its sons. We assume for that the leaves in the hierarchy of partitions must all satisfy the constraint to ensure a non-void feasible set.

In LMCC the minimal cut is unique, even when the energy is not strictly h-increasing. This authorizes the use of sup and inf composed energies. The comparison between LME and LMCC frameworks is instructive. One can also notice that the assumption 3 of the theorem 6 turns out to be weaker than the corresponding notion of in Theorem 5, as it involves an inequality only. Both models show that Lagrangians still work for lattices of cuts, and not just only on numerical lattices of energies. An interesting feature is that Theorem 6 applies for infinite partitions of E, as soon as the number of classes is locally finite. This covers the "remote sensing type" of situations, where the zone under study is incomparably smaller than the total extension of the scene. The LMCC works since it avoids the constraint approximation problem.

7   Class Constrained Minimization (CCM)

We now study a constraint applied purely to classes. This restricts the constraint function to be defined now on the classes and no more on the partitions. This section treats firstly the case of hierarchies and then that of braids, and develops an alternative method for constrained optimization, which does not resort to Lagrangians.

The hierarchy H under study is supposed to be finite. Provide the classes $S \in \mathcal{S}$ of the hierarchy H with an energy ω_∂ and fix a constraint value C. Introduce now an objective energy ω_φ that is h-increasing and singular. We can now set the following problem:

Problem 5.   Find the cut(s) of H of smallest energy $\omega_\varphi(\pi)$, such that all classes in these cuts satisfy the constraint $\omega_\partial(S) \leq C$.

The method consists in generating a new hierarchy H' where the minimization of ω_φ is no longer conditioned. Let $\mathcal{A}(C)$ stand for the family of the cuts π of H whose energies of all classes are $\leq C$:

$$\pi \in \mathcal{A}(C) \quad \Leftrightarrow \quad \{S \sqsubseteq \pi \Rightarrow \omega_\partial(S) \leq C\}. \tag{12}$$

Obviously, the problem is feasible if and only if $\mathcal{A}(C)$ is not empty. Since the family $\mathcal{A}(C)$ is closed under the refinement infimum, it admits a smallest element π_0:

$$\pi_0 = \wedge\{\pi, \pi \in \mathcal{A}(C)\} \tag{13}$$

The classes of π_0 can be interpreted as the set of leaves of a new hierarchy H', identical to H above and on π_0, but where all classes below π_0 are removed (see Figure 3). The cuts π of H' are exactly those of H that satisfy the constraint $\omega_\partial(\pi) \leq C$. The problem now reduces to find the minimal cut of H' w.r.t. ω_φ, a question that we already know how to treat. As the minimization is understood in the ω_φ-energetic lattice $\Pi(\omega_\varphi, H')$ relative to H', we have to suppose ω_φ singular and h-increasing, and we can state:

Proposition 1. *When ω_φ is a singular and h-increasing energy, then the minimal cut π_φ^* in the ω_φ-energetic lattice $\Pi(\omega_\varphi, H')$ is also a cut of smallest ω_φ energy in $\Pi(\omega_\varphi, H)$ whose all classes S^* satisfy the cost constraint $\omega_\partial(S^*) \leq C$.*

The result is important. It grants the existence and the uniqueness of the minimal cut π_φ^* under very large conditions: no prerequisite is needed for ω_∂, and uniquely singularity and h-increasingness for ω_φ. Note that the cost C need not be constant. Equivalence (12) holds on each class separately. C may vary through the space, or according to the level i in the hierarchy. When the energy ω_φ is also increasing w.r.t. the refinement of the cuts (e.g. the usual version of Mumford-Shah objective energy), i.e. when:

$$\pi_1 \leq \pi_2 \quad \Rightarrow \quad \omega_\varphi(\pi_1) \leq \omega_\varphi(\pi_2), \tag{14}$$

then the minimal cut π_φ^* coincides with π_0, since $\pi_0 \leq \pi \Rightarrow \omega_\varphi(\pi_0) = \wedge\{\omega_\varphi(\pi), \pi \in \mathcal{A}(C)\} = \omega_\varphi(\pi_\varphi^*)$. It remains to build up the hierarchy H' i.e. to find the leaves π_0. Suppose now that ω_∂ is inf-modular. Let

$$\omega_\partial(S) \leq \wedge\{\omega_\partial(T), T \text{ son of } S\}, \tag{15}$$

i.e. the energy ω_∂ of class S is smaller or equal to the smallest energy of the sons of S. Such class inf-modularity acts on classes and no longer on p.p. as in Rel.(6), but both are equivalent. The partition π_0 is obtained at the end of the scan, i.e. in one pass. A toy example is given in Figure 3. W.r.t. the ω_∂-energetic lattice $\Pi_{\omega_\partial, H'}$, the cut π_0 turns out to be a maximum.

8   Conclusion

We began this theoretical paper by using the Everett's theorem to show that λ-cuts in case of Guigues [6], and optimal prunings of Salembier [11] provide only an upper-bound on minimal objective energy. This was explicated further by the dependence of the constraint function values on the Lagrangian multiplier, and also the possibility of the non-existence of multipliers for certain constraint values and vice versa. The constraint function values were shown to be lattice structured and not varying continuously. This motivated the use of an energetic lattice framework. This gave us two ways to enforce of a constraint:

Numerical Minimization: Lagrangian Minimization by Energy (LME) model enforces a numerical constraint on energy, without referring to the partition structure. The Energetic lattice was used to generalize the Lagrangian model in LME, when one works in the space of partitions from a braid.

Lattice structured Minimization: Lagrange Minimization by Cut-Constraints (LMCC) and Class Constrained Minimization (CCM) models enforce the constraint in the form a partition. This does not involve any numerical constraint function, but one that is driven or evaluated on the energetic lattice.

References

1. Bach, F.: Learning with submodular functions: A convex optimization perspective. Foundations and Trends in Machine Learning 6(2-3), 145–373 (2013)
2. Ballester, C., Caselles, V., Igual, L.: Level lines selection with variational models for segmentation and encoding. Journal of Mathematical Imaging and Vision 27(1), 5–27 (2007)
3. Boyd, S., Vandenberghe, L.: Convex Optimization. Cambridge University Press, New York (2004)
4. Breiman, L., Friedman, J.H., Olshen, R.A., Stone, C.J.: Classification and Regression Trees. Wadsworth (1984)
5. Everett, H.: Generalized Lagrange multiplier method for solving problems of optimum allocation of resources. Operations Research 11(3), 399–417 (1963)
6. Guigues, L., Cocquerez, J.P., Le Men, H.: Scale-sets image analysis. International Journal of Computer Vision 68(3), 289–317 (2006)
7. Kiran, B.R., Serra, J.: Global-local optimizations by hierarchical cuts and climbing energies. Pattern Recognition 47(1), 12 (2014)
8. Kiran, B.R.: Energetic Lattice based optimization. Ph.D. thesis, Université Paris-Est, LIGM-A3SI, ESIEE (2014)
9. Kiran, B.R., Serra, J.: Braids of partitions. In: Benediktsson, J.A., Chanussot, J., Najman, L., Talbot, H. (eds.) ISMM 2015. LNCS, vol. 9082, pp. 217–228. Springer, Heidelberg (2015)
10. Ronse, C.: Partial partitions, partial connections and connective segmentation. J. Math. Imaging Vis. 32(2), 97–125 (2008)
11. Salembier, P., Garrido, L.: Binary partition tree as an efficient representation for image processing, segmentation, and information retrieval. IEEE Trans. on Image Processing 9(4), 561–576 (2000)
12. Shoham, Y., Gersho, A.: Efficient bit allocation for an arbitrary set of quantizers. IEEE Transactions on Acoustics, Speech & Signal Processing 36(9), 1445–1453 (1988)

Segmentation of Multimodal Images
Based on Hierarchies of Partitions

Guillaume Tochon[1(✉)], Mauro Dalla Mura[1], and Jocelyn Chanussot[1,2]

[1] GIPSA-lab, Grenoble Institute of Technology, Saint Martin d'Hères, Grenoble,
France
{guillaume.tochon,mauro.dalla-mura,
jocelyn.chanusshot}@gipsa-lab.grenoble-inp.fr
[2] Department of Electrical and Computer Engeneering, University of Iceland,
Reykjavik, Iceland

Abstract. Hierarchies of partitions are widely used in the context of
image segmentation, but when it comes to multimodal images, the fusion
of multiple hierarchies remains a challenge. Recently, braids of partitions
have been proposed as a possible solution to this issue, but have never
been implemented in a practical case. In this paper, we propose a new
methodology to achieve multimodal segmentation based on this notion of
braids of partitions. We apply this new method in a practical example,
namely the segmentation of hyperspectral and LiDAR data. Obtained
results confirm the potential of the proposed method.

Keywords: Image segmentation · Multimodal image · Hierarchy of
partitions · Energy minimization

1 Introduction

Multimodality is nowadays increasingly used in signal and image processing.
In fact, multimodal data (i.e., data of a physical phenomenon collected from
different sensors/locations, each of these showing a particular aspect of this phenomenon) allow to take advantage of both the correlation and complementarity
between each mode (i.e., data collected by one particular sensor) to better understand the underlying physical phenomenon of the source. However, there is a
good number of challenges that still must be faced in order to fully exploit the nature of multimodal data [8]. One talks in particular of multimodal images when
several images of the same scene have been acquired by different sensors. This
multimodality phenomenon occurs in several fields of image processing, such as
medical imaging [1] or remote sensing [3,4]. However, the design of adapted tools
to process multimodal images remains a challenge, notably due to the diverse
physical meanings and contents of images produced by all possible imaging sensors. Image segmentation is a particular process that would surely benefit from
the development of such multimodal tools, since it aims at partitioning an image into regions that "make sense" with respect to some underlying goal. The
segmentation of a multimodal image should benefit from the complementarity

© Springer International Publishing Switzerland 2015
J.A. Benediktsson et al. (Eds.): ISMM 2015, LNCS 9082, pp. 241–252, 2015.
DOI: 10.1007/978-3-319-18720-4_21

of its modes to ensure a more accurate delineation of its regions, in particular when those regions share similar features in one mode but not in the other ones.

Image segmentation constitutes an ill-posed problem since a given image can often be properly segmented at various levels of detail, and the precise level to choose depends on the underlying application (an optimal level might not exist). A potential solution to this intrinsic multiscale nature issue is to use a hierarchy of segmentations, which organizes in its structure all the potential scales of interest in a nested way. The hierarchy can be built once for a given image regardless of the application, and its level of exploration can then be tuned afterwards to produce the desired segmentation [13]. In [5] for example, this tuning relies on some energy minimization process over all the possible segmentations that can be extracted from the hierarchy. The *optimal* scale thus depends on the definition of the energy. However, handling the case of a multimodal image (and thus of several hierarchies) still remains an open question. Recently, the concept of braids of partitions has been introduced [6] as a potential tool to tackle this issue. We define in this paper a strategy of energy minimization for segmenting hierarchies of segmentations issued from different modalities, based on this concept of braids of partitions.

In Section 2, we summarize the works of [5, 7] and [6] about energy minimization over hierarchies and braids of partitions, respectively. In Section 3, we introduce a new methodology to achieve multimodal segmentation, based on energy minimization over braids of partitions. Section 4 features the application of the proposed methodology in a practical case, namely the joint segmentation of hyperspectral and LiDAR data, and presents some results. Conclusion and future work are drawn in Section 5.

2   Segmentation by Energy Minimization

We first define the notations used throughout the paper, before quickly recalling the notions of energy minimization over hierarchies and braids. The words segmentation and partition are used interchangeably in the following.

2.1   Definitions and Notations

Let $\mathcal{I} : E \rightarrow V$, $E \subseteq \mathbb{Z}^2$, $V \subseteq \mathbb{R}^n$, be a generic image, of elements (pixels) $\mathbf{x}_i \in E$. A partition of E, denoted π, is a collection of regions $\{\mathcal{R}_i \subseteq E\}$ (also called classes) of E such that $\mathcal{R}_i \cap \mathcal{R}_{j \neq i} = \emptyset$ and $\bigcup_i \mathcal{R}_i = E$. The set of all possible partitions of E is denoted Π_E. For any two partitions $\pi_i, \pi_j \in \Pi_E$, $\pi_i \leq \pi_j$ when for each region $\mathcal{R}_i \in \pi_i$, there exists a region $\mathcal{R}_j \in \pi_j$ such that $\mathcal{R}_i \subseteq \mathcal{R}_j$. π_i is said to refine π_j in such case. Π_E is a complete lattice for the refinement ordering $\leq$. Minimizing some energy function over Π_E requires first the definition of a regional energy, i.e., a function $\mathcal{E}$ that maps any region $\mathcal{R} \subseteq E$ to $\mathbb{R}^+$, and the definition of some operator $\mathfrak{D}$ (such as $\sum$, $\prod$ or $\bigvee$ for instance) to express the energy of a partition as a composition of the energies of its regions:

$$\mathcal{E}(\pi = \{\mathcal{R}_i\}) = \underset{\mathcal{R}_i \in \pi}{\mathfrak{D}} \, \mathcal{E}(\mathcal{R}_i). \tag{1}$$

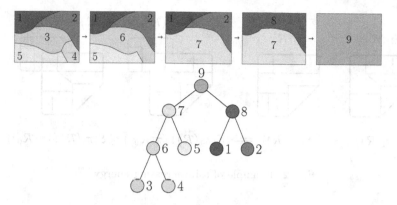

Fig. 1. Example of hierarchy of partitions (binary partition tree [14])

Well-known methods to perform image segmentation by energy minimization non-exhaustively include the Mumford-Shah functional [10], graph cuts [2] or Markov random fields [9]. However, finding the optimal partition that minimizes a given energy remains a difficult task, mainly due to the huge cardinality of Π_E (a 5×5 image can be partitioned in more than 4.6×10^{18} different ways). Hierachies, by restraining the space of possible partitions, are an appealing tool to minimize the energy on.

2.2   Minimization over a Hierarchy

A hierarchy of segmentations of E is a collection $H = \{\mathcal{R} \subseteq E\}$ such that $\emptyset \notin H$, $E \in H$ and $\forall \mathcal{R}_i, \mathcal{R}_j \in H, \mathcal{R}_i \cap \mathcal{R}_j \in \{\emptyset, \mathcal{R}_i, \mathcal{R}_j\}$. In other words, any two regions belonging to a hierarchy are either disjoint or nested. The most common way to obtain a hierarchical decomposition of an image is to start from an initial partition π_0 and to iteratively merge its regions until the whole image support is reached [12,14], resulting in a sequence of partitions $\pi_0 \leq \pi_1 \leq \cdots \leq \pi_n = \{E\}$, as displayed in Figure 1. Regions of π_0 are called *leaves*, $\pi_n = \{E\}$ is called the *root* of the hierarchy, and each non leaf node $\mathcal{R}$ contains a set of $S(\mathcal{R})$ children nodes. A *cut* of H is a partition π of E whose regions belong to H. The set of all cuts of a hierarchy H built over the image $\mathcal{I}$ is denoted $\Pi_E(H)$, and is a sub-lattice of Π_E. $H(\mathcal{R})$ denotes the sub-hierarchy of H rooted at $\mathcal{R}$. Any cut of the sub-hierarchy $H(\mathcal{R})$ is called a *partial partition* of $\mathcal{R}$ following [11], and is denoted $\pi(\mathcal{R})$. The cut of H that is minimal (i.e., optimal) with respect to the energy $\mathcal{E}$ is defined as:

$$\pi^\star = \underset{\pi \in \Pi_E(H)}{\mathrm{argmin}}\ \mathcal{E}(\pi) \qquad (2)$$

Assumptions on $\mathcal{E}$ under which it is easy to retrieve the minimal cut $\pi^\star$ have been studied in [5] in the context of separable energies (i.e., $\mathcal{E}(\pi) = \sum_{\mathcal{R} \in \pi} \mathcal{E}(\mathcal{R})$) and later generalized in [7] to wider classes of composition laws $\mathfrak{D}$, namely *h-increasing energies*. An energy $\mathcal{E}$ is said to be h-increasing when given any two

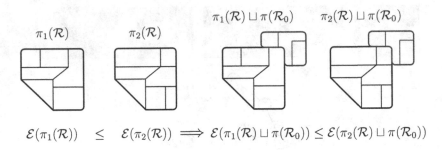

$$\mathcal{E}(\pi_1(\mathcal{R})) \quad \leq \quad \mathcal{E}(\pi_2(\mathcal{R})) \implies \mathcal{E}(\pi_1(\mathcal{R}) \sqcup \pi(\mathcal{R}_0)) \leq \mathcal{E}(\pi_2(\mathcal{R}) \sqcup \pi(\mathcal{R}_0))$$

Fig. 2. Example of a h-increasing energy

$\mathcal{R}, \mathcal{R}_0 \in H$ disjoint, given partial partitions $\pi_1(\mathcal{R})$, $\pi_2(\mathcal{R})$ and $\pi(\mathcal{R}_0)$, then $\mathcal{E}(\pi_1(\mathcal{R})) \leq \mathcal{E}(\pi_2(\mathcal{R})) \Rightarrow \mathcal{E}(\pi_1(\mathcal{R})\sqcup\pi(\mathcal{R}_0)) \leq \mathcal{E}(\pi_2(\mathcal{R})\sqcup\pi(\mathcal{R}_0))$, with $\sqcup$ denoting disjoint union (concatenation). An example of h-increasing energy is depicted in Figure 2. In that case, it is possible to find the minimal cut of H by solving for each node $\mathcal{R}$ the following dynamic program:

$$\mathcal{E}^\star(\mathcal{R}) = \min\left\{\mathcal{E}(\mathcal{R}), \underset{r \in S(\mathcal{R})}{\mathcal{D}} \mathcal{E}(\pi^\star(r))\right\} \tag{3}$$

$$\pi^\star(\mathcal{R}) = \begin{cases} \{\mathcal{R}\} & \text{if } \mathcal{E}(\mathcal{R}) \leq \underset{r \in S(\mathcal{R})}{\mathcal{D}} \mathcal{E}(\pi^\star(r)) \\ \underset{r \in S(\mathcal{R})}{\bigsqcup} \pi^\star(r) & \text{otherwise} \end{cases} \tag{4}$$

The optimal cut of $\mathcal{R}$ is given by comparing the energy of $\mathcal{R}$ and the energy of the disjoint union of the optimal cuts of its children, and by picking the smallest of the two. The optimal cut of the whole hierarchy is the one the root node, and is reached by scanning all nodes in the hierarchy in one ascending pass [5].

Energies in the literature often depend in practice on a positive real-valued parameter λ that acts as a trade-off between simplicity (i.e., favoring under-segmentation) and a good data fitting of the segmentation (i.e., leading to over-segmentation). These energies $\mathcal{E}_\lambda$ generate sequences of optimal cuts $\{\pi_\lambda^\star\}$ in turn indexed by this parameter λ. The behavior of $\pi_\lambda^\star$ with respect to λ has been studied in [7], which introduced in particular the property of *scale-increasingness*: $\mathcal{E}_\lambda$ is scale-increasing if for any $\mathcal{R} \in H$, any of its partial partition $\pi(\mathcal{R})$, and any $0 \leq \lambda_1 \leq \lambda_2$, $\mathcal{E}_{\lambda_1}(\mathcal{R}) \leq \mathcal{E}_{\lambda_1}(\pi(\mathcal{R})) \Rightarrow \mathcal{E}_{\lambda_2}(\mathcal{R}) \leq \mathcal{E}_{\lambda_2}(\pi(\mathcal{R}))$.

In the case where the energy is h-increasing for any λ and scale-increasing with respect to λ, the family $\{\pi_\lambda^\star\}$ of optimal cuts is hierarchically organized, that is

$$\lambda_1 \leq \lambda_2 \Rightarrow \pi_{\lambda_1}^\star \leq \pi_{\lambda_2}^\star. \tag{5}$$

In such case, it is possible to transform some hierarchy H into an optimal version $H^\star$, composed of all the optimal cuts $\pi_\lambda^\star$ of H when λ spans $\mathbb{R}^+$. In practice, the energy $\mathcal{E}_\lambda$ is seen as a function of λ, and (3) is conducted over

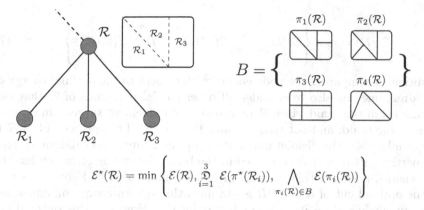

$$\mathcal{E}^*(\mathcal{R}) = \min \left\{ \mathcal{E}(\mathcal{R}), \overset{3}{\underset{i=1}{\square}} \mathcal{E}(\pi^*(\mathcal{R}_i)), \bigwedge_{\pi_i(\mathcal{R}) \in B} \mathcal{E}(\pi_i(\mathcal{R})) \right\}$$

Fig. 3. Illustration of a step of the dynamic program (7) applied to a braid structure: one has to choose between $\{\mathcal{R}\}$, $\bigsqcup \pi^*(\mathcal{R}_i)$ or any other $\pi_i(\mathcal{R}) \in B$. Note however that $\mathcal{R} \neq E$, otherwise B would not be a braid since $\pi_3(\mathcal{R}) \vee \pi_4(\mathcal{R}) = \mathcal{R}$.

the space of such functions. The output of the dynamic program is no longer some optimal cut for a given value of λ, but some partition of $\mathbb{R}^+$ into intervals $[0, \lambda_1[\cup [\lambda_1, \lambda_2[\cup \cdots \cup [\lambda_p, +\infty[$ where all λ values within a given interval $[\lambda_i, \lambda_{i+1}[$ are leading to the same optimal cut $\pi_{\lambda_i}^*$. The reader is referred to [5] for more practical implementation details.

2.3 Minimization over a Braid

Braids of partitions have been recently introduced in [6] as a potential tool to combine multiple hierarchies and thus tackle segmentation of multimodal images, but these have not been yet investigated in practice in multimodal data fusion. Braids of partitions are defined as follows: a family of partitions $B = \{\pi_i\}$ is called a *braid* whenever there exists some hierarchy H_m, called *monitor hierarchy*, such that:

$$\forall \pi_i, \pi_j \in B, \pi_i \vee \pi_{j \neq i} \in \Pi_E(H_m) \backslash \{E\} \tag{6}$$

where $\pi_i \vee \pi_j$ denotes the refinement supremum, i.e. the smallest partition that is refined by both π_i and π_j. In other words, a braid is a family of partitions such that the refinement suprema of any pair of different partitions of the family are hierarchically organized, even though the partitions composing the braid might not be. For this reason, braids of partitions are more general than hierarchies of partitions: while hierarchies are braids, the converse is not necessarily true. It is also worth noting that the refinement supremum of any two partitions must differ from the whole image $\{E\}$ in (6). Otherwise, any family of arbitrary partitions would form a braid with $\{E\}$ as a supremum, thus loosing any interesting structure. The optimal cut of a braid of partitions is reached by solving the dynamic program (3) for every node $\mathcal{R}$ of the monitor hierarchy H_m, with a slight modification:

$$\mathcal{E}^\star(\mathcal{R}) = \min\left\{ \mathcal{E}(\mathcal{R}), \underset{r \in S(\mathcal{R})}{\mathfrak{D}}\, \mathcal{E}(\pi^\star(r)), \bigwedge_{\pi_i(\mathcal{R}) \in B} \mathcal{E}(\pi_i(\mathcal{R})) \right\} \qquad (7)$$

In addition to comparing the node energy with respect to the optimal energy of its children, one has also to consider all other partial partitions of $\mathcal{R}$ that can be contained in the braid, since $\mathcal{R}$ represents the refinement supremum of some regions in the braid, and not those regions themselves. The optimal cut of $\mathcal{R}$ is then given by $\{\mathcal{R}\}$, the disjoint union of the optimal cuts of its children or some other partial partition of $\mathcal{R}$ contained in the braid, depending on which has the lowest energy. A step of this dynamic program is illustrated by Figure 3. Notice that the optimal cut of a braid B is obtained through an energy minimization procedure conducted on its monitor hierarchy H_m. However, this optimal cut may be composed of regions that are solely contained in the braid and therefore not supported by nodes of the monitor hierarchy (it would be the case in the example depicted by Figure 3 if $\pi_4(\mathcal{R})$ were for instance chosen to be the optimal cut of $\mathcal{R}$).

3   Proposed Methodology

3.1   Generation of a Braid from Multiple Hierarchies

The refinement supremum of two cuts of a hierarchy remains a cut of this hierarchy. For this reason, it is straightforward to compose a braid with cuts coming from the same hierarchy since any family of such cuts is a braid. It also implies in that case that the regions composing the corresponding monitor hierarchy are a subset of the regions composing the initial hierarchy. However, this guarantee is lost when one wants to compose a braid from cuts coming from multiple hierarchies: all those cuts must be sufficiently related to ensure that all their pairwise refinement suprema are hierarchically organized. As an example, let $B = \{H_1 = \{\pi_1^1 \geq \pi_1^2\}, H_2 = \{\pi_2^1 \geq \pi_2^2\}\}$ be some family of partitions composed of two supposedly independent hierarchies H_1 and H_2, both composed of two ordered cuts. B being a braid implies that all pairwise refinements suprema are hierarchically organized. In particular, this must be true for $\pi_1^1 \vee \pi_1^2 = \pi_1^1$ and $\pi_2^1 \vee \pi_2^2 = \pi_2^1$, which were initally assumed to come from independent hierarchies. Thus, the partitions composing B cannot be chosen arbitrarily. This leads us to introduce the property of *h-equivalence* (h standing here for *hierarchical*): two partitions π_a and π_b are said to be h-equivalent, and one notes $\pi_a \overset{h}{\simeq} \pi_b$ if and only if $\forall \mathcal{R}_a \in \pi_a, \forall \mathcal{R}_b \in \pi_b, \mathcal{R}_a \cap \mathcal{R}_b \in \{\emptyset, \mathcal{R}_a, \mathcal{R}_b\}$. In other words, π_a and π_b may not be globally comparable, but they are locally comparable (for instance, $\pi_1(\mathcal{R})$ and $\pi_2(\mathcal{R})$ of Figure 3 are not globally comparable, but they locally are). In particular, given a hierarchy H, $\forall \pi_1, \pi_2 \in \Pi_E(H), \pi_1 \overset{h}{\simeq} \pi_2$: all cuts of a hierarchy are h-equivalent. $\overset{h}{\simeq}$ is a tolerance relation: it is reflexive and symmetric, but not transitive. Given some hierarchy H and a partition $\pi_* \in \Pi_E$, we denote by $H \overset{h}{\simeq} \pi_*$

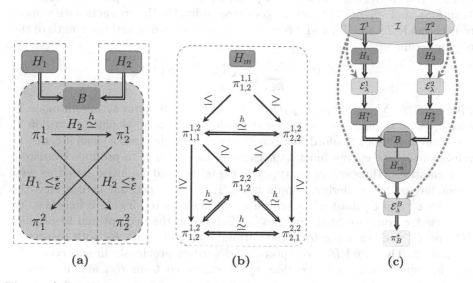

Fig. 4. a) Possible composition of a braid B with cuts from two hierarchies H_1 and H_2, b) cuts of the corresponding monitor hierarchy H_m, and c) workflow of proposed multimodal segmentation methodology.

the set of cuts of H that are h-equivalent to π_*. Obviously, $H \overset{h}{\simeq} \pi_* \subseteq \Pi_E(H)$ with equality if and only if $\pi_* \in \Pi_E(H)$. Provided some hierarchy H, some h-increasing and scale-increasing energy $\mathcal{E}_\lambda$ and some partition $\pi_* \in \Pi_E$, we also define $H \leq^\star_\mathcal{E} \pi_* = \{\pi \in \Pi_E(H^\star)$ s.t $\pi \leq \pi_*\}$ as the set of optimal cuts of H with respect to $\mathcal{E}_\lambda$ that are refinements of π_*. Following, it is possible to compose a braid B with cuts extracted from two hierarchies H_1 and H_2 using these two relations, as depicted in figure 4a: Given $\pi_1^1 \in \Pi_E(H_1)$, take some $\pi_2^1 \in H_2 \overset{h}{\simeq} \pi_1^1$. Then, π_1^2 and π_2^2 are taken in $H_1 \leq^\star_\mathcal{E} \pi_2^1$ and $H_2 \leq^\star_\mathcal{E} \pi_1^1$, respectively. In practice, we choose $\pi_2^1 = \bigvee\{H_2 \overset{h}{\simeq} \pi_1^1 \backslash \{E\}\}$, $\pi_1^2 = \bigvee\{H_1 \leq^\star_\mathcal{E} \pi_2^1\}$ and $\pi_2^2 = \bigvee\{H_2 \leq^\star_\mathcal{E} \pi_1^1\}$. Under this configuration, it is guaranteed that $B = \{\pi_1^1, \pi_1^2, \pi_2^1, \pi_2^2\}$ forms a braid with monitor hierarchy H_m whose cuts $\pi_{i,j}^{k,l} = \pi_i^k \vee \pi_j^l$ are organized as displayed by figure 4b. Other configurations for the composition of B may work as well.

3.2   Methodology

We now propose a methodology to perform multimodal image segmentation, using the previously introduced concept of braids of partitions to fuse the output of several hierarchies. The proposed method is illustrated by the workflow in figure 4c, detailed step by step in the following. Let $\mathcal{I} = \{\mathcal{I}^1, \mathcal{I}^2\}$ be a multimodal image, assumed to be composed of two modes $\mathcal{I}^1$ and $\mathcal{I}^2$ having the same spatial support E, for a matter of clarity (the extension to a greater number of modes follows the same scheme). First, two hierarchies H_1 and H_2 are built on $\mathcal{I}^1$

and $\mathcal{I}^2$, respectively. Two energies $\mathcal{E}_\lambda^1$ and $\mathcal{E}_\lambda^2$ are defined as piecewise constant Mumford-Shah energies [10] whose goodness-of-fit (GOF) term acts with respect to each mode $\mathcal{I}^1$ and $\mathcal{I}^2$, and whose regularization term is half the length of the region perimeter:

$$\mathcal{E}_\lambda^i(\pi) = \sum_{R \in \pi} \left(\Xi_i(\mathcal{R}) + \frac{\lambda}{2}|\partial \mathcal{R}| \right) \tag{8}$$

with $\Xi_i(\mathcal{R}) = \sum_{\mathbf{x} \in \mathcal{R}} \|\mathcal{I}^i(\mathbf{x}) - \boldsymbol{\mu}_i(\mathcal{R})\|_2^2$ being the GOF term acting on mode $\mathcal{I}^i$ and $\boldsymbol{\mu}_i(\mathcal{R})$ is the mean value/vector in mode $\mathcal{I}^i$ of pixels belonging to region $\mathcal{R}$. Piecewise constant Mumford-Shah energies are a popular choice when it comes to minimizing some energy function because of their ability to produce consistent segmentations. However, other types of energies could be investigated as well, depending on the underlying application. The only constraint here is that the energies $\mathcal{E}_\lambda^1$ and $\mathcal{E}_\lambda^2$ must be h-increasing and scale-increasing. It is known to be the case for Mumford-Shah energies [7]. Following, the two optimal hierarchies H_1^* and H_2^* are generated from the optimal cuts of H_1 and H_2 with respect to $\mathcal{E}_\lambda^1$ and $\mathcal{E}_\lambda^2$. The braid B is composed as described previously in subsection 3.1 and by figure 4a: a first partition $\pi_1^{1\star}$ is extracted from H_1^*, and is used to extract two partitions $\pi_2^{1\star}$ and $\pi_2^{2\star}$ from H_2^* following the relations $\overset{h}{\simeq}$ and $\leq_\mathcal{E}^\star$, respectively. A second partition $\pi_1^{2\star}$ is finally extracted from H_1^* using $\leq_\mathcal{E}^\star$ and $\pi_2^{1\star}$. Eventually, B is composed of 4 partitions $\{\pi_1^{1\star}, \pi_1^{2\star}, \pi_2^{1\star}, \pi_2^{2\star}\}$ extracted from the two hierarchies H_1^* and H_2^*, and the braid structure is guaranteed, allowing to construct the monitor hierarchy H_m. A last energy term $\mathcal{E}_\lambda^B$ is defined, relying on both modes of the multimodal image $\mathcal{I}$:

$$\mathcal{E}_\lambda^B(\pi) = \sum_{\mathcal{R} \in \pi} \left(\max \left(\frac{\Xi_1(\mathcal{R})}{\Xi_1(\mathcal{I}^1)}, \frac{\Xi_2(\mathcal{R})}{\Xi_2(\mathcal{I}^2)} \right) + \frac{\lambda}{2}|\partial \mathcal{R}| \right) \tag{9}$$

The GOF term of each region $\mathcal{R}$ is now defined as the maximum with respect to both modes of the normalized GOFs. The normalization allows both GOF terms to be in the same dynamical range. $\mathcal{E}_\lambda^B$ is also a h-increasing and scale-increasing energy. Its minimization over H_m and B following the dynamic program (7) gives some optimal segmentation π_B^* of $\mathcal{I}$, which should contain salient regions shared by both modes as well as regions exclusively expressed by $\mathcal{I}^1$ and $\mathcal{I}^2$.

4   Results

4.1   Conducted Experiments

We apply the proposed methodology on the multimodal data set described in [4] composed of a hypespectral (HS) image $\mathcal{I}^1$ of 144 spectral bands evenly spaced between 380 nm and 1050 nm, and a LiDAR-derived digital surface model (DSM) $\mathcal{I}^2$, with the same ground-sampling distance of 2.5 m. Data were acquired over the University of Houston campus. The study site features an urban area with several buildings of various heights and made of different materials, some parking

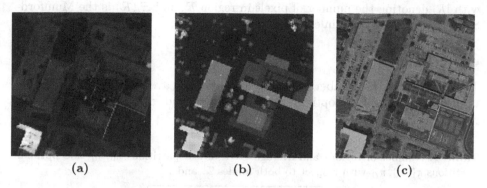

(a)   (b)   (c)

Fig. 5. a) RGB composition of the hyperspectral image, b) corresponding LiDAR-derived DSM, c) very-high resolution RGB image of the same site

lots, tennis courts, roads and some portions of grass and trees. A RGB composition of the hyperspectral image is displayed in figure 5a, and the corresponding LiDAR-derived DSM is shown in figure 5b. It is also shown for visualization purpose a very-high resolution RGB image of the scene in figure 5c[1].

Hierarchies H_1 and H_2 are obtained by building two binary partition trees [14] on $\mathcal{I}^1$ and $\mathcal{I}^2$ with standard parameters (mean spectrum and spectral angle for the region model and merging criterion of the HS mode, mean value and Euclidean distance for the DSM). Both hierarchies are built on the same inital partition π_0, obtained as the refinement infimum of two mean shift clustering procedures conducted on the RGB composition of the HS and on the DSM. The braid B is constructed following the procedure exposed in figure 4a: $\pi_1^{1\star}$ is the first cut extracted from $H_1^\star$ and contains around 125 regions. It is used to extract $\pi_2^{1\star}$ and $\pi_2^{2\star}$ from $H_2^\star$, which comprise 342 and 349 regions, respectively. Finally, $\pi_1^{2\star}$ is extracted from $H_1^\star$ using $\pi_2^{1\star}$ and contains 379 regions. The four partitions composing B generate $\binom{4}{2} = 6$ cuts of the monitor hierarchy H_m, which is built by re-organizing those cuts in a hierarchical manner. The leaf partition of H_m, denoted π_0^B, is obtained as $\bigwedge\{\pi_i \vee \pi_{j\neq i}, \pi_i, \pi_j \in B\}$. Finally, the minimization of $\mathcal{E}_\lambda^B$ over H_m, following (7), is conducted with λ being empirically set to 5.10^{-5}, and produces an optimal segmentation $\pi_B^\star$ of the braid composed of 302 regions. To evaluate the improvements brought by the braid structure, we propse to extract from $H_1^\star$ and $H_2^\star$ the two optimal cuts $\pi_1^\star$ and $\pi_2^\star$ that have the same (or a close) number of regions as $\pi_B^\star$ (in practice, $\pi_1^\star$ and $\pi_2^\star$ have 301 and 302 regions, respectively). This should allow a fair visual comparison since all three partitions should feature regions of similar scales. In addition, we compute for the partitions $\pi_0, \pi_0^B, \pi_1^\star, \pi_2^\star$ and $\pi_B^\star$ their average GOF with respect to both modes $\mathcal{I}^1$ and $\mathcal{I}^2$ as follows:

$$\epsilon(\pi|\mathcal{I}^i) = \frac{1}{|E|} \sum_{\mathcal{R}\in\pi} |\mathcal{R}| \times \Xi_i(\mathcal{R}) \tag{10}$$

---

[1] https://goo.gl/maps/VVXE6

with $|\mathcal{R}|$ denoting the number of pixels in region $\mathcal{R}$, and $\Xi_i(\mathcal{R})$ is the Mumford-Shah GOF term defined in equation (8).

4.2 Results

Table 1 presents the number of regions as well as the average GOF of leaf partitions π_0 and π_0^B, and of optimal partitions $\pi_1^\star, \pi_2^\star$ and $\pi_B^\star$ with respect to both

Table 1. Number of regions and average GOF of leaf partitions π_0, π_0^B and optimal partitions $\pi_1^\star, \pi_2^\star, \pi_B^\star$ with respect to both modes $\mathcal{I}^1$ and $\mathcal{I}^2$

	π_0	π_0^B	$\pi_1^\star$	$\pi_2^\star$	$\pi_B^\star$
$|\pi|$	416	354	301	302	302
$\epsilon(\pi|\mathcal{I}^1)$	13.2	16.6	16.0	57.2	19.8
$\epsilon(\pi|\mathcal{I}^2)$	262.2	297.7	611.7	413.0	358.8

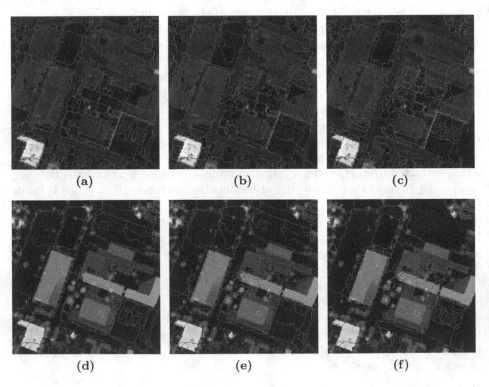

Fig. 6. Top row: optimal partitions a) $\pi_B^\star$, b) $\pi_1^\star$ (optimal with respect to the HS image) and c) $\pi_2^\star$ (optimal with respect to the DSM) superimposed over the HS image. Bottom row: optimal partitions d) $\pi_B^\star$, e) $\pi_1^\star$ and f) $\pi_2^\star$ superimposed over the DSM image.

modes $\mathcal{I}^1$ and $\mathcal{I}^2$. Its analysis demonstrates the interest of the proposed methodology using the braid structure. One can indeed remark, not surprinsigly, that $\pi_1^\star$ and $\pi_2^\star$ score a low average GOF value with respect to their corresponding mode, but a greater average GOF with respect to the complementary mode. On the other hand, $\pi_B^\star$ outperforms $\pi_1^\star$ with respect to $\mathcal{I}^2$ while scoring a similar value for $\mathcal{I}^1$, and outperforms $\pi_2^\star$ both with respect to $\mathcal{I}^1$ and $\mathcal{I}^2$. Thus, $\pi_B^\star$ better fits both modes of the multimodal image at the same time. In addition, it contains fewer regions than π_0 and π_0^B while not increasing the average GOF too much. Therefore, $\pi_B^\star$ decreases over-segmentation compared to the two leaf partitions while maintaining comparable GOF values. Figure 6 shows the optimal partitions $\pi_B^\star$, $\pi_1^\star$ and $\pi_2^\star$ superimposed over the RGB composition of the HS image (top row, from figure 6a to 6c) and over the DSM image (bottom row, from figure 6d to 6f). The qualitative analysis of figure 6 leads to similar conclusions : while $\pi_1^\star$ tends to under-segment regions featuring the same spectral properties but not the same elevation (typically, the buildings in the center of the scene), those regions are correctly segmented in $\pi_B^\star$. Similary, regions at the same elevation are often under-segmented in $\pi_2^\star$ even if they are made of different materials (parking lots, roads and grass for instance) but correctly delineated in $\pi_B^\star$. This demonstrates how the construction of the braid and associated monitor hierarchy as well as the following energy minimization were able to fuse the information contained in both modes to produce a more accurate segmentation of the multimodal image.

5 Conclusion

In conclusion, we presented in this article a new method to perform multimodal segmentation, based on the hierarchical minimization of some energy function. In particular, we used the recently introduced concept of braids of partitions and associated monitor hierarchies and we adapted to them the dynamic program procedure conducted to perform energy minimization over hierarchies. The proposed framework was investigated over a multimodal image composed of a hyperspectral and a LiDAR mode. Results demonstrated, quantitatively and qualitatively, the ability of the proposed approach to produce a segmentation that not only retains salient regions shared by both modes, but also regions appearing in only one mode of the multimodal image.

Future work include a deeper investigation on the way to compose a braid with cuts coming from several hierarchies, and a more thorough assessment analysis of the improvements brought by the proposed methodology.

Acknowledgement. This work was partially funded through the ERC CHESS project, ERC-12-AdG-320684-CHESS.

References

1. Bießmann, F., Plis, S., Meinecke, F., Eichele, T., Muller, K.: Analysis of multimodal neuroimaging data. QIEEE Reviews in Biomedical Engineering 4, 26–58 (2011)

2. Boykov, Y., Veksler, O., Zabih, R.: Fast approximate energy minimization via graph cuts. IEEE Transactions on Pattern Analysis and Machine Intelligence 23(11), 1222–1239 (2001)

3. Dalla Mura, M., Prasad, S., Pacifici, F., Gamba, P., Chanussot, J.: Challenges and opportunities of multimodality and data fusion in remote sensing. In: 2013 Proceedings of the 22nd European Signal Processing Conference (EUSIPCO), pp. 106–110. IEEE (2014)

4. Debes, C., Merentitis, A., Heremans, R., Hahn, J., Frangiadakis, N., van Kasteren, T., Bellens, W.L.R., Pizurica, A., Gautama, S., Philips, W., Prasad, S., Du, Q., Pacifici, F.: Hyperspectral and lidar data fusion: Outcome of the 2013 grss data fusion contest. IEEE Journal of Selected Topics in Applied Earth Observations and Remote Sensing 7(6), 2405–2418 (2014)

5. Guigues, L., Cocquerez, J., Le Men, H.: Scale-sets image analysis. International Journal of Computer Vision 68(3), 289–317 (2006)

6. Kiran, B.: Energetic-Lattice based optimization. PhD thesis, Université Paris-Est, Paris (October 2014)

7. Kiran, B., Serra, J.: Global–local optimizations by hierarchical cuts and climbing energies. Pattern Recognition 47(1), 12–24 (2014)

8. Lahat, D., Adaly, T., Jutten, C.: Challenges in multimodal data fusion. In: 2013 Proceedings of the 22nd European Signal Processing Conference (EUSIPCO), pp. 101–105. IEEE (2014)

9. Li, S.: Markov random field modeling in computer vision. Springer-Verlag New York, Inc. (1995)

10. Mumford, D., Shah, J.: Optimal approximations by piecewise smooth functions and associated variational problems. Communications on Pure and Applied Mathematics 42(5), 577–685 (1989)

11. Ronse, C.: Partial partitions, partial connections and connective segmentation. Journal of Mathematical Imaging and Vision 32(2), 97–125 (2008)

12. Soille, P.: Constrained connectivity for hierarchical image partitioning and simplification. IEEE Transactions on Pattern Analysis and Machine Intelligence 30(7), 1132–1145 (2008)

13. Tarabalka, Y., Tilton, J., Benediktsson, J., Chanussot, J.: A marker-based approach for the automated selection of a single segmentation from a hierarchical set of image segmentations. IEEE Journal of Selected Topics in Applied Earth Observations and Remote Sensing 5(1), 262–272 (2012)

14. Valero, S., Salembier, P., Chanussot, J.: Hyperspectral image representation and processing with binary partition trees. IEEE Transactions on Image Processing 22(4), 1430–1443 (2013)

Multi-image Segmentation: A Collaborative Approach Based on Binary Partition Trees

Jimmy Francky Randrianasoa[1(✉)], Camille Kurtz[2], Éric Desjardin[1],
and Nicolas Passat[1]

[1] Université de Reims Champagne-Ardenne, CReSTIC, France
jimmy.randrianasoa@univ-reims.fr
[2] Université Paris-Descartes, LIPADE, France

Abstract. Image segmentation is generally performed in a "one image, one algorithm" paradigm. However, it is sometimes required to consider several images of a same scene, or to carry out several (or several occurrences of a same) algorithm(s) to fully capture relevant information. To solve the induced segmentation fusion issues, various strategies have been already investigated for allowing a consensus between several segmentation outputs. This article proposes a contribution to segmentation fusion, with a specific focus on the "n images" part of the paradigm. Its main originality is to act on the segmentation research space, *i.e.*, to work at an earlier stage than standard segmentation fusion approaches. To this end, an algorithmic framework is developed to build a binary partition tree in a collaborative fashion, from several images, thus allowing to obtain a unified hierarchical segmentation space. This framework is, in particular, designed to embed consensus policies inherited from the machine learning domain. Application examples proposed in remote sensing emphasise the potential usefulness of our approach for satellite image processing.

Keywords: Segmentation fusion · Morphological hierarchies · Multi-image · Collaborative strategies · Binary partition tree · Remote sensing

1 Introduction

In image processing / analysis, segmentation is a crucial task. The concept of segmentation is also quite generic, in terms of semantics (from low-level definition of homogeneous areas to high-level extraction of specific objects), in terms of definition (object versus background, or total partition of the image support), and in terms of algorithmics.

The principal invariant of segmentation is the "one image, one algorithm" paradigm. Indeed, for a given application, a specific algorithm is generally chosen (or designed) with respect to its adequacy with the considered image processing / analysis problem; parametrized with respect to the physical / semantic properties of the targeted images; and then applied (once) on a given image, or on each image of a given dataset.

Since segmentation is an ill-posed problem, it is plain that the result of a segmentation algorithm applied on an image cannot be completely satisfactory. Based on this

This research was partially funded by the French *Agence Nationale de la Recherche* (Grant Agreements ANR-10-BLAN-0205 and ANR-12-MONU-0001).

J.A. Benediktsson et al. (Eds.): ISMM 2015, LNCS 9082, pp. 253–264, 2015.
DOI: 10.1007/978-3-319-18720-4_22

assertion, it is sometimes relevant to relax the "one image, one algorithm" paradigm, either by operating one algorithm on several images of a same scene – in order to enrich / improve the information provided as input – or by applying several algorithms, or the same algorithm with several parameter sets – in order to enrich / improve the information provided as output.

In the literature, these "n images, one algorithm" and "one image, n algorithms" paradigms are dealt with under the common terminology of *segmentation fusion*. Most of the time, the strategy consists of computing n segmentation maps, and to develop a consensus strategy to gather and unify these n results into a single, assuming that an accurate global result will emerge from a plurality of less accurate ones.

This fusion strategy has also been proposed – and intensively studied – in the research field of machine learning, especially in clustering [1]. Intrinsically, the problems are the same, with the only – but important – difference that image segmentation, by contrast with general data clustering, implies the handling of the spatial organisation of the processed data.

In this article, we propose a new approach for segmentation fusion, that focuses on the "n images, one algorithm" paradigm. The relevance and novelty of this approach, that lies in the framework of morphological hierarchies and connected operators [2], derives from the early stage where the fusion occurs. By contrast with most segmentation fusion approaches, we do not intend to fuse several segmentation maps obtained from several images, but we directly develop a consensus when defining the research space, *i.e.*, during the construction of a hierarchy that models the n images.

This strategy, that acts at the research space level, induces two important side effects. First, it allows us to rely on hierarchical approaches for segmentation, and thus to propose a rich and versatile segmentation framework, that can be easily instantiated according to the application field. Second, by operating the fusion on the internal data-structures involved in the algorithmic construction of the hierarchy, instead of spatial regions of the segmentation maps, we can directly benefit from all the strategies devoted to "non-spatial" data fusion, previously proposed by the machine learning community.

Our approach relies on Binary Partition Trees (BPTs) [3] as hierarchical model. This choice is motivated both by the possibility to tune a BPT construction, by contrast with other tree structures more deterministically deriving from the images (namely, component-trees or trees of shapes). It is also justified by the frequent and successful application of BPTs in remote sensing, where the use of multiple images of a same scene is frequent, such as the use of fusion strategies for clustering issues.

This article is organized as follows. In Section 2, a brief overview of related works is proposed, describing previous segmentation fusion strategies, and applications of BPTs and hierarchies for analysing multiple remote sensing images. In Section 3, the general data-structure / algorithmic framework for building a BPT from several images is described, and various consensus strategies for instantiating this framework are discussed. In Section 4, experiments performed on remote sensing images, in order to illustrate the relevance and usefulness of BPT segmentation from several images, by considering two application cases, namely mono-date multi-imaging on urban areas, and multi-date imaging on agriculture areas. In Section 5, a discussion concludes this article by emphasising perspectives and potential further works.

2   Related Works

2.1   Segmentation Fusion

Segmentation fusion consists of establishing a consensus between several segmentation maps. The taxonomy of segmentation fusion is directly mapped on that of segmentation.

On the one hand, segmentation can be viewed as a process that aims to extract one structure (the object) versus the remainder of the image (the background). Typical examples of such segmentation strategies are deformable models, graph-cuts, *etc.* In this context, segmentation fusion can be interpreted as a geometrical problem of shape or contour averaging / interpolation (see, *e.g.*, [4,5]). This subfamily of segmentation partition fusion methods, mainly used in medical imaging, is out of the scope of our study.

On the other hand, segmentation can be viewed as a process that aims to define a partition of the whole image, in order to extract meaningful homogeneous areas. Examples of such segmentation strategies are watersheds, split-and-merge and, more generally, connected operators. Clustering methods also enter in this category, with the difference that they generally do not take into account the spatial organisation of the image data.

Following a machine learning vision, some methods handled segmentation fusion via clustering ensemble [1]; a comparative study can be found in [6]. It was also proposed to interpret the information gathered at each pixel in the various segmentation maps as feature vectors then involved in optimization procedures. In particular, a two-stage K-MEANS approach was considered in [7], or probabilistic frameworks in [8,9].

Other approaches explicitly took into account the spatial organisation of the segmentation maps. In pioneering works [10], images of a same scene obtained from various modalities were merged and optimized based on edge information. Later, statistical analysis of the co-occurrence probability of neighbour pixels was considered to improve the accuracy of a partition from several versions with slightly disturbed borders [11].

More recently, connected operators were also involved. Stochastic watersheds [12] were introduced as a solution for improving the robustness of marker-based watersheds, by fusing the results obtained from different initial seeds. This approach further inspired a stochastic minimum spanning forest approach [13] for segmentation / classification of hyperspectral remote sensing images. Random walkers [14] were also considered for segmenting a graph generated from the information derived from the degree of accordance between different segmentation maps.

2.2   Morphological Hierarchies, Multi-images and Remote Sensing

Morphological hierarchies associated to connected operators [2] have been successfully involved in image segmentation in the "one image, one algorithm" paradigm. In the associated tree-structures, the nodes model homogeneous regions in the image whereas the edges represent their inclusion relations.

Classical trees, as component-trees [15] or trees of shapes [16], allow us to perform hierarchical segmentation by fusion of flat zones. These structures provide as output partial partitions of an image with tunable levels of details. However, they strongly rely on the image intensity, which is not compliant with the specificities of satellite images.

A first solution to deal with this issue relies on the constrained connectivity [17]. The connectivity relation generates a partition of the image domain; fine to coarse partition hierarchies are then produced by varying a threshold value associated with each connectivity constraint. In a different manner, the BPT [3] reflects a (chosen) similarity measure between neighbouring regions, and models the hierarchy between these regions. The BPTs were used to segment various types of satellite images [18,19]. However, objects in satellite images appear often too much heterogeneous to be correctly segmented from a single image. It appears then relevant to consider n images of a same scene to enrich the data space and improve the capability of segmentation hierarchies.

In this context, efforts were conducted to extend morphological hierarchies for handling n images. In [20] an extension of the BPT model was proposed to deal with multiresolution satellite images by considering one hierarchy per resolution image. In a similar vein, an approach based on multiple morphological hierarchies was developed in [21] to segment a multispectral image. The originality of this approach was to build independently one hierarchy per radiometric band and combine them to select meaningful segments, optimizing a mixed spectral / connectivity measure.

These recent works show the interest of considering a multi-image paradigm to enhance the hierarchical segmentation of complex structures from remote sensing images. Multi-image fusion segmentation methods take advantage of the complementarity of available data, and can be adapted for morphological hierarchies by interpreting the "n images, one algorithm" as "n images, one hierarchy". By performing collaborations where the fusions occur relatively to the different image contents, it is then possible to gradually build a unique consensual hierarchy, which can be used to detect complex patterns while avoiding (spectral, semantic, ...) noise appearing in a single image.

Based on these considerations, we propose in the next section a hierarchical collaborative segmentation method, extending the BPT to deal with multi-images. By contrast with classical segmentation fusion approaches, we do not intend to fuse several segmentation maps, but we directly develop a consensus during the construction of the hierarchy that models the n images. To build this consensual hierarchy, we propose different algorithmic consensus strategies (inspired from ensemble clustering strategies previously developed by the machine learning community) that seek widespread agreement among the content of different images of a same scene.

3   Building a Binary Partition Tree from Several Images

We first recall the BPT construction algorithm. In particular, we focus on the data-structure point of view, which is the cornerstone of our contribution. Then, we describe our generalization of this algorithm to deal with several images. Various families of consensus strategies for instantiating this framework are finally discussed.

3.1   The Standard BPT Construction

Summary of the Algorithm [3]. A BPT is a hierarchical representation of an image. More precisely, it is a binary tree, whose each node is a connected region. Each of these nodes is either a leaf – then corresponding to an "elementary" region – or an internal node, modelling the union of the regions of its two children nodes. The root is the node corresponding to the support of the image. Practically, a BPT is built from its leaves – provided by an initial partition of the image support – to its root, in a bottom-up fashion, by iteratively choosing and merging two adjacent regions which minimize a merging criterion that reflects, *e.g.*, the spectral and / or geometrical likenesses of the regions.

Structural Description of the Algorithm. An image is a function $I : \Omega \to V$ that associates to each point x of the finite set Ω a value $I(x)$ of the set V. A crucial point for building a BPT – and more generally for performing connected filtering – is to take into account the structure of Ω. More precisely, it is mandatory to model the fact that two points x and y of Ω are neighbours. This is done by defining an adjacency A_Ω, *i.e.*, an irreflexive, symmetric, binary relation on Ω. In other words, (Ω, A_Ω) is a graph that models the structure of the image space.

Let us now consider an initial partition $\mathcal{L}$ of Ω. (Each node $L \subseteq \Omega$ of $\mathcal{L}$ is generally assumed to be connected with respect to A_Ω.) This partition $\mathcal{L}$ defines the set of the leaves of the BPT we are going to build (*e.g.*, $\mathcal{L}$ can be the set of the image flat zones).

For any partition $\mathcal{P}$ of Ω (and in particular for $\mathcal{L}$) we can define an adjacency inherited from that of Ω. More precisely, we say that two distinct nodes $N_1, N_2 \in \mathcal{P}$ are adjacent if there exist $x_1 \in N_1$ and $x_2 \in N_2$ such that (x_1, x_2) is an edge of A_Ω, *i.e.*, x_1 and x_2 are adjacent in (Ω, A_Ω). This new adjacency relation $A_\mathcal{P}$ is also irreflexive and symmetric. In the case of $\mathcal{L}$, it allows us to define a graph $\mathfrak{G}_\mathcal{L} = (\mathcal{L}, A_\mathcal{L})$ that models the structure of the partition of the image I.

The BPT is the data-structure that describes the progressive collapse of $\mathfrak{G}_\mathcal{L}$ onto the trivial graph $(\Omega, \emptyset)$. This process consists of defining a sequence $(\mathfrak{G}_i = (N_i, A_{N_i}))_{i=0}^n$ (with $n = |\mathcal{L}| - 1$) as follows. First, we set $\mathfrak{G}_0 = \mathfrak{G}_\mathcal{L}$. Then, for each i from 1 to n, we choose the two nodes N_{i-1} and N'_{i-1} of $\mathfrak{G}_{i-1}$ linked by the edge $(N_{i-1}, N'_{i-1}) \in A_{N_{i-1}}$ that minimizes a merging criterion, and we define $\mathfrak{G}_i$ such that $N_i = N_{i-1} \setminus \{N_{i-1}, N'_{i-1}\} \cup \{N_{i-1} \cup N'_{i-1}\}$; in other words, we replace these two nodes by their union. The adjacency A_{N_i} is defined accordingly from $A_{N_{i-1}}$: we remove the edge (N_{i-1}, N'_{i-1}), and we replace each edge (N_{i-1}, N''_{i-1}) and / or (N'_{i-1}, N''_{i-1}) by an edge $(N_{i-1} \cup N'_{i-1}, N''_{i-1})$ (in particular, two former edges may be fused into a single).

From a structural point of view, the BPT $\mathfrak{T}$ is the Hasse diagram of the partially ordered set $(\bigcup_{i=0}^n N_i, \subseteq)$. From an algorithmic point of view, $\mathfrak{T}$ is built in parallel to the progressive collapse from $\mathfrak{G}_0$ to $\mathfrak{G}_n$; in other words, $\mathfrak{T}$ stores the node fusion history. More precisely, we define a sequence $(\mathfrak{T}_i)_{i=0}^n$ as follows. We set $\mathfrak{T}_0 = (N_0, \emptyset) = (\mathcal{L}, \emptyset)$. Then, for each i from 1 to n, we build $\mathfrak{T}_i$ from $\mathfrak{T}_{i-1}$ by adding the new node $N_{i-1} \cup N'_{i-1}$, and the two edges $(N_{i-1} \cup N'_{i-1}, N_{i-1})$ and $(N_{i-1} \cup N'_{i-1}, N'_{i-1})$. The BPT $\mathfrak{T}$ is finally defined as $\mathfrak{T}_n$.

Remark. The classical – image and application-oriented – description of the BPT construction algorithm considers as input: the image I (*i.e.*, the geometrical embedding of Ω, and the value associated to each point of Ω); a region model, that allows us to "describe" the nodes; and a merging criterion, that allows us to quantify the homogeneousness of nodes before and after a putative fusion. These information are important from an applicative point of view. However, from an algorithmic point of view, their only use is to define a valuation on the edges that allows us to *choose* which nodes to fuse at any given step. In the sequel, we will then consider – without loss of correctness – that a BPT is fully defined by only two input information: (1) the graph $\mathfrak{G}_{\mathcal{L}} = (\mathcal{L}, A_{\mathcal{L}})$ that models the initial partition of the image; (2) a valuation function $W : (2^{\Omega})^2 \times V^{\Omega} \to \mathbb{R}$ that allows us to choose, at each step of the process, the next pair of nodes to be merged.

Data-Structures. The above description of the BPT construction algorithm implies to define and update, during the whole process, several data-structures, namely: the graph $\mathfrak{G}$, that allows us to know what nodes remain to be merged and what are their adjacency links; and the tree $\mathcal{T}$ that is progressively built. In order to efficiently compute the valuation W, it is also important to associate each node of $\mathfrak{G}$ to the corresponding part of the image I, e.g., via a mapping between $\mathfrak{G}$ (actually, $\mathcal{N}$) and Ω.

The last – but not least – required data-structure is a sorted list $\mathcal{W}$ that gathers the valuations of each remaining edge of $\mathfrak{G}$. This list contains the information that will authorise, at each of the n iterative steps of the process, to choose the couple of nodes to be merged. This choice is made in constant time $O(1)$, since $\mathcal{W}$ is sorted. After the merging operation, $\mathcal{W}$ has to be updated: (1) to remove the edge between the two nodes; (2) to update the edges affected by the merging operation; and (3) to re-order these updated edges. Operation (1) is carried out in constant time $O(1)$. Operation (2) is carried out in $O(\alpha.T_W)$, where T_W is the cost of the computation of W for an edge, and α is the number of neighbours of the merged nodes (α is generally bounded by a low constant value). Operation (3) is carried out in $O(\alpha. \log_2 |\mathcal{W}|)$.

3.2   Generalizing the BPT Construction to Several Images

Let us now consider $k > 1$ images I_j, instead of one. Assuming that they correspond to a same scene, we consider – up to resampling – that all are defined on the same support. We then have a set of images $\{I_j : \Omega \to V_j\}_{j=1}^k$. The purpose is now to build a BPT from these k images, by generalizing the algorithmic framework described in Section 3.1. A step of this algorithmic process is illustrated in Figure 1.

Structural Evolutions. As stated above, we first need a graph that models the initial partition $\mathcal{L}$ of the image(s). Since all the I_j share the same support Ω, such a graph can still be obtained easily, either by subdividing Ω into one-point singleton sets – the induced graph $\mathfrak{G}_{\mathcal{L}}$ is then isomorphic to (Ω, A_{Ω}) – or by considering flat zones, e.g., maximal connected sets of constant value with respect to the Cartesian space $\Pi_{j=1}^k V_j$.

The k images I_j share the same support, but they take their values in different sets V_j. As a consequence, following the standard BPT construction paradigm, each of them is associated with a specific valuation function $W_j : (2^{\Omega})^2 \times V_j^{\Omega} \to \mathbb{R}$ that is defined in particular with respect to the value set V_j.

From a data-structure point of view, the generalized BPT construction algorithm will still handle one graph $\mathfrak{G}$, that will be progressively collapsed; and one tree $\mathcal{T}$ that will be built to finally provide the BPT. A unique mapping between $\mathcal{N}$ and Ω will still allow to have access to the values of a node for the k images. The main difference now lies

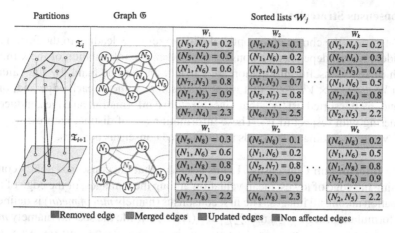

| Partitions | Graph $\mathfrak{G}$ | Sorted lists $\mathcal{W}_j$ | | |

Fig. 1. One step of the building of a BPT from k images. Left: the partition of Ω before and after the fusion of two nodes. Center: the associated graph $\mathfrak{G}$, before and after the fusion of N_3 and N_4, forming the new node N_8. The red edge is removed. The blue and orange edges are updated, *e.g.*, (N_1, N_3) becomes (N_1, N_8); the orange are merged by pairs, *e.g.*, (N_7, N_3) and (N_7, N_4) become (N_7, N_8). The green edges are not affected. Right: the k lists $\mathcal{W}_j$ each corresponding to an image. The red cells are removed, as the edge (N_3, N_4) is suppressed; this edge had been chosen according to a given consensus policy, due to its "high" position in the k lists. The scores of blue and orange cells are updated with respect to N_8; the orange cells are merged by pairs. The positions of the blue and orange cells are updated with respect to their new scores. The scores of the green cells are not affected.

into the fact that each function W_j induces – in first approximation – a specific sorted list $\mathcal{W}_j$ to gather the valuation of the remaining edges of $\mathfrak{G}$, with respect to W_j.

Algorithmic Consequences. From an algorithmic point of view, each iteration of the construction process preserves the same structure. An edge is chosen and the two incident nodes of the graph are merged. This operation leads to update the nodes and edges of $\mathfrak{G}$, and adds a new node plus two edges in $\mathcal{T}$. The main differences are that: (*i*) several sorted lists then have to be updated instead of one; and (*ii*) the choice of the optimal edge has to be made with respect to the information carried by these k sorted lists instead of only one, for a standard BPT.

From a computational point of view, choosing the edge to remove is no longer a constant time operation, but will depend on the way information are used and compared. Afterwards, operations (1–3) described in the standard BPT construction algorithm, for the sorted list maintenance, have to be duplicated for each list. These operations are then carried out in $O(k)$, $O(k.\alpha.T_{W_\star})$ and $O(k.\alpha.\log_2 |\mathcal{W}_\star|)$, respectively.

However, this initial generalization of the BPT construction algorithm can be refined by studying more precisely the policies that are considered to choose an edge, with respect to the information carried by the W_j valuation functions and / or the $\mathcal{W}_j$ sorted lists.

3.3   Consensus Strategies

At each iteration, the choice of the optimal edge to remove, leading to the fusion of its two incident nodes, depends on a consensus between the information of the k images. For each image I_j, useful information are carried, on the one hand, by the valuation function $W_j : (2^\Omega)^2 \times V^\Omega \to \mathbb{R}$ that gives an *absolute* value to each edge and, on the other hand, by the sorted list $\mathcal{W}_j$, that gives a *relative* information on edges, induced by their ordering with respect to W_j. These information are of distinct natures; we study their relevance according to the kinds of considered consensus policies.

Absolute Information Consensus. Let us consider that the consensus policy consists of choosing the edge of lowest mean valuation among the k images, or the edge of minimal valuation among all images. The first consensus (namely *min of mean*) is defined by a linear formulation: $\arg_{(N,N')\in\mathcal{N}} \min \sum_{j=1}^k W_j((N, N'))$, while the second (namely *min of min*) is defined by a non-linear formulation: $\arg_{(N,N')\in\mathcal{N}} \min \min_{j=1}^k W_j((N, N'))$. However, in both cases the decision is made by considering the absolute information carried by the edges. In other words, it is sufficient to know the k values of each point of Ω with respect to the images I_j. Then, the k sorted lists $\mathcal{W}_j$ are useless, and a single sorted list $\mathcal{W}$ that contains the information of these – linear or non-linear – formulations is indeed sufficient. The construction of a BPT from k images is then equivalent to that from one image defined as $I : \Omega \to \Pi_{j=1}^k V_j$.

Relative Local Information Consensus. Let us now consider that the consensus policy consists of choosing the edge that is the most often in first position in the k sorted lists $\mathcal{W}_j$, or the most frequently present in the $r \ll |\mathcal{W}_\star|$ first positions in the k sorted lists $\mathcal{W}_j$. These consensus (namely, *majority vote* and *most frequent*, potentially weighted) policies do not act on the absolute valuations of the edges, but on their relative positions in the lists. In such case, it is then mandatory to maintain k sorted lists. However, the decision process does not require to explicitly access the whole lists, but it can be restricted to the first (or the first r) element(s) of each, leading to a *local* decision process.

Relative Global Information Consensus. Let us finally consider that the consensus policy consists of choosing the edge that has the best global ranking among the k sorted lists $\mathcal{W}_j$. Such consensus (*e.g.*, *best average*, or *best median ranking*) policy, also acts on the relative positions of the edges in the lists. By contrast with the above case, the decision process requires to explicitly access the whole content of all these lists, leading to a *global* decision process of high computation cost. (Such cost may be reduced by maintaining, in favourable cases, a $(k + 1)$-th list that summarises the global information, and / or by adopting heuristic strategies that update the lists only after a given number of steps).

Algorithmic and Structural Consequences. The choice of a consensus strategy is strongly application-dependent. As a consequence, it is important to consider a trade-off between the structural and computational cost of the approach versus the benefits in terms of results accuracy. In particular, these costs are summarized in Table 1.

Table 1. Space and time cost of the BPT construction for various families of consensus policies. For the sake of readability, r, α and $T_{W_\star}$, which are practically bounded by low constant values have been omitted here.

Consensus policies	# $\mathcal{W}_\star$	Edge choice	Edge removal	Edges update	Edges sorting
Absolute information	1	$\Theta(1)$	$\Theta(1)$	$\Theta(1)$	$\Theta(\log_2 |\mathcal{W}_\star|)$
Relative local inf.	k	$\Omega(k)$	$\Theta(k)$	$\Theta(k)$	$\Theta(k.\log_2 |\mathcal{W}_\star|)$
Relative global inf.	k	$\Omega(k.|\mathcal{W}_\star|)$	$\Theta(k)$	$\Theta(k)$	$\Theta(k.\log_2 |\mathcal{W}_\star|)$

4 Experiments

To experiment our framework, two applications have been considered in the context of remote sensing: one-time, one-sensor, several (noisy) images, to assess the ability to retrieve information despite image degradation; and multi-time, one-sensor, one image per date, to assess the ability to capture time-independent and redundant information.

The BPT construction and segmentation approaches were voluntarily chosen as very simple, in order to avoid any bias related to these choices, thus better focusing on the actual structural effects of multi-image BPT versus standard BPT.

At this stage, these experiments have to be considered as toy-examples, since neither quantitative validation nor fine parameter tuning were carried out. Our purpose is mainly to give the intuition of potential uses of such BPTs in the field of remote sensing.

4.1 Urban Noisy Images

Data. The dataset used here was sensed over the town of Strasbourg (France). The original sample (Figure 2(a)) is an urban image (1024×1024 pixels) acquired by the PLÉIADES satellite in 2012 (courtesy LIVE, UMR CNRS 7263). It is a pansharpened image at a spatial resolution of 60 cm with four bands (NIR, R, G, B). From this image, a series of 7 noisy images was generated by adding Gaussian and speckle noise (Figure 2(b)).

Method and Results. The BPTs are built from the trivial partition $\mathcal{L}$ composed by all singleton sets, *i.e.*, one pixel per region. The valuation function $W_\star : (2^\Omega)^2 \times V^\Omega \to \mathbb{R}$ is defined as the increase of the ranges of the intensity values (for each radiometric band), potentially induced by the fusion of the incident regions. In the case of multi-images, the relative local information consensus policy *most-frequent*, weighted according to the position of the edges within the lists is applied for the first 10% of the lists $\mathcal{W}_\star$.

The non-noisy image of Figure 2(a) is first segmented by extracting a user-defined horizontal cut from its "standard" BPT (Figure 2(c)). According to a same number of regions, a cut is then selected from the multi-images BPT leading to a comparable segmentation result (Figure 2(d)).

From these figures, we observe that the results obtained from noisy images are slightly degraded, but of comparable quality, with respect to Figure 2(c). This tends to confirm

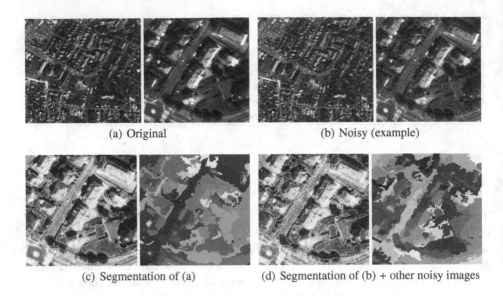

(a) Original   (b) Noisy (example)

(c) Segmentation of (a)   (d) Segmentation of (b) + other noisy images

Fig. 2. (a) Initial image, PLÉIADES, 2012; and a zoomed sample, 200×200 pixels. (b) An example of noisy image generated from (a), Gaussian ($\sigma = 10\%$) and speckle noise (5%), and a zoomed sample, 200×200 pixels. (c) Segmentation from the mono-image BPT of (a). (d) Segmentation from the multi-image BPT of (b) plus 6 other noisy images.

the ability of the multi-image BPT-based segmentation to generate accurate results, by discriminating relevant information from noise thanks to the consensus operated between the various images, even in the case of low signal-to-noise ratio.

4.2   Agricultural Image Time Series

Data. The dataset used here is a time series of agricultural images (1000×1000 pixels) of an area located near Toulouse (France). Images were acquired by the FORMOSAT-2 satellite over the 2007 cultural year, see Figure 3(a–c). They were ortho-rectified and have a spatial resolution of 8 m, with four spectral bands (NIR, R, G, B).

Method and Results. The BPTs are built from the same partition $\mathcal{L}$ and valuation function $W_\star : (2^\Omega)^2 \times V^\Omega \to \mathbb{R}$ as in Section 4.1. They are also segmented in the same way.

The segmented results, depicted in Figure 3(e), provide regions that are not the same as those obtained from a standard mono-image BPT, computed from one of the images of the series (Figure 3(d)). On the one hand, some segmentation effects deriving from semantic noise in mono-image segmentation are sometimes corrected by the redundant information obtained from multi-images. On the other hand, the multi-image BPT focuses on time-specific details that are only accessible via a temporal analysis, providing a potentially useful tool for such data.

(a) May   (b) July   (c) August   (d) Seg. mono-im. (e) Seg. multi-im.

Fig. 3. (a–c) 3 samples (200 × 200 pixels) of the image time series, FORMOSAT-2, 2007. (d) Segmentation obtained from the mono-image BPT from (a). (e) Segmentation obtained from the multi-image BPT from (a–c).

5 Conclusion

This article has presented a data-structure / algorithmic framework and different fusion consensus strategies for building a unique BPT from several images. This contribution is, to the best of our knowledge, the first attempt to handle segmentation fusion in the framework of (morphological) hierarchies, in the "*n* images, one algorithm" paradigm.

The experiments carried out on satellite multi-image datasets have shown that the quality of the induced morphological hierarchies are sufficient to further perform improved segmentation, *e.g.*, from noisy or multi-temporal images of a same scene. The consensus strategies considered in this study remain, however, mostly basic (most-frequent, majority vote, *etc.*). Integrating higher-level consensus may then allow us to improve the quality of the hierarchies and the induced segmentation.

In the case of mono-date images, the fusion decisions underlying to the consensus strategies could also be guided by semantic information. Recent advances concerning the hierarchical modelling of such semantic information, in the context of classified remote sensing data [22], may facilitate such approaches.

In the case of multi-date images, we may handle the land-cover evolutions of the observed territories by considering an adequate region model and merging criterion [23]. Such spatio-temporal information could be used to follow local consensus between images leading to hyper-trees where the branches model local temporal fusion decisions.

In a next issue, the results obtained with this method will be fully assessed by quantitative comparisons (using datasets provided by different sensors) and compared to the results produced by other hierarchical and fusion-based segmentation methods.

References

1. Topchy, A., Jain, A.K., Punch, W.: Clustering ensembles: Models of consensus and weak partitions. IEEE TPAMI 27, 1866–1881 (2005)
2. Salembier, P., Wilkinson, M.H.F.: Connected operators: A review of region-based morphological image processing techniques. IEEE SPM 26, 136–157 (2009)
3. Salembier, P., Garrido, L.: Binary partition tree as an efficient representation for image processing, segmentation, and information retrieval. IEEE TIP 9, 561–576 (2000)
4. Rohlfing, T., Maurer Jr., C.R.: Shape-based averaging. IEEE TIP 16, 153–161 (2007)
5. Vidal, J., Crespo, J., Maojo, V.: A shape interpolation technique based on inclusion relationships and median sets. IVC 25, 1530–1542 (2007)
6. Franek, L., Abdala, D.D., Vega-Pons, S., Jiang, X.: Image segmentation fusion using general ensemble clustering methods. In: Kimmel, R., Klette, R., Sugimoto, A. (eds.) ACCV 2010, Part IV. LNCS, vol. 6495, pp. 373–384. Springer, Heidelberg (2011)
7. Mignotte, M.: Segmentation by fusion of histogram-based K-means clusters in different color spaces. IEEE TIP 17, 780–787 (2008)
8. Calderero, F., Eugenio, F., Marcello, J., Marqués, F.: Multispectral cooperative partition sequence fusion for joint classification and hierarchical segmentation. IEEE GRSL 9, 1012–1016 (2012)
9. Wang, H., Zhang, Y., Nie, R., Yang, Y., Peng, B., Li, T.: Bayesian image segmentation fusion. KBS 71, 162–168 (2014)
10. Chu, C.C., Aggarwal, J.K.: The integration of image segmentation maps using region and edge information. IEEE TPAMI 15, 72–89 (1993)
11. Cho, K., Meer, P.: Image segmentation from consensus information. CVIU 68, 72–89 (1997)
12. Angulo, J., Jeulin, D.: Stochastic watershed segmentation. In: ISMM, pp. 265–276 (2007)
13. Bernard, K., Tarabalka, Y., Angulo, J., Chanussot, J., Benediktsson, J.A.: Spectral-spatial classification of hyperspectral data based on a stochastic minimum spanning forest approach. IEEE TIP 21, 2008–2021 (2012)
14. Wattuya, P., Rothaus, K., Praßni, J.S., Jiang, X.: A random walker based approach to combining multiple segmentations. In: ICPR, pp. 1–4 (2008)
15. Salembier, P., Oliveras, A., Garrido, L.: Antiextensive connected operators for image and sequence processing. IEEE TIP 7, 555–570 (1998)
16. Monasse, P., Guichard, F.: Scale-space from a level lines tree. JVCIR 11, 224–236 (2000)
17. Soille, P.: Constrained connectivity for hierarchical image decomposition and simplification. IEEE TPAMI 30, 1132–1145 (2008)
18. Vilaplana, V., Marques, F., Salembier, P.: Binary partition trees for object detection. IEEE TIP 17, 2201–2216 (2008)
19. Benediktsson, J.A., Bruzzone, L., Chanussot, J., Dalla Mura, M., Salembier, P., Valero, S.: Hierarchical analysis of remote sensing data: Morphological attribute profiles and binary partition trees. In: Soille, P., Pesaresi, M., Ouzounis, G.K. (eds.) ISMM 2011. LNCS, vol. 6671, pp. 306–319. Springer, Heidelberg (2011)
20. Kurtz, C., Passat, N., Gançarski, P., Puissant, A.: Extraction of complex patterns from multiresolution remote sensing images: A hierarchical top-down methodology. PR 45, 685–706 (2012)
21. Akcay, H.G., Aksoy, S.: Automatic detection of geospatial objects using multiple hierarchical segmentations. IEEE TGRS 46, 2097–2111 (2008)
22. Kurtz, C., Naegel, B., Passat, N.: Connected filtering based on multivalued component-trees. IEEE TIP 23, 5152–5164 (2014)
23. Alonso-González, A., López-Martínez, C., Salembier, P.: PolSAR time series processing with binary partition trees. IEEE TGRS 52, 3553–3567 (2014)

Scale-Space Representation Based on Levelings Through Hierarchies of Level Sets

Wonder A.L. Alves[1,2], Alexandre Morimitsu[1], and Ronaldo F. Hashimoto[1(✉)]

[1] Department of Computer Science, Institute of Mathematics and Statistics,
University of São Paulo, São Paulo, Brazil
{wonder,alem,ronaldo}@ime.usp.br
[2] Department of Informatics, Nove de Julho University, São Paulo, Brazil

Abstract. This paper presents new theoretical contributions on scale-space representations based on levelings through hierarchies of level sets, i.e., component trees and tree of shapes. Firstly, we prove that reconstructions of pruned trees (component trees and tree of shapes) are levelings. After that, we present a new and fast algorithm for computing the reconstruction based on marker images from component trees. Finally, we show how to build morphological scale-spaces based on levelings through the reconstructions of successive pruning operations (whether based on increasing attributes or marker images).

Keywords: Scale-space · Levelings · Component trees · Tree of shapes

1 Introduction

As we know, an operator in Mathematical Morphology (MM) can be seen as a mapping between complete lattices. In particular, mappings on the set of all gray level images $\mathcal{F}(\mathcal{D})$ defined on domain $\mathcal{D} \subset \mathbb{Z}^2$ and codomain $\mathbb{K} = \{0, 1, ..., K\}$ are of special interest in MM and they are called *image operators*. Furthermore, when ψ enlarges the partition of the space created by the flat zones, it is called *connected operator* [1]. They represent a wide class of operators in which F. Meyer [2, 3, 4] extensively studied their specializations. One of these specializations, known as *levelings*, are powerful simplifying filters that preserve order, do not create new structures (regional extrema and contours) and their values are enclosed by values of a neighborhood of pixels (see Def. 1).

Definition 1 (F. Meyer [3, 2]). *An operator $\psi : \mathcal{F}(\mathcal{D}) \to \mathcal{F}(\mathcal{D})$ is said to be leveling, if and only if, for any $f \in \mathcal{F}(\mathcal{D})$ the following relation is valid for all pairs of adjacent pixels, i.e., $\forall (p, q) \in \mathcal{A}$,*

$$[\psi(f)](p) > [\psi(f)](q) \Rightarrow f(p) \geq [\psi(f)](p) \ and \ [\psi(f)](q) \geq f(q).$$

where $\mathcal{A}$ is a adjacency relation[1] on $\mathcal{D}$.

---

[1] An adjacency relation $\mathcal{A}$ on $\mathcal{D}$ is a binary relation on pixels of $\mathcal{D}$. Thus, $(p, q) \in \mathcal{A}$ if and only if p is an adjacent of q. 4 or 8-adjacency are common examples of adjacencies relation on $\mathcal{D}$.

© Springer International Publishing Switzerland 2015
J.A. Benediktsson et al. (Eds.): ISMM 2015, LNCS 9082, pp. 265–276, 2015.
DOI: 10.1007/978-3-319-18720-4_23

From the definition of a class of operators, it is possible to build a binary relation $\mathcal{R}$ on $\mathcal{F}(\mathcal{D})$ as follows: $(f, g) \in \mathcal{R}$ if and only if there exists ψ in this class such that $g = \psi(f)$. Thus, the definition of levelings can be seen as a binary relation $\mathcal{R}_{leveling}$ on $\mathcal{F}(\mathcal{D})$. So, we say that g is leveling of f if and only if $(f, g) \in \mathcal{R}_{leveling}$. In [2], F. Meyer, shows that $\mathcal{R}_{leveling}$ is reflexive and transitive and if we ignore the constant images then $\mathcal{R}_{leveling}$ is anti-symmetric, i.e., $\mathcal{R}_{leveling}$ is a order relation. With the help of this order relation, the levelings can be nested to create a scale-space decomposition of an image $f \in \mathcal{F}(\mathcal{D})$ in the form of a series of levelings $(g_0 = f, g_1, ..., g_n)$, where g_k is leveling of g_{k-1} and as a consequence of transitivity, g_k is also a leveling of each image g_j, for $j < k$ [4, 5]. Thus, a morphological scale-space is generated with the following features: simplification, causality and fidelity [4, 5]. For example, this scale-space can be created with the help from a traditional algorithm Λ to construct levelings that takes as arguments two images: an input image $f \in \mathcal{F}(\mathcal{D})$ and a marker image $h \in \mathcal{F}(\mathcal{D})$. It modifies h in such a way that it becomes a leveling of f. Thus, we will say that $g = \Lambda(f, h)$ is a leveling of f, obtained from the marker h [3, 4]. With this algorithm it is possible to construct a morphological scale-space based on levelings from any family of markers $(h_1, h_2, ..., h_n)$ with the following chaining: $g_1 = \Lambda(f, h_1), g_2 = \Lambda(g_1, h_2), ..., g_n = \Lambda(g_{n-1}, h_n)$ [4].

In this work, we follow a different approach for construction of scale-spaces. Our approach consists of representing an image through a tree based on hierarchies of level sets (i.e., component tree and tree of shapes) and from this tree is proved that reconstruction of pruned trees are levelings. Despite these facts are knowns and/or mentioned by several authors [6, 7, 8, 9, 10, 11] this is the first study that presents a formal proof on the perspective of trees which reconstruction of pruned trees are levelings. In addition, we present a new fast algorithm for computing the reconstruction by dilation (or erosion) which is faster than the algorithm by Luc Vincent [12]. Finally, we show how to construct a morphological scale-space based on levelings through the reconstructions of successive pruning operations (whether based on increasing attributes or marker images).

The remainder of this paper is structured as follows. Section 2 briefly recalls some definitions and properties of image representation by tree structures. In Section 3, we provide the first original result of this work where establishes theoretical links between reconstruction pruned trees with levelings. In Section 4, we associate reconstructions of pruned trees with several morphological operators based on marker images. In Section 5, we present constructions of morphological scale-space either based on increasing attributes or marker images. In Section 6, we show an application of scale-space based on levelings to construct residual operators. Finally, Section 7 concludes this work.

2   Theoretical Background

For any $\lambda \in \mathbb{K} = \{0, 1, ..., K\}$, we define $\mathcal{X}_{\downarrow}^{\lambda}(f) = \{p \in \mathcal{D} : f(p) < \lambda\}$ and $\mathcal{X}_{\lambda}^{\uparrow}(f) = \{p \in \mathcal{D} : f(p) \geq \lambda\}$ as the *lower* and *upper level sets* at value λ from an image $f \in \mathcal{F}(\mathcal{D})$, respectively. These level sets are nested, i.e., $\mathcal{X}_{\downarrow}^{1}(f) \subseteq$

$\mathcal{X}_\downarrow^2(f) \subseteq \dots \subseteq \mathcal{X}_\downarrow^K(f)$ and $\mathcal{X}_K^\uparrow(f) \subseteq \mathcal{X}_{K-1}^\uparrow(f) \subseteq \dots \subseteq \mathcal{X}_0^\uparrow(f)$. Thus, the image f can be reconstructed using either the family of lower or upper sets, i.e., $\forall x \in \mathcal{D}$, $f(x) = \inf\{\lambda - 1 : x \in \mathcal{X}_\downarrow^\lambda(f)\} = \sup\{\lambda : x \in \mathcal{X}_\downarrow^\uparrow(f)\}$. From these sets, we define two other sets $\mathcal{L}(f)$ and $\mathcal{U}(f)$ composed by the connected components (CCs) of the lower and upper level sets of f, i.e., $\mathcal{L}(f) = \{C \in CC_4(\mathcal{X}_\downarrow^\lambda(f)) : \lambda \in \mathbb{K}\}$ and $\mathcal{U}(f) = \{C \in CC_8(\mathcal{X}_\lambda^\uparrow(f)) : \lambda \in \mathbb{K}\}$, where $CC_4(X)$ and $CC_8(X)$ are sets of 4 and 8 connected CCs of X, respectively. Then, the ordered pairs consisting of the CCs of the lower and upper level sets and the usual inclusion set relation, i.e., $(\mathcal{L}(f), \subseteq)$ and $(\mathcal{U}(f), \subseteq)$, induce two dual trees [13]. They can be represented by a non-redundant data structures known as min-tree and max-tree. Combining this pair of dual trees, min-tree and max-tree, into a single tree, we have the tree of shapes [13]. Then, let $\mathcal{P}(\mathcal{D})$ denote the *powerset* of $\mathcal{D}$ and let $sat : \mathcal{P}(\mathcal{D}) \to \mathcal{P}(\mathcal{D})$ be the operator of saturation [13] (or filling holes), i.e.,

Definition 2. *We call* holes *of $A \in \mathcal{L}(f) \cup \mathcal{U}(f)$ the CCs of $\mathcal{D} \setminus A$. Thus, we call* internal holes *of A, denoted by $Int(A)$, the CCs of $\mathcal{D} \setminus A$ that are subsets of $sat(A)$. Likewise, we call* exterior hole *of A, denoted by $Ext(A)$, the set $\mathcal{D} \setminus sat(A)$. Thus, $sat(A) = \cup\{H \in Int(A)\} \cup A$, where the unions are disjoint.*

Note that, the complement of a CC A is in $Ext(A) \cup Int(A)$ and if $H \in Int(A)$ then $sat(H) \subseteq sat(A)$. Moreover, if $A \in \mathcal{L}(f)$ with internal holes, then elements of $Int(A)$ are CCs of $\mathcal{U}(f)$ (or vice versa, $A \in \mathcal{U}(f) \Rightarrow Int(A) \subseteq \mathcal{L}(f)$). Now, let $\mathcal{SAT}_L(f) = \{sat(C) : C \in \mathcal{L}(f)\}$ and $\mathcal{SAT}_U(f) = \{sat(C) : C \in \mathcal{U}(f)\}$ be the family of CCs of the lower and upper level sets, respectively, with holes filled and consider $\mathcal{SAT}(f) = \mathcal{SAT}_L(f) \cup \mathcal{SAT}_U(f)$. The elements of $\mathcal{SAT}(f)$, called *shapes*, are nested by an inclusion relation and thus the pair $(\mathcal{SAT}(f), \subseteq)$, induces the tree of shapes [13]. The tree of shapes, such as component trees, is a complete representation of an image which can be represented by a compact and non-redundant data structure [14] so that a pixel $p \in \mathcal{D}$ which is associated with the smallest shape or CC of the tree containing it, by the parenthood relationship, is also associated to all the ancestors shapes. Then, we denote by $SC(\mathcal{T}, p)$ the smallest shape or CC containing p in a tree $\mathcal{T}$.

Extended Trees: In this work, we also consider the extended versions of these trees, i.e., the trees containing all the possible components of an image, defined as follows: Let $Ext(\mathcal{L}(f)) = \{(C, \mu) \in \mathcal{L}(f) \times \mathbb{K} : C \in CC_4(\mathcal{X}_\downarrow^\mu(f))\}$ and $Ext(\mathcal{U}(f)) = \{(C, \mu) \in \mathcal{U}(f) \times \mathbb{K} : C \in CC_8(\mathcal{X}_\mu^\uparrow(f))\}$ the set of all possible CCs of lower and upper level sets, respectively. Consider $\sqsubseteq$ a relation on $Ext(\mathcal{L}(f))$ (resp. $Ext(\mathcal{U}(f))$), i.e., $\forall (A, i), (B, j) \in Ext(\mathcal{L}(f)), (A, i) \sqsubseteq (B, j) \Leftrightarrow A \subseteq B$ and $i \leq j$ (resp. $\forall (A, i), (B, j) \in Ext(\mathcal{U}(f)), (A, i) \sqsubseteq (B, j) \Leftrightarrow A \subseteq B$ and $i \geq j$). Although, we can similarly build $Ext(\mathcal{SAT}(f))$, in this paper, we only use extended versions of max-tree and min-tree. Therefore, $(Ext(\mathcal{L}(f)), \sqsubseteq)$, $(Ext(\mathcal{U}(f)), \sqsubseteq)$ and $(Ext(\mathcal{SAT}(f)), \sqsubseteq)$ are the version extended of trees $(\mathcal{L}(f), \subseteq)$, $(\mathcal{U}(f), \subseteq)$ and $(\mathcal{SAT}(f), \subseteq)$, respectively. Fig. 1 is an example of these trees for a given image f. Note that, the smallest shape or CC of a tree $\mathcal{T}_f$ containing a $p \in \mathcal{D}$ in $Ext(\mathcal{T}_f)$ is the node $(SC(\mathcal{T}_f, p), f(p)) \in Ext(\mathcal{T}_f)$.

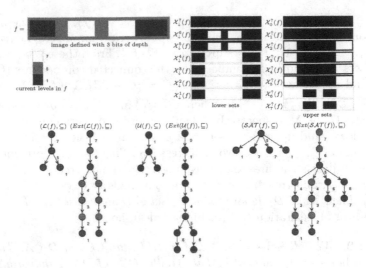

Fig. 1. An example of the construction of the extended trees

2.1 Image Reconstruction and Pruning Operation

As we have seen, an image can be reconstructed from its level sets. Now, we will show how to reconstruct an image f given a tree $\mathcal{T}_f$ (min-tree, max-tree or tree of shapes). This leads us to define the functions $level_{\mathcal{L}} : \mathcal{L}(f) \to \mathbb{K}$, $level_{\mathcal{U}} : \mathcal{U}(f) \to \mathbb{K}$ and $level_{\mathcal{SAT}} : \mathcal{SAT}(f) \to \mathbb{K}$ as follows $level_{\mathcal{L}}(C) = \min\{\lambda - 1 : C \in \mathcal{CC}_4(\mathcal{X}_{\downarrow}^{\lambda}(f)), \lambda \in \mathbb{K}\}$, $level_{\mathcal{U}}(C) = \max\{\lambda : C \in \mathcal{CC}_8(\mathcal{X}_{\lambda}^{\uparrow}(f)), \lambda \in \mathbb{K}\}$ and $level_{\mathcal{SAT}}(C) = f(y)$ such that $y \in \arg\max\{|\mathcal{SC}(\mathcal{T}_f, x)| : x \in C\}$. For the sake of simpler notation, from now on, the subscript $\mathcal{T}_f$ will be dropped from the level function when it is clear from context. Obviously, the function $level$ for $Ext(\mathcal{L}(f))$ and $Ext(\mathcal{U}(f))$ is simply $level_{Ext(\mathcal{L})}(C) = \min\{\lambda : (C, \lambda) \in Ext(\mathcal{L}(f))\}$ and $level_{Ext(\mathcal{U})}(C) = \max\{\lambda : (C, \lambda) \in Ext(\mathcal{U}(f))\}$, respectively. Using this function, it is possible to prove that an image $f \in \mathcal{F}(\mathcal{D})$ can be reconstructed from a tree $\mathcal{T}_f$ as follows: $\forall x \in \mathcal{D}, f(x) = level(\mathcal{SC}(\mathcal{T}_f, x))$. In such a case, we write: $f = Rec(\mathcal{T}_f)$. In particular, if f is obtained by $Rec((\mathcal{L}(f), \subseteq))$ (resp. $Rec((\mathcal{U}(f), \subseteq))$, and $Rec((\mathcal{SAT}(f), \subseteq)))$ then we call this operation *lower* (resp. *upper* and *shape*) *reconstruction*.

Now, the following definition (Def. 3) gives the conditions for a pruning operation of a tree $(\mathcal{T}, \preceq)$.

Definition 3. *We say that $(\mathcal{T}', \preceq)$ is obtained by a pruning operation of a tree $(\mathcal{T}, \preceq)$ if and only if, $\mathcal{T}' \subseteq \mathcal{T}$, for any $X \in \mathcal{T}'$, $\nexists Y \in (\mathcal{T} \setminus \mathcal{T}')$ such that $X \preceq Y$. In such a case, we write $\mathcal{T}' = \mathcal{P}_{runing}(\mathcal{T})$.*

Following this definition, if $\mathcal{T}_f$ is a tree of an image $f \in \mathcal{F}(\mathcal{D})$, then we say $\mathcal{T}_g$ is the pruned version of $\mathcal{T}_f$ if and only if $\mathcal{T}_g = \mathcal{P}_{runing}(\mathcal{T}_f)$. Also, one can easily see that $\mathcal{T}_g \subseteq \mathcal{T}_f$ and $\mathcal{T}_g$ is still a tree. In addition, since the nodes of $\mathcal{T}_f$ and $\mathcal{T}_g$ are nested by the order relation, it can be proved that, $p \in \mathcal{D}, \mathcal{SC}(\mathcal{T}_f, p) \subseteq \mathcal{SC}(\mathcal{T}_g, p)$.

3   Links Between Reconstruction of Pruned Trees and Levelings

Once the pruning operation and the reconstruction of pruned trees are established, we can relate the reconstruction of pruned trees with levelings. We begin observing that, if $\mathcal{T}_g$ is obtained by a pruning operation of a max-tree (resp. min-tree) $\mathcal{T}_f$, then, $\forall p \in \mathcal{D}$, $level(SC(\mathcal{T}_f, p)) \geq level(SC(\mathcal{T}_g, p))$ (resp. $level(SC(\mathcal{T}_f, p)) \leq level(SC(\mathcal{T}_g, p)))$, thanks to the well-defined ordering of the level sets. Thus, $Rec(\mathcal{T}_f) \geq Rec(\mathcal{T}_g)$ (resp. $Rec(\mathcal{T}_f) \leq Rec(\mathcal{T}_g)$). This property shows that upper (resp. lower) reconstructions are anti-extensive (resp. extensive). Now, we state a simple property, given by Prop. 1, thanks to the well-defined ordering of the level sets.

Proposition 1. *Let $(\mathcal{U}(f), \subseteq)$ be the max-tree (resp. min-tree $(\mathcal{L}(f), \subseteq)$) of an image f. Let $(p, q) \in \mathcal{A}$. Then, $f(p) > f(q)$ (resp. $f(p) < f(q)$) if and only if $SC(\mathcal{U}(f), p) \subset SC(\mathcal{U}(f), q)$ (resp. $SC(\mathcal{L}(f), p) \subset SC(\mathcal{L}(f), q))$.*

This fact shows that, if $\mathcal{T}_g$ is obtained by a pruning operation of a max-tree (resp. min-tree) $\mathcal{T}_f$, then $g = Rec(\mathcal{T}_g)$ is a leveling of $f = Rec(\mathcal{T}_f)$, since, for any $(p, q) \in \mathcal{A}$, the following condition holds: $g(p) > g(q) \Rightarrow f(p) \geq g(p) > g(q) = f(q)$ (resp. $g(p) > g(q) \Rightarrow f(p) = g(p) > g(q) \geq f(q)$). Furthermore, if we consider the extended version of the max-tree (resp. min-tree) then there is an equivalence between upper (resp. lower) reconstruction and anti-extensive (resp. extensive) levelings.

Theorem 1. *Anti-extensive (resp. extensive) levelings and upper (resp. lower) reconstructions are equivalent.*

Proof. Let $f \in \mathcal{F}(\mathcal{D})$ be an image and $\mathcal{T}_f = (Ext(\mathcal{L}(f)), \subseteq)$ the extended version of the max-tree of f. Thus, we have: $g \in \mathcal{F}(\mathcal{D})$ is an upper reconstruction of f

$\Longleftrightarrow g = Rec(\mathcal{T}_g)$ such that $\mathcal{T}_g = Pruning(\mathcal{T}_f)$.

$\Longleftrightarrow g \leq f$ and $\forall (p, q) \in \mathcal{A}$,
$$\begin{cases} \text{either } SC(\mathcal{T}_f, p) \sqsubseteq SC(\mathcal{T}_f, q) \sqsubseteq SC(\mathcal{T}_g, p) \sqsubseteq SC(\mathcal{T}_g, q) \\ \text{or }\quad SC(\mathcal{T}_f, p) \sqsubseteq SC(\mathcal{T}_g, p) \sqsubseteq SC(\mathcal{T}_f, q) = SC(\mathcal{T}_g, q) \end{cases}$$

$\Longleftrightarrow g \leq f$ and $\forall (p, q) \in \mathcal{A}$, $SC(\mathcal{T}_g, p) \sqsubset SC(\mathcal{T}_g, q)$
$$\Rightarrow \begin{cases} SC(\mathcal{T}_f, p) \sqsubseteq SC(\mathcal{T}_g, p) \\ \text{and} \\ SC(\mathcal{T}_g, q) = SC(\mathcal{T}_f, q) \end{cases}$$

$\Longleftrightarrow g \leq f$ and $\forall (p, q) \in \mathcal{A}$, $level(SC(\mathcal{T}_g, p)) > level(SC(\mathcal{T}_g, q))$
$$\Rightarrow \begin{cases} level(SC(\mathcal{T}_f, p)) \geq level(SC(\mathcal{T}_g, p)) \\ \text{and} \\ level(SC(\mathcal{T}_g, q)) = level(SC(\mathcal{T}_f, q)) \end{cases}$$

$\Longleftrightarrow g \leq f$ and $\forall (p, q) \in \mathcal{A}$, $g(p) > g(q) \Rightarrow f(p) \geq g(p)$ and $g(q) = f(q)$

$\Longleftrightarrow g$ is anti-extensive leveling of f.

The proof for extensive levelings and lower reconstruction follows similarly.

Now, to establish links between shape reconstructions and levelings, it is necessary to know relations between neighboring pixels in the nodes of the tree. In this sense, Propositions 2, 3, 4 and 5 help us understand how the neighboring pixels are related in the tree. Thus, the Prop. 2 is a corollary of Theo. 2.16 given in [15], the Prop. 3 is a directly consequence of Prop. 1 and the Prop. 5 is a direct consequence of Prop. 4.

Proposition 2. *Let $(\mathcal{SAT}(f), \subseteq)$ be the tree of shapes of an image f. If $(p, q) \in \mathcal{A}$ such that $f(p) \neq f(q)$ then $\mathcal{SC}(\mathcal{SAT}(f), p)$ and $\mathcal{SC}(\mathcal{SAT}(f), q)$ are comparable or disjoint.*

Proposition 3. *Let $(\mathcal{SAT}(f), \subseteq)$ be the tree of shapes of an image f. If $(p, q) \in \mathcal{A}$ such that $\mathcal{SC}(\mathcal{SAT}(f), p) \subset \mathcal{SC}(\mathcal{SAT}(f), q)$ and both $\mathcal{SC}(\mathcal{SAT}(f), p)$ and $\mathcal{SC}(\mathcal{SAT}(f), q)$ belong to $\mathcal{SAT}_\mathcal{U}(f)$ (resp. $\mathcal{SAT}_\mathcal{L}(f)$), then $f(p) > f(q)$ (resp. $f(p) < f(q)$).*

Proposition 4. *Let $A \in \mathcal{L}(f) \cup \mathcal{U}(f)$ such that $sat(A) \in \mathcal{SAT}(f)$. If $B \in Int(A)$ and $(p, q) \in \mathcal{A}$ such that $p \in B$ and $q \notin B$, then $q \in sat(A)$.*

Proof. Suppose, by contradiction, $q \notin sat(A)$. Then q belongs to the complement of $sat(A)$, i.e., $q \in (\mathcal{D} \setminus sat(A)) = Ext(A) \subseteq (\mathcal{D} \setminus A)$. As $Int(A)$ contains the CCs of $(\mathcal{D} \setminus A)$ included in $sat(A)$ and $B \in Int(A)$, we have that both B and $Ext(A)$ are CCs of $(\mathcal{D} \setminus A)$. With that fact in mind, and, since $p \in B$, $q \in Ext(A)$, and $(p, q) \in \mathcal{A}$, we have that $Ext(A) = B$. But, this is a contradiction, since $q \notin B$. Therefore, $q \in sat(A)$.

Corollary 1. *Let $A, B \in \mathcal{SAT}(f)$ such that $B \subset A$, $A \in \mathcal{SAT}_\mathcal{U}(f)$ and $B \in \mathcal{SAT}_\mathcal{L}(f)$ (resp. $A \in \mathcal{SAT}_\mathcal{L}(f)$ and $B \in \mathcal{SAT}_\mathcal{U}(f)$). If $(p, q) \in \mathcal{A}$ such that $p \in B$ and $q \notin B$, then $q \in A$.*

Proposition 5. *Let $(p, q) \in \mathcal{A}$ and let $\mathcal{SC}(\mathcal{SAT}(f), p)$ and $\mathcal{SC}(\mathcal{SAT}(f), q)$ be elements of $\mathcal{SAT}_\mathcal{U}(f)$ (resp. $\mathcal{SAT}_\mathcal{L}(f)$). If $X \in \mathcal{SAT}(f)$ such that $\mathcal{SC}(\mathcal{SAT}(f), p) \subset X \subset \mathcal{SC}(\mathcal{SAT}(f), q)$, then $X \in \mathcal{SAT}_\mathcal{U}(f)$ (resp. $X \in \mathcal{SAT}_\mathcal{L}(f)$).*

Proof. Suppose, by contradiction, $X \notin \mathcal{SAT}_\mathcal{U}(f)$. Thus, $X \in \mathcal{SAT}_\mathcal{L}(f)$ since $X \in \mathcal{SAT}(f)$. Then, thanks to Corol. 1, it follows that $q \in X$, since $\mathcal{SC}(\mathcal{SAT}(f), p) \in \mathcal{SAT}_\mathcal{U}(f)$ and $\mathcal{SC}(\mathcal{SAT}(f), p) \subset X$. So we have a contradiction, since $X \subset \mathcal{SC}(\mathcal{SAT}(f), q)$ and $\mathcal{SC}(\mathcal{SAT}(f), q)$ is the smallest shape containing q. Therefore, $X \in \mathcal{SAT}_\mathcal{U}(f)$.

Theorem 2. *Shape reconstructions are levelings.*

Proof. Let $f \in \mathcal{F}(D)$ be an image and $\mathcal{T}_f$ be the tree of shapes of f. Then, $g \in \mathcal{F}(D)$ is a shape reconstruction of f if and only if $g = Rec(\mathcal{T}_g)$ such that $\mathcal{T}_g = \mathcal{P}runing(\mathcal{T}_f)$. To prove that g is leveling of f, we just need to check if, $\forall (p, q) \in \mathcal{A}$, the definition of leveling holds, that is, $g(p) > g(q) \Rightarrow f(p) \geq g(p)$ and $g(q) \geq f(q)$.

Let us consider two cases, where $f(p) = f(q)$ and $f(p) \neq f(q)$. In the first case, g meets the definition of leveling by vacuity. In the second case, we have $\mathcal{SC}(\mathcal{T}_f, p) \neq \mathcal{SC}(\mathcal{T}_f, q)$ and the pruning in $\mathcal{T}_f$, which generates $\mathcal{T}_g$, can: (1) preserve both nodes; or (2) eliminate both nodes; or (3) eliminate one of the nodes. See in Fig. 2 the illustrations of pruning settings.

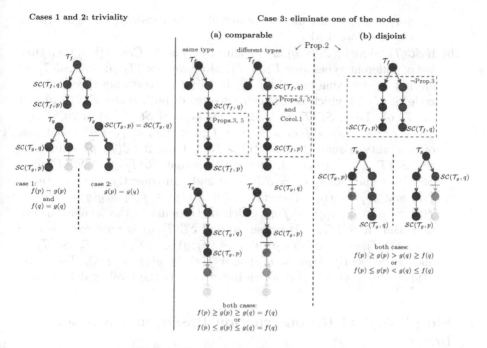

Fig. 2. Illustrations of pruning settings

1. If both nodes are preserved, then $f(p) = g(p)$ and $f(q) = g(q)$ that meets the definition of leveling (see Fig. 2 - Case 1);
2. If both nodes are eliminated and comparable, then $g(p) = g(q)$, and thus the definition of leveling is valid by vacuity (see Fig. 2 - Case 2). In case they are eliminated and not comparable, see case 3(b);
3. If only one of the two nodes is eliminated, then, thanks to Prop. 2, $SC(\mathcal{T}_f, p)$ and $SC(\mathcal{T}_f, q)$ are either comparable or disjoint (see Fig. 2 - Case 3).

   (a) If $SC(\mathcal{T}_f, p)$ and $SC(\mathcal{T}_f, q)$ are comparable (see Fig. 2 - Case 3(a)), then suppose without loss of generality that $SC(\mathcal{T}_f, p) \subset SC(\mathcal{T}_f, q)$. Thus, $SC(\mathcal{T}_f, p) \neq SC(\mathcal{T}_g, p)$ and $SC(\mathcal{T}_f, q) = SC(\mathcal{T}_g, q)$ and consequently $g(q) = f(q)$.

   – If $SC(\mathcal{T}_f, p)$ and $SC(\mathcal{T}_f, q)$ belong to $SAT_\mathcal{U}(f)$ (resp. $SAT_\mathcal{L}(f)$) then thanks to Prop. 5, follows that $SC(\mathcal{T}_g, p) \in SAT_\mathcal{U}(f)$. Thus, thanks to Prop. 3, follows that $f(p) \geq g(p) \geq g(q) = f(q)$ (resp. $f(p) \leq g(p) \leq g(q) = f(q)$), that meets the definition of leveling (see left tree of Fig. 2 - Case 3(a));

   – If $SC(\mathcal{T}_f, p)$ and $SC(\mathcal{T}_f, q)$ are of different types, then thanks to Corollary 1, it follows that $\forall (r, s) \in \mathcal{A}$ such that $r \in SC(\mathcal{T}_f, p)$ and $s \notin SC(\mathcal{T}_f, p)$ follows that $s \in SC(\mathcal{T}_f, q)$. Thus, $SC(\mathcal{T}_f, p) \subset SC(\mathcal{T}_f, s) \subseteq SC(\mathcal{T}_f, q)$ and consequently either $f(p) < f(s) \leq f(q) \Rightarrow f(p) \leq g(p) \leq g(q) = f(q)$ or $f(p) > f(s) \geq f(q) \Rightarrow f(p) \geq g(p) \geq$

$g(q) = f(q)$, that meets the definition of leveling (see right tree of Fig. 2 - Case 3(a)).

(b) If $SC(\mathcal{T}_f, p)$ and $SC(\mathcal{T}_f, q)$ are disjoint (see Fig. 2 - Case 3(b)), then they are of different types (see Prop. 3). Moreover, $SC(\mathcal{T}_f, p)$ and $SC(\mathcal{T}_f, q)$ are not in the same branch of $\mathcal{T}_f$. In this case, certainly there exists a node $SC(\mathcal{T}_f, r)$ which is common ancestor of both nodes $SC(\mathcal{T}_f, p)$ and $SC(\mathcal{T}_f, q)$. Thus, $SC(\mathcal{T}_f, r)$ is the same type of $SC(\mathcal{T}_f, p)$ or $SC(\mathcal{T}_f, q)$. Then, either $f(p) > f(r) > f(q)$ or $f(p) < f(r) < f(q)$. Without loss of generality, assume $f(p) > f(r) > f(q)$. Thus, if $SC(\mathcal{T}_f, p)$ is removed and $SC(\mathcal{T}_f, q)$ is preserved, then we have that $SC(\mathcal{T}_f, p) \subset SC(\mathcal{T}_g, p) \subseteq SC(\mathcal{T}_f, r)$ and $SC(\mathcal{T}_g, q) = SC(\mathcal{T}_f, q)$ and consequently $f(p) > g(p) \geq f(r)$ and $g(q) = f(q)$. Therefore, $f(p) > g(p) \geq f(r) > g(q) = f(q) \Rightarrow f(p) > g(p) > g(q) = f(q)$ which in turn meets the leveling definition. But, if $SC(\mathcal{T}_f, p)$ is preserved and $SC(\mathcal{T}_f, q)$ is removed, then we have $SC(\mathcal{T}_g, p) = SC(\mathcal{T}_f, p)$ and $SC(\mathcal{T}_f, q) \subset SC(\mathcal{T}_g, q) \subseteq SC(\mathcal{T}_f, r)$ and consequently $f(p) = g(p)$ and $f(r) \geq g(q) > f(q)$. Therefore, $f(p) = g(p) > g(q) > f(q)$ which in turn meets the leveling definition.

4   Morphological Reconstruction Based on a Marker Image

In this section, present a new and fast algorithm for morphological reconstructions based on a marker image by reconstruction of pruned trees. In fact, pruning strategy based on marker image is not well explored in the literature. The strategy is to use the marker image to determine the place of pruning in extended versions of max-trees and min-trees.

The reconstruction operator is relatively simple in binary case, which consists in extracting the CCs of an binary image $X \subseteq \mathcal{D}$ which are marked by another binary image $M \subseteq X$. The binary images M and X are respectively called marker and mask. Then, the reconstruction $\rho_B(X, M)$ of mask X from marker M is the union of all the CCs of X which contain at least one pixel of M, i.e., $\rho_B(X, M) = \{C \in CC(X) : C \cap M \neq \emptyset\}$ [12].

To extend the reconstruction operator to grayscale images, we recall that any increasing operator defined for binary images can be extended to grayscale images through threshold decomposition [16]. Thus, given a mask image $f \in \mathcal{F}(\mathcal{D})$ and a marker image $g \in \mathcal{F}(\mathcal{D})$ such that $g \leq f$, we have:

$$\forall p \in \mathcal{D}, [\rho(f, g)](p) = \sup\{\mu \in \mathbb{K} : p \in \rho_B(\mathcal{X}_\mu^\uparrow(f), \mathcal{X}_\mu^\uparrow(g))\}$$

$$\iff [\rho(f, g)](p) = \sup\{\mu \in \mathbb{K} : p \in \{C \in CC(\mathcal{X}_\mu^\uparrow(f)) : C \cap \mathcal{X}_\mu^\uparrow(g) \neq \emptyset\}\}$$

$$\iff [\rho(f, g)](p) = \sup\{\mu \in \mathbb{K} : SC(\mathcal{U}(f), p) \cap \mathcal{X}_\mu^\uparrow(g) \neq \emptyset\}.$$

Note that, to construct $\rho(f, g)$ for $p \in \mathcal{D}$, we need to find the smallest set $\mathcal{X}_\mu^\uparrow(g)$ such that $\mathcal{X}_\mu^\uparrow(g) \cap SC(\mathcal{U}(f), p) \neq \emptyset$. Fortunately, this can be expressed as a pruning operation in the extended version of $(\mathcal{U}(f), \subseteq)$ as follows: remove (resp. preserve) all nodes $(C, \mu) \in Ext(\mathcal{U}(f))$, if and only if, there exists a pixel $p \in C$ such that $\mu > g(p)$ (resp. $g(p) \geq \mu$). Therefore, $\rho(f, g) = Rec((\mathcal{T}_\mathcal{U}, \subseteq))$ such that $\mathcal{T}_\mathcal{U} = \{(C, \mu) \in Ext(\mathcal{U}(f)) : \bigvee_{p \in C} g(p) \geq \mu\}$.

Following these ideas, we present the Algorithm 1 to computes the reconstruction by dilation. This algorithm makes use of a priority queue to process the pixels of the marker image in an orderly manner and so we do not reprocess the nodes that already were visited.

Algorithm 1. Compute the reconstruction by dilation

```
 1  Image reconstruction(Max-tree 𝒯_f, Image marker g) begin
 2  |   Initialize priority queue 𝒬
 3  |   foreach C ∈ Ext(𝒯_f) do  remove[C] = true  foreach p ∈ 𝒟 such that g(p) ≤ f(p) do
    |     add (𝒬, p, g(p))
 4  |   while 𝒬 is not empty do
 5  |   |   p = removeMaxPriority(𝒬);
 6  |   |   (C, μ) = SC(Ext(𝒯_f), p)
 7  |   |   if C was not processed then
 8  |   |   |   while remove[C] is true AND μ > g(p) do
 9  |   |   |   |   C = parent (C)
10  |   |   |   remove[C] = false
11  |   |   |   while parent(C) is not null AND remove[parent(C)] is true do
12  |   |   |   |   remove[parent(C)] = false
13  |   |   |   |   C = parent(C)

    |   /* reconstruction of pruned tree, i.e., Rec(Pruning(𝒯_f)) such that
    |      Pruning(𝒯_f) = {(C, μ) ∈ Ext(𝒯_f) : remove[C] = false}                      */
14  |   Initialize queue 𝒬_fifo
15  |   enqueue(𝒬_fifo, root(Ext(𝒯_f)))
16  |   while 𝒬_fifo is not empty do
17  |   |   (C, μ) = dequeue(𝒬_fifo)
18  |   |   if remove[C] is false then  ∀p ∈ C, f(p) = μ  foreach S ∈ children(C) do
    |   |     enqueue(𝒬_fifo, S)

    |   /* The image f = ρ(f, g) is the result of reconstruction by dilation of mask image
    |      f using the marker image g.                                                   */
19  |   return f
```

The reconstruction using upper level sets can also be defined through the geodesic dilation of f with respect to g iterated until stability. In this respect, a traditional fast algorithm for computing reconstruction by dilation has been proposed by Luc Vincent [12]. Thus, we show in Fig. 3 a graphic for a comparison simples.

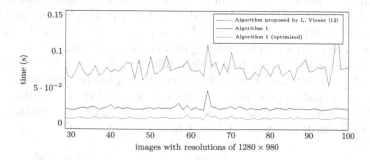

Fig. 3. Comparison of computation time. In this comparison each image of ICDAR dataset [17] is tested with 10 different marker images and the average value is plotted.

Of course, by duality, one can obtain the reconstruction by erosion ρ^* using similarly lower level sets. These two morphological reconstructions techniques are at basis of numerous operators, such as: opening and closing by reconstruction, top-hat by reconstruction, h-basins, h-domes and others [3, 12].

5   Scale-Space Representation Through a Sequence of Reconstructions of Successive Prunings

It is already known that some operators can be obtained by reconstructions of pruned trees, as for example: attribute opening (resp. closing), grain filters, and others. From the previous section, we showed that some operators based on marker images also can be obtained by pruning operation such as opening by reconstruction, top-hat by reconstruction, h-basins and others. Taking advantage of this property, we will show in this section a way to build scale-space representation by a sequence of successive prunings.

Let $\mathcal{T}_f$ be the tree (max-tree, min-tree or tree of shapes) that represents an image f. Since levelings can be nested to create a space-scale decomposition of an image, by Theo. 1 and 2, we have that $Rec(\mathcal{P}runing(\mathcal{P}runing(\mathcal{T}_f)))$ is a leveling of $Rec(\mathcal{P}runing(\mathcal{T}_f))$ and $Rec(\mathcal{P}runing(\mathcal{T}_f))$ is a leveling of $Rec(\mathcal{T}_f) = f$. Then, by transitivity, we also have that $Rec(\mathcal{P}runing(\mathcal{P}runing(\mathcal{T}_f)))$ is a leveling of $Rec(\mathcal{T}_f)$. This shows that the tree generates a family of levelings that further simplifies the image f, thus constituting a morphological space-scale and this leads us to Prop. 6.

Proposition 6. *Let $\mathcal{T}_f$ be the tree (max-tree, min-tree or tree of shapes) that represents an image f. Then, the sequence of reconstructions of successive prunings $(g_0 = Rec(\mathcal{T}_f),\ g_1 = Rec(\mathcal{P}runing(\mathcal{T}_f)),\ g_2 = Rec(\mathcal{P}runing(\mathcal{P}runing(\mathcal{T}_f))),\ ...,\ g_n = Rec(\mathcal{P}runing(...(\mathcal{P}runing(\mathcal{P}runing(\mathcal{T}_f)))))\ )$ is a space-scale of levelings such that g_k is a leveling of g_l for all $0 \leq l \leq k \leq n$.*

Thus, we can build through successive pruning: (1) scale-space based on attributes from increasing criteria on attributes and so generate scale-space of opening, closing and grain filter by attribute (or extinction values) and others; (2) scale-space based on marker images from a family of markers and so generate scale-space of reconstruction by opening and closing, top-hat by reconstruction, h-basins, h-domes and others. In addition, following F. Meyer [2], from Eq. 1 is possible define the self-dual reconstruction combining the reconstruction by dilation and erosion, and so generate scale-space of self-dual reconstruction. In fact, different families of markers may be used to generate a morphological scale-space based on levelings as shown by F. Meyer in [4].

$$\forall x \in \mathcal{D}, [\nu(f,g)](x) = \begin{cases} [\rho(f, g \wedge f)](x) & \text{, if } g(x) < f(x), \\ [\rho^*(f, g \vee f)](x) & \text{, if } g(x) > f(x), \\ f(x) & \text{, otherwise.} \end{cases} \quad (1)$$

Based on this idea, Fig. 4 presents some images of a scale-space generated with marker images produced by alternate sequential filtering.

Fig. 4. Some images of a scale-space generated with markers produced by alternate sequential filtering

6 Application Example

In many application in Image Processing and Analysis, the objects of interest which must be detected, measured, segmented, or recognized in an image are, in general case, not in a fixed but in many scales. For such situations, several multi-scale operators have been developed over the last few decades. In this sense, this section briefly illustrates (see Fig. 5) the application of some residual operators defined on an scale-space based on levelings [18, 19, 20]. They are: ultimate attribute opening (UAO) (resp. closing (UAC)) [20] and ultimate grain filters (UGF) [19]. They belong to a larger class of residual operators that we call ultimate levelings and defined from a indexed family of levelings $\{\psi_i : i \in I\}$ such that $i, j \in I$, $i \leq j \Rightarrow \psi_j$ is a leveling of ψ_i. Thus, the an ultimate leveling is defined by $\mathcal{R}_\theta(f) = \mathcal{R}_\theta^+(f) \vee \mathcal{R}_\theta^-(f)$ where $\mathcal{R}_\theta^+(f) = \sup\{r_i^+(f) : r_i^+(f) = [\psi_i(f) - \psi_{i+1}(f) \vee 0]\}$ and $\mathcal{R}_\theta^-(f) = \sup\{r_i^-(f) : r_i^-(f) = [\psi_{i+1}(f) - \psi_i(f) \vee 0]\}$. They can be implemented efficiently through of a max-tree, min-tree or tree of shapes [21, 19].

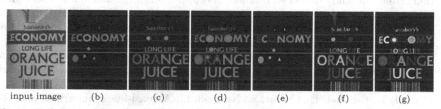

input image   (b)   (c)   (d)   (e)   (f)   (g)

Fig. 5. Example of extraction of contrast and segmentation using UAO (b) and (e), UAC (c) and (f), and UGF (d) and (g)

7 Conclusion

In this work, we have presented scale-space representations of an image based on levelings through hierarchies of level sets (component trees and tree of shapes). For that, we first proved the main result of this paper in Section 3 that reconstructions of pruned trees are levelings. After that, in Section 4 we present a new and fast algorithm for computing the reconstruction based on marker images from component trees. Finally, in Section 5 we show how to build morphological scale-spaces based on levelings through the reconstructions of successive pruning operations (whether based on increasing attributes or marker images).

Acknowledgements. We would like to thank the financial support from CAPES, CNPq, and FAPESP (grant #2011/50761-2).

References

[1] Salembier, P., Serra, J.: Flat zones filtering, connected operators, and filters by reconstruction. IEEE Transactions on Image Processing 4, 1153–1160 (1995)

[2] Meyer, F.: From connected operators to levelings. In: Mathematica Morphology and its Applications to Image and Signal Processing, pp. 191–198 (1998)

[3] Meyer, F.: The levelings. In: Mathematical Morphology and its Applications to Image and Signal Processing, pp. 199–206 (1998)

[4] Meyer, F., Maragos, P.: Nonlinear scale-space representation with morphological levelings. J. Vis. Commun. Image Represent. 11(2), 245–265 (2000)

[5] Meyer, F., Maragos, P.: Morphological scale-space representation with levelings. In: Nielsen, M., Johansen, P., Fogh Olsen, O., Weickert, J. (eds.) Scale-Space 1999. LNCS, vol. 1682, pp. 187–198. Springer, Heidelberg (1999)

[6] Salembier, P., Oliveras, A., Garrido, L.: Anti-extensive connected operators for image and sequence processing. IEEE Trans. on Image Processing 7, 555–570 (1998)

[7] Caselles, V., Monasse, P.: Grain filters. J. M. Imaging Vision 17(3), 249–270 (2002)

[8] Xu, Y., Geraud, T., Najman, L.: Morphological filtering in shape spaces: Applications using tree-based image representations. In: 2012 21st International Conference on Pattern Recognition (ICPR), pp. 485–488 (2012)

[9] Dalla Mura, M., Benediktsson, J.A., Bruzzone, L.: Self-dual attribute profiles for the analysis of remote sensing images. In: Soille, P., Pesaresi, M., Ouzounis, G.K. (eds.) ISMM 2011. LNCS, vol. 6671, pp. 320–330. Springer, Heidelberg (2011)

[10] Salembier, P., Wilkinson, M.: Connected operators. IEEE Signal Processing Magazine 26(6), 136–157 (2009)

[11] Terol-Villalobos, I.R., Vargas-Vázquez, D.: Openings and closings with reconstruction criteria: a study of a class of lower and upper levelings. Journal of Electronic Imaging 14(1), 013006–013006–11 (2005), doi:10.1117/1.1866149

[12] Vincent, L.: Morphological grayscale reconstruction in image analysis: applications and efficient algorithms. IEEE Trans. on Image Processing 2(2), 176–201 (1993)

[13] Caselles, V., Meinhardt, E., Monasse, P.: Constructing the tree of shapes of an image by fusion of the trees of connected components of upper and lower level sets. Positivity 12(1), 55–73 (2008)

[14] Géraud, T., Carlinet, E., Crozet, S., Najman, L.: A quasi-linear algorithm to compute the tree of shapes of nd images. In: Hendriks, C.L.L., Borgefors, G., Strand, R. (eds.) ISMM 2013. LNCS, vol. 7883, pp. 98–110. Springer, Heidelberg (2013)

[15] Caselles, V., Monasse, P.: Geometric Description of Images As Topographic Maps, 1st edn. Springer Publishing Company, Incorporated (2009)

[16] Maragos, P., Ziff, R.: Threshold superposition in morphological image analysis systems. IEEE Trans. on Pattern Analysis and Machine Intel. 12, 498–504 (1990)

[17] Lucas, S.M., Panaretos, A., Sosa, L., Tang, A., Wong, S., Young, R.: Icdar 2003 robust reading competitions. ICDAR 2003 2, 682 (2003)

[18] Alves, W., Morimitsu, A., Castro, J., Hashimoto, R.: Extraction of numerical residues in families of levelings. In: 2013 26th SIBGRAPI - Conference on Graphics, Patterns and Images (SIBGRAPI), pp. 349–356 (2013)

[19] Alves, W.A.L., Hashimoto, R.: Ultimate grain filter. In: 2014 IEEE International Conference on Image Processing (ICIP), Paris, France, pp. 2953–2957 (2014)

[20] Marcotegui, B., Hernández, J., Retornaz, T.: Ultimate opening and gradual transitions. In: Soille, P., Pesaresi, M., Ouzounis, G.K. (eds.) ISMM 2011. LNCS, vol. 6671, pp. 166–177. Springer, Heidelberg (2011)

[21] Fabrizio, J., Marcotegui, B.: Fast implementation of the ultimate opening. In: Proc. 9th International Symposium on Mathematical Morphology, pp. 272–281 (2009)

Cluster Based Vector Attribute Filtering

Fred N. Kiwanuka[1(✉)] and Michael H.F. Wilkinson[2]

[1] College of Computing and Information Sciences, Makerere University, P.O. Box 7062
Kampala, Uganda
kiwanoah@gmail.com
[2] Institute for Mathematics and Computing Science, University of Groningen, P.O. Box 407,
9700 Groningen AK, The Netherlands
m.h.f.wilkinson@rug.nl

Abstract. Morphological attribute filters operate on images based on properties
or attributes of connected components. Until recently, attribute filtering was based
on a single global threshold on a scalar property to remove or retain objects.
A single threshold struggles in case no single property or attribute value has a
suitable, usually multi-modal, distribution. Vector-attribute filtering allows better description of characteristic features for 2D images. In this paper, we apply
vector-attribute filtering to 3D and incorporate unsupervised pattern recognition,
where connected components are classified based on the similarity of feature vectors. Using a single attribute allows multi-thresholding for attribute filters where
more than two classes of structures of interest can be selected. In vector-attribute
filters automatic clustering avoids the need for either setting very many attribute
thresholds, or finding suitable class prototypes in 3D and setting a dissimilarity
threshold. Explorative visualization reduces to visualizing and selecting relevant
clusters. We show that the performance of these new filters is better than those of
regular attribute filters in enhancement of objects in medical images.

Keywords: Image enhancement · Object detection · Attribute Filters · Connected
Operators · Max Tree · Clustering

1 Introduction

The field of connected mathematical morphology has contributed a wide range of operators to image processing. Efficient techniques and algorithms have been developed
for extracting image components that are useful in the representation and description
of shapes. In many applications, an important task is to extract particular regions of an
image while preserving as much of the contour information as possible. This is the main
aim of connected filters [15, 16], a strictly edge preserving class of operators in mathematical morphology. These operators act by merging flat zones given some criteria, and
filter an image without introducing new contours.

An important sub-class of connected filters are attribute filters [2, 14]. They allow
filtering based on the properties or *attributes* of connected components in the image.
Examples of attribute filters include attribute openings, closings, thickenings, and thinnings [2, 14, 19]. Despite the development of many types of attributes, in their current
format, attribute filters have two drawbacks. First, the attributes used are often a single

© Springer International Publishing Switzerland 2015
J.A. Benediktsson et al. (Eds.): ISMM 2015, LNCS 9082, pp. 277–288, 2015.
DOI: 10.1007/978-3-319-18720-4_24

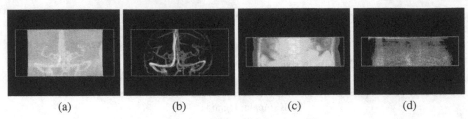

(a)   (b)   (c)   (d)

Fig. 1. Attribute Filtering: (a)original:*angiolarge* (b)filtered with non-comp(λ=3.2) (c)original:*kidney-stone* (d)filtered with non-comp(λ=7.8)

scalar value describing either size or shape properties of connected components. This works well if the desired structures can be separated easily from undesired structures, especially when attributes can be found of high discriminative power [5,6,12,19]. However, in many cases, objects of different classes are not easily discriminated by a single shape number. An example of this is shown in Fig. 1. The *non-compactness* is a known robust attribute filter and easily performs well on relatively low-clutter volumes, such as the (angiolarge) data set in Fig. 1(b). However, on a more noisy image with more clutter, like kidney-stone, the filter completely fails, as shown in Fig. 1(d). This could imply that perhaps a single threshold or attribute is deficient.

This is the reason why vector-attribute filtering [11,17,18] was proposed. This allows a better description of characteristic features such as size and shape of the objects in the image. These features have been studied for synthetic images of characters and 2D dermatological images, and are based on dissimilarity measures such as Euclidean or Mahalanobis distance in feature space. Components that are similar to a set of reference shapes can be preserved or removed. This has been applied when a priori knowledge of a suitable reference shape is known. In many image filtering tasks such as medical images a priori knowledge of a given object is not readily available. In this research we develop 3D vector-attribute filters which do not rely on reference shapes.

In this research, we propose improving the robustness and the versatility of attribute filters by developing vector-attribute filters in which features are selected or rejected based on feature vectors, just as in [11,17,18], rather than a single property. Unlike the previous work, we apply this to 3D medical volume data sets, in which the selection of reference shapes is far more difficult than in the 2D case, where we can simply delineate features of interest on the screen. Therefore, we want to develop a method of interactive attribute filtering which does not need prior knowledge of ideal target shapes, and preferably requires minimal a-priori setting of parameters. To achieve this we adapt unsupervised pattern recognition approaches, where object classes are learned based on the clustering of attribute vectors. We demonstrate the capacity of these approaches on 3D biomedical images for both size and shape based attributes.

The paper is organized as follows. The theory of vector-attribute filters is covered in Section 2. In Section 3, a description of attribute cluster filter computation and implementation for vector-attribute filtering is built. While Section 4, presents experimental results obtained for the vector-attributes in 3D medical image enhancement with a comparison to other methods are presented. A discussion of results is also in this section. We give concluding remarks in Section 5.

2   Vector-Attribute Filtering

In the binary case, attribute filters [2], retain those connected components of an image, which meet certain criteria. After computing the connected components, some property or attribute of each component is computed. A threshold is usually applied to these attributes to determine which components are retained and which removed. Thus, the criterion Λ, usually has the form

$$\Lambda(C) = Attr(C) \geq \lambda \tag{1}$$

with C the connected component, $Attr(C)$ some real-valued attribute of C and λ the attribute threshold. For grey scale image f, we compute these attributes for the connected components of threshold sets $X_h(f)$, defined as

$$X_h(f) = \{x \in E | f(x) \geq h\}. \tag{2}$$

Urbach et al. [17] replaced the single attribute by a feature vector of dimensionality D. Rather than setting D thresholds, they based the criterion on dissimilarity to a reference vector r, ideally obtained from some reference shape.

They define a multi-variate attribute thinning $\Phi^{\{\Lambda_i\}}(X)$ with scalar attributes τ_i and their corresponding criteria $\{\Lambda_i\}$, with $1 \leq i \leq D$, such that connected components are preserved if they satisfy at least one of the criteria $\{\Lambda_i(C)\}$ and are removed otherwise:

$$\Phi^{\{\Lambda_i\}}(X) = \bigcup_{i=1}^{D} \Phi^{\Lambda_i}(X) \quad \text{with} \quad \Lambda_i(C) \equiv \tau_i(C) \geq \lambda_i. \tag{3}$$

with λ_i are the attribute thresholds.

The set of scalar attributes τ_i can also be considered as a single vector-attribute $\tau = \{\tau_1, \tau_2, \dots, \tau_D\}$, in which case a vector-attribute thinning is needed with a criterion:

$$\Lambda_\lambda^\tau \equiv \exists i : \tau_i(C) \geq \lambda_i \quad \text{for } 1 \leq i \leq D. \tag{4}$$

with λ the attribute threshold vector. Urbach et al. [17] then proceed to a more useful criterion defined as

$$\Lambda_{r,\epsilon}^\tau(C) \equiv d(\tau(C), r) \geq \epsilon \tag{5}$$

where d is some dissimilarity measure, r is a reference vector, and ϵ is a dissimilarity threshold. This replaces D parameters with just a single value ϵ, but adds the need for a reference vector. A binary vector-attribute thinning $\Phi_{r,\epsilon}^\tau(X)$, with D-dimensional vectors removes the connected components of a binary image X whose vector-attributes differ more than a given quantity from a reference vector $r \in \Upsilon$.

Alternatively [11] suggested

$$\Lambda_{r,\epsilon}^\tau(C) \equiv d(\tau(C), r) \leq \epsilon \tag{6}$$

which is essentially the complement (but not quite) of the form of [17]. Therefore, this preserves all objects with attribute vectors sufficiently similar to the reference, rejecting all others. We will work with this latter form in the remainder of the paper.

While it is possible to compose reference shapes in the 2D case of letters from a known font [17], in 3D it becomes much harder. Therefore, it could be useful to consider approaches that do not need these reference vectors a priori.

The approach here derives some inspiration from [18]. This paper introduces the notion of *context attributes* of components. Context attributes describe how a component relates to other components in the image. Alignment, distance, and similarities in size, shape, and orientation between the individual components are used to determine which components belong to the same object. Contextual filter preserves only those components which visually appear to belong to a certain group of similar components. Urbach [18] only considers the binary case, and focuses mainly on spatial relations such as proximity and alignment. Here we move to grey scale and volume data, and focus exclusively on similarities in terms of (vector-)attributes.

Another related approach is that of Xu et al. [21], who propose a method for filtering Max-Trees with non-increasing attribute, by building a Max-Tree of a Max-Tree. This method cannot deal with vector attributes, however. We will therefore not perform a comparison. A similar approach in [13], was published after submission of our initial work [7], and that a comparison would be of interest in future work. It proposes a hierarchical Markovian unsupervised algorithm in order to classify the nodes of the traditional Max-Tree to handle multivariate attributes.

2.1   The Clustering Approach

Here we follow a different approach. Ideally, we would like to select attributes in such a way that vectors belonging to different categories of objects occupy compact and disjoint regions in D-dimensional attribute vector space. Thus, given a suitable set of attributes or features, we could automatically organize the huge number of connected components of all threshold sets into a much smaller number of groups by automatic clustering. Instead of painstakingly setting reference shape and correct distance threshold, the user now inspects a limited number of clusters.

Let $\mathcal{C} = \{C_1, C_2, \ldots, C_N\}$, be set of connected components of image X where $\tau(C_i) \in \mathbb{R}^D$ denotes the associated attribute vector. As in [17] τ is the vector attribute function.

Any clustering partitions $\mathbb{R}^D$ into k sets. We denote the partition classes $P_j \subset \mathbb{R}^D$, $j = 1, 2, \ldots, k$. Every vector $x \in \mathbb{R}^D$ lies in one or more partition classes. The cluster criterion Λ_j becomes

$$\Lambda_j(C) = (\tau(C) \in P_j) \tag{7}$$

i.e. it returns true if the attribute vector of C lies in partition P_j. Replacing the usual criterion (1) in attribute filters with (7), we can draw up the *attribute cluster filter* ψ_{Λ_j}

$$\psi_{\Lambda_j}(X) = \bigcup_{x \in X} \psi^{\Lambda_j}(\Gamma_x(X)). \tag{8}$$

It is trivial to show that ψ_{Λ_j} adheres to all the properties of vector-attribute filters.

3   Attribute Cluster Filter Computation and Implementation

In this section we describe vector-attribute filtering pattern classications in brief detail.

1. **Feature selection**: A large number of both size and shape attributes for filtering in 3D is now available [5, 6, 12, 19, 20]. These attributes enhance the ability of connected filters to select structures of interest for different imaging modalities. Ideally in selecting the attributes we require these attributes to distinguish patterns belonging to different clusters and be less immune to noise. Currently this is done manually. For efficient computation of the attributes, we utilize the Max-Tree [14]. The Max-Tree is a data structure that was designed for morphological attribute filtering. The filtering process is separated into four stages: build, compute attributes, filter and restitution. It is this filtering process that we change in our research, rather than decision being based on attribute signature of the connected component, the decision is based on the feature vector and class of the component determines whether to be removed or retained.

2. **Clustering**: Clustering is ubiquitous and a wealth of clustering algorithms have been developed to solve different problems in various fields. There is no clustering algorithm that can be universally used to solve all problems [8]. In this research we explore four well researched clustering algorithms: k-means [10], fuzzy c-means (FCM) [1], Vector quantization [9] and Mean Shift [3, 4]. In the final results, we eliminated Vector quantization because it's performance was very similar to k-means but slower.

3. **Implementation**: We implemented vector-attributes for 3D grey-scale attribute filtering in the MTdemo package [19]. This uses the Max-Tree [14] data structure to compute and visualize volumetric data. In [19], Max-Tree construction and attribute computation are separated, allowing computation of different attributes without complete re-building of the tree. However, filtering is based on a single attribute property in this structure, and the notion of filtering is hard-coded as comparison to a threshold λ. To accommodate vector-attribute computation and attribute cluster filtering a number of changes are included in the structure: more notably, *NodeSelector* abstract class, which enables us to use any type of filtering criteria other than attribute signature *clusterID* an extra field per node is required, this field stores the cluster label of the node, *ComputeAttributes()* function used in the vector-attributes computation class to compute any number of attributes and store the attributes for each node in an auxiliary array index or vector.

4   Results and Discussion

To demonstrate the performance of attribute-cluster filtering, we ran tests on a number of 3D medical images of different modalities, courtesy of http://www.volvis.org and the Department of Radiology and Medical Imaging, University General Hospital of Alexandroupolis, Greece. In the following, we often use just a single attribute, in order to compare more reasonably with classical attribute filters based on attribute thresholds. Due to difficulties in quantifying filtering results in these cases, we first perform a more quantitative test on a printed document.

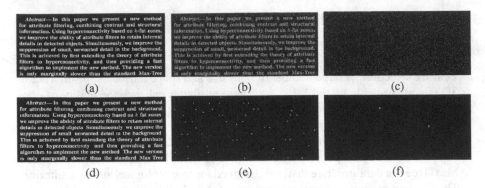

Fig. 2. Mean shift attribute cluster filtering of document: (a) Original (b) corrupted (c,d,e,f) the different clusters

Document Validation: We begin this section by demonstrating how attribute cluster filters are computed through a simple example of a document that has been corrupted with salt and pepper noise of density 0.3 as shown in Fig. 2. We then apply an attribute cluster filter using the volume attribute (equivalent to area filtering in 2-D) to the document which yields four foreground clusters: (i) the cluster containing the alphabet characters, (ii) the noise, (iii and iv) the punctuation marks. A close inspection shows that this is a valid decom- position of the original document. All clusters are shown in Fig. 2. To provide a quantitative measure of the quality of these methods in image filtering we used the document and filtered the document using the area attribute. Performance analysis of attribute-cluster filtering is carried out using universal quality index (UQI), which models image distortion as a combination of loss of correlation, luminance distortion, and contrast distortion. The regular attribute filter using a manually selected area-threshold achieves a UQI index of 0.91 while the attribute-cluster filter achieves a UQI index of 0.89, both with respect to the original, uncorrupted document. Thus the automatic thresholds chosen by k-means clustering yield a result very close to the manual method.

4.1   MRI and CT Scan Performance

The angiolarge: In this experiment we compared the performance of the attribute cluster filter against the conventional attribute filters. In their current format, attribute filtering for 3d medical images based on size-based attributes perform very poorly, they not only fail in enhancing blood vessels but also amplify noise. This is illustrated in Fig. 3 (b), a *volume* attribute ($\lambda = 9000$) simply amplifies noise on this data set. This also applies for all size based attributes like *surface area, X-extent, Y-extent, Z-extent*. However, when attribute cluster filter is applied on the *volume* attribute, the performance of this attribute is seen in Fig. 3(c). The noise is not only suppressed but the blood vessels are clearly enhanced. In this case the k-means clustering was used with the number of clusters $k = 11$. The result presents one of the 11 clusters computed for the data set but this was selected as a basis of comparison because in this particular cluster more blood vessels were retained. The other clusters can be availed on request. The performance of

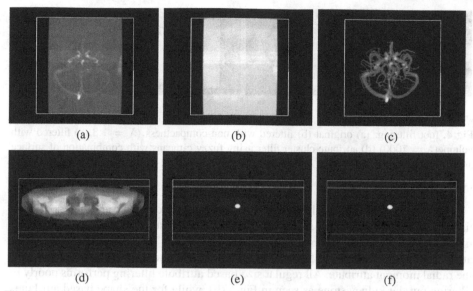

Fig. 3. (a) original; (b) filtered with volume($\lambda = 9000$); (c) attribute cluster filter by the same attribute using k-means with $k = 8$; (d) original; (e) filtered with radial moment ($\beta = 5, \lambda = 0.00256$); (f) attribute cluster filter by the same attribute using k-means with $k = 17$

the other size based attributes is also improved by using attribute cluster filtering. It's important to note that the performance improvement is irrespective of the increase in the dimensionality of the vector-attributes used. We tested this up to 6 attributes. The other clustering techniques are also able to achieve this result but at higher computational time. Shape based attributes like *non-compactness, radial moment* always perform well on this data set even with simple attribute thresholding.

The prostate-stone: On this data set, attribute filters in their current format are able to isolate the stone but they are never successful in suppressing the noise Fig. 3(e). The problem has been eradicated in [5] [6] by filtering using 2 attributes successively. First, a *non-compactness, or sphericity or radial moment* is applied to the data set to obtain result shown in Fig. 3(e), then a *volume or surface area* is applied to remove the remaining noise. However, using attribute cluster filtering, the result in Fig. 3(f) is obtained in a single step. This result was obtained using k-means and the *non-compactness* attribute with $k = 17$. Higher dimension of the vector up to 5 attributes was capable of isolating the stone in a single step but at a higher computational cost. The other clustering techniques are also able to achieve this result with mean shift clustering using 3 attributes while fuzzy c-means a single attribute but at higher computational cost as compared to k-means.

The human foot: Attribute filters normally struggle to suppress noise on this data set like seen in Fig. 4. In Fig. 4(b), the *non-compactness* attribute is able to enhance the bones but with noise still visible, while Fig. 4(c) the *surface area* attribute like all other size based attributes simply amplifies noise. However, attribute cluster filter using a combination of *surface area [6], surface area [12] and volume* perfectly enhances the bones and suppresses noise as seen in Fig. 4(d). All the clustering techniques perform well on this data set.

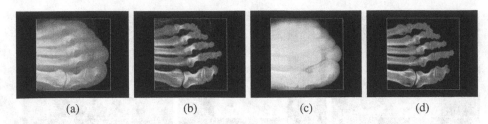

Fig. 4. foot filtering: (a) original (b) filtered with non-compactness ($\lambda = 1.3$ (c) filtered with volume($\lambda = 7000$) (d) attribute cluster filter using fuzzy c-means with combination of surface area, volume, surface area $k = 8$

The kidney-stone: The kidney-stone data set is more complex and has poor soft-tissue contrast, low SNR and shading effect. The kidney-stone data set is very hard to filter for most attributes when done on a single attribute. The performance of regular attributes on this data set is shown in Fig. 5(b) for volume attribute and Fig. 5(c) for the radial moment attribute. All regular size based attribute filtering performs poorly in filtering out the kidney stone as seen in Fig. 5(b), while for the shape based attributes only radial moment does a relative good job but still struggles to suppress noise as shown in Fig. 5(c). To suppress the noise the surface area or volume attributes are applied to Fig. 5(c) like in [5]. However, with attribute cluster filtering using mean shift algorithm and 5 size based attributes in a single step the kidney-stone is isolated without any noise as shown in Fig. 5(d). Attribute cluster filtering also reveals that there are more than one stone Fig. 5(f). This attribute filtering method is able to clearly enhance bony like structures like part of the spinal cord Fig. 5(e). The k-means and fuzzy c-Means fail to isolate the kidney stone but succeed in revealing the spinal cord.

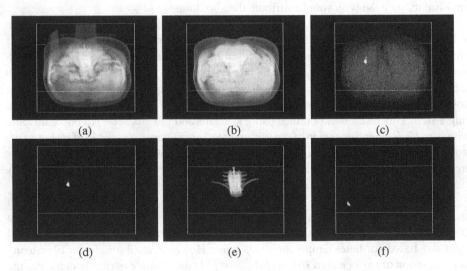

Fig. 5. kidney-stone filtering: (a) original; (b) filtered with volume ($\lambda=7400$); (c) filtered with radial moment ($\beta=3$, $\lambda=0.124$); (d,e,f) the result of attribute cluster filter using the mean shift

The female Chest: From Fig. 6 the performance of regular attribute filter is seen in Fig. 6(e), the radial moment(β) attribute is able to enhance the skeleton but other unwanted tissue still remains. However, an attribute cluster filter of any combination of size attributes not only enhances the skeleton without leaving unwanted tissue but also enhances the heart though faintly seen as in Fig. 6(f). All clustering techniques attain this result. In this experiment volume and non-compactness attributes were used with fuzzy c-means $k = 9$.

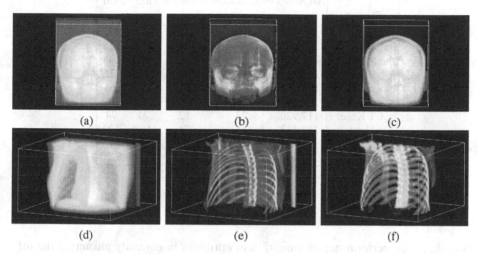

(a)   (b)   (c)

(d)   (e)   (f)

Fig. 6. Female-chest/Head Filtering: Left column (a) original head (c) filtered with non-compactness ($\lambda = 2.6$) (e) attribute cluster filter using k-means ($k = 9$); Right column (b) original chest (d) filtered with radial moment ($\beta = 3, \lambda = 0.034$); (f) attribute cluster filter using fuzzy c-means ($k = 12$)

The human Head: From Fig. 6 the performance of regular attribute filter is seen in Fig. 6(b), the non-compactness attribute enhances the head but other unwanted tissue still remains. However, an attribute cluster filter combination of any number of size attributes gives the result seen in Fig. 6(c). All noise is suppressed.

Computational Timings: Using a standard Core 2 Duo E8400 at 2.0 GHz, 2GB RAM machine we ran timings for the computation of each algorithm for attribute cluster filter-ing up to 6 attributes for different medical images of varying sizes and gray scale levels. The timings include the computation of the attributes and the clusters. The number of clusters computed was determined by the mean shift method because in this algorithm the number is not explicitly pre-determined thus we used it as the reference point. We consider the following attributes: *volume, surface area [6], surface area [12], X-extent, Y-extent and Z-extent*, all these attributes are size based. From Table 1, the results of the various clustering methods are shown. K-means algorithm is by far the faster algorithm as expected others follow interchangeably. The computation of shape based attributes is slower by an average factor of three, this is because of the floating values of shape descriptors. But overall the computational complexity of these methods is good even for large data sets, for instance, even for very large data sets like mrt16_angio with 1,554,454 nodes for 6 attributes for $k = 23$: it takes 17 seconds for size based attributes and 58 seconds for shape attributes. This is faster than most users can select an optimal setting for a single attribute.

Table 1. Cluster Computing Time(seconds)

Dataset	No.Nodes	No.Clusters	K-means	Fuzzy	MSF
angiolarge	361,463	10	3.02	170	172
mrt16_angio	1,554,454	7	12.1	819	525
mrt16_angio2	419,454	11	5.74	389	148
foot	279,513	11	5.50	64.0	62.9
prostate-stone	124,477	9	1.58	18.2	26.0
Kidney-stone	387,462	17	17.6	149	202
CT-Knee	117,920	23	12	71	74
Head	752,333	9	24	168	123
Chest	85,414	15	5.2	30.6	24.7
Document	290,446	3	0.11	0.40	8.20

4.2 Discussion

To evaluate the performance of the different attributes in correctly clustering the different data sets various combinations of the attributes were used. The performance of the combination of size based attributes was good on data sets that involved separating hard tissue from soft tissue that is the CT scans *foot, chest knee, kidney-stone*. This is in part due to the fact that these structures are brighter, and therefore are peaks in the Max-Tree. The combination of shape based attributes performed better on enhancing and noise suppression on data sets that exclusively were made of soft tissue that is the MRI *angiolarge, mrt16_angio, mrt16_angio2* . In clustering, size attributes dominate over shape. On the *angiolarge* data set the *non-compactness* normally does a good job in terms of vessel filtering and noise suppression, but *volume* always performs poorly. Volume has larger values as compared to non-compactness, and therefore volume decides the clustering results and thus the filtering looks more like volume attribute filter. This has to do with use of the Minkowski metric where the largest scaled features dominate others. This could be solved by normalizing the volume attributes but in this research we did not explore this. Experiments also show that with this kind of mixture, more structures seem to be retained when a pre-filter is first applied to very difficult data sets such as time of flight.

The clustering of scalar attributes (i.e. $d = 1$) using a suitable number of clusters for all attributes and most data sets gives very good results as compared to regular attribute filtering irrespective of whether it is a shape or size based attribute. A further increment in the number of attributes to more than 6 reveals little or no changes in performance for both size and shape based attributes. This could be due to the distance used in the clustering process as the similarity measure. Normalization or relevance learning could be used to combine features in a better way. The mean-shift algorithm has particular difficulties, where the performance in high dimensions degrades rapidly, possibly through the sparseness of data space.

Overall, all three clustering algorithms succeed on a number of data sets. By far k-means performs much better than the others while the mean shift looks promising with a number of great results. Perhaps the major concern of these clustering techniques is how to determine the *optimal* number of clusters. The mean shift determines this automatically but it has so many parameters that need to be set.

5 Conclusion

In this paper, we presented methods for computing attribute-cluster filtering in 3-D using unsupervised pattern classification where image or volume features are selected or rejected based on feature vectors rather than a single property. We have shown that the performance of attribute-cluster filters is better than those of regular attribute filters in enhancing structures in medical images and noise suppression in most cases. These filters show a lot of flexibility in selecting features of interest. Though the implementation of these techniques is not very sophisticated, their computational load is already acceptable. Algorithmic advances could improve this further. Attribute cluster filtering using k-means is fastest and gives good results. The fuzzy c-means is slower but allows us to have flexibility in deciding cluster membership especially in images without clear boundaries which is prevalent in medical images. The mean-shift method though slow as expected, performs well using standard kernel estimation profiles. Through its automatic cluster learning and unique image decomposition exhibits a lot of potential for further investigation.

In future work, we will study the behavior of attribute cluster filters for higher dimensional (≥ 10) attribute vectors. The current metric used in the clustering process as the similarity measure is not suitable in high dimensional space and certainly needs rescaling to better combine attributes with very different ranges. Thus, we intend to look at other metrics. Dimensionality reduction techniques are also required, in part to reduce the actual clustering times, but also because it is clear that not all the attributes contribute equally to the separation of the data.

References

1. Bezdek, J.C.: Pattern Recognition with Fuzzy Objective Function Algorithms. Kluwer Academic Publishers, Norwell (1981)
2. Breen, E.J., Jones, R.: Attribute openings, thinnings and granulometries. Comp. Vis. Image Understand. 64(3), 377–389 (1996)
3. Comaniciu, D., Meer, P.: Mean shift: A robust approach toward feature space analysis. IEEE Trans. Pattern Anal. Mach. Intell. 24(5), 603–619 (2002)
4. Fukunaga, K., Hostetler, L.D.: Estimation of the gradient of a density function with applications in pattern recognition. IEEE Trans. Inform. Theor. IT-21, 32–40 (1975)
5. Kiwanuka, F.N., Wilkinson, M.H.F.: Radial Moment Invariants for Attribute Filtering in 3D. In: Kthe, U., Montanvert, A., Soille, P. (eds.) Proc. Workshop on Applications of Discrete Geometry and Mathematical Morphology (WADGMM), pp. 37–41 (2010)

6. Kiwanuka, F.N., Ouzounis, G.K., Wilkinson, M.H.: Surface-area-based attribute filtering in 3d. In: Wilkinson, M.H.F., Roerdink, J.B.T.M. (eds.) ISMM 2009. LNCS, vol. 5720, pp. 70–81. Springer, Heidelberg (2009)
7. Kiwanuka, F.N., Wilkinson, M.H.F.: Cluster-based vector-attribute filtering for ct and mri enhancement. In: ICPR, pp. 3112–3115. IEEE (2012)
8. Kleinberg, J.: An impossibility theorem for clustering, pp. 446–453. MIT Press (2002)
9. Linde, Y., Buzo, A., Gray, R.: An algorithm for vector quantizer design. IEEE Transactions on Communications 28(1), 84–95 (1990)
10. Macqueen, J.B.: Some methods of classification and analysis of multivariate observations. In: Proceedings of the Fifth Berkeley Symposium on Mathematical Statistics and Probability, pp. 281–297 (1967)
11. Naegel, B., Passat, N., Boch, N., Kocher, M.: Segmentation using vector-attribute filters: Methodology and application to dermatological imaging. In: Proc. Int. Symp. Math. Morphology, ISMM 2007, pp. 239–250 (2007)
12. Ouzounis, G.K., Giannakopoulos, S., Simopoulos, C.E., Wilkinson, M.H.F.: Robust extraction of urinary stones from ct data using attribute filters. Proc. Int. Conf. Image Proc. (2009) (submitted)
13. Perret, B., Collet, C.: Connected image processing with multivariate attributes: An unsupervised markovian classification approach. Computer Vision and Image Understanding 133, 1–14 (2015)
14. Salembier, P., Oliveras, A., Garrido, L.: Anti-extensive connected operators for image and sequence processing. IEEE Trans. Image Proc. 7, 555–570 (1998)
15. Salembier, P., Serra, J.: Flat zones filtering, connected operators, and filters by reconstruction. IEEE Trans. Image Proc. 4, 1153–1160 (1995)
16. Salembier, P., Wilkinson, M.H.F.: Connected operators: A review of region-based morphological image processing techniques. IEEE Signal Processing Magazine 26(6), 136–157 (2009)
17. Urbach, E.R., Boersma, N.J., Wilkinson, M.H.F.: Vector-attribute filters. In: Mathematical Morphology: 40 Years On, Proc. Int. Symp. Math. Morphology (ISMM) 2005, April 18-20, pp. 95–104. Paris (2005)
18. Urbach, E.: Contextual image filtering. In: 24th International Conference on Image and Vision Computing New Zealand, IVCNZ 2009, pp. 299–303 (November 2009)
19. Westenberg, M.A., Roerdink, J.B.T.M., Wilkinson, M.H.F.: Volumetric attribute filtering and interactive visualization using the max-tree representation. IEEE Trans. Image Proc. 16, 2943–2952 (2007)
20. Wilkinson, M.H.F., Westenberg, M.A.: Shape preserving filament enhancement filtering. In: Niessen, W.J., Viergever, M.A. (eds.) MICCAI 2001. LNCS, vol. 2208, pp. 770–777. Springer, Heidelberg (2001)
21. Xu, Y., Géraud, T., Najman, L.: Morphological filtering in shape spaces: Applications using tree-based image representations. CoRR abs/1204.4758 (2012)

Intelligent Object Detection Using Trees

Erik R. Urbach[✉]

University of Florence, Florence, Italy
Erik.R.Urbach@ieee.org

Abstract. In this paper a method is proposed for detection and local-isation of objects in images using connected operators. Existing meth-ods typically use a moving window to detect objects, which means that an image needs to be scanned at each pixel location for each possible scale and orientation of the object of interest which makes such meth-ods computationally expensive. Some of those methods have made some improvements in computational efficiency but they still rely on a mov-ing window. Use of connected operators for efficiently detecting objects has typically been limited to objects consisting of a single connected region (either based on simple or more generalized connectivities). The proposed method uses component trees to efficiently detect and locate objects in an image. These objects can consist of many segments that are not necessarily connected. The computational efficiency of the connected operators is maintained as objects of interest of all scales and orienta-tions are detected using two component trees constructed from the input image without using any moving window.

Keywords: Object detection · Component trees · Connected filters · Trainable filters · Vector attributes · AdaBoost · Contextual filters

1 Introduction

Commonly, object detection methods [14] use a moving window to scan for ob-jects in a given input image. This can be computationally expensive as the object detection method, e.g., template matching, needs to be applied at (nearly) every pixel location in the image and for every possible orientation and scale. Much work has been done on developing more efficient algorithms for window-based object detection. A very popular approach by Viola and Jones [19] uses Haar-like feature types that are very fast to compute and decision processes that on aver-age decide quickly to accept or reject a window. A drawback of the Viola-Jones approach is the limitation in the use of 24x24 training set images. Although larger training set images could be used, this would render the training pro-cedure infeasible due to excessive memory and computing time requirements. Although images of human faces (the original purpose of Viola-Jones method) can be scaled to 24x24 without losing essential image details or lead to problems with the fixed (square) image ratio, neither of these would be true in general for all object detection applications. Similarly, extending their approach to 3-D

© Springer International Publishing Switzerland 2015
J.A. Benediktsson et al. (Eds.): ISMM 2015, LNCS 9082, pp. 289–300, 2015.
DOI: 10.1007/978-3-319-18720-4_25

images would generally lead to unacceptable computational and memory requirements for the training process. Various variants of the original Viola-Jones method have been proposed that are more efficient [1, 23] with detecting objects at multiple orientations. While these are more complex than the original Viola-Jones approach they still use local scanning windows.

The method proposed here computes features that are invariant under translation, scaling, and rotation which are computed from the whole image at once instead of using a local window. The features are computed using connected operators [8, 11], which are a commonly used set of tools from the field of mathematical morphology [13] for filtering [20], segmentation [12], and classification [15, 16] of features in 2-D and 3-D images. Existing approaches of using connected operators for object detection can in general be distinguished into two categories, namely: feature segmentation and image classification. In the former, the purpose is to automatically segment image features in an image. These image features usually consist of only one connected region (whether defined using conventional or more sophisticated connectivities such as second-generation connectedness [10]). Examples of this approach are filtering blood vessels in medical images [20] and segmenting cracks in 3-D images of shale rock samples [18]. The purpose of the latter (i.e., image classification) is not to locate or segment objects of interests in images but just to distinguish different kinds of images. Many approaches in this category use a classifier with feature vectors where each feature vector summarises key information about the image (or of a predetermined region mask thereof) it represents. Intensity, size or shape histograms are commonly used to construct feature vectors. These size or shape histograms can be computed by using operators with structuring elements [2] or by using connected operators [15]. A straightforward use of pattern spectra for object detection is to use them with a moving window and obtain a pattern spectra-based variety of the Viola-Jones method which will also have the same limitations.

A different approach is needed for detection of more complicated objects such as those in Fig. 1 or human faces that consist of multiple components with certain spatial relationships between them in the midst of many other components that vary between images. In the case of a human face, one might be able to use conventional connected operators to detect face components such as eye pupils but as many objects in nature have features that are similar in appearance to these pupils a method is needed that can detect objects based on the co-occurrence and the spatial interrelationships of multiple significant features in the object of interest while ignoring irrelevant features such as a beard, glasses, or even shadows.

An approach referred to as a contextual image filter [17] has been proposed which uses connected operators to segment regions of components that share certain spatial relationships. There the aim was to segment regions of components that form patterns whose spatial relationships can be described by explicit rules. In contrast, the aim of the proposed method is to use connected operators to efficiently detect (or segment) objects consisting of multiple segments where the detector is formed by training it with a set of labeled example images without any a priori information other than the example images.

dots forming letter A   traffic sign   Martian crater

Fig. 1. Examples of three kinds of objects, each consisting of multiple components

A major benefit of the proposed method (besides speed) over other trainable object detection methods is that the training set can be much smaller as the method does not need to be trained for different scales or orientations and can even handle some small variations in shape or location among the segments between different instances of the same object.

Neither the proposed method nor the aforementioned contextual image filter should be confused with the term "context object detection" [5] as used increasingly in computer vision. There the aim is to increase the object detection performance by using information that is not part of the object itself. Such information could be present elsewhere in the image (e.g., the presence of a street in an image makes it more likely that certain objects will be cars) or could be non-image information related to the image (e.g., image captions). In the proposed method the only information used by the method are the intensity values and positions of the pixels belonging to a possible object of interest.

For the implementation of this method, the Max-tree algorithm [11] was used to construct and process the component trees. The *Max-tree* uses a tree-based representation of a gray-scale image f where node N_h^k represents peak component P_h^k in the image, for which holds:

$$P_{h2}^{k2} \subset P_{h1}^{k1} \iff h2 > h1 \wedge N_{h1}^{k1} \text{ is an ancestor of } N_{h2}^{k2} \tag{1}$$

and

$$P_{h1}^{k1} = P_{h2}^{k2} \iff k1 = k2 \wedge h1 = h2. \tag{2}$$

A *peak component* $P_h^k(f)$, where k runs over some index set I_h^f, is defined as the kth connected component of the threshold set $\mathcal{T}_h(f)$ of image f which is defined as

$$\mathcal{T}_h(f) = \{x \in \mathbf{M} | f(x) \geq h\}, \tag{3}$$

with image domain $\mathbf{M} \subset \mathbb{R}^n$ (in this paper $n = 2$).

In this paper, peak components will be simply referred to as *components*, For a detailed discussion on component trees in general and Max-trees in particular the reader is referred to [3, 9, 11].

The computational cost for detecting objects using the proposed method is dominated by the time needed to construct the component trees [4]. The number of nodes in the component trees do form a minor influence on the total computation time. For the training process, the computational costs are, similar to the Viola-Jones method, dominated by the training algorithm of AdaBoost [6]. While the computational and memory requirements of the Viola-Jones method increase dramatically with even minor increases in the size or the number of dimensions of the training set images, the computational and memory requirements of the proposed method are only influenced by the number of components that need to be processed. A major focus in this paper is to reduce the total set of components to only a very small subset that will be used for training the AdaBoost classifier.

The main contribution of this work is the development of a multi-component object detection using component trees that can detect objects without using a moving local window. Other contributions presented here are automatically estimating the orientation and scale of a detected object (compared with the training examples) and trainable vector-attribute filters using AdaBoost. The proposed method will be discussed here for 2-D gray-scale images only but can be easily adapted to higher dimensions [20, 22].

2   Method

The idea behind the proposed method is to automatically decompose the (complex) objects of interest (e.g., human faces, traffic signs, craters) into multiple simple shapes. Connected operators are used to segment these simple shapes with each component representing a simple shape. AdaBoost [6] is used for training and detecting simple shapes in a way similar to how Viola and Jones [19] used it to train and detect objects. AdaBoost was chosen here for two reasons: i) it is one of the most commonly used classification algorithms in object detection applications, and ii) it allows one to investigate why or how it decides to accept or reject a test sample (a simple shape).

An implementation of the proposed method would consist of two programs: one to train the classifier and one to use that classifier to detect objects in image. Below, the former will be discussed first, followed by a description of the latter.

2.1   Training the Object Detection Classifier

The proposed object detection classifier takes as input a training set T_{N_t} consisting of N_p positive and N_n negative images. The images for this set do not need to be scaled to the same size and do not need to be normalized but it will be assumed that i) all positive examples are aligned to the same orientation and ii) that all positive examples have been cropped in about the same way. These two conditions are added to make the training process faster and for the sake of simplicity. They do not affect the performance of the trained classifier.

The complete training process can be summarized to the following steps:

1. Construct component trees for each image of the training set [11]
2. Pre-process the trees of the images to reduce the number of nodes
3. Construct shape classifiers that each can detect a simple shape
4. Construct the object classifier by combining the simple classifiers

2.2   Constructing the Max-Trees

For each of the $N_t = N_p + N_n$ images of the training set a component tree is constructed. For the results discussed in this paper, the popular Max-tree algorithm [11] was used, but in essence any algorithm that produces a connected tree-based representation with the ability to compute strict-shape attributes [15] is suitable. An approach that might offer improvements over the Max/Min-trees used here uses energy estimators with trees of shapes [21]. A very recent review of various component tree algorithms is presented in [4]. Strict-shape attributes [15] refers here to attributes that are invariant under translation, scale, and rotation. As Max-trees are used to process bright image features, Min-trees are needed to handle the dark image details. These Min-trees were implemented by inverting the images of the training set.

For each of N_t Max-trees and N_t Min-trees the following attributes were computed: Area, Centroid, Bounding Box, Ratio and Inverse Ratio of the length of both axes, non-compactness and Hu's set of 7 moment invariant. All but the first three of these are strict-shape attributes. The Centroid, the Ratio, the Non-compactness, and Hu's moment invariant attributes can be computed efficiently in a Max- or Min-tree from the central moments [7]. The Non-compactness is essentially the same as Hu's first moment invariant. The Centroid consists of two values: the mean x- and the mean y-coordinates of a component. The Bounding Box is represented by four values, representing the minimum and maximum x- and y-coordinates of the pixels in a component.

2.3   Pre-processing

In a later step simple classifiers will be trained to detect the various features present in the positive training examples. Although the 'constructing simple classifiers' step could be done on all the components found in the positive examples of the training set, this would cost much extra computing time that can be avoided by using pre-processing to remove irrelevant components.

In this pre-processing step, components are removed that can never be part of the object of interest. In the current implementation, nodes are removed if either their Area $\leq \lambda_A$ or if they touch any of the boundaries of the image. This first condition ensures that features deemed too small to be significant for the training process are removed (note that λ_A could either be in pixels or as a percentage of the image size). The second condition ensures that all components that are considered in the training process are fully represented in the image without parts being clipped off (and thus changing their shapes). A reason for adding this condition was that some training sets that consist of a large round object (such as craters or round traffic signs) the cropped training images all contain the same rounded triangular corner shapes, which due to their frequency (appearing in each image) would appear significant to the classifier. Adding this condition is a simple and reasonable way of removing these clipped shapes. Fig. 2 shows an example of this pre-processing, where the number of components (nodes) in the input and inverted input images are 2063 and 1953 respectively. After pre-processing this is reduced to 565 and 549 nodes respectively.

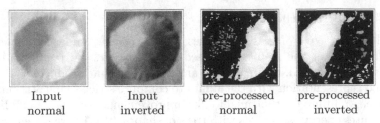

| Input | Input | pre-processed | pre-processed |
| normal | inverted | normal | inverted |

Fig. 2. Pre-processing the original and inverted input (training) image of a crater

2.4 Training Simple-shape Classifiers

The next step in the procedure to construct the object detection classifier is to train classifiers that can identify components with similar shapes in the set of positive training images. A straightforward approach would be to manually segment and label components and supply these to a classification algorithm which has been done previously for detecting craters in images of Mars [16]. As manual segmenting and labelling of suitable shapes is generally cumbersome and time-consuming, an automatic procedure is presented and used here for the proposed method.

This automatic procedure has as input the set of components of the positive training images that have remained after pre-processing. The idea is to create N_C training sets where each training set contains N_p components from the positive training images (using one component per pre-processed image) and all components from the negative training images after pre-processing. Each component is labeled with the label of the image it belongs to. Let $S_i^{P,\text{pre}}$ be the set of components in the ith pre-processed positive training image and let $S_i^{N,\text{pre}}$ be the set of components in the ith pre-processed negative training image, then the number N_C of training sets for training the simple-shape features is:

$$N_C = \prod_{i=0}^{N_p-1} ||S_i^{P,\text{pre}}||, \tag{4}$$

where $||S_i^{P,\text{pre}}||$ denotes the number of components in $S_i^{P,\text{pre}}$.

Let $C_i^{\text{pre,pos}}(n)$ be a function that yields the n'th component of $S_i^{P,\text{pre}}$ then the training set $T_{\mathbf{v}}^{shape}$ for the simple shape classifier can be defined as:

$$T_{\mathbf{v}}^{\text{shape}} = T_{\mathbf{v}}^{\text{shape},P} \cup T^{\text{shape},N}, \tag{5}$$

which is the union of the set of components of the negative examples

$$T^{\text{shape},N} = \bigcup_{i=0}^{N_n} S_i^{N,\text{pre}} \tag{6}$$

and a set of N_p components of the positive examples

$$T_{\mathbf{v}}^{\text{shape},P} = \bigcup_{i=0}^{N_p} C_i^{\text{pre,pos}}(v_n), \tag{7}$$

where $\mathbf{v} = \{v_0, ..., v_{N_p-1}\}$ denotes a vector of N_p component indices with index v_i defined as $0 \leq v_i < ||S_i^{P,\text{pre}}||$ and $0 \leq i < N_p$.

The procedure to construct simple-shape classifiers is then to compute a set of feature vectors from each of the N_C training sets $T_\mathbf{v}^{shape}$. For each component a feature vector is computed consisting of the strict-shape attributes that were discussed in 2.2 leading to feature vectors containing only 10 values, which is remarkably short compared with the more than the 160,000 feature values used by the Viola-Jones method. Furthermore, the number of components used is considerable higher than the number of training images but still relatively small due to the pre-processing step. This means that training one simple-shape classifier will be very fast but much computing time would be needed to compute all the classifiers for the N_C possible training sets.

For only a very small number of these N_C training sets will classifiers have to be made if the following conditions are met for all of the positive training images:

- The shapes that are significant for detecting the object do not vary much in shape or in size and position relative to image size.
- Each positive training image contains one object that has been cropped in a consistent way around the object.
- Each positive training image can have a different size but the aspect ratio should be roughly the same.
- The object of interest in each positive training image has the same orientation w.r.t. the image axes.

Note that these conditions only apply within one set of training images and not between different sets. Furthermore, no conditions apply to the negative training images except that they should not contain the object of interest.

The last three conditions can be met easily for any object detection application as these conditions affect how the training set is created by cropping and rotating and do not restrict the object itself in any way. By contrast, the first condition seems to affect which kinds of objects could be detected with the proposed method. However, this condition just means that characteristic features of the object should look similar and have similar position and size relative to the size of the training image. In the case of detecting human faces it would for example mean that in each positive image of the training set the characteristic shapes, e.g., eye pupils, are at about the same location with about the same size and with similar shape properties. This seems reasonable as one would not expect that in each image the eye pupils would appear at very different positions in the face, or change much in relative size or shape.

If these conditions are used then instead of training N_C classifiers, classifiers will be trained for only a very small fraction of the N_C sets. Let $\text{Match}(C_a, C_b)$ be a boolean function that yields true if and only if component C_a and C_b have similar size and position relative within the positive training images (a and b respectively) they are part of, and yields false otherwise. The similarity of size and shape can be obtained by computing the ratio R_A and the distance D_C between respectively the Area A and Centroid (C_x, C_y) attribute values of these components. The Match function yield true if $D_C \leq \lambda_D$ and $1/\lambda_R \leq R_A(C_a, C_b) \leq \lambda_R$ using parameters λ_D and λ_R. The number of components after pre-processing of the crater image shown in Fig. 2 was for the normal and inverted input images 565 and 549 nodes respectively. The total number of shape training sets that were computed from two positive (one of which was shown) and two negative training images was 1656. This relatively high number of sets was caused by the large number of components that represent only a few actual images features. As a tree-representation was used, this high number can be easily reduced by removing all components that have ancestor or descendant nodes with similar size and shape.

Since the shape classifiers only have to learn simple shapes from short feature vectors with strong descriptive power, the number of weak classifiers that AdaBoost needs to train for each simple shape classifier can be kept low. The training time should therefore remain feasible for very difficult kinds of object where large training sets might be needed.

2.5   Constructing the Object Classifier

Each of the simple-shape classifiers trained in the previous step, can be evaluated using the training set images. How well a simple-shape classifier will perform will depend on the (positive) components it has been trained on and how significant the shape is for detecting the object of interest. Although all of the positive components a classifier was trained on are located in the same area of each training image, this does not guarantee that all of them represent the same object feature. For example, a classifier might have been trained on a set of components from human face images where some of these components represent the pupil and some perhaps a small part of a pair of glasses that was in the same area of the image. It is to be expected that a classifier trained on only eye pupils will be better at detecting pupils then one that is trained on a set consisting of some eye pupils and some other object features. It is to be expected therefore that better classifiers will detect more shapes in the positive and fewer in the negative training images. The pseudocode for constructing the object classifier Z is shown in Algorithm 1. The function MeetsTarget(T) yields true iff object classifier T achieves a certain detection performance (based on true and false positives) w.r.t. the training set or to an evaluation set of images. The output of the object classifier is a set of n shape classifiers $\{Z_0, ..., Z_{n-1}\}$ and list of n sets of shapes $\{A_0, ..., A_{n-1}\}$ where each set A_i contains exactly those shapes that were detected by shape classifier Z_i in each and every positive image of the training set.

2.6   Detecting Objects

The process for detecting objects in a test image g using object classifier Z consists of the following steps:

1. Component trees are constructed for image g and its inverse.
2. The trees are pre-processed using the same operators and parameters as was done for the training process.

3. Two sets of feature vectors (normal and inverted image with one vector for each component) is created using the same attribute types as during the training process.
4. Apply the shape detectors (that were trained using AdaBoost) on the corresponding set of feature vectors (classifiers trained on components from "normal" images vs. classifiers trained using components from inverted images).
5. Each shape in the list of shapes $\{A_0, ..., A_{n-1}\}$ needs to match (based on shape and spatial relationships with other shapes) with a component in the sets of shapes produced at the previous step. If each shape from $\{A_0, ..., A_{n-1}\}$ can be matched with a certain subset O from the test image, then O represents the object of interest.
6. Remove the components representing O from the set of components from the test image
7. Repeat the previous two steps until no more objects are found.

Algorithm 1. Algorithm for constructing the object classifier Z

$Z \leftarrow \emptyset$ {Object classifier Z is a set of shape classifiers}
$A \leftarrow \emptyset$ {A is the set of object features detected in each of the positive images}
while not MeetsTarget(T) **do**
    $Z_{\text{best}} \leftarrow \emptyset$
    $D_{\text{best}} \leftarrow 0$
    for all $\mathbf{v} \in V$ **do**
        $Z_s \leftarrow$ TrainShapeClassifier($\mathbf{v}$)
        for all $i = 0$ **to** N_P **do**
            $S_i \leftarrow$ DetectShapes($Z_s, T_{\mathbf{v}}^{\text{shape},P}$)
            $S \leftarrow S \cap S_i$
        end for
        if $||S|| > D_{\text{best}}$ **then**
            $Z_{\text{best}} \leftarrow Z_s$
            $D_{\text{best}} \leftarrow ||S||$
        end if
    end for
    $Z \leftarrow Z \cup \{Z_{\text{best}}\}$
    $A \leftarrow A \cup S$
end while

---

3 Experiments

The method was evaluated using two kinds of objects: a dotted letter A (representative of objects consisting of common shapes where the uniqueness of the object is determined by the spatial relationships between these simple shapes) and Martian impact craters (representative of more real-world applications). The training sets and results of these are shown in Fig. 3. Detected objects are marked with rectangular outlines. Pixels within these outlines were slightly

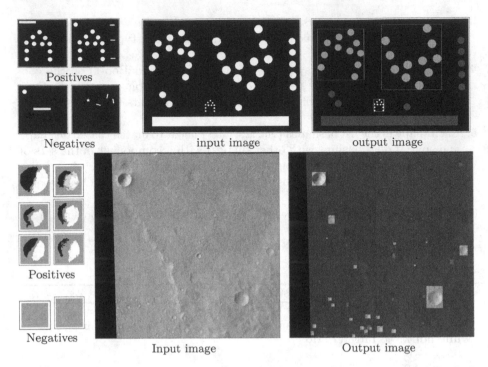

Fig. 3. Detecting dotted letter A objects (top) and craters in a 2000x2000 pixels section of HRSC image h0905_0000 using simplified training set (bottom)

darkened if that object's orientation did not match with the orientation of the training examples. Pixels not belonging to detected objects have darkened pixel values. The 3 objects with different orientations and scales in the dotted letter A test were correctly detected without false detections. The number of objects detected as craters was 55 of which 25 were actual craters. The false detections can be distinguished by their deviating orientations as all true craters in this image have the same orientation. The time needed to process and detect craters in the 2000x2000 pixel image was 3.15s on a 2.2GHz Intel Core i7 processor (2720QM) using a single thread using an inefficient implementation of the method.

The scaling and rotation needed to transform an object O_R in the training set to a detected object O_D can be estimated using the vector between the pair of components of the object that are the furthest apart. Let $\mathbf{v}_R$ and $\mathbf{v}_D$ be such vectors of the training and detected objects respectively, then the rotation is the difference between the directional angles of $\mathbf{v}_R$ and $\mathbf{v}_D$ and the scaling is $d(\mathbf{v}_D)/d(\mathbf{v}_R)$ with d being the Euclidean distance function. The estimated transformations of the detected dotted A objects are:

Object	Max.dist. O_R	Max.dist. O_D	Scaling Actual	Scaling Estimated	Rotation Actual	Rotation Estimated
1	127.95px	128.03px	1.00×	1.00060×	15°	15.0°
2	127.95px	31.96px	0.25×	0.24979×	163°	0.1°
3	127.95px	159.91px	1.25×	1.24973×	0°	163.0°

4   Conclusions

A method for detecting objects using connected operators was presented. Contrary to commonly-used object detection methods such as Viola-Jones, the proposed method does not use a local window to scan the image for objects but instead uses a tree representation of the image to efficiently locate and detect objects. Furthermore, by computing feature vectors using attributes that are invariant under translation, scaling, and rotation, the resulting classifier can detect objects at any orientation or scale without expensive extra computations. The features used are more descriptive and powerful than the simple Haar-like features used by methods such as Viola-Jones which should ease the training of the classifiers. The proposed object detector can handle some variation in the sizes, shapes and spatial relationships of the simple shapes. All of this together should allow object detectors to be trained using a fraction of the huge training sets needed to properly train methods such as Viola-Jones. Finally, the proposed method can be easily extended for 3-D or 4-D images without incurring excessive use of memory or computing time. Hu's 2-D moment invariants [7] that are currently used could be replaced with 3-D moment invariants [22].

Currently only shape and size information is used to detect objects. Although it means the proposed method is contrast invariant, object detection results might be improved by adding gray-scale information. The benefits of using context-based energy estimators with contrast-invariant tree of shapes [21] is being studied. More work needs to be done on evaluating the proposed method on training sets large enough to produce results comparable or better than Viola-Jones on real-world applications. Furthermore, although the connected operators used produce features that are more descriptive and generally less influenced by noise, they do have a problem with gaps or narrow connections caused by noise or occlusions. One solution would be to use second-generation connectivities [10] to handle these issues. The feature types used by Viola-Jones essentially do not permit any change in the appearance of an object. Their approach demonstrates that given large enough training sets and training time, AdaBoost can learn to handle all the variations. This is also to be expected for the proposed method.

References

1. Ali, K., Fleuret, F., Hasler, D., Fua, P.: A real-time deformable detector. IEEE Trans. Patt. Anal. Mach. Intell. 34(2), 225–239 (2012)
2. Batman, S., Dougherty, E.R.: Size distributions for multivariate morphological granulometries: texture classification and statistical properties. Optical Engineering 36, 1518–1529 (1997)

3. Breen, E.J., Jones, R.: Attribute openings, thinnings and granulometries. Comp. Vis. Image Understand. 64, 377–389 (1996)
4. Carlinet, E., Geraud, T.: A comparative review of component tree computation algorithms. IEEE Trans. Image Proc. 23(9), 3885–3895 (2014)
5. Chen, Q., Song, Z., Dong, J., Huang, Z., Hua, Y., Yan, S.: Contextualizing object detection and classification. IEEE Trans. Patt. Anal. Mach. Intell. 37(1), 13–27 (2015)
6. Freund, Y., Schapire, R.E.: A decision-theoretic generalization of on-line learning and an application to boosting. Journal of Computer and System Sciences 55(1), 119–139 (1997)
7. Gonzalez, R.C., Woods, R.E.: Digital Image Processing, 2nd edn. Addison-Wesley (2002)
8. Heijmans, H.J.A.M.: Connected morphological operators for binary images. Comput. Vis. Image Understand. 73, 99–120 (1999)
9. Najman, L., Couprie, M.: Building the component tree in quasi-linear time. IEEE Trans. Image Proc. 15(11), 3531–3539 (2006)
10. Ouzounis, G.K., Wilkinson, M.H.F.: Mask-based second generation connectivity and attribute filters. IEEE Trans. Patt. Anal. Mach. Intell. 29(2), 990–1004 (2007)
11. Salembier, P., Oliveras, A., Garrido, L.: Anti-extensive connected operators for image and sequence processing. IEEE Trans. Image Proc. 7, 555–570 (1998)
12. Sofou, A., Tzafestas, C., Maragos, P.: Segmentation of soilsection images using connected operators. In: Int. Conf. Image Proc. 2001, pp. 1087–1090 (2001)
13. Soille, P.: Morphological Image Analysis: Principles and Applications, 2nd edn. Springer, New York (2002)
14. Szeliski, R.: Computer Vision: Algorithms and Applications, 1st edn. Springer-Verlag New York, Inc., New York (2010)
15. Urbach, E.R., Roerdink, J.B.T.M., Wilkinson, M.H.F.: Connected shape-size pattern spectra for rotation and scale-invariant classification of gray-scale images. IEEE Trans. Patt. Anal. Mach. Intell. 29(2), 272–285 (2007)
16. Urbach, E.R., Stepinski, T.F.: Automatic detection of sub-km craters in high resolution planetary images. Planetary and Space Science 57, 880–887 (2009)
17. Urbach, E.R.: Contextual image filtering. In: 24th International Conference Image and Vision Computing New Zealand, pp. 299–303 (November 2009)
18. Urbach, E.R., Pervukhina, M., Bischof, L.: Segmentation of cracks in shale rock. In: Soille, P., Pesaresi, M., Ouzounis, G.K. (eds.) ISMM 2011. LNCS, vol. 6671, pp. 451–460. Springer, Heidelberg (2011)
19. Viola, P., Jones, M.J.: Robust real-time face detection. International Journal of Computer Vision 57(2), 137–154 (2004)
20. Westenberg, M.A., Roerdink, J.B.T.M., Wilkinson, M.H.F.: Volumetric attribute filtering and interactive visualization using the max-tree representation. IEEE Trans. Image Proc. 16(12), 2943–2952 (2007)
21. Xu, Y., Geraud, T., Najman, L.: Context-based energy estimator: Application to object segmentation on the tree of shapes. In: Proc. Int. Conf. Image Proc. 2012, pp. 1577–1580 (September 2012)
22. Yang, B., Flusser, J., Suk, T.: 3d rotation invariants of gaussian-hermite moments. Pattern Recogn. Lett. 54, 18–26 (2015)
23. Zimmermann, K., Hurych, D., Svoboda, T.: Non-rigid object detection with local interleaved sequential alignment (lisa). IEEE Trans. Patt. Anal. Mach. Intell. 36(4), 731–743 (2014)

GraphBPT: An Efficient Hierarchical Data Structure for Image Representation and Probabilistic Inference

Abdullah Al-Dujaili[1,2], François Merciol[1], and Sébastien Lefèvre[1(✉)]

[1] IRISA, University of Bretagne-Sud, Vannes, France
{francois.merciol,sebastien.lefevre}@irisa.fr
[2] Nanyang Technological University, School of Computer Engineering, Singapore
aldujail001@e.ntu.edu.sg

Abstract. This paper presents GraphBPT, a tool for hierarchical representation of images based on binary partition trees. It relies on a new BPT construction algorithm that have interesting tuning properties. Besides, access to image pixels from the tree is achieved efficiently with data compression techniques, and a textual representation of BPT is also provided for interoperability. Finally, we illustrate how the proposed tool takes benefit from probabilistic inference techniques by empowering the BPT with its equivalent factor graph. The relevance of GraphBPT is illustrated in the context of image segmentation.

Keywords: Image processing · Hierarchical segmentation · Binary partition tree · Compression · Probabilistic inference

1 Introduction

A strong interest in the recent decades has been developed towards realizing machines that can perceive and understand their surroundings. However, computer vision is still facing a lot of challenges even with high-performance computing systems. One of these challenges is how to deal with the input of these machines. Typically, the input to computer vision is of images in their pixel-based rectangular representation, whereas the output is associated with actions or decisions. Clearly, what kind of output or performance is desired from such a system imposes a set of constraints on the visual data representation. A representation for a storage-efficient system is not the same as for high-accuracy systems.

In these early approaches for analyzing visual data, pixels were treated independently [1]. This proved to be successful for a while but the increase in image resolution due to the advancing sensors technology created the need for a model that considers spatial relationships. This led to context-based models such as superpixels, edge-based, and segmentation-based representations, that brought up the object-based paradigm [2]. Although such models show an advantage in several applications [3], it still experienced uncertainty in defining what a context is due to factors like scale, context inter- and intra-variance. Consequently,

© Springer International Publishing Switzerland 2015
J.A. Benediktsson et al. (Eds.): ISMM 2015, LNCS 9082, pp. 301–312, 2015.
DOI: 10.1007/978-3-319-18720-4_26

the concept of hierarchical representation was adopted and it has proven useful in analyzing images [4–6]. The acyclic nature of some of these representations makes many of the growing machine-learning techniques exact and efficient. For instance, the belief propagation algorithm for probabilistic inference is exact on tree-graphical models [7].

In this paper, we focus on the binary partition trees (BPTs), a special case of hierarchical representations that allows for greater flexibility than many other morphological trees. We elaborate on this representation and introduce three complementary contributions to the state-of-the-art:

1. An efficient implementation of BPT construction algorithm. The implementation offers a flexible framework to specify and control the way a BPT is built. The code is freely available[1] under LGPL license[2].
2. A textual representation of the BPT which makes it portable across different programming environments and platforms.
3. A demonstration of empowering BPTs with probabilistic inference.

These contributions aim to support the dissemination of the BPT (and more generally hierarchical representations) to solve computer vision problems.

The paper organization is as follows. We first recall in Sec. 2 the concept of BPT and introduce in Sec. 3 an efficient algorithm for its computation. We then propose in Sec. 4 a compact and portable representation of BPTs through textual files, with efficient access to image data. Sec. 5 presents how BPT can be combined with the framework of probabilistic inference, with an illustrative application in image segmentation. The last section is dedicated to conclusion and future directions.

2   Binary Partition Tree (BPT)

There exist several hierarchical representations that come with useful properties, e.g. min- and max-trees [8], component trees [9], or trees of shapes [10] that all aim to extract regional extrema of the image. However, such regions sometimes do not correspond to objects in the scene. On the other hand, the nodes of a binary partition tree (BPT) are potential candidates for objects as BPT is able to couple image regions based on their similarities.

BPT was introduced in [11, 12] as a structured representation of the regions that can be obtained from an initial partition using binary mergings. In other words, it is an agglomerative hierarchical clustering of image pixels (see Figs. 3, 4, and 5 for an illustration with a color image, its associated tree, and the nested partitions, respectively). Image filtering, segmentation, object detection and recognition based on BPT were demonstrated e.g. in [12–16].

The basic framework of constructing a BPT is simple and straightforward: starting with the image pixels as the initial regions, a BPT is constructed by successively merging the most similar pairs of regions until all regions are merged into a single region [17]. The process is governed by the following factors [11]:

---

[1] http://www-obelix.irisa.fr/software
[2] GNU Lesser General Public License, FSF, https://www.gnu.org/licenses/lgpl.html.

- **Region Model** defines how each region is represented based on its characteristics, e.g. color, shape, texture, etc.
- **Merging Criterion** defines a score of merging two regions. It is a function of their **region model**.
- **Merging Order** defines the rules to guide the merging process based on **merging criterion**.

There is no unique choice for these various parameters. However, a commonly adopted strategy is to represent each region by its average color in a given color space, and to first merge two regions that either have models similar one to each other, or similar to the model of the region built from their possible union [14]. Furthermore, Vilaplana et al. [14] also explore how to take into account edge and contrast information in the merging process through advanced merging criteria. Let us note that, similarly to the underlying BPT model, the methodology proposed in this paper is generic, i.e. it can be apply to various region models and merging criteria.

3   BPT Construction

Based on the strategy proposed for building a BPT in [17], Valero et al. [16] described an algorithm for constructing BPTs and presented a detailed analysis on its complexity. For an $(N = m * n)$-pixel image and assuming 4-connectivity, the complexity can be expressed as the following:

$$O_{BPT}(N) = N * O_{leaf} + (4 * N) * (O_{edge} + O_{insert}) + (N - 1) * O_{merge} \quad (1)$$

where $O_{leaf}, O_{edge}, O_{insert} = O(log_2 N)$ are the complexity costs of building a leaf node, building an edge between two nodes, inserting an edge into the priority queue, respectively. $O_{merge} = O_{parent} + a * (O_{edge} + O_{insert}) + b * O_{pop}$ is the complexity of merging two nodes; with O_{parent} being the cost for constructing their parent node, and O_{pop} being the cost of poping an edge off the priority queue. Scale factors a and b correspond respectively to the number of parent node's neighbors and the number of two nodes' neighbors. Here, we describe an optimized algorithm that lowers some of the terms in Eq. (1).

3.1   Proposed Algorithm

First, the leaf nodes are computed from the image N pixels. Their edges are also built based on the 4-connectivity scheme. Each edge is built once, so the term $(4 * N)$ becomes $(2 * N - m - n)$. Moreover, instead of inserting all the edges for a node, we insert only the most light edge (corresponding to the most similar neighboring pixels) into the queue; all other edges are irrelevant for the node of interest. Nevertheless, if it gets merged into a new node, then all of its neighbours are going to be considered even those whose connecting edges are not in the queue. This considerably reduces the priority queue size as only one insert per node is carried out, whilst the image support is fully considered.

Algorithm 1. Proposed Algorithm for BPT Construction

Input	: An image I of N pixels
Variables:	Min-Priority Queue PQ,
	A set of nodes V representing the binary partition tree nodes,
	A set of edges E connecting neighboring nodes
Output	: Binary partition tree of the image BPT

1 **foreach** *pixel p of the image I* **do**
2     $u \leftarrow$ BuildLeafNode (p)
3     $BPT \leftarrow$ UpdateBPT (u)
4   **foreach** *neighboring pixel q of p* **do**
5        $v \leftarrow$ leaf node of q
6        $E_{uv} \leftarrow$ UpdateNeighborhoodEdges (u,v)
7   PQ.insert($E_{uv}.smallestEdge$)

8 **for** $i \leftarrow 1$ **to** $N - 1$ **do**
9   **do**
10       $e \leftarrow PQ$.getHighestPriority()
11   **while**(e.nodes() have no parents)
12     $u, v \leftarrow e$.nodes()
13     $w \leftarrow$ BuildParentNode (u, v)
14     $BPT \leftarrow$ UpdateBPT (w)
15     $E_w \leftarrow$ UpdateNeighborhoodEdges $(w$, neighbors of u, neighbors of $v)$
16   PQ.insert($E_w.smallestEdge$)

17 **return** BPT

Consequently, the number of insertions and pops is decreased. Nodes are merged successively in $N - 1$ steps. In each of the merging steps, an edge is taken off the queue, one edge (the most light one) is inserted due to the new neighborhood formed; while edges of the merged nodes in the queue are invalidated. We do not bother about removing the invalidated edges from the priority queue. Instead, whenever a merging step is done, we pop edges from the priority queue and merge on the first valid popped edge. This on average, brings the factor b down to a b'. At the $N - 1$ step, the BPT root is computed and the BPT construction is complete. The optimized algorithm is listed in Alg. 1.

3.2 Efficiency Evaluation

The new algorithm comes with the following complexity:

$$O_{BPT'}(N) = N*O_{leaf} + (2*N - m - n)*O_{edge} + N*O_{insert} + (N-1)*O_{merge'} \quad (2)$$

with

$$O_{merge'} = O_{parent} + a*O_{edge} + O_{insert} + b'*O_{pop}. \quad (3)$$

Table 1. Performance statistics for a subset of MSRA images (120,000 pixels)

Performance statistics	CPU time (in seconds)
Minimum	1.499
Maximum	66.277
Standard Deviation	3.788
Mean	2.534
Mode	1.663

Let $a_{average}$ and $b'_{average}$ be the average estimation of a and b', respectively. With $O_{pop} = O(log_2(N))$ and O_{edge} being a constant operation, the complexity can be approximated as:

$$O_{BPT'}^{average}(N) \approx O(N * a_{average}) + O(N * b'_{average} * log_2(N)). \qquad (4)$$

A close look on b' shows that it can have an average estimation of ≈ 1 because in the beginning the priority queue has N edges and each of the $N-1$ merging steps adds a single edge and pops one valid edge. Hence the average number of popped invalid edges $b'_{average}$ is $\frac{2N-1}{N-1} - 1 \approx 1$. Similar analysis can be conducted on a, leading to an average number being a fraction of N, with a peak of $\frac{2}{3}N$ in the worst case (i.e. a thin elongated region).

Besides theoretical analysis, we also evaluated efficiency through runtime measurement. Previous implementations of BPT in the literature reported an execution time of 1.5 seconds for a 25,344-pixel image on a 200 MHz processor [11] and 1.03 seconds on a 1.87 GHz processor for the same image size [14]. Alg. 1 (coded in Java) reported an execution time of 0.282 seconds on a 2.2 GHz for the same image size.

A further analysis was conducted on MSRA's 5,000 images [18]. We ran our code with no tuning of BPT parameters and choosing the spectral similarity as the region model. The execution time varies from 0.6 seconds for a 36,630-pixel color image to 3.8 minutes for a 154,400-pixel color image. Although the execution time is greatly affected by the number of pixels N, it is as well affected by the image content. Indeed, image content determines the weights of nodes edges which consequently affect the performance of both priority queue operations and image regions compression. Table 1 shows the execution time statistics for 1238 color images of 120,000 pixels using a 2.2 GHz processor. As a future work, we plan to study extensively the effects of the similarity measures and provide a benchmark for comparing various implementations of BPT.

3.3   BPT Tuning

As already indicated, the BPT model is attractive due to its flexibility. We keep such property in our tool by providing a set of tunable parameters to fit the application needs (e.g. in object detection, regions of compact geometry are usually

Table 2. BPT construction parameters

Parameter	Range	Description
Number of Nodes	$[1, 2N-1]$	Specifies the number of nodes the BPT should have. *It can also be set as a fraction of the total number of nodes.*
Number of levels	$[1, N]$	Specifies how many levels the BPT should have.
Node Size	$[1, N]$	Specifies how many pixels a node should at least contain. *It can also be set as a fraction of the total number of pixels.*
Similarity Measure Weights	$[0, 1]^k$	Specifies the contribution of the k individual features to the overall similarity measure between two nodes.
Node Orientation	$up, down$	Specifies whether the nodes levels are assigned in a top-down or a bottom-up manner.

more preferred over others which might not be the case for a segmentation-based application). Currently, these parameters can be set from the source files. As a future work, we intend to add a friendly interface to the tool for setting them. Table 2 lists these parameters. Some of them are related and might override each other. For instance, the number of nodes and levels for a tree are quite related (a binary tree of l levels, has at most $2^l - 1$ nodes). Controlling BPT's number of nodes, number of levels, and their sizes helps in bringing images of different scale or size to a normalized representation under their BPTs. The tree construction relies on a similarity between nodes, that is computed here as a linear combination of k similarity measures. Such measures as well as their contribution to the overall similarity measure can be tuned as well. Let us note that we consider in this paper three measures dealing respectively with color, spatial, and geometric information. As BPTs could have irregular structure (e.g. leaf nodes can have different distances from the root), we provide a parameter, *node orientation*, that assigns the level of each node based on its distance from either the root level (level 1) or the leaf nodes level (specified by the parameter, *number of levels*). In other words, each node can be assigned to a level either in a top-down or a bottom-up order.

4   BPT Indexing and Management

4.1   Textual Representation

To make our BPT implementation portable and readable across different programming environments and platforms, we worked out a compact textual representation of the BPT. The labels: $0, \ldots, N-1$ are assigned to the nodes of an N-BPT in a depth-first order from right to left as shown in Fig. 1.

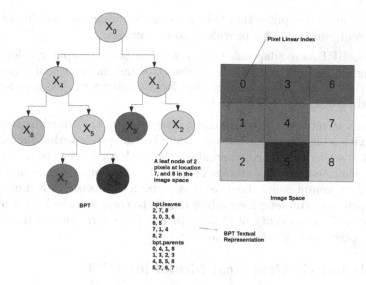

Fig. 1. BPT textual representation

Each leaf node l is represented with a single line of comma-separated values (csv) as the following: $[A_l, B_l]$ where A_l is the node label, and B_l is the set of node's pixels linear indices. On the other hand, each parent node p is represented by the line: $[C_p, D_p, E_p, F_p]$ where C_p is the node label, D_p, E_p are the children nodes labels in a descending order, and F_p is the greatest label among descendant nodes labels. Such a textual representation allows to build BPTs in a top-down approach directly. Besides, indexing p's image region is nothing but the pixels union of leaf nodes whose labels A_l intersect with the interval $[E_p, F_p]$.

As each node is represented with a csv line, it is easy to add any other feature/attribute to its textual representation. For instance, the number of descendant nodes can be appended for each node to help in drawing the tree. Our tool automatically produces two text files named as *bpt.leaves* and *bpt.parents* which can be read readily to retrieve the BPT structure. Along with the tool, we have provided MATLAB functions for accessing these files and producing the BPT structure in a MATLAB environment.

4.2 Region Indexing

Given a node, it is sometimes needed to access the corresponding region in the image space. Usually, the leaf nodes would have the direct access to the image pixels. As a result, graph traversal is the technique commonly used to traverse from a node of interest to its descendant leaf nodes, and hence accessing the corresponding image region. To avoid traversing the tree, and provide a direct access to a node's image region, a bounding box is created for each node that covers all the pixels included in the corresponding image region, as well as some neighboring pixels (i.e. pixels that belong to the bounding box but do not belong to the node of interest). This provides two advantages:

1. It makes BPT more adaptable to grid/window-based computer vision models and paradigms such as kernel descriptors [19] and convolutional networks [4].
2. It provides a constant time operation for accessing a relaxed representation (bounding box) of a specific node.

Nevertheless, to retrieve the node's exact image region, the bounding box can be provided with a bitmap whose bits correspond to the pixels within the bounding box. A bit is set to 1 if the corresponding pixel belongs to the node, or 0 otherwise. The additional memory storage incurred by these bitmaps can significantly be reduced by compressing them using a suitable compression technique. The tool currently uses run-length encoding (RLE) to compress the BPT's bitmaps. This adds to the complexity of O_{parent} in Eq. (3) a term that is linear in the number of pixels in newly-formed node.

5   Probabilistic Graphical Model for BPT

Some of the problems in the domain of computer vision such as object detection and recognition are naturally *ill-posed* in a way that it is very difficult to determine with absolute certainty their exact solutions. In these settings, probabilistic graphical models become very handy in not only providing a single solution but a probability distribution over all feasible solutions [20]. Therefore, instead of the conventional methods that analyse each node as a separate entity for example in object detection and recognition problems [13], treating the BPT as an entire structure by encoding the relationships between its nodes is elegantly done using a probabilistic graphical model. Here, we focus on representing BPTs with a discrete factor graph with a conditional distribution. The reader is referred to [20] for an introduction to these models. The practical interest of such a connection between morphological representations and probabilistic inference will be illustrated in the context of image segmentation.

5.1   Inducing a BPT's Factor Graph

A BPT can be defined as the tuple (X, E) where X is the set of measurements/observations nodes (color, shape, or other features) that correspond to the BPT nodes and E is the set of edges connecting the nodes and hence X. X can be considered as the set of input variables that are always available. On the other side, an output variable is provided for each node and collectively denoted as Y. Their values represent the solution to the problem of interest. For instance, in object detection, we can have a binary variable per node to denote whether it corresponds to a sought object or not. The factor graph captures the interaction among these variables by introducing a set of factor nodes $\mathcal{F}$. These factors can be seen as potential functions that assign scores to the output variables assignments and are application-dependent. Let $\mathcal{V} = X \cup Y$, then the factor graph is the tuple $(\mathcal{V}, \mathcal{F}, \mathcal{E})$ where $\mathcal{E} \subseteq \mathcal{V} \times \mathcal{F}$. Figure 2 shows how a factor graph is induced from a BPT.

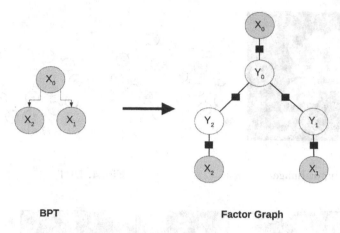

BPT   **Factor Graph**

Fig. 2. Inducing a factor graph from a 3-BPT

5.2   Probabilistic Inference on BPT

Once the factor graph is built, probabilistic inference is a straightforward process. We integrated libDAI [21], a free and open source C++ library for performing probabilistic inference. As the generated factor graph is of a tree structure, inference is efficient and exact. For an output variable domain of $\mathcal{L}$, the inference complexity is of $O(|Y||\mathcal{L}|^2)$. We recall that performing an exact inference on a general network is NP-hard [22, 23].

5.3   Application

We demonstrate here how the tool can be used for one of the most common problems in computer vision, namely image segmentation (into foreground and background). Given an image, we would like to segment out the object of interest using BPT. In other words, we are interested in labelling BPT nodes and hence image regions with either foreground and background class. As a first step, the BPT is built from the initial image. Figure 4 shows BPT built for the image in Fig. 3. As we are targeting objects of homogeneous texture, the contribution of color information to the similarity measure is set to be the highest. As already indicated, other sources of information (e.g. edge, spatial, geometric, or complex features) could be considered for the similarity measure depending on the application context.

Conversely from previous approaches in analyzing BPTs, we deal with them holistically by performing probabilistic inference on their induced factor graphs. Put it mathematically, a node x_i in the BPT X has a label variable $y_i \in \{0, 1\}$, and $Y = \{y_i\}$ is the set of X's nodes label variables. Our goal is to find the highest probable joint assignment of Y given X:

$$Y^* = \underset{Y \in \{0,1\}^{|Y|}}{\arg\max} \; P(Y|X) \tag{5}$$

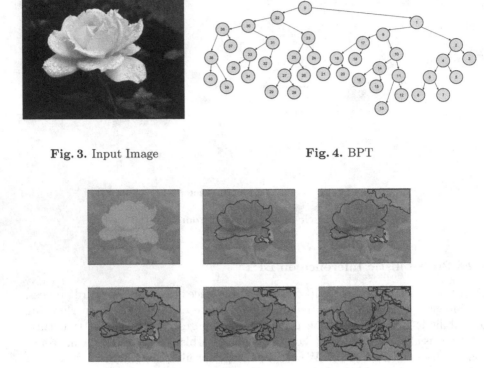

Fig. 3. Input Image   **Fig. 4.** BPT

Fig. 5. BPT Nodes Labels

where $P(Y|X)$ is nothing but the normalized product of the BPT's factor graph factors:

$$P(Y|X) = \frac{1}{Z(\mathcal{F})} \prod_{f \in \mathcal{F}: y_f \in Y, x_f \in X} \phi_f(y_f; x_f) \tag{6}$$

with Z the normalizing function for a proper probability distribution and ϕ_f the potential function for the factor node f [20]. A crucial aspect of exploiting the power of factor graphs is how carefully factors (potential functions) are designed. These factors can be either hand-crafted or learned using well-established machine learning techniques to suit more complex problems. For the sake of demonstration, we have designed the factors based on nodes color information. Figure 5 shows the labelling of BPT nodes across its six levels with transparent green assigned to background nodes and transparent red assigned to foreground ones.

6   Conclusion

This paper presented an efficient tool for building and managing binary partition trees as hierarchical representations of images. It relies on a new algorithm that allows for efficient BPT construction, while still offering several control parameters to guide the construction process. Besides, we also introduced an indexing scheme based on compressed bit maps of the nodes regions. With an additional manageable storage cost, it avoids the recursive graph traversals that is usually needed for accessing all pixels belonging to a BPT node. Furthermore, empowering the BPT with probabilistic inference features is made available by inducing the corresponding factor graph of the BPT. As the induced factor graph is as well acyclic, probabilistic inferences are exact and efficient. These complementary contributions, gathered in a publicly available tool, support the BPT as a tunable model that can be combined with recent machine learning paradigms to solve various computer vision problems.

We have provided here only a very limited comparison with existing works [11, 14]. In order to better assess the relevance of our findings, we plan now to perform a deeper experimental evaluation of our contributions (both the computational cost of the construction algorithm and the memory cost of the proposed data structure) and to compare them with more recent works, e.g. [24]. Besides, we are considering to apply the proposed probabilistic framework to various problems faced in computer vision, e.g. object recognition or image classification. To do so, we will also need to explore various similarity measures and their impact on the performance of the resulting BPT model.

References

1. Hussain, M., Chen, D., Cheng, A., Wei, H., Stanley, D.: Change detection from remotely sensed images: From pixel-based to object-based approaches. ISPRS Journal of Photogrammetry and Remote Sensing 80, 91–106 (2013)
2. Blaschke, T., Lang, S., Lorup, E., Strobl, J., Zeil, P.: Object-oriented image processing in an integrated GIS/remote sensing environment and perspectives for environmental applications. Environmental Information for Planning, Politics and the Public 2, 555–570 (2000)
3. Walter, V.: Object-based classification of remote sensing data for change detection. ISPRS Journal of Photogrammetry and Remote Sensing 58(3–4), 225–238 (2004)
4. Farabet, C., Couprie, C., Najman, L., LeCun, Y.: Learning hierarchical features for scene labeling. IEEE Transactions on Pattern Analysis and Machine Intelligence 35(8), 1915–1929 (2013)
5. Lefvre, S., Chapel, L., Merciol, F.: Hyperspectral image classification from multiscale description with constrained connectivity and metric learning. In: Proceedings of the 6th International Workshop on Hyperspectral Image and Signal Processing: Evolution in Remote Sensing, WHISPERS 2014 (2014)
6. Valero, S., Salembier, P., Chanussot, J.: Hyperspectral image representation and processing with binary partition trees. IEEE Transactions on Image Processing 22(4), 1430–1443 (2013)
7. Koller, D., Friedman, N.: Probabilistic Graphical Models: Principles and Techniques - Adaptive Computation and Machine Learning. The MIT Press (2009)
8. Salembier, P., Oliveras, A., Garrido, L.: Antiextensive connected operators for image and sequence processing. IEEE Transactions on Image Processing 7(4), 555–570 (1998)

9. Jones, R.: Component trees for image filtering and segmentation. In: IEEE Workshop on Nonlinear Signal and Image Processing, NSIP (1997)
10. Monasse, P., Guichard, F.: Scale-space from a level lines tree. Journal of Visual Communication and Image Representation 11(2), 224–236 (2000)
11. Garrido, L., Salembier, P., Garcia, D.: Extensive operators in partition lattices for image sequence analysis. Signal Processing 66(2), 157–180 (1998)
12. Salembier, P., Garrido, L.: Binary partition tree as an efficient representation for image processing, segmentation, and information retrieval. IEEE Transactions on Image Processing 9(4), 561–576 (2000)
13. Salerno, O., Pardàs, M., Vilaplana, V., Marqués, F.: Object recognition based on binary partition trees. In: International Conference on Image Processing, vol. 2, pp. 929–932. IEEE (2004)
14. Vilaplana, V., Marques, F., Salembier, P.: Binary partition trees for object detection. IEEE Transactions on Image Processing 17(11), 2201–2216 (2008)
15. Giró-i Nieto, X.: Part-Based Object Retrieval With Binary Partition Trees. PhD thesis, Universitat Politècnica de Catalunya, UPC (2012)
16. Valero, S., Salembier, P., Chanussot, J.: Hyperspectral image representation and processing with binary partition trees. IEEE Transactions on Image Processing 22(4), 1430–1443 (2013)
17. Garrido, L.: Hierarchical Region Based Processing of Images and Video Sequences: Application to Filtering, Segmentation and Information Retrieval. PhD thesis, Universitat Politècnica de Catalunya, UPC (2002)
18. Liu, T., Yuan, Z., Sun, J., Wang, J., Zheng, N., Tang, X., Shum, H.Y.: Learning to detect a salient object. IEEE Transactions on Pattern Analysis and Machine Intelligence 33(2), 353–367 (2011)
19. Bo, L., Lai, K., Ren, X., Fox, D.: Object recognition with hierarchical kernel descriptors. In: IEEE Conference on Computer Vision and Pattern Recognition (CVPR), pp. 1729–1736. IEEE (2011)
20. Nowozin, S., Lampert, C.H.: Structured learning and prediction in computer vision. Foundations and Trends in Computer Graphics and Vision 6(3–4), 185–365 (2011)
21. Mooij, J.M.: libDAI: A free and open source C++ library for discrete approximate inference in graphical models. Journal of Machine Learning Research 11, 2169–2173 (2010)
22. McAuley, J., de Campos, T., Csurka, G., Perronnin, F.: Hierarchical image-region labeling via structured learning. In: Proceedings of the British Machine Vision Conference, pp. 49.1–49.11. BMVA Press (2009)
23. Cooper, G.F.: The computational complexity of probabilistic inference using bayesian belief networks. Artificial Intelligence 42(2), 393–405 (1990)
24. Najman, L., Cousty, J., Perret, B.: Playing with kruskal: Algorithms for morphological trees in edge-weighted graphs. In: International Symposium on Mathematical Morphology, pp. 135–146 (2013)

Probabilistic Hierarchical Morphological Segmentation of Textures

Dominique Jeulin[✉]

Centre de Morphologie Mathématique,
MINES Paristech, PSL Research University,
35 rue Saint-Honoré, 77300 Fontainebleau, France
dominique.jeulin@mines-paristech.fr

Abstract. A general methodology is introduced for texture segmentation in binary, scalar, or multispectral images. Textural information is obtained from morphological operations of images. Starting from a fine partition of the image in regions, hierarchical segmentations are designed in a probabilistic framework by means of probabilistic distances conveying the textural information, and of random markers accounting for the morphological content of the regions and of their spatial arrangement.

Keywords: Texture segmentation · Morphological operations · Random markers · Probabilistic segmentation · Poisson process

1 Introduction

In many cases images contain regions with different textures, rather than objects on a background or regions with homogeneous grey level easily segmented by thresholding.

Automatic texture extraction is required in different areas such as for instance industrial control [6] or remote sensing [20, 21]. In these two last cases, in every pixel of images multivariate information is available, like results of morphological transformations applied to grey level or to binary images [5, 6], or like the wavelength response of a sensor in multispectral images [20, 21]. In this context a typical approach of segmentation makes use of pixels classification by means of multivariate image analysis [5, 6], sometimes combined with a watershed segmentation based on some multivariate gradient [20, 21].

In what follows, we introduce a hierarchical probabilistic segmentation of textures based on multivariate morphological information available on every pixel.

2 Morphological Texture Descriptors

We will consider images as domains $\mathcal{D}$ in the n dimensional space $\mathbb{R}^n$. Every pixel x is described by a set of morphological parameters or transformations building a vector with dimension p in the parameter space $\mathbb{R}^p$. Many types of transformations can be used. From experience, some standard families of morphological

J.A. Benediktsson et al. (Eds.): ISMM 2015, LNCS 9082, pp. 313–324, 2015.
DOI: 10.1007/978-3-319-18720-4_27

transformations Ψ [14, 23], performed on an initial image, are efficient as texture descriptors [5, 6]: dilations $\delta(\rho)$, erosions $\varepsilon(\rho)$, openings $\gamma(\rho)$ or closings $\varphi(\rho)$ by convex structuring elements with size ρ. These operations are as well defined for binary images as for scalar grey level images. Efficient texture descriptors, from the point of view of pixels classification, are increments of transformations with respect to the size ρ. Thus for a binary image A, vectors of description are obtained for each type of transformation, with components $I_\alpha(x)$ where I_α is the indicator function of the set $\Psi(A, \rho_\alpha) \triangle \Psi(A, \rho_{\alpha-1})$, $\triangle$ being the set difference, and α ranging from 1 to s, with $\rho_0 = 0$. For a grey level image $Y(x)$, the components are given by the increments $Z_a(x) = |\Psi(Y(x), \rho_\alpha) - \Psi(Y(x), \rho_{\alpha-1})|$. When Ψ is an opening or a closing, the increments $Z_a(x)$ provide a granulometric spectrum, as used in various domains: binary textures [5, 24], rough surfaces [2], satellite imagery [22], to mention a few. In some specific situations, the components $I_\alpha(x)$ or $Z_a(x)$ are averaged in a local window $K(x)$ around x, to provide local granulometries [5, 6, 10] or the output of linear filters, like curvelet transform [5, 6]. These descriptors are easily extended in a marginal way to the components of multispectral images.

In [18], oriented textures are extracted with information on local orientation deduced from the gradient vector.

3   Texture Classification

The morphological descriptors generate a vector field on the domain $\mathcal{D}$, from which a classification of pixels in the various textures present in the image can be looked for. For this, a partition in classes C_β must be built in the high dimensional parameter space. A convenient methodology is based on multivariate factor analysis to reduce the dimension of the data and to remove noise: in [20, 21] use is made of Factor Correspondence analysis FCA, well suited to positive data, like multi spectral images or like probability distributions as encountered in granulometric spectra; for heterogeneous data, Principal Component Analysis PCA can be a good method to produce the dimensional reduction [6]. Each of these analysis makes use of some specific distance in the parameter space, which we will denote $\|Z(x_1) - Z(x_2)\|$ for the descriptors of the two pixels x_1 and x_2. Various distances can be chosen (for instance the chi-squared between distributions in the case of FCA). We will not discuss this choice here, which is highly application dependent.

A classification of pixels is then made in the parameter space or in its reduced version, after keeping the most prominent factors. This classification can be unsupervised, using random germs in the K-means algorithm or a hierarchical classification, as described in the book [3]. When the textures are documented by a set of representative pixels, supervised statistical learning methods can be implemented for later classification (see an extensive presentation in the book [11]). In [5] a Linear Discriminant Analysis LDA is used. It follows a PCA in [6].

4   Probabilistic Texture Segmentation

In this section we assume that in every pixel x in the image embedded in $\mathbb{R}^n$, multivariate information is available (like multispectral data, or transformed images as described in section 2), stored in a vector $Z(x)$ with components $Z_\alpha(x)$. For any pair of pixels x_1 and x_2, a multivariate distance $\|Z(x_1) - Z(x_2)\|$ is defined in the parameter space $\mathbb{R}^p$.

4.1   Watershed Texture Segmentation

Considering points y in the neighborhood $B(x)$ of point x, a multivariate gradient can be defined as [19–21]:

$$\text{grad}\,(Z(x)) = \vee_{y \in B(x)} \|Z(x) - Z(y)\| - \wedge_{y \in B(x)} \|Z(x) - Z(y)\|$$

The gradient image can be used as the starting point of the segmentation of the domain $\mathcal{D}$ in homogeneous regions A_i. In fact it is expected that a texture sensitive gradient will provide weak values in homogeneous regions, and high values on the boundary A_{ij} between two regions A_i and A_j. A separation of the domain $\mathcal{D}$ in homogeneous connected regions A_i is obtained by the construction of the watershed of the gradient image from markers generated by the minima of the gradient, as initially defined for scalar images [4, 16] and later extensively used for multispectral images [19–21].

The main drawback of the watershed segmentation is its sensitivity to noise, resulting in systematic oversegmentation of the image. This is alleviated by means of a careful choice of markers, driven by some local content, like for instant chosen from a multivariate classification in [21].

Another approach, the stochastic watershed [1], makes use of random markers replacing the usual markers, enabling us to estimate a local probability of boundaries at each point $x \in A_{ij}$. The main idea is to evaluate the strength of contours by their probability, estimated from Monte Carlo simulations in a first step, as developed in the scalar [1] and in the multispectral cases [20, 21]. Simulations can be replaced by a direct calculation of the probability of contour for each boundary A_{ij} between adjacent regions A_i and A_j [13, 17]. This was successfully applied for 3D multiscale segmentation of granular media using point markers [8] or oriented Poisson lines markers [9].

4.2   Probabilistic Hierarchical Segmentation

In what follows, we design a new probabilistic segmentation obtained by a hierarchical merging of regions from a fine partition of a domain $\mathcal{D}$ in regions A_i. This initial partition can be obtained in a first step from the watershed of a gradient image, or from some classification of pixels. Given two regions A_i and A_j, not necessarily connected or even adjacent, we will estimate for various criteria the probability p_{ij}:

$$p_{ij} = P\{A_i \text{ and } A_j \text{ contain different textures}\} \tag{1}$$

The probability p_{ij} will play the same role as a gradient (or a distance) between regions A_i and A_j. In a hierarchical approach, a progressive aggregation of regions is performed, starting from lower values of p_{ij} and updating the probability after fusion of regions containing similar textures. This approach was proposed for the case of random markers [13] and implemented in an iterative segmentation [8]. Using a probabilistic framework makes easier the combination of different criteria for the segmentation, as illustrated later.

In the context of texture classification, the probability $1 - p_{ij} = P\{A_i$ and A_j contain the same texture$\}$ is also a similarity index between regions A_i and A_j.

Probabilistic Distance. Consider two points x_1 and x_2 in a domain $\mathcal{D}$, and the multivariate distance $\|Z(x_1) - Z(x_2)\|$. When the two points are located randomly in $\mathcal{D}$, $\|Z(x_1) - Z(x_2)\|$ becomes a random variable, characterized by its cumulative distribution function $P\{\|Z(x_1) - Z(x_2)\| \geq d\} = T(x_1, x_2, d)$. We have the following property:

Proposition 1. *For any $d > 0$, the distribution function $T(x_1, x_2, d)$ is a distance in $\mathcal{D}$.*

Proof. We have $T(x_1, x_1, d) = 0$ and $T(x_1, x_2, d) = T(x_2, x_1, d)$. T satisfies the triangular inequality: for any triple (x_1, x_2, x_3),

$$\|Z(x_1) - Z(x_2)\| \leq \|Z(x_1) - Z(x_3)\| + \|Z(x_2) - Z(x_3)\|.$$

Therefore

$$\|Z(x_1) - Z(x_2)\| \geq d \Longrightarrow \|Z(x_1) - Z(x_3)\| + \|Z(x_2) - Z(x_3)\| \geq d$$

and

$$T(x_1, x_2, d) \leq P\{\|Z(x_1) - Z(x_3)\| + \|Z(x_2) - Z(x_3)\| \geq d\}$$
$$\leq T(x_1, x_3, d) + T(x_2, x_3, d).$$

Definition 1. *Consider two regions A_i and A_j in $\mathcal{D}$, and two independent random points $x_i \in A_i, x_j \in A_j$. For any $d > 0$, the probability $P(A_i, A_j, d) = P\{\|Z(x_i) - Z(x_j)\| \geq d\}$ defines a probabilistic distance between A_i and A_j.*

By construction, $P(A_i, A_j, d)$ is a pseudo-distance, since we have not for every d, $P(A_i, A_i, d) = 0$. For any triple $(A_i \subset \mathcal{D}, A_j \subset \mathcal{D}, A_k \subset \mathcal{D})$ and $(x_i \in A_i, x_j \in A_j, x_k \in A_k)$ we have $P(A_i, A_j, d) \leq P(A_i, A_k, d) + P(A_j, A_k, d)$ as a result of the triangular inequality satisfied by $T(x_i, x_j, d)$.

The probabilistic distance can also be used between classes C_β obtained for a partition in the parameter space as a result of a classification. In that case we will define for two classes C_α and C_β the probability

$$p_{\alpha\beta} = P\{\|Z(x_\alpha) - Z(x_\beta)\| \geq d\}$$

estimated for independent random points $x_\alpha \in C_\alpha$ and $x_\beta \in C_\beta$. For a classification of textures in homogeneous classes, we expect that the diagonal of the matrix P, with elements $p_{\alpha\beta}$, is close to 0. This can drive the choice of the threshold d, based on the data used for the classification.

Remark 1. For any pair of regions A_i and A_j in $\mathcal{D}$, where the proportions of pixels belonging to class C_α are p_α^i and p_α^j respectively, and for independent uniform random points $x_i \in A_i, x_j \in A_j$ we have $P\{x_i \in C_\alpha, x_j \in C_\beta\} = p_\alpha^i p_\beta^j$. In that case we get $P(A_i, A_j, d) = \sum_{\alpha,\beta} p_\alpha^i p_\beta^j \, p_{\alpha\beta}$. The calculation of the probabilistic distance between A_i and A_j is made faster after a preliminary storage of the probability matrix P.

Probabilistic Distance and Hierarchical Segmentation. As mentioned before, we can make use of the probabilistic distance $P(A_i, A_j, d)$ to build a hierarchical segmentation, starting from the lowest probability. Let $P(A_i, A_j, d) < P(A_i, A_k, d)$ and $P(A_i, A_j, d) < P(A_j, A_k, d), \forall k \neq i, k \neq j$. By merging regions A_i and A_j with measures $|A_i|$ and $|A_j|$ (for instance area in $\mathbb{R}^2$ and volume in $\mathbb{R}^3$), we generate a new region $A_l = A_i \cup A_j$. For any k we get $P(A_k, A_l, d) = \frac{|A_i|}{|A_l|} P(A_k, A_i, d) + \frac{|A_j|}{|A_l|} P(A_k, A_j, d)$. Therefore we have

$$P(A_k, A_i, d) \vee P(A_k, A_j, d) \geq P(A_k, A_l, d) \geq P(A_k, A_i, d) \wedge P(A_k, A_j, d)$$
$$> P(A_i, A_j, d).$$

The probabilistic distance increases when merging two classes, so that it can be used as an index in the hierarchy. All remaining values $P(A_k, A_l, d)$ are updated after fusion of two regions, and the process can be iterated. Indeed, $\partial(A_i, A_j) = Inf\{p, A_i \text{ and } A_j \text{ are included in the same region } A_l\}$ is equivalent to the diameter of the smallest region of the hierarchy containing A_i and A_j, which satisfies the ultrametric inequality required to generate a hierarchy [3]. Alternatively we can use the probabilistic distance involved in every level of merging, to generate an ultrametric distance used to build the hierarchy [7]. The segmentation involved with the probabilistic distance is unsupervised in the general case.

Remark 2. In the context of segmentation, a partition of the domain $\mathcal{D}$ is obtained by considering all subdomains obtained when cutting the hierarchy at a given level (probability) p. The choice of the threshold p can be driven by the results of the preliminary classification of pixels, using the values of the elements of the matrix P, or by the number of subdomains, that should correspond to the number of textures present in $\mathcal{D}$. An estimation of this number can be derived from the spectral analysis of the matrix of the graph Laplacian derived from the matrix with elements $1 - P(A_i, A_j, d)$, which is an adjacency matrix [11].

Remark 3. The proposed hierarchy generates non necessarily connected subdomains, since no adjacency conditions is imposed in the choice of regions to be merged. This condition can be required, as is made for the construction of watersheds, by restricting the use of the probabilistic distances to adjacent regions. In addition to connectedness of segmented regions, it reduces the cost of calculations by limiting the number of pairs, instead of considering the full cross-product $A_i \times A_j$. Intermediate constructions can involve the probabilistic distances of iterated adjacent regions. This approach can be followed in the first steps of the segmentation, to reduce the number of regions, and released for the remaining steps of the process, in order to allow for the extraction of non connected regions with the same texture.

Remark 4. By integrating the probability $P(A_i, A_j, d)$ with respect to the threshold d, we obtain the average distance $\|Z(x_i) - Z(x_j)\|$ for independent random points $x_i \in A_i, x_j \in A_j$. This average can be used in a hierarchy where the aggregation is made according to the average distance criterion [3].

Combination of Probabilistic Segmentations. It can be useful to enrich the probabilistic distance by other probability distributions concerning the comparison of the content of two regions, in order to combine them for the segmentation. We will have to restrict the choice of probability distributions on the cross-product $A_i \times A_j$ according to the following definition.

Definition 2. *A probability $P(A_i, A_j)$ is said to be increasing with respect to the fusion of regions, when it satisfies $P(A_k, A_l) \geq P(A_k, A_i) \wedge P(A_k, A_j)$ for any i, j, k, with $A_l = A_i \cup A_j$.*

As shown before, the probabilistic distance satisfies the property given in definition 2. Other probability distributions with the same property will be introduced later.

The property given in definition 2 is satisfied when $P(A_i, A_j)$ is increasing with respect to $\subset$, which means that $P(A_k, A_l) \geq P(A_k, A_i)$ when $A_i \subset A_l$. However, this is not a necessary condition.

We start from two probabilistic segmentations, based on separate aggregation conditions, involving the probability of separation of regions A_i and A_j, $P^1(A_i, A_j)$ and $P^2(A_i, A_j)$. P^1 and P^2 are assumed to own the fusion property of definition 2. These probabilities can be combined according to different rules. For instance:

1. probabilistic independence: $P(A_i, A_j) = P^1(A_i, A_j)P^2(A_i, A_j)$
2. more reliable event: $P(A_i, A_j) = P^1(A_i, A_j) \vee P^2(A_i, A_j)$
3. least reliable event: $P(A_i, A_j) = P^1(A_i, A_j) \wedge P^2(A_i, A_j)$
4. ponderation between the two events (with probabilities λ_1 and λ_2: $P(A_i, A_j) = \lambda_1 P^1(A_i, A_j) + \lambda_2 P^2(A_i, A_j)$
5. any combination $P(A_i, A_j) = \Phi(P_1, P_2)(A_i, A_j)$, where P is a probability.

These rules are easily extended to more than two conditions of aggregation. We have the following result.

Proposition 2. *The previous rules of combination of the probability of separation of regions satisfy the property given in definition 2.*

Proof. We start from $P(A_i, A_j) = \Phi(P_1, P_2)(A_i, A_j)$. As before, consider $A_l = A_i \cup A_j$ and the condition: $P(A_i, A_j) < P(A_i, A_k)$ and $P(A_i, A_j) < P(A_j, A_k)$, $\forall k \neq i, k \neq j$. We have for any region A_k

$$P(A_k, A_l) = \frac{|A_i|}{|A_l|}P(A_k, A_i) + \frac{|A_j|}{|A_l|}P(A_k, A_j).$$

and

$$P(A_k, A_i) \vee P(A_k, A_j) \leq P(A_k, A_l) \geq P(A_k, A_i) \wedge P(A_k, A_j) > P(A_i, A_j)$$

Local Probability Distributions. The regions of the fine partition (or obtained after some steps of aggregation) can be characterized by some local probability distributions.

Probabilistic classification. If pixels x_i in region A_i can be attributed to various classes of textures C_α by a probabilistic classification, the probability $p_\alpha^i = P\{x_i \in C_\alpha\}$ can be used as a probabilistic descriptor of A_i. Considering now independent uniform random points $x_i \in A_i$ and $x_j \in A_j$, the probability (1) can be written

$$P(A_i, A_j) = 1 - P\{x_i \text{ and } x_j \text{ belong to the same texture}\} = 1 - p_{ij} \qquad (2)$$
$$= 1 - \sum_\alpha p_\alpha^i p_\alpha^j$$

Noting $I_\alpha(x)$ the indicator function of class C_α, and $I(x)$ the vector with components $I_\alpha(x)$, we have $P\{\|I_\alpha(x_i) - I_\alpha(x_j)\| = 0\} = P\{x_i \in C_\alpha, x_j \in C_\alpha\} = p_\alpha^i p_\alpha^j$ and therefore $P(A_i, A_j) = P\{\|I(x_i) - I(x_j)\| > 0\}$, so that $P(A_i, A_j)$ defined by 2 is a probabilistic distance corresponding to definition 1. It satisfies the property given in definition 2. This can be checked as follows:using $A_l = A_i \cup A_j$ we obtain

$$p_\alpha^l = \frac{|A_i|}{|A_l|} p_\alpha^i + \frac{|A_j|}{|A_l|} p_\alpha^j$$

and

$$1 - P(A_k, A_l) = \sum_\alpha p_\alpha^k p_\alpha^l = \frac{|A_i|}{|A_l|} \sum_\alpha p_\alpha^i p_\alpha^k + \frac{|A_j|}{|A_l|} \sum_\alpha p_\alpha^j p_\alpha^k$$
$$= \frac{|A_i|}{|A_l|} (1 - P(A_i, A_k)) + \frac{|A_j|}{|A_l|} (1 - P(A_j, A_k))$$

Therefore we get

$$P(A_k, A_l) = \frac{|A_i|}{|A_l|} P(A_i, A_k) + \frac{|A_j|}{|A_l|} P(A_j, A_k)$$

and

$$P(A_k, A_i) \vee P(A_k, A_j) \geq P(A_k, A_l) \geq P(A_k, A_i) \wedge P(A_k, A_j)$$

Local granulometric spectrum. As indicated in section 2, granulometric information can be provided after transformation of the image by morphological opening or closing operations. A local granulometric spectrum can be obtained in each region A_i by averaging the components $I_\alpha(x_i)$ or $Z_\alpha(x_i)$ over A_i. After normalization, we obtain local granulometric spectra in classes of sizes C_α where for size α, $p_\alpha^i = P\{x_i \in C_\alpha\}$. This local classification with respect to size can be introduced in the probability (2) to build a hierarchy. We can alternatively combine different granulometries, like opening and closing, for instance by linear combination as proposed in section 4.2, to generate a composite hierarchy.

Use of Random Markers. Following the approach proposed for the stochastic watershed [1], we can use random markers to randomly select regions for which the previous probabilistic segmentation will be performed. Doing this, some morphological content on the regions of the hierarchy and on their location in $\mathcal{D}$ is accounted for, in addition to the previous probabilistic textural information. The aim of this section is to calculate the probability $P_R(A_i, A_j)$ of selection of two regions (A_i, A_j) by random markers. We follow the results introduced in [13].

Reminder on random allocation of germs. We use m random points (or germs) x_k with independent uniform distribution in the domain $\mathcal{D}$ containing r regions. The probability p_i for a germ to fall in A_i is given by

$$p_i = \frac{|A_i|}{|\mathcal{D}|}, \text{ with } \sum_{i=1}^{i=r} p_i = 1.$$

By construction, the allocation of germs in the regions of a partition follows a multinomial distribution (N_i being the random number of germs in A_i). An interesting case is asymptotically obtained when $|\mathcal{D}| \to \infty$ and $m \to \infty$, with $\frac{m}{|\mathcal{D}|} \to \theta$. For these conditions, the multinomial distribution converges towards the multivariate Poisson distribution, the random numbers $N_1, N_2, ..., N_r$ being independent Poisson random variables with intensities $\theta_i = \theta |A_i|$ and $P\{N_i = 0\} = \exp -\theta_i$

Calculation of the probability $P_R(A_i, A_j)$ for point markers. Random markers are used to select regions of a partition by reconstruction. With this process, the reconstructed regions for any realization of the random germs are left intact, while regions without germs are merged. Considering many realizations of the germs, we can compute the probability $P_R(A_i, A_j)$ for the two regions to remain separate.

Proposition 3. *For m independent uniformly distributed random germs, the probability $P_R(A_i, A_j)$ for the two regions A_i, A_j to remain separate is given by:*

$$P_R(A_i, A_j) = 1 - (1 - p_i - p_j)^m \tag{3}$$

Proof. The pair (A_i, A_j) is merged $\Longleftrightarrow \{N_i = 0 \text{ and } N_j = 0\} \Longleftrightarrow N_i + N_j = 0 \Longleftrightarrow N(A_i \cup A_j) = 0$.

Working on images, the probabilities $P_R(A_i, A_j)$ computed for all pairs (A_i, A_j) are easily ranked in increasing order. A hierarchical fusion of regions is obtained by starting with the lowest probability $P_R(A_i, A_j)$. After fusion of two regions with $A_l = A_i \cup A_j$ the probabilities $P_R(A_k, A_l)$ are updated. The pair (A_k, A_l) is merged $\iff \{N_k = 0 \text{ and } N_l = 0\} \iff N_k + N_l = 0 \iff N(A_k \cup A_l) = 0$ $\iff N(A_i \cup A_j \cup A_k) = 0$. We get:

$$P_R(A_k, A_l) = 1 - (1 - p_i - p_j - p_k)^m > P_R(A_i, A_j)$$

and the probability $P_R(A_k, A_l)$ is increasing with respect to the fusion of regions as in definition 2.

In general no conditions of connectivity or of adjacency of regions are required for the fusion process. It is easy to force the connectivity by working on connected components of regions, or to limit the fusion to adjacent regions.

The random germs can be generated by a Poisson point process.

Proposition 4. *For Poisson point germs with intensity θ, the probability $P_R(A_i, A_j)$ for the two regions A_i, A_j to remain separate is given by:*

$$P_R(A_i, A_j) = 1 - \exp\left[-\theta\left(|A_i| + |A_j|\right)\right] \tag{4}$$

and $P_R(A_k, A_l)$ is increasing with respect to the fusion of regions as in definition 2.

The morphological content in the probabilities (3, 4) only depends on the Lebesgue measure (area in $\mathbb{R}^2$ and volume in $\mathbb{R}^3$) of regions. It increases with the measure of regions, larger regions resisting more to fusion. For a pair of regions, $P_R(A_i, A_j)$ is maximal when $|A_i| = |A_j|$, so that the random markers hierarchy tends to generate by fusion regions with homogeneous sizes, the regions with lower measure disappearing first.

Calculation of the probability $P_R(A_i, A_j)$ for Poisson lines and Poisson flats markers. It can be interesting to obtain other ponderations of regions with a probabilistic meaning, like the perimeter in $\mathbb{R}^2$ or the surface area in $\mathbb{R}^3$. Restricting to the Poisson case, it is easy to make this extension, provided use is made of appropriate markers. For this purpose, we will consider now isotropic Poisson lines in $\mathbb{R}^2$, isotropic Poisson planes and Poisson lines in $\mathbb{R}^3$[12, 15]. Oriented Poisson lines in $\mathbb{R}^3$ were used as markers in the context of the stochastic watershed (and so with another type of probability), and applied to the segmentation of granular structures [9].

Proposition 5. *Consider stationary isotropic Poisson lines with intensity λ as random markers in $\mathbb{R}^2$. The probability $P_R(A_i, A_j)$ for the two regions A_i, A_j to remain separate, when $A_i \cup A_j$ is a connected set, is given by:*

$$P_R(A_i, A_j) = 1 - \exp\left[-\lambda L(\mathcal{C}(A_i \cup A_j))\right] \tag{5}$$

where L is the perimeter and $\mathcal{C}(A_i \cup A_j)$ is the convex hull of $A_i \cup A_j$.

Proposition 6. *Consider stationary isotropic Poisson lines with intensity λ as random markers in $\mathbb{R}^3$. The probability $P_R(A_i, A_j)$ for the two regions A_i, A_j*

with surface areas $S(A_i)$ and $S(A_i)$ to remain separate, when $A_i \cup A_j$ is a connected set, is given by:

$$P_R(A_i, A_j) = 1 - \exp\left[-\lambda \frac{\pi}{4} S(\mathcal{C}(A_i \cup A_j))\right] \tag{6}$$

For random markers in $\mathbb{R}^3$ made of stationary isotropic Poisson planes with intensity λ, the probability $P_R(A_i, A_j)$ for the two regions A_i, A_j with integrals of mean curvature $\mathcal{A}(A_i)$ and $\mathcal{A}(A_j)$ to remain separate, when $A_i \cup A_j$ is a connected set, is given by:

$$P_R(A_i, A_j) = 1 - \exp\left[-\lambda \mathcal{A}(\mathcal{C}(A_i \cup A_j))\right] \tag{7}$$

It is possible to combine various types of Poisson markers (points and lines in $\mathbb{R}^2$, points, planes and lines in $\mathbb{R}^3$) with their own intensities. For instance in $\mathbb{R}^2$ we obtain, when $A_i \cup A_j$ is a connected set,

$$P_R(A_i, A_j) = 1 - \exp\left[-\left\{\theta\left(|A_i| + |A_j|\right) + \lambda L\left(\mathcal{C}(A_i \cup A_j)\right)\right\}\right] \tag{8}$$

where a ponderation by the area and the perimeter of the regions acts for the segmentation. Similarly in $\mathbb{R}^3$ is introduced a ponderation of the volume, and the surface area and integral of mean curvature of $\mathcal{C}(A_i \cup A_j)$ in the process of segmentation.

Calculation of the probability $P_R(A_i, A_j)$ for compact markers. Further morphological information on the regions can be accounted for when introducing compact random markers (not necessarily connected). In the process of selection of regions of a partition by reconstruction, point markers are replaced by a compact grain A' located on Poisson points, and generating a Boolean model A. We have:

Proposition 7. *For compact markers A' generating a Boolean model with intensity θ, the probability $P_R(A_i, A_j)$ for the two regions A_i, A_j to remain separate is given by:*

$$P_R(A_i, A_j) = 1 - \exp\left[-\theta\left|(\check{A}_i \oplus A') \cup (\check{A}_j \oplus A')\right|\right] \tag{9}$$
$$= 1 - \exp\left[-\theta\left(|\check{A}_i \oplus A'| + |\check{A}_j \oplus A'| - |(\check{A}_i \oplus A') \cap (\check{A}_j \oplus A')|\right)\right]$$

Proof. The pair (A_i, A_j) is merged $\iff A_i \cup A_j$ is outside the Boolean model with primary grain A'. The expression (9) is the Choquet capacity $T(K)$ of the Boolean model when $K = A_i \cup A_j$.

The compact markers can be random sets (for instance spheres wit a random radius). In that case, the measures $||$ are replaced by their mathematical expectations with respect to the random set A'. Using for A' a ball with radius ρ, $P_R(A_i, A_j)$ increases until a constant value when the distance between A_i and A_j increases from 0 to 2ρ: the probability to merge two regions is higher when their distance is lower.

Combination of textural and of morphological information. We can now combine the use of random markers, conveying morphological content on the partition and on its evolution in the hierarchy, to the previous textural content (probabilistic distance, or local probability information. For instance, we can decide to merge two regions when they are not reconstructed by markers (with a marker dependent probability $1 - P_R(A_i, A_j)$) and the textures they enclose are similar (with a probability $1 - P(A_i, A_j, d)$). In this context the probability p_{ij} (1) becomes

$$P(A_i, A_j) = \tag{10}$$
$$\Phi(P_R, P_d)(A_i, A_j) = P_R(A_i, A_j) + P(A_i, A_j, d) - P_R(A_i, A_j)P(A_i, A_j, d)$$

By construction, this composite probability is increasing with respect to the fusion of regions as in definition 2, and will generate a hierarchy for the segmentation, by updating each terms of (10) according to the previous rules.

5 Conclusion

The probabilistic hierarchical segmentation tools introduced in this work are flexible enough to handle various types of textures (scalar or multivariate) and their spatial distribution, by progressively merging regions of a fine partition. Combining appropriate morphological operations and texture classification, proved to be efficient in previous studies [2, 5, 6, 10, 19–22, 24], supervised or unsupervised texture segmentation can be implemented. A probabilistic distance between regions is defined, from which hierarchies bearing statistical information on texture are built. Additionally, morphological information on the regions of the fine partition and of the hierarchy can be accounted for in the process, through the use of random markers.

References

1. Angulo, J., Jeulin, D.: Stochastic watershed segmentation. In: Banon, G., Barrera, J., Braga-Neto, U. (eds.) Proc. ISMM 2007, 8th International Symposium on Mathematical Morphology, Rio de Janeiro, Brazil, October 10-13, pp. 265–276 (2007) ISBN 978-85-17-00032-4
2. Aubert, A., Jeulin, D.: Classification morphologique de surfaces rugueuses, Revue de Métallurgie - CIT/Sience et Génie des Matériaux, pp. 253–262 (February 2000)
3. Benzecri, J.P.: L'analyse des données, TIB (3) § 3-4, pp. 133–149; TIB (4) §2.3, pp. 180–183, Dunod (1973)
4. Beucher, S., Lantuéjoul, C.: Use of watersheds in contour detection. In: International Workshop on Image Processing, Real-time Edge and Motion Detection (1979)
5. Cord, A., Jeulin, D., Bach, F.: Segmentation of random textures by morphological and linear operators. In: Banon, G., Barrera, J., Braga-Neto, U. (eds.) Proc. ISMM 2007, 8th International Symposium on Mathematical Morphology, Rio de Janeiro, Brazil, October 10-13, pp. 387–398 (2007) ISBN 978-85-17-00032-4

6. Cord, A., Bach, F., Jeulin, D.: Texture classification by statistical learning from morphological image processing. Application to metallic surfaces. Journal of Microscopy 239, 159–166 (2010)

7. Duda, R.O., Hart, P.E.: Pattern recognition and scene analysis, pp. 236–237. Wiley, New York (1973)

8. Gillibert, L., Jeulin, D.: Stochastic Multiscale Segmentation Constrained by Image Content. In: Soille, P., Pesaresi, M., Ouzounis, G.K. (eds.) ISMM 2011. LNCS, vol. 6671, pp. 132–142. Springer, Heidelberg (2011)

9. Gillibert, L., Peyrega, C., Jeulin, D., Guipont, V., Jeandin, M.: 3D Multiscale Segmentation and Morphological Analysis of X-ray Microtomography from Cold-sprayed Coatings. Journal of Microscopy 248(pt. 2), 187–199 (2012)

10. Gratin, C., Vitria, J., Moreso, F., Seron, D.: Texture classification using neural networks and local granulometries. In: Serra, J., Soile, P. (eds.) Mathematical Morphology and its Applications to Image Processing, pp. 309–316. Kluwer Academic Publishers (1994)

11. Hastie, T., Tibshirani, R., Friedman, J.: The Elements of Statistical Learning, Data Mining, Inference, and Prediction, pp. 544–547. Springer (2001)

12. Modèles, J.D.: Morphologiques de Structures Aléatoires et de Changement d'Echelle. Thèse de Doctorat d'Etat ès Sciences Physiques, Université de Caen (Avril 25, 1991)

13. Jeulin D.: Remarques sur la segmentation probabiliste, N-10/08/MM, Internal report, CMM, Mines ParisTech (September 2008)

14. Matheron, G.: Eléments pour une théorie des milieux poreux, Masson, Paris (1967)

15. Matheron, G.: Random sets and integral geometry. Wiley (1975)

16. Meyer, F., Beucher, S.: Morphological segmentation. Journal of Visual Communication and Image Representation 1(1), 21–46 (1990)

17. Meyer, F., Stawiaski, J.: A stochastic evaluation of the contour strength. In: Goesele, M., Roth, S., Kuijper, A., Schiele, B., Schindler, K. (eds.) Pattern Recognition. LNCS, vol. 6376, pp. 513–522. Springer, Heidelberg (2010)

18. Jeulin, D., Moreaud, M.: Segmentation of 2D and 3D textures from estimates of the local orientation. Image Analysis and Stereology 27, 183–192 (2008)

19. Noyel, G., Angulo, J., Jeulin, D.: Morphological segmentation of hyperspectral images. Image Analysis and Stereology 26, 101–109 (2007)

20. Noyel, G., Angulo, J., Jeulin, D.: Classification-driven stochastic watershed. Application to multispectral segmentation. In: Proc. IS&T's Fourth European Conference on Color in Graphics, Imaging and Vision (CGIV 2008), Terrassa - Barcelona, Spain, June 9-13, pp. 471–476 (2008)

21. Noyel, G., Angulo, J., Jeulin, D.: A new spatio-spectral morphological segmentation for multispectral remote sensing images. International Journal of Remote Sensing 31(22), 5895–5920 (2010)

22. Pesaresi, M., Benediktsson, J.: A new approach for the morphological segmentation of high resolution satellite imagery. Geoscience and Remote Sensing 39(2), 309–320 (2001)

23. Serra, J.: Image Analysis and Mathematical Morphology, vol. I. Academic Press, London (1982)

24. Sivakumar, K., Goutsias, J.: Monte Carlo estimation of morphological granulometric discrete size distributions. In: Serra, J., Soille, P. (eds.) Mathematical Morphology and its Applications to Image Processing, pp. 233–240. Kluwer Academic Publishers (1994)

The Waterfall Hierarchy on Weighted Graphs

Fernand Meyer[✉]

CMM – Centre de Morphologie Mathématique,
MINES ParisTech – PSL Research University,
35 rue St Honoré, 77300 Fontainebleau, France
fernand.meyer@mines-paristech.fr

Abstract. We study and present two new algorithms for constructing the waterfall hierarchy of a topographic surface. The first models a topographic surface as a flooding graph, each node representing a lake filling a catchment basin up to its lowest pass point ; each edge representing such a pass point. The second algorithm produces the waterfall partition in one pass through the edges of a minimum spanning tree of the region adjacency graph associated to a topographic surface.

Keywords: Waterfall · Watershed · Edge and node weighted graphs · 1 Pass waterfall algorithm

1 Introduction

The waterfall hierarchy represents the nested structure of the catchment basins of a topographic surface [3]. Consider a series of increasing floodings of a topographic surface. The associated watershed partitions form a hierarchy, as each catchment basin of a higher flooding is the union of catchment basins of a lower flooding of the surface. The waterfall hierarchy is obtained when a topographic surface is submitted to the highest flooding possible without any overflow from one catchment basin to a neighboring bassin ; each basin is then filled up to its lowest pass leading to a neighboring basin. This flooding level is thus independent of the shape, size or depth of the basins. As a result is the waterfall hierarchy invariant under the change of scale, orientation or contrast of the images.

Fig.1 presents a topographic surface with 3 successive levels of waterfall flooding. The associated waterfall partitions are indicated below each flooded surface.

The waterfall hierarchy unfolds the nested structure of the catchment basins of a topographic surface. From one level to the next of the waterfall hierarchy, a number of catchment basins merge, revealing a new scale, until a last flooding completely covers the domain. In general does the waterfall hierarchy possess only a low number of hierarchical levels. The lowest levels provide highly simplified partitions compared to the initial oversegmentation produced by the watershed. The waterfall hierarchy has been used with success for segmenting not only gray tone images [11], but also color and texture images [19]. It has been used not only on images but also on probability distribution functions used for the stochastic watershed of multipectral images [2] or on triangular meshes [9]. The domain of application is broad going from multimedia images [22] to SAR imagery [20].

© Springer International Publishing Switzerland 2015
J.A. Benediktsson et al. (Eds.): ISMM 2015, LNCS 9082, pp. 325–336, 2015.
DOI: 10.1007/978-3-319-18720-4_28

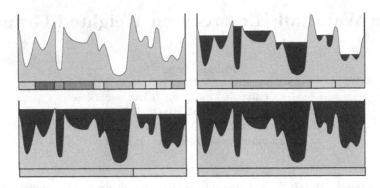

Fig. 1. 4 levels of the waterfall hierarchy. The watershed partition is presented below each topographic surface. Each topographic surface has been flooded up to the lowest pass point of the catchment basins associated to the preceding surface.

After a rapid discussion of the classical constructions of the waterfall hierarchy, we present two new methods having each its merits.

The first method models the topographic surface as a particular graph, perfectly representing the waterfall flooding, where all lakes are full, at the verge of overflow. The nodes of the graph represent the full lakes and their weight equals the altitude of the lake. A drop of water added to a full lake provokes an overflow into a neighboring lake, called exhaust lake of the first one. An edge links each lake with its exhaust lakes ; its weight is equal to the altitude of the lake which overflows through this edge. Such node and edge weighted graphs are called flooding graphs. On such a graph the trajectory of a drop of water is the same, whether one considers the edge weights alone or the node weights alone. Efficient, precise and fast algorithms based on the node weights only permit to construct the watershed partitions [16,17,18].

The second method constructs a complete waterfall hierarchy in one pass through the edges of the minimum spanning tree of the region adjacency graph

2   The Waterfall Hierarchy: State of the Art

2.1   An Algorithm Based on Successive Image Floodings

The waterfall hierarchy has been defined by Serge Beucher for grey tone images [3]. A first algorithm works at the image level [4]. Each catchment basin is flooded up to its lowest pass point. As the pass points are regional minima of the watershed line, the solution consists in the following steps:

 * constructing the watershed line of the topographic surface f, producing the level 1 of the waterfall hierarchy.

 * constructing a ceiling function equal to f on the watershed line and ∞ elsewhere

 * constructing the highest flooding of f under this ceiling function

A new topographic surface is produced, which is then submitted to the same process, producing the next level of the waterfall hierarchy.

Discussion: Using the watershed line is an advantage, as the ceiling function is easily constructed. It also is a disadvantage, as the watershed line has no meaning in terms of image content. It "takes place" to the detriment of thin and narrow structures which cannot be represented. It makes the construction of a hierarchy more difficult. Below we present a method where each level of the waterfall hierarchy is a partition, without watershed line separating neighboring tiles.

2.2   A Watershed Algorithm on the Region Adjacency Graph

A second algorithm [5] constructs the catchment basins of the topographic surface at the pixel level. They constitute the finest level of the waterfall hierarchy. The higher levels of the waterfall hierarchy are then constructed on the region adjacency graph, whose nodes are the catchment basins, whose edges represent the pass points between neighboring basins, weighted by the altitude of the pass point. B.Marcotegui et al use the watershed partition of this graph by constructing a minimum spanning forest rooted in its regional minima [13]. The edge inside each tree are then contracted within the MST, producing a new tree on which the same process is repeated. This has been the first watershed algorithm on edge weighted graphs. More recent alternative algorithms could be used for the construction of the watershed on edge weighted graphs [8,10,17].

Discussion: The minimum spanning tree of the region adjacency graph is not unique and with a unlucky choice, the results could be poor. This is due to the fact that the watershed algorithms on an edge weighted graph are myopic and evaluate the steepness of the flooding on the immediate neighboring edges of each node. Below we present a watershed algorithm in which the steepness of the flooding is better taken into account.

3   Construction of the Waterfall Hierarchy, Thanks to the Flooding Graph

3.1   Reminders on Node and/or Edge Weighted Graphs

A *non oriented graph* $G = [N, E]$ contains a set N of vertices or nodes and a set E of edges ; an edge being a pair of vertices. The nodes are designated with small letters: p, q, r...The edge linking the nodes p and q is designated by e_{pq}.

A *path*, π, is a sequence of vertices and edges, interweaved in the following way: π starts with a vertex, say p, followed by an edge e_{pq}, incident to p, followed by the other endpoint q of e_{pq}, and so on.

Edges and/or nodes may be weighted. Denote by $\mathcal{F}_e$ and $\mathcal{F}_n$ the sets of non negative weight functions on the edges and on the nodes respectively. The function $\eta \in \mathcal{F}_e$ takes the value η_{pq} on the edge e_{pq}, and the graph holding only edge weights is designated by $G(\eta, nil)$. The function $\nu \in \mathcal{F}_n$ takes the weight

ν_p on the node p and the graph holding only these node weights is designated by $G(nil, \nu)$. If both nodes and edges are weighted we write $G(\eta, \nu)$.

Images defined on a grid may also be modelled by such a graph : the pixels of the image become the nodes of the graph, with the same weight ; neighboring pixels are linked by an unweighted edge.

We define two operators between the node and edge weights:

* an operator δ_{en} associating to the function $\nu \in \mathcal{F}_n$ the function $\delta_{en}\nu \in \mathcal{F}_e$ taking the value $(\delta_{en}\nu)_{pq} = \nu_p \vee \nu_q$ on the edge e_{pq}.
* an operator ε_{ne} associating to the function $\eta \in \mathcal{F}_e$ the function $\varepsilon_{ne}\eta \in \mathcal{F}_n$ taking the value $(\varepsilon_{ne}\eta)_p = \bigwedge_{(s \text{ neighbors of } p)} \eta_{ps}$ on the node p.

The pair of operators $(\delta_{en}, \varepsilon_{ne})$ form an adjunction:

$$\forall \eta \in \mathcal{F}_e, \forall \nu \in \mathcal{F}_n : \delta_{en}\nu < \eta \Leftrightarrow \nu < \varepsilon_{ne}\eta.$$ It follows that ([21]):
* δ_{en} is a dilation from $\mathcal{F}_n$ into $\mathcal{F}_e$
* ε_{ne} is an erosion from $\mathcal{F}_e$ into $\mathcal{F}_n$

This section is based on the flooding graph, whose nodes represent the full lakes of a topographic surface when the flooding of each basin reaches its lowest pass points . We first rigorously establish the altitude of a full lake. The flooding graph is completed by adding the edges linking each lake with its exhaust lakes. The flooding graph contains the necessary and sufficient information for constructing the watershed partition or the waterfall hierarchy of a topographic surface.

3.2   The Lowest Pass Point of a Catchment Basin

Consider a watershed partition π of a topographic surface, verifying the following condition : if s and p are two neighboring nodes belonging to the catchment basins of two distinct minima m_1 and m_2, then there exists a non increasing path going from s to m_1 and a non increasing path from p to m_2.

Our aim is to figure out the level of the full lakes on such a topographic surface, given this watershed partition.

Consider a catchment basin CB_1 filled by a lake up to its lowest pass point, leading to an unflooded neighboring basin CB_2. The regional minima of CB_1 and CB_2 are respectively m_1 and m_2.

Consider among all paths linking the minimum m_1 with the minimum m_2, a path whose highest altitude is the lowest. Such a path crosses the pass point between both basins CB_1 and CB_2. The highest altitude of the path is equal to the altitude of the pass point.

Consider the altitude of the flooded relief g along one such path ϖ (there may be several of them). The altitude is constant and equal to the altitude of the full lake covering m_1 until it reaches the last node with this altitude, say p ; the next node q having a lower altitude. But $g_p > g_q \Rightarrow g_p = f_p$, an implication which characterizes floodings [14] indicating that the node p is dry and has the altitude of the full lake. Three cases are to be considered:

* $p \in CB_1$ and $q \in CB_2$. Then $f_p = g_p > g_q \geq f_q$ and the altitude of the lake is $f_p \vee f_q = f_p$

* $p \in CB_2$ and $q \in CB_2$ and the last node preceding p on the path ϖ is a node s belonging to CB_1, then $f_s \leq g_s = g_p = f_p$. The altitude of the lake is $f_s \vee f_p = f_p$

* $p \in CB_2$ and $q \in CB_2$ and the last pixel of the path ϖ belonging to CB_1 is a pixel s, followed by a series of pixels $t_1, t_2, ...p$ belonging to CB_2. All these pixels belong to the full lake : $g_s = g_{t_1} = ... = g_p$. On the other hand $t_1, t_2, ...p$ constitutes a non ascending path for the function $f : f_{t_1} \geq f_{t_1} \geq ... \geq f_p$. As $g_p = g_{t_1} \geq f_{t_1} \geq f_p = g_p$, all internal inequalities are in fact equalities and $f_{t_1} = g_p$. Here again the level of the lake is $f_s \vee f_{t_1}$ for two pixels belonging one to CB_1 and the other to CB_2.

This gives a clue for determining the altitude of the full lake which will cover the minimum m_1. One considers all couples of neighboring pixels which form the boundary of CB_1, with $p \in CB_1$ and $q \notin CB_1$. The altitude of the full lake is then the smallest value taken by $f_p \vee f_q$: $\lambda(CB_1) = \bigwedge\limits_{\substack{p \in CB_1 \text{ and } q \notin CB_1 \\ p,q \text{ neighbors}}} f_p \vee f_q$.

Each exhaust passpoint of the catchment basin $CB1$ towards a basin $CB2$ is thus a pair of nodes $p \in CB_1$ and $q \in CB_2$ verifying $f_p \vee f_q = \lambda(CB_1)$. We say that $CB2$ is an exhaust basin of $CB1$, as each additional drop of water falling in the full lake occupying $CB1$ provokes an overflow into $CB2$.

3.3   Modelling the Catchment Basins of a Topographic Surface as a Flooding Graph

Modeling a Topographic Surface as a Flooding Graph. Consider a topographic surface in which each catchment basin has been flooded up to its exhaust pass point. We create a graph $G = [N, E]$ in which each node n_i represents a full lake occupying a catchment basin, with a weight ν_i equal to the altitude of this lake. An edge e_{ij} with the same weight $\eta_{ij} = \nu_i$ links this node n_i with the node n_j if the catchment basin CB_j is an exhaust catchment basin of CB_i. We say that e_{ij} is an exhaust edge of n_i. We call this node and edge weighted graph a flooding graph of f, as it models the highest flooding of f without overflow.

Properties of a flooding graph. As each catchment basin has one or several lowest pass points, each node n_k of $G = [N, E]$ has one or several exhaust edges. All other adjacent edges of n_k correspond to higher pass points of CB_k and represent the exhaust edges of neighboring catchment basins. Hence each exhaust edge of a node has the lowest weight among all adjacent edges of this node: $\nu = \varepsilon_{ne}\eta$. In particular, if a node n_i is linked with n_k through an exhaust edge e_{ik} of n_k, we have $\nu_k = \eta_{ik} \geq \nu_i$. It follows that each node is linked by an exhaust edge with its neighboring nodes whose weight is lower or equal ; and the weight of each edge is equal to the highest weight of its extremities : $\eta_{ik} = \nu_i \vee \nu_k$. Hence $\eta = \delta_{en}\nu$.

A graph where the node and edge weights are coupled by $\nu = \varepsilon_{ne}\eta$ and $\eta = \delta_{en}\nu$ is called a flooding graph [18].

Flooding paths and regional minima. Adding a drop of water to a full lake provokes an overflow through an exhaust pass point into a neighboring full lake
with a lower or equal altitude, which itself provokes an overflow and so on until,
ultimately, a regional minimum lake is reached. Such a trajectory of a drop of
water is modelled on the flooding graph as a flooding path.

Definition 1. *A path* $p - e_{pq} - q - e_{qs} - s...$ *is a flooding path if each node except
the last one is followed by one of its exhaust edges.*

By construction we then have $\nu_p = \eta_{pq} \geq \nu_q = \eta_{qs} \geq \nu_s$. Considering only
the node weights, the path will be a flooding path if $\nu_p \geq \nu_q \geq \nu_s$. Considering
only the edge weights it is a flooding path if each edge is an exhaust edge of the
preceding node. A drop of water following a flooding path ultimately is trapped
in a regional minimum. It arrives to a node whose exhaust lead to adjacent
nodes with the same weight. There is no possibility to reach a deeper node as
each neighboring node of a regional minimum has a higher weight and is linked
with the regional minimum by an exhaust edge also with a higher weight.

The catchment zone of a regional minimum is the set of nodes linked with
this minimum through a flooding path. Minimum and flooding path being the
same whether one considers the node weights only or the edge weights only.

Having the same flooding paths and the same regional minima, both graphs
also have the same catchment zones.

Theorem 1. *For a flooding graph* $G(\eta, \nu)$*, the flooding paths, regional minima
and catchment zones of the graphs* $G(nil, \nu)$ *and* $G(\eta, nil)$ *are identical.*

An Order Relation Between Flooding Paths. A drop of water following a
flooding path reaching a regional minimum node p can quit this node through
one of its exhaust edges and reach another node q with the same altitude. From
this node, it may then come back. Going hence and forth between p and q, the
flooding path is thus artificially prolongated as a path of infinite length. A **lexicographic preorder relation** compares the infinite paths $\pi = (p_1, p_2, ...p_k, ...)$
and $\chi = (q_1, q_2, ...q_k, ...)$:

* $\pi \prec \chi$ if $\nu_{p_1} < \nu_{q_1}$ or there exists t such that $\begin{array}{l} \forall l < t : \nu_{p_l} = \nu_{q_l} \\ \nu_{p_t} < \nu_{q_t} \end{array}$

* $\pi \preceq \chi$ if $\pi \prec \chi$ or if $\forall l : \nu_{p_l} = \nu_{q_l}$.

The preorder relation $\preceq$ is total, as it permits to compare all paths with the
same origin p. The lowest of them are called the *steepest* paths of origin p.

Constructing the Steepest Watershed Partition. A node may be linked by
flooding paths with 2 or more distinct regional minima. In this case the watershed
zones overlap. If one considers only the *steepest* paths, this will rarely happen,
as the weights all along the paths should be absolutely identical. In particular,
steepest paths reaching regional minima with different altitude necessarily have
a distinct steepness. One obtains like that highly accurate watershed partitions,

whereas the classical algorithms, being myopic as they use only the adjacent edges of each node, pick one solution out of many.

The classical algorithm for constructing a watershed partition in which the flooding of a node weighted graph is governed by a hierarchical queue [12] propagates the labels of the minima along the steepest flooding paths of the graph. We start detecting and labeling the node regional minima.

Initialisation: Create a hierarchical queue HQ. Put the outer boundary nodes of the regional minima in the queue corresponding to their weight. Assign to each regional minimum a distinct label

Repeat until the HQ is empty:

Extract the node p with the highest priority from the queue.

For each node q without label, such that (p, q) neighbors:

* $label(q) = label(p)$
* put q in the queue with priority n_q

At the end of the process, we obtain a watershed partition Π_1 of the nodes. By merging all basins corresponding to nodes belonging to the same tile of Π_1, we obtain a new partition of the domain D. The boundaries of this partition are a subset of the boundaries of the initial watershed partition of f.

The complete process is repeated: creating the flooding graph of the new partition, constructing the watershed partition Π_2 on the nodes etc.

The Flooding Graph of a Region Adjacency Graph. An alternative solution consists in constructing the region neighborhood graph $G = [N, E]$ associated to the initial watershed partition. The subsequent processing is done on this graph and illustrated in fig.2. The flooding graph G' is constructed first: for each node p of G, one retains in G' only its exhausts edges, i.e. the adjacent edges with minimal weight. The node p gets the same weight as its exhaust edges. The watershed partition of G' is constructed as above, producing a number of connected components. Contracting the inside edges of each component within the graph G, produces a new edge weighted graph, to which the same process may be applied.

4   The Waterfall Partition in One Run

We now compute the waterfall hierarchy associated to an arbitrary edge weighted tree T with edge weights $\eta > 0$. In the context of morphological segmentation, T is the minimum spanning tree of the region adjacency graph associated to the topographic surface. But the developments below make sense for any edge weighted tree. The waterfall hierarchy will be expressed as another tree Θ, having the same structure, i.e. nodes and edges as the tree T but not the same weights. We associate to the tree Θ the ultrametric distance χ : the distance between two nodes n_1 and n_2 is equal to the highest edge on the unique path inside Θ linking these two nodes.

Initially all edges are assigned a weight equal to ∞. Hence the ultrametric distance between any two node also is equal to ∞.

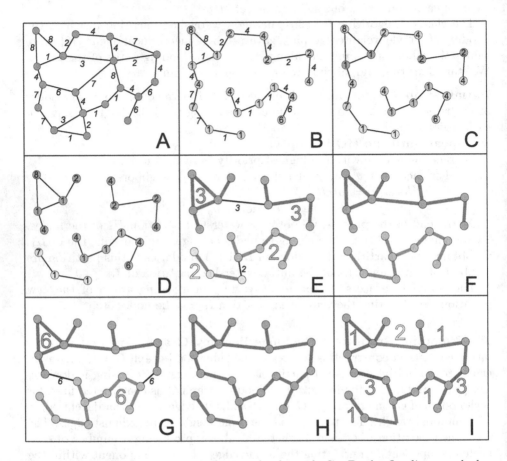

Fig. 2. A: the region adjacency edge weighted graph G ; B: the flooding graph, by retaining for each node its exhaust edges ; C: assigning to each node the weight of its exhaust edges and skipping the edge weights ; D: watershed partition on the nodes obtained by the hierarchical queue algorithm ; E: contraction of all edges which belong to the same tile of the watershed partition These contracted edges are indicated as bold lines. At the same time, construction of the associated flooding graph: one retains for each node obtained by contraction its lowest adjacent edges within the graph G. Each node gets the weight of its exhaust edge ; F: the watershed partition associated to the previous node weighted graph ; G: new contraction of edges, and associated flooding graph. Each node gets the weight of its exhaust edge. There is only one regional minimum ; H: the associated watershed partition has only one region, which marks the end of the process ; I: the waterfall hierarchy of all edges which have been used as exhaust edges : each edge is indexed by the time when it has been introduced in a flooding graph.

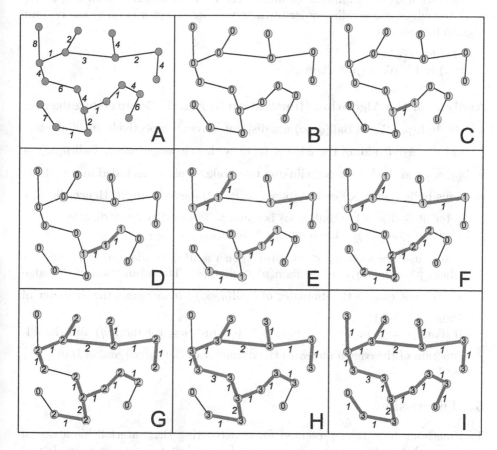

Fig. 3. A: Minimum spanning tree ; B: Initially Θ has only nodes, with a diameter 0 ; C: The first edge is introduced in Θ with a weight 1. Its extremities have a weight 1 ; D,E: Additional edges are introduced in Θ with a weight 1. Their extremities have a weight 1 ; F: First edge linking two subtrees of Θ with diameter 1. This edge gets a waterfall weight 2, and the diameter of its subtree is 1 ; G: Introduction of new edges with a waterfall level of 1. The diameters of the subtrees is 2 ; H: Introduction of an edge linking two subtrees with a diameter 2. The resulting edge gets a weight equal to 3 and its subtree a diameter 3 ; I: After introduction of the last edge, all edges got their waterfall index.

We define the open ball $\overset{\circ}{\mathrm{Ball}}(p, \lambda) = \{q \mid \chi(p, q) < \lambda\}$ and $\mathrm{diam}(p)$ its diameter, i.e. the weight of the highest edge linking two nodes of $\overset{\circ}{\mathrm{Ball}}(p, \infty)$. Initially no such edge is present and we set $\mathrm{diam}(p) = 0$.

We start with the minimum spanning tree T of the neighborhood graph. We treat the edges in ascending order of their weights η. Let $u = (p, q)$ the current edge to be processed:

- $\Theta = \Theta \cup \{u\}$
- $\theta(u) = 1 + (\mathrm{diam}(p) \wedge \mathrm{diam}(q))$

Analysis of the Algorithm (Illustration in Fig.3). Before adding the edge u to Θ, $\overset{\circ}{\mathrm{Ball}}(p, \infty)$ and $\overset{\circ}{\mathrm{Ball}}(q, \infty)$ are disjoint, having respectively the diameters d_p and d_q. After adding the edge u to Θ with the weight $\theta(u)$, $\overset{\circ}{\mathrm{Ball}}(p, \infty) = \overset{\circ}{\mathrm{Ball}}(q, \infty)$ and the largest edge linking two nodes is either included in one of the previous balls $\overset{\circ}{\mathrm{Ball}}(p, \infty)$ or $\overset{\circ}{\mathrm{Ball}}(p, \infty)$ or it is the edge u itself. Hence the new diameter of $\overset{\circ}{\mathrm{Ball}}(p, \infty) = \overset{\circ}{\mathrm{Ball}}(q, \infty)$ is equal to $\theta(u) \vee \mathrm{diam}(p) \vee \mathrm{diam}(q)$.

Let us analyze $\theta(u) = 1 + (\mathrm{diam}(p) \wedge \mathrm{diam}(q))$:

- if for instance $\mathrm{diam}(p) < \mathrm{diam}(q)$ then $\mathrm{diam}(p) \leq \mathrm{diam}(q) - 1$ and $1 + (\mathrm{diam}(p) \wedge \mathrm{diam}(q)) = 1 + \mathrm{diam}(p) \leq \mathrm{diam}(q)$. The adjunction of the edge u does not change the diameter of $\overset{\circ}{\mathrm{Ball}}(q, \infty)$; it increases the diameter of $\overset{\circ}{\mathrm{Ball}}(p, \infty)$ by 1.
- if $\mathrm{diam}(p) = \mathrm{diam}(q)$, then $\theta(u) = 1 + \mathrm{diam}(p) = 1 + \mathrm{diam}(q)$ and the adjunction of the edge u increases the diameter of $\overset{\circ}{\mathrm{Ball}}(p, \infty)$ and of $\overset{\circ}{\mathrm{Ball}}(q, \infty)$ by 1.

5   Discussion

Two methods have been presented for constructing the waterfall hierarchy of a topographic surface. The second is the fastest but relies upon a number of arbitrary choices: choice of a minimum tree among many of the regiona adjacency graph. The edges of the tree are processed in the order of increasing values ; among edges with the same weight, there are again many arbitrary choices. The structure of the waterfall hierarchy will remain the same, but the boundaries of the partitions forming each level will differ. The first method is the most precise as it relies on a watershed algorithm which reduces to the outmost the arbitrary choices between alternative solutions [16,17,18] For instance, if the regional minima have distinct weights, the watershed is unique .

6   Conclusion

The waterfall hierarchy of a grey tone image is illustrated by the waterfall saliency of its contours in figure <ref>theowfal</ref>.

The waterfall hierarchy of a grey tone image is illustrated by the waterfall saliency of its contours in the following figure.

Initial image and the saliency of its waterfall contours

The waterfall hierarchy permits a structural analysis of a topographic surface as it explores the nested structure of the catchment basins, without any consideration of depth or size. It is in fact a purely ordinal process, as shown by the last algorithm presented above, which increases by one the waterfall level each time one encounters a deeper nested level.

The waterfall hierarchy has been invented, with floodings on a topographic surface in mind. It has then be, conceptually and algorithmically, extended to node or edge weighted graphs. It may be extremely useful for structuring large graphs representing complex interconnections in many domains, beyond the world of image processing.

Sometimes, there are only very few nested structures, and one may wish enriching the watershed hierarchy. B.Marcotegui and S.Beucher have shown how to prevent some regions from mergings. This domain of research merits further investigations.

References

1. Angulo, J., Serra, J.: Color segmentation by ordered mergings. In: Image Processing, ICIP 2003, vol. 2, pp. 121–125. IEEE (2003)
2. Angulo, J., Velasco-Forero, S., Chanussot, J.: Multiscale stochastic watershed for unsupervised hyperspectral image segmentation. In: Geoscience and Remote Sensing Symposium, IGARSS 2009, vol. 3, pp. 93–96. IEEE (2009)
3. Beucher, S.: Segmentation d'Images et Morphologie Mathématique. PhD thesis, E.N.S. des Mines de Paris (1990)
4. Beucher, S.: Watershed, hierarchical segmentation and waterfall algorithm. In: Mathematical Morphology and its Applications to Image Processing, pp. 69–76. Springer (1994)
5. Marcotegui, B., Beucher, S.: Fast implementation of waterfalls based on graphs. In: ISMM 2005: Mathematical Morphology and its Applications to Signal Processing, pp. 177–186 (2005)

6. Beucher, S.: Geodesic reconstruction, saddle zones and hierarchical segmentation. Image Anal. Stereol. 20(2), 137–141 (2001)

7. Beucher, S., Marcotegui, B., et al.: P algorithm, a dramatic enhancement of the waterfall transformation (2009)

8. Cousty, J., Bertrand, G., Najman, L., Couprie, M.: Watershed cuts: Minimum spanning forests and the drop of water principle. IEEE Transactions on Pattern Analysis and Machine Intelligence 31, 1362–1374 (2009)

9. Delest, S., Bone, R., Cardot, H.: Fast segmentation of triangular meshes using waterfall. In: VIIP, vol. 6, pp. 308–312 (2006)

10. Golodetz, S.M., Nicholls, C., Voiculescu, I.D., Cameron, S.A.: Two tree-based methods for the waterfall. Pattern Recognition 47(10), 3276–3292 (2014)

11. Hanbury, A., Marcotegui, B.: Waterfall segmentation of complex scenes. In: Narayanan, P.J., Nayar, S.K., Shum, H.-Y. (eds.) ACCV 2006. LNCS, vol. 3851, pp. 888–897. Springer, Heidelberg (2006)

12. Meyer, F.: Un algorithme optimal de ligne de partage des eaux. Actes duu, 847–859 (1991)

13. Meyer, F.: Minimum spanning forests for morphological segmentation. In: Procs. of the Second International Conference on Mathematical Morphology and its Applications to Image Processing, pp. 77–84 (1994)

14. Meyer, F.: Flooding and segmentation. In: Goutsias, J., Vincent, L., Bloomberg, D.S. (eds.) Mathematical Morphology and its Applications to Image and Signal Processing. Computational Imaging and Vision, vol. 18, pp. 189–198 (2002)

15. Meyer, F., Stawiaski, J.: Morphology on graphs and minimum spanning trees. In: Wilkinson, M.H.F., Roerdink, J.B.T.M. (eds.) ISMM 2009. LNCS, vol. 5720, pp. 161–170. Springer, Heidelberg (2009)

16. Meyer, F.: The steepest watershed: from graphs to images. arXiv preprint arXiv:1204.2134 (2012)

17. Meyer, F.: Watersheds, waterfalls, on edge or node weighted graphs. arXiv preprint arXiv:1204.2837 (2012)

18. Meyer, F.: Watersheds on weighted graphs. Pattern Recognition Letters (2014)

19. Ogor, B., Haese-coat, V., Ronsin, J.: Sar image segmentation by mathematical morphology and texture analysis. In: Geoscience and Remote Sensing Symposium, IGARSS 1996. 'Remote Sensing for a Sustainable Future.', International, pp. 1:717–719 (1996)

20. Shafarenko, L., Petrou, M., Kittler, J.: Automatic watershed segmentation of randomly textured color images. IEEE Transactions on Image Processing 6(11), 1530–1544 (1997)

21. Serra, J. (ed.): Image Analysis and Mathematical Morphology. II: Theoretical Advances. Academic Press, London (1988)

22. Stottinger, J., Banova, J., Ponitz, T., Sebe, N., Hanbury, A.: Translating journalists' requirements into features for image search. In: 15th International Conference on Virtual Systems and Multimedia, VSMM 2009, pp. 149–153. IEEE (2009)

Color, Multivalued and Orientation Fields

N-ary Mathematical Morphology

Emmanuel Chevallier[1]([⊠]), Augustin Chevallier[2], and Jesús Angulo[1]

[1] CMM-Centre de Morphologie Mathématique,
MINES ParisTech, PSL-Research University, Paris, France
[2] Ecole Normale Supérieur de Cachan, Cachan, France
`emmanuel.chevallier@mines-paristech.fr`

Abstract. Mathematical morphology on binary images can be fully described by set theory. However, it is not sufficient to formulate mathematical morphology for grey scale images. This type of images requires the introduction of the notion of partial order of grey levels, together with the definition of sup and inf operators. More generally, mathematical morphology is now described within the context of the lattice theory. For a few decades, attempts are made to use mathematical morphology on multivariate images, such as color images, mainly based on the notion of vector order. However, none of these attempts has given fully satisfying results. Instead of aiming directly at the multivariate case we propose an extension of mathematical morphology to an intermediary situation: images composed of a finite number of independent unordered labels.

Keywords: Mathematical morphology · Labeled images · Image filtering

1 Introduction

A key idea of mathematical morphology is the extension and the reduction of the surface of the different objects of an image over their neighbors. This idea leads naturally to the two basic morphological operators in binary images.

Binary Images. In such images, there are only two kinds of objects: black or white objects. Two dual and adjoint operators have been defined: the erosion and the dilation. The erosion extends the black objects over the white objects, the dilation extend the white objects over the black objects. Formally, a binary image can be seen as a support set Ω, and X a subset of Ω. Let B be a subset of Ω called the structuring element. We assume that Ω disposes of a translation operation. The erosion $\varepsilon_B(X)$ and the dilation $\delta_B(X)$ of X according to a structuring element are defined as follows [5,7]:

$$\varepsilon_B(X) = \bigcap_{y \in B} X_{-y} = \{p \in \Omega \ : \ B_p \subset X\} = \{x \ : \ \forall p \in \check{B}, x \in X_p\}, \qquad (1)$$

$$\delta_B(X) = \bigcup_{y \in B} X_y = \{x + y \ : \ x \in X, y \in B\} = \{p \in \Omega \ : \ X \cap \check{B}_p \neq \emptyset\}. \quad (2)$$

where $\check{X} = \{-x \ : \ x \in X\}$ is the transpose of X (or symmetrical set with respect to the origin O) and $X_p = \{x + p \ : \ x \in X\}$ the translate of X by p. For the

J.A. Benediktsson et al. (Eds.): ISMM 2015, LNCS 9082, pp. 339–350, 2015.
DOI: 10.1007/978-3-319-18720-4_29

sake of simplicity, we limit the rest of our notation to symmetric structuring elements: $B = \check{B}$.

Grey-scale Images. With the apparition of grey-scale images, mathematical morphology was reformulated in terms of inf and sup convolution where the kernel is the structuring element B [7]. An image is now considered as a function I defined as

$$I : \begin{cases} \Omega \to V \\ p \mapsto I(p) \end{cases}$$

where V is the set of grey-levels, which can be generally assumed as a subset of the real line $V \subset \mathbb{R}$. Grey-scale flat erosion and dilation of I by structuring element B are now defined as follows:

$$\varepsilon_B(I)(p) = \inf_{q \in B_p} \{f(q)\}, \quad \delta_B(I)(p) = \sup_{q \in B_p} \{f(q)\}.$$

In this framework, each grey-level is not fully considered as an independent label (i.e., a different category) but simply as an intermediary level between black and white. This point of view is actually justified when interesting objects of the images are local extrema.

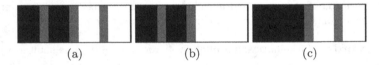

(a)   (b)   (c)

Fig. 1. Grey-level morphological processing: (a) original image I, (b) closing $\varphi_B(I)$, (c) opening $\gamma_B(I)$

Let us see what happens in the situation depicted in Fig. 1. It corresponds to process a rather simple grey-level by a closing and an opening. We recall that the closing of I by B is the composition of a dilation followed by a erosion; i.e., $\varphi_B(I) = \varepsilon_B(\delta_B(I))$. The opening is just the dual operator; i.e., $\gamma_B(I) = \delta_B(\varepsilon_B(I))$. Closing (resp. opening) is appropriate to remove dark (resp. bright) structures smaller than the structuring element B. This behavior is based on the fact that the dilation "reduces" dark structures by B while the erosion "restores" the dark structures which are still present.

In the current example, it is not possible to remove the central grey spot using erosion and dilation with B larger than the spot size. This grey spot is not considered as an interesting object in itself but simply as an intermediary value between the black object and the white object. If this assumption is often coherent, this is not always the case.

Let consider the grey-scale image in Fig. 2(a). In this image, each grey level has the same semantic level: each represents a different component, sometimes called a phase. However, in the morphological grey-scale framework, the grey is processed as an intermediary level. It is possible to replace each grey level by a color, see Fig. 2(b). We would like then to process both images using the same approach.

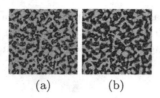

(a)   (b)   (c)

Fig. 2. (a) A three independent grey-scale image and (b) the same image where the grey values has been replaced by colors. (c) Example of multivariate (color) image.

Multivariate Images. For multivariate images, no canonical framework has yet appeared. Most processing consist in endowing the structure with a partial order relationship. The structure has to be a complete lattice in order to define erosion and dilation in terms of inf and sup. The notion of order induces the notion of intermediary level as in the precedent framework. However, the notion of intermediary levels often leads to non-intuitive situations, see example in Fig. 2(c). As the red usually has a real signification in terms of a particular class of objects, it is very natural to try to remove the red spot, which is not possible using generic classical morphology. The more the image has a complex semantic structure, such as a color image, the more it is difficult to find a lattice structure which makes every interesting object an extremum.

Aim and Paper Organization. Historically, mathematical morphology has been generalized from binary to grey scale, and then to multi-variate images. However, the gap between grey-scale and color is much more significant than the gap between binary and grey-scale. As we discussed previously, the grey-scale structure only enables to define intermediary colors between two references. This structure is obviously too weak to describe color information. Note that to simplify the vocabulary, we use the notion of color for any non-scalar valuation of the pixels on the image.

Before extending mathematical morphology to color images, we might want to define a coherent approach for mathematical morphology with n independent unordered colors, without considering them as intermediary levels. This is the aim of this paper. Then only, we might try in the future to define mathematical morphology for the full color space. The difference between the frameworks can be interpreted in term of a change of metric on the value space:

- grey-scale framework: $\forall (i,j), \; d(color_i, color_j) = |i - j|$;
- $n-$ary framework: $\forall (i,j), \; d(color_i, color_j) = 1$.

Our paper is not the first to consider the problems of classical mathematical morphology for images composed of independents labels. Authors of [1] and [2] have very similar motivations but the development we propose is different. In contrast to operators proposed in [1], [2] or labelled openings from [3], we are interested in filling gaps left by anti-extensive operators. We note that the theory of morphological operators for partitions [9] and hierarchies of partitions [6] is not compatible with our framework.

The rest of the paper is organized as follows. A proposition of n-ary morphological operators and a study of their theoretical properties in Sections 2 and 3. Some applications to image filtering are discussed in Section 4. Section 5 of conclusions closes the paper.

2 n-ary Morphological Operators

Let us come back to the key idea of mathematical morphology is to reduce and extend objects over their neighbors. In the case of binary images, two operations where introduced: the erosion extends the black over the white and the dilation extends the white over the black. In a general way, we would like to allow to reduce and extend the surface of each label of object. This makes four theoretic operations in the binary case, reduced to two in practice due to the coincidence of certain operations: reducing the black is the same as extending the white and conversely. This duality is one of the basic principle of binary morphology.

2.1   Dilation and Erosion of Label i

Let I be an n-ary image defined as

$$I : \begin{cases} \Omega \to \{1, 2, \cdots, n\} \\ p \mapsto I(p) \end{cases}$$

In the $n-$ary case, it seems natural to try to introduce the corresponding pair $(\varepsilon_i, \delta_i)$ of operators for each label i. Erosion ε_i is the operator that reduces the surface of the objets of label i, and dilation δ_i the operator that extends the label i. Above $n > 2$, we unfortunately lose the duality between operations, such that the number of elementary operators is then equal to $2n$. Let us formulate more precisely these operators.

The *dilation of label i on image I by structuring element B* presents no difficulty:

$$\delta_i(I; B)(x) = \begin{cases} I(x) & \text{if } \forall p \in B_x, I(p) \neq i \\ i & \text{if } \exists p \in B_x, I(p) = i \end{cases} \tag{3}$$

$\delta_i(I; B)$ extends objects of label i over their neighbors. The case of the erosion presents more theoretical difficulties. Indeed, if we want to reduce the objects of label i, we need to decide how to fill the gaps after the reduction.

Let us first define the erosion for pixels where there are no ambiguities. Thus the *erosion of label i on image I by structuring element B* is given by

$$\varepsilon_i(I; B)(x) = \begin{cases} I(x) & \text{if } I(x) \neq i \\ i & \text{if } \forall p \in B_x, I(p) = i \\ \theta(x, I) & \text{otherwise} \end{cases} \tag{4}$$

We will address later definition of $\theta(x, I)$. Sections 2.2 and 2.3 are independent of θ. Although the image is a partition of Ω the proposed framework differs from [2] and [9].

2.2 Opening and Closing of Label i

Once the dilation and erosion have been defined, we can introduce by composition of these two operators the *opening and the closing on I by B of label i* respectively as

$$\gamma_i(I; B) = \delta_i \circ \varepsilon_i = \delta_i \left(\varepsilon_i(I; B); B \right), \tag{5}$$
$$\varphi_i(I; B) = \varepsilon_i \circ \delta_i = \varepsilon_i \left(\delta_i(I; B); B \right). \tag{6}$$

Let us set a few notations used in the following. If ϕ is an operator, let ϕ^k be $\phi \circ .... \circ \phi$ the iteration of ϕ, k times. Let $\phi_{|A}$ be the restriction of ϕ to the subset A. Let us set $E_i^I = I^{-1}(i)$. To simplify, $\mathbf{1}_{E_i^I}$ will be noted $\mathbf{1}_i^I$.

We have the following property of stability.

Proposition 1. *Opening and closing of label i are idempotent operators, i.e.,*

$$\gamma_i(I; B) = \gamma_i^2(I; B),$$
$$\varphi_i(I; B) = \varphi_i^2(I; B).$$

Proof. Since the binary opening is idempotent, one has $E_i^{\gamma_i(I)} = E_i^{\gamma_i^2(I)}$. Furthermore we have that $E_j^{\gamma_i(I)} \subset E_j^{\gamma_i^2(I)}$, for all $j \neq i$. Since sets $(E_i)_i$ from a partition of the support space, necessarily $E_j^{\gamma_i(I)} = E_j^{\gamma_i^2(I)}$, $\forall j$. Indeed, if all the elements of a partition are extensive, then they all remain stable. Then $\gamma_i = \gamma_i^2$.

Properties cannot directly be transported by duality, as in binary morphology, however the property remains true for the closing. We first show the binary property $\varepsilon \delta \varepsilon = \varepsilon$. The binary erosion and opening can be written as

$$\varepsilon_B(X) = \cup_{B_x \subset X} \{x\}, \quad \text{and} \quad \gamma_B(X) = \cup_{B_x \subset X} B_x.$$

Then $\varepsilon(\gamma(X)) = \cup_{B_x \subset \gamma(X)} \{x\}$. Since $\{B_x \subset \gamma(X)\} = \{B_x \subset \cup_{B_x \subset X} B_x\} = \{B_x \subset X\}$, then $\varepsilon(\gamma(X)) = \varepsilon(X)$. Thus, $\varepsilon \delta \varepsilon = \varepsilon$ and by duality, $\delta \varepsilon \delta = \delta$. Then $E_i^{\delta_i} = E_i^{\delta_i \varepsilon_i \delta_i}$. It can be shown that $E_j^{\delta_i} \subset E_j^{\delta_i \varepsilon_i \delta_i}$ for all $j \neq i$. Using the same reasoning as in the proof for the opening, we have that for all j, $E_j^{\delta_i} = E_j^{\delta_i \varepsilon_i \delta_i}$. In other words, $\delta_i \varepsilon_i \delta_i = \delta_i$. Thus $\varepsilon_i \delta_i \varepsilon_i \delta_i = \varepsilon_i \delta_i$, or equivalently $\varphi_i = \varphi_i^2$.

2.3 Composed n-ary Filters

We can now try to define label filters from the openings and the closings of label i. In binary morphology, the simplest filters are of the following form: $\gamma \circ \phi$ and $\phi \circ \gamma$. In the n-ary framework, with $n = 2$, they can be rewritten as

$$\gamma_1 \circ \gamma_2 = \phi_2 \circ \phi_1, \quad \text{and} \quad \gamma_2 \circ \gamma_1 = \phi_1 \circ \phi_2.$$

The opening removes peaks smaller than the structuring element and the closing removes holes, which are dual notions in binary morphology. However, peaks and holes are no longer a dual notion in n-ary morphology with $n > 2$. Fig. 3

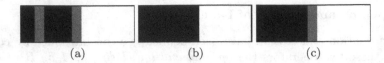

Fig. 3. Opening and closing on a 3-ary image: (a) original image I, (b) opening of red color $\gamma_{red}(I; B)$, (c) closing of black color $\varphi_{black}(I; B)$

illustrates the difference between openings and closings on a 3-ary image: three colors, black, white and red. The structuring element is a square whose size is half of the width of the red line. Removing the red line using φ_{black} requires a structuring element twice bigger than with γ_{red}.

As a good candidate to filter out small object of a labeled image I, independently of the label of the objects, we introduce the operator ψ, named *composed n-ary filter by structuring element B*, defined as

$$\psi(I; B) = \gamma_n(I; B) \circ \gamma_{n-1}(I; B) \circ \cdots \circ \gamma_1(I; B). \tag{7}$$

Unfortunately on the contrary to $\gamma \circ \phi$ in binary morphology, ψ is generally not idempotent. Worst, the sequence ψ^k do not necessarily converge. However we still have a stability property for relevant objects. Let us be more precise.

Proposition 2. *Let Ω be a finite set. Given a structuring element B, the interior with respect to B of the composed n-ary filter $\psi(I; B)$ converges for any image I, i.e.,*

$$\forall i, \varepsilon(E_i^{\psi^k}) \ converges \ .$$

Proof. Since $\varepsilon = \varepsilon \circ \delta\varepsilon$, $\forall i$, $\varepsilon(E_i^{\psi^k}) = \varepsilon(E_i^{\gamma_i \circ \psi^k})$. Furthermore, since $\varepsilon(E_i^{\psi^k}) \subset \varepsilon(E_i^{\gamma_j \circ \psi^k})$, we have that $\forall i, \varepsilon(E_i^{\psi^k}) \subset \varepsilon(E_i^{\psi^{k+1}})$. Since Ω is a finite set, $\varepsilon(E_i^{\psi^k})$ converges.

This property ensures that the variations between ψ^k and ψ^{k+1} do not affect the interior of objects and is only limited to boundaries. Nevertheless, as we shown in section 4, ψ^k is almost always stable after a few iterations.

2.4   n-ary Geodesic Reconstruction

The binary reconstruction can be transposed in the n-ary framework as follows. The proposition of reconstruction is similar to the one proposed in [2]. Given two labeled images R and M, for each label i,

- Perform a binarisation of the reference R and the marker M between i and $\complement i$, which correspond respectively to binary images X_i and Y_i.
- Compute $\gamma^{rec}(X_i; Y_i)$, that is the binary geodesic reconstruction of the marker in the reference.

Then, the *n-ary geodesic reconstruction of the reference R by the marker M* is given by

$$\gamma^{\text{rec}}(R; M)(x) = \begin{cases} i & \text{if } x \in \gamma^{\text{rec}}(X_i; Y_i) \\ M(x) & \text{if } \forall i, \ x \notin \gamma^{\text{rec}}(X_i; Y_i) \end{cases}$$

Fig.4 illustrates the difference between classical geodesic reconstruction and the proposed n-ary reconstruction. For the classical reconstruction, the 3-label image is simply viewed as a grey-scale image.

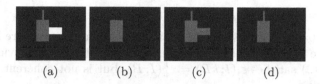

(a)   (b)   (c)   (d)

Fig. 4. Geodesic reconstruction of a 3-ary image: (a) reference image R, (b) marker image M, (c) classical grey-scale reconstruction, (d) n-ary reconstruction $\gamma^{\text{rec}}(R; M)$

The aim of this definition is to symmetrize labels. In Fig.4 (d), the grey object is considered as an object in itself. The proposed reconstruction is a connected operator in the sense of [10]

3   On the Choice of an Erosion of Label i

Before any application, we need to come back to the erosion problem. More precisely, we need to define a consistent rule to fill the space created by the erosion operation.

First of all, we note that the definition in Eq. (8) of the erosion of label i ε_i does not indicate how to behave on the following set:

$$A = \{x \mid I(x) = i \text{ and } \exists p \in B_x \text{ such that } I(p) \neq i\}.$$

For points $x \in A$ we have to decide by which label to replace label i and therefore to define ε_i on A, i.e.,

$$\varepsilon_i(I; B)(x) = \begin{cases} I(x) & \text{if } I(x) \neq i \\ i & \text{if } \forall p \in B_x, I(p) = i \\ ? & \text{if } x \in A \end{cases} \tag{8}$$

Many alternatives are possible. Two criteria have to be taken into account: (i) the direct coherence in terms of image processing, and (ii) the number of morphological properties verified by the erosion, such as $\varepsilon_i(I; kB) = \varepsilon_i^k(I; B)$ where $kB = \{kx \mid x \in B\}$ (i.e., homothetic of size k). Let us consider in particular the three following rules for $x \in A$:

1. **Fixed-label Erosion:** Erosion always fills the gaps with label 1 (or any other fixed label):
$$\varepsilon_i(I; B)(x) = 1. \tag{9}$$

2. **Majority-based Erosion:** Erosion takes the value of the major label different from i in the structuring element B:

$$\varepsilon_i(I;B)(x) = min(\arg\max_{j\neq i}(\text{Card}\{p \in B_x | I(p) = j\})). \tag{10}$$

3. **Distance-based Erosion:** Erosion replaces label i by the closest label on the support space Ω:

$$\varepsilon_i(I;B)(x) = min(\arg\min_{j\neq i} d_x^j). \tag{11}$$

where $d_x^j = \inf\{\|x - p\|_\Omega \mid p \in \Omega, I(p) = j\}$

The majority-based erosion (10) and distance-based erosion (11) are initially not defined in case of equality. Hence the apparition of the min. Obviously, fixed-label erosion (9) satisfies $\varepsilon_i(I;kB) = \varepsilon_i^k(I;B)$, but is not coherent in terms of image processing.

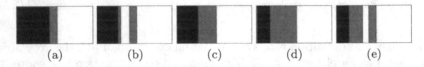

(a) (b) (c) (d) (e)

Fig. 5. Comparison of erosions of black color: (a) original image I, (b) erosion $\varepsilon_{black}(I;B)$ using majority-based formulation (10), (c) erosion $\varepsilon_{black}(I;B)$ using distance-based formulation (11). Let l be the width of the red line, the structuring element B is a square whose size is now between l and $2l$. (d) iterated black erosion $\varepsilon_{black}^2(I;B)$ using majority-based formulation, (e) erosion $\varepsilon_{black}(I;2B)$ using majority-based formulation.

Majority-based erosion and distance-based erosion both look potentially interesting in terms of image processing; however, as shown in the basic example of Fig. 5, majority-based erosion (to compare Fig. 5(b) to Fig. 5(c)) can produce unexpected results. The same example, now Fig. 5(d) and Fig. 5(e), shows that majority-based erosion do not satisfies $\varepsilon_i(I;kB) = \varepsilon_i^k(I;B)$.

Let us formalize the iterative behavior of the distance-based erosion by the following result on isotropic structuring elements.

Proposition 3. *Let (Ω, d) be a compact geodesic space and I a n-ary image on Ω. For any $R_1, R_2 > 0$ the distance-based erosion of label i satisfies:*

$$\varepsilon_i(I;B_{R_1+R_2}) = \varepsilon_i(I;B_{R_2}) \circ \varepsilon_i(I;B_{R_1}).$$

where B_R is the open ball of radius R.

Proof. For the sake of notation, let $X = E_i^I$ and $X_j = E_j^I$, for $j \neq i$. Let $X' = \{x \in X | d(x, \complement X) \geq R_1\}$, X' is the binary eroded of X. Let $proj_X(a) = \{b | b \in \overline{\complement X}, d(a,b) = d(a, \complement X)\}$ with $\overline{A}$ the closure of A. Let $\mathcal{P} = \{X_j\}$ and for $a \in \Omega, I_a^{\mathcal{P}} = \{j | \exists b \in proj_X(a) \cap \overline{X_j}\}$. Let $X_j' = \{a \notin X' | proj_X(a) \cap \overline{X_j} \neq \emptyset\}$ and $\mathcal{P}' = \{X_j'\}$. Using the property of the min function, $min(A \cup B) = min(\{min(A), min(B)\})$, it can be shown that $I_a^{\mathcal{P}} = I_a^{\mathcal{P}'}$ for all $a \in X'$ implies proposition 3.

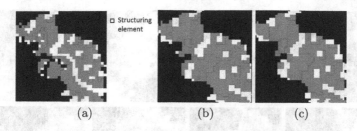

Fig. 6. Morphological filtering of a 4-ary image: (a) original color image I and structuring element B, (b) color operator $\gamma_B \circ \varphi_B$ using a color order (c) color operator $\varphi_B \circ \gamma_B$ using same order.

- Let $j \in I_a^{\mathcal{P}}$. Thus, $\exists b \in proj_X(a) \cap \overline{X_j}$. Let γ be a geodesic joining a and b with $\gamma(0) = b$ and $\gamma(d(a,b)) = a$. Let $c = \gamma(R_1)$. For all $x \in \gamma([0, R_1[)$, $x \notin X'$ and $b \in proj_X(x)$. Hence $x \in X_j'$ and $c \in \overline{X_j'}$. The triangle inequality gives $c \in proj_{X'}(a)$. Indeed, assuming that $c \notin proj_{X'}(a)$ easily leads to $b \notin b \in proj_X(x)$. Thus, $j \in I_a^{\mathcal{P}'}$.
- Let $j \in I_a^{\mathcal{P}'}$. Thus, $\exists c \in proj_{X'}(a) \cap \overline{X_j'}$. By definition of the closure, $\exists c_n \to c, c_n \in X_j'. c_n \in X_j' \Rightarrow (\exists d_n \in proj_X(c_n), d_n \in \overline{X_j})$. The compact assumption enables us to consider that $d_n \to d$ (it is at least valid for a sub-sequence). $d \in \overline{X_j}$. By continuity, $d(d,c) = lim(d_n, c_n) = lim(\complement X, c_n) = d(\complement X, c)$. Thus, $d \in proj_X(c)$. Proposition 4 tells us that $d \in proj_X(a)$. Hence $j \in I_a^{\mathcal{P}}$.

Thus, $I_a^{\mathcal{P}} = I_a^{\mathcal{P}'}$.

Proposition 4. *Using the notation introduced in the demonstration of proposition 3:*

$$\forall a \in X', \forall c \in proj_{X'}(a), \forall d \in proj_X(c), d \in proj_X(a).$$

Proof. Let $a \in X', c \in proj_{X'}(a), d \in proj_X(c)$.

- Since X' is closed, $\complement X'$ is open and the existence of geodesics implies that $c \notin \complement X'$. Hence $c \in X'$ and $d(c, \complement X) \geq R_1$. Since c is a projection on $\complement X'$, we have $c \in \overline{\complement X'}$, $d(c, \complement X) \leq R_1$. Hence $d(c, \complement X) = R_1$ and $d(c, d) = R_1$.
- Let $b \in proj_X(a)$ and γ a geodesic such that $\gamma(0) = a$ and $\gamma(d(a,b)) = b$. Let $c' = \gamma(sup\{t|\gamma(t) \in X'\})$. Since X' is closed, $c' \in X'$. We have $d(c', \complement X) \geq R_1$. Since $c' \in \overline{\complement X'}$, $d(a, c') \geq d(a, \complement X') = d(a, c)$. Thus, $d(a, b) = d(a, c') + d(c', b) \geq d(a, c) + d(c', \complement X) = d(a, c) + d(c', \complement X) \geq d(a, c) + R_1 = d(a, c) + d(c, d) \geq d(a, d)$. Hence $d \in proj_X(a)$.

Note that in a compact subset of a vector space, since proposition 3 is valid for any norm, the property holds for any convex structuring element. Note also that 3 is based on a metric definition of the erosion. For vector spaces norms, this erosion is identical to the translation based erosion. According to this discussion, in all what follows we adopt distance-based erosion.

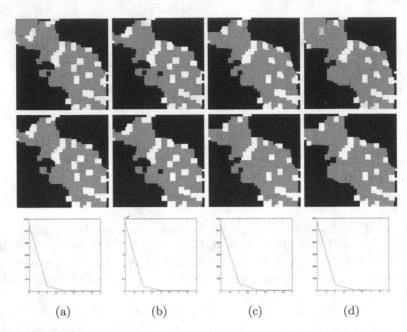

(a)   (b)   (c)   (d)

Fig. 7. n-ary processing of Fig. 6 (a). first row: composed n-ary filter $\psi(I; B)$, second row: iterated composed n-ary filter $\psi^k(I; B)$ until convergence, third row: convergence speed w.r.t. to k. Column (a) $\psi(I; B) = \gamma_1 \circ \gamma_2 \circ \gamma_3 \circ \gamma_4$, (b) $\psi(I; B) = \gamma_2 \circ \gamma_1 \circ \gamma_3 \circ \gamma_4$, (c) $\psi(I; B) = \gamma_3 \circ \gamma_4 \circ \gamma_1 \circ \gamma_2$, (d) $\psi(I; B) = \gamma_4 \circ \gamma_3 \circ \gamma_2 \circ \gamma_1$.

4   Applications to Image Filtering

In the first case study, depicted in Fig. 6 and Fig. 7, we consider the behavior and interest of the composed n-ary filter $\psi(I; B)$. The aim is to filter out objects smaller than the structuring element B of the 4-ary labeled image Fig. 6(a). Results in Fig. 6(b) and (c) are respectively the classical color operator $\gamma_B \circ \varphi_B$ and $\varphi_B \circ \gamma_B$ based on an arbitrary order. Alternative composed n-ary filter $\psi(I; B)$, which correspond to different permutations of the composition of openings of label i. On the one hand, note that in all the cases, the iterated filter converges rather fast to a stable (idempotent) result ant that the difference between the first iteration and the final result are rather similar. On the other hand, the different permutations produce different results, however in all the cases, the small objects seems better removed than in the case of the classical color order operators.

The second example, given in Fig. 8 attempts to regularize the 3-ary image, by removing small objects without deforming the contours of the remaining objects. More precisely, Fig. 8(a) represents the electron microscopy image of a ceramic eutectic, with three different phases after segmentation. The filtering process is composed of two steps: morphological size filter followed by geodesic reconstruction. We compare the result of filtering the color image according to two pipelines: (i) color total order framework, Fig. 8(b), where the filter is an opening by reconstruction composed with a closing by reconstruction; (ii) n-ary

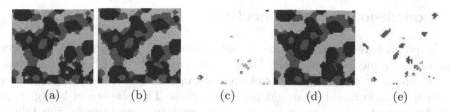

Fig. 8. Image size-regularization: (a) original image I, (b) classical order-based filtering, (c) residue between (a) and (b), (d) n-ary based filtering, (e) residue between (a) and (d). See the text for details.

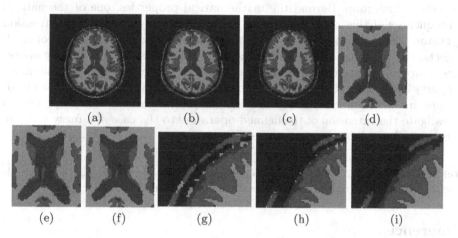

Fig. 9. Image size-regularization: (a) original image I, (b) classical order-based filtering, (c) n-ary based filtering, (d) zoom-in (a), (e) zoom-in (b), (f) zoom-in (c), (g) zoom-in (a), (h) zoom-in (b),(i) zoom-in (c). See the text for details.

framework, Fig. 8(d), where the marker is a n-ary filter $\psi(I; B)$ followed by a n-ary geodesic reconstruction. In the case of the color ordering, black and blue are extreme color whereas red is the intermediary color. As we can observe, using the order-based approach all red objects that lay between black and blue objects are not extracted. Both n-ary and color total order frameworks give the same results for the blue grains. This corresponds to the fact that, in the color order, blue is an extreme color whereas red is an intermediary. Therefore the n-ary framework provides a more symmetric processing of all the colors.

The last example is a classification image from the brain. Fig. 9 (a) is a result of a classification where the red represents the grey matter, the green represents the white matter and the blue represents the cerebrospinal fluid. The processing is same as for the second example. The miss-classified white matter around the brain and some miss-classified grey matter spots around the cerebrospinal fluid are successfully removed by the n-ary framework, whereas they remain after the classical processing.

5   Conclusions and Perspectives

We proposed here an approach to extend mathematical morphology to images composed of n independent levels. The approach presents two key particularities: first, the increase of the number of elementary operators, second, the absence of a notion of background or an indeterminate class. The absence of background or indeterminate class transforms the problem into a problem of gap filling. We prove that some of the elementary properties of standard morphological operators are preserved, such as the idempotence of openings and closings per label. Despite its quasi experimental validity, the main lost property is the granulometric semigroup. Beyond the mathematical properties, one of the natural consequences of this n-ary framework is the definition of a new reconstruction operator. The main application of the proposed operators is the filtering of small objects, the presented examples demonstrate the relevance of the n-ary operators. Our immediate research will focus on the recovery of the "granulometry property". In a second step, we plan to extend our applications to other classification images such as classified satellite images. Then in a third step, we will investigate the extension of the defined operators to the case of a fuzzy mixture between independent unordered labels.

Acknowledgement. We would like to thank Jasper van de Gronde, for the multiple discussions and constructive suggestions during our academic visits.

References

1. Busch, C., Eberle, M.: Morphological Operations for Color-Coded Images. Proc. of the EUROGRAPHICS 1995 14(3), C193–C204 (1995)
2. Ronse, C., Agnus, V.: Morphology on label images: Flat-type operators and connections. Journal of Mathematical Imaging and Vision 22(2-3), 283–307 (2005)
3. Hanbury, A., Serra, J.: Morphological operators on the unit circle. IEEE Trans. Image Processing 10(12), 1842–1850 (2001)
4. Heijmans, H.J.A.M.: Morphological image operators. Academic Press, Boston (1994)
5. Matheron, G.: Random Sets and Integral Geometry. Wiley, New York (1975)
6. Meyer, F.: Adjunctions on the lattice of hierarchies. HAL, hal-00566714, 24p (2011)
7. Serra, J.: Image Analysis and Mathematical Morphology. Academic Press, London (1982)
8. Soille, P.: Morphological Image Analysis. Springer, Berlin (1999)
9. Ronse, C.: Ordering Partial Partition for Image Segmentation and Filtering: Merging, Creating and Inflating Blocks. Journal of Mathematical Imaging and Vision 49(1), 202–233 (2014)
10. Salembier, P., Serra, J.: Flat Zones Filtering, Connected Operators, and Filters by reconstruction. IEEE Trans. on Images Processing 4(8), 1153–1160 (2014)

Sponges for Generalized Morphology

Jasper J. van de Gronde[✉] and Jos B.T.M. Roerdink

Johann Bernoulli Institute for Mathematics and Computer Science,
University of Groningen, P.O. Box 407, 9700 Groningen AK, The Netherlands
j.j.van.de.gronde@rug.nl

Abstract. Mathematical morphology has traditionally been grounded
in lattice theory. For non-scalar data lattices often prove too restrictive,
however. In this paper we present a more general alternative, sponges,
that still allows useful definitions of various properties and concepts
from morphological theory. It turns out that some of the existing work
on "pseudo-morphology" for non-scalar data can in fact be considered
"proper" mathematical morphology in this new framework, while other
work cannot, and that this correlates with how useful/intuitive some of
the resulting operators are.

Keywords: Mathematical morphology · Pseudo-morphology · Weakly
associative lattices · Sponges

1 Introduction

Lattice theory has brought mathematical morphology very far when it comes to
processing binary and greyscale images. However, for vector- and tensor-valued
images lattice theory appears to be overly restrictive [5, 12, 20, 27]: vectors and
tensors simply do not seem to naturally fit a lattice structure. For example, we
cannot have a lattice that is compatible with a vector space structure while also
behaving in a rotationally invariant manner [20]. And having a lattice that can
deal with periodic structures is equally impossible (due to it being based on an
order relation).

Some attempts have been made to still apply mathematical morphology to
vector- and tensor-valued images by letting go of the lattice structure while still
having something resembling the infimum and supremum operations [1, 3, 6, 7,
10, 11, 17, 30]. However, over the years, mathematical morphology has developed
a host of concepts that (in their usual formulation) rely on a lattice structure.
Take away the lattice structure and all these concepts make very little sense any
more. For example, a dilation is defined as an operator that commutes with the
supremum of the lattice. And some pseudo-morphological operators can indeed
lead to unintuitive (and undesired) behaviour [19].

In this paper we introduce a novel theoretical framework which generalizes
lattices, while retaining some crucial properties of infima and suprema. This

This research is funded by the Netherlands Organisation for Scientific Research
(NWO), project no. 612.001.001.

© Springer International Publishing Switzerland 2015
J.A. Benediktsson et al. (Eds.): ISMM 2015, LNCS 9082, pp. 351–362, 2015.
DOI: 10.1007/978-3-319-18720-4_30

more flexible structure, which we will call a "sponge", is inspired by the vector
levelings of Zanoguera and Meyer [30] and the tensor dilations/erosions by Burgeth et al. [6, 9, 11], and is effectively a variant of what is known as a "weakly
associative lattice" or "trellis" [14, 15, 28]. Its relations to lattices and the earlier generalizations are discussed and examples are shown of existing and new
methods that are not interpretable in a traditional lattice theoretic framework,
but that do lead to sponges. The method proposed by Burgeth et al. [6], on the
other hand, is shown not to lead to a sponge (consistent with some of the issues
with this method). We hope this new framework will be useful in guiding future
developments in non-scalar morphology, and that it will provide more insight
into the properties of operators based on such schemes.

2 Related Work

If we step away from lattices, what options do we have? Some attempts at developing specific methods that still behave much like a traditional lattice include
non-separable vector levelings by Zanoguera and Meyer [30], morphology for
hyperbolic-valued images by Angulo and Velasco-Forero [3], and the Loewner-
order based operations by Burgeth et al. [6]. All these methods support the
concepts of upper and lower bounds, as well as some sort of join and meet (infimum and supremum), but do not rely on a lattice structure. The framework
presented in this work will be shown to encompass some of these methods, but
not all (in the latter case this can be linked to some issues with the method).

Below we will present a generalization of a partial order that will be called an
oriented set, as a starting point for our generalization of a lattice. An oriented set
is so named because it can also be considered an oriented graph[1] and vice versa.
Also, if all elements in some subset of an oriented set are comparable, this subset
can be called a tournament (analogous to a chain). This structure was already
used, under different names, as the basis for a subtly different generalization of
a lattice: a weakly associative lattice (WAL), trellis, or T-lattice [14, 15, 28].

Based on oriented sets we will introduce a generalization of a lattice called a
sponge, which supports (partial) join and meet operations on sets of elements.
A sponge is a lattice if the orientation is transitive *and* the join and meet are
defined for all pairs (as a consequence of being a lattice they must then also be
defined for all finite sets). If the latter condition does not hold the result would
still be a partially ordered set with a join and meet defined for all finite subsets
that have an upper/lower bound (which is a bit more specific than the concept of
a partial lattice used by [18, Def. 12]). On the other hand, a weakly associative
lattice [14, 15, 28] is defined in *almost* the exact same way as a sponge. The
difference is that a weakly associative lattice requires the join and meet of *every*
pair of elements to be defined, while not guaranteeing that the existence of an
upper/lower bound implies the existence of the join/meet of a (finite) set of

---

[1] Fried and Grätzer [16] called an oriented graph a directed graph.

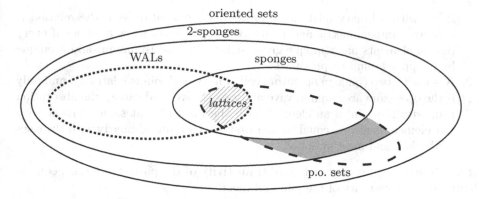

Fig. 1. An Euler diagram of the oriented and ordered structures discussed in the text. The shaded area is empty, but all other areas are not. The set of all lattices is the (hatched) intersection of WALs and p.o. sets.

elements [13]. The concept of a *partial* weakly associative lattice seems to be no more powerful than that of an oriented set [16, Lemma 1][2].

Imagine a variant of a sponge where the join and meet only need to be defined for all pairs (rather than all finite sets) with upper/lower bounds. If (against our better judgement, since the concept is more general than a sponge) we call such structures 2-sponges, then (as captured in Fig. 1)

- WALs, sponges and partially ordered sets generalize lattices,
- there are WALs that are not sponges (and vice versa),
- a WAL that is also a partial order is a lattice (and thus also a sponge),
- a partially ordered set that is also a 2-sponge is a sponge,
- there are partially ordered sets that are not 2-sponges (and vice versa),
- 2-sponges are (strictly) more general than both WALs and sponges,
- oriented sets are (strictly) more general than 2-sponges and partial orders.

Since many morphological operators and concepts are based on joins and meets of sets, sponges provide a much more natural framework for generalized morphology than WALs. Also, WALs require the join and meet to be defined for *all* pairs of elements, and all our examples violate this property. Partial WALs on the other hand provide too few guarantees to really be useful. As a consequence, we believe sponges are the right choice in the current context.

3   Sponges

In order to define sponges, it is useful to first quickly recap the two main ways of defining lattices (we will provide analogous definitions for sponges):

---

[2] The definition of a partial weakly associative lattice is a little vague, but it seems clear that at least any oriented set gives rise to a partial weakly associative lattice.

1. A set with a binary relation is considered a partial order if the relation is reflexive, antisymmetric and transitive. A partial order is a lattice if every pair of elements has a unique greatest lower bound (infimum) and a unique least upper bound (supremum).
2. A set with two (binary) operators called meet and join is a lattice if and only if the operators are commutative and associative, and satisfy the absorption property: the join of an element with the meet of that same element with any element is always equal to that first element, and similarly with the roles of the join and meet swapped.

Roughly speaking, sponges let go of transitivity of the partial order, or, equivalently, the associativity of the join and meet.

3.1 Sponges as Oriented Sets

We define a *(partially) oriented set*[3] S to be a set with a binary relation '$\preceq$' – a (partial) orientation[4] – that has the following two properties:

reflexivity: $a \preceq a$ for all $a \in S$, and
antisymmetry: $a \preceq b$ and $b \preceq a \implies a = b$.

We also write $P \preceq Q$ for subsets P and Q of S if and only if $[\forall a \in P, b \in Q : a \preceq b]$. For reasons of simplicity, we *will* say that a is less than or equal to b (or a lower bound of b) if $a \preceq b$, even though the relation need not be transitive. If the orientation relation is total (in the sense that all elements are comparable), the set is called totally oriented and the relation is a total orientation (or tournament).

We now define a sponge as an oriented set in which there exists a supremum/infimum for every non-empty *finite* subset of S which has at least one common upper/lower bound. Here a supremum a of a subset P of S is defined as an element in S such that $P \preceq \{a\}$ and $a \preceq b$ for all b such that $P \preceq \{b\}$; the infimum is defined analogously. Note that antisymmetry guarantees that if a supremum/infimum exists, it is unique.[5]

3.2 Algebraic Definition of Sponges

Analogous to the algebraic definition of a lattice, we now define a sponge as a set S with partial functions J (join) and M (meet) defined on non-empty *finite* subsets of the set S, satisfying the properties (with $a, b \in S$ and P a finite subset of S)

idempotence: $M(\{a\}) = a$,
absorption: if $M(P)$ is defined, then $[\forall a \in P : J(\{a, M(P)\}) = a]$,
part preservation: $[\forall a \in P : M(\{a, b\}) = b] \implies M(\{M(P), b\}) = b$,

---

[3] Fried [13] called an oriented set a partial tournament.
[4] Rachůnek [26] called an orientation a semi-order, while Skala [28] and Fried and Grätzer [16] called it a pseudo-order.
[5] In fact, Fried [13] already showed that in any orientation the set of "least upper bounds" of a set is either empty or a set of just one element.

and the same properties with J substituted for M and vice versa. Since J and M are operators on (sub)sets, they preserve the commutativity of lattice-based joins and meets, but not necessarily their associativity. In a lattice $\mathcal{L}$, idempotence follows from absorption: $a \wedge a = a \wedge (a \vee (a \wedge b)) = a$ for all $a, b \in \mathcal{L}$. In a sponge, on the other hand, the join and meet need *not* be defined for all pairs of elements in the sponge, and this argument breaks down (but we still need it, so it is included as a separate property). Part preservation[6] is essentially "half" of associativity, in the sense that if the implication was replaced by a logical equivalence, J and M would be associative. In some cases we wish to write down $M(P)$ or $J(P)$ without worrying about whether or not it is actually defined for the set P. We then consider M or J to return a special value when the result is undefined. This value propagates much like a NaN: if it is part of the input of M or J, then the output takes on this "undefined" value as well.

It is important to note if a (finite) subset P of a sponge has a common lower (upper) bound b, the premise of part preservation is true, and P must then have a meet (join), or the left-hand side of the conclusion would be undefined.

From now on, we will omit braces around explicitly enumerated sets whenever this need not lead to any confusion (as this greatly enhances readability). So we will write $M(a, b)$ and $P \preceq a$ rather than $M(\{a, b\})$ and $P \preceq \{a\}$.

3.3 Equivalence of Definitions

We now proceed to show that both definitions above are, in fact, equivalent. For example, part preservation can be interpreted as: $b \preceq P$ implies $b \preceq M(P)$.

Theorem 1. *An oriented set-based sponge gives rise to an algebraic sponge, in which the partial functions J and M recover precisely the suprema and infima in the oriented set.*

Proof. Since the supremum is unique whenever it is defined, we can construct a partial function J that gives the supremum of a (finite) set of elements; we construct M analogously. Due to reflexivity the resulting J and M must be idempotent. Part preservation also follows, as by definition any upper bound of a set of elements is an upper bound of the supremum of those elements (note that by definition, if there is an upper bound, there must also be a supremum).

To see that our candidate sponge also satisfies the absorption laws, suppose that the set P has a common lower bound, so its infimum $\inf(P)$ is defined. By definition, $a \preceq a$ as well as $\inf(P) \preceq a$ for any $a \in P$. Since the two elements share an upper bound (a), the supremum of a and $\inf(P)$ must be defined. Again by definition, we must have that $a \preceq \sup(a, \inf(P))$, but also that $\sup(a, \inf(P)) \preceq a$ (since a is an upper bound for all of the arguments). Due to the antisymmetry of the orientation $\sup(a, \inf(P))$ must thus equal a. Since the same can be done in the dual situation, the J and M induced by '$\preceq$' must give rise to an algebraic sponge, in which J and M recover the suprema and infima in the oriented set. $\square$

---

[6] The name of this property was taken from the analogous property on binary joins/meets given by Skala [28], and presumably refers to the meet (join) preserving all joint lower (upper) bounds ("parts").

Theorem 2. *An algebraic sponge gives rise to an oriented-set-based sponge, such that a finite set has a supremum (infimum) if and only if J (M) of the set is defined, and if it is, J (M) gives the supremum (infimum).*

Proof. We define $a \preceq b$ if and only if $M(a,b) = a$. Note that the absorption laws guarantee that it does not matter whether we base the relation $\preceq$ on M or on J, as they imply that $M(a,b) = a \implies J(a,b) = J(M(a,b),b) = b$ (the dual statement follows analogously).

Since M is idempotent, the induced relation '$\preceq$' must be reflexive. Also, as M is a (partial) function, $a \preceq b$ and $a \neq b$ together imply $b \npreceq a$ (a function cannot take on two values at the same time). In other words: the relation '$\preceq$' is antisymmetric, and we can thus conclude that '$\preceq$' is an orientation.

We will now show that every finite set with a common upper (lower) bound (according to '$\preceq$') has a supremum (infimum) if and only if J (M) is defined for that set, and that if it exists, the supremum (infimum) is given by J (M). Due to the absorption and part preservation properties, the join provides every finite set that has a common upper bound with a supremum. Thus, the relation '$\preceq$' induced by J and/or M is a sponge. Furthermore, we cannot have any finite subsets for which J (M) is not defined but a supremum (infimum) does exist, as J and M must be defined for *all* finite subsets with a common upper/lower bound (due to the part preservation property). This concludes our proof. □

3.4   Further Properties

Lemma 1. *For any finite set of finite subsets $P_1, P_2, \ldots$ of a sponge, we have $M(P) \preceq M(M(P_1), M(P_2), \ldots)$, with $P = P_1 \cup P_2 \cup \ldots$, assuming $M(P)$ exists (and similarly for joins).*

Proof. We have $M(P) \preceq P$ (absorption). Now, since $P_i \subseteq P$, $M(P)$ is a lower bound of all elements of P_i (for any i), and thus of $M(P_i)$ as well (part preservation). The lemma now follows from another application of part preservation (since we have established that $M(P)$ is a lower bound of all the P_i). □

In a sponge the join of a finite set exists if (and only if) there is a common upper bound of that set. A conditionally complete sponge guarantees that this is true for *all* non-empty sets (finite or otherwise) that have at least one common upper bound, and similarly for the meet. A complete sponge would be a sponge for which all sets are guaranteed to have a join and a meet. All of the examples given in Section 4 are conditionally complete, and most of them have a smallest element (so all non-empty meets exist). We expect conditionally complete sponges with a least element to play a role analogous to that of complete lattices in traditional morphological theory. In such a sponge the meet of any set is well-defined, as is the join of any set with a common upper bound, making something similar to a structural opening well-defined (see Section 5).

Analogous to semilattices, we can define semisponges: a meet-semisponge is an oriented set such that any finite set with a lower bound has an infimum (a join-semisponge can be defined analogously). We can consider a meet-semisponge to have an operator M (the meet) that gives the infimum of a set. As the infimum is defined as the unique lower bound that is an upper bound of all lower bounds, M would still satisfy the part preservation property, as well as a modified form of the absorption property: if $M(P)$ is defined, then $[\forall a \in P : M(a, M(P)) = M(P)]$.

Theorem 3. *If S is a conditionally complete meet-semisponge it is a conditionally complete sponge (by duality the same holds for a conditionally complete join-semisponge).*

Proof. If S is a conditionally complete meet-semisponge, this means that the meet is defined for all (non-empty) sets that have a lower bound. We can now define J as giving the meet of the set of all upper bounds of a given set. For any non-empty set with a common upper bound, the set of all (common) upper bounds is again non-empty, and bounded from below by the original set, so its infimum is well-defined. As a consequence, if this construction turns S into a sponge, then this sponge is conditionally complete. Due to part preservation, the meet of the set of all upper bounds of a set is still an upper bound of the original set, and due to the absorption property it must also be a lower bound of all the upper bounds of the original set. In other words, J can indeed be interpreted as giving the supremum of any (non-empty) subset of S with an upper bound. We can thus conclude that S is a conditionally complete sponge.   □

If a conditionally complete meet-semisponge has a least element we can give meaning to the meet of *all* non-empty subsets, as well as the join of the empty set (which would give the least element).

Another important property of a sponge is whether it is acyclic or not. In a lattice we cannot have any cycles whatsoever (due to the combination of transitivity and antisymmetry of the underlying partial order). In a sponge we can have cycles (an example will be given below), but not having cycles allows us to define a partial order or preorder on a sponge, making it easier to talk about the convergence of sequences. It may also be interesting to consider *locally* acyclic sponges, whose restriction to any set of lower/upper bounds is acyclic.

Yet another interesting property that sponges can have is that the meet of a set of lower bounds is still a lower bound (and dually for upper bounds/joins). It is possible to construct sponges that do not have this property, but the examples given in Section 4 do all have this property. An immediate consequence is that in these sponges the join and meet preserve *both* upper and lower bounds: $a \preceq P$ and $P \preceq b$ imply $a \preceq M(P)$ and $M(P) \preceq b$ (and similarly for the join).

4   Examples

4.1   Inner Product and Hyperbolic Sponges

Inspired by the vector levelings developed by Zanoguera and Meyer [30], we can consider a vector $\mathbf{a}$ in some Hilbert space as "less" than (or equal to) another vector $\mathbf{b}$ if and only if $\mathbf{a} \cdot (\mathbf{b} - \mathbf{a}) \geq 0$. This does not give rise to a partial order,

Fig. 2. Computing the meet and join of the points a and b in the 2D (Euclidean) inner product sponge. The line segment between a and b is the convex hull of those two points. Each point has an associated circle enclosing all of its lower bounds (the dark dot represents the origin). The thick, dashed *lines* show the boundaries of the half-spaces of upper bounds for a and b. The shaded area below the meet is the intersection of the lower bounds of a and b, and is a subset of the set of lower bounds of the meet, consistent with part preservation.

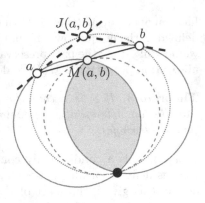

or even a preorder, as it is not transitive. However, it can be checked that it does give an orientation, and we will show that it even gives rise to a sponge (Fig. 2 illustrates the orientation and a meet and join in toy example).

The relation $\mathbf{a} \preceq \mathbf{b} \iff \mathbf{a} \cdot (\mathbf{b} - \mathbf{a}) \geq 0$ implies that the set of upper bounds of some element $\mathbf{a}$ is the half-space defined by $\mathbf{a} \cdot \mathbf{b} \geq \|\mathbf{a}\|^2$. We now define the meet of a set of elements as the element closest to the origin in/on the closed convex hull of the set. If the convex hull includes the origin, this is the origin itself (and in this case there is indeed no other lower bound of the entire set). If the origin is outside the convex hull, the meet is still well-defined (minimization of a strictly convex function over a convex set) and must lie on the boundary of the convex hull. It is possible to see that the original points must thus be upper bounds of the meet. Also, since the meet is in the closed convex hull of the original points, and the set of upper bounds of any element is closed and convex, any element which was a lower bound of all of the original points must still be a lower bound of the meet. Based on Theorem 3, we can now conclude that – based on the meet described above – we have a conditionally complete sponge with the origin as its least element.

The "geodesic ordering" on the Poincaré upper-half plane developed by Angulo and Velasco-Forero [3] can be shown to form a conditionally complete sponge along much the same lines as the inner product sponge.

4.2   Angle Sponge

Another problem area for the lattice formalism is that of periodic values, like angles. Several solutions [2, 4, 21, 25, 29] have been proposed to deal with angles, but none of them really deal with the inherent periodicity of angles. This is not by accident: it is impossible to have a periodic lattice.

Interestingly, we *can* create a periodic sponge: consider an angle a to be less than an angle b (both considered to be in the interval $[0, 2\pi)$) if and only if $b - \pi < a \leq b$ or $b - \pi < a - 2\pi \leq b$. In other words: a is less than b if and only if a is less than 180°clockwise from b.

It is clear that the above gives an orientation (the relation is reflexive and antisymmetric). Furthermore, if a set of angles has a common upper bound, all angles must lie on some arc of less than 180 degrees, and there must be a unique supremum (similarly for the infimum). We can thus conclude that we have defined a (periodic) conditionally complete sponge on angles.

4.3   A Non-sponge: The Loewner Order

The Loewner order [6] considers a (symmetric) matrix $\mathbf{A}$ less than or equal to another (symmetric) matrix $\mathbf{B}$ if the difference $\mathbf{B} - \mathbf{A}$ is positive semidefinite. This is a partial order compatible with the vector space structure of (symmetric) matrices, but it does not give rise to a lattice, or even a sponge. Any join/meet based on the Loewner order cannot satisfy both the absorption property and part preservation at the same time (we gave an example where part preservation breaks down in previous work [19]). As a partial fix, Burgeth et al. originally [9] computed the meet as the matrix inverse of the join of the inverses, so at least positive semidefiniteness would be preserved. However, matrix inversion does not reverse the order, and this still does not solve the problem that no upper bound of two matrices can be a lower bound of all common upper bounds.

In later work Burgeth et al. compute both the join and meet in a way that is compatible with the Loewner order [7, 8], but as a result they have to be careful not to get values outside the original range. The resulting structure is likely to be a so-called χ-lattice [23], but it is not yet clear how important this is from a morphological perspective. Based on the arguments presented by Pasternak et al. [24] and some preliminary experimentation, we expect it could be interesting to simply use the inner product sponge directly on the vector space of symmetric tensors (which would still preserve positive semidefiniteness).

5   Operators

One of the main advantages of the lattice-theoretical framework is that it allows us to classify operators into various categories based on certain lattice-related properties, and that these classes often have useful and intuitive interpretations. Although it remains to be seen to what extent existing classes carry over to the sponge case, here we show that at least one crucial property is preserved when we directly translate so-called "structural openings" (and in the process, that we can reason about such things for sponges in general).

We can try to translate structural dilations and erosions on images defined on a (translation-invariant) domain E to the sponge case. We then get a dilation-like operator defined by $\delta_A(f)(x) = J(\{f(y) \mid x \in A_y\})$ and an erosion-like operator $\varepsilon_A(f)(x) = M(\{f(y) \mid y \in A_x\})$ (where $x \in E$ and A_y is taken to be the structuring element translated by y). These operators need not commute with taking the join or meet, respectively, nor do they need to satisfy $\delta_A(f) = M(\{g \mid f \le \varepsilon_A(g)\})$ like in a complete lattice [22, Prop. 3.14]. It is an open question whether there exist different definitions that recover a bit more of the traditional properties (while remaining compatible with the lattice case). Nevertheless, we can use these operators to define an operator that behaves a bit like a lattice-based structural opening (and is a structural opening if the sponge is a lattice):

$$\gamma(f)(x) = J(\{M(\{f(z) \mid z \in A_y\}) \mid x \in A_y\}).$$

It is immediately obvious that the resulting operator is still guaranteed to be anti-extensive (due to each of the $M(\{f(z) \mid z \in A_y\})$ being a lower bound of $f(x)$), something which is not guaranteed by Loewner-based operators [19]. The operator may no longer be idempotent though. Increasingness is also potentially violated, but this is mostly due to the meet and join not necessarily being increasing in a sponge, so it may make sense to look for a different property. In any case, the above shows that for any image, the set of lower bounds that can be written as a dilation of some other image is non-empty, so it should be possible to implement some sort of projection onto this set, giving an operator that would clearly be anti-extensive *and* idempotent.

6   Conclusion

There is a need for mathematical morphology beyond what can be done with lattices. To this end, we have proposed a novel algebraic structure (closely related to the notion of a weakly associative lattice). This structure generalizes lattices not by forgoing (unique) meets and joins, but rather by letting go of having a (transitive) order. It preserves the absorption property though, as well as a property called "part preservation". These properties are important in the intuitive interpretation of joins and meets: absorption guarantees that the meet of a set is a lower bound of the set, while part preservation guarantees that a meet is truely the "greatest" lower bound, in the sense that it is an upper bound of all other lower bounds.

To demonstrate the potential relevance of this new framework, we give several examples of sponges, some of which were actually defined in earlier work (but not recognized as fitting into some general framework). One of our example sponges operates on a periodic space (something which is utterly impossible with a lattice). One advantage of recognizing these as sponges rather than ad-hoc constructions is that sponges guarantee relatively intuitive interpretations of joins and meets, and allow us to recover at least some properties of familiar operators like structural openings. Also, sponges allow us to reason about the structure itself. For example: we can recognize conditionally complete semisponges and see that any conditionally complete semisponge is a full (conditionally complete) sponge. It may also be possible to give characterizations of certain classes of sponges, as has been done for weakly associative lattices [26].

In terms of future work, the field is wide open: are sponges the "right" generalization of lattices? What kinds of sponges give what opportunities? What classes of operators can be defined? The list of open questions is endless. Also, it would be interesting to see if more interesting sponges can be constructed or found in the literature. For example, one may wonder whether some of the more algebraic methods given by Angulo [1], Burgeth et al. [10] give rise to sponges.

Acknowledgements. We would like to thank Jesús Angulo and Fernand Meyer for an inspiring discussion on vector levelings. We would also like to thank Emmanuel Chevallier for numerous discussions on the properties and possible applications of sponges, as well as recognizing that the two definitions of sponges are equivalent.

References

[1] Angulo, J.: Supremum/Infimum and Nonlinear Averaging of Positive Definite Symmetric Matrices. In: Nielsen, F., Bhatia, R. (eds.) Matrix Information Geometry, pp. 3–33. Springer, Heidelberg (2013)

[2] Angulo, J., Lefèvre, S., Lezoray, O.: Color Representation and Processing in Polar Color Spaces. In: Fernandez-Maloigne, C., Robert-Inacio, F., Macaire, L. (eds.) Digital Color Imaging, chap. 1, pp. 1–40. John Wiley & Sons, Inc., Hoboken (2012)

[3] Angulo, J., Velasco-Forero, S.: Complete Lattice Structure of Poincaré Upper-Half Plane and Mathematical Morphology for Hyperbolic-Valued Images. In: Nielsen, F., Barbaresco, F. (eds.) GSI 2013. LNCS, vol. 8085, pp. 535–542. Springer, Heidelberg (2013)

[4] Aptoula, E., Lefèvre, S.: On the morphological processing of hue. Image Vis. Comput. 27(9), 1394–1401 (2009)

[5] Astola, J., Haavisto, P., Neuvo, Y.: Vector median filters. Proc. IEEE 78(4), 678–689 (1990)

[6] Burgeth, B., Bruhn, A., Didas, S., Weickert, J., Welk, M.: Morphology for matrix data: Ordering versus PDE-based approach. Image Vis. Comput. 25(4), 496–511 (2007)

[7] Burgeth, B., Kleefeld, A.: Morphology for Color Images via Loewner Order for Matrix Fields. In: Hendriks, C.L.L., Borgefors, G., Strand, R. (eds.) ISMM 2013. LNCS, vol. 7883, pp. 243–254. Springer, Heidelberg (2013)

[8] Burgeth, B., Kleefeld, A.: An approach to color-morphology based on Einstein addition and Loewner order. Pattern Recognit. Lett. 47, 29–39 (2014)

[9] Burgeth, B., Papenberg, N., Bruhn, A., Welk, M., Feddern, C., Weickert, J.: Morphology for Higher-Dimensional Tensor Data Via Loewner Ordering. In: Ronse, C., Najman, L., Decencière, E. (eds.) Mathematical Morphology: 40 Years On, Computational Imaging and Vision, vol. 30, pp. 407–416. Springer, Netherlands (2005)

[10] Burgeth, B., Welk, M., Feddern, C., Weickert, J.: Morphological Operations on Matrix-Valued Images. In: Pajdla, T., Matas, J(G.) (eds.) ECCV 2004. LNCS, vol. 3024, pp. 155–167. Springer, Heidelberg (2004)

[11] Burgeth, B., Welk, M., Feddern, C., Weickert, J.: Mathematical Morphology on Tensor Data Using the Loewner Ordering. In: Weickert, J., Hagen, H. (eds.) Visualization and Processing of Tensor Fields. Math. Vis, pp. 357–368. Springer, Heidelberg (2006)

[12] Chevallier, E., Angulo, J.: Optimized total order for mathematical morphology on metric spaces. Tech. rep., Centre de Morphologie Mathématique, MINES ParisTech (2014)

[13] Fried, E.: Weakly associative lattices with join and meet of several elements. Ann. Univ. Sci. Budapest. Eötvös Sect. Math. 16, 93–98 (1973)

[14] Fried, E., Grätzer, G.: A nonassociative extension of the class of distributive lattices. Pacific J. Math. 49(1), 59–78 (1973)

[15] Fried, E., Grätzer, G.: Some examples of weakly associative lattices. Colloq. Math. 27, 215–221 (1973)

[16] Fried, E., Grätzer, G.: Partial and free weakly associative lattices. Houston J. Math. 2(4), 501–512 (1976)

[17] Gomila, C., Meyer, F.: Levelings in vector spaces. In: Int. Conf. Image Proc., vol. 2, pp. 929–933. IEEE (1999)

[18] Grätzer, G.: General Lattice Theory, 2nd edn. Birkhäuser Verlag (2003)

[19] van de Gronde, J.J., Roerdink, J.B.T.M.: Frames, the Loewner order and eigendecomposition for morphological operators on tensor fields. Pattern Recognit. Lett. 47, 40–49 (2014)

[20] van de Gronde, J.J., Roerdink, J.B.T.M.: Group-Invariant Colour Morphology Based on Frames. IEEE Trans. Image Process. 23(3), 1276–1288 (2014)

[21] Hanbury, A.G., Serra, J.: Morphological operators on the unit circle. IEEE Trans. Image Process. 10(12), 1842–1850 (2001)

[22] Heijmans, H.J.A.M.: Morphological image operators. Academic Press (1994)

[23] Leutola, K., Nieminen, J.: Posets and generalized lattices. Algebra Universalis 16(1), 344–354 (1983)

[24] Pasternak, O., Sochen, N., Basser, P.J.: The effect of metric selection on the analysis of diffusion tensor MRI data. NeuroImage 49(3), 2190–2204 (2010)

[25] Peters, R.A.: Mathematical morphology for angle-valued images. Proceedings of the SPIE 3026, 84–94 (1997)

[26] Rachůnek, J.: Semi-ordered groups. Sborník prací Přírodovědecké fakulty University Palackého v Olomouci. Matematika 18(1), 5–20 (1979)

[27] Serra, J.: Anamorphoses and function lattices. In: Dougherty, E.R., Gader, P.D., Serra, J.C. (eds.) Image Algebra and Morphological Image Processing IV, vol. 2030, pp. 2–11. SPIE Proceedings (1993)

[28] Skala, H.L.: Trellis theory. Algebra Universalis 1(1), 218–233 (1971)

[29] Tobar, M.C., Platero, C., González, P.M., Asensio, G.: Mathematical morphology in the *HSI* colour space. In: Martí, J., Benedí, J.M., Mendonça, A.M., Serrat, J. (eds.) IbPRIA 2007. LNCS, vol. 4478, pp. 467–474. Springer, Heidelberg (2007)

[30] Zanoguera, F., Meyer, F.: On the implementation of non-separable vector levelings. In: Talbot, H., Beare, R. (eds.) Mathematical Morphology, p. 369. CSIRO Publishing (2002)

A Color Tree of Shapes with Illustrations on Filtering, Simplification, and Segmentation

Edwin Carlinet[1,2(✉)] and Thierry Géraud[1]

[1] EPITA Research and Development Laboratory (LRDE), Kremline-Bicêtre, France
{firstname.lastname}@lrde.epita.fr
[2] LIGM, Équipe A3SI, ESIEE, Université Paris-Est, Créteil, France

Abstract. The Tree of Shapes (ToS) is a morphological tree that provides a high-level, hierarchical, self-dual, and contrast invariant representation of images, suitable for many image processing tasks. When dealing with color images, one cannot use the ToS because its definition is ill-formed on multivariate data. Common workarounds such as marginal processing, or imposing a total order on data are not satisfactory and yield many problems (color artifacts, loss of invariances, etc.) In this paper, we highlight the need for a self-dual and contrast invariant representation of color images and we provide a method that builds a single ToS by merging the shapes computed marginally, while guarantying the most important properties of the ToS. This method does not try to impose an arbitrary total ordering on values but uses only the inclusion relationship between shapes. Eventually, we show the relevance of our method and our structure through some illustrations on filtering, image simplification, and interactive segmentation.

Keywords: Tree of shapes · Hierarchical representation · Color image processing · Connected operators

1 Introduction

The Tree of Shapes (ToS) [6] is a hierarchical representation of an image in terms of the inclusion of its level lines. Its powerfulness lies in the number of properties verified by this structure. First, it is a morphological representation based on the inclusion of the connected components of an image at the different levels of thresholding. As such, a basic filtering of this tree is a connected filter, i.e., an operator that does not move the contours of the objects but either keep or remove some of them [16]. In addition, not only is it invariant by contrast changes on the whole image but it is also robust to local changes of contrast [2]. This property is very desirable in many computer vision applications where we face the problem of illumination changes, e.g., for scene matching, object recognition... In Fig. 1c, we show this invariance by simulating a change of illumination directly in the ToS so we do have the exact same tree representation as for the original image in Fig. 1a. Third, besides being contrast change invariant, the ToS is also a self-dual representation of the image. This feature is fundamental in a context where structures may appear both on a brighter

© Springer International Publishing Switzerland 2015
J.A. Benediktsson et al. (Eds.): ISMM 2015, LNCS 9082, pp. 363–374, 2015.
DOI: 10.1007/978-3-319-18720-4_31

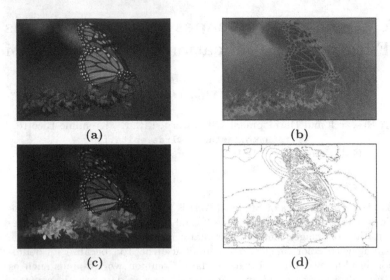

Fig. 1. On the need for contrast change/inversion invariance. (b) and (c) have been generated from the original image. We have changed and/or inverted the contrast of each channel for (b) and we have changed locally the contrast for (c) (to simulate a change of illumination). The three images (a), (b), and (c) give the same CToS whose level lines are shown in (d).

background or on a lighter one. Therefore, self-dual operators are particularly well adapted to process images without prior knowledge on their contents. While many morphological operators try to be self-dual by combining extensive and anti-extensive operators (for instance, the Alternating Sequential Filters), many of them actually depend on the processing order (i.e., on which filter comes first). Self-dual operators have the ability to deal with both dark and light objects in a *symmetric* way [12,17] (see Fig. 1b).

Despite all these wonderful features, the ToS is still widely under-exploited even if some authors have already been using it successfully for image processing and computer vision applications. In [9,21,22], an energy optimization approach is performed on the ToS for image simplification and image segmentation, while Cao et al. [2] rely on an *a-contrario* approach to select meaningful level-lines. Some other applications include blood vessel segmentation [21], scene matching extending the Maximally Stable Extremal Regions (MSER) through the Maximally Stable Shapes [6], image registration [6], and so on.

While the ToS is well-defined on grayscale images, it is getting more complicated with multivariate data. Indeed, like in the case of the min and max-trees, the ToS relies on an ordering relation of values which has to be total. If it is not, the definition based on lower and upper threshold sets yield components that may overlap and the tree of inclusion does not exist. To overcome this problem, most authors have focused on defining a total order on multivariate data. However, from our point of view, the most important feature of morphological trees lies in the inclusion of components/shapes. As a consequence, this paper introduces a novel approach which does not intend to build a total order, but

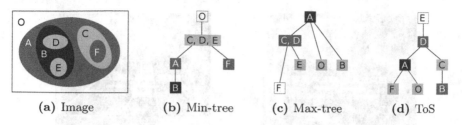

Fig. 2. An image (a) and its morphological component trees (b) to (d)

tries to build up a set of non-overlapping shapes from an arbitrary set of shapes using the inclusion relation only.

The paper is organized as follows. In Section 2, we remind the basis about the ToS and its formal definition. We also review some classical approaches to handle multivariate data with morphological trees. In Section 3, we introduce a new method to build a Color Tree of Shapes (CToS) by merging the shapes issued from marginal ToS's. In Section 4, we show some illustrations to highlight the potential of the CToS and its versatility.

2   Background

2.1   The Tree of Shapes

Let an image $u : \Omega \to E$ defined on a domain Ω and taking values on a set E embedded with an ordering relation $\le$. Let, $[u < \lambda]$ (resp. $[u > \lambda]$) with $\lambda \in \mathbb{R}$ be a threshold set of u (also called respectively lower cut and upper cut) defined as $[u < \lambda] = \{x \in \Omega, u(x) < \lambda\}$. We note $\mathcal{CC}(X), X \in \mathcal{P}(\Omega)$ the set of connected components of X. If $\le$ is a total relation, any two connected components $X, Y \in \mathcal{CC}([u < \lambda])$ are either disjoint or nested. The set $\mathcal{CC}([u < \lambda])$ endowed with the inclusion relation forms a tree called the *min-tree* and its dual tree, defined on upper cuts, is called the *max-tree* (see Figs. 2b and 2c). Given the hole-filling operator $\mathcal{H}$, we call a *shape* any element of $\mathcal{S} = \{\mathcal{H}(\Gamma), \Gamma \in \mathcal{CC}([u < \lambda])\}_\lambda \cup \{\mathcal{H}(\Gamma), \Gamma \in \mathcal{CC}([u > \lambda])\}_\lambda$. If $\le$ is total, any two shapes are either disjoint or nested, hence the cover of $(\mathcal{S}, \subseteq)$ forms a tree called the Tree of Shapes (ToS) (see Fig. 2d). In the rest of the paper, we implicitly consider the cover of $(\mathcal{S}, \subseteq)$ while writing $(\mathcal{S}, \subseteq)$ only. The level lines of u are the contours of the shapes. Using the image representation in [11], one can ensure each level line is an isolevel closed curve given that $\le$ is a total order. Actually, the "totality" requirement about $\le$ comes from the definition of the level lines in terms of contours of lower or upper sets. Note that the ToS encodes the shapes inclusion but also the level lines inclusion that are the contours of the shapes. Without loss of generality, we will consider $E = \mathbb{R}^n$ throughout this paper, and we will note u for scalar images ($n = 1$) and $\mathbf{u}$ for multivariate ones ($n > 1$).

2.2   The Color Problem: Common Solutions and Related Works

The previous definitions of level lines (in terms of iso-level curves and as contour of shapes) are both ill-formed when dealing with partial orders. Indeed, iso-level

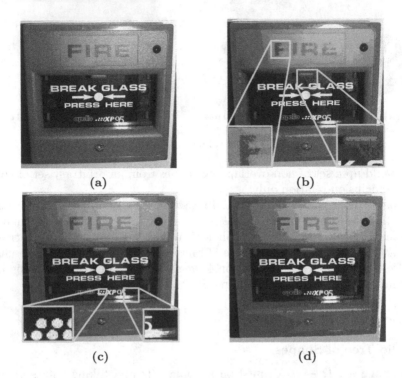

(a)   (b)

(c)   (d)

Fig. 3. Simplification issues with "classical" color image processing. (b) shows the simplification on the luminance of the original image (198 regions) issuing leakage problems because it does not allow to retrieve the whole geometric information. (c) shows the marginal processing (123 + 141 + 136 regions) that introduces false colors. (d) is the simplification with our method (158 regions). It retrieves correctly the main contents of the image while preventing the introduction of false colors.

sets in colors do not form closed curves and the shapes issued from lower and uppers cuts may intersect without being nested, i.e., $(\mathcal{S}, \subseteq)$ is a graph.

An unacceptable but widely used workaround for color image processing is to get rid of colors and to process a gray-level version of the color image. This workaround makes sense if we pretend that the geometric information is mainly held by the luminance [5]. However, many images exist where edges are only available in the color space (e.g. in document or synthetic images), emphazing that the chrominance holds some geometric information as well (see Fig. 3b).

Another commonly used solution is processing the image channel-wise and finally recombine the results. Marginal processing is subject to the well-known false color problem: it usually creates new colors that were not in the original images. False colors may or may not be a problem in itself (e.g. if the false colors are perceptually close to the original ones) but for image simplification it may produce undesirable color artifacts as shown in Fig. 3c. Also marginal processing leads to several trees (each of them representing a single channel the image) whereas we aim at producing a single representation of the image.

Since the pitfall of overlapping shapes is due to the partial ordering of colors, some authors tend to impose an "arbitrary" total ordering or total pre-ordering on values. They differ in the fact that a node may get associated with several colors. The way of ordering a multivariate space has been widely studied to extend gray-scale morphological operators. A non-extensive review of classical way of ordering values can be found in [1]. Also more advanced strategies have been designed to build a more "sensitive" total ordering that depends on the image contents (see for example [19,20,13,8]).

Another approach introduced by [15] uses directly the partial ordering of values and manipulates the underlying structure, which is a graph. The component-graph is still at a development level but has shown promising results for filtering tasks [14]. However, the component-graph faces an algorithmic complexity issue that compels the authors to perform the filtering locally. Thus, it is currently not suitable if we want to have a single representation of the whole image.

In [4], we introduced an approach where instead of trying to impose a total ordering on values, we compute marginally the ToS's and merge them into a single tree. The merge decision does not rely on values anymore but rather on some properties computed in a shape space. However, the merging procedure proposed in that paper shows a loss of "coherence" by merging unrelated shapes together. In [3], inspired by the work of [15], we proposed the Graph of Shapes (GoS) which merges the marginal ToS's into a single structure in an efficient way. We showed that this structure has a strong potential compared to the methods that impose a total order on values. Yet, the method builds a graph that prevents from using tools from the component tree framework (filtering, object detection, segmentation methods, etc.) and complicates the processing. The work presented here can be seen as a continuation of the ideas introduced in [4] and [3] since the GoS is used as an intermediate representation to extract a single tree from the marginal ToS's.

3   Merging the Trees of Shapes

3.1   Overview and Properties

Let us first relax the definition of a *shape*. A *shape* X is a connected component of Ω without holes (i.e., such that $\mathcal{H}(X) = X$). Given a family of shape sets, namely $\mathcal{M} = \{\mathcal{S}_1, \mathcal{S}_2, \ldots, \mathcal{S}_n\}$, where each element $(\mathcal{S}_i, \subseteq)$ forms a tree, we note $\mathcal{S} = \bigcup \mathcal{S}_i$ the primary shape set. Note that $(\mathcal{S}, \subseteq)$ generally does not form a tree but a graph since shapes may overlap. We aim at defining a new set of shapes $\boldsymbol{S}$ such that any two shapes are either nested or disjoint. We do not impose the constraint $\boldsymbol{S} \subseteq \mathcal{S}$. In other words, we do allow the method to build new shapes that were not in the primary shape set. We note $T : \Omega^{\mathbb{R}^n} \to (\mathcal{P}(\mathcal{P}(\Omega)), \subseteq)$ the process that builds a ToS $(\boldsymbol{S}(\mathbf{u}), \subseteq)$ from an image $\mathbf{u} \in (\mathbb{R}^n)^\Omega$.

A transformation ψ is said contrast change invariant if given a strictly increasing function $g : \mathbb{R} \to \mathbb{R}$, $g(\psi(u)) = \psi(g(u))$. Moreover, the transformation is said self-dual if it is invariant w.r.t. the complementation i.e. $\complement(\psi(u)) = \psi(\complement(u))$ (for images with scalar values $\complement(u) = -u$). When ψ is both self-dual and contrast

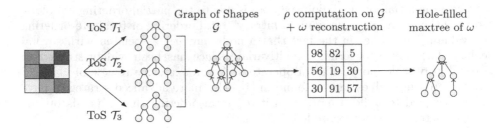

Fig. 4. Scheme of the proposed method to compute a Color Tree of Shapes (CToS)

change invariant, then for any strictly monotonic function G (i.e., either strictly increasing or decreasing), we have $G(\psi(u)) = \psi(G(u))$. The ToS is actually a support for many self-dual morphological operators and a representation T is said self-dual and morphological if $T(G(u)) = T(u)$.

More formally, we want the method T to produce $T(\mathbf{u}) = (\mathbf{S}(\mathbf{u}), \subseteq)$ having the following properties:

(P1) Domain covering $\left(\bigcup_{X \in S(\mathbf{u})} X \right) = \Omega$

(a point belongs to one shape at least)

(P2) Tree structure $\forall X, Y \in \mathbf{S}(\mathbf{u})$, either $X \cap Y = \emptyset$ or $X \subseteq Y$ or $Y \subseteq X$

(any two shapes are either nested or disjoint)

(P3) Scalar ToS equivalence. If $\mathcal{M} = \{\mathcal{S}_1\}$ then $\mathbf{S}(\mathbf{u}) = \mathcal{S}_1$ (for scalar images, the tree built by the method is equivalent to the gray-level ToS).

(P4) If a shape $X \in \mathbf{S}(\mathbf{u})$ verifies:

$$\forall Y \neq X \in S, \ X \cap Y = \emptyset \text{ or } X \subset Y \text{ or } Y \subset X$$

then $X \in \mathbf{S}(\mathbf{u})$ (any shape that does not overlap with any other shape should exist in the final shape set).

(P5) Marginal contrast change/inversion invariance.

Let us consider $\mathbf{G}(\mathbf{u}) = (G_1(u_1), G_2(u_2), \ldots, G_n(u_n))$, where G_i is a strictly monotonic function, then T shall be invariant by marginal inversion/change of contrast, that is, $T(\mathbf{G}(\mathbf{u})) = T(\mathbf{u})$.

3.2   Method Description

The method we propose is a 4-steps process (see Fig. 4). First, we start with computing the marginal ToS's $\mathcal{T}_1, \mathcal{T}_2, \ldots, \mathcal{T}_n$ of $\mathbf{u}$ associated with the shape sets $\mathcal{S}_1$, $\mathcal{S}_2, \ldots \mathcal{S}_n$ that give a primary shape set $\mathcal{S} = \bigcup \mathcal{S}_i$. The multiple trees provide a representation of the original image and $\mathbf{u}$ can be reconstructed marginally from them. However, handling several trees is not straightforward and they lack some important information: how the shapes of one tree are related (w.r.t the inclusion) to the shapes of the other trees. The graph $\mathcal{G}$, the cover of $(S, \subseteq)$, is nothing more than these trees merged in a unique structure that adds the inclusion relation that was missing previously. As a consequence, $\mathcal{G}$ is "richer" than $\{\mathcal{T}_1, \ldots, \mathcal{T}_n\}$ and because the transformation from $\{\mathcal{T}_1, \ldots, \mathcal{T}_n\}$ to $\mathcal{G}$ is re-

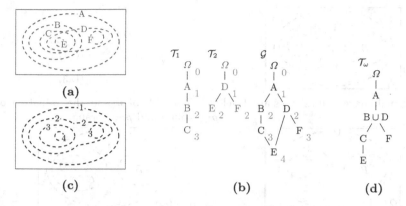

Fig. 5. The method illustrated on an example. (a) A 2-channels image **u** and its shapes (resp. in green and red). (b) The marginal ToS's $\mathcal{T}_1$, $\mathcal{T}_2$ and the GoS. The depth appears in gray near the nodes. (c) ω image built from $\mathcal{G}$. (d) The max-tree $\mathcal{T}_\omega$ of ω.

versible, $\mathcal{G}$ is a complete representation of **u** (i.e. **u** can be reconstructed from $\mathcal{G}$). Moreover, $\mathcal{G}$ is also a self-dual, contrast invariant representation of **u**.

The second part of the method tries to extract a tree from $\mathcal{G}$ verifying the constraints (P1) to (P5). The major difficulty of this task is to get from $\mathcal{G}$ a new set of shapes that do not overlap. The first observation is that for any decreasing attribute $\rho : \mathcal{P}(\Omega) \to \mathbb{R}$ (i.e. $\forall A, B \in \mathcal{S}, A \subset B \Rightarrow \rho(A) > \rho(B)$), then $(\mathcal{S}, \subset)$ is isomorphic to $(S, \mathcal{R})$ where $A \mathcal{R} B \Leftrightarrow \rho(A) > \rho(B)$ and $A \cap B \neq \emptyset$. This just means that the inclusion relationship between shapes that we want to preserve can be expressed in terms of a simple ordering relation on $\mathbb{R}$ with the values on a decreasing attribute. An example of such an attribute is the *depth* where the *depth* of a shape in $\mathcal{G}$ stands for the length of the longest path of a shape A from the root. Consider now the image $\omega(x) = \max_{x \in X, X \in \mathcal{S}} \rho(x)$. ω is an image that associates for each point x, the depth of the deepest shape containing x (see Figs. 5b and 5c). The latter may form component with holes so we consider $S(\mathbf{u}) = \mathcal{H}(\mathbb{C})$ and $(S(\mathbf{u}), \subseteq)$ as the final ToS $\mathcal{T}_\omega$ (see Fig. 5d). Note that in the case where $(\mathcal{S}, \subset)$ is already a tree, we thus have $\mathbb{C} = \{\mathcal{CC}([\omega \geq h]), h \in \mathbb{R}\} = \mathcal{S}$. In other words, the max-tree of w reconstructed from ρ valuated on a tree $\mathcal{T}$ yields the same tree (property **P3**) and more generally, if a shape A do not overlap any other shape, it belongs to $\mathcal{CC}([\omega \geq h])$ (property **P4**). From a computational standpoint, the most expensive part is the graph computation which is $O(n^2.H.N)$, where n is the number of channels to merge, H the maximal depth of the trees, and N the number of pixels. Indeed, for each shape of one tree, we need to track its smallest including shape in the other trees which is basically an incremental least common ancestor attribute computation. The other steps are either linear or quasi-linear with the number of pixels. In the next section, we explain why we choose ρ to be the *depth* in $\mathcal{G}$.

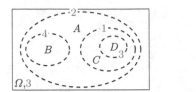

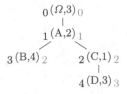

(a) Original image u, its shapes and level lines. | (b) The ToS of u and the valuation of ρ_{TV} (blue) and ρ_{CV} (orange).

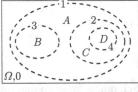

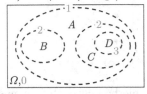

(c) The level lines of ω_{TV} | (d) The level lines of ω_{CV}

Fig. 6. Equivalence between the level lines of a gray-level image u and the level lines of the distance maps ω_{TV} and ω_{CV}

3.3 Computing the Inclusion Map

The 4th step of the method involves the choice of an attribute to be computed over the GoS $\mathcal{G}$. This is a critical step since it decides which shapes are going to be merged or removed. In Section 3.2, the *depth* is used as the attribute to merge the shapes, yet without justification. We now explain the rationale for this choice. Consider the distance measure between two points (p, p') in Ω:

$$d_{\text{TV}}(p, p') = \min_{C_{pp'}} \int_0^1 |\nabla u(C_{pp'}(t)).\dot{C}_{pp'}(t)|.dt, \qquad (1)$$

where $C_{pp'}(t)$ is a path in Ω from p to p'. Equation (1) is actually the minimum total variation among the paths from p to p'. This measure has been used by [10] for segmenting where the ToS helps to compute efficiently the level set distance. Let $\omega_{\text{TV}}(x) = d_{\text{TV}}(\partial\Omega, x)$ the Total Variation distance map from the border. It can be computed using the ToS by summing the variations from the root to the nodes. Then, instead of considering the tree $\mathcal{T}$ of u level lines, one can consider the max-tree $\mathcal{T}_\omega$ of equidistant lines. Both are equivalent in gray-level.

The problem with the Total Variation metric lies in that it depends on u, i.e., ω_{TV} is no contrast invariant. A contrast invariant counterpart would be to only count the number of variation (CV), i.e., the minimum number of level lines to traverse to get to p. Algorithmically speaking, building ω_{CV} consists in computing the depth attribute $\rho_{\text{CV}}(A) = |\{S \in \mathcal{S} \mid A \subset S\}|$ and reconstructing $\omega_{\text{CV}}(x) = \max_{X \in \mathcal{S},\, x \in X} \rho_{\text{CV}}(X)$. This process is shown on Fig. 6. Based on the equivalence between level lines and equidistant lines for scalar images, we would want to build such a distance map for color images as well. Once again, the idea is to count the number of marginal level lines to traverse. More formally:

$$\rho(A) = \max_{\phi \in [\Omega \rightsquigarrow A]} |\phi| \quad \text{and} \quad \omega_{\text{CV}}(x) = \max_{X \in \mathcal{S},\, x \in X} \rho(X)$$

(a) Original   (b) Depth image ω (c) Reconstruction from $\mathcal{T}_\omega$

(d) Grain filters of sizes 10, 100 and 500

Fig. 7. Grain filters

where $[\Omega \rightsquigarrow A)$ stands for the set of paths from the root to A in $\mathcal{G}$. $\omega_{\mathrm{CV}}(x)$ actually counts the number of marginal level lines (*that are nested*) along the path from the border to the deepest shape that contains x. ρ can be computed efficiently from $\mathcal{G}$ using a basic shortest path algorithm.

4   Illustrations

4.1   Grain Filters

A grain filter [7] is an operator that removes the regions of the image which are local extrema and whose area is below a given threshold. Using the ToS, a grain filter is thus simply a pruning removing the nodes which do not pass the size criterion. Grain filters allow to reveal the "correctness" of the tree in the sense that a small grain size should filter out what we perceive as noise or details while an high grain size should show the main objects and the structure of the image. In Fig. 7, we show the inclusion map ω computed using our method and the image reconstructed from the max-tree $\mathcal{T}_\omega$. The reconstruction consists in computing for each node the average color of the pixels it contains and then, assigning this value to the pixels. Because $\mathcal{T}_\omega$ is not a reversible representation of **u**, the latter cannot be recovered from $\mathcal{T}_\omega$, however the reconstruction is close to the original. This illustration is rather a "structure validation" experience and does not aim at getting the best filtering results. In particular, we are in the same case of preorder-based filters, where a node may be associated with many color values. More advanced reconstruction strategies can be found in [18] that limit the restitution artifacts. In Fig. 7d, we have applied size-increasing grain filters that eliminate details in a "sensitive" way, and provide a reconstruction with few color artifacts that validate the structure organization of our tree.

(a) 112 / 288k level lines selected   **(b)** 1600 / 120k level lines selected

Fig. 8. Document simplification. Original images on top and simplified images below.

4.2   Image Simplification

To illustrate the ability of the CToS to represent the main structures of the images, we tested the tree against image simplification. This assessment uses the method proposed by [22] that minimizes the Mumford-Shah cartoon model constrained by the tree topology. More formally, we have to select a subset of shapes $\mathcal{S}' \subset \mathcal{S}$ that minimizes the energy:

$$E(\mathcal{S}') = \sum_{S \in \mathcal{S}'} \sum_{x \in S \,|\, S_x = S} ||f(x) - \bar{f}(S)||_2^2 + \lambda |\partial S|,$$

where S_x denotes the smallest shape containing x, $\bar{f}(S)$ is the average color of the region and $|\partial S|$ the length of the shape boundary. In [22], the authors use a greedy algorithm that removes the level lines sorted by meaningfulness until the energy does not decrease anymore.

Figure 8 illustrates the need for contrast inversion invariance in the case of document restoration. Here, the important point is that the CToS is able to retrieve low-contrasted letters even in the presence of "show-through". Since we use a segmentation energy, we do not pretend that it is the perfect solution for document binarization, however since the documents are largely simplified while keeping all the objects of interest, it may serve as a pre-processing for a more specific binarization method.

4.3   Interactive Object Segmentation

In [10], the authors introduced a method for interactive image segmentation using the level set representation of the image. We extend basically the same idea to the CToS. Given a set of markers B and F (both in $\mathcal{P}(E)$), where B stands for the background class $\mathcal{B}$ and F for the foreground class $\mathcal{F}$, we aim at classifying all the other pixels to one of these classes. We then use the Nearest

Fig. 9. Interactive segmentation using the CToS. original images with the markers on the top line and the segmentation below.

Neighbor classifier where the distance between two points x and y is the minimal total variation along all the all paths from x to y (see Eq. (1)). The ToS allows a fast computation of the distance between any two points x and y by summing up the variations along the paths of S_x and S_y to their least common ancestors. As a consequence, instead of working at the pixel level, the classification can be done equivalently with the ToS by computing the influence zones of the shapes having a marker pixel using the tree topology. With the CToS, a node may contain pixels of different colors, so we consider that the distance between a shape and its parent is simply the L^2-distance of their average color (in RGB, or better in the La*b* space).

A strong advantage of the method is its ability to recover large regions of interest with very few markers (see Fig. 9) whereas many other methods using statistics require larger markers for a better learning accuracy. We did not show the results using the ToS computed on the luminance only but the same problems (so the same remarks) stand as for the simplification.

5   Conclusion and Perspectives

We have presented a method to extend the ToS on multivariate images. Contrary to standard approaches, our CToS does not rely on any choice of multivariate total ordering but is only based on the inclusion relationship of the shapes and outputs a tree which is marginally both self-dual and contrast change invariant. In this paper, we have tried to highlight why those properties are important for image processing and computer vision tasks. Eventually, we have shown the versatility and the potential of our representation. As perspectives, we will focus on some other kinds of multivariate data such as hyperspectral satellite images or multimodal medical images to validate the contributions of our approach to the processing of such data. Moreover, we also plan to compare the CToS to other hierarchical representations such as hierarchies of partitions (quasi-flat zones hierarchy, binary partition trees...) to further study the pros and cons of our method.

References

1. Aptoula, E., Lefèvre, S.: A comparative study on multivariate mathematical morphology. Pattern Recognition 40(11), 2914–2929 (2007)
2. Cao, F., Musé, P., Sur, F.: Extracting meaningful curves from images. Journal of Mathematical Imaging and Vision 22(2-3), 159–181 (2005)
3. Carlinet, E., Géraud, T.: Getting a morphological tree of shapes for multivariate images: Paths, traps and pitfalls. In: Proc. of ICIP, France, pp. 615–619 (2014)
4. Carlinet, E., Géraud, T.: A morphological tree of shapes for color images, Sweden, pp. 1132–1137 (August 2014)
5. Caselles, V., Coll, B., Morel, J.M.: Geometry and color in natural images 16(2), 89–105 (2002)
6. Caselles, V., Monasse, P.: Geometric Description of Images as Topographic Maps. Lecture Notes in Mathematics, vol. 1984. Springer (2009)
7. Caselles, V., Monasse, P.: Grain filters 17(3), 249–270 (November 2002)
8. Chevallier, E., Angulo, J.: Image adapted total ordering for mathematical morphology on multivariate image, pp. 2943–2947, Paris, France (October 2014)
9. Dibos, F., Koepfler, G.: Total variation minimization by the Fast Level Sets Transform. In: IEEE Workshop on Variational and Level Set Methods in Computer Vision, pp. 179–185. IEEE Computer Society (2001)
10. Dubrovina, A., Hershkovitz, R., Kimmel, R.: Image editing using level set trees, pp. 4442–4446, Paris, France (October 2014)
11. Géraud, T., Carlinet, E., Crozet, S., Najman, L.: A quasi-linear algorithm to compute the tree of shapes of n-D images. In: Hendriks, C.L.L., Borgefors, G., Strand, R. (eds.) ISMM 2013. LNCS, vol. 7883, pp. 98–110. Springer, Heidelberg (2013)
12. Heijmans, H.J.A.M.: Self-dual morphological operators and filters 6(1), 15–36 (1996)
13. Lezoray, O., Charrier, C., Elmoataz, A., et al.: Rank transformation and manifold learning for multivariate mathematical morphology, vol. 1, pp. 35–39 (2009)
14. Naegel, B., Passat, N.: Towards connected filtering based on component-graphs. In: Hendriks, C.L.L., Borgefors, G., Strand, R. (eds.) ISMM 2013. LNCS, vol. 7883, pp. 353–364. Springer, Heidelberg (2013)
15. Passat, N., Naegel, B.: An extension of component-trees to partial orders, pp. 3933–3936. IEEE Press (2009)
16. Salembier, P., Serra, J.: Flat zones filtering, connected operators, and filters by reconstruction 4(8), 1153–1160 (1995)
17. Soille, P.: Beyond self-duality in morphological image analysis. Image and Vision Computing 23(2), 249–257 (2005)
18. Tushabe, F., Wilkinson, M.H.F.: Color processing using max-trees: A comparison on image compression. In: Proc. of ICSAI, pp. 1374–1380. IEEE (2012)
19. Velasco-Forero, S., Angulo, J.: Supervised ordering in R_p: Application to morphological processing of hyperspectral images 20(11), 3301–3308 (2011)
20. Velasco-Forero, S., Angulo, J.: Random projection depth for multivariate mathematical morphology. IEEE Journal of Selected Topics in Signal Processing 6(7), 753–763 (2012)
21. Xu, Y.: Tree-based shape spaces: Definition and applications in image processing and computer vision. Ph.D. thesis, Université Paris-Est (December 2013)
22. Xu, Y., Géraud, T., Najman, L.: Salient level lines selection using the Mumford-Shah functional. In: Proc. of ICIP, pp. 1227–1231 (2013)

Elementary Morphological Operations on the Spherical CIELab Quantale

Marcos Eduardo Valle[1](✉) and Raul Ambrozio Valente[2]

[1] Department of Applied Mathematics, University of Campinas, Campinas – SP, Brazil
[2] Ariam Equipamentos Metalúrgicos, Londrina – PR, Brazil
valle@ime.unicamp.br

Abstract. Mathematical morphology is a theory with applications in image and signal processing and analysis. This paper presents a quantale-based approach to color morphology based on the CIELab color space with spherical coordinates. The novel morphological operations take into account the perceptual difference between color elements by using a distance-based ordering scheme. Furthermore, the novel approach allows the use of non-flat structuring elements. Although the paper focuses on dilations and erosions, many other morphological operations can be obtained by combining these two elementary operations. An illustrative example reveals that non-flat dilations and erosions may preserve more features of a natural color image than their corresponding flat operations.

Keywords: Mathematical morphology · Quantale · Color image processing and analysis · CIELab color system · Distance-based total ordering

1 Introduction

Mathematical morphology (MM) is a theory with many applications in image and signal processing and analysis [11,18,29]. From the theoretical point of view, MM can be very well defined in a mathematical structure called complete lattice [18,23,27]. A complete lattice is a partially ordered set in which any subset has both a supremum and an infimum. Since the requirement is a partial ordering with well defined extrema operations, complete lattices allows for the development of morphological operations for multivalued data such as color images [3]. Precisely, multivalued MM attracted the attention of many researchers since the 1990s [8,9,12,22,32]. In general terms, the researches on multivalued MM focused on appropriate ordering schemes. A comprehensive discussion on several approaches for multivalued MM, including color MM, can be found in [3]. In particular, total orderings such as the conditional ordering schemes have been widely used in multivariate MM partially because they prevent the appearance of "false colors" [4,28]. For instance, [17] introduced a conditional ordering on the CIELab space for color MM. Also, [25] proposed an ordering scheme based on the distance to a reference color followed by a lexicographical ordering used to resolve ambiguities. Some distance-based ordering schemes have also been proposed by other

---

This work was supported in part by CNPq and FAPESP under grants nos. 305486/2014-4 and 2013/12310-4, respectively.

© Springer International Publishing Switzerland 2015
J.A. Benediktsson et al. (Eds.): ISMM 2015, LNCS 9082, pp. 375–386, 2015.
DOI: 10.1007/978-3-319-18720-4_32

prominent researchers including Angulo [2], Aptoula and Lefèvre [5], and De Witte at al. [34]. Recent developments in multivalued MM include operations that are invariant to a certain group of transformations [13,15], operations derived from Einstein addition and Loewner ordering scheme [7,14], and probabilistic pseudo-morphology [10].

Despite the suitability of complete lattices, many important approaches to MM are defined in richer mathematical structures [19,24,30]. For instance, the famous umbra approach to gray-scale MM is defined in a mathematical structure obtained by enriching a complete lattice with an order-preserving group operation [19,31]. Also, some researchers advocate that many morphological operations can be defined in a structure called *quantale* [24,30]. In few words, a quantale is a complete lattice endowed with an associative binary operation that commutes with the supremum operation [20]. From an algebraic point of view, the quantale framework comprises many approaches – such as the umbra approach – but it is included in the general complete lattice framework.

Recently, Valle et al. introduced a quantale based on the CIELab color space with spherical coordinates, referred to as the *spherical CIELab quantale* [33]. Although the spherical CIELab quantale has been successfully applied for the development of a class of associative memories, it has not been used as the mathematical background for color MM yet. The purpose of this paper is to provide support and some insights into the elementary operations of color MM defined on the spherical CIELab quantale.

The paper is organized as follows. Section 2 provides the basic concepts on MM, including the complete lattice and quantale-based frameworks. The spherical CIELab quantale as well as the operations of dilation and erosion are discussed in Section 3. An illustrative example of the dilation and the erosion of a natural color image is given in this section. The paper finishes with concluding remarks in Section 4.

2   Mathematical Morphology on Complete Lattices and Quantales

First of all, a complete lattice $\mathcal{L}$ is a partially ordered set in which any subset $X \subseteq \mathcal{L}$ has both a sumpremum and an infimum, denoted respectively by $\bigvee X \in \mathcal{L}$ and $\bigwedge X \in \mathcal{L}$ [6]. From an algebraic point of view, the two elementary operations of MM are defined as follows using the adjunction relationship [18]:

Definition 1 (Adjunction, Dilation, and Erosion). *Given complete lattices $\mathcal{L}$ and $\mathcal{M}$, we say that $\varepsilon : \mathcal{L} \to \mathcal{M}$ and $\delta : \mathcal{M} \to \mathcal{L}$ form an adjunction between $\mathcal{L}$ and $\mathcal{M}$ if the following equivalence holds true for $x \in \mathcal{L}$ and $y \in \mathcal{M}$:*

$$\delta(y) \leq x \iff y \leq \varepsilon(x). \tag{1}$$

In this case, δ and ε are called respectively a dilation and an erosion.

Many other operations of MM are obtained by combining dilations and erosions [11,29]. For example, an opening and a closing, denoted respectively by $\gamma : \mathcal{L} \to \mathcal{L}$ and $\phi : \mathcal{M} \to \mathcal{M}$, are obtained by the compositions

$$\gamma = \delta \circ \varepsilon \quad \text{and} \quad \phi = \varepsilon \circ \delta. \tag{2}$$

The operations of opening and closing are used, for instance, in granulometries as well as for the removal of noise [18,29].

Besides the complete lattice, many approaches to MM are defined in richer mathe-matical structures such as *quantales* [30,24]. A quantale, denoted in this paper by the triple $(Q, \leq, \cdot)$, is the algebraic structure in which the set Q, ordered by the relation "$\leq$", is a complete lattice and the binary operation "$\cdot$" commutes with the supremum operation in both arguments. In other words, the following equations hold true for any $q \in Q$ and $X \subseteq Q$:

$$q \cdot \left(\bigvee X \right) = \bigvee_{x \in X} (q \cdot x) \quad \text{and} \quad \left(\bigvee X \right) \cdot q = \bigvee_{x \in X} (x \cdot q). \qquad (3)$$

The binary operation "$\cdot$" is often referred to as the *multiplication* of the quantale. We say that $(Q, \leq, \cdot)$ is a *commutative quantale* if "$\cdot$" is commutative. Similarly, we have a *unital quantale* if "$\cdot$" has an identity, that is, if there exist $e \in Q$ such that $e \cdot q = q \cdot e = q$ for all $q \in Q$. The spherical CIELab quantale presented in the following section is an instance of commutative unital quantale.

In a quantale $(Q, \leq, \cdot)$, the multiplication is always residuated [24]. In particular, there exists a binary operation "$/$", called the *left residuum* of "$\cdot$", such that

$$x \cdot y \leq z \iff x \leq z/y, \quad \forall x, y, z \in Q. \qquad (4)$$

Furthermore, the left residuum is uniquely determined by the equation

$$y/x = \bigvee \{z \in Q : z \cdot x \leq y\}, \quad \forall x, y \in Q. \qquad (5)$$

In the following, we review how the elementary operations of MM are defined using a quantale [24,30]. For simplicity, we shall restrict our attention to images $f : X \to Q$, where $(Q, \leq, \cdot)$ is a commutative unital quantale and X is a subset of either $\mathbb{R}^d$ or $\mathbb{Z}^d$. The set of all images $f : X \to Q$ is denoted by Q^X. Note that Q^X is a complete lattice under the natural ordering: $f \leq g$ if and only if $f(x) \leq g(x)$ for all $x \in X$.

Such as the classical approaches to MM, the quantale-based dilation and erosion of an image $f \in Q^X$ are defined in terms of a structuring element (SE) $s \in Q^Y$, where the domain Y is also a subset of either $\mathbb{R}^d$ or $\mathbb{Z}^d$. The SE s is used to extract relevant information about the shape and form of objects in the probed image f. Formally, the two elementary operations of MM are defined as follows on a commutative unital quantale:

Definition 2 (Quantale-based Dilation and Erosion). *Let $(Q, \leq, \cdot)$ denote a commutative unital quantale. The quantale-based dilation of $f \in Q^X$ by an SE $s \in Q^Y$, denoted by $\delta_s^Q(f)$, is the image given by*

$$\delta_s^Q(f)(x) = \bigvee_{y \in Y, x-y \in X} (f(x-y) \cdot s(y)), \quad \forall x \in X. \qquad (6)$$

Dually, the quantale-based erosion of an image $f \in Q^X$ by an SE $s \in Q^Y$, denoted by $\varepsilon_s^Q(f)$, is defined as follows where "$/$" denotes the residuum of the multiplication "$\cdot$":

$$\varepsilon_s^Q(f)(x) = \bigwedge_{y \in Y, x+y \in X} (f(x+y)/s(y)), \quad \forall x \in X. \qquad (7)$$

Remark 1. Let $(Q, \leq, \cdot)$ be a quantale and $f \in Q^X$ an image. The spatial translation of an image f by $y \in X$, denoted by f_y, is the image defined by $f_y(x) = f(x - y)$ for all $x \in X$. The pixel value translation of f by α, denoted by $\alpha \cdot f$, is the image given by $(\alpha \cdot f)(x) = \alpha \cdot f(x)$, for all $x \in X$. We say that an image operator $\psi : Q^X \to Q^X$ is invariant under spatial translations if $\psi(f_y) = [\psi(f)]_y$ for all $f \in Q^X$ and $y \in X$. Similarly, ψ is invariant under pixel value translations if $\psi(\alpha \cdot f) = \alpha \cdot \psi(f)$, for all $f \in Q^X$ and $\alpha \in Q$. Now, Maragos showed that a dilation $\delta : Q^X \to Q^X$ is invariant under both spatial and pixel value translations if and only if it is given by (6) [19]. Therefore, the quantale-based approach is used whenever one imposes that the dilation is invariant under both spatial and pixel value translations.

We would like to conclude this section recalling that many important approaches, including the widely used umbra and some fuzzy-based approaches [21,31], belong to the quantale-based framework. Further examples of morphological operations on quantales can be found in [24,30].

3   Mathematical Morphology on the Spherical CIELab Quantale

In this paper, a color image is a function $f : X \to C$, where the domain is $X = \mathbb{R}^d$ or $X = \mathbb{Z}^d$ and C, usually a subset of $\bar{\mathbb{R}}^3$, $\bar{\mathbb{R}} = \mathbb{R} \cup \{+\infty, -\infty\}$, is the color space. There are many color spaces in the literature but, in this paper, we restrict our attention to the RGB and CIELab spaces, denoted respectively by C_{RGB} and C_{Lab} [1].

The RGB color space is based on the *tristimulus theory* of vision in which a color is decomposed into the primitives: *red* (R), *green* (G), and *blue* (B) [1]. Geometrically, this color space is represented by the cube $C_{RGB} = [0,1] \times [0,1] \times [0,1]$, whose axes correspond to the intensities in each primitive. In the RGB space, a certain color $c = (c_R, c_G, c_B)$ is a point in or inside the cube C_{RGB}. The origin corresponds to "black" while the edge $(1,1,1)$ represents "white".

Although the RGB color space is widely used by imaging devices as well as for color image representation and processing, it is not a perceptually uniform color space [1]. Specifically, it is claimed that the Euclidean distance between two elements $c, c' \in C_{RGB}$ hardly resembles the perceptual difference between the two colors. As a consequence, the RGB color space is not advisable for applications in which the visual perception of the colors is important. A perceptually uniform color space, such as the CIELab color space, is recommended in such applications [1].

In the CIELab color space $C_{Lab} \subseteq [0, 100] \times \mathbb{R} \times \mathbb{R}$, the components of a color element $c = (c_L, c_a, c_b)$ have the following connotations:

- The component c_L models the lightness.
- The component c_a indicates the green-red position.
- The component c_b yields the blue-yellow position.

The value $c_c = \sqrt{c_a^2 + c_b^2}$, referred to as *chroma*, measures the colorfulness of c with respect to white. Also, the angle $c_h = \tan^{-1}(c_b/c_a) \in (-\pi, \pi]$ is called *hue* of the color c [35]. Gray-scale elements are in the line segment $c_L = [0, 100]$, $c_a = c_b = 0$. The reader interested in the details of the conversion between the RGB and CIELab color spaces is invited to consult [1].

3.1 The Spherical CIELab Complete Lattice

A color element $c = (c_L, c_a, c_b) \in C_{Lab}$ can be expressed as $(c_\rho, c_\phi, c_\theta)$ using spherical coordinates centered at a fixed color reference $r = (r_L, r_a, r_b) \in C_{Lab}$. Recall that the spherical coordinates c_ρ (radius), c_ϕ (elevation), and c_θ (azimuth) are given by

$$c_\rho = \|c - r\|_2, \text{ with } c_\rho \in \mathbb{R}_{\geq 0} = \{x \in \mathbb{R} : x \geq 0\}, \tag{8}$$

$$c_\phi = \tan^{-1}\left(\frac{c_L - r_L}{\sqrt{(c_a - r_a)^2 + (c_b - r_b)^2}} \right), \text{ with } c_\phi \in \Phi = [-\pi/2, \pi/2], \tag{9}$$

$$c_\theta = \tan^{-1}\left(\frac{c_b - r_b}{c_a - r_a} \right), \text{ with } c_\theta \in \Theta = (-\pi, \pi]. \tag{10}$$

From now on, the symbol $\mathcal{S}_r = \mathbb{R}_{\geq 0} \times \Phi \times \Theta$ denotes the CIELab color space in spherical coordinates centered at a reference color r.

Let us now enrich the spherical CIELab system $\mathcal{S}_r$ with a total ordering scheme. Given two elements $c = (c_\rho, c_\phi, c_\theta) \in \mathcal{S}_r$ and $c' = (c'_\rho, c'_\phi, c'_\theta) \in \mathcal{S}_r$, we define

$$c \leqslant_{\mathcal{S}_r} c' \iff \begin{cases} c_\rho > c'_\rho, \text{ or} \\ c_\rho = c'_\rho, \text{ and } c_\phi \prec c'_\phi, \text{ or} \\ c_\rho = c'_\rho, c_\phi = c'_\phi, \text{ and } c_\theta \preceq c'_\theta, \end{cases} \tag{11}$$

where ">" denotes the usual greater than ordering on $\mathbb{R}$, "$\prec$" is the ordering given by

$$x \prec y \Leftrightarrow \begin{cases} |x| < |y|, \text{ or} \\ |x| = |y| \text{ and } x < y, \end{cases} \tag{12}$$

and $x \preceq y$ if and only if $x \prec y$ or $x = y$.

The total ordering given by (11) can be interpreted as follows: The greater of the two elements c and c' is the color closer (in the Euclidean distance sense) to the reference r, which corresponds to the origin $(0, 0, 0)$ in the spherical CIELab system $\mathcal{S}_r$. In case both c and c' have the same distance to r, the greater is the one with larger elevation, which corresponds to the one whose chroma is closer to the chroma of the reference r. Finally, the last condition in (11) avoids ambiguities and makes "$\leqslant_{\mathcal{S}_r}$" a total ordering scheme.

Although the greatest element of $\mathcal{S}_r$ is the origin, the algebraic structure $(\mathcal{S}_r, \leqslant_{\mathcal{S}_r})$ is not a complete lattice because it does not have a least element. Nevertheless, we can circumvent this problem by introducing an artificial point $\perp = (+\infty, 0, 0)$ as the least element of $\mathcal{S}_r$. Now, the set $\bar{\mathcal{S}}_r = \mathcal{S}_r \cup \{\perp\}$ is a complete lattice under the total ordering "$\leqslant_{\mathcal{S}_r}$" given by (11). The artificial element $\perp$ plays a role similar to the infinities $+\infty$ and $-\infty$ on the extended real numbers $\bar{\mathbb{R}}$.

3.2 The Spherical CIELab Quantale

Let us now endow the complete lattice $(\bar{\mathcal{S}}_r, \leqslant_{\mathcal{S}_r})$ with a binary operation "$\cdot_{\mathcal{S}_r}$" such that the algebraic structure $(\bar{\mathcal{S}}_r, \leqslant_{\mathcal{S}_r}, \cdot_{\mathcal{S}_r})$ is a commutative unital quantale.

Given two elements $c = (c_\rho, c_\phi, c_\theta) \in \bar{\mathcal{S}}_r$ and $c' = (c'_\rho, c'_\phi, c'_\theta) \in \bar{\mathcal{S}}_r$, the multiplication $c \cdot_{\mathcal{S}_r} c'$ is defined by

$$c \cdot_{\mathcal{S}_r} c' = (c_\rho \times' c'_\rho, c_\phi \curlywedge c'_\phi, c_\theta \curlywedge c'_\theta), \tag{13}$$

where the binary operation "$\curlywedge$" is given by

$$x \curlywedge y = \begin{cases} x, & \text{if } x \preceq y, \\ y, & \text{otherwise.} \end{cases} \tag{14}$$

The operation "$\times'$" coincides with the usual multiplication on $\mathbb{R}_{\geqslant 0}$ and it is extended to the infinity by means of the equations $(+\infty) \times' x = x \times' (+\infty) = +\infty$ for all $x \in \bar{\mathbb{R}}_{\geqslant 0} = \mathbb{R}_{\geqslant 0} \cup \{+\infty\}$.

The identity of "$\cdot_{\mathcal{S}_r}$" is the element $e = (1, \frac{\pi}{2}, \pi) \in \bar{\mathcal{S}}_r$. Furthermore, the left residuum of the multiplication is the binary operation "$/_{\mathcal{S}_r}$" given by the following equation for all $c, c' \in \bar{\mathcal{S}}_r$:

$$c /_{\mathcal{S}_r} c' = (c_\rho /' c'_\rho, c_\phi /^\Phi c'_\phi, c_\theta /^\Theta c'_\theta), \tag{15}$$

where the symbols "$/'$", "$/^\Phi$", and "$/^\Theta$" denote the binary operations defined as follows for appropriate elements x and y and "$\div$" denotes the usual division of real numbers:

$$y /' x = \begin{cases} 0, & \text{if } x = +\infty, \\ 0, & \text{if } x = 0,\ y = 0, \\ +\infty, & \text{if } x = 0,\ y > 0, \\ y \div x, & \text{otherwise,} \end{cases} \tag{16}$$

$$y /^\Phi x = \begin{cases} \frac{\pi}{2}, & \text{if } x \preceq y, \\ y, & \text{otherwise,} \end{cases} \quad \text{and} \quad y /^\Theta x = \begin{cases} \pi, & \text{if } x \preceq y, \\ y, & \text{otherwise.} \end{cases} \tag{17}$$

3.3 Dilation and Erosion on the Spherical CIELab

According to Definition 2, the elementary operations of MM on the spherical CIELab quantale $(\bar{\mathcal{S}}_r, \leqslant_{\mathcal{S}_r}, \cdot_{\mathcal{S}_r})$ are defined as follows:

Definition 3 (Spherical CIELab Dilation and Erosion). *The spherical CIELab dilation of $f \in \bar{\mathcal{S}}_r^X$ by an $s \in \bar{\mathcal{S}}_r^Y$, denoted by $\delta_s^{\mathcal{S}_r}(f)$, is the color image given by*

$$\delta_s^{\mathcal{S}_r}(f)(x) = \bigvee_{y \in Y, x - y \in X} \left(f(x - y) \cdot_{\mathcal{S}_r} s(y) \right), \quad \forall x \in X. \tag{18}$$

Dually, the spherical CIELab erosion of a color image $f \in \bar{\mathcal{S}}_r^X$ by an SE $s \in \bar{\mathcal{S}}_r^Y$, denoted by $\varepsilon_s^{\mathcal{S}_r}(f)$, is the color image computed as follows

$$\varepsilon_s^{\mathcal{S}_r}(f)(x) = \bigwedge_{y \in Y, x + y \in X} \left(f(x + y) /_{\mathcal{S}_r} s(y) \right), \quad \forall x \in X, \tag{19}$$

where "$/_{\mathcal{S}_r}$" denotes the left residuum given by (15).

Fig. 1. Original color image f of size 768×512

Remark 2. Since the dilation is based on the supremum, the ordering "$\leqslant_{S_r}$" given by (11) yields an operator that enlarges the objects having a color close to the reference. Dually, the erosion shrinks the objects which have a color similar to the reference because of the infimum operation. Furthermore, the lattice structure given by (11) is very similar to the one proposed by Angulo in the CIElab space with the Euclidean distance [2]. Specifically, the two approaches differ in the lexicographical cascade after the first comparison.

3.4 Example with a Natural Color Image

Consider the color image $f \in \mathcal{C}_{RGB}^X$ depicted in Fig. 3.4, whose domain $X \subseteq \mathbb{Z}^2$ corresponds to a grid of size 768×512. This natural color image, provided by Bruce Lindbloom[1], have been used previously by Gronde and Roerdink [13] as well as Burgeth and Kleefeld [7].

First, Fig. 2 shows the spherical CIElab elementary operations $\delta_p^{\mathcal{S}^{\text{white}}}(f)$, $\delta_s^{\mathcal{S}^{\text{white}}}(f)$, $\varepsilon_p^{\mathcal{S}^{\text{white}}}(f)$, and $\varepsilon_s^{\mathcal{S}^{\text{white}}}(f)$, where the reference corresponds to white. Similarly, Fig. 3 presents $\delta_p^{\mathcal{S}^{\text{yellow}}}(f)$, $\delta_s^{\mathcal{S}^{\text{yellow}}}(f)$, $\varepsilon_p^{\mathcal{S}^{\text{yellow}}}(f)$, and $\varepsilon_s^{\mathcal{S}^{\text{yellow}}}(f)$ obtained by considering yellow as the reference color. Here, we used the SEs $p, s \in \bar{\mathcal{S}}_r^Y$ defined as follows

$$ p(y) = e = \left(1, \frac{\pi}{2}, \pi\right) \quad \text{and} \quad s(y) = \left(1 + \frac{\sqrt{2}}{10}\|y\|, \frac{\pi}{2}, \pi\right), \quad \forall y \in Y, \quad (20) $$

where $Y \subseteq \mathbb{Z}^2$ is the 21×21 square grid centered at the origin $(0, 0)$. Note that the SE p is flat and symmetric. Also, for any $x \in X$, we have

$$ \delta_p^{\mathcal{S}_r}(f)(x) = \bigvee_{y \in Y, x-y \in X} f(x - y) \quad \text{and} \quad \varepsilon_p^{\mathcal{S}_r}(f)(x) = \bigwedge_{y \in Y, x+y \in X} f(x + y). \quad (21) $$

---

[1] Available at: http://www.brucelindbloom.com/.

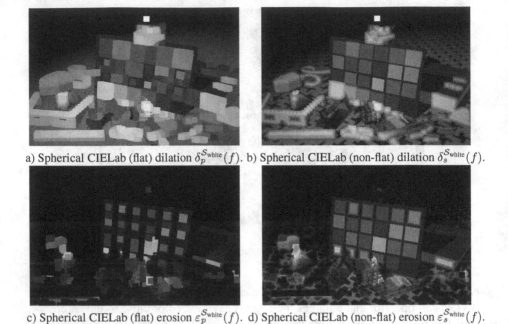

a) Spherical CIELab (flat) dilation $\delta_p^{\mathcal{S}_{\text{white}}}(f)$. b) Spherical CIELab (non-flat) dilation $\delta_s^{\mathcal{S}_{\text{white}}}(f)$.

c) Spherical CIELab (flat) erosion $\varepsilon_p^{\mathcal{S}_{\text{white}}}(f)$. d) Spherical CIELab (non-flat) erosion $\varepsilon_s^{\mathcal{S}_{\text{white}}}(f)$.

Fig. 2. Spherical CIELab elementary operations using white as the reference color

In contrast, the SE s is non-flat (but it is symmetric). For instance, we have $s(0,0) = (1, \pi/2, \pi)$ in the origin and $s(\pm10, \pm10) = (3, \pi/2, \pi)$ in the four corners of Y. For comparison purposes, Fig. 4 presents color images obtained from the Euclidean distance-based approach of Angulo in the CIELab space [2]. Specifically, $\delta_Y^{\mathcal{A}_{\text{white}}}(f)$ denotes the dilation of the image f by an SE Y using white as the reference color. Similarly, $\varepsilon_Y^{\mathcal{A}_{\text{yellow}}}(f)$ denotes the erosion of f by Y using yellow as reference.

The images in the first row of Fig. 2 show that the spherical CIElab dilations $\delta_p^{\mathcal{S}_{\text{white}}}$ and $\delta_s^{\mathcal{S}_{\text{white}}}$, obtained considering white as the reference color, expanded whiter objects of f. Similarly, the first row of Fig. 3 reveals that yellowish objects of f have been expanded by the spherical CIElab dilations $\delta_p^{\mathcal{S}_{\text{yellow}}}$ and $\delta_s^{\mathcal{S}_{\text{yellow}}}$ obtained by using yellow as the reference. Dually, the spherical CIElab erosions $\varepsilon_p^{\mathcal{S}_{\text{white}}}$ and $\varepsilon_s^{\mathcal{S}_{\text{white}}}$ expanded dark objects of the original color image f. Similarly, $\varepsilon_p^{\mathcal{S}_{\text{yellow}}}$ and $\varepsilon_s^{\mathcal{S}_{\text{yellow}}}$ expanded blueish objects. Recall that blue is opposite to yellow in the yellow-blue axis in the CIELab color space. Therefore, in some sense, we can control the objects that we intent to expand or shrink by tuning the reference color. The color reference can be chosen according to some *a priory* knowledge about the relevant or irrelevant information on the image. Alternatively, a machine algorithm may be devised to select appropriate color references.

Note that the dilated image $\delta_Y^{\mathcal{A}_{\text{white}}}(f)$ obtained by the approach of Angulo is very similar to the spherical CIELab dilation $\delta_p^{\mathcal{S}_{\text{white}}}(f)$, where the latter is computed using the flat SE p. Also, $\varepsilon_Y^{\mathcal{A}_{\text{yellow}}}(f)$ resembles $\varepsilon_p^{\mathcal{S}_{\text{yellow}}}(f)$. Indeed, the elementary operations in the distance-based approach of Angulo are similar to (21) but using an ordering scheme which first compares the distance to the reference color followed by the usual

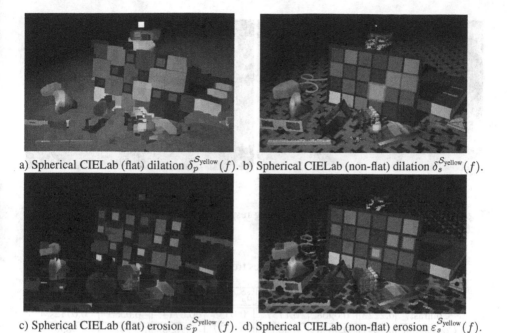

a) Spherical CIELab (flat) dilation $\delta_p^{\mathcal{S}_{\text{yellow}}}(f)$. b) Spherical CIELab (non-flat) dilation $\delta_s^{\mathcal{S}_{\text{yellow}}}(f)$.

c) Spherical CIELab (flat) erosion $\varepsilon_p^{\mathcal{S}_{\text{yellow}}}(f)$. d) Spherical CIELab (non-flat) erosion $\varepsilon_s^{\mathcal{S}_{\text{yellow}}}(f)$.

Fig. 3. Spherical CIELab elementary operations using yellow as the reference color

lexicographical cascade on the color components. Since the maximum and minimum is often determined by the first comparison (the distance to the reference or the radius in the spherical CIElab), the spherical CIELab and the approach of Angulo usually yield similar outcomes. The advantage of the spherical CIELab approach consists in the use of non-flat structuring elements.

Finally, note that some features of the original image f, such as the black patterns on the ground, cannot be found on the dilated images $\delta_p^{\mathcal{S}_{\text{white}}}(f)$ and $\delta_p^{\mathcal{S}_{\text{yellow}}}(f)$ obtained by considering the flat SE p. In contrast, the dilated images $\delta_s^{\mathcal{S}_{\text{white}}}(f)$ and $\delta_s^{\mathcal{S}_{\text{yellow}}}(f)$, obtained by considering the non-flat SE s, preserved many features of f. Quantitatively, Table 1 provides some texture statistics obtained from the normalized gray-level co-occurrence matrix (GLCM) of the luminance (in the NTSC color space) of the probed image and its dilations [1]. Note that $\delta_p^{\mathcal{S}_{\text{white}}}(f)$ and $\delta_Y^{\mathcal{A}_{\text{white}}}(f)$, as well as $\delta_p^{\mathcal{S}_{\text{yellow}}}(f)$ and $\delta_Y^{\mathcal{A}_{\text{yellow}}}(f)$, have the same statistic values. This remark confirms that the flat spherical CIELab approach yielded images very similar to the ones produced by the approach of Angulo. Also, note that the non-flat approach yielded statistic values between those produced by f and the corresponding flat dilation. Hence, the non-flat approach preserved more texture characteristics than the non-flat approach. Analogously, the eroded images $\varepsilon_s^{\mathcal{S}_{\text{white}}}(f)$ and $\varepsilon_s^{\mathcal{S}_{\text{yellow}}}(f)$, obtained by considering the non-flat SE s, preserved features of f which cannot be found in the eroded images $\varepsilon_p^{\mathcal{S}_{\text{white}}}(f)$ and $\varepsilon_p^{\mathcal{S}_{\text{yellow}}}(f)$, determined using the flat SE p. In view of these facts, the non-flat approach may prove advantageous in applications such as image segmentation and texture classification.

a) Distance-based dilation $\delta_Y^{\mathcal{A}^{\text{white}}}(f)$.   b) Distance-based dilation $\delta_Y^{\mathcal{A}^{\text{yellow}}}(f)$.

Fig. 4. Dilation $\delta_Y^{\mathcal{A}^{\text{white}}}$ and erosion $\varepsilon_Y^{\mathcal{A}^{\text{white}}}$ using respectively white and yellow as the reference color

Table 1. Second-order texture statistics from the original and dilated images

Image	f	$\delta_s^{\mathcal{S}^{\text{white}}}(f)$	$\delta_s^{\mathcal{S}^{\text{yellow}}}(f)$	$\delta_p^{\mathcal{S}^{\text{white}}}(f)$	$\delta_p^{\mathcal{S}^{\text{yellow}}}(f)$	$\delta_Y^{\mathcal{A}^{\text{white}}}(f)$	$\delta_Y^{\mathcal{A}^{\text{yellow}}}(f)$
Contrast	0.2223	0.0968	0.1757	0.0555	0.0714	0.0555	0.0712
Correlation	0.9541	0.9838	0.9633	0.9927	0.9858	0.9927	0.9858
Energy	0.1983	0.1600	0.1779	0.1475	0.1827	0.1475	0.1826
Homogeneity	0.9563	0.9640	0.9557	0.9856	0.9864	0.9856	0.9864
		(non-flat approach)		(flat approach)		(approach of Angulo)	

4   Concluding Remarks

Mathematical morphology (MM) is a theory with many applications in image and signal processing and analysis [26,29]. From a theoretical point of view, MM can be very well conducted in a mathematical structure called complete lattices [23,27]. The complete lattice framework, which relies primary on a partial ordering with well defined extrema operations, contributed for the development of many morphological approaches to color images [2,3,4,16,17]. Notwithstanding, some researchers such as Russo and Stell argued that some widely used morphological approaches are defined in a richer mathematical structure called quantale [24,30]. Furthermore, the quantale-based approach is appropriate if one requires dilations invariant under both vertical and horizontal translation.

A quantale based on the CIELab color space with spherical coordinates have been introduced recently by Valle and Sussner [33]. In general terms, the spherical CIELab quantale is obtained by enriching the CIELab space with a distance-based total ordering and a binary operation called multiplication. The total ordering scheme is partially motivated by the works of [2], [17], and [35]. The multiplication is given by the product of the radii and the minimum between either the elevation or the azimuth angles. Although the spherical CIELab quantale have been effectively used for the development of a class of morphological associative memories, the performance of morphological operations defined in such mathematical structure have not been investigated previously. This paper illustrate the effect of dilation and erosion using a natural color image. In particular, we observed that the elementary operations defined using a non-flat structuring element (SE) may retain more details of an image than an approach based on flat SE. We believe that the novel approach has potential application in problems such as detection of the boundaries of objects in color images. We would like to alert, however, that the SE as well as the reference color must be carefully chosen for certain applications.

References

1. Acharya, T., Ray, A.: Image Processing: Principles and Applications. John Wiley and Sons, Hoboken (2005)
2. Angulo, J.: Morphological colour operators in totally ordered lattices based on distances: Application to image filtering, enhancement and analysis. Computer Vision and Image Understanding 107(1-2), 56–73 (July-August 2007) (special issue on color image processing)
3. Aptoula, E., Lefèvre, S.: A Comparative Study on Multivariate Mathematical Morphology. Pattern Recognition 40(11), 2914–2929 (2007)
4. Aptoula, E., Lefèvre, S.: On Lexicographical Ordering in Multivariate Mathematical Morphology. Pattern Recognition Letters 29(2), 109–118 (2008)
5. Aptoula, E., Lefèvre, S.: On the morphological processing of hue. Image and Vision Computing 27(9), 1394–1401 (2009)
6. Birkhoff, G.: Lattice Theory, 3rd edn. American Mathematical Society, Providence (1993)
7. Burgeth, B., Kleefeld, A.: An approach to color-morphology based on Einstein addition and Loewner order. Pattern Recognition Letters 47, 29–39 (2014)
8. Chanussot, J., Lambert, P.: Total ordering based on space filling curves for multivalued morphology. In: Proceedings of the Fourth International Symposium on Mathematical Morphology and its Applications to Image and Signal Processing, ISMM 1998, pp. 51–58. Kluwer Academic Publishers, Norwell (1998)
9. Comer, M.L., Delp, E.J.: Morphological operations for color image processing. Journal of Electronic Imaging 8(3), 279–289 (1999)
10. Căliman, A., Ivanovici, M., Richard, N.: Probabilistic pseudo-morphology for grayscale and color images. Pattern Recognition 47(2), 721–735 (2014)
11. Dougherty, E.R., Lotufo, R.A.: Hands-on Morphological Image Processing. SPIE Press (July 2003)
12. Goutsias, J., Heijmans, H.J.A.M., Sivakumar, K.: Morphological Operators for Image Sequences. Computer Vision and Image Understanding 62, 326–346 (1995)
13. van de Gronde, J., Roerdink, J.: Group-invariant frames for colour morphology. In: Hendriks, C.L.L., Borgefors, G., Strand, R. (eds.) ISMM 2013. LNCS, vol. 7883, pp. 267–278. Springer, Heidelberg (2013)
14. van de Gronde, J., Roerdink, J.: Frames, the Loewner order and eigendecomposition for morphological operators on tensor fields. Pattern Recognition Letters 47, 40–49 (2014)
15. van de Gronde, J., Roerdink, J.: Group-invariant colour morphology based on frames. IEEE Transactions on Image Processing 23(3), 1276–1288 (2014)
16. Hanbury, A., Serra, J.: Morphological Operators on the Unit Circle. IEEE Transactions on Image Processing 10, 1842–1850 (2001)
17. Hanbury, A., Serra, J.: Mathematical Morphology in the CIELAB Space. Image Analysis and Stereology 21, 201–206 (2002)
18. Heijmans, H.J.A.M.: Mathematical Morphology: A Modern Approach in Image Processing Based on Algebra and Geometry. SIAM Review 37(1), 1–36 (1995)
19. Maragos, P.: Lattice Image Processing: A Unification of Morphological and Fuzzy Algebraic Systems. Journal of Mathematical Imaging and Vision 22(2-3), 333–353 (2005)

20. Mulvey, C.J.: &. Rend. Circ. Mat. Palermo 12, 99–104 (1986)
21. Nachtegael, M., Kerre, E.E.: Connections between binary, gray-scale and fuzzy mathematical morphologies. Fuzzy Sets and Systems 124(1), 73–85 (2001)
22. Peters II, R.A.: Mathematical Morphology for Angle-valued images. In: Dougherty, E.R., Astola, J.T. (eds.) Proceedings of the SPIE. Nonlinear Image Processing III, vol. 3026, pp. 84–94 (February 1997)
23. Ronse, C.: Why Mathematical Morphology Needs Complete Lattices. Signal Processing 21(2), 129–154 (1990)
24. Russo, C.: Quantale Modules and their Operators, with Applications. Journal of Logic and Computation 20(4), 917–946 (2010)
25. Sartor, L.J., Weeks, A.R.: Morphological operations on color images. Journal of Electronic Imaging 10(2), 548–559 (2001)
26. Serra, J.: Image Analysis and Mathematical Morphology. Academic Press, London (1982)
27. Serra, J.: Image Analysis and Mathematical Morphology. Theoretical Advances, vol. 2. Academic Press, New York (1988)
28. Serra, J.: The "false colour" problem. In: Wilkinson, M.H.F., Roerdink, J.B.T.M. (eds.) ISMM 2009. LNCS, vol. 5720, pp. 13–23. Springer, Heidelberg (2009)
29. Soille, P.: Morphological Image Analysis. Springer, Berlin (1999)
30. Stell, J.G.: Why mathematical morphology needs quantales. In: Wilkinson, M., Roerdink, J. (eds.) Abstract book of the 9th International Symposium on Mathematical Morphology (ISMM 2009), pp. 13–16. University of Groningen, The Netherlands (August 2009)
31. Sussner, P., Valle, M.E.: Classification of Fuzzy Mathematical Morphologies Based on Concepts of Inclusion Measure and Duality. Journal of Mathematical Imaging and Vision 32(2), 139–159 (2008)
32. Talbot, H., Evans, C., Jones, R.: Complete ordering and multivariate mathematical morphology. In: Proceedings of the Fourth International Symposium on Mathematical Morphology and its Applications to Image and Signal Processing, ISMM 1998, pp. 27–34. Kluwer Academic Publishers, Norwell (1998)
33. Valle, M.E., Sussner, P.: Quantale-based autoassociative memories with an application to the storage of color images. Pattern Recognition Letters 34(14), 1589–1601 (2013)
34. Witte, V., Schulte, S., Nachtegael, M., Weken, D., Kerre, E.: Vector Morphological Operators for Colour Images. In: Kamel, M.S., Campilho, A.C. (eds.) ICIAR 2005. LNCS, vol. 3656, pp. 667–675. Springer, Heidelberg (2005)
35. Witte, V., Schulte, S., Nachtegael, M., Mélange, T., Kerre, E.: A Lattice-Based Approach to Mathematical Morphology for Greyscale and Colour Images. In: Kaburlasos, V., Ritter, G. (eds.) Computational Intelligence Based on Lattice Theory. SCI, vol. 67, pp. 129–148. Springer, Heidelberg (2007)

Spectral Ordering Assessment
Using Spectral Median Filters

Hilda Deborah[1,2]([✉]), Noël Richard[1], and Jon Yngve Hardeberg[2]

[1] Laboratory XLIM-SIC UMR CNRS 7252, University of Poitiers, Poitiers, France
hildad@hig.no, richard@sic.univ-poitiers.fr
[2] The Norwegian Colour & Visual Computing Laboratory,
Gjøvik University College, Gjøvik, Norway
jon.hardeberg@hig.no

Abstract. Distance-based mathematical morphology offers a promising opportunity to develop a metrological spectral image processing framework. Within this objective, a suitable spectral ordering relation is required and it must be validated by metrological means, e.g. accuracy, bias, uncertainty, etc. In this work we address the questions of suitable ordering relation and its uncertainty for the specific case of hyperspectral images. Median filter is shown to be a suitable tool for the assessment of spectral ordering uncertainty. Several spectral ordering relations are provided and the performances of spectral median filters based on the aforementioned ordering relations are compared.

Keywords: Spectral ordering · Median filters · Mathematical morphology

1 Introduction

Taking the benefits of its discriminating power that goes beyond color or grayscale imaging, more and more application fields explore the possibilities offered by hyperspectral imaging [7,10,16]. Unfortunately, the advances in image processing does not quite follow the imaging technology. Regardless of the vast amount of data contained in a hyperspectral image, most of the processing requires either dimensionality reduction or band selection so as to make processing with the existing operations possible [17,19]. Considering such amount of data, using data reduction approaches is indeed understandable. However, such approaches negate the initial purpose of using hyperspectral images, i.e. to improve accuracy by acquiring and processing more information. With that reason, the focus of this work is to process hyperspectral data in vector and full-band approach.

Hyperspectral data are not basic mathematical objects, they are physical measures of energy or percentage of energy across the wavelengths. Due to the mathematical and physical nature of hyperspectral images, there is no valid addition, multiplication by scalar value, and subtraction operations available for such data. Fortunately, as it is possible to define and validate distance measure between hyperspectral data [8], it is also possible to define ordering relation based on distance function [12].

© Springer International Publishing Switzerland 2015
J.A. Benediktsson et al. (Eds.): ISMM 2015, LNCS 9082, pp. 387–397, 2015.
DOI: 10.1007/978-3-319-18720-4_33

The article is organized as follows. First, we consider in Section 2 the questions regarding the ordering of spectral data, and then we propose some basic ordering relations linked to the particular structure of a spectrum. Then we develop the questions around the quality assessment of ordering relation for spectral data in Section 3. The result and discussion of using spectral median filters to assess the quality of spectral ordering relation is provided in Section 4.

2   Hyperspectral Ordering

Ordering relation is the core of many important image processing tools, e.g. mathematical morphology and rank order filters. While natural order exists in scalar domain [5,11], ordering in vector or multivariate case is ambiguous and many different approaches of multivariate ordering exist [2,4,18]. Furthermore, when constructing an ordering, the nature of the data at hand needs to be taken into account as opposed to only considering them as mathematical objects. Hyperspectral data is defined a sequence of energy measured on spectrally contiguous channels where each channel is defined with a certain bandwidth. This basic definition of hyperspectral data then leads to the digitization of a continuous energy spectrum across the wavelengths and, as a consequence, a spectrum is neither a vector in Euclidean space nor a distribution [8]. In this work, the studied hyperspectral data are represented in terms of spectral reflectance and were acquired in the visible and near-infrared spectral range. In reflectance space, each measure is a ratio of the acquired energy divided by the illuminant energy of the corresponding spectral band. Nevertheless, the purpose can be directly extended to other spectral ranges or radiance space.

By respecting the physical definition of a spectrum and considering the ordering constructions proposed by Barnett [4] and further extended to multivariate images by Aptoula [2], several spectral ordering relations are provided in the followings. Given an arbitrary spectrum as a function of wavelength or spectral band λ_p, $S_i = \{S_i(\lambda_p),\ p \subset [0, n-1]\}$, marginal or component-wise ordering approach is as shown in Eq. 1. The main idea is to order a spectrum based on the values of each of its component and to combine logically the result. Due to such construction, this ordering is not able to order all spectra, e.g. when $S_1(\lambda_{p-1}) < S_2(\lambda_{p-1})$ and $S_1(\lambda_p) > S_2(\lambda_p)$. Moreover, this partial ordering is not able to preserve correlations that exist between neighboring spectral band.

$$S_1 \preceq S_2 \Leftrightarrow \begin{cases} S_1(\lambda_1) \leq S_2(\lambda_1) \text{ and} \\ S_1(\lambda_2) \leq S_2(\lambda_2) \text{ and} \\ \quad \cdots \qquad\qquad \text{and} \\ S_1(\lambda_n) \leq S_2(\lambda_n) \end{cases} \tag{1}$$

Lexicographic ordering is an example of conditional ordering approach; it is a total order. In this approach the ordering is based on conditions or prioritizations applied on each spectral channel, see Eq. 2 where priorities are given to shorter wavelengths and C is an arbitrary constant number. Prioritization over a hyperspectral image with hundreds of channels is challenging since in most of the cases the discriminating channels are unknown, unlike in a color image of typically three channels. Moreover, this construction states that the order of a spectrum is determined by the order of few of its channels, while the rest of the channels will be considered to be almost negligible.

$$S_1 \preceq S_2 \Leftrightarrow \sum_{p=0}^{n-1} C^{n-p} \cdot S_1(\lambda_p) \leq \sum_{p=0}^{n-1} C^{n-p} \cdot S_2(\lambda_p) \tag{2}$$

Considering that a spectrum is an energy, the ordering relation in Eq. 3 is proposed. The underlying hypothesis to this point of view states that a spectrum is considered as 'larger' than another if its total amount of energy is bigger. With this ordering relation, a white spectrum is always 'larger' than a black spectrum. This ordering is in a direct equivalence to morphology based on intensity images.

$$S_1 \preceq S_2 \Leftrightarrow \int_{\lambda} S_1(\lambda) \, d\lambda \leq \int_{\lambda} S_2(\lambda) \, d\lambda \tag{3}$$

The previous ordering relation is limited to energy point of view and implicitly defines the convergence of spectral extrema toward the black and white spectra. In the context of color image processing, the notion of distance function has been used to address several image processing tasks [1,14,20]. Using distance function, two spectra can be ordered by their distance relative to a reference spectrum S_{ref} (Eq. 4). If a white spectrum is used as the reference, the behavior obtained from Eq. 4 will be similar to that of Eq. 3 where a spectrum is considered as an energy, i.e. larger spectrum is the one having more energy.

$$S_1 \preceq S_2 \Leftrightarrow d(S_1, S_{ref}) \geq d(S_2, S_{ref}) \tag{4}$$

The two previous ordering relations do not satisfy total ordering property as well as not allowing to fully control the extrema in a morphological process according to certain application goals. Ledoux et al. [13] proposed the Convergent Colour Mathematical Morphology (CCMM) by combining two reference coordinates to control the convergence of the color set toward selected infimum and supremum. This total ordering relation was shown to be suitable for color vector processing such as texture discrimination or pattern detection in a spatio-chromatic template. A preliminary work on hyperspectral images demonstrated that the purpose was extendable to spectral domain [12]. CCMM defines two convergence coordinates $O^{+\infty}$ and $O^{-\infty}$ in order to compute the ordering relation for the maximum and minimum extraction in a set of spectra (Eq. 5).

$$S_1 \succeq S_2 \Leftrightarrow d(S_1, O^{+\infty}) \leq d(S_2, O^{+\infty})$$
$$S_1 \preceq S_2 \Leftrightarrow d(S_1, O^{-\infty}) \leq d(S_2, O^{-\infty}) \tag{5}$$

The ordering relation of CCMM (Eq. 5) does not satisfy the idempotency property of opening and closing transforms and a construction in which the two distances are expressed in a single relationship is required [9]. Thus, we propose an ordering that uses the ratio between the distance to the reference coordinate for the maximum $O^{+\infty}$ and for the minimum $O^{-\infty}$ (Eq. 6). With this construction, the maximum in set of spectra will be the one closest to $O^{+\infty}$. If several spectra are at the same distance to $O^{+\infty}$, the maximum is the one closest to $O^{-\infty}$ coordinate; this construction ensures idempotency property.

$$S_1 \preceq S_2 \Leftrightarrow \frac{d(S_1, O^{-\infty})}{d(S_1, O^{+\infty})} \leq \frac{d(S_2, O^{-\infty})}{d(S_2, O^{+\infty})} \tag{6}$$

The ordering relation in Eq. 6 does not guarantee a total order as different color or spectral coordinates can produce the same ratio. Thus, additional constraints are required in order to produce a total order, i.e. using a conditional construction as in Eq. 7. In this case, the first part of the relation is identical to that of Eq. 6 and the additional constraint is proportional to a spectral angle.

$$S_1 \preceq S_2 \Leftrightarrow \begin{cases} R_0(S_1) \leq R_0(S_2) & \text{or} \\ R_0(S_1) = R_0(S_2) & \text{and} \quad R_1(S_1) \leq R_1(S_2) \end{cases}$$

$$\text{with } R_0(S) = \frac{d(S, O^{-\infty})}{d(S, O^{+\infty})}, \; R_1(S) = 2\frac{d(S, O^{-\infty})}{d(O^{-\infty}, O^{+\infty})} \tag{7}$$

For n-dimensional data such as hyperspectral images, experiments show that the second relation of Eq. 7 is almost never used. Spectral complexity, acquisition noise, and other factors affecting a spectral data allow to work only with the ordering relation shown in Eq. 6. Such behaviour can be expressed through the mathematical conjecture in Eq. 8, i.e. the probability of obtaining more than one spectral coordinate as the maximum (or minimum) at level R_i of a conditional ordering relation, k being the number of spectral channels.

$$prob\left(\#\left\{\bigcap_{i=1}^{k-1} Sd_i\right\} > 1\right) > prob\left(\#\left\{\bigcap_{i=1}^{k} Sd_i\right\} > 1\right) \tag{8}$$

3   Spectral Ordering Accuracy Assessment

3.1   Uncertainty in Color Ordering

Given a sequence of ordered colors, where each two colors are slightly different by a certain color gradient, any arbitrary ordering relation must be able to retain the initial order of this sequence of colors [15], see Eq. 9. This hypothesis enables a way to assess the uncertainty of color ordering relations. A randomly generated sequence of colors can be obtained by, first, randomly selecting extremal colors and, second, iteratively adding a certain amount of color gradient to the extremal colors so as to obtain the in-between colors. This color ordering uncertainty test is however too restrictive for spectral images. As illustrated by metamerism phenomenon, reducing spectral data into color data might yield identical colors despite the different spectral signatures. And eventually, reducing spectral ordering uncertainty test into color domain loses the meaning of spectral image processing, i.e. to improve accuracy.

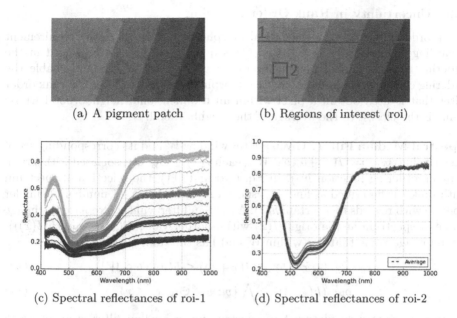

(a) A pigment patch   (b) Regions of interest (roi)

(c) Spectral reflectances of roi-1   (d) Spectral reflectances of roi-2

Fig. 1. A pigment patch was acquired using a hyperspectral scanner and its color image was constructed using color matching function. The obtained spectral reflectance signals illustrate how global extrema extraction in roi-1 is a less complex problem than obtaining spectral ordering in a small region such as in roi-2 where the spectra are similar to the average spectrum of this region.

$$\exists! C_m \mid C_m = \bigwedge_{i:0...p} C_i = \bigwedge_{i:0...p} R(C_i)$$
$$\text{and } R(C_{i-1}) > R(C_i), \forall i \leq m \tag{9}$$
$$R(C_{i+1}) > R(C_i), \forall i \geq m$$

In order to assess spectral ordering uncertainty, rather a sequence of colors, a sequence of spectra should be generated and the physical definition of a spectrum must be taken into account. Furthermore, a set of spectral images where its ground truth is available must be obtained so as to have a set of data that is close to reality. Unfortunately, there is no such dataset. And even if there is, the problems of ordering will not be encountered when the global extrema, i.e. maximum and minimum, are to be extracted from the dataset. Such is, in fact, less complex than when working with a small-sized structuring element on a region where each of the spectra are similar to the average of this region, where consequently accuracy should be at its maximum. To summarize, spectral ordering is less likely to face a problem at the level of generated uncertainty. Challenges will be encountered at texture feature level where uncertainties are integrated in the data in multiscale construction, see Fig. 1. Finally, the uncertainty of a spectral ordering relation must be assessed within the distribution of input spectral data.

3.2 Uncertainty in Rank Order Filtering

Rank order filters and mathematical morphology share the same requirement regarding a valid ordering relation. Consequently, several criteria used in the ordering construction of nonlinear filters can be used to partially enable the ordering construction in mathematical morphology. Median filter is a rank order filter that is used obtain a pixel within an n-dimensional neighborhood whose value is the closest to the average of the neighborhood.

Spectral Median Filter. Given a filter window W and its corresponding set of image values $\mathfrak{S}_W = \{I(x_0+b), b \in W\}$, each $I(x) \in \mathfrak{S}_W$ is associated with a rank r by an ordering function $h(\cdot)$. In Eq. (10) and (11), the selection of spectrum with rank r is described as finding a spectrum which has $r-1$ number of smaller spectra with regards to a certain $h(v)$. A median filter will then replace the image value or spectrum at the origin $I(x)$ with another value $F(x) = \rho_{W,r}(h(I(x)))$, i.e. an image value that lies within W and has the rank $r = \frac{n_W - 1}{2}$.

$$\mathfrak{c}_v = \# \left\{ I(x + b) : h(I(x + b)) \leq h(v), \forall b \in W \right\} \tag{10}$$

$$\rho_{W,r}(h(I(x))) = \bigwedge \left\{ v : v \in \mathfrak{S}_W, \mathfrak{c}_v \leq r \right\} \tag{11}$$

As the primary requirement of constructing a median filter is an ordering relation, many of the spectral ordering relations provided in Section 2 can be used to construct different spectral median filters. As mentioned before, a lexicographic ordering overestimates the impact of a certain band. Therefore a spectral median filter with a lexicographic ordering O_{lex} which gives priority to shorter wavelengths (Eq. 2) will be compared to an ordering which is based on only the first component of the spectral channels $O_{\lambda1}$, i.e. the wavelength at 414.62 nm. Other ordering relations to be employed to construct spectral median filters are the ordering based on the sum of energy O_{esum} (Eq. 3), the distance-based ordering with white reflectance spectrum as the reference O_{dw} (Eq. 4), the ordering based on ratio O_{ratio} (Eq. 6), and the conditional ordering based on ratio and angle information O_{cra} (Eq. 7).

Spectral Median Filters Quality Assessment. Deborah et al. proposed several quality assessment protocols for spectral image processing algorithms [6], where experimental results were provided for the assessment of spectral distance and spectral median filters. In this work, the previously proposed spectral median filters quality assessment protocol is modified so as to be able to evaluate the quality of ordering relations embedded in the filters.

Several pigment patches as shown in Fig. 2 were acquired using a hyperspectral scanner and will be used as the input data to Vector Median Filters and the different spectral median filters that are enabled by the different spectral ordering relations. Vector Median Filter or VMF is a multivariate median filter that is based on aggregate distance. And since VMF was shown to be the most accurate median filter for several classes of random distribution [3], the result of VMF will be used as the benchmark of this quality assessment. All the results of other spectral median filters will be evaluated against this benchmark.

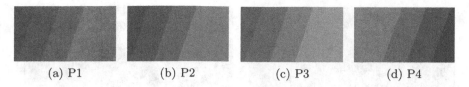

| (a) P1 | (b) P2 | (c) P3 | (d) P4 |

Fig. 2. Four pigment patches acquired by a hyperspectral scanner

4   Result and Discussion

Table 1 shows image differences between the result of VMF and the 6 spectral median filters. The comparison criteria is pixelwise root mean square error. The results obtained by lexicographic ordering O_{lex} are similar to those obtained by $O_{\lambda 1}$. Such results are as expected, due to the fact that lexicographic ordering is a conditional ordering with the highest priority given to $\lambda 1$. These results explain that the other channels are rarely used in the ordering process and therefore is also explained by the mathematical conjecture in Eq. 8. Hence, there is no significant difference between the results of O_{lex} and $O_{\lambda 1}$. In the case of the pigment patches shown in Fig. 2, better results could indeed be obtained if the lexicographic approach gives more priority to the dominant wavelength of the respective color pigments. Nevertheless, these results for the color pigments illustrate the limitation of lexicographic approach, i.e. when the discriminating spectral channel is unknown and therefore prioritization is given to the wrong channels, the processing result becomes unreliable.

The better performances of O_{esum}, O_{dw}, O_{ratio}, and O_{cra} compared to those of O_{lex} and $O_{\lambda 1}$ are due to the global approach of the four ordering relations. O_{dw}, O_{ratio}, and O_{cra} consider the whole dimension of spectra and especially the inner correlation that exists between neighboring spectral channels due to the use of distance function. For O_{esum}, even though it still considers the whole dimension of spectra by reducing a spectrum into a total amount of energy,

Table 1. Average and standard deviation of pixel differences between the results of VMF and other spectral median filters, computed using root mean square error. The more similar performance to that of VMF is considered to be the better ordering relation, hence the smaller values. Significant differences are shown by O_{lex} and $O_{\lambda 1}$, while the other ordering approaches perform similarly.

Ordering	P1		P2		P3		P4	
relation	Mean	Std	Mean	Std	Mean	Std	Mean	Std
O_{lex}	8.076	7.009	7.408	10.818	6.963	8.448	7.072	5.033
$O_{\lambda 1}$	7.982	6.890	7.380	10.831	6.929	8.397	7.057	5.049
O_{esum}	4.574	2.980	4.240	2.872	4.317	2.532	4.262	2.864
O_{dw}	4.539	3.131	4.303	2.886	4.305	2.543	4.187	2.832
O_{ratio}	4.596	3.022	4.306	3.088	4.332	2.537	4.297	2.881
O_{cra}	4.596	3.022	4.306	3.088	4.332	2.537	4.297	2.881

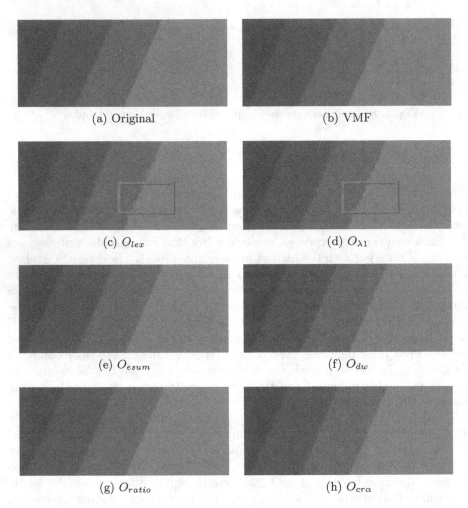

(a) Original

(b) VMF

(c) O_{lex}

(d) $O_{\lambda 1}$

(e) O_{esum}

(f) O_{dw}

(g) O_{ratio}

(h) $O_{cr\alpha}$

Fig. 3. Results of applying VMF and several spectral median filters with different ordering relations on pigment P2. It can be observed in the regions inside the blue squares that the spectral median filters that are based on ordering relations O_{lex} and $O_{\lambda 1}$ are not able to maintain the edges that exist between two color shades.

it is not considering the inner correlation. In the table the results of O_{ratio} and $O_{cr\alpha}$ are identical. As previously explained for O_{lex} and $O_{\lambda 1}$, $O_{cr\alpha}$ is also a conditional ordering and its second level of condition is almost never used. Again, the mathematical conjecture in Eq. 8 explains this behavior.

While Table 1 only provides the performance of the different spectral median filters globally, Fig. 3 and 4 provides an idea of how the different filters perform locally in the spatial context. One of the desired capabilities of a median filter is to preserve edges from shapes or local gradient and to remove additive noise. In the two figures, original images of pigment P2 and P3 are provided as well

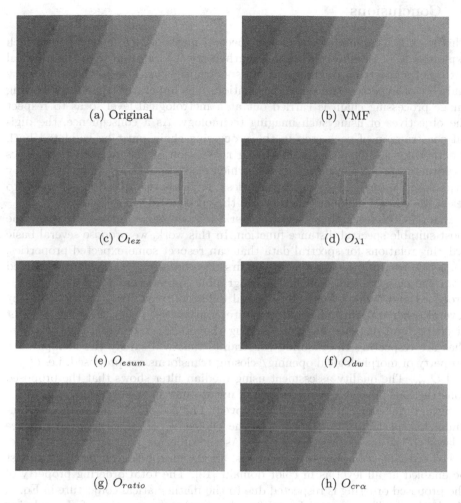

(a) Original   (b) VMF

(c) O_{lex} (d) $O_{\lambda 1}$

(e) O_{esum} (f) O_{dw}

(g) O_{ratio} (h) $O_{cr\alpha}$

Fig. 4. Results of applying VMF and several spectral median filters with different ordering relations on pigment P3. Similar to the case of P2 in Fig. 3, the spectral median filters that are based on ordering relations O_{lex} and $O_{\lambda 1}$ are also not able to maintain the edges that exist between two color shades, see regions inside the blue squares.

as the result of the benchmark, i.e. VMF, and the other spectral median filters. In this figure we can observe that O_{lex} and $O_{\lambda 1}$ are not able to retain the edges that exist between two color shades. Finally, if we are to choose between the 6 spectral ordering relations provided in this work, the choice are left with O_{ratio} and $O_{cr\alpha}$, since O_{esum} and O_{dw} do not satisfy total ordering property.

5   Conclusions

This work is a preliminary study on spectral mathematical morphology which explores the possibilities of relating mathematical formalism with the physical nature of a spectrum originating from spectral sensors. Due to the complexity and cost of hyperspectral image acquisition, it is imperative that the following image processing chain is carried out at a metrological level so as to respect the objectives of using such imaging technology. As a consequence, the digital construction of each step in the processing chain must be validated both theoretically and experimentally. Using the notion of distance function allows constructing spectral ordering scheme which respects the constraints induced by hyperspectral image domain. Nevertheless, it yields two questions, i.e. how to select the correct distance function and the correct ordering scheme.

In a previous research [8], several criteria were developed in order to select the most suitable spectral distance function. In this work, we propose several basic ordering relations for spectral data that can respect some expected properties, i.e. total order and idempotency. The first two ordering relations are extended from the classical marginal and lexicographic approaches. Then a relation is proposed and defined from the physical definition of a spectrum, followed by several other ordering relations that are constructed using the total amount of energy and distance function. Among the proposed ordering relations, two that are respecting the required constraints that will satisfy the idempotency property of morphological opening/ closing transforms are proposed, i.e. O_{ratio} and $O_{cr\alpha}$. The quality assessment using median filter shows that the proposed constructions present a reduced level of uncertainty in the ordering.

Several questions must be still improved. The notion of maximum and minimum for spectral data is related to the choice of convergence spectra, hence related to the image processing goals. As the ordering relation is based on reduced ordering using distance functions, all the morphological properties must be enabled at all level as in color domain [13]. The total ordering property of the proposed ordering is respected due to the mathematical conjecture in Eq. 8. While the duality property is satisfied due to the construction of the ordering relation. The next challenge will then be to prove the idempotency for opening and closing operation so as to reach the advanced morphological filters and processing.

References

1. Angulo, J.: Morphological colour operators in totally ordered lattices based on distances: Application to image filtering, enhancement and analysis. Computer Vision and Image Understanding: Special issue on color image processing 107(1–2), 56–73 (2007)
2. Aptoula, E., Lefèvre, S.: A comparative study on multivariate mathematical morphology. Pattern Recognition 40(11), 2914–2929 (2007)
3. Astola, J., Haavisto, P., Neuvo, Y.: Vector median filters. Proceedings of the IEEE 78(4), 678–689 (1990)

4. Barnett, V.: The ordering of multivariate data. Journal of the Royal Statistical Society. Series A (General) 139(3), 318–355 (1976)
5. Bell, C.B., Haller, H.S.: Bivariate symmetry tests: Parametric and nonparametric. The Annals of Mathematical Statistics 40(1), 259–269 (1969)
6. Deborah, H., Richard, N., Hardeberg, J.Y.: On the quality evaluation of spectral image processing algorithms. In: 10th International Conference on Signal Image Technology & Internet Systems (SITIS) (November 2014)
7. Deborah, H., George, S., Hardeberg, J.Y.: Pigment mapping of the scream (1893) based on hyperspectral imaging. In: Elmoataz, A., Lezoray, O., Nouboud, F., Mammass, D. (eds.) ICISP 2014. LNCS, vol. 8509, pp. 247–256. Springer, Heidelberg (2014)
8. Deborah, H., Richard, N., Hardeberg, J.Y.: A comprehensive evaluation on spectral distance functions and metrics for hyperspectral image processing. IEEE Journal of Selected Topics in Applied Earth Observations and Remote Sensing 8 (to appear, 2015)
9. Goutali, R., Richard, N., Ledoux, A., Ellouze, N.: Problématique de l'idempotence pour les images couleurs et multi-valuées. TAIMA Proceedings – Atelier Traitement et Analyse des Images, Méthodes et Applications (TAIMA) (May 2013)
10. Karaca, A.C., Erturk, A., Gullu, M.K., Elmas, M., Erturk, S.: Plastic waste sorting using infrared hyperspectral imaging system. In: 2013 21st Signal Processing and Communications Applications Conference (SIU), pp. 1–4 (April 2013)
11. Kendall, M.G.: Discrimination and classification. In: Krishnaiah, P.R. (ed.) Multivariate Analysis, pp. 165–185. Academic Press, New York (1966)
12. Ledoux, A., Richard, N., Capelle-Laizé, A.S., Deborah, H., Fernandez-Maloigne, C.: Toward a full-band texture features for spectral images. In: IEEE International Conference on Image Processing (ICIP) (2014)
13. Ledoux, A., Richard, N., Capelle-Laizé, A.: The fractal estimator: A validation criterion for the colour mathematical morphology. In: 6th European Conference on Colour in Graphics, Imaging, and Vision, CGIV 2012, pp. 206–210. IS&T - The Society for Imaging Science and Technology (May 2012)
14. Ledoux, A., Richard, N., Capelle-Laizé, A.-S., Fernandez-Maloigne, C.: Perceptual color hit-or-miss transform: application to dermatological image processing. Signal, Image and Video Processing, 1–11 (2013)
15. Ledoux, A., Richard, N., Capelle-Laizé, A.S., et al.: Limitations et comparaisons d'ordonnancement utilisant des distances couleur. TAIMA Proceedings – Atelier Traitement et Analyse des Images, Méthodes et Applications (TAIMA) (2011)
16. Mehrubeoglu, M., Teng, M.Y., Savage, M., Rafalski, A., Zimba, P.: Hyperspectral imaging and analysis of mixed algae species in liquid media. In: IEEE International Conference on Imaging Systems and Techniques (IST), pp. 421–424 (2012)
17. Qian, S.E., Chen, G.: A new nonlinear dimensionality reduction method with application to hyperspectral image analysis. In: IEEE International Geoscience and Remote Sensing Symposium, IGARSS 2007, pp. 270–273 (July 2007)
18. Velasco-Forero, S., Angulo, J.: Supervised ordering in $\mathbb{R}^p$: Application to morphological processing of hyperspectral images. IEEE Transactions on Image Processing 20(11), 3301–3308 (2011)
19. Yang, H., Du, Q., Chen, G.: Unsupervised hyperspectral band selection using graphics processing units. IEEE Journal of Selected Topics in Applied Earth Observations and Remote Sensing 4(3), 660–668 (2011)
20. Yeh, C.W., Pycock, D.: Similarity colour morphology. In: 2013 5th Computer Science and Electronic Engineering Conference (CEEC), pp. 71–76 (September 2013)

Non-adaptive and Amoeba Quantile Filters
for Colour Images

Martin Welk[1(⊠)], Andreas Kleefeld[2], and Michael Breuß[2]

[1] Department of Biomedical Computer Science and Technology, University of Health Sciences,
Medical Informatics and Technology (UMIT),
Eduard-Wallnöfer-Zentrum 1, 6060 Hall/Tyrol, Austria
martin.welk@umit.at
[2] Faculty of Mathematics, Natural Sciences and Computer Science, Brandenburg Technical
University Cottbus-Senftenberg,
03046 Cottbus, Germany
{kleefeld,breuss}@tu-cottbus.de

Abstract. Quantile filters, or rank-order filters, are local image filters which assign quantiles of intensities of the input image within neighbourhoods as output
image values. Combining a multivariate quantile definition developed in matrix-
valued morphology with a recently introduced mapping between the RGB colour
space and the space of symmetric 2×2 matrices, we state a class of colour image
quantile filters, along with a class of morphological gradient filters derived from
these. Using amoeba structuring elements, we devise image-adaptive versions of
both filter classes. Experiments demonstrate the favourable properties of the filters.

Keywords: Quantile · Rank-order filter · Color image · Matrix field · Amoebas

1 Introduction

The core of mathematical morphology is formed by the study of grey-value image filters
that are equivariant under automorphisms of the image plane and under monotonically
increasing transformations of the intensities, see e.g. [15]. Equivariance means that the
filtering step and the respective transform of the data commute: Transforming the filter
result by one of the mentioned transforms yields the same result as if the same transform
had been applied to all input values, and the filter applied to the so transformed data.

Equivariance under monotonically increasing grey-value maps is often called *morphological invariance*. This axiomatic definition of morphological filtering, following
[15], includes the most fundamental morphological operations, dilation and erosion,
and numerous filters composed of these, but also further filters like the median filter.

Median filtering has been established since Tukey's work [19] as a simple and robust
denoising filter for (univariate) signals and images with favourable structure-preserving
properties. Since the median of a set of data is equivariant with respect to arbitrary
monotonically increasing intensity transformations, the median filter is morpholog-
ically invariant. The same equivariance with regard to monotonous transformations
holds for arbitrary α-quantiles, giving rise to α-quantile filters (also known as rank-
order filters) as a class of morphological filters that nicely interpolate between erosion
($\alpha = 0$), median filter ($\alpha = 1/2$) and dilation ($\alpha = 1$).

© Springer International Publishing Switzerland 2015
J.A. Benediktsson et al. (Eds.): ISMM 2015, LNCS 9082, pp. 398–409, 2015.
DOI: 10.1007/978-3-319-18720-4_34

Adaptive morphology and amoebas. Like other local image filters, median filtering can be understood as the combination of two steps: first, a sliding-window *selection* step, and second, the *aggregation* of the so selected input values. For median filtering, aggregation is done by taking the median; other local filters use different aggregation procedures, such as maximum for dilation, etc. Changing the selection rule, away from a fixed shape sliding window towards spatially adaptive neighbourhoods, provides a means to increase the sensitivity of these filters to important image structures; such approaches are summarised as *adaptive morphology*. One class of such adaptive neighbourhoods are *morphological amoebas* as introduced by Lerallut et al. [12,13]. In their construction, one combines spatial distance in the image domain with the intensity contrast into an image-adaptive *amoeba metric*. Structuring elements called *amoebas* are then defined as neighbourhoods of prescribed radius in this amoeba metric. By the construction of the amoeba metric, these neighbourhoods adapt sensitively to image structures.

On the theoretical side, amoeba filters for scalar-valued images have been investigated further in [22,23], especially by relating space-continuous versions of them to image filters based on partial differential equations (PDEs). Put very short, it is proven there that amoeba median filtering is an approximation of the self-snakes PDE [17] where the specific choice of the amoeba metric translates into the choice of the edge-stopping function in the self-snakes PDE. Amoeba dilation and erosion filters as well as α-quantile filters with $\alpha \neq 1/2$ are shown to approximate Hamilton–Jacobi PDEs for front propagation with different image-dependent speed functions. These results generalise known facts about non-adaptive filters, namely that median filtering approximates (mean) curvature motion [8], and dilation and erosion are related to Hamilton–Jacobi equations with constant speed functions.

An interesting filter derived from morphological dilation and erosion is the (self-dual) morphological gradient, or Beucher gradient [16], defined as the difference between dilation and erosion of the input image with the same structuring element. It provides an approximation to the gradient magnitude $|\nabla u|$ of the input image u, which is also consistent with the previously mentioned approximation of Hamilton–Jacobi equations by dilation and erosion. Note that the morphological gradient is not morphologically invariant as it depends on grey-value differences.

The interpolation between dilation and erosion afforded by quantiles motivates to consider also the difference between the $(1/2 + \alpha)$-quantile and the $(1/2 - \alpha)$-quantile of the same image with the same structuring element as a morphological gradient operator. This is further supported by the above-mentioned relation between quantile filters and Hamilton–Jacobi equations, from which it is evident that such a filter, too, approximates $|\nabla u|$ up to some scaling factor. Already [16] implies this possibility by defining a gradient operator as the difference of an extensive and an anti-extensive operator. We will call morphological gradients established in this way *quantile gradients*.

Multivariate Morphological Filters. Due to the favourable robustness and structure preservation of the classical median filter, interest in median filtering procedures for multivariate data developed soon in the image processing community. Nowadays, median filtering for multivariate images is mostly based on the multivariate generalisation of the median concept that is known in the statistical literature as *spatial median* or L^1

median. The L^1 median, going back to [9,20] has been applied to colour images [18] as well as to diffusion tensor images [24,25] where pixel values are symmetric matrices.

Also for morphological dilation and erosion, multivariate counterparts have been developed. Unlike the median, the concepts of supremum and infimum of data values that underlie dilation and erosion require reference to some ordering on the data. For instance, dilation and erosion for diffusion tensor data have been established in [5] using the *Loewner order* [14] of symmetric matrices in connection with the non-strict total ordering relation given by the traces of matrices.

In [3], this well-understood framework for matrix-valued morphology has been used to establish a concept of colour morphology. To this end, RGB colour images were transformed via an intermediate HCL (hue–chroma–luminance) colour space to matrix-valued images, such that the matrix-valued supremum from [5] could be used to define colour dilation; analogously for erosion. Recently, [11] used the same colour–matrix translation for colour image median filtering.

Multivariate α-quantile filters that generalise the L^1 median concept were considered in [6] and more recently in the case of matrix-valued images in [25]. These approaches differ in how they handle the inherent directionality of the quantile concept. In [6] it is pointed out that the parameter α of a scalar-valued α-quantile can be rescaled to $2\alpha - 1 \in [-1, 1]$ and then describes a direction and amplitude of deviation of the quantile from the median within the input distribution. Thus their quantile concept for n-dimensional data uses a multidimensional parameter from the unit ball in $\mathbb{R}^n$ in place of α. In contrast, [25] use the magnitude of input matrices as a natural direction of preference to allow for a one-dimensional parameter α as in the scalar-valued case. Our present work, too, builds on filtering matrix-valued data, and the magnitude of matrix values represents the luminance of the underlying colour values which again constitutes a natural preferred direction. Thus we follow here the quantile definition in [25].

To construct from matrix-valued α-quantile filters also quantile gradients is straightforward. However, the application to colour images imposes an additional hurdle if gradient values are to be represented as colour values for convenient visualisation. For this purpose, [4] suggests to use *Einstein co-subtraction*, which we will also do here.

One more word of care needs to be said. Although the multivariate morphological filters in general, and their matrix-valued versions in particular, mimick numerous properties of scalar-valued morphology, important differences remain. Not only must one abandon the property of scalar-valued median filter, dilation, and erosion to yield always data values from the input data set; also the PDE limit relationships break down to some extent. As demonstrated in [22], L^1-median filtering of multivariate data yields a PDE limit that appears practically unmanageable due to the inconvenient structure of the PDE and its coefficient functions that involve elliptic integrals.

Structure of the Paper. Section 2 collects the definitions of matrix-valued morphological filters that are used in the sequel. Section 3 describes the transformation between colour images and matrix fields that is used afterwards to obtain colour image filters from matrix-valued ones. In Section 4 we recall the construction of amoeba metrics and amoeba structuring elements, and their adaptation to the multivariate setting under consideration. Section 5 presents experiments to demonstrate the effect of our filters. A short summary and outlook is presented in Section 6.

2   Matrix-Valued Morphological Filters

Median Filter. Notice first that the median of scalar-valued data $x_1, \ldots, x_k \in \mathbb{R}$ is

$$\mu := \operatorname*{argmin}_{x \in \mathbb{R}} \sum_{i=1}^{k} |x - x_i| . \tag{1}$$

(This argmin is set-valued for even k, which is commonly disambiguated in some way. As this set-valuedness disappears in the multivariate case except for degenerate situations, we do not further discuss it here.) Generalising this observation, the L^1 median [20] of a set of points $x_1, \ldots, x_k$ in the Euclidean space $\mathbb{R}^n$ is defined to be the point x that minimises the sum of Euclidean distances to the given data points, i.e.

$$\mu := \operatorname*{argmin}_{x \in \mathbb{R}^n} \sum_{i=1}^{k} \|x - x_i\| . \tag{2}$$

For data from the set $\mathrm{Sym}(n)$ of symmetric $n \times n$ matrices, the Euclidean norm $\|x - y\|$ in (2) is naturally replaced with the Frobenius norm $\|X - Y\|_{\mathrm{F}}$ of the matrix $X - Y$, i.e. the square root of the sum of its squared entries. Other matrix norms can be used instead, such as the nuclear (or trace) norm (the sum $\sum_{i=1}^{n} |\lambda_i|$ of the moduli of eigenvalues λ_i of $X - Y$) or the spectral norm (the maximum of the $|\lambda_i|$), see [25].

It should be noticed that the median concept, by referring to a *central* value of the data distribution, does not make use of the *direction* of the ordering relation. Therefore also the multivariate median concepts do not require an ordering on the data, and are therefore equivariant under Euclidean rotations of $\mathbb{R}^n$.

Dilation and Erosion. The Loewner order [14] is a half-order $\preccurlyeq$ for symmetric matrices in which $X \preccurlyeq Y$ is defined to hold for matrices X, Y if and only if $Y - X$ is positive semidefinite. For a set of data values $\mathcal{X} := X_1, \ldots, X_k$, the Loewner order defines a convex set of upper bound matrices,

$$\mathcal{U}(\mathcal{X}) := \{X \in \mathrm{Sym}(n) \mid X_i \preccurlyeq X, \ i = 1, \ldots, k\} . \tag{3}$$

To distinguish within this set a unique supremum of $\mathcal{X}$, an additional total ordering relation is required. As proposed by [5] the non-strict ordering relation that compares matrices by their trace can serve this purpose. The supremum of $\mathcal{X}$ is then defined as minimal element of $\mathcal{U}(\mathcal{X})$ with regard to the trace order,

$$\mathbf{Sup}(\mathcal{X}) := Y \quad \text{such that} \quad \mathrm{trace}(Y) \leq \mathrm{trace}(X) \quad \forall X \in \mathcal{U}(\mathcal{X}) . \tag{4}$$

Dilation for matrix-valued images is then achieved by combining selection via a suitable structuring element with aggregation by the supremum operation (4).

Quantiles. Scalar-valued α-quantiles can be described analogously to (1) by replacing the modulus $|x - x_i|$ with $f_\alpha(x - x_i)$ where $f_\alpha(z) := |z| + (1 - 2\alpha)z$. This motivates to define matrix-valued quantiles of a set $\mathcal{X}$ of symmetric matrices [25] as

$$Q_\alpha(\mathcal{X}) := \operatorname*{argmin}_{X \in \mathrm{Sym}(n)} \sum_{i=1}^{k} \|F_\alpha(X - X_i)\| \tag{5}$$

where $F_\alpha : \mathrm{Sym}(n) \to \mathrm{Sym}(n)$ is the matrix-valued generalisation of f_α; given a symmetric matrix Y with spectral decomposition $Y = Q \operatorname{diag}(\lambda_1, \dots, \lambda_n) Q^{\mathrm{T}}$, it is obtained via $F_\alpha(Y) := Q \operatorname{diag}(f_\alpha(\lambda_1), \dots, f_\alpha(\lambda_n)) Q^{\mathrm{T}}$.

Limit cases of quantiles. One is interested in the limit cases $\alpha \to 0$, $\alpha \to 1$ of matrix quantiles. The matrix-valued quantile definition (5) is based on the Frobenius norm. It is easy to see that for $\alpha \to 1$, the α-quantile of a set $\mathcal{X} = \{X_1, \dots, X_k\}$ of symmetric matrices converges to an element Q_1 of $\mathcal{U}(\mathcal{X})$. As a result of the minimisation condition in (5), Q_1 will be the (unique) extremal point of $\mathcal{U}(\mathcal{X})$ for which the Frobenius norm of $Q_1 - \frac{1}{k}(X_1 + \dots + X_k)$ becomes minimal. A rigorous proof of this fact will be included in a forthcoming paper. We point out that in a large variety of cases Q_1 coincides with the supremum (4), but there exist cases in which the two differ.

3   Translating Between Colours and Symmetric Matrices

We start by recalling the conversion procedure from intensity triples (r, g, b) in RGB colour space to symmetric 2×2 matrices $A \in \mathrm{Sym}(2)$ as introduced in [3] and used in [11]. By pixelwise application, an RGB image u is then transformed into a matrix field F of equal dimensions.

The conversion from [3] is a two-step procedure. Each RGB triple (r, g, b) is first mapped non-linearly into a (slightly modified) HCL colour space. The second step is a Euclidean isometry from the HCL space into the space $\mathrm{Sym}(2)$.

To begin with the first step, let an intensity triple (r, g, b) with $0 \le r, g, b \le 1$ be given. Using the abbreviations

$$M := \max\{r, g, b\}, \qquad\qquad m := \min\{r, g, b\}, \qquad (6)$$

we compute hue $h \in [0, 1)$, chroma $c \in [0, 1]$, and luminance $l \in [-1, 1]$ as

$$c := M - m, \quad l := M + m - 1, \quad h := \begin{cases} \frac{1}{6}(g - b)/M \text{ modulo } 1, & M = r, \\ \frac{1}{6}(b - r)/M + \frac{1}{3}, & M = g, \quad (7) \\ \frac{1}{6}(r - g)/M + \frac{2}{3}, & M = b. \end{cases}$$

Except for a rescaling of the luminance, this corresponds to Algorithm 8.6.3 from [1]. These values represent colours in a cylindrical coordinate system, with c as radial, $2\pi h$ as angular, and l as axial coordinate. The gamut of RGB colours represented by the cube $[0, 1]^3$ is thereby bijectively mapped onto the bi-cone Γ given by $c + |l| \le 1$.

For the second step, we transform the cylindrical coordinates (c, h, l) to Cartesian coordinates by $x = c \cos(2\pi h)$, $y = c \sin(2\pi h)$, $z = l$, and further to symmetric matrices $A \in \mathrm{Sym}(2)$ via

$$A = \frac{\sqrt{2}}{2} \begin{pmatrix} z - y & x \\ x & z + y \end{pmatrix}. \qquad (8)$$

Note that (8) defines an isometry between the Euclidean space $\mathbb{R}^3$ and the space $\mathrm{Sym}(2)$ with the metric $d(A, B) := \|A - B\|_{\mathrm{F}}$. The set of all matrices A which correspond to points of the bi-cone Γ is therefore itself a bi-cone in $\mathrm{Sym}(2)$ which in the following will be identified with Γ.

From symmetric matrices to RGB triples. Concerning the converse transform, i.e. from matrices to RGB triples, it is straightforward to invert the previously described transform, compare [3].

However, in the context of quantile and gradient computation, additional difficulties arise. First, as [4] points out, the matrix supremum (4) and the corresponding infimum do not necessarily lie within the convex hull of the input data. Even for input matrices from the bi-cone Γ, the supremum is only guaranteed to belong to the unit ball $\mathcal{B}$ given by $l^2 + c^2 \leq 1$. The same is true for the α-quantiles (5) as soon as $\alpha \neq 1/2$.

Following [4], we use therefore the inverse of the map Θ from [4, eq. (5)] to map quantiles from $\mathcal{B}$ back to Γ before transforming them back to the RGB colour space. Written in terms of the (h, c, l) colour space, the inverse map reads as

$$\Theta^{-1}(h, c, l) = (h, c/\kappa, l/\kappa) \tag{9}$$

where κ is the solution of $\kappa^{\nu+1} - \kappa^{\nu} - (c + l)^{\nu}(c + l - \sqrt{c^2 + l^2})/\sqrt{c^2 + l^2} = 0$ in the interval $\kappa \in [1, \sqrt{2}]$, with a constant ν that we fix to 10 as proposed in [4].

Second, the difference of matrices from $\mathcal{B}$ obviously needs not to belong to $\mathcal{B}$. Therefore the morphological gradient in [4] is not defined via standard subtraction of supremum and infimum but instead by a so-called Einstein co-subtraction $\boxminus$ (similar to a relativistic subtraction of velocities) to ensure that the difference is within the ball $\mathcal{B}$. For symmetric matrices $A, B \in \mathcal{B}$, the Einstein co-subtraction is defined as [4, Sec. 5]

$$A \boxminus B := \frac{2C}{1 + \|C\|_{\mathrm{F}}^2} \quad \text{where} \quad C := \frac{\sqrt{1 - \|B\|_{\mathrm{F}}^2}\, A - \sqrt{1 - \|A\|_{\mathrm{F}}^2}\, B}{\sqrt{1 - \|B\|_{\mathrm{F}}^2} + \sqrt{1 - \|A\|_{\mathrm{F}}^2}} . \tag{10}$$

Following this approach, we define the quantile-based gradient $D_\alpha(\mathcal{X})$ as

$$D_\alpha(\mathcal{X}) := Q_{1/2+\alpha}(\mathcal{X}) \boxminus Q_{1/2-\alpha}(\mathcal{X}) . \tag{11}$$

Since $D_\alpha(\mathcal{X})$ belongs to $\mathcal{B}$, it can be mapped to Γ via Θ^{-1} from (9) and finally be represented in the RGB colour space, with grey ($r = g = b = 1/2$) as neutral value. Like their scalar-valued counterpart, colour morphological gradients can be expected to be useful for edge detection.

4   Amoebas

In order to extend our previously defined colour quantile and quantile gradient filters into adaptive morphological filters, we use amoebas as structuring elements. This section is therefore devoted to the construction of amoebas using spatial distance and image contrast (tonal distance). The construction presented here basically follows [11] where Lerallut et al.'s original amoeba framework [12] was adapted to symmetric matrices as data values. However, unlike in [11,12], where spatial and tonal information were combined via an L^1 sum, and spatial distance measurement itself is based on 4-neighbourhoods, we use for our quantile and quantile gradient filters in this work an L^2 spatial-tonal sum with spatial 8-neighbourhoods as in [22, Sec. 4.3]. The latter are preferred for their better approximation of Euclidean distance in the image plane; note however that continuous-scale arc length could be approximated even better by more sophisticated approaches, see [2,7,10]. Regarding alternatives for how the spatial and tonal distances could be combined, see also [21] (in univariate formulation).

Let a matrix field F over a discrete image domain Ω be given, such that $F_i \in$ Sym(2) denotes the data value assigned to the pixel location $i \in \Omega$. Let (x_i, y_i) be the spatial coordinates of pixel i. We introduce an *amoeba metric* $d_\mathcal{A}$ for pairs (i, i') of adjacent pixels (where adjacency can be horizontal, vertical, or diagonal, thus 8-neighbourhoods are used) by

$$d_\mathcal{A}(i, i') := \sqrt{(x_i - x_{i'})^2 + (y_i - y_{i'})^2 + \beta^2 \|F_i - F_{i'}\|_F^2} , \tag{12}$$

which is an L^2 sum of the Euclidean distance of i and i' in the image plane, and the Frobenius distance of their data values weighted with $\beta > 0$.

To construct a structuring element around some given pixel $i_0 \in \Omega$, we consider paths $P = (i_0, i_1, \ldots, i_k)$ starting at the given i_0 with $i_j \in \Omega$ such that each two subsequent pixels i_j, i_{j+1} are adjacent in Ω horizontally, vertically or diagonally (thus, 8-neighbourhoods are used). We measure the length of such a path P using the amoeba metric introduced above as

$$L(P) := \sum_{j=0}^{k-1} d_\mathcal{A}(i_j, i_{j+1}) . \tag{13}$$

A pixel $i^* \in \Omega$ is included in the amoeba structuring element around i_0 if and only if there exists some k and a path P starting at i_0 and ending at $i_k = i^*$ with $L(P) \leq \varrho$.

This procedure is repeated for each pixel i_0 to generate a complete set of structuring elements for the given matrix field. The amoeba construction has two free parameters: the amoeba radius $\varrho > 0$ and the contrast scale $\beta > 0$.

Note that the path length $L(P)$ equals the Euclidean length of P in constant image regions but the more data variation is met along P, the more $L(P)$ exceeds the Euclidean path length. As a consequence, structuring elements adapt to image structures, extending preferredly towards locations with similar data values, but avoiding to cross strong contrast edges.

5   Experiments

In this section we demonstrate the effect of our matrix-based colour quantiles and quantile gradients using two test images, see Figure 1(a) and Figure 4(a). From the nature of the filters in question, it is expected that α-quantile filters for $\alpha = 0 \ldots 1$ provide a gradual transition from erosion via median to dilation. Quantile gradients are expected to highlight colour edges with high sensitivity to colour differences, making them usable as a building block for colour image edge detection. Up to some α-dependent scaling colour quantile gradients should yield similar results as the classical Beucher gradient, but possibly with increased robustness. Amoeba versions of both filter classes are expected to improve the sharp edge preservation over their non-adaptive counterparts.

Figure 1(b)–(f) show the results of α-quantile filtering with a non-adaptive structuring element applied on the first test image, where α is varied in the range 0.1 to 0.9.

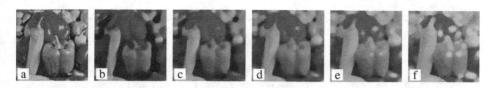

Fig. 1. Colour α-quantile filtering with a non-adaptive disc-shaped structuring element of radius 5. **Left to right: (a)** Test image, 128×128 pixels. – **(b)** Colour quantile, $\alpha = 0.1$. – **(c)** $\alpha = 0.3$. – **(d)** $\alpha = 0.5$ (median). – **(e)** $\alpha = 0.7$. – **(f)** $\alpha = 0.9$.

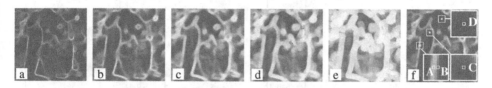

Fig. 2. Gradient filtering of the test image from Figure 1(a) with a non-adaptive disc-shaped structuring element of radius 5. **Left to right: (a)** Quantile gradient, $\alpha = 0.1$. – **(b)** Quantile gradient, $\alpha = 0.2$. – **(c)** Quantile gradient, $\alpha = 0.3$. – **(d)** Quantile gradient, $\alpha = 0.4$. – **(e)** Beucher gradient based on the dilation and erosion operators from [3]. – **(f)** Same as (b) with four locations A–D marked for more detailed analysis.

The gradual transition between an erosion-like and dilation-like behaviour is evident. Colour tones are preserved in a visually appealing way across the parameter range.

In Figure 2(a)–(d) we display quantile gradients D_α for the same test image and α from $0.1 \ldots 0.4$. The series is completed with the Beucher gradient computed from dilation and erosion according to [3] (although this is not the exact limit case of D_α, as pointed out at the end of Section 2). Remember that grey ($r = g = b = 1/2$) represents a zero matrix, thus also zero gradient. Colours brighter than that correspond to matrices of larger trace. By the construction of the morphological (quantile or standard Beucher) gradients, this is naturally the case for all its values. Furthermore, it is evident that despite using the same structuring element across the series (a)–(e), the quantile gradients with smaller α are not just gradients with reduced contrast; instead, they are sharper than those for larger α or the standard Beucher gradient. This is consistent with the fact proven in the univariate case [22] that α-quantiles approximate the same continuous process as dilation or erosion but at a reduced speed; thus using quantiles with α closer to $1/2$ in computing the gradient has a similar effect as a smaller structuring element but without the increased noise sensitivity of such a smaller structuring element.

Figure 2(f) marks four locations in the image domain: Pixel A is located on an edge as evident from the gradient data; pixel B is located slightly off the same edge; pixel C is placed in an area where colours vary slowly but do not feature sharp edges; pixel D resides in an almost flat area.

For these four locations, Figure 3 illustrates the dependency of their respective colour quantiles Q_α (with the same structuring element as in the preceding experiments) on α. The plots in the upper row represent the coordinates x, y, z of the intermediate colour space as functions of α. For the edge pixel A, the resulting curves have essentially a

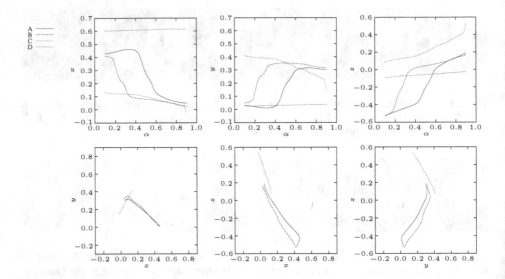

Fig. 3. Colour quantiles of the image from Fig. 1(a) as functions of α for four exemplary points A–D marked in Fig. 2(f). **Top row:** Cartesian coordinates x, y, z in the intermediate colour space as functions of α for $\alpha \in [0.1, 0.9]$. – **Bottom row:** Projections of the colour quantile curve for $\alpha \in [0.1, 0.9]$ from the x-y-z space to the x-y, x-z and y-z planes.

sigmoid shape centred at the median ($\alpha = 0.5$). For the nearby pixel B, similar curve shapes are observed but the inflection point of the sigmoid is shifted to $\alpha \approx 0.2$. Simultaneous consideration of quantiles across a suitable range of α could therefore be used for precise localisation of edges. For pixel C, it is visible that its quantiles vary almost linearly with α, while for pixel D they are essentially constant, as could be expected. The bottom row of Figure 3 shows the projections of the trajectories of Q_α for pixels A–D in the (x, y, z) space to its three coordinate planes. The almost linear shape of the curves confirms that colour quantile gradients D_α for different α differ mainly in amplitude but not in their direction in colour space. It is also evident that the variation of quantiles, and thus the gradient, is largest on and nearby the edge, smaller for the slope region around C and very small in the homogeneous region at D.

Figure 4 demonstrates non-adaptive quantiles and gradients on a second test image.

Turning to adaptive filtering using the amoeba framework, we show in Figure 5 quantile filtering of the same test image as in Figure 1 but replacing the non-adaptive structuring element of radius 5 with amoebas of radius $\varrho = 5$. Thus, the same structuring elements as in the non-adaptive case result in homogeneous regions, while the filter effect is attenuated where contrasts prevail. Frames (a)–(d) of Figure 5 show quantiles for α in the range 0.1 to 0.9 and amoeba contrast scale parameter $\beta = 10$, while (e) and (f) vary the contrast parameter. As expected, amoeba filtering gives sharper results than non-adaptive filters but this effect goes away when β is chosen smaller (e). Increasing β to 30, see Figure 5(f), does not result in much additional sharpness of the result.

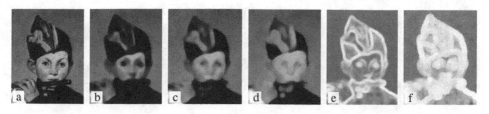

Fig. 4. Non-adaptive α-quantile and gradient filtering with a disc-shaped structuring element of radius 5. **Left to right: (a)** Test image, 131×173 pixels. – **(b)** Quantile, $\alpha = 0.2$. – **(c)** Quantile, $\alpha = 0.5$ (median). – **(d)** Quantile, $\alpha = 0.8$. – **(e)** Quantile gradient, $\alpha = 0.3$. – **(f)** Beucher gradient based on the dilation and erosion operators from [3].

Fig. 5. Colour amoeba α-quantile filtering of the test image from Figure 1(a) with structuring element radius $\varrho = 5$. **Left to right: (a)** $\alpha = 0.1$, contrast scale $\beta = 10$. – **(b)** $\alpha = 0.3$, $\beta = 10$. – **(c)** $\alpha = 0.7$, $\beta = 10$. – **(d)** $\alpha = 0.9$, $\beta = 10$. – **(e)** $\alpha = 0.7$, $\beta = 3$. – **(f)** $\alpha = 0.7$, $\beta = 30$.

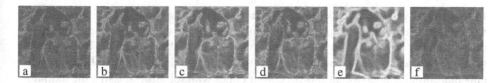

Fig. 6. Colour amoeba gradient filtering of the test image from Figure 1(a) with structuring element radius $\varrho = 5$. **Left to right: (a)** Quantile gradient, $\alpha = 0.2$, $\beta = 10$. – **(b)** Quantile gradient, $\alpha = 0.3$, $\beta = 10$. – **(c)** Quantile gradient, $\alpha = 0.4$, $\beta = 10$. – **(d)** Beucher gradient using the dilation and erosion from [3], $\beta = 10$. – **(e)** Quantile gradient, $\alpha = 0.4$, $\beta = 3$. – **(f)** Quantile gradient, $\alpha = 0.4$, $\beta = 30$.

Figure 6 shows colour quantile and Beucher gradients of the same image. Again, (a)–(d) use the same contrast scale $\beta = 10$ to demonstrate how the amplitude of the gradient image increases from small to larger α and up to the Beucher gradient. Note how in all cases the amoeba method achieves sharper localisation of edges compared to the non-adaptive approach. In (e) and (f) variation of β is shown. Again, $\beta = 3$ appears too small for the amoeba procedure to take substantial effect. In contrast, $\beta = 30$ suppresses the filter at edges so much that edges almost cannot be detected in the filtered image while the moderate contrasts within smooth regions survive.

Finally, Figure 7 demonstrates the amoeba quantile and gradient filtering on the second test image, with similar results as in Figures 5 and 6. Note that unlike in the non-adaptive case, see Figure 4, where the dilating or eroding effect of quantiles is significant the two amoeba quantile filtering results in Figure 7 (a), (b) keep contours

Fig. 7. Amoeba α-quantile and gradient filtering of the test image from Figure 4(a) with structuring element radius $\varrho = 5$. **Left to right: (a)** Quantile, $\alpha = 0.2$, $\beta = 10$. – **(b)** Quantile, $\alpha = 0.8$, $\beta = 10$. – **(c)** Quantile gradient, $\alpha = 0.3$, $\beta = 10$. – **(d)** Beucher gradient based on the dilation and erosion operators from [3], $\beta = 10$. – **(e)** Quantile gradient, $\alpha = 0.3$, $\beta = 3$. – **(f)** Quantile gradient, $\alpha = 0.3$, $\beta = 30$.

fairly well in place, while at the same time some smoothing together with a darkening ($\alpha = 0.2$) or brightening ($\alpha = 0.8$) takes place. Given that also the colour tones are fairly well preserved, the amoeba quantile filters lend themselves as robust operators for adjusting image brightness.

6   Summary and Outlook

In this paper we have extended the work from [3,4,11] on the application of matrix-valued morphology to colour image processing. Using the matrix-valued quantile definition from [25] we have provided colour quantile filters that interpolate between erosion, median, and dilation, and can be used to obtain a variant of morphological gradients that combine good localisation of colour edges with robustness. Using the morphological amoeba framework [11,12,13,23] we have formulated image-adaptive versions of quantile and quantile gradient filters with favourable edge-preserving properties.

Ongoing research is directed at further theoretical analysis, including the interaction between structuring element radius, quantile parameter α, and the amoeba adaptivity in gradient computation, as well as the limit relation between quantile filters and dilation/erosion. Also, the numerics of multivariate quantile computation will be a subject of future work. On the application side, the use of quantile gradient filtering in edge detection and image segmentation will be of interest but also the suitability of amoeba quantile filters for image brightness adjustment.

References

1. Agoston, M.K.: Computer Graphics and Geometric Modeling: Implementation and Algorithms. Springer, London (2005)
2. Borgefors, G.: Distance transformations in digital images. Computer Vision, Graphics and Image Processing 34, 344–371 (1986)
3. Burgeth, B., Kleefeld, A.: Morphology for color images via Loewner order for matrix fields. In: Luengo Hendriks, C.L., Borgefors, G., Strand, R. (eds.) ISMM 2013. LNCS, vol. 7883, pp. 243–254. Springer, Heidelberg (2013)

4. Burgeth, B., Kleefeld, A.: An approach to color-morphology based on Einstein addition and Loewner order. Pattern Recognition Letters 47, 29–39 (2014)
5. Burgeth, B., Welk, M., Feddern, C., Weickert, J.: Mathematical morphology on tensor data using the Loewner ordering. In: Weickert, J., Hagen, H. (eds.) Visualization and Processing of Tensor Fields, pp. 357–368. Springer, Berlin (2006)
6. Chaudhuri, P.: On a geometric notion of quantiles for multivariate data. Journal of the American Statistical Association 91(434), 862–872 (1996)
7. Fabbri, R., Da F. Costa, L., Torelli, J.C., Bruno, O.M.: 2D Euclidean distance transform algorithms: A comparative survey. ACM Computing Surveys, 40(1), art. 2 (2008)
8. Guichard, F., Morel, J.-M.: Partial differential equations and image iterative filtering. In: Duff, I.S., Watson, G.A. (eds.) The State of the Art in Numerical Analysis. IMA Conference Series (New Series), vol. 63, pp. 525–562. Clarendon Press, Oxford (1997)
9. Hayford, J.F.: What is the center of an area, or the center of a population? Journal of the American Statistical Association 8(58), 47–58 (1902)
10. Ikonen, L., Toivanen, P.: Shortest routes on varying height surfaces using gray-level distance transforms. Image and Vision Computing 23(2), 133–141 (2005)
11. Kleefeld, A., Breuß, M., Welk, M., Burgeth, B.: Adaptive filters for color images: Median filtering and its extensions. In: Trémeau, A., Schettini, R., Tominaga, S. (eds.) CCIW 2015. LNCS, vol. 9016, pp. 149–158. Springer, Heidelberg (2015)
12. Lerallut, R., Decencière, É., Meyer, F.: Image processing using morphological amoebas. In: Ronse, C., Najman, L., Decencière, E. (eds.) Mathematical Morphology: 40 Years On. Computational Imaging and Vision, vol. 30, pp. 13–22. Springer, Dordrecht (2005)
13. Lerallut, R., Decencière, É., Meyer, F.: Image filtering using morphological amoebas. Image and Vision Computing 25(4), 395–404 (2007)
14. Löwner, K.: Über monotone Matrixfunktionen. Mathematische Zeitschrift 38, 177–216 (1934)
15. Meyer, F., Maragos, P.: Nonlinear scale-space representation with morphological levelings. Journal of Visual Communication and Image Representation 11, 245–265 (2000)
16. Rivest, J.-F., Soille, P., Beucher, S.: Morphological gradients. Journal of Electronic Imaging 2(4), 326–336 (1993)
17. Sapiro, G.: Vector (self) snakes: a geometric framework for color, texture and multiscale image segmentation. In: Proc. 1996 IEEE International Conference on Image Processing, Lausanne, Switzerland, vol. 1, pp. 817–820 (September 1996)
18. Spence, C., Fancourt, C.: An iterative method for vector median filtering. In: Proc. of 2007 IEEE International Conference on Image Processing, vol. 5, pp. 265–268 (2007)
19. Tukey, J.W.: Exploratory Data Analysis. Addison–Wesley, Menlo Park (1971)
20. Weber, A.: Über den Standort der Industrien. Mohr, Tübingen (1909)
21. Welk, M.: Amoeba techniques for shape and texture analysis. Technical Report cs:1411.3285, arXiv.org (2014)
22. Welk, M., Breuß, M.: Morphological amoebas and partial differential equations. In: Hawkes, P.W. (ed.) Advances in Imaging and Electron Physics, vol. 185, pp. 139–212. Elsevier Academic Press (2014)
23. Welk, M., Breuß, M., Vogel, O.: Morphological amoebas are self-snakes. Journal of Mathematical Imaging and Vision 39, 87–99 (2011)
24. Welk, M., Feddern, C., Burgeth, B., Weickert, J.: Median filtering of tensor-valued images. In: Michaelis, B., Krell, G. (eds.) DAGM 2003. LNCS, vol. 2781, pp. 17–24. Springer, Heidelberg (2003)
25. Welk, M., Weickert, J., Becker, F., Schnörr, C., Feddern, C., Burgeth, B.: Median and related local filters for tensor-valued images. Signal Processing 87, 291–308 (2007)

Ordering on the Probability Simplex of Endmembers for Hyperspectral Morphological Image Processing

Gianni Franchi and Jesús Angulo[✉]

MINES ParisTech, PSL-Research University,
CMM-Centre de Morphologie Mathématique, Paris, France
gianni.franchi@mines-paristech.fr, Jesus.Angulo@ensmp.fr

Abstract. A hyperspectral image can be represented as a set of materials called endmembers, where each pixel corresponds to a mixture of several of these materials. More precisely pixels are described by the quantity of each material, this quantity is often called abundance and is positive and of sum equal to one. This leads to the characterization of a hyperspectral image as a set of points in a probability simplex. The geometry of the simplex has been particularly studied in the theory of quantum information, giving rise to different notions of distances and interesting preorders. In this paper, we present total orders based on theory of the ordering on the simplex. Thanks to this theory, we can give a physical interpretation of our orders.

Keywords: Hyperspectral image · Mathematical morphology · Learning an order · Quantum information

1 Introduction

Hyperspectral images, which represent a natural extension of conventional optical images, can reconstruct the spectral profiles through the acquisition of hundreds of narrow spectral bands, generally covering the entire optical spectral range. Thus, at each pixel, there is a vector which corresponds to the spectrum of reflected light. For a long time the processes associated with this type of images were limited to treatments where each pixel was considered just as a vector independently of its location on the image domain. Subsequently techniques to account for spatial information were developed [10] [12]. Between these techniques, mathematical morphology has been also used [13][19]. Adding spatial information in the treatment of hyperspectral images greatly improve tasks such as classification.

Mathematical morphology is a non-linear methodology for image processing based on a pair of adjoint and dual operators, dilation and erosion, used to compute sup/inf-convolutions in local neighborhoods. Therefore the extension of morphological operators to multivariate images, and in particular to hyperspectral images, requires the introduction of appropriate vector ordering strategies.

© Springer International Publishing Switzerland 2015
J.A. Benediktsson et al. (Eds.): ISMM 2015, LNCS 9082, pp. 410–421, 2015.
DOI: 10.1007/978-3-319-18720-4_35

Lots of research done has aimed at developing orders in multivariate images. First, we found techniques based on marginal order [15]: one applies a morphological operator at each channel of the hyperspectral image (or each eigenimage after PCA)[12], without taking care of the vector nature of spectral information. Then there are other approaches based on learning a vector order using classification techniques [17] or nonparametric statistical tools [18]. Supervised learning an order [17] involves to introduce a prior spectral background and foreground. This happens if, for example, one knows that the vegetation areas are the object of interest, and does not care of image structures from other materials: then prior vegetation spectra would be the foreground, and the spectra of other materials the background. But, sometimes, this prior configuration is unknown. There might be two possible solutions. One approach proposed in [18] consists in computing an order based on the "distance to central value", obtained by the statistical depth function. Another solution builds an order based on distances to pure materials on the image [2]: more a spectrum is a mixture of materials less its position in the order is. In hyperspectral image processing, these pure materials are called endmembers. The goal of this paper is to investigate order techniques in a physical representation similar to the one proposed in [2]. From a geometric viewpoint, the space where the spectral values are represented is not an Euclidean space, but a simplex, which is also useful to manipulate discrete probability distributions. In our case, we will consider several alternatives, which can be related to notions as majorization, stochastic dominance, distances and divergence between probability distributions, etc.

2   Hyperspectral Image Linear Model

From a mathematical viewpoint, a hyperspectral image is considered as a function f defined as

$$f : \begin{cases} E \to \mathbb{R}^D \\ x \mapsto v_i \end{cases}$$

where D is the number of spectral bands and E is the image domain (support space of pixels. This multivariate image can be also seen as set of D grey-scale images which correspond to the spectral bands, i.e., $f(x) = (f_1(x), \cdots, f_D(x))$, $f_j : E \to \mathbb{R}$.

Unmixing on Hyperspectral Images. Spectral unmixing [6,7,8] is an ill-posed inverse problem. The idea is to find both the physical pure materials, also known as endmembers, present in the scene and their corresponding abundances. By abundance we mean the quantity of each material at each pixel. However, in hyperspectral imaging, because of the low resolution of these images, a pixel may correspond to a large surface, where several different physical materials are present: these materials can be mixed and may affect the unmixing algorithms. A model often used by its simplicity is the linear one. In this model, ground is considered flat such that when the sun's rays reach the study area, they are directly reflected to the sensor. It receives, from a convex area, a linear combination of all the materials present in the area. A strong and extreme assumption which is often made is that one can find the endmembers from the hyperspectral image itself, because they are already present in the image, or because the geometric problem is not very complex.

Working on this paradigm, many different algorithms have been proposed to compute the endmembers [8]. We have used in our examples N-FINDR [21]. Once the endmembers are obtained, the corresponding abundance should be computed at each pixel, which consequently yields an abundance image for each endmember, i.e., $\alpha_r(x) : E \to \mathbb{R}_+$, $1 \leq r \leq R$, where $R \in \mathbb{N}$ is the number of endmembers. We have used a fully constrained least squared on each pixel [8], where one imposes that each pixel is a linear nonnegative and convex combination of endmembers. We just discuss below the geometric interpretation of such constrained regression.

Due to the fact that in general the number of endmembers is significatively lower than the number of spectral bands, i.e., $R \ll D$, working on the abundance maps $\{\alpha_j(x)\}_{1 \leq j \leq R}$ instead the original spectral bands $\{f_j(x)\}_{1 \leq j \leq D}$ is a way to counterpart the curse of dimensionality [4], which designates the phenomena that occur when data are studied in a large dimensionality space. In summary, thanks to unmixing techniques, the hyperspectral image is embedded into a lower dimension space.

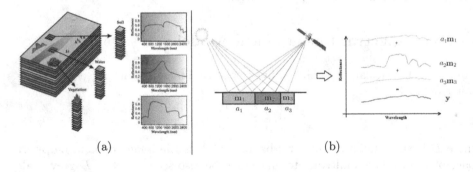

(a)   (b)

Fig. 1. (a) Representation of a hyperspectral image, (b) Zoom in the linear mixing model. Figures borrowed from [14].

Geometrical Viewpoint of Unmixing. Let us consider in Fig. 1(a) the typical diagram to illustrate spectral unmixing. To fix the ideas, we can consider that the image is composed of water, soil, and vegetation. Finding the endmembers on this image is therefore a way to find the spectra of these three pure materials. Under the linear model, each pixel of the image can be written as a positive combination of the different endmembers m_1, m_2 and m_3, see Figure 1(b), with a_1, a_2 and a_3 being the abundances. Thus, if we consider the spectrum at a pixel as the vector $v_i \in \mathbb{R}^D$, then, it can be written as

$$v_i = \sum_{r=1}^{R} a_{r,i} m_r + n_i, \tag{1}$$

where $\{m_r\}_{r=1}^{R}$ represent the set of R endmembers, $a_{r,i}$ the abundance at vector i of each endmember r, and n_i an additive noise. This last term can be neglected.

Let us consider a geometric interpretation of the representation by endmembers [11]. Spectra correspond to a set of points $v_i \in \mathbb{R}^D$, $1 \leq i \leq |E|$. Since each v_i represents a physical spectrum, the v_i are nonnegative data, thus every point lies in the positive orthant $\mathcal{P}^D$ of $\mathbb{R}^D$. Therefore the endmembers should be also nonnegative vectors. The endmembers basis $\Phi = \{m_r\}_{1 \leq r \leq R}$, $m_r \in \mathbb{R}^D$ generates a simplicial cone Γ_Φ containing the data and which lies in $\mathcal{P}^D$:

$$\Gamma_\Phi = \{v : v = \sum_{r=1}^{R} a_r m_r, \ a_r \geq 0\}. \tag{2}$$

Hence, the extraction of endmembers can be seen as finding the simplicial cone containing the data. In general, for a given set of vectors there are many possible simplicial cones containing the vectors. A way to reduce the number of possible representations consist in restrict the nonnegative coefficients a_r to be a convex combination such that $\sum_{r=1}^{R} a_r = 1$. By using this additional constrain, it is guaranteed to work on a $(R-1)$-simplex. We remind that the unit n-simplex is the subset of $\mathbb{R}^{n+1}$ given by

$$\Delta^n = \{(s_1, \cdots, s_{n+1})^t \in \mathbb{R}^{n+1} : \sum_{k-1}^{n+1} s_k = 1 \text{ and } s_k \geq 0\}. \tag{3}$$

where the $n+1$ vertices of a regular tetrahedron. In our case, the vertices are the endmembers $\{m_r\}_{1 \leq r \leq R}$, but we can work geometrically on the canonical simplex Δ^{R-1}. In summary, the abundances $(a_{1,i}, \cdots, a_{R,i})^t$ are just the barycentric coordinates of vector i in the Δ^{R-1}. In the case of the hyperspectral model studied in Fig. 1 with just 3 endmembers, its abundances lie in the triangle represented in Fig. 2(a).

3   Ordering on the Simplex of Endmembers

Let us consider that the hyperspectral image of spectral pixels $v_i \in \mathbb{R}^D$ is represented by their abundances, together with the set of endmembers $\{m_r\}_{1 \leq r \leq R}$. Let us consider that we have two spectral vectors v_i and v_j whose coordinates in the simplex are $(a_{1,i}, \cdots, a_{R,i})^t$ and $(a_{1,j}, \cdots, a_{R,j})^t$. We address in this section the question of how can we order this pair of vectors.

3.1   Lexicographic Abundance Order

An easy way to define a total order between v_i and v_j is based on a lexicographic order of their barycentric coordinates. To avoid an arbitrary order between the endmembers, we propose to order them according to their norm $\|m_r\|$,

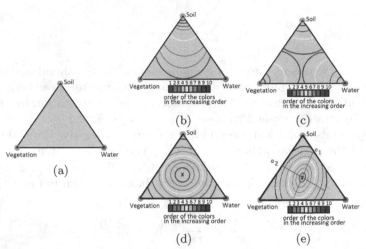

Fig. 2. (a) Simplex representation of an hyperspectral image with three endmembers. (b) First iso-order lines for $\leq_{\text{lex}}$. (c) First iso-order lines for $\leq_{\text{maj}}$. (d) Iso-order lines for $\leq_{d_{L^2}}$. (e) Iso-order lines for $\leq_{d_M}$.

i.e., $m_r \preceq m_s \Leftrightarrow \|m_r\| \leq \|m_s\|$. Then a permutation τ according to this order of endmembers is applied on the abundance coordinates, so v_i and v_j are represented by $(a_{\tau^{-1}(1),i}, \ldots, a_{\tau^{-1}(R),i})^t$ and $(a_{\tau^{-1}(1),j}, \ldots, a_{\tau^{-1}(R),j})^t$. Finally, we define the lexicographic abundance order $\leq_{\text{lex}}$ as

$$
v_i \leq_{\text{lex}} v_j \Leftrightarrow
\begin{cases}
a_{\tau^{-1}(1),i} < a_{\tau^{-1}(1),j} \text{ or} \\
a_{\tau^{-1}(1),i} = a_{\tau^{-1}(1),j} \text{ and } a_{\tau^{-1}(2),i} < a_{\tau^{-1}(2),j} \text{ or} \\
\vdots \\
a_{\tau^{-1}(1),i} = a_{\tau^{-1}(1),j} \text{ and } \ldots \text{ and } a_{\tau^{-1}(R),i} \leq a_{\tau^{-1}(R),j}
\end{cases}
\tag{4}
$$

Fig. 2(b) gives an example of first iso-order lines, where the soil is the dominant material (i.e., largest endmember in the sense of its norm). By first iso-order lines we mean those which correspond to the first condition in (4): $a_{\tau^{-1}(1),i} \leq a_{\tau^{-1}(1),j}$.

We note that this order has two important weaknesses. First, all the information of the hyperspectral images is not taking into account, since the first coordinate dominates all the others. Second, we need to order the materials. Nevertheless it is easy to use and to apply.

3.2   Foreground Abundance Order: Use of Majorization

We consider now a partial ordering based on the position of vectors in the simplex with respect to the so-called foreground. More precisely the foreground here corresponds to the set of endmembers.

The proposed approach is based on the notion of majorization [9], which is a technique for ordering vectors of same sum. Let us consider two vectors $\mathbf{c} = (c_1, \ldots, c_n) \in \mathbb{R}^n$ and $\mathbf{d} = (d_1, \ldots, d_n) \in \mathbb{R}^n$, then we say that C weakly majorizes D, written $\mathbf{c} \succ_w \mathbf{d}$, if and only if

$$\begin{cases} \sum_{i=1}^{k} c_i^{\downarrow} \geq \sum_{i=1}^{k} d_i^{\downarrow} \ , \ \forall k \in [1,n] \\ \sum_{i=1}^{n} c_i = \sum_{i=1}^{n} d_i \end{cases} \tag{5}$$

where $c_i^{\downarrow}$ and $d_i^{\downarrow}$ represent respectively the coordinates of C and D sorted in descending order. Majorization is not a partial order, since $\mathbf{c} \succ \mathbf{d}$ and $\mathbf{d} \succ \mathbf{c}$ do not imply $\mathbf{c} = \mathbf{d}$, it only implies that the components of each vector are equal, but not necessarily in the same order.

We propose a majorization-like partial order adapted to the abundances. Similarly to the majorization, a permutation τ_i of the coordinates of the vectors v_i in the simplex is applied such that they are sorted in descending order. The majorization-like order $\leq_{\mathrm{maj}}$ is given defined as

$$v_i \leq_{\mathrm{maj}} v_j \Leftrightarrow \begin{cases} a_{\tau_i^{-1}(1),i} < a_{\tau_j^{-1}(1),j} \ \text{or} \\ a_{\tau_i^{-1}(1),i} = a_{\tau_j^{-1}(1),j} \ \text{and} \ a_{\tau_i^{-1}(2),i} < a_{\tau_j^{-1}(2),j} \ \text{or} \\ \vdots \\ a_{\tau_i^{-1}(1),i} = a_{\tau_j^{-1}(1),j} \ \text{and} \ \dots \ \text{and} \ a_{\tau_i^{-1}(R),i} \leq a_{\tau_j^{-1}(R),j} \end{cases} \tag{6}$$

See in Fig. 2(c), the corresponding first iso-order lines. An important advantage of this order is the fact that there is no need for an order between the different materials. However, as one may notice, it is not possible to find a order-related metric that would follow the geometry of the simplex.

Partial order based on distances to reference vectors is classical in color morphology [1]. The partial order $\leq_{\mathrm{maj}}$ is also related to the approach of hyperspectral order proposed in [2]. The fundamental difference is that the present order is based on abundances (barycentric coordinates in the $R - 1$-simplex) whereas in [2], the partial order is based on distances in $\mathbb{R}^D$ between v_i and each endmember m_r (we note that computing distances in a high dimensional space is a tricky issue).

3.3 Foreground Abundance Order: Use of Stochastic Dominance

The stochastic dominance is a partial ordering between probability distributions [16]. The term is used in decision theory to say if a distribution can be considered "superior" to another one. We first introduce the second order stochastic dominance [16]. Let us consider two random variables X and Y of respective cumulative distribution F_X and F_Y, then we say that X is second-order stochastically dominant over Y if and only if

$$\int_{-\infty}^{c} F_X(w)dw \leq \int_{-\infty}^{c} F_Y(w)dw \ , \ \forall c \in \mathbb{R} \tag{7}$$

In our case, the abundance coordinates $(a_{1,i}, \dots, a_{R,i})^t$ of each vector v_i can be seen as a discrete probability distribution, since the sum of the coordinates is equal to one. However, to be able to calculate the cumulative distribution of the vector, we need an order between the coordinates. We use the order of the endmembers and the corresponding permutation τ introduced for the lexicographic abundance order. So v_i is represented by $(a_{\tau^{-1}(1),i}, \ldots, a_{\tau^{-1}(R),i})^t$, then we calculate the discrete cumulative distribution from v_i, noted by $(b_{1,i}, \ldots, b_{R,i})^t$:

$$(a_{1,i}, \ldots, a_{R,i})^t \mapsto (b_{1,i}, \ldots, b_{R,i})^t, \text{ where } b_{j,i} = \sum_{k=1}^{j} a_{\tau^{-1}(k),i}.$$

Stochastic dominance gives rise to a partial order, which induces a supremum $\vee_{dom}$ between vectors v_i and v_j whose barycentric coordinates are $(a_{\tau(1),\vee}, \ldots, a_{\tau(R),\vee})^t$ such that

$$(a_{1,\vee}, \ldots, a_{R,\vee})^t \to v_1 \vee_{dom} v_2,$$

where

$$a_{1,\vee} = b_{1,i} \vee b_{1,j},$$
$$a_{2,\vee} = b_{2,i} \vee b_{2,j} - b_{1,i} \vee b_{1,j},$$
$$\ldots$$
$$a_{R,\vee} = b_{R,i} \vee b_{R,j} - b_{R-1,i} \vee b_{R-1,j}.$$

Supremum $\vee_{dom}$ introduces "false abundances", in the sense that they are not present in the original set of vectors, however, the corresponding spectra are obtained from the endmembers and therefore the spectra are physically plausible. The infimum is obtained dually.

3.4 Background Abundance Order: Use of Distance/Divergence

Previous orders focuss mainly on the relationship between the vectors in the simplex and its vertices, which correspond in a certain way to the foreground of the hyperspectral image (the pure pixels).

We propose now to think in a dual paradigm. Basically, the idea is to have an order given by how far the vectors are from the background. The background correspond to the situation where the material are totally mixed: the vector which has the same quantity of materials or in geometric terms, the center of the simplex $c \in \mathbb{R}^R$, i.e., in barycentric coordinates $c = (1/R, \ldots, 1/R)^t$. Then, the partial order between two vectors v_i and v_j will depend on the distance between c and each vector. We adopted the convention that further a point is from the center c, higher its position in the order is. Given a distance $d(\cdot, \cdot)$, we have the corresponding background partial order $\leq_{d_{L^2}}$:

$$v_i \leq_d v_j \Leftrightarrow d\left(c, (a_{1,i}, \ldots, a_{R,i})^t\right) \leq d\left(c, (a_{1,j}, \ldots, a_{R,j})^t\right). \tag{8}$$

In case of equality, we just use the lexicographic abundance order to have a total order, similarly to [1]. We have considered several distance metrics, for instance the L^1 and L^2 Minkowski norm (see in Fig. 2(d) the iso-order lines for $\leq_{d_{L^2}}$. We have also considered the interest of the Mahalanobis distance $d_M(\cdot, \cdot)$, i.e., given two random vectors $\mathbf{c}$ and $\mathbf{d}$ following the same distribution with covariance matrix Σ, it is given by

$$d_M(\mathbf{c}, \mathbf{d}) = \sqrt{(\mathbf{c} - \mathbf{d})^t \Sigma^{-1} (\mathbf{c} - \mathbf{d})}.$$

In our case, we have defined Σ as the empirical covariance matrix from the endmembers $\{m_r\}_{1 \leq r \leq R}$. We represented the first iso-order lines for $\leq_{d_M}$ in Fig. 2(e). As one may notice, Mahalanobis distance involves elliptical lines whose directions of principal axes e_1 and e_2 are the two eigenvectors of Σ. Hence the geometry is deformed according to the correlations between the endmembers.

However, neither the L^p metric nor the Mahalanobis distance do not follow the intrinsic geometry of the simplex. For this purpose, we have studied the interest of the Kullback-Leibler divergence and its generalization, the Rényi divergence [3,5]. Given two discrete probability measures P and Q, their similarity can be computed by the (non-symmetric) Kullback-Leibler divergence defined as

$$D_{KL}(P\|Q) = \sum_i P(i) \log \frac{P(i)}{Q(i)}. \tag{9}$$

Rényi divergence of order q, $q > 0$ of a distribution P from a distribution Q is given

$$D_q(P\|Q) = \frac{1}{q-1} \log \sum_i P(i)^q Q(i)^{1-q}. \tag{10}$$

Parameter q allows the introduction to a new family of divergences, for instance $D_{q \to 1}(P\|Q) = D_{KL}(P\|Q)$. The corresponding partial order is obtained as:

$$v_i \leq_{D_q} v_j \Leftrightarrow D_q\left(c, (a_{1,i}, \ldots, a_{R,i})^t\right) \leq D_q\left(c, (a_{1,j}, \ldots, a_{R,j})^t\right). \tag{11}$$

Fig. 3 illustrates the iso-order lines in the simplex for different values of q. We can notice in particular that the case $q = 1/2$ follows rather well the geometry of the simplex. Finally, we note that Rényi divergence and majorization are related [20].

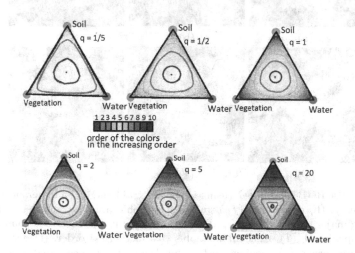

Fig. 3. Iso-order lines for partial order $\leq_{D_q}$ based on for Rényi divergence of parameter q

4   Application to Hyperspectral Image Processing

Given a hyperspectral image f, represented by its endmembers and the abundance maps:

$$f(x) = \sum_{r=1}^{R} \alpha_r(x) m_r,$$

the previous partial orders can be used to compute supremum and infimum needed for dilation and erosion of the abundance images. That is the vector order on the simplex is used on multivariate image $\{\alpha_r(x)\}_{1 \leq r \leq R}$, to compute for instance the operator $\phi : E \times \Delta^{R-1} \to E \times \Delta^{R-1}$. By abuse of notation, we write $\phi(\alpha_r)$ the r-abundance map obtained by operator ϕ. Then, the processed hyperspectral image by operator ϕ, noted $\phi(f)$ is obtained as

$$\phi(f)(x) = \sum_{r=1}^{R} \phi(\alpha_r)(x) m_r.$$

Fig. 4 gives a comparative example of the opening $\gamma_B(f)$ of a hyperspectral image f using various of the discussed orders.

To judge the effectiveness of a particular morphological operator is not easy. It depends mainly on the application. We have decided to evaluate the alternative

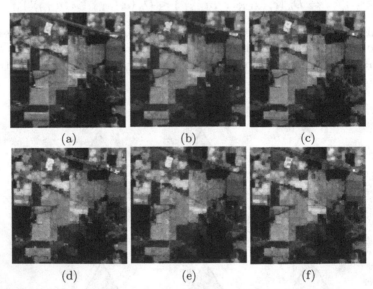

(a)   (b)   (c)

(d)   (e)   (f)

Fig. 4. (a) False RGB color image (using three spectral bands) of a hyperspectral image Indian Pines f, (b) opening $\gamma_B(f)$ using lexicographic abundance order, (c) opening $\gamma_B(f)$ using majorization-like order, (d) opening $\gamma_B(f)$ using stochastic dominance-like order, (e) opening $\gamma_B(f)$ using Mahalanobis distance-based order, (f) opening $\gamma_B(f)$ using L^1-based order, (g) opening $\gamma_B(f)$ using Rényi divergence-based order, $q = 1$.

orders for hyperspectral images in the context of image regularization and its interest to improve spectral classification. Basically, we compare the result of the supervised spectral classification obtained on the original hyperspectral image or the hyperspectral image filtered by a sequential filter according to one of the presented orders in the simplex.

This empirical study has been made on two images conventionally used in hyperspectral image processing domain: i) the Pavia image, which represents the campus of Pavia university (urban scene), of size 610×340 pixels and $D = 103$ spectral bands; ii) the Indian Pines image, test site in North-western Indiana composed for two thirds of agriculture, and one-third of forest, of 145×145 pixels and $D = 224$ spectral bands. We fixed the number of endmembers: $R = 9$ for Pavia image and $R = 10$ for the Indian Pines image. The endmembers are computed using N-FINDR [21] and the abondances by fully-constrained linear regression. Finally, for the different orders, a sequential filter $\gamma_B \varphi_B \gamma_B(f)$ is computed, where the structuring element B is a square size 3. The supervised classification is done using the least square SVM [22] algorithm with a RBF kernel as learning technique. Result of classification for both images are summarized in Tables 1 and 2.

Table 1. Comparison of result of classification on the Indian Pine hyperspectral image

Order	Overall Accuracy with RBF kernel	Kappa statistic with RBF kernel
without morpho. processing	60	0.57
lexicographic abundance order	61	0.59
majorization-like order	60	0.57
stochastic dominance-like order	71	0.69
Background order L^1	66	0.64
Background order L^2	61	0.59
Background order L^∞	58	0.55
Background order Mahalanobis distance,	64	0.62
Background order Rényi divergence $q = 0.2$,	68	0.66
Background order Rényi divergence $q = 1$,	69	0.67
Background order Rényi divergence $q^* = 4.65$,	74	0.72

Table 2. Comparison of result of classification on the Pavia hyperspectral image

Order	Overall Accuracy with RBF kernel	Kappa statistic with RBF kernel
without morpho. processing	88	0.87
lexicographic abundance order	87	0.86
majorization like-order	88	0.86
stochastic dominance like-order	90	0.89
Background order L^1	84	0.82
Background order L^2	85	0.83
Background order L^∞	87	0.85
Background order Mahalanobis distance,	88	0.86
Background order Rényi divergence $q = 0.2$,	85	0.83
Background order Rényi divergence $q = 1$,	81	0.79
Background order Rényi divergence $q^* = 1.74$,	93	0.92

We do not claim that this is the best way to improve spectral classification, however this study highlights the impact of the order used in the morphological operators, which involves in certain case on the choice of the metric. We observe that, by optimizing the parameter q of Rényi divergence, noted q^*, it is possible to significantly improve the classification score.

5   Conclusions

We have proposed different kinds of partial orders based on the an endmember representation of the hyperspectral images. They are just useful to compute morphological operators for images represented by the linear mixing model. From a mathematical morphology viewpoint, orders considered here can be used for other data lying in the simplex. This is the case for instance of images where at each pixel a discrete probability distribution is given.

In the experimental section, we have illustrate the potential interest of corresponding morphological operators for a spatial regularization before the classification. However, any image processing task tackled with morphological operators (scale-space decomposition, image enhancement, etc.) can be extended to hyperspectral images using the present framework.

References

1. Angulo, J.: Morphological colour operators in totally ordered lattices based on distances: Application to image filtering, enhancement and analysis. Computer Vision and Image Understanding 107(1), 56–73 (2007)
2. Aptoula, E., Courty, N., Lefevre, S.: An end-member based ordering relation for the morphological description of hyperspectral images. In: Proc. of the IEEE ICIP 2014 (2014)
3. Bengtsson, I., Zyczkowski, K.: Geometry of quantum states: an introduction to quantum entanglement. Cambridge University Press (2006)
4. Bellman, R.E.: Adaptive control processes. Princeton University Press, New Jersey (1961)
5. Bromiley, P.A., Thacker, N.A., Bouhova-Thacker, E.: Shannon entropy, Renyi entropy, and information. Statistics and Inf. Series (2004-004) (2004)
6. Chang, C.I.: Further results on relationship between spectral unmixing and subspace projection. IEEE Trans. on Geoscience and Remote Sensing 36(3), 1030–1032 (1998)
7. Chang, C.I. (ed.): Hyperspectral data exploitation: theory and applications. John Wiley & Sons (2007)
8. Keshava, N.: A survey of spectral unmixing algorithms. Lincoln Laboratory Journal 14(1), 55–78 (2003)
9. Marshall, A.W., Olkin, I., Arnold, B.C.: Inequalities: Theory of Majorization and Its Applications: Theory of Majorization and Its Applications. Springer (2010)
10. Mohan, A., Sapiro, G., Bosch, E.: Spatially coherent nonlinear dimensionality reduction and segmentation of hyperspectral images. IEEE Geoscience and Remote Sensing Letters 4(2), 206–210 (2007)

11. Donoho, D., Stodden, V.: When does non-negative matrix factorization give a correct decomposition into parts? In: Proc. of Advances in Neural Information Processing Systems, vol. 16. MIT Press (2003)
12. Fauvel, M., Benediktsson, J.A., Chanussot, J., Sveinsson, J.R.: Spectral and spatial classification of hyperspectral data using SVMs and morphological profiles. IEEE Trans. on Geoscience and Remote Sensing 46(11), 3804–3814 (2008)
13. Fauvel, M., Tarabalka, Y., Benediktsson, J.A., Chanussot, J., Tilton, J.C.: Advances in spectral-spatial classification of hyperspectral images. Proceedings of the IEEE 101(3), 652–675 (2013)
14. Altmann, Y.: Nonlinear unmixing of hyperspectral images, Ph.D. thesis manuscript, Université de Toulouse (France), HAL, tel-00945513 (2013)
15. Serra, J.: Anamorphoses and Function Lattices (Multivalued morphology). In: Dougherty, E. (ed.) Mathematical Morphology in Image Processing, ch. 13. Marcel Dekker (1993)
16. Shaked, M., Shanthikumar, J.G.: Stochastic Orders and their Applications. Associated Press (1994)
17. Velasco-Forero, S., Angulo, J.: Supervised ordering in $\mathbb{R}^n$: Application to morphological processing of hyperspectral images. IEEE Transactions on Image Processing 20(11), 3301–3308 (2011)
18. Velasco-Forero, S., Angulo, J.: Random projection depth for multivariate mathematical morphology. IEEE Journal of Selected Topics in Signal Processing 6(7), 753–763 (2012)
19. Velasco-Forero, S., Angulo, J.: Classification of hyperspectral images by tensor modeling and additive morphological decomposition. Pattern Recognition 46(2), 566–577 (2013)
20. van Erven, T., Harremoës, P.: Rényi divergence and majorization. In: Proc of the IEEE International Symposium on Information Theory, ISIT (2010)
21. Winter, M.E.: N-FINDR: an algorithm for fast autonomous spectral end-member determination in hyperspectral data. In: Proc. of SPIE Image Spectrometry V. SPIE, vol. 3753, pp. 266–277 (1999)
22. Camps-Valls, G., Bruzzone, L.: Kernel-based methods for hyperspectral image classification. IEEE Transactions on Geoscience and Remote Sensing 43(6), 1351–1362 (2005)

Bagging Stochastic Watershed on Natural Color Image Segmentation

Gianni Franchi and Jesús Angulo[✉]

CMM-Centre de Morphologie Mathématique, MINES ParisTech, PSL-Research
University, Paris, France
gianni.franchi@mines-paristech.fr, Jesus.Angulo@ensmp.fr

Abstract. The stochastic watershed is a probabilistic segmentation approach which estimates the probability density of contours of the image from a given gradient. In complex images, the stochastic watershed can enhance insignificant contours. To partially address this drawback, we introduce here a fully unsupervised multi-scale approach including bagging. Re-sampling and bagging is a classical stochastic approach to improve the estimation. We have assessed the performance, and compared to other version of stochastic watershed, using the Berkeley segmentation database.

Keywords: Unsupervised image segmentation · Stochastic watershed · Berkeley segmentation database

1 Introduction

The goal of image segmentation is to find a simplified representation such that each pixel of the image belongs to a connected class, thus a partition of the image into disjoint classes is obtained. One can also talk of region detection. We would like to emphasize from the beginning the difference between this type of technique and edge detection. Indeed the contours obtained by edge detection are often disconnected and may not necessarily correspond to closed areas.

In this paper, we introduce a new variant of the stochastic watershed [2] to detect regions and also to calculate the probability of contours of the edges of a given image. Inspired by the milestone segmentation paradigm of Arbelaez *et al.* [4] and in order to improve the results, we will combine some of their tools, like the probability boundary gradient or the spectral probability gradient, with the morphological segmentation obtained by the stochastic watershed. The work that we present can be applied to both gray-scale and color images. Our main assumption here is that the approach should be fully unsupervised. In order to quantitatively evaluate the performance of our contributions, we have used the benchmark available in the Berkeley Segmentation Database (BSD) [11].

Besides the classical contour detection techniques, such as the Sobel filter, the Canny edge detector or the use of a morphological gradient, based exclusively on local information, an innovative and powerful solution is the globalized probability of boundary detector (gPb) [4]. In this case, global information is added

© Springer International Publishing Switzerland 2015
J.A. Benediktsson et al. (Eds.): ISMM 2015, LNCS 9082, pp. 422–433, 2015.
DOI: 10.1007/978-3-319-18720-4_36

to the local one thanks to a multi-scale filter and also by means of a notion of texture gradient. However, supervised learning has been used to optimize gPb which could limit its interest for other type of images than the natural color images. We note also that in [4] the gPb can be the input for a ultrametric contour map computation, which leads to a hierarchical representation of the image into closed regions. This approach is somehow equivalent to the use of watershed transform on a gradient since a hierarchy of segmentation is obtained [13,14].

The interest of the stochastic watershed is to be able to estimate the probability density of contours of the image from a given gradient [2,14,15]. However, in complex images as the ones of the BSD, the stochastic watershed enhances insignificant contours. Dealing with this problem has been the object of recent work, see [5] or [9]. We propose here a multi-scale solution including bagging. The interest of a pyramidal representation is widely considered in image processing, including for instance state-of-the-art techniques such as convolutional networks [8]. Re-sampling and bagging [7] is a classical stochastic approach to improve the estimation.

2 Background

2.1 Basics on Watershed Transform for Image Segmentation

Mathematical morphology operators are non-linear transforms based on the spatial structure of the image that need a complete lattice structure, i.e. an ordering relation between the different values of the image. This is the case of erosion/dilation, opening/closing, etc. but also for the watershed transform (WT) [6,12,16,13], which is a a well known segmentation technique. Consider that we are working in a gray scale image, and that all image values are partially ordered, so it is possible to represent the image as a topographic relief. WT starts by flooding from the local minima until each catchment basin is totally flooded. At this point, a natural barrier between two catchment basins is detected, this ridge represents the contour between two regions. To segment an image, the WT is often applied to the gradient of the image, since the barrier between regions will correspond to local maxima of the gradient. Hence, in watershed segmentation the choice of the gradient is of great importance: it should represent significant edges of the image, and minimizes the presence of secondary edges. In the case of a color image, there are many alternatives to define the topographic relief function (i.e., the gradient) used for the WT.

Our practical motivation involves to identify a relevant gradient that maximizes the F-measure on the Berkeley database, being also compatible with watershed transforms (involving mainly local information). Hence, we have tested different gradients on the BSD, in particular some morphological color gradients [1], as well as the probability boundary and the globally probability boundary [4]. Following our different tests, we favor the probability boundary gradient, because it produces excellent results, being also unsupervised.

2.2 Probability Boundary Gradient (Pb)

Let us recall how the Pb gradient is computed [4]. First, the color image, which can be seen as a third order tensor, is converted in a fourth order tensor, where the first three channels correspond to the CIE Lab color space, and a fourth channel based on texture information is added. Then, on each of the first three channels, an oriented gradient of histogram $G(x, \theta)$ is calculated. To calculate the oriented gradient at angle θ for each pixel x of the image, a circular disk is centred at this pixel and split by a diameter at angle θ, see Fig. 1. The histograms of each half-disk are computed, and their difference according to the χ^2 distance represents the magnitude of the gradient at θ. In the case of the texture channel, first, a texton is assigned to each pixel, which is a 17 dimensional vector containing texture descriptor information. The set of vectors in then clustered using K-means, the result is a gray-scale image of integer value in $[1, K]$. Second, an oriented gradient of histogram is computed from this image. The four oriented gradients are linearly combined, i.e., $Pb(x, \theta) = \frac{1}{4} \sum_i^4 G_i(x, \theta)$, with i the number of channels, where $G_i(x, \theta)$ measures the oriented gradient in position x of channel i. Finally, the Pb gradient at x is obtained as the maximum for all orientations: $Pb(x) = \max_\theta \{Pb(x, \theta)\}$.

In [4], it was also considered the so-called globalized probability of boundary (gPb), which consists in computing the Pb at different scales, then added together, and finally, an enhancement by combining this gradient with the so-called spectral gradient.

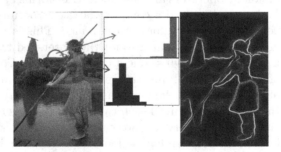

Fig. 1. Computing oriented gradient of histograms for a given orientation. Figure borrowed from [4].

2.3 From Marked Watershed to Stochastic Watershed Using Graph Model

Stochastic watershed (SW) was pioneered as a MonteCarlo approach based on simulations [2]. This is also the paradigm used for computing the examples given in this paper. However, in order to describe the SW, we follow the formulation introduced in [14], which considers a graph image representation of the WT and leads to an explicit computation of the probabilities. A similar paradigm has been also used in [10] for efficient implementations.

First let us consider the partitions formed by the catchment basins resulting from the WT. It is possible to work with the graph resulting from this segmentation called the Region Adjacency Graph (RAG) where each node is a catchment basin of the image, and where edges are weighted by a dissimilarity between regions. In particular, the weight of the edge between two nodes is the minimum level of water λ needed so that the two catchment basins are fused. Another way to see it is to say that the weight between two nodes is the lowest pass point on the gradient image between adjacent regions. Let us now explain how to compute a marked watershed from this graph [12,13]. First, thanks to some prior knowledge, markers are selected and play the role of sources from which the topographic surface will be flooded. Their flow is such that they create lakes. Another important point is that these markers represent the only local minimum of the gradient. Then on this new image we proceed at a WT. As previously, a region is associated to each local minimum. Another view of this problem is the RAG point of view. First, the Minimum Spanning Tree (MST) of the RAG is calculated, then between any two nodes, there exists a unique path on the MST and the weight of the largest edge along this path is equal to the flooding ultrametric distance between these nodes. For example, if one puts three markers on the RAG then, to have the final partition, we just have to cut the two highest edges of the MST. This operation will produce a forest where each subtree spans a region of the domain. We can easily see that if one uses $n > 1$ markers, we would cut $n - 1$ edges in order to produce n trees. However markers must be chosen accurately, since the final partition depend on it. Since one may not have a priori information about the marker positions, the rationale behind the SW is to use randomly placed markers on the MST of the RAG. By using random markers, it is possible to build a density function of the contours. Let us consider an example of image with 20 catchment basins whose MST is represented in Fig 2(a). Then if one wants to calculate the probability of the boundary between the catchment basins corresponding to the nodes e and f, we will write this probability $\mathcal{P}(e, f)$. One has first to cut all the edges above the value of the edge $e - f$, so that we get a Minimum Spanning Forest represented on Fig 2(b). Then let us consider the tree containing node e and the one containing node f that are respectively represented in purple and blue on Fig 2(b), and denoted by T_e and T_f respectively. The edge $e - f$ is cut if and only if during the process where the markers are placed randomly, there is at least one marker on T_e and at least one marker on T_f.

Let us consider the node f, and let us write $\mathcal{P}(f)$ the probability that one marker is placed on f such that $\sum_{\nu \in V} \mathcal{P}(\nu) = 1$ where V is the set of all the nodes of the image, and ν is a node. Then the probability that one marker is placed on a tree T is:

$$\mathcal{P}(T) = \mathcal{P}(\bigcup_{\nu \in T} \nu) = \sum_{\nu \in T} \mathcal{P}(\nu) \qquad (1)$$

One can express $\mathcal{P}(e, f)$ as the probability that there is at least one marker out of N markers on T_e and one on T_f, but one can also reformulate it by taking the opposite event. Then we have $\mathcal{P}(e, f)$ is equal to the opposite that there is

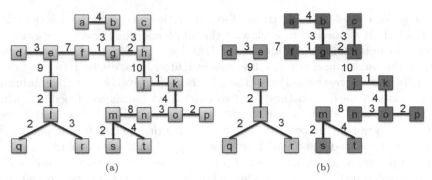

Fig. 2. (a) the Minimum Spanning Tree of on image, (b) the Minimum Spanning Forest of the same image.

no markers on T_e or on T_f. The probability that out of N markers there is no markers on a tree T is $\mathcal{P}(\bar{T}) = (1 - \mathcal{P}(T))^N$. Thus, we have:

$$\mathcal{P}(e, f) = 1 - (1 - \mathcal{P}(T_e))^N - (1 - \mathcal{P}(T_f))^N + (1 - \mathcal{P}(T_e \bigcup T_f))^N \qquad (2)$$

There are different ways to express the probability that one marker is placed on a node, a classical one developed in [2] is to consider that the seed are chosen uniformly, in this case we consider the surfacic stochastic watershed. Since the probability of the tree will depend mainly on it surface. One can also consider other criteria, it is proposed for instance in [14] a volumic stochastic watershed, where the probability of the tree will depend on the surface$\times\lambda$, where λ is the minimum altitude to flood the nodes of this tree. For other variants, see [15].

We use here a variant of the stochastic watershed, where the probability of each pixel depends on value of the function to be flood, typically the gradient. First the gradient is rescaled, such that the sum of all the pixels of the gradient is equal to one, then the probability of a node is equal to the sum of all the pixel of the node. This kind of probability map improves the result with respect to the classical (surface) SW. However this may not be sufficient, that is the motivation to improve our result thanks to bagging.

3   Bagging Stochastic Watershed

3.1   Bootstrap and Bagging

In digital images, one has to deal with a discrete sampling of an observation of the real word. Thus image processing techniques are dependent on this sampling. A solution to improve the evaluation of image estimators consists in using bootstrap [7]. Bootstrap techniques involves building a new world, the so-called "Bootstrap world" which is supposed to be an equivalent of the real world. Then, by generating multiple sampling of this "Bootstrap world", one speaks of resampling, a new estimator is obtained by aggregating the results of the estimator on each of these samples. That is the reason to call the approach bootstrap aggregating or bagging. In our case, the "Bootstrap image" is the original one, and we have to produce different samples from this image, and the estimator is the probability density of contours obtained from the SW.

3.2   Bagging Stochastic Watershed (BSW)

We can immediately note that bagging has a cost: a loss of image resolution. However this is not necessary a drawback in natural images. Indeed, most of digital cameras produces nowadays color images which are clearly over-resolved in terms of number of pixels with respect to the optical resolution. For instance, the typical image size of for a smartphone is 960×1280 pixels.

SW combined at different scales has been already the object of previous research. In [3], it was proposed to improve the result of SW working with seeds that are not just points, but dilated points, i.e., disk of random diameter, see also [14] for the corresponding computation. By doing that, we change the scale of the study.

Moreover if one has a look at Fig. 3(b), it is possible to see that the human ground truth only focuses on most significant and salient boundaries. This ground truth represents the average of the boundaries drawn by about 5 subjects. Typically, texture information is ignored. In addition, the selected contours are not always based on local contrast. Human visual system (HVS) integrates also some high level semantic information in the selection of boundaries. It is well known that multi-scale image representation is part of the HVS pipeline and consequently algorithms based on it are justified. This is the case for instance of convolutional neural networks.

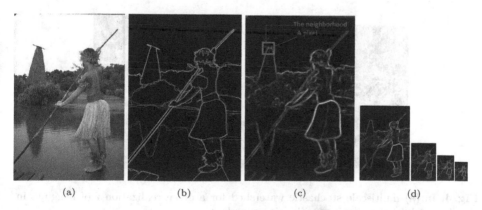

(a)   (b)   (c)   (d)

Fig. 3. (a) Example of image from the BSD [11], (b) corresponding ground truth [11], (c) the original Pb gradient, (d) a multi-scale representation based on sampling by mean value computation.

In the present approach, the multiresolution framework for SW includes also the notion of bagging, which is also potentially useful in case of "noisy data". Let us denote by g the "Bootstrap image gradient" and by $\mathcal{P}_{SW}(g)$ the probability density function (pdf) of contours obtained from a gradient g. The bagging stochastic watershed (BSW) is computed as follows.

- **Multi-scale representation by resampling:** Given g, at each iteration i, the image g is resampled in a squared grid of size fixed : $n_j \times n_j$. This resampling is done by selecting 50% of pixels in a $n_j \times n_j$ neighbourhood, which are used to estimate the mean value. The corresponding downsampled by averaging gradient is denoted $g_i^{\downarrow j}$. The step is applied for $S = 4$ scales, such that $n_1 = 2 \times 2$, $n_2 = 4 \times 4$, $n_3 = 6 \times 6$ and $n_4 = 8 \times 8$. An example is illustrated in Fig. 3(d). We denote $g_i^{\downarrow 0} = g$.
- **Multi-scale SW:** For each realization i of resampled gradient, compute the SW at each scale j: $\mathcal{P}_{SW}(g_i^{\downarrow j})$, see example in Fig. 4.
- **Multiple realizations and bagging:** Resampling procedure is iterated N times for each of the S scales. Bagging involves to combine all the pdf of contours. Note that an upsampling step, here we chose the bilinear interpolation, is required to combine at the original resolution. Thus, the bagging stochastic watershed is computed as

$$\mathcal{P}_{BSW}(g) = (SN)^{-1} \sum_{i=1}^{N} \sum_{j=0}^{S} w_j \left[\mathcal{P}_{SW}(g_i^{\downarrow j})\right]^{\uparrow j}. \tag{3}$$

We have empirically fixed the following scale weights: $w_1 = 0.6$, $w_2 = 0.2$, $w_3 = 0.1$, $w_4 = 0.1$. The weights can be chosen according to the pattern spectrum of the image.

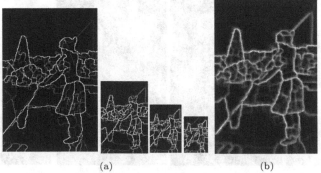

(a)   (b)

Fig. 4. In (a) multiscale stochastic watershed for a given realization i of bagging, in (b) the result of the bagging stochastic watershed

3.3   Improved Accuracy of BSW

Before discussing the results, let us prove that by aggregating the different multi-scale replicates of SW the accuracy of this estimator is theoretically improved.

Let us write $\mathcal{B} = \{b\}$ the set of boundaries b of an image, and $P(b)$ the true probability boundary of $b \in \mathcal{B}$. Let us consider $\mathcal{P}_{SW}(b \in \mathcal{B}, g)$ the estimation of the probability boundary by means of the SW from gradient g, and $\mathcal{P}_{BSW}(b \in \mathcal{B}, g)$ the estimation using the BSW. We note by $\mathcal{L} = \{g^{\downarrow}\}$ the set of resampled gradients where the SW is applied. Then, we have:

$$\mathcal{P}_{BSW}(b \in \mathcal{B}, g) = \mathbb{E}_{\mathcal{L}}[\mathcal{P}_{SW}(b \in \mathcal{B}, g^{\downarrow})]. \tag{4}$$

We define the quadratic error of estimation of the probability boundary on $g^{\downarrow}$ as:

$$\mathrm{E}_{\mathcal{P}_{SW}} = \mathbb{E}_{\mathcal{B}}[P(b) - \mathcal{P}_{SW}(b \in \mathcal{B}, g^{\downarrow})]^2. \tag{5}$$

The expectation of this error on $\mathcal{L}$ is given by $\mathbb{E}_{\mathcal{L}}\mathbb{E}_{\mathcal{B}}[P(b) - \mathcal{P}_{SW}(b \in \mathcal{B}, g^{\downarrow})]^2$, which can be rewritten as

$$\mathbb{E}_{\mathcal{B}}(P(b)^2) - 2\mathbb{E}_{\mathcal{B}}[P(b)\mathbb{E}_{\mathcal{L}}[\mathcal{P}_{SW}(b \in \mathcal{B}, g^{\downarrow})]] + \mathbb{E}_{\mathcal{B}}[\mathbb{E}_{\mathcal{L}}[\mathcal{P}_{SW}(b \in \mathcal{B}, g^{\downarrow})]^2].$$

Now, by using the fact that $\mathbb{E}(z^2) \geq (\mathbb{E}(z))^2$, we obtain:

$$\mathbb{E}_{\mathcal{L}}\mathbb{E}_{\mathcal{B}}[P(b) - \mathcal{P}_{SW}(b \in \mathcal{B}, g^{\downarrow})]^2 \geq \mathbb{E}_{\mathcal{B}}[P(b) - \mathcal{P}_{BSW}(b \in \mathcal{B}, g)]^2 = \mathrm{E}_{\mathcal{P}_{BSW}}. \tag{6}$$

Therefore, we note that the average error of $\mathcal{P}_{SW}$ is higher than the error of $\mathcal{P}_{BSW}$. In conclusion, by aggregating the SW we decrease the error of estimation of the probability boundaries, but we also decrease the image resolution.

4   Results on BSD

Evaluation of the BSW has been done in the BSD, containing 200 natural images and their ground truths (with contours manually segmented by 5 different subjects. We have compared the results of Pb and gPb from [4], a simple morphological color gradient (G) [1], the stochastic watershed (SW) [2], the improved stochastic watershed (ISW) as introduced in [5] and the proposed bagging stochastic watershed (BSW). We have also include a last result which corresponds to a pdf obtained by multiplying the BSW with the spectral probability gradient proposed also in [4], this combination is named the spectral bagging stochastic watershed (SBSW).

Since the result of the algorithm is an image of probability of contours, by selecting different thresholds, one has a totally different set of contours. Similarly to [4], at each level of the threshold, the F-measure is calculated: F-measure $= (2 \times \text{Precision} \times \text{Recall})/(\text{Precision} + \text{Recall})$, showing a tradeoff between precision and recall. The main advantage of the F-measure is that, contrary to the ROC curve, does not depend on the true negatives, which may turn the results to be less discriminative. Table 1 summarizes the results for different methods. The

Table 1. (a) Comparison of F-measure (obtained by thresholding the pdf) on BSD. (b) Comparison of C-measure (obtained by hierarchy based on contrast) on BSD.

		Pb	gPb	G	SW	ISW	BSW	SBSW
(a)	ODS	0.64	0.69	0.25	0.59	0.66	0.63	0.65
	OIS	0.66	0.70	0.26	0.62	0.69	0.64	0.67

		Pb	gPb	SW	ISW	BSW	SBSW
(b)	ODS	0.62	0.66	0.57	0.63	0.62	0.63
	OIS	0.65	0.70	0.58	0.67	0.65	0.66
	mean(C-measure)	0.19	0.071	0.54	0.19	0.39	0.01

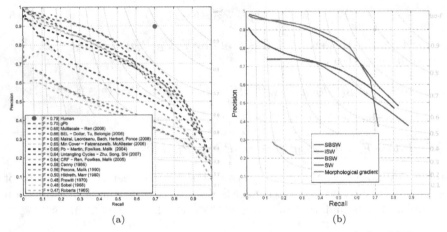

(a)   (b)

Fig. 5. (a) F-measure of different algorithms, source of this plot [17] . (b) F-measure of algorithms compared on this paper.

curves of the F-measure are given in Fig. 5. As in [4], two features are computed from the F-measure: the ODS which is the optimal scale for the entire data set, and the OIS which is the optimal scale per image. We note the ISW and the here proposed BSW improves the results produced by the SW, with comparable performances. However, none of the morphological approaches are better than the gPb, which we remind is based on a learning procedure.

Thresholding and calculating F-measure of the pdf produce the optimal set of boundaries, but not the optimal regions. Instead of using a simple threshold on the pdf, we propose to compute a watershed transform after applying a h-reconstruction on the pdf. That produces a segmentation into closed regions and the choice of h is related to the contrast of probability. When this approach is obtained for different values of h we obtain a hierarchy of segmentations [13], related also to the so-called ultrametric contour map in [4]. Now, the F-measure can be computed at each value of the hierarchy h. In addition, we propose to compute the following contrast feature:

$$C\text{-measure} = (\sum_h h \times F\text{-measure}(S_h))/\sum_h h \qquad (7)$$

where S_h is the set of closed contours obtained from the h-reconstruction watershed. The C-measure 7 represents the fact that one does not know what is the optimal value h^*, so the C-measure, that we develop is a mean over different values of h. On Table 1.(b), for each algorithm we have computed the mean of the C-measures over the 200 images. As one can notice, using this feature, the BSW and the ISW seem to be the best compromise in terms of contrast/F-measure.

Fig. 6 provides also a comparison for the current image of the pdf of contours as well as two segmentations from each pdf, obtained by watershed combined with h-reconstruction ($h = 0.1$ and $h = 0.25$).

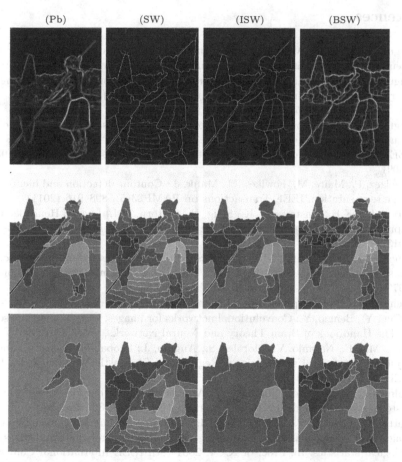

Fig. 6. First row, pdf of contours; second row, segmentation by keeping just the contours with more than 10% of contrast; third row, more than 25% of contrast. First column, Pb; second column, classical stochastic watershed (SW); third column, improved SW; fourth column bagging SW.

5   Conclusion

Stochastic watershed is a probabilistic segmentation approach which can be integrated into a bagging paradigm to improve estimation of probability of contours. We have in particular considered a multi-scale approach of bagging stochastic watershed which at this point is fully unsupervised. In order to assess the performance in the BSD, we have used the F-measure and new criterion called the C-measure.

The combination of multi-scale bagging stochastic watershed with segmentation learning architectures based for instance in convolutional networks should be considered in future research.

References

1. Angulo, J., Serra, J.: Modelling and Segmentation of Colour Images in Polar Representations. Image Vision and Computing 25(4), 475–495 (2007)
2. Angulo, J., Jeulin, D.: Stochastic watershed segmentation. In: Proc. of the 8th International Symposium on Mathematical Morphology (ISMM 2007), pp. 265–276. MCT/INPE (2007)
3. Angulo, J., Velasco-Forero, S., Chanussot, J.: Multiscale stochastic watershed for unsupervised hyperspectral image segmentation. In: Proc. of IEEE IGARSS 2009 (IEEE International Geoscience & Remote Sensing Symposium), vol. III, pp. 93–96 (2009)
4. Arbelaez, P., Maire, M., Fowlkes, C., Malik, J.: Contour detection and hierarchical image segmentation. IEEE Transactions on PAMI 33(5), 898–916 (2011)
5. Bernander, K.B., Gustavsson, K., Selig, B., Sintorn, I.M., Luengo Hendriks, C.L.: Improving the stochastic watershed. Pattern Recognition Letters 34(9), 993–1000 (2013)
6. Beucher, S., Lantuéjoul, C.: Use of watershed in contour detection. In: Proc. of Int. Worshop Image Processing, Real-Time Edge and Motion Detection/Estimation (1979)
7. Breiman, L.: Bagging predictors. Machine Learning 24(2), 123–140 (1996)
8. LeCun, Y., Bengio, Y.: Convolutional networks for images, speech, and time series. In: The Handbook of Brain Theory and Neural Networks, vol. 3361 (1995)
9. López-Mir, F., Naranjo, V., Morales, S., Angulo, J.: Probability Density Function of Object Contours Using Regional Regularized Stochastic Watershed. In: Proc. IEEE ICIP 2014 (2014)
10. Malmberg, F., Luengo Hendriks, C.L.: An efficient algorithm for exact evaluation of stochastic watersheds. Pattern Recognition Letters 47, 80–84 (2014)
11. Martin, D., Fowlkes, C., Tal, D., Malik, J.: A database of human segmented natural images and its application to evaluating segmentation algorithms and measuring ecological statistics. In: Proc. of ICCV 2001 (8th IEEE International Conference on Computer Vision), vol. 2, pp. 416–423 (2001)
12. Meyer, F., Beucher, S.: Morphological Segmentation. Journal of Mathematical Imaging and Vision 1(1), 21–46 (2012)
13. Meyer, F.: An overview of morphological segmentation. International Journal of Pattern Recognition and Artificial Intelligence 15(07), 1089–1118 (2001)

14. Meyer, F., Stawiaski, J.: A stochastic evaluation of the contour strength. In: Goesele, M., Roth, S., Kuijper, A., Schiele, B., Schindler, K. (eds.) Pattern Recognition. LNCS, vol. 6376, pp. 513–522. Springer, Heidelberg (2010)
15. Meyer, F.: Stochastic Watershed Hierarchies. In: Proc. of ICAPR 2015 (2015)
16. Roerdink, J.B.T.M., Meijster, A.: The Watershed Transform: Definitions, Algorithms and Parallelization Strategies. Fundamenta Informaticae 41(1-2), 187–228 (2000)
17. http://www.eecs.berkeley.edu/Research/Projects/CS/vision/grouping/resources.html

Binary Partition Trees-Based Spectral-Spatial Permutation Ordering

Miguel Ángel Veganzones[1(✉)], Mauro Dalla Mura[2], Guillaume Tochon[2], and Jocelyn Chanussot[2,3]

[1] GIPSA-lab, CNRS, Saint Martin d'Hères, Grenoble, France
miguel-angel.veganzones@gipsa-lab.fr
[2] GIPSA-lab, Grenoble-INP, Saint Martin d'Hères, Grenoble, France
[3] Faculty of Electrical and Computer Engineering, University of Iceland, Reykjavik, Iceland
mauro.dalla-mura@gipsa-lab.fr, guillaume.tochon@gipsa-lab.fr, jocelyn.chanussot@gipsa-lab.fr

Abstract. Mathematical Morphology (MM) is founded on the mathematical branch of Lattice Theory. Morphological operations can be described as mappings between complete lattices, and complete lattices are a type of partially-ordered sets (*poset*). Thus, the most elementary requirement to define morphological operators on a data domain is to establish an ordering of the data. MM has been very successful defining image operators and filters for binary and gray-scale images, where it can take advantage of the natural ordering of the sets $\{0, 1\}$ and $\mathbb{R}$. For multivariate data, *i.e.* RGB or hyperspectral images, there is no natural ordering. Thus, other orderings such as reduced orderings (R-orderings) have been proposed. Anyway, all these orderings are based solely on sorting the spectral set of values. Here, we propose to define an ordering based on both, the spectral and the spatial information, by means of a binary partition tree (BPT) representation of images. The proposed ordering aims to find a permutation of the pixel indexes, that is, a sorting of the pixels arrangement in the data matrix. Morphological operations using the proposed ordering are able to enlarge (shrink) spatial structures independently of their spectral values, as far as the spatial structures are encoded in the BPT representation. We provide examples of potential use of the proposed ordering using binary and RGB images.

Keywords: Mathematical morphology · Lattice theory · Ordering · Spectral spatial analysis · Binary partition trees · Permutations

1 Introduction

Mathematical Morphology (MM) [15,10,16,7,18] has been very successful in defining image operators and filters for binary and gray-scale images. A very appealing characteristic of MM is that its mathematical support is well known. Lattice Theory [5,8,9] gives the most general theoretical background for MM [13,11]. Basically, morphological operations can be described as mappings between complete lattices. From now on, we denote complete lattices by the symbols

© Springer International Publishing Switzerland 2015
J.A. Benediktsson et al. (Eds.): ISMM 2015, LNCS 9082, pp. 434–445, 2015.
DOI: 10.1007/978-3-319-18720-4_37

$\mathbb{L}$ and $\mathbb{M}$. For every subset $Y \subseteq \mathbb{L}$ an *erosion* is a mapping $\varepsilon : \mathbb{L} \rightarrow \mathbb{M}$ that commutes with the infimum operation, $\varepsilon \left(\bigwedge Y \right) = \bigwedge_{y \in Y} \varepsilon \left(y \right)$. Similarly, a *dilation* is a mapping $\delta : \mathbb{L} \rightarrow \mathbb{M}$ that commutes with the supremum operation, $\delta \left(\bigvee Y \right) = \bigvee_{y \in Y} \delta \left(y \right)$. An *anti-erosion* operator, $\bar{\varepsilon} : \mathbb{L} \rightarrow \mathbb{M}$, and an *anti-dilation* operator, $\bar{\delta} : \mathbb{L} \rightarrow \mathbb{M}$, are defined as mappings holding $\bar{\varepsilon} \left(\bigwedge Y \right) = \bigvee_{y \in Y} \bar{\varepsilon} \left(y \right)$ and $\bar{\delta} \left(\bigvee Y \right) = \bigwedge_{y \in Y} \bar{\delta} \left(y \right)$, respectively. Any mapping Ψ between complete lattices $\mathbb{L}$ and $\mathbb{M}$ can be expressed in terms of supremums and infimums of these four morphological operators [3].

Binary and gray scale images are complete lattices given the natural order of the binary set, $\{0, 1\}$, and the positive integers, $\mathbb{Z}^{+}$, respectively (or the set of real vectors, $\mathbb{R}$, in general). The extension of MM to multivariate images is not straightforward since pixels are (high dimensional) vectors without an intrinsic natural total order. There are different strategies to build up an order from multivariate data [4,12,17,2]. The Marginal ordering (M-ordering) corresponds to univariate orderings realized on every component of the given vectors, so the M-ordering is also called *component-wise ordering*. The M-ordering allows the use of MM but all the between components information is ignored, since each spectral band is independently processed. Furthermore, the use of a M-ordering results in the apparition of vectors which are not present in the original image, what is called the *false color problem*.

To avoid the false color problem, the Conditional ordering (C-ordering) establishes a priority between the vectors marginal components. The vectors components are ranked and sequentially selected according to this rank. *Lexicographical ordering* is the most known example of C-ordering. The C-ordering is useful when a natural priority exists among vector components, which is not often the case for multi- and hyperspectral images. Even when this is suitable, the use of C-orderings as the lexicographical ordering, yields to the use of only a few components dismissing the information contained in the ones left. One way to define a multivariate ordering that makes use of all the components and interdependencies is by constructing a surjective mapping into a lattice $h : \mathbb{R}^{n} \rightarrow \mathbb{L}$ so that we can assume an ordering induced by this mapping. This is the so called Reduced ordering (R-ordering). Some authors [1,24] define R-orderings on the basis of a supervised classifier trained with some pixel values. Discriminant function values and class a posteriori probabilities provide the surjective mapping h.

1.1 Motivation

All the previous orderings, that allow to define complete lattices, and thereof, to apply mathematical operators over binary, gray-scale or multivariate images, are based on the ordering of the set of spectral values only. No spatial information of the pixels arrangement is employed to define the ordering. If we assume that the false color problem is undesirable, defining an ordering of the pixel values could be reduced to find a permutation of the pixels arrangement. Suppose, the pixels are arbitrarily arranged in matrix form, $\mathbf{X} \in \mathbb{R}^{N \times M}$, where N denotes the number of pixels and M the number of spectral bands. Then, an ordering of the pixels could be defined by a permutation of the rows of the data matrix, $\mathbf{X}$:

$$\forall \mathbf{x}_\alpha, \mathbf{x}_\beta \in \mathbb{R}^M, \ \mathbf{x}_\alpha \leq_\pi \mathbf{x}_\beta \Leftrightarrow \pi_\alpha \leq \pi_\beta, \tag{1}$$

where π denotes a permutation of the row indexes, $\{1, \ldots, N\}$.

The C- and R-orderings can be understood as mappings from the spectral domain to complete lattices that provide a permutation, π, of the data matrix solely based on spectral information. However, redefining the ordering problem as finding the right permutation of the arrangement of the pixels in the data matrix allows to introduce spatial information into the underneath mapping in a natural way. Thus, we aim to define mappings from a combination of the spectral and the spatial domains to complete lattices, and to use the permutation-based definition of an ordering to provide such mappings.

1.2 Contribution

Here, we propose to define an ordering using permutations obtained by sorting the pixels using both, the spectral values of the pixels an their spatial locations. In particular, we propose to use the spatial structures encoded in a binary partition tree (BPT) representation of images [14]. We show that the ordering of the leaf nodes defines an arrangement of the pixels. Thus, obtaining an ordering is reduced to obtain a permutation of the leaf nodes. The image spectral-spatial information is encoded in a BPT representation. The BPT construction provides an arbitrary arrangement of the BPT leaves which we consider as the reference pixels arrangement. Then, we find a permutation of the leaves in base to the spectral information of the nodes and the spatial structures encoded in the BPT hierarchy. This defines a permutation of the pixels arrangement that raises an ordering of the data samples. We show the potential use of this novel BPT-based spectral-spatial permutation ordering, and we discuss on further extensions of the proposed ideas.

2   Binary Partition Trees

Hierarchical segmentation algorithms have proved to be very valuable to explore and exploit the spatial content of images by providing a hierarchy of segmentations working at different scales. The BPT is a hierarchical region-based representation of an image in a tree structure [14]. Recently, some authors have proposed the use of the Binary Partition Tree (BPT) to handle very high dimensional images such as hyperspectral images [21,22,20,23].

In the BPT representation, the leaf nodes correspond to an initial partition of the image, which can be the individual pixels, or a coarser segmentation map (superpixels). From this initial partition, an iterative bottom-up region merging algorithm is applied until only one region remains. This last region represents the whole image and corresponds to the root node. All the nodes between the leaves and the root result of the merging of two adjacent children regions. An example of BPT is displayed in Fig. 1. If the initial partition contains N leaf nodes, the BPT representation contains $2N - 1$ nodes.

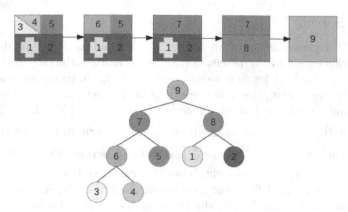

Fig. 1. Construction of the Binary Partition Tree (BPT)

Two notions are of prime importance when defining a BPT: i) the *region model* which specifies how a region $\mathcal{R}$ is modelled, and ii) the *merging criterion* which is a dissimilarity measure between the region models of any two regions $\mathcal{R}_\alpha$ and $\mathcal{R}_\beta$. Each merging iteration involves the search of the two neighbouring regions which achieve the lowest pair-wise dissimilarity among all the pairs of neighbouring regions in the current segmentation map. Those two regions are consequently merged. The first-order parametric model, defined by the component-wise average of the pixels in the region, is a widely used region model. Given the first-order parametric model, the Euclidean or the spectral angle distance (SAD) between the averaged values of two adjacent regions, are the most employed merging criteria for RGB and hyperspectral images, respectively. The building of a BPT may suffer from small and meaningless regions resulting in a spatially unbalanced tree. To overcome this limitation, a priority term is included in the merging criterion that forces those regions smaller than a given percentage of the average region size to be merged first [6,19].

3 BPT-based Spectral-spatial Permutation Ordering

Given a BPT representation of an image, the leaves are sorted from left to right following the hierarchical structure in an arbitrary way, normally selected according to implementation aspects. In other words, in the construction of the tree, there is no ordering between sibling nodes. In this work, we propose to impose an ordering among nodes in the tree. Specifically, the ordering is done on the leaves of the tree. Given an image with N leaf nodes, *i.e.* with N pixels, there are $N-1$ merging operations needed to build the BPT. Each of the merging operations implies a decision about the sorting of nodes that are being merged. Thus, there are 2^{N-1} possible permutations of the BPT leaves. Next, we propose a criterion to select one among all the possible permutations of the leaves using the hierarchical spatial structures encoded in the BPT representation and to use such permutation to define a BPT-based spectral-spatial permutation ordering of the image pixels.

3.1 BPT Sorting

Given a BPT representation of an image, $\mathbf{X}$, denoted as $\mathcal{T}$, let us define the reference arrangement of the pixels, $(\mathbf{x}_1, \ldots, \mathbf{x}_N)$, as the arbitrary arrangement of the image pixels. Also, let us associate each pixel, $\mathbf{x}_\alpha$, to its corresponding leaf node, $\mathcal{R}_\alpha$. By convention, the nodes resulting of the merging process take identification numbers consecutive to the ones of the leaves, $\{N+1, \ldots, 2N-1\}$. Let $\boldsymbol{\Pi} = \left\{\boldsymbol{\pi}^{(k)}\right\}_{k=1}^{2^{N-1}}$ be the set of all the possible permutations of the leaves of $\mathcal{T}$ where each $\boldsymbol{\pi}^{(k)} = \left[\pi_1^{(k)}, \ldots, \pi_N^{(k)}\right]$, denotes a permutation that sorts the N reference leaves from left-to-right. Thus, for each pixel, $\mathbf{x}_\alpha$, $1 \le \alpha \le N$, the permutation index, π_α, defines the location of the leaf node, $\mathcal{R}_\alpha$, in the sorted BPT, here on denoted as $\mathcal{T}_\pi$. We aim to define an educated criterion to select an optimal permutation of the BPT, $\boldsymbol{\pi}^* \in \boldsymbol{\Pi}$.

Consider the BPT representation shown in Fig. 1. Assume that the pixels identification numbering corresponds to an arbitrary arrangement of the pixels of the image, $(\mathbf{x}_1, \mathbf{x}_2, \mathbf{x}_3, \mathbf{x}_4, \mathbf{x}_5)$. The BPT construction gives an arbitrary permutation of the leaf nodes (the pixels), denoted as $\boldsymbol{\pi}^{(1)} = [4, 5, 1, 2, 3]$, which defines the ordering $\mathbf{x}_3 \le \mathbf{x}_4 \le \mathbf{x}_5 \le \mathbf{x}_1 \le \mathbf{x}_2$. The ordering can be visually interpreted in the BPT representation as an arrangement of the BPT leaves from left to right (see Fig. 2(a)). For instance, the permutation index, $\pi_1^{(1)} = 4$, means that the first leaf node, $\mathcal{R}_1$, identifying the pixel, $\mathbf{x}_1$, is located in the fourth position of the sorted BPT counting from the left to the right. Now, by educatedly sorting the BPT leaves we obtain a second permutation, $\boldsymbol{\pi}^{(2)} = [5, 4, 2, 1, 3]$, which defines the ordering $\mathbf{x}_4 \le \mathbf{x}_3 \le \mathbf{x}_5 \le \mathbf{x}_2 \le \mathbf{x}_1$ (see Fig. 2(b)).

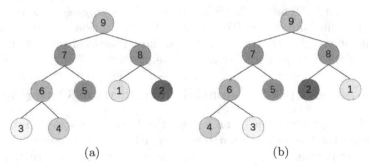

(a)   (b)

Fig. 2. Two examples of possible permutations of the BPT leaves encoding the same hierarchical structure: (a) $\boldsymbol{\pi}^1 = [4, 5, 1, 2, 3]$; and (b) $\boldsymbol{\pi}^2 = [5, 4, 2, 1, 3]$

Following, we propose as a criterion to select a permutation, $\boldsymbol{\pi}^*$, among the set of all possible permutations of the leaves, $\boldsymbol{\Pi}$, of a given BPT. The proposed methodology follows a top-down approach where at each node, the child node with minimum dissimilarity to the parent node is set to the left of the branch division. The pseudo-code of this method that we have termed as the *minimum dissimilarity top-down BPT sorting* is given in Alg. 1. The proposed method recursively sorts the BPT representation starting from the root node. The recursion goes down by sorting the children of the node in turn, and then calling himself to sort each of the sub-trees defined by each of the sorted child nodes. The recursion stops when the node in turn is a leaf node. Once the BPT have been sorted, the leaves form a permutation of the original reference BPT.

Algorithm 1. Minimum dissimilarity top-down BPT sorting pseudo-code

---

1. $\mathcal{T} \leftarrow \mathrm{BPT}\,(\mathbf{X})$ ▷ BPT construction
2. $\mathcal{R}_0 \leftarrow \mathrm{root}\,(\mathcal{T})$,   ▷ Top-down initialization
3. $\mathcal{T}_\pi \leftarrow \mathrm{sortBPT}\,(\mathcal{R}_0)$ ▷ Recursive sorting algorithm
4. $\{\mathcal{R}_{\pi_i}\}_{i=1}^{N} \leftarrow \mathrm{leaves}\,(\mathcal{T}_\pi)$ ▷ Leaf nodes of the sorted BPT
5. **return** π

function sortBPT $(\mathcal{R})$
  $[\mathcal{R}_l, \mathcal{R}_r] \leftarrow \mathrm{children}\,(\mathcal{R})$ ▷ Children nodes (left and right)
  if $\mathcal{R}_l = \emptyset$ **or** $\mathcal{R}_r = \emptyset$ **then**   ▷ Stop condition
   break
  end
  if $d\,(\mathcal{R}, \mathcal{R}_r) < d\,(\mathcal{R}, \mathcal{R}_l)$ **then**   ▷ Compare region models
   $\mathcal{R}_{\mathrm{tmp}} \leftarrow \mathcal{R}_l$ ▷ Exchange subtrees
   $\mathcal{R}_l \leftarrow \mathcal{R}_r$
   $\mathcal{R}_r \leftarrow \mathcal{R}_{\mathrm{tmp}}$
  end
  sortBPT $(\mathcal{R}_l)$ ▷ Sort left subtree
  sortBPT $(\mathcal{R}_r)$ ▷ Sort right subtree
end

3.2 BPT-based Spectral-Spatial Ordering

Given a pixel, $\mathbf{x}$, and its neighbouring pixels, $\mathcal{N}_\mathcal{S}\,(\mathbf{x})$, according to some structural element $\mathcal{S}$, an erosion (dilation), using the proposed BPT-based permutation ordering, will change the given pixel value by the value of the pixel in the neighbourhood identified by the leftmost (rightmost) leaf of the sorted BPT. Since the BPT encodes a hierarchy of spatial structures, eroding (dilating) an image shrinks the pixels values to those spectral values of spatial structures encoded to the left (right) of the BPT. A consequence is that an erosion will enlarge the spatial structures encoded to the leftmost part of the spatial hierarchies, while a dilation will enlarge those spatial structures to the rightmost part. In all cases, the enlarging (shrinking) works inside the subtrees that are involved according to the structure element.

4 Examples

In order to show the potential use of the proposed spectral-spatial π-ordering defined on base to the sorting of the BPT leaves, we have applied basic morphological operations to the logo of the ISMM conference in binary and RGB versions, and also on a grayscale and RGB image from the COREL dataset.

Fig. 3. Application of the proposed BPT-based ordering methodology to the logo of the ISMM conference: (a) original binary logo, (b) original RGB logo

4.1 ISMM Logos

Fig. 3 shows the binary and RGB versions of the ISMM logo. In Fig. 4 we compare the result of applying erosion and dilation operations using a 11 × 11 square structural element, given a conventional component-wise ordering and the proposed π-ordering from a BPT representation of the images obtained using the first-order parametric model and the Euclidean distance. Figs. 4(a-d) show the results obtained by the conventional component-wise ordering, while Figs. 4(e-h) depicts the results obtained by the proposed π-ordering.

For the binary logo, the conventional ordering results in shrinking the white areas and enlarging the black ones for the erosion operation, and the contrary for the dilation, as it was expected. However, the morphological operations using the proposed π-ordering enlarge or reduce the geometrical structures of the logo independently of the foreground and background colors. That is, it works directly on the foreground and background structures independently of the binary encoding used to represent them in each case. For instance, in Fig. 4(e), an erosion was applied according to the BPT-based π-ordering. Both foreground objects were enlarged independently of their colors. From the point of view of a conventional use of the erosion/dilation operators in binary images, the proposed erosion operated as a conventional dilation on the left part of the image and as conventional erosion on the right part. But this behaviour results of a single morphological operation and it is given by the spectral and spatial nature of the proposed ordering.

The results using the RGB logo could be explained using the same reasoning than for the binary logo. Here, we were interested in highlight that the proposed π-ordering naturally operates on multivariate data. The conventional component-wise ordering can not handle the multivariate RGB information properly, modifying the colors of the geometrical structures, while the proposed π-ordering is able to return the expected result, *i.e.*, all foreground objects are consistently processed regardless their color.

4.2 Real Image

Next, we provide a comparison using a real image from the COREL database. In all cases, we make use of a 11 × 11 square structural element. Fig. 5 depicts the gray-scale and RGB versions of the image. Fig. 6 depicts the results obtained

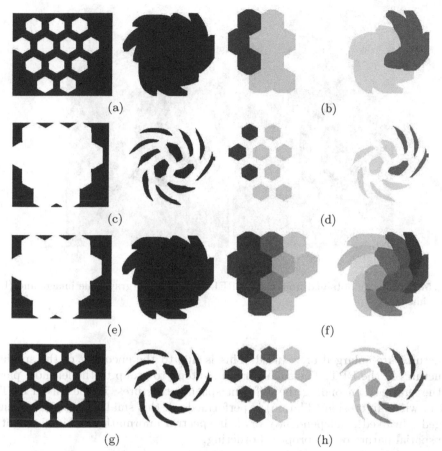

Fig. 4. Application of the erosion and dilation operations to the ISMM logos using a square 11×11 structural element: (a) conventional erosion of the binary logo, (b) component-wise erosion of the RGB logo, (c) conventional dilation of the binary logo, (d) component-wise dilation of the RGB logo, (e) proposed erosion of the binary logo, (f) proposed erosion of the RGB logo, (g) proposed dilation of the binary logo, and (h) proposed dilation of the RGB logo.

using conventional gray-scale and RGB component-wise morphological operations compared to the proposed π-ordering. The potential use of the proposed π-ordering can be appreciated by looking at the background. This is composed of sand with several scattered dark and bright spots, probably corresponding to small rocks, salt grains and other small particles in the sand. Using conventional erosion and dilation morphological operators it is not possible to get rid of both dark and bright structures, while using the proposed methodology, either both

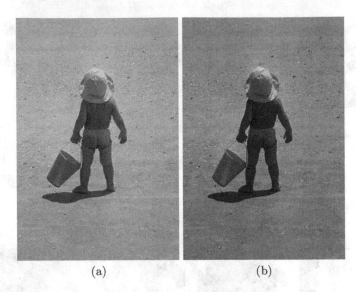

(a)   (b)

Fig. 5. Real image obtained from the COREL database: (a) gray-scale image, and (b) RGB image.

structures are enlarged or reduced. This is due to the encoding of the spatial structures in the BPT. The small structures tend to group to the leftmost part of the subtrees, encoding the different spatial structures, in the sorted BPT. Then, when an erosion (dilation) is performed, all this small structures are enlarged (shrinked), independently of their spectral information. This highlights the spatial nature of the proposed ordering.

5 Discussion

We would like to point-out some aspects and open discussions of the proposed methodology we are working on:

- We have proposed a definition of a permutation ordering, which relies in a pixel to pixel mapping, instead of a spectral value to a real value mapping. We have implemented this permutation ordering taking advantage of the capability of a BPT representation to encode the spectral-spatial information of an image.
- We have focused on a BPT-based definition of the proposed π-ordering. However, the same underlined methodology could be defined for any structure that encodes a hierarchy of partitions, or that at least, in the case of trees, their leaves form a partition of the data.

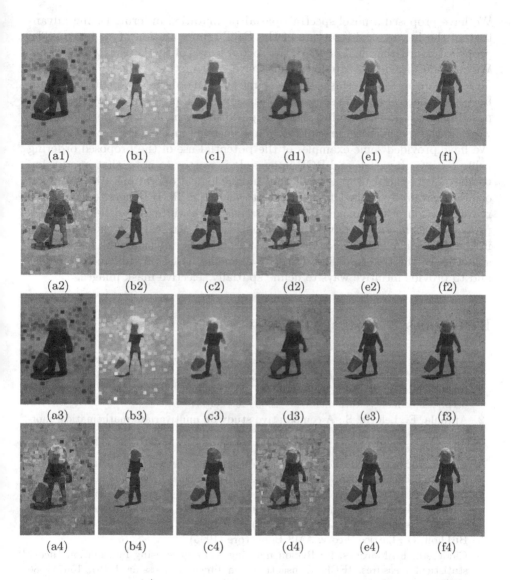

Fig. 6. Comparison of the conventional gray-scale (row 1) and the conventional RGB component-wise ordering (row 3) morphological operations, with respect to the proposed BPT-based spatial-spectral π-ordering (rows 2 and 4) obtained using a square 11×11 structural element: (a) erosion, (b) dilation, (c) closing, (d) opening, (e) closing by reconstruction, and (f) opening by reconstruction

6   Conclusions

We have proposed a novel spectral-spatial permutation ordering taking advantage of the hierarchical structure of the BPT representation of an image. The proposed π-ordering, makes use of the sorted leaves of the BPT representation. We have also provided a top-down recursive algorithm to sort the BPT leaves. As far as the authors know, this is the first time that an ordering taking into account both, the spectral and the spatial information of the pixels of an image, is proposed. The proposed ordering allows to perform morphological operations that work on the spatial structures of the image encoded in the BPT representation. We have provided some examples of the potential use of the proposed ordering using binary, gray-scale and RGB images. It is worthy to note, that in addition to the novel spatial properties of the morphological operations, the proposed ordering naturally deals with multivariate data, $i.e.$ RGB images, once a BPT representation of the image is given. Further work will explore the theoretical properties of the proposed ordering and potential new research avenues provided in the discussion section. For instance, the possibility of using the hierarchical structure of the tree to define structure elements based on sub-tree properties and to provide intuitive ways to define spatially selective mask images.

References

1. Angulo, J.: Morphological colour operators in totally ordered lattices based on distances: Application to image filtering, enhancement and analysis. Comput. Vis. Image Underst. 107(1-2), 56–73 (2007),
   http://dl.acm.org/citation.cfm?id=1265986.1266116
2. Aptoula, E., Lefevre, S.: A comparative study on multivariate mathematical morphology. Pattern Recogn 40(11), 2914–2929 (2007),
   http://dl.acm.org/citation.cfm?id=1274191.1274319
3. Banon, G., Barrera, J.: Decomposition of mappings between complete lattices by mathematical morphology, part i. general lattices. Signal Processing 30(3), 299–327 (1993), http://www.sciencedirect.com/science/article/pii/0165168493900153
4. Barnett: The ordering of multivariate data. Journal of The Royal Statistical Society Series A General 139(3), 318–355 (1976)
5. Birkhoff, G.: Lattice theory. AMS Bookstore (1995)
6. Calderero, F., Marques, F.: Region merging techniques using information theory statistical measures. IEEE Transactions on Image Processing 19(6), 1567–1586 (2010)
7. Goutsias, J., Heijmans, H.: Mathematical Morphology. IOS Press (January 2000)
8. Gratzer, G.: General Lattice Theory, 2nd edn. Birkhäuser, Basel (2003)
9. Gratzer, G.: Lattice Theory: Foundation, 1st edn. Springer, Basel (2011)
10. Haralick, R., Sternberg, S., Zhuang, X.: Image analysis using mathematical morphology. IEEE Transactions on Pattern Analysis and Machine Intelligence PAMI-9(4), 532–550 (1987)
11. Hawkes, P.W., Heijmans, H.J.A.M., Kazan, B.: Morphological Image Operators. Academic Press (December 1993)

12. Pitas, I., Tsakalides, P.: Multivariate ordering in color image filtering. IEEE Transactions on Circuits and Systems for Video Technology 1(3), 247– 259, 295–296 (1991)
13. Ronse, C.: Why mathematical morphology needs complete lattices. Signal Processing 21(2), 129–154 (1990)
14. Salembier, P., Garrido, L.: Binary partition tree as an efficient representation for image processing, segmentation, and information retrieval. IEEE Transactions on Image Processing 9(4), 561–576 (2000)
15. Serra, J.: Image Analysis and Mathematical Morphology, vol. 1. Image Analysis & Mathematical Morphology Series). Academic Press (February 1984)
16. Serra, J.: Image Analysis and Mathematical Morphology, Vol. 2: Theoretical Advances, 1st edn. Academic Press (February 1988)
17. Serra, J.: Anamorphoses and function lattices. In: Proceedings of SPIE, vol. 2030, pp. 2–11 (1993)
18. Soille, P.: Morphological Image Analysis: Principles and Applications, 2nd edn. Springer (2004)
19. Tochon, G., Feret, J., Martin, R.E., Tupayachi, R., Chanussot, J., Asner, G.P.: Binary partition tree as a hyperspectral segmentation tool for tropical rainforests. In: 2012 IEEE International Geoscience and Remote Sensing Symposium (IGARSS), pp. 6368–6371 (2012)
20. Tochon, G., Féret, J.B., Valero, S., Martin, R.E., Knapp, D.E., Salembier, P., Chanussot, J., Asner, G.P.: On the use of binary partition trees for the tree crown segmentation of tropical rainforest hyperspectral images. Remote Sensing of Environment 159, 318–331 (2015)
21. Valero, S., Salembier, P., Chanussot, J.: Comparison of merging orders and pruning strategies for binary partition tree in hyperspectral data. In: 2010 17th IEEE International Conference on Image Processing (ICIP), pp. 2565–2568. IEEE (2010)
22. Valero, S., Salembier, P., Chanussot, J.: Hyperspectral image representation and processing with binary partition trees. IEEE Transactions on Image Processing 22(4), 1430–1443 (2013)
23. Veganzones, M., Tochon, G., Dalla-Mura, M., Plaza, A., Chanussot, J.: Hyperspectral image segmentation using a new spectral unmixing-based binary partition tree representation. IEEE Transactions on Image Processing 23(8), 3574–3589 (2014)
24. Velasco-Forero, S., Angulo, J.: Supervised ordering in R^p: Application to morphological processing of hyperspectral images. IEEE Transactions on Image Processing 20(11), 3301–3308 (2011)

Shape-Based Analysis on Component-Graphs for Multivalued Image Processing

Éloïse Grossiord[1,4(✉)], Benoît Naegel[2], Hugues Talbot[1], Nicolas Passat[3], and Laurent Najman[1]

[1] Université Paris-Est, ESIEE-Paris, LIGM, CNRS, France
[2] Université de Strasbourg, ICube, CNRS, France
[3] Université de Reims Champagne-Ardenne, CReSTIC, France
[4] KeoSys, Nantes, France

Abstract. The extension of mathematical morphology to multivalued images is an important issue. This is particularly true in the context of connected operators based on morphological hierarchies, which aim to provide efficient image filtering and segmentation tools in various application fields, e.g. (bio)medical imaging, remote sensing, or astronomy. In this article, we propose a preliminary study that describes how two notions recently introduced for connected filtering, namely component-graphs (that extend component-trees from a spectral point of view) and shaping (that extend component-trees from a conceptual point of view) can be associated for the effective processing of multivalued images. Structural, algorithmic and experimental developments are proposed. This study opens the way to new paradigms for connected filtering based on hierarchies.

Keywords: Connected filtering, morphological hierarchies, component-graph, component-tree, shaping, multivalued images, medical imaging.

1 Introduction

Connected operators have been intensively studied for the last twenty years in the framework of mathematical morphology [1]. In this context, operators based on hierarchical image models (i.e., trees) have been the object of several structural, algorithmic and methodological developments [2], in order to tackle specific issues associated to various application fields.

In the meantime, mathematical morphology – first defined on binary, and then on grey-level images [3] – progressively extended its framework and tools to the case of multivalued images [4], with a strong focus on colour imaging, but also with contributions in label, multimodal, multi- and hyperspectral imaging.

At the convergence of both issues, a question naturally arises: *How can we perform connected filtering on multivalued images based on morphological hierarchies?* Two principal answers were given to this question. The first defined morphological trees by

This research was partially funded by the French *Agence Nationale de la Recherche* (Grant Agreement ANR-10-BLAN-0205) and the *Programme d'Investissements d'Avenir* (LabEx Bézout, ANR-10-LABX-58).

© Springer International Publishing Switzerland 2015
J.A. Benediktsson et al. (Eds.): ISMM 2015, LNCS 9082, pp. 446–457, 2015.
DOI: 10.1007/978-3-319-18720-4_38

considering a simplifying metric (e.g., a saliency measure for hierarchical watersheds [5], a merging order for partition trees [6,7], or via hyperconnections [8]) in their construction process. The second simplified the multivalued space of images *a priori*, to retrieve tractable totally ordered values, e.g., by marginal or vectorial policies [4].

The latter strategy allows us to rely on morphological trees specifically designed for grey-level images, namely component-trees [9] and trees of shapes [10]. The simplification of multivalued space however induces a loss of information. To cope with this problem, efforts were conducted to extend these data-structures to such complex spaces. Nevertheless, preserving a tree structure still requires a final simplification [11], or restrictive constraints on the value space [12]. Indeed, a true extension of such hierarchies to multivalued spaces necessarily leads to a data-structure that is no longer a tree, but a directed acyclic graph. This is in particular the case for the notion of component-graph [13], that extends the component-tree.

The higher richness and structural complexity of the component-graph, with respect to the component-tree, induces algorithmic issues when considering the classical antiextensive filtering process developed in [9,14]. This is in particular the case for handling the spatial complexity [15], pruning policies and image reconstruction [16].

Recently, a new notion of *shaping* [17] was introduced as an efficient way to improve the framework of anti-extensive filtering of [9,14], by considering a two-layer component-tree for grey-level image processing [18,19].

The key-idea of this article is to consider that the paradigm of shaping can be used not only to build a *tree on a tree*, but also a *tree on a graph*. This may allow us to associate the shaping and component-graph in a common framework that takes advantage of both notions, for developing connected operators on multivalued images. Beyond this simple idea, some practical issues remain to be dealt with.

After briefly summarizing recent works on component-trees, in Section 2, we define the minimal set of definitions required to make this article self-contained. To this end, Section 3 describes the notions of component-tree and component-graph in a unified graph-based formalism. Section 4 discusses the principal advantages and issues raised by coupling shaping and component-graph, for applications on multivalued images. Then, Section 5 proposes some algorithmic solutions to handle node selection and the two steps of reconstruction. An illustration in the field of 3D medical imaging is proposed in Section 6, in order to show the potentiality of this approach. A discussion concludes this article in Section 7, by emphasising the various ways to develop this framework, by extension to other kinds of hierarchies or to richer attribute spaces.

2   Related Work on Component-Trees

The component-tree is a compact, information lossless, hierarchical model for greylevel images. Indeed, it can represent an image in a mixed spatial / spectral space where basic operations can be interpreted in terms of image processing. In particular, filtering and segmentation [9,14,20] can easily be carried out by simply selecting nodes, leading to connected operators. The versatility of the component-tree structure also has led to many other image applications, such as retrieval [21], classification [22], visualisation [23], or document binarisation [24].

The efficiency of the component-tree first relies on its low computation cost. In this context, many efforts were conducted to build component-trees in quasi-linear time, in sequential [9,25] and distributed ways [26]. (The reader is refered to [27] for a recent survey.) The success of component-trees also relies on the development of efficient algorithmic processes for node selection. To cope with filtering and segmentation issues, two main approaches were developed. The first consists of minimizing an energy globally defined over the tree nodes, leading to define an optimal cut [28], that can be interpreted as a segmentation of the underlying image. This approach is the basis for carrying interactive segmentation [29]. The second consists of determining locally the nodes that should be preserved or discarded, based on attribute values [30]. This approach is formalized as an anti-extensive filtering framework [9,14] – recalled in Section 4 – that constitutes the methodological basis of the present work.

The two main limitations of the component-tree are (1) structural: it is heavily constrained by the topological structure of the image; and (2) spectral: it is limited to grey-level (i.e., totally ordered) value images. Structural extensions of the component-tree have been proposed in [31] to deal with ordered families of connectivities, leading to component-hypertrees, and in [32] to handle images defined as valued directed graphs, leading to directed acyclic graphs (DAGs) structured over a tree. Spectral extensions were first considered by exploring marginal approaches for colour image handling [33]. Then, actual extensions of component-trees to partially-ordered value images were pioneered in [34] and further formalized in [13]. Except in specific cases where the values are themselves hierarchically organized [12], the induced data-structure, namely a component-graph, is no longer a tree, but a DAG. The antiextensive framework proposed for component-tree filtering remains valid in theory, but algorithmic issues have to be dealt with both for node selection and image reconstruction [15,16].

3   Background Notions

We now recall some basic notions on graphs. They will allow us to describe the component-trees and component-graphs in a simple and unified formalism, and to discuss, in Sections 4 and 5, how to carry out shaping on component-graphs to handle multivalued images.

3.1   Vertex-Valued Graphs

A graph $\mathcal{G}$ is a couple $(\Gamma, \frown)$ where Γ is a nonempty finite set, and $\frown$ is a binary relation on Γ. The elements of Γ are called vertices. If two vertices x, y of Γ satisfy $x \frown y$, we say that they are adjacent; any such couple (x, y) is called an edge. A subgraph $\mathcal{G}'$ of $\mathcal{G}$ is a graph $(\Gamma', \frown)$ such that Γ' is a subset of Γ, equipped with the restriction of $\frown$ to Γ'.

We consider irreflexive graphs, i.e., we never have $x \frown x$. We also consider non-directed graphs, i.e., $x \frown y \Leftrightarrow y \frown x$; the edges (x, y) and (y, x) are then the same.

In $\mathcal{G}$, a path between two vertices x and y is defined as a sequence of distinct vertices of $\mathcal{G}$ from x to y such that any two successive vertices are adjacent. If this path exists and is unique for any two vertices of the graph, then the graph is a tree. The connected components of $\mathcal{G}$ are the maximal sets of vertices that can be linked by a path. The set of all these connected components is noted $C[\mathcal{G}]$; it is a partition of Γ.

Let $\mathcal{F} : \Gamma \to \mathbb{V}$ be a function such that $\mathbb{V}$ is canonically equipped with an order relation $\leq$. The triple $(\mathcal{G}, \mathbb{V}, \mathcal{F})$ is called a (vertex-)valued graph. We now define the notions of component-tree and component-graph based on this notion of valued graph.

3.2 Component-Tree [9]

Let $(\mathcal{G}, \mathbb{V}, \mathcal{F})$ be a valued graph. We assume that $\leq$ is a total order on $\mathbb{V}$, and that $\mathcal{G}$ is connected, i.e., $C[\mathcal{G}] = \{\Gamma\}$ contains a unique connected component. Since Γ is finite, so is the set $\mathcal{F}(\Gamma) = \{\mathcal{F}(x) \mid x \in \Gamma\} \subseteq \mathbb{V}$. Without loss of generality, we can assume that $\mathbb{V} = \mathcal{F}(\Gamma)$ and is then finite. In particular, $(\mathbb{V}, \leq)$ admits a minimum, noted $\perp$.

For any $v \in \mathbb{V}$, we define the threshold set $\Gamma_v = \{x \in \Gamma \mid v \leq \mathcal{F}(x)\}$. Any such threshold set induces a subgraph $\mathcal{G}_v = (\Gamma_v, \frown)$ of $\mathcal{G}$. For any $v, v' \in \mathbb{V}$ we have $v \leq v' \Leftrightarrow \Gamma_{v'} \subseteq \Gamma_v$. In addition, for any connected component $X_{v'}$ of $C[\mathcal{G}_{v'}]$, there exists a (unique) connected component X_v of $C[\mathcal{G}_v]$ such that $X_{v'} \subseteq X_v$.

The component-tree of $(\mathcal{G}, \mathbb{V}, \mathcal{F})$, noted $C\mathcal{T}$, is the Hasse diagram of the partially ordered set $(\Psi, \subseteq)$, where $\Psi = \bigcup_{v \in \mathbb{V}} C[\mathcal{G}_v]$ is the set of all the connected components of the subgraphs $\mathcal{G}_v$ obtained by successive thresholdings of $\mathcal{G}$.

As suggested by its denomination, the component-tree has a tree structure. Its vertices are also called nodes. Among them, the largest is the maximum for the Hasse diagram, namely the set Γ, obtained as the unique connected component of $\mathcal{G} = \mathcal{G}_{\perp}$; it is the root of the tree. On the opposite side, the leaves are the minimal elements of the Hasse diagram, i.e., the nodes of Ψ that do not strictly include any other nodes.

For image processing purposes, each node of $C\mathcal{T}$ generally stores a value: either an energy (for global optimization) or an attribute (for local selection); this value is most often real. In both cases, this valuation is modeled by a function $\mathcal{V} : \Psi \to \mathbb{R}$. In other words, such enriched component-tree can be interpreted as a valued graph $(C\mathcal{T}, \mathbb{R}, \mathcal{V})$.

3.3 Component-Graph [13]

Let $(\mathcal{G}, \mathbb{V}, \mathcal{F})$ be a valued graph. We still assume that $\mathbb{V} = \mathcal{F}(\Gamma)$ is finite and that $(\mathbb{V}, \leq)$ admits a minimum, noted $\perp$. The graph $\mathcal{G}$ also remains connected, but we no longer assume that $\leq$ is a total order on $\mathbb{V}$.

We extend the notion of connected component in the following way: for any $X \in C[\mathcal{G}_v]$, the couple $K = (X, v)$ is called a valued connected component. We note $\Theta = \bigcup_{v \in \mathbb{V}} C[\mathcal{G}_v] \times \{v\}$ the set of all valued connected components of $\mathcal{G}$, with respect to its successive thresholds. From the order relation $\leq$ and the inclusion relation $\subseteq$, we define[1] the order relation $\trianglelefteq$ on Θ as $(X_1, v_1) \trianglelefteq (X_2, v_2) \Leftrightarrow (X_1 \subset X_2) \vee (X_1 = X_2 \wedge v_2 \leq v_1)$, which intuitively mixes the inclusion and value orders in a lexicographic way.

The component-graph $C\mathcal{G}$ of the valued graph $(\mathcal{G}, \mathbb{V}, \mathcal{F})$ is the Hasse diagram of the partially ordered set $(\Theta, \trianglelefteq)$ (Figure 1). It does not necessarily have a tree structure.

---

[1] Practically, when $\leq$ is a total order, the component-graph and the component-tree are isomorphic. Consequently, it would make sense to also consider the valued connected components and the order $\trianglelefteq$ for building the component-tree, as the threshold value that leads to the generation of a connected component is useful for image modeling and reconstruction, see Equation (1).

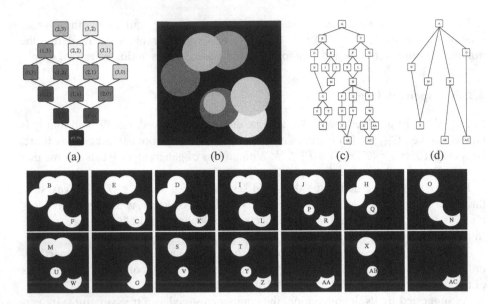

Fig. 1. (a) The Hasse diagram of the ordered set $(\mathbb{V}, \leq)$, where $\mathbb{V} \subset \mathbb{N}^2$ is equipped with the canonical order relation $\leq$. For the sake of readability, each value of $\mathbb{V}$ is associated to an arbitrary colour. (b) A multivalued image, viewed as a valued graph $(\mathcal{G}, \mathbb{V}, \mathcal{F})$ where $\mathcal{G}$ is a part of $\mathbb{Z}^2$ equipped with the standard 4-adjacency relation. (c) The component-graph $\mathcal{CG}$ associated to $(\mathcal{G}, \mathbb{V}, \mathcal{F})$. (d) A simplified version of the component-graph (see Section 5). Second and third rows: thresholded images obtained from (b). Each (valued) connected component is represented by a letter: A, B, C, etc. The same letters identify these components in (c,d).

This derives from the fact that two connected components can meet without inclusion. The component-graph has a largest node that is the maximum for the Hasse diagram, namely the set $\varGamma$; it is the root of the graph. Its leaves are the minimal elements of the Hasse diagram. The main difference is that several paths may exist between two nodes.

As for component-trees, each node of $\mathcal{CG}$ can contain an attribute value and this valuation can also be interpreted as a function $\mathcal{A} : \varTheta \rightarrow \mathbb{R}$. Then, such enriched component-graph is also interpreted as a valued graph $(\mathcal{CG}, \mathbb{R}, \mathcal{A})$.

4   Shape-Space Analysis of Multivalued Images: Concept

A (discrete) image is a mapping $\mathcal{I}$ from a finite spatial domain $\varOmega$ to a value space $\mathbb{V}$. To develop connected operators, it is necessary to handle the structure of $\varOmega$, i.e., to know the adjacency between its points, leading to a graph $\mathcal{S}$. In addition, to develop morphological hierarchies such as component-trees and component-graphs, it is required to know the order $\leq$ on $\mathbb{V}$. An image is then modeled as a valued graph $(\mathcal{S}, \mathbb{V}, \mathcal{I})$.

4.1 Antiextensive Filtering With the Component-Tree

The component-tree and the component-graph are image lossless models. More precisely, the mapping $\mathcal{I}$ can be fully recovered from the (de)composition formula

$$\mathcal{I} = \bigvee_{(X,v)\in\Theta} C_{(X,v)} = \bigvee_{X\in\Psi} C_{(X,v)} \tag{1}$$

where $C_{(X,v)} : \Omega \to \mathbb{V}$ is the cylinder function of support X and value v, that maps $x \in X$ onto v and $x \notin X$ onto $\bot$.

In the case of component-trees (i.e., for grey-level images, i.e., when $\leq$ is a total order), this formula leads to a well-defined image for Ψ, but also for any subset $\widehat{\Psi} \subseteq \Psi$. This consideration led to the proposal of an anti-extensive filtering framework [9,14] that basically consists of three successive steps:

 (*i*) construction of the component-tree $\mathcal{CT}$ associated to the image;
 (*ii*) reduction of the component-tree by selection of nodes $\widehat{\Psi} \subseteq \Psi$; and
(*iii*) reconstruction of the result image $\widehat{\mathcal{I}} \leq \mathcal{I}$ from the reduced component-tree $\widehat{\mathcal{CT}}$.

Step (*i*) is carried out from a wide range of available component-tree construction methods, while Step (*iii*) is straightforward from Equation (1). The core of the process is Step (*ii*) that is dealt with by considering attribute values carried by each node of the component-tree, namely the valuation $\mathcal{V} : \Psi \to \mathbb{R}$, that guides the decision of preserving or discarding a node (together with pruning policies whenever $\mathcal{V}$ is not increasing, see [9,14]).

4.2 Coupling Shaping and Component-Graphs

Anti-Extensive Filtering with the Component-Graph. On the one hand, we can extend the above anti-extensive filtering approach to images taking their values in any value space $\mathbb{V}$, without the assumption that $\leq$ is a total order. Instead of a component-tree, we then have to consider a component-graph. This allows us to process any image in the same framework as initially proposed in [9,14]. Nevertheless, it raises two difficulties: Step (*ii*) is now more complex, as the standard pruning policies have to be adapted for dealing with non-linear bottom-up or top-down node parsing; and Step (*iii*) can be an ill-posed problem, depending on the nature of the order $\leq$ and the preserved nodes $\widehat{\Theta}$.

Anti-Extensive Filtering in the Shape-Space [17]. On the other hand, the paradigm of shaping proposes to perform anti-extensive filtering based on a double layer of component-trees, i.e., on the component-tree of the component-tree of the image. The inner component-tree is seen as an image whose points are the nodes, and grey-level values are the attributes. It is then possible to process any grey-level image in the framework initially proposed in [9,14], by performing node selection in a data-structure that is no longer defined at the image level, but at a higher semantic level. This allows us also to define increasing attribute values on the outer component-tree, and to perform real-time, threshold-based node selection. The main limitation of this framework is that it considers a tree as intermediate data-structure, thus limiting its use to grey-level images.

From "A Tree on a Tree" to "a Tree on a Graph". The formalism of valued graphs sheds light on the common structure of images, component-trees and component-graphs. As a side effect, it emphasises the fact that the inner layer of shape-space filtering only requires a graph, and not necessarily a tree. The cornerstone of this work is to consider that the initial shaping paradigm of a "tree on a tree", can be generalised to a "tree on a graph". This simple idea, summarized by Diagram (2), allows us – in theory – to process any image via a shape-based filtering.

$$
\begin{array}{ccccc}
(\mathcal{S}, \mathbb{V}, \mathcal{I}) & \xrightarrow{\ (i)\ } & (\mathcal{CG}, \mathbb{R}, \mathcal{A}) & \xrightarrow{\ (i)\ } & (\mathcal{CT}, \mathbb{R}, \mathcal{V}) \\
\downarrow & & & & \downarrow{\scriptstyle(ii)} \\
(\mathcal{S}, \mathbb{V}, \widehat{\mathcal{I}}) & \xleftarrow{\ (iii)\ } & (\widehat{\mathcal{CG}}, \mathbb{R}, \mathcal{A}_{|\widehat{\Theta}}) & \xleftarrow{\ (iii)\ } & (\widehat{\mathcal{CT}}, \mathbb{R}, \mathcal{V}_{|\widehat{\Psi}})
\end{array}
\tag{2}
$$

Based on the above remarks, this approach has the following virtues:

- it avoids the complex selection of nodes directly in the component-graph, since this task is indirectly carried out on the outer-layer component-tree;
- it extends the initial shaping approach beyond grey-level images;
- it inherits the good properties of shape-space filtering from increasing criteria.

Nevertheless, behind this simple idea, and its intrinsic advantages, some algorithmic issues remain to be considered, in particular for the two reconstruction steps (*iii*), from the component-tree to the component-graph, and then to the image. In Section 5, we propose some solutions to these issues in the case of multivalued images, that are defined as combinations of several grey-level images, opening the way to applications in multimodal / multispectral imaging.

5   Shape-Space Analysis of Multivalued Images: Algorithmics

The formalism of component-graphs handles valued graphs $(\mathcal{S}, \mathbb{V}, \mathcal{I})$ where $\leq$ can be any order. We focus here on the case of multivalued images where $\mathbb{V}$ is composed of k spectral bands $\mathbb{V}_i$, each equipped with a total order. In particular, we consider the canonical partial order $\leq$ on $\mathbb{V}$ defined by $(v_i)_{i=1}^k \leq (w_i)_{i=1}^k \Leftrightarrow \forall i \in [1, k], v_i \leq w_i$.

Component-Graph Construction. In [13], several variants of component-graphs were introduced, in particular to simplify $\mathcal{CG}$ by considering smaller subsets of Θ. In the first part of Step (*i*), that builds $\mathcal{CG}$ from $(\mathcal{S}, \mathbb{V}, \mathcal{I})$, we chose to consider the lightest version of component-graph (Figure 1(d)), i.e., the one that represents only the nodes which actually contribute to the construction of the image according to Equation (1), defined as $\ddot{\Theta} = \{(X, v) \in \Theta \mid \exists x \in X, v = \mathcal{F}(x)\}$. (For the sake of simplicity, we will now note $\ddot{\Theta}$ as Θ.) This choice is motivated by the fact that such component-graphs are sufficient to process images defined in the above value space. From a complexity point of view, its construction has a lower time cost compared to the other variants of component-graphs. Moreover, its spatial complexity is in the same order as that of the initial image support Ω. The component-graph $\mathcal{CG}$ is built from the algorithm proposed in [15].

Component-Graph Valuation. At this stage, an attribute can be associated to each node of Θ, in the component-graph $\mathcal{C}\mathcal{G}$. We chose to consider here an attribute taking its values in $\mathbb{R}$, namely a set where all values are comparable. While alternative choices are possible (see Section 7), we assume here that a valuation $\mathcal{A} : \Theta \to \mathbb{R}$ is indeed sufficient to accurately filter the nodes, while authorising the design of a tree structure at the second layer. The criteria potentially modeled by $\mathcal{A}$ for each node $K = (X, v) \in \Theta$ can depend on: (1) spectral properties (then, we practically have $\mathcal{A} : \mathbb{V} \to \mathbb{R}$); (2) geometric properties (then, we practically have $\mathcal{A} : 2^{\Omega} \to \mathbb{R}$); (3) structural properties (then, $\mathcal{A}(K)$ depends on the relationships of K within $\mathcal{C}\mathcal{G}$); or a combination of some of these three classes. The structure of the chosen version of $\mathcal{C}\mathcal{G}$ is relatively light, and a criterion of type (3) would be weakly relevant. For building the component-tree of the outer layer, only geometric criteria are considered here. This choice is coherent with the paradigm of shaping, and also motivated by the fact that the spectral handling of images can be carried out at the inner layer, either before or after the shaping stage.

Component-Tree Construction and Pruning. From the valued graph $(\mathcal{C}\mathcal{G}, \mathbb{R}, \mathcal{A})$ associated to the component-graph, a shape-based component-tree can now be defined. Assuming that the relevant values of $\mathcal{A}$ are the highest, two policies can be considered to build $\mathcal{C}\mathcal{T}$: either as a min-tree or a max-tree. In the first case, the nodes of interest will be located near the root; in the second, they will be located near the leaves. We chose here to consider the max-tree case, that allowed us to select the relevant nodes by only preserving the distal parts, i.e., the branches of the tree. Practically, each node $Y \in \mathcal{Y}$ of the component-tree $\mathcal{C}\mathcal{T}$ is a connected component gathering nodes of a subgraph of $\mathcal{C}\mathcal{G}$, for a given threshold value with respect to $\mathcal{A}$. This threshold value then constitutes the valuation of this node, and thus defines $\mathcal{V}$. Following the above criteria classification, the valuation $\mathcal{V}$ – that is however directly obtained from a valuation of class (2) – is now a valuation of class (1) in the shape-space. In addition, it defines a monotonic (here, decreasing) criterion, allowing for an easy selection by thresholding, and avoiding the use of any specific pruning policies.

Component-Graph Filtering. A "standard" component-tree – defined from a grey-level image – contains nodes which represent connected components of points of the image, obtained at a given threshold value. In contrast, the component-tree $\mathcal{C}\mathcal{T}$ defined at the outer layer of the shape-space model – defined from the valued graph $(\mathcal{C}\mathcal{G}, \mathbb{R}, \mathcal{A})$ – contains nodes that are connected components of Θ which are themselves connected components of Ω. Such node $Y \in \mathcal{Y}$ is then defined as a set $\{K_i = (X_i, v_i)\}_{i=1}^{k} \subseteq \Theta$, with $k \geq 1$. Each node $K_i \in Y$ is either included in another node $K_j \in Y$, or is a maximal element in Y with respect to the $\trianglelefteq$ relation. When dealing with geometric criteria, only these latter nodes, that contribute to define the support $\bigcup_{i=1}^{k} X_i$ of Y in Ω are of actual interest. In other words, if Y is preserved in $\widehat{\mathcal{Y}}$, only these nodes should be preserved, both spatially and spectrally in the filtered image. We note $\widehat{Y} \subseteq Y$ the subset of Y formed by such nodes. The other nodes of Y are not taken into account; however, this is not a problem, as any node $K \in \Theta$ belongs to $\widehat{Y}$ for at least one $Y \in \mathcal{Y}$. Consequently, it may be preserved based on the chosen geometric criterion, via this node. The main difference between the initially proposed shaping paradigm ("a tree on a tree") and the present one ("a tree on a graph"), is that the first defines any $\widehat{Y}$ as a singleton set $\{K\}$, while the second can now associate several – overlapping – nodes of Θ into a same $\widehat{Y}$, since some values of $\mathbb{V}$ may be non-comparable.

Image Filtering. As we deal with multivalued images, the space $(\mathbb{V}, \leq)$ is structured as a lattice. Based on this hypothesis, two strategies can be used to reconstruct the filtered image $\widehat{\mathcal{I}}$. The first – that considers each band of $\mathbb{V}$ with a same degree of relevance – consists of assigning the value v – defined as the infimum of all the v_i – to the reduced node $\widehat{Y}$, associated to each node $Y \in \widehat{\mathcal{Y}}$ and thus to each node $K_i = (X_i, v_i) \in \widehat{Y}$. This policy is justified by the fact that the node Y has been preserved with respect to a geometrical attribute computed for the union of all the supports X_i of the K_i; in such conditions, the least common threshold value associated to all these nodes should be considered. However, a given node $K \in \Theta$ may belong to $\widehat{Y_j}$, for several nodes $Y_j \in \widehat{\mathcal{Y}}$. In that case, the value assigned to K should be defined as the supremum of all these values. This policy is justified by the fact that a node $Y \in \mathcal{Y}$, defined as the union of several nodes of Θ, should not lose its geometry in the filtered image. The reconstruction of the filtered image can then be formalized as follows.

$$\widehat{\mathcal{I}} = \bigvee_{Y \in \widehat{\mathcal{Y}}} C_{(\bigcup_{(X,v) \in \widehat{Y}} X, \bigwedge_{(X,v) \in \widehat{Y}} v)} \tag{3}$$

The second strategy – that gives priority to one or several given band(s) versus others – consists of applying the first strategy on a strict subspace of $\mathbb{V}$ that corresponds to specific spectral bands. In the case where only one band is considered, the reconstructed image is a grey-level one, and the supremum and infimum on $\mathbb{V}$ considered above are simply replaced by the maximum and minimum in the considered band.

6   Application Example: PET / CT Image Filtering

Component-trees have been involved in the development of various tools devoted to process 3D medical images where the structures of interest have locally extremal values. This is the case for angiographic imaging [35,36] where high signal corresponds to flowing blood, and nuclear imaging [19] where it corresponds to high metabolic (often tumoral) activity. We illustrate the potential usefulness of our framework in this latter application field, by filtering coupled Positron Emission Tomography (PET) and X-ray Computed Tomography (CT) images. These experiments remain to be clinically confirmed. As a consequence, they only constitute an illustrative proof of concept for our framework, and not an actual medical image analysis tool.

PET images (Figure 2(a)) showing metabolic activity, are classically associated to morphological CT images (Figure 2(b)) for visualizing the anatomy. Such coupled images provide complementary information. It is pertinent to process them as a unique bivalued image in order to more accurately extract the lesions and their activity.

In contrast to PET images, where the canonical order $\leq$ on $\mathbb{R}$ captures the semantics of metabolic activity, this order is – partially – meaningless with respect to the Hounsfield scale in CT. Consequently, we apply a non-injective mapping on CT images, in order to

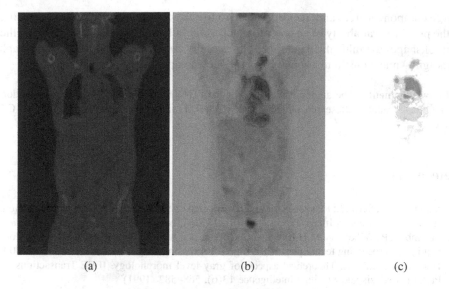

(a)   (b)   (c)

Fig. 2. Coupled CT (a) and PET (b) images. (b) Ground-truth of the lesions, in purple. (c) Multivalued shape-based filtering from (a+b), visualized in the PET value space.

associate the lowest values to tissues of extremal (low, e.g., water and blood; and high, e.g., bones) intensities. The order $\leq$ on $\mathbb{R}$ for the resulting image associates the least values in the CT data to tissues which are more likely to induce false positives in PET.

The value space is subsampled to 256 values for both PET and CT, leading to a space $\mathbb{V}$ of 65536 distinct values. The criterion considered for filtering is the compactness factor [19] defined as the ratio between the extremal eigenvalues of the matrix of inertia. The filtered image is reconstructed in $\mathbb{V}$, following the first proposed policy, see Equation (3). Results of the process are exemplified in Figure 2(c), emphasising a good discrimination of lesions versus false positives, and a satisfactory spatial accuracy.

7   Conclusion

By coupling the two recently introduced notions of shaping and component-graph, this work opens the way to the development of new connected operators based on morphological hierarchies, and devoted to process images taking their values in rich spaces. The conceptual and algorithmic results presented here for handling multivalued images, constitute a first step toward such developments.

This preliminary study dealt with scalar attributes on the component-graph. Considering vectorial attributes [37] may enrich the potential of this framework, with the counterpart of having to cope with a component-graph at the outer layer, leading to "a graph on a tree" for grey-level imaging, and "a graph on a graph" in the most general cases.

Another limitation was to consider images where each value band was equipped with a "natural" or "semantic" order. A way to partially relax this constraint could be to merge component-trees and trees of shapes, to use the optimal data-structure according to the putative availability of an order for each value subspace. Recent extensions of the trees of shapes to multivalued images [11] – actually connected to the component-graph paradigm – may constitute a sound basis for such an approach.

Acknowledgements. The authors thank M. Meignan (Hôpitaux Universitaires Henri-Mondor, Lymphoma Academic Research Organisation, Créteil, France), for providing the PET / CT images.

References

1. Salembier, P., Serra, J.: Flat zones filtering, connected operators, and filters by reconstruction. IEEE Transactions on Image Processing 4(8), 1153–1160 (1995)
2. Salembier, P., Wilkinson, M.H.F.: Connected operators: A review of region-based morphological image processing techniques. IEEE Signal Processing Magazine 26(6), 136–157 (2009)
3. Heijmans, H.J.A.M.: Theoretical aspects of gray-level morphology. IEEE Transactions on Pattern Analysis and Machine Intelligence 13(6), 568–582 (1991)
4. Aptoula, E., Lefèvre, S.: A comparative study on multivariate mathematical morphology. Pattern Recognition 40(11), 2914–2929 (2007)
5. Najman, L., Schmitt, M.: Geodesic saliency of watershed contours and hierarchical segmentation. IEEE Transactions on Pattern Analysis and Machine Intelligence 18(12), 1163–1173 (1996)
6. Salembier, P., Garrido, L.: Binary partition tree as an efficient representation for image processing, segmentation, and information retrieval. IEEE Transactions on Image Processing 9(4), 561–576 (2000)
7. Soille, P.: Constrained connectivity for hierarchical image decomposition and simplification. IEEE Transactions on Pattern Analysis and Machine Intelligence 30(7), 1132–1145 (2008)
8. Perret, B., Lefèvre, S., Collet, C., Slezak, É.: Hyperconnections and hierarchical representations for grayscale and multiband image processing. IEEE Transactions on Image Processing 21(1), 14–27 (2012)
9. Salembier, P., Oliveras, A., Garrido, L.: Antiextensive connected operators for image and sequence processing. IEEE Transactions on Image Processing 7(4), 555–570 (1998)
10. Monasse, P., Guichard, F.: Scale-space from a level lines tree. Journal of Visual Communication and Image Representation 11(2), 224–236 (2000)
11. Carlinet, E., Géraud, T.: A morphological tree of shapes for color images. In: Proc. of the ICPR, pp. 1132–1137 (2014)
12. Kurtz, C., Naegel, B., Passat, N.: Connected filtering based on multivalued component-trees. IEEE Transactions on Image Processing 23(12), 5152–5164 (2014)
13. Passat, N., Naegel, B.: Component-trees and multivalued images: Structural properties. Journal of Mathematical Imaging and Vision 49(1), 37–50 (2014)
14. Jones, R.: Connected filtering and segmentation using component trees. Computer Vision and Image Understanding 75(3), 215–228 (1999)
15. Naegel, B., Passat, N.: Colour image filtering with component-graphs. In: Proc. of the ICPR, pp. 1621–1626 (2014)
16. Naegel, B., Passat, N.: Toward connected filtering based on component-graphs. In: Hendriks, C.L.L., Borgefors, G., Strand, R. (eds.) ISMM 2013. LNCS, vol. 7883, pp. 353–364. Springer, Heidelberg (2013)
17. Xu, Y., Géraud, T., Najman, L.: Morphological filtering in shape spaces: Applications using tree-based image representations. In: Proc. of the ICPR, pp. 485–488 (2012)

18. Xu, Y., Géraud, T., Najman, L.: Two applications of shape-based morphology: Blood vessels segmentation and a generalization of constrained connectivity. In: Hendriks, C.L.L., Borgefors, G., Strand, R. (eds.) ISMM 2013. LNCS, vol. 7883, pp. 390–401. Springer, Heidelberg (2013)

19. Grossiord, É., Talbot, H., Passat, N., Meignan, M., Tervé, P., Najman, L.: Hierarchies and shape-space for PET image segmentation. In: Proc. of the ISBI, pp. 1118–1121 (2015)

20. Ouzounis, G.K., Wilkinson, M.H.F.: Mask-based second-generation connectivity and attribute filters. IEEE Transactions on Pattern Analysis and Machine Intelligence 29(6), 990–1004 (2007)

21. Alajlan, N., Kamel, M.S., Freeman, G.H.: Geometry-based image retrieval in binary image databases. IEEE Transactions on Pattern Analysis and Machine Intelligence 30(6), 1003–1013 (2008)

22. Urbach, E.R., Roerdink, J.B.T.M., Wilkinson, M.H.F.: Connected shape-size pattern spectra for rotation and scale-invariant classification of gray-scale images. IEEE Transactions on Pattern Analysis and Machine Intelligence 29(2), 272–285 (2007)

23. Westenberg, M.A., Roerdink, J.B.T.M., Wilkinson, M.H.F.: Volumetric attribute filtering and interactive visualization using the max-tree representation. IEEE Transactions on Image Processing 16(12), 2943–2952 (2007)

24. Naegel, B., Wendling, L.: A document binarization method based on connected operators. Pattern Recognition Letters 31(11), 1251–1259 (2010)

25. Najman, L., Couprie, M.: Building the component tree in quasi-linear time. IEEE Transactions on Image Processing 15(11), 3531–3539 (2006)

26. Wilkinson, M.H.F., Gao, H., Hesselink, W.H., Jonker, J.E., Meijster, A.: Concurrent computation of attribute filters on shared memory parallel machines. IEEE Transactions on Pattern Analysis and Machine Intelligence 30(10), 1800–1813 (2008)

27. Carlinet, E., Géraud, T.: A comparative review of component tree computation algorithms. IEEE Transactions on Image Processing 23(9), 3885–3895 (2014)

28. Guigues, L., Cocquerez, J.P., Le Men, H.: Scale-sets image analysis. International Journal of Computer Vision 68(3), 289–317 (2006)

29. Passat, N., Naegel, B., Rousseau, F., Koob, M., Dietemann, J.L.: Interactive segmentation based on component-trees. Pattern Recognition 44(10–11), 2539–2554 (2011)

30. Breen, E.J., Jones, R.: Attribute openings, thinnings, and granulometries. Computer Vision and Image Understanding 64(3), 377–389 (1996)

31. Passat, N., Naegel, B.: Component-hypertrees for image segmentation. In: Soille, P., Pesaresi, M., Ouzounis, G.K. (eds.) ISMM 2011. LNCS, vol. 6671, pp. 284–295. Springer, Heidelberg (2011)

32. Perret, B., Cousty, J., Tankyevych, O., Talbot, H., Passat, N.: Directed connected operators: Asymmetric hierarchies for image filtering and segmentation. IEEE Transactions on Pattern Analysis and Machine Intelligence, doi:10.1109/TPAMI.2014.2366145

33. Naegel, B., Passat, N.: Component-trees and multivalued images: A comparative study. In: Wilkinson, M.H.F., Roerdink, J.B.T.M. (eds.) ISMM 2009. LNCS, vol. 5720, pp. 261–271. Springer, Heidelberg (2009)

34. Passat, N., Naegel, B.: An extension of component-trees to partial orders. In: Proc. of the ICIP, pp. 3981–3984 (2009)

35. Wilkinson, M.H.F., Westenberg, M.A.: Shape preserving filament enhancement filtering. In: Niessen, W.J., Viergever, M.A. (eds.) MICCAI 2001. LNCS, vol. 2208, pp. 770–777. Springer, Heidelberg (2001)

36. Dufour, A., Tankyevych, O., Naegel, B., Talbot, H., Ronse, C., Baruthio, J., Dokládal, P., Passat, N.: Filtering and segmentation of 3D angiographic data: Advances based on mathematical morphology. Medical Image Analysis 17(2), 147–164 (2013)

37. Urbach, E.R., Boersma, N.J., Wilkinson, M.H.F.: Vector attribute filters. In: Proc. of the ISMM. Computational Imaging and Vision, vol. 30, pp. 95–104. Springer (2005)

Elementary Morphology
for SO(2)- and SO(3)-Orientation Fields

Andreas Kleefeld[1]([⊠]), Anke Meyer-Baese[2], and Bernhard Burgeth[3]

[1] Faculty of Mathematics, Natural Sciences and Computer Science,
Brandenburg Technical University Cottbus - Senftenberg,
03046, Cottbus, Germany
kleefeld@tu-cottbus.de

[2] Department of Scientific Computing, Florida State University,
Tallahassee, Florida, USA
ameyerbaese@fsu.edu

[3] Saarland University,
Faculty of Mathematics and Computer Science,
66041 Saarbrücken, Germany
burgeth@math.uni-sb.de

Abstract. In this article, techniques to process fields of special orthogonal matrices by means of mathematical morphology are proposed. Since the group structure of SO(2)- resp. SO(3)-fields is not suitable to establish useful notions of infimum and supremum, they are transformed into scalar resp. SYM(2)-fields utilizing a matrix-valued version of the Cayley transform. However, for symmetric 2×2 matrices, that is for SYM(2)-fields, elementary morphological operations are available, see [6] and [7]. Several examples and numerical results are presented to show the merits and the limitations of this novel approach. Additionally, the series of transformations utilized in the proposed methods open up possibilities to visualize SO(2)- and SO(3)-fields.

Keywords: Orientation field · Tensor field · Loewner order · Lie group · Lie algebra · Cayley transform · Symmetric matrix · Erosion · Dilation · Morphological Laplacian · Beucher gradient

1 Introduction

Mathematical morphology came into existence with the work of Matheron and Serra [12,13] in the late sixties. The research on morphology over the last 5 decades has provided us with an abundance of tools and techniques to process successfully scalar-valued images for applications ranging from medical imaging to geological sciences [11,14,16]. Recent years have seen an increasing interest in the development of morphological tools for non-scalar images; this means, images with values in sets of vectors, i.e. colors, see for example [8] and [9] for a more historic account, and [2] and [15] for background in order theory. For a still up-to-date survey for morphology on color images the reader is referred to [1]

© Springer International Publishing Switzerland 2015
J.A. Benediktsson et al. (Eds.): ISMM 2015, LNCS 9082, pp. 458–469, 2015.
DOI: 10.1007/978-3-319-18720-4_39

and the extensive list of literature cited therein. In this context we mention [6] and especially, due to its general framework, [17] as well. Images with symmetric 2×2- or 3×3-matrices as function values, matrix fields for short, have been considered in [3,4,7] to mention a few. The spectral decomposition of fields of 3×3 symmetric matrices (considering them as second order generalization of real numbers, diagonal matrices being the first order generalization) $S = ODO^\top$, where O is an orthonormal matrix, leads naturally to SO(3)-fields. There are numerous fields of applications for symmetric matrices, ranging from medical imaging, civil engineering or image processing itself. Processing these fields (not only with morphological methods!) is important, hence it is in demand. However, there are unwanted dissipative numerical effects stemming from the mutual influence of directional information (captured in the eigenvectors in O) and size-information (captured in the eigenvalues in D), see [5]. The obvious way out of this dilemma is to treat them separately. This is a first motivation for the topic of this article. Image processing based on Lie groups (and their quotients), coding orientation, is a current research topic with applications in neuro and retinal imaging, see [10] and the literature cited therein. The approach presented here is somewhat complementary to [10], since the Lie group appears in the co-domain of the image, and not in its domain. We see possible applications of the proposed method, for example, in civil engineering where data coming from stress tests or simulations of mechanical components have to be analyzed.

In this article, we are attempting to provide basic morphological operations for orientation fields. Here, n-dimensional orientation fields may be considered as mappings $f : \Omega \longmapsto SO(n)$ from a two- or three-dimensional image domain Ω into the set $SO(n)$ of $n \times n$-orthonormal matrices with determinant 1. In other words: as SO(n)-valued images. In the sequel, we will refer to them as SO(2)-, resp., SO(3)-fields. SO(n) forms a group with respect to matrix multiplication, a so-called Lie group. As such it is related via the matrix-valued exponential map to its Lie algebra, which is well-known to be the set SKEW(n) of all skew-symmetric real $n \times n$ matrices. However, we will use the so-called Cayley transform as an easy to compute substitute for the exponential resp. its inverse, the logarithmic map.

The key idea is to take advantage of the linear structure of the space SKEW(n). In particular, since at the current state we are solely dealing with $n = 2$ and $n = 3$, we have that SKEW(2) and SKEW(3) are isometric to $\mathbb{R}$ resp., $\mathbb{R}^3$. Hence we can rewrite one-to-one a SO(2)-, resp. SO(3)-field as a scalar image resp. a vector field with values in $\mathbb{R}^3$ or as a color image with values in a suitable color space. Hence, the gray-scale image associated to a SO(2)-field can be subjected to a morphological operation and the result is transformed back into a SO(2)-field. In principle the same procedure applies in the case of SO(3)-field, however, the range of the corresponding vector field is a subset of $\mathbb{R}^3$ which does not allow for a total order. As a remedy we use another transformation step: an isometry between $\mathbb{R}^3$ and the set SYM(2) of all real symmetric 2×2-matrices M satisfying $M = M^\top$, with $\top$ denoting transposition. Now, SYM(2) has a somewhat natural order, the Loewner order, which indeed allows for elementary morphological operations, as pointed out, for example, in [4] and [6].

The article has the following structure: in Section 2, we introduce the Cayley transform and some of its properties including the conversion between SO(3)-fields and SYM(2)-fields. In addition the visualization of such fields is discussed followed by some exemplary calculations of suprema and infima of two SO(3)-matrices using the proposed technique based on the Loewner order. We report on experimental results, mainly employing morphological derivatives, that can serve as a proof-of-concept in Section 3. Section 4 containing a summary and a short outlook followed by a list of references concludes the article.

2 From Orientation to Symmetric Matrices via Cayley Transform

2.1 Cayley Transform

There is an interesting, less known connection between the Lie group SO(n) and its Lie algebra SKEW(n). The Cayley transform well-known from *complex analysis* and *hyperbolic geometry* has an obvious generalization to matrices by means of

$$\text{cay}(S) = (I - S)(I + S)^{-1} =: \frac{I - S}{I + S},$$

where I stands for the $n \times n$-unit matrix. Note that $I - S$ and $I + S$ commute. It is not difficult to see that cay is its own inverse: cay = cay^{-1}. We consider this a clear advantage over the usual exp-log-relation between SO(n) and SKEW(n). However, the Cayley transform is defined only for matrices S, whose set of eigenvalues SPEC(S) does not contain -1. Hence we are dealing in the sequel mainly with SO(n)$^* := \{S \in$ SO(n) $| -1 \notin$ SPEC(S)$\}$ in the cases $n = 2, 3$. How the Cayley transform is exploited, is demonstrated explicitly in the simple case of SO(2)-fields in the next subsection as a preparation for the more involved situation of SO(3)-fields.

2.2 SO(2)-orientation Fields

Let us first consider SO(2)-fields, that is images with values in the set of all rotations with angle ϕ around the origin in the two-dimensional Euclidean space,

$$\text{SO}(2) = \left\{ \begin{pmatrix} \cos(\phi) & -\sin(\phi) \\ \sin(\phi) & \cos(\phi) \end{pmatrix} \middle| \phi \in [-\pi, \pi) \right\}.$$

A short calculation gives

$$\text{cay} \begin{pmatrix} \cos(\phi) & -\sin(\phi) \\ \sin(\phi) & \cos(\phi) \end{pmatrix} = \begin{pmatrix} 0 & \tan(\frac{\phi}{2}) \\ -\tan(\frac{\phi}{2}) & 0 \end{pmatrix}$$

with $\phi \in (-\pi, \pi)$. Hence cay is a bijection from SO(2)*, to its Lie algebra, the linear space SKEW(2) $= \left\{ \begin{pmatrix} 0 & s \\ -s & 0 \end{pmatrix} \middle| s \in \mathbb{R} \right\}$. The mapping

$$\kappa : \begin{pmatrix} 0 & s \\ -s & 0 \end{pmatrix} \longmapsto \frac{s}{\sqrt{2}}$$

is an isometry between $(\text{SKEW}(2), \|\cdot\|_F)$ equipped with the Frobenius norm $\|\cdot\|_F$ and the set of real numbers $\mathbb{R}$ with the absolute value as norm. In the sequel, we will employ the notation $\widetilde{\text{cay}} := \kappa \circ \text{cay}$. For the inverse $\widetilde{\text{cay}}^{-1} = \text{cay} \circ \kappa^{-1}$ one finds for $s \in \mathbb{R}$ that indeed

$$\widetilde{\text{cay}}^{-1}(s) = \begin{pmatrix} \cos(\phi(s)) & -\sin(\phi(s)) \\ \sin(\phi(s)) & \cos(\phi(s)) \end{pmatrix}$$

with $s/\sqrt{2} = \tan(\phi(s)/2)$, or equivalently $\phi(s) = 2 \cdot \arctan(s/\sqrt{2}) \in [-\pi, \pi)$. Since $\mathbb{R}$ resp. $[-\pi, \pi)$ are totally ordered sets with a strictly monotone function $s \mapsto \phi(s)$ the above reasoning has the agreeable interpretation that a rotation in SO(2) is the larger, the larger the angle of rotation is.

Visualization of SO(2)-fields. In effect $\widetilde{\text{cay}}$ provides a one-to-one correspondence between SO(2)-fields and regular scalar images with values in the interval $[-\pi, \pi)$ or $[0, 1)$ after a trivial rescaling. However, for the latter ones the full apparatus of image processing is at our disposal. Now, an orientation field can be $\widetilde{\text{cay}}$-transformed, the resulting scalar image may be processed with the desired tool, and the outcome $\widetilde{\text{cay}}^{-1}$-transformed back into a (modified) orientation field. It becomes apparent from the example above that the morphological operations on a SO(2)-field boil down to standard gray-scale morphology on images with values in $[-\pi, \pi)$ or $[0, 1)$. We will depict SO(2)-fields both as gray-scale images and as 2D-unit-vector fields, where the angle between the unit vectors and the positive x-axis represents the angle of rotation. Due to the simplicity and straightforwardness of the underlying idea we report on the results of only a few experiments with some basic morphological operations such as erosion, dilation, Beucher gradient, and morphological Laplacian in the experimental section.

2.3 SO(3)-orientation Fields

SO(3)-fields are images with values in the set SO(3) of all rotations with angle ϕ around an axis through the origin of the three-dimensional Euclidean space $\mathbb{R}^3$. The matrix

$$R(u, v, w, \theta) := \begin{pmatrix} u^2(1 - \cos(\theta)) + \cos(\theta) & uv(1 - \cos(\theta)) - w\sin(\theta) & uw(1 - \cos(\theta)) + v\sin(\theta) \\ uv(1 - \cos(\theta)) + w\sin(\theta) & v^2(1 - \cos(\theta)) + \cos(\theta) & vw(1 - \cos(\theta)) - u\sin(\theta) \\ uw(1 - \cos(\theta)) - v\sin(\theta) & vw(1 - \cos(\theta)) + u\sin(\theta) & w^2(1 - \cos(\theta)) + \cos(\theta) \end{pmatrix}$$

is a typical element from SO(3) describing a rotation around the axis determined by the **unit vector** $(u, v, w) \in \mathbb{R}^3$ and with $\theta \in [0, \pi)$ as angle of rotation. Indeed, $\theta \in [0, \pi)$ suffices, since a rotation with angle θ around $(u, v, w)^T$ coincides with a rotation with angle $-\theta$ around $(-u, -v, -w)^T$. The corresponding Lie algebra is

$$\text{SKEW}(3) = \left\{ \begin{pmatrix} 0 & x & -y \\ -x & 0 & z \\ y & -z & 0 \end{pmatrix} \middle| \; x, y, z \in \mathbb{R} \right\}$$

equipped, for example, with the well-known Frobeniusnorm $\| \cdot \|$:

$$\left\| \begin{pmatrix} 0 & x & -y \\ -x & 0 & z \\ y & -z & 0 \end{pmatrix} \right\| = \sqrt{2} \cdot \sqrt{x^2 + y^2 + z^2} \,.$$

Similar to the case $n = 2$ we consider the isometry κ

$$\kappa \,:\, \text{SKEW}(3) \longleftrightarrow \mathbb{R}^3, \quad \begin{pmatrix} 0 & z & -y \\ -z & 0 & x \\ y & -x & 0 \end{pmatrix} \longleftrightarrow \sqrt{2} \cdot \begin{pmatrix} x \\ y \\ z \end{pmatrix} \,.$$

The Cayley transform is a bijection between $\text{SO}(3)^*$ and $\text{SKEW}(3)$. Using the notation $\widetilde{\text{cay}} := \kappa \circ \text{cay}$, a lengthy calculation reveals for $\theta \in (0, \pi)$ that

$$\widetilde{\text{cay}}(R(u,v,w,\theta)) = \kappa \underbrace{\left(\frac{\sin(\theta)}{1 + \cos\theta} \cdot \begin{pmatrix} 0 & w & -v \\ -w & 0 & u \\ v & -u & 0 \end{pmatrix} \right)}_{\in\, \text{SKEW}(3)} = \underbrace{\sqrt{2} \tan\left(\frac{\theta}{2}\right) \cdot \begin{pmatrix} u \\ v \\ w \end{pmatrix}}_{\in\, \mathbb{R}^3} \,.$$

In contrast to the case $n = 2$, now $\widetilde{\text{cay}}$ provides a one-to-one correspondence between $\text{SO}(3)$-fields and vector fields, that is, images with values in $\mathbb{R}^3$. However, there is no total order on $\mathbb{R}^3$. In order to devise some useful morphological operations for such vector fields, one can refer to a partial order at best. In this article we opt for the partial order that is defined by the cone $\mathcal{C} := \{(x, y, z) \in \mathbb{R}^3 \mid x^2 + y^2 \le z^2\}$. The reason for this choice is the isometry between $\mathbb{R}^3$ and the set $\text{SYM}(2)$ of symmetric 2×2-matrices given by

$$g \,:\, \mathbb{R}^3 \longrightarrow \text{SYM}(2), \quad \begin{pmatrix} u \\ v \\ w \end{pmatrix} \longmapsto \frac{1}{\sqrt{2}} \begin{pmatrix} w - v & u \\ u & w + v \end{pmatrix} \,.$$

Especially, g maps $\mathcal{C}$ onto $\text{SYM}^+(2)$, the cone of positive semi-definite symmetric 2×2-matrices; that is $g(\mathcal{C}) = \text{SYM}^+(2)$ holds. Summarizing the above considerations we have established a bijection

$$g \circ \widetilde{\text{cay}} \,:\, \begin{cases} \text{SO}(3)^* \longrightarrow \text{SYM}(2) \\ R(u, v, w, \theta) \longmapsto \tan\left(\frac{\theta}{2}\right) \begin{pmatrix} w - v & u \\ u & w + v \end{pmatrix}, \end{cases} \tag{1}$$

with unit vector $(u, v, w)^T$ and angle of rotation $\theta \in [0, \pi)$. Using the inverse

$$g^{-1} : \begin{pmatrix} a & b \\ b & c \end{pmatrix} \longmapsto \frac{1}{\sqrt{2}} \begin{pmatrix} 2b \\ c - a \\ c + a \end{pmatrix}$$

one can show that the inverse mapping $\text{cay} \circ \kappa^{-1} \circ g^{-1} : \text{SYM}(2) \longrightarrow \text{SO}(3)^*$ is formally given by

$$\begin{pmatrix} a & b \\ b & c \end{pmatrix} \longmapsto \frac{1}{c^2 + a^2 + 2(1 + b^2)} \begin{pmatrix} 2 + 2b^2 - c^2 - a^2 & 2(bc - c - a - ba) & 2(bc + ba + c - a) \\ 2(c + a + bc - ba) & -2ca + 2 - 2b^2 & -4b + c^2 - a^2 \\ 2(bc - c + a + ba) & 4b + c^2 - a^2 & 2ca - 2b^2 + 2 \end{pmatrix} \quad (2)$$

In [7] a framework for elementary morphological operations for fields of symmetric matrices is proposed. As before in the case $n = 2$ we can employ the morphological frame work for fields of 2×2-symmetric matrices as follows:

1. the SO(3)-field is transformed into a SYM(2)-field,
2. the morphological operation available for SYM(2)-fields is applied,
3. the processed field is transformed back into a SO(3)-field.

A straightforward calculation proves that the right hand side of (2) is indeed an element of SO(3). However, it is possible to infer the unit vector (u, v, w) determining the axis of rotation and the angle θ of rotation in the representation $R(u, v, w, \theta)$ directly from given symmetric 2×2-matrix: Keeping in mind the formula (1) we calculate for a given non-zero matrix $A = \begin{pmatrix} a & b \\ b & c \end{pmatrix}$ the components $u^* := b$, $v^* := \frac{1}{2}(c - a)$, $w^* := \frac{1}{2}(c + a)$ of a vector (u^*, v^*, w^*), which in its normalized version $(u, v, w)^\top$ (with respect to the Euclidean norm $\| \cdot \|_2$) determines the oriented axis of rotation. Furthermore, we obtain $\theta = 2 \cdot \arctan(\|(u^*, v^*, w^*)^\top\|_2) \in (0, \pi)$ as the angle of rotation. For the basic operations of dilation and erosion the notion of infimum and supremum is decisive. As pointed out in [7] the supremum resp. infimum of two matrices $A, B \in \text{SYM}(2)$ with respect to this partial order can be calculated. With the above correspondences a reckoning of the supremum resp. infimum of two orthogonal matrices is within our reach.

Example 1. The matrices

$$A := \begin{pmatrix} 1/2 & -\sqrt{3}/2 & 0 \\ \sqrt{3}/2 & 1/2 & 0 \\ 0 & 0 & 1 \end{pmatrix} \quad \text{resp.} \quad B := \begin{pmatrix} -1/2 & -\sqrt{3}/2 & 0 \\ \sqrt{3}/2 & -1/2 & 0 \\ 0 & 0 & 1 \end{pmatrix}$$

are orthogonal matrices representing rotations around the x_3-axis with angles $\frac{\pi}{3}$ resp. $\frac{2\pi}{3}$ in a $x_1 - x_2 - x_3$-coordinate system. The corresponding symmetric 2×2-matrices read

$$a := \begin{pmatrix} \sqrt{3}/3 & 0 \\ 0 & \sqrt{3}/3 \end{pmatrix} \quad \text{resp.} \quad b := \begin{pmatrix} \sqrt{3} & 0 \\ 0 & \sqrt{3} \end{pmatrix}.$$

Using the functional calculus advocated, for instance in [7], a straightforward calculation leads to

$$\sup(a,b) = \frac{1}{2}(a+b+|a-b|) = b \quad \text{resp.} \quad \inf(a,b) = \frac{1}{2}(a+b-|a-b|) = a\,,$$

and hence $\sup(A,B) = B$ and $\inf(A,B) = A$, as expected.

Example 2. The matrices

$$A := \begin{pmatrix} 2/3 & -1/3 & 2/3 \\ 2/3 & 2/3 & -1/3 \\ -1/3 & 2/3 & 2/3 \end{pmatrix} \quad \text{resp.} \quad B := \begin{pmatrix} 2/3 & 2/3 & -1/3 \\ -1/3 & 2/3 & 2/3 \\ 2/3 & -1/3 & 2/3 \end{pmatrix}$$

are describing a rotations with angle $\pi/3$ around the axes $\frac{1}{\sqrt{3}} \cdot (1,1,1)^\top$ resp.
$\frac{-1}{\sqrt{3}} \cdot (1,1,1)^\top$. The associated matrices $a, b \in \mathrm{SYM}(2)$ are

$$a := \begin{pmatrix} 0 & 1/3 \\ 1/3 & 2/3 \end{pmatrix} \quad \text{resp.} \quad b := \begin{pmatrix} 0 & -1/3 \\ -1/3 & -2/3 \end{pmatrix}$$

yielding matrices

$$\inf(a,b) = \begin{pmatrix} -\sqrt{2}/6 & -\sqrt{2}/6 \\ -\sqrt{2}/6 & -\sqrt{2}/2 \end{pmatrix} \quad \text{resp.} \quad \sup(a,b) = \begin{pmatrix} \sqrt{2}/6 & \sqrt{2}/6 \\ \sqrt{2}/6 & \sqrt{2}/2 \end{pmatrix}.$$

They correspond to orthogonal matrices

$$\inf(A,B) := \begin{pmatrix} 7/12 & 1/12+\sqrt{2}/2 & 1/6-\sqrt{2}/4 \\ 1/12-\sqrt{2}/2 & 7/12 & 1/6+\sqrt{2}/4 \\ 1/6+\sqrt{2}/4 & 1/6-\sqrt{2}/4 & 5/6 \end{pmatrix} \quad \text{resp.}$$

$$\sup(A,B) := \begin{pmatrix} 7/12 & 1/12-\sqrt{2}/2 & 1/6+\sqrt{2}/4 \\ 1/12+\sqrt{2}/2 & 7/12 & 1/6-\sqrt{2}/4 \\ 1/6-\sqrt{2}/4 & 1/6+\sqrt{2}/4 & 5/6 \end{pmatrix}$$

describing rotations with angle $\pi/3$ around the axes $(-\sqrt{6}/6, -\sqrt{6}/6, -\sqrt{6}/3)^\top$
resp. $(\sqrt{6}/6, \sqrt{6}/6, \sqrt{6}/3)^\top$.

Visualization of $\mathrm{SO}(3)$-Fields. A two-dimensional $\mathrm{SO}(3)$-field is uniquely
determined by the angle of rotation and the oriented axis of rotation given by a
3D unit vector in each pixel. Hence, such a 2D-$\mathrm{SO}(3)$-field can be visualized by
a 2D-vector field with values in the unit sphere $S^2 \subset \mathbb{R}^3$ representing the axis
in conjunction with a scalar image representing the angle. We refrained from
depicting a 2D-$\mathrm{SO}(3)$-field by means of a 2D-vector field with values in $\mathbb{R}^3$, the
length of the vector coding the angle, since this type of visualization leads to
cluttering and a large amount of ambiguities. However, we can visualize $\mathrm{SO}(3)$-
fields as color images by performing the following steps.

1. The Cayley-transform of an element in $\mathrm{SO}(3)$ is a matrix

$$\begin{pmatrix} 0 & x & -y \\ -x & 0 & z \\ y & -z & 0 \end{pmatrix} \in \mathrm{SKEW}(3)$$

2. Normalize $(x,y,z)^\top$: $(u,v,w)^\top := (x,y,z)^\top / \sqrt{x^2+y^2+z^2}$.

3. Determine the spherical coordinates $\phi \in [0, 2\pi)$ and $\xi \in [0, \pi]$ of $(u, v, w)^\top$ ($u = \sin(\xi)\cos(\phi)$, $v = \sin(\xi)\sin(\phi)$, and $\cos(\xi)$), and in addition, the angle of rotation $\theta \in [0, \pi)$ via $\sqrt{x^2 + y^2 + z^2} = \tan(\theta/2)$.

4. Normalize ϕ, ξ, θ yielding $\hat{\phi}$, $\hat{\xi}$, $\hat{\theta} \in [0, 1]$.

5. Assign $\hat{\phi}$ to the *hue*-, $\hat{\xi}$ to the *saturation*-, and $\hat{\theta}$ to the *lightness*-value of of a pixel in a *hsl*-color image.

6. Transform the resulting *hsl*-image to a *rgb*-image.

The 2D-SO(3)-field can be visualized as an *rgb*-color image. Vice-versa, a *rgb*-image gives rise to a 2D-SO(3)-field, since these steps can be reversed. In the experimental section we will display the results of some morphological operations in both modalities: A 2D-SO(3)-field visualized as a *rgb*-image and as a normalized vector field in conjunction with the angle of rotation as a gray-scale image.

3   Experimental Results

The first experimental example is concerned with SO(2)-fields which in essence is nothing but a scalar image with values in $[0, 1)$ after a trivial rescaling of $[-\pi, \pi)$. Hence, the SO(2)-field in turn stems from a gray-scale version of the image *peppers*. The original 512×512 color image is taken from the *SIPI Image Database* and downsampled to the resolution 32×32 and converted to a gray-valued image. Next, the scalar image is subjected to a morphological operation, where we restrict ourselves to Beucher gradient and morphological Lapalacian. The results are depicted both as a processed gray-valued image and again by a simple rescaling into angles of rotation around the origin depicted as rotated normalized arrows in each pixel of the images as shown in Fig. 1. As we can see in Figures 1(a) and 1(d) the structure of the original image is clearly visible in the SO(2) field. Homogeneous regions can easily be identified. This is even more apparent for the Beucher gradient as shown in Figures 1(b) and 1(e) and the morphological Laplacian as shown in Figures 1(c) and 1(f). Most of the regions for the morphological Laplacian are very dark which correspond to vectors pointing south-west. Next, we process the colored image *peppers* downsampled to the size 32×32 which is shown in Figure 2(a). In Figures 2(b) and 2(c) we show the application of dilation and erosion applied to the original image shown in Figure 2(a) by means of [6] using a 3×3 squared structuring element denoted by $\mathrm{SE_{square}}$. We convert those three color images to matrix fields consisting of 2×2 symmetric matrices. Those fields are further processed into SO(3) fields as presented at the end of in Section 2. Precisely, we plot the scaled angles of rotation as a gray-valued image as shown in Figures 2(d), 2(e), and 2(f), respectively. The unit vectors are depicted as quiver plots given in Figures 2(g), 2(h), and 2(i), respectively. Note that we provided a top-view of the field, although the arrows are three dimensional vectors. Those give the best possible view for the reader, but any view can be used to extract meaningful information from the quiver plots. Interestingly, homogeneous areas are pictured as black regions in the scaled gray-valued angle plot 2(f). Also homogeneous regions are clearly

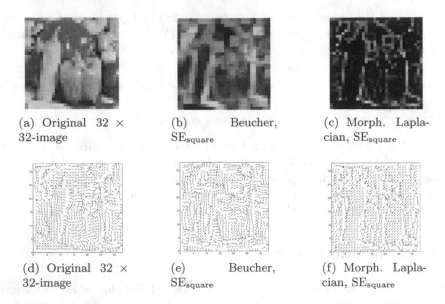

(a) Original 32 × 32-image

(b)   Beucher, SE_square

(c) Morph. Laplacian, SE_square

(d) Original 32 × 32-image

(e)   Beucher, SE_square

(f) Morph. Laplacian, SE_square

Fig. 1. Top row: the angle of rotation θ is represented as a gray-valued image. Bottom row: angle of rotation θ captured by the rotated versions of the unit vector "$\longrightarrow$". First column: original image. Second and third column: output of the Beucher gradient, respectively, the morphological Laplacian with a structuring element SE_square applied to a SO(2)-field originating from a gray-scale version of *peppers*.

visible in the quiver plot 2(i). Additionally, the structure of the pepper can be recognized. We also applied dilation and erosion to the color image *peppers* of size 32×32 given in Figure 2(a) using a 3×3 structuring element denoted by SE_square. The corresponding results are shown as a color image in Figures 3(a) and 3(d), respectively. The used structuring element is clearly visible in those two pictures. Additionally, we provide the processed matrix fields consisting of 2×2 symmetric matrices to corresponding SO(3) fields. The scaled angles of rotation converted into a gray-valued images are shown in Figures 3(b) and 3(e), respectively. Also here, the structuring element is clearly visible. It is noted that the images are dark-grayish. Furthermore, we show the unit vectors depicted as quiver plots in Figures 3(c) and 3(f), respectively. Again, a top-view is provided for the same reason as before. As expected, several different colored regions are clearly distinguishable through different regions of vectors pointing in the same direction. Additonally, we present the SO(3) visualization of the downsampled color image *parrot* of size 192×128. The *parrot* image of size 768×512 is taken from the website http://www.lucnix.be/v/BEST+OF/ara_ararauna.html. The parrot image is visualized in Figure 4(a). The scaled angles of rotation converted into a gray-valued image is shown in Figure 4(d). The unit vectors are depicted as quiver plots as shown in Figure 4(c). Again, the parrot is clearly visible both in the gray-valued scaled θ-plot 4(d) and in the quiver plot 4(c). Finally,

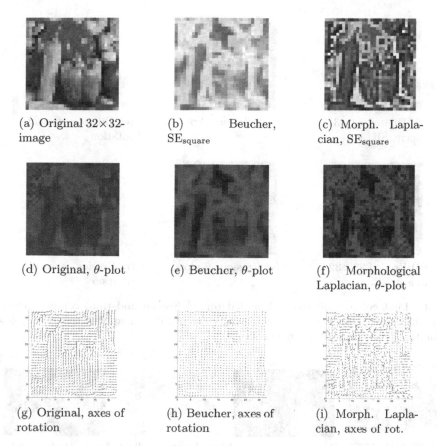

(a) Original 32×32-image

(b)   Beucher, SE$_{\text{square}}$

(c) Morph. Laplacian, SE$_{\text{square}}$

(d) Original, θ-plot

(e) Beucher, θ-plot

(f) Morphological Laplacian, θ-plot

(g) Original, axes of rotation

(h) Beucher, axes of rotation

(i) Morph. Laplacian, axes of rot.

Fig. 2. Beucher and morphological Laplacian with SE$_{\text{square}}$ applied to a SO(3)-field originating from *peppers*

we assume that we have an SO(3)-field. Here we take the SO(3)-field obtained through the conversion of the *parrot* image. Finally, we use the visualization process explained in Section 2. This visualization process of the SO(3)-field is shown in Figure 4(b). Interestingly, we obtain an image that has a strong magenta tinge. It is noteworthy that we did not encounter numerical instabilities due to the Cayley transform in all of our experiments.

Finally, we remark the following: the choice of the order cone $\mathcal{C} = \{(x, y, z) \in \mathbb{R}^3 \mid x^2 + y^2 \leq z^2\}$ clearly favours rotations about the z-axis in $\mathbb{R}^3$ as "large" elements. Other order cones are justified as well, for example, one could rotate the cone $\mathcal{C}$, so the new cone $\mathcal{C}^*$ has a vector $m \neq (0,0,1)^\top$ defining its center line. If $M \in SO(3)$ is a matrix mapping $(0,0,1)^\top$ to m this can be achieved computationally by using the slightly modified Cayley transform $\text{cay}_{M^\top} := \text{cay} \circ \mu$, where μ is given by $\mu_{M^\top} : v \longmapsto M^\top v$ with inverse $\mu_M : w \longmapsto Mw$. The corresponding inverse modified Cayley transform cay_M^{-1} then reads $\text{cay}_M^{-1} = \mu_M \circ \text{cay}^{-1}$. This subject will be part of future research.

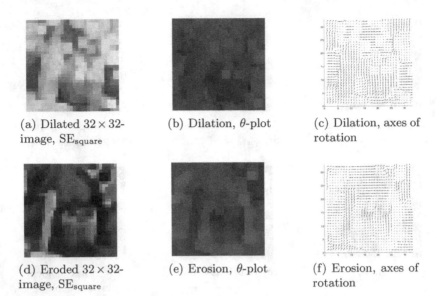

(a) Dilated 32×32-image, SE_{square}

(b) Dilation, θ-plot

(c) Dilation, axes of rotation

(d) Eroded 32×32-image, SE_{square}

(e) Erosion, θ-plot

(f) Erosion, axes of rotation

Fig. 3. First column: dilated and eroded image. Second and third column: output of dilation, resp., erosion with a structuring element SE_{square} applied to a $SO(3)$-field originating from the color image *peppers*.

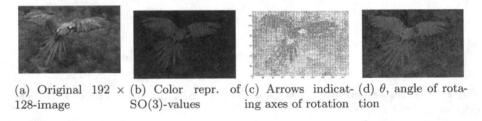

(a) Original 192×128-image

(b) Color repr. of $SO(3)$-values

(c) Arrows indicating axes of rotation

(d) θ, angle of rotation

Fig. 4. Visualization of $SO(3)$-fields. First: original *parrot* image. Second: Alternative color coding in the *hsl*-model. Third: quiver plot depicting axes of rotation. Fourth: θ-values as gray-scale image.

4   Conclusion

It is not obvious how the group structure of $SO(3)$ with respect to matrix multiplication gives directly rise to the notions of supremum and infimmum that might pave the way to elementary morphological operations on $SO(3)$-fields. We circumvent this difficulty by transforming the $SO(3)$-field into a $SYM(2)$-field making a detour to $SKEW(3)$ and $\mathbb{R}^3$ by means of the matrix-valued Cayley transform. For $SYM(2)$-field elementary morphological operations are at our disposal using the Loewner order. Since all the mappings used above are reversible, we have performed a morphological operation on the original $SO(3)$-field. In future research we will employ ordering cones different from $\mathcal{C}$, the one used in this article. Currently, we are investigating various transformations of SO(3)-fields into fields of symmetric matrices for which morphological processes are available. Additionally, new techniques to visualize SO(3)-fields will be a subject of our future research efforts.

References

1. Aptoula, E., Lefèvre, S.: A comparative study on multivariate mathematical morphology. Pattern Recognition 40(11), 2914–2929 (2007)
2. Barnett, V.: The ordering of multivariate data. Journal of the Statistical Society A139(3), 318–355 (1976)
3. Brox, T., Weickert, J.: A TV flow based local scale measure for texture discrimination. In: Pajdla, T., Matas, J(G.) (eds.) ECCV 2004. LNCS, vol. 3022, pp. 578–590. Springer, Heidelberg (2004)
4. Burgeth, B., Bruhn, A., Papenberg, N., Welk, M., Weickert, J.: Mathematical morphology for tensor data induced by the Loewner ordering in higher dimensions. Signal Processing 87(2), 277–290 (2007)
5. Burgeth, B., Didas, S., Florack, L.M.J., Weickert, J.: A generic approach to the filtering of matrix fields with singular PDEs. In: Sgallari, F., Murli, A., Paragios, N. (eds.) SSVM 2007. LNCS, vol. 4485, pp. 556–567. Springer, Heidelberg (2007)
6. Burgeth, B., Kleefeld, A.: An approach to color-morphology based on Einstein addition and Loewner order. Pattern Recognition Letters 47, 29–39 (2014)
7. Burgeth, B., Papenberg, N., Bruhn, A., Welk, M., Feddern, C., Weickert, J.: Morphology for higher-dimensional tensor data via Loewner ordering. In: Ronse, C., Najman, L., Decencière, E. (eds.) Mathematical Morphology: 40 Years On. Computational Imaging and Vision, vol. 30, pp. 407–418. Springer, Dordrecht (2005)
8. Comer, M.L., Delp, E.J.: Morphological operations for color image processing. Journal of Electronic Imaging 8(3), 279–289 (1999)
9. Goutsias, J., Heijmans, H.J.A.M., Sivakumar, K.: Morphological operators for image sequences. Computer Vision and Image Understanding 62, 326–346 (1995)
10. Dela Haije, T.C.J., Duits, R., Tax, C.M.W.: Sharpening fibers in diffusion weighted mri via erosion. In: Westin, C.-F., Vilanova, A., Burgeth, B. (eds.) Visualization and Processing of Tensors and Higher Order Descriptors for Multi-Valued Data, pp. 97–126. Springer, Berlin (2014)
11. Heijmans, H.J.A.M.: Morphological Image Operators. Academic Press, Boston (1994)
12. Matheron, G.: Eléments pour une théorie des milieux poreux. Masson, Paris (1967)
13. Serra, J.: Echantillonnage et estimation des phénomènes de transition minier. PhD thesis, University of Nancy, France (1967)
14. Serra, J.: Image Analysis and Mathematical Morphology, vol. 1 and 2. Academic Press, London (1982, 1988)
15. Serra, J.: Anamorphoses and function lattices (multivalued morphology). In: Dougherty, E.R. (ed.) Mathematical Morphology in Image Processing, pp. 483–523. Marcel Dekker, New York (1993)
16. Soille, P.: Morphological Image Analysis, 2nd edn. Springer, Berlin (2003)
17. van de Gronde, J.J., Roerdink, J.B.T.M.: Group-invariant frames for colour morphology. In: Hendriks, C.L.L., Borgefors, G., Strand, R. (eds.) ISMM 2013. LNCS, vol. 7883, pp. 267–278. Springer, Heidelberg (2013)

Optimization, Differential Calculus
and Probabilities

When Convex Analysis Meets Mathematical Morphology on Graphs

Laurent Najman[✉], Jean-Christophe Pesquet, and Hugues Talbot

Laboratoire d'Informatique Gaspard Monge – CNRS UMR 8049, Université Paris-Est
77454 Marne la Vallée Cedex 2, Paris, France
first.last@univ-paris-est.fr

Abstract. In recent years, variational methods, *i.e.*, the formulation of problems under optimization forms, have had a great deal of success in image processing. This may be accounted for by their good performance and versatility. Conversely, mathematical morphology (MM) is a widely recognized methodology for solving a wide array of image processing-related tasks. It thus appears useful and timely to build bridges between these two fields. In this article, we propose a variational approach to implement the four basic, structuring element-based operators of MM: dilation, erosion, opening, and closing. We rely on discrete calculus and convex analysis for our formulation. We show that we are able to propose a variety of continuously varying operators in between the dual extremes, *i.e.*, between erosions and dilation; and perhaps more interestingly between openings and closings. This paves the way to the use of morphological operators in a number of new applications.

Keywords: Optimization · Convex analysis · Discrete calculus · Graphs

1 Introduction

In recent years, variational methods have received a great deal of attention in the image processing community, mainly due to their success in addressing a wide range of signal and image processing tasks [24] (*e.g.*, denoising, restoration, reconstruction, impainting, segmentation,...). Concurrently, mathematical morphology (MM) is widely recognized as a fundamental methodology for solving image processing problems [23]. It thus appears useful to build bridges between these two fields. A first step in this direction may consist of looking for variational formulations of the essential structuring-element based operators of morphological processing, *i.e*, the dilation, erosion, opening, and closing operators.

Several formulations can be found in the literature. Since the dilation, for instance, is in some way similar to a propagation, it seems natural to express it as a propagation or a diffusion PDE. The first proposal for a PDE-based formulation of morphology was independently proposed by Alvarez *et al.* [1,2], Brockett and Maragos [4], and Boomgaard and Smeulders [27]. In [19], this approach is studied from a more geometrical point of view. In [10], standard derivatives are

© Springer International Publishing Switzerland 2015
J.A. Benediktsson et al. (Eds.): ISMM 2015, LNCS 9082, pp. 473–484, 2015.
DOI: 10.1007/978-3-319-18720-4_40

replaced by metric-based differential calculus. In [17], Maragos proposes a PDE-based formalism based on the slope transform. The slope transform is studied in details, including its relationship with convex analysis in [14]. We also mention [3,25], where binary dilations and erosions are modeled as curve evolution with constant normal speed. An overview of PDE-based, contrast-invariant image processing, including morphology operators, is given in [13]. Many references could be cited here, not limited to dilation and erosions; for example, the equivalents of connected operators are proposed in PDE form in [21]. Obviously all these approaches need to be discretized for practical implementations, typically with finite differences. All of these share a similar goal: they seek for implementing the notion of morphological operator in the continuous domain. The discrete structure used to specifying the operators is not seen as important.

In contrast, some formulations exploit the discrete nature of image data. For instance, an algebraic analysis of the theory of differential operators [5] leads, in the discrete case, to the discrete exterior calculus [9,8]. In [12], Grady proposes a graph-based, variational implementation, with applications to imaging. The variational discrete calculus is expanded in [6]. A similar approach is followed in [11]. An overview of the various graph-based variational approaches is given in [16]. In all of these, the graph is not a discretization artifact, it remains a core feature, allowing for more flexibility and arguably better data fidelity.

In the discrete setting, a natural framework for mathematical morphology is precisely that of graphs. Pioneering papers are in [28,15]. Many works have been proposed since, for instance [20,7]. The framework has been extended far beyond the basic operators, the interested reader can refer to the recent survey given in [22].

To the best of our knowledge, a variational framework for implementing the morphological operators on graphs has not been proposed. Maragos [18] was the first to propose a variational approach to study the basic and the connected operators of MM, but in a continuous setting. A graph-based PDE approach to MM is proposed in [26], however the approach is not variational. In this work, we follow an alternative, conceptually simpler way, merging variational formulation and graphs, based on discrete calculus and convex analysis.

The graph-based variational framework for MM developped in this paper, allows us to overcome some of the main difficulties of the continuous setting. In particular, in the continuous domain, it has been not possible so far to propose a proper differential definition of the combined operators: openings and closings; a major difficulty being the idempotence of these operators. In contrast, we show in this paper that the discrete nature of the graph allows a complete definition of these operators. In future work, such a variational formulation of morphological operators could be used as asymmetric, non self-dual, regularisers, something difficult (and thus unusual) in standard applications of the variational framework.

The organization of the paper is the following. By designing objective functions using judiciously chosen support functions, we show that a fully convex formulation of dilation, erosion, dilation, and opening is possible. More precisely, we establish that these morphological operators correspond to asymptotic forms of the solutions to classes of convex optimization problems. As a side result, by considering a non-asymptotic regime, a broader set of solutions can be obtained which allow us to generate, in a continuous manner, intermediate behaviours between dilation/erosion and opening/closing. Our proposal for a graph-based variational formulation of the erosion and dilation operators is exposed in section 2, while the one for opening and closing operators is developed in section 3. Due to space constraints, the proofs of the results in this paper will be provided in an extended version.

Notation. Let $G = (V, E)$ be a valued directed reflexive graph. The cardinality of the vertex set V is assumed to be equal to $n \in \mathbb{N}^*$. In the case of images, the vertices reduce to pixels. An edge joining vertex $v_i \in V$ to vertex $v_j \in V$ with $(i, j) \in \{1, \ldots, n\}^2$, $i \neq j$, is denoted by $e_{i,j} \in E$. For every $i \in \{1, \ldots, n\}$, a weight $y_i \in \mathbb{R}$ is associated with each vertex v_i, and we introduce the set of neighboring indices of i, $N_i = \{j \in \{1, \ldots, n\} \mid e_{i,j} \in E\}$, which will be assumed to be nonempty. The reciprocal neighborhood of N_i is $\check{N}_i = \{j \in \{1, \ldots, n\} \mid i \in N_j\}$. Since G is reflexive, $(\forall i \in \{1, \ldots, n\})\, i \in N_i \cap \check{N}_i$.

2   A Unifying Variational Formulation of Erosion, Dilation and Median

For every $i \in \{1, \ldots, n\}$, let U_i be a nonempty subset of V (although it may be any subset, we will be mainly interested in the case when it corresponds to a neighborhood or reciprocal neighborhood of node i). We will first consider the following convex minimization problem:

$$\underset{x=(x_i)_{1 \leqslant i \leqslant n} \in \mathbb{R}^n}{\text{minimize}} \quad \sum_{i=1}^{n} \sum_{j \in U_i} \varphi_i\big(\sigma_{\Omega_i}(x_i - y_j - w_{i,j})\big) \tag{2.1}$$

where $(w_{i,j})_{j \in U_i} \in \mathbb{R}^{|U_i|}$ is a vector of shift parameters, $\varphi_i \colon [0, +\infty[\to \mathbb{R}$ is a (strictly) increasing lower-semicontinuous convex function, Ω_i is a closed real interval, and σ_{Ω_i} denotes the support function of Ω_i, which is defined as

$$(\forall v \in \mathbb{R}) \qquad \sigma_{\Omega_i}(v) = \sup_{\xi \in \Omega_i} v\xi$$

$$= \begin{cases} \alpha_i^- v & \text{if } v < 0 \\ 0 & \text{if } v = 0 \\ \alpha_i^+ v & \text{if } v > 0 \end{cases} \tag{2.2}$$

with $\alpha_i^- = \inf \Omega_i$ and $\alpha_i^+ = \sup \Omega_i$.

Any solution to the multivariate optimization problem (2.1) is a vector of node weights which can be viewed as the result of a generally nonlinear processing applied to the original vector of node weights $y = (y_i)_{1 \leqslant i \leqslant n}$. The result is dependent on the choice of the intervals $(\Omega_i)_{1 \leqslant i \leqslant n}$ and our goal will be to better

X understand this dependence. In order to guarantee the existence of a solution to problem (2.1), it will be assumed that:

Assumption 1. *For every* $i \in \{1, \ldots, n\}$, $-\infty \leqslant \alpha_i^- < 0 < \alpha_i^+ \leqslant +\infty$ *with* $(\alpha_i^-, \alpha_i^+) \neq (-\infty, +\infty)$.

Note that α_i^- (resp. α_i^+) is allowed to be equal to $-\infty$ (resp. $+\infty$), which will turn out to be useful in the rest of our discussion.

A first result concerning the solution to this optimization problem is as follows:

Proposition 1. *Suppose that Assumption 1 holds. Then Problem* (2.1) *admits a solution.*
Let $i \in \{1, \ldots, n\}$ *and let* $\hat{x}_i$ *be the* i-*th component of such a solution* $\hat{x}$. *We have the following properties:*

(i) *If* $\alpha_i^+ = +\infty$, *then*

$$\hat{x}_i = \min \{y_j + w_{i,j} \mid j \in U_i\}. \tag{2.3}$$

(ii) *If* $\alpha_i^- = -\infty$, *then*

$$\hat{x}_i = \max \{y_j + w_{i,j} \mid j \in U_i\}. \tag{2.4}$$

(iii) *If* φ_i *is strictly convex, then* $\hat{x}_i$ *also is uniquely defined.*

Remark 1. We see that standard morphological mathematical operators are recovered as specific solutions to Problem (2.1):

- If $\alpha_i^+ = +\infty$ and $U_i = N_i$ with $i \in \{1, \ldots, n\}$, then (2.3) corresponds to an erosion with (spatially variant) structuring element:

$$(\forall (i,j) \in \{1, \ldots, n\}) \quad \tilde{w}_{i,j} = \begin{cases} -w_{i,j} & \text{if } j \in N_i \\ -\infty & \text{otherwise.} \end{cases} \tag{2.5}$$

- If $\alpha_i^- = -\infty$ and $U_i = N_i$ with $i \in \{1, \ldots, n\}$, then (2.4) corresponds to a dilation with (spatially variant) structuring element:

$$(\forall (i,j) \in \{1, \ldots, n\}) \quad \tilde{w}_{i,j} = \begin{cases} w_{i,j} & \text{if } j \in N_i \\ -\infty & \text{otherwise.} \end{cases} \tag{2.6}$$

By making $-\alpha_i^+/\alpha_i^-$ vary, intermediate behaviours between an erosion and a dilation can be obtained. In general however, one must resort to numerical methods to find a solution to Problem (2.1).

We now mention a specific choice of functions $(\varphi_i)_{1 \leqslant i \leqslant n}$, for which minimizers can be expressed in closed forms.

Proposition 2. *Suppose that Assumption 1 holds. Let $i \in \{1,\ldots,n\}$ and let $\widehat{x}_i$ be the i-th component of the optimal solution to Problem (2.1) where $\varphi_i \colon [0,+\infty[\mapsto \mathbb{R} \colon \xi \mapsto \xi^2$. Let us denote by $(z_{i,j})_{1 \leqslant j \leqslant |U_i|}$ the coefficients $(y_j + w_{i,j})_{j \in U_i}$ which have been re-indexed in increasing order, and let us set $z_{i,0} = -\infty$ and $z_{i,|U_i|+1} = +\infty$. If $(\alpha_i^-, \alpha_i^+) \in \mathbb{R}^2$, then $\widehat{x}_i$ is the unique value in $[z_{i,j_i-1}, z_{i,j_i}]$ with $j_i \in \{1,\ldots,|U_i|+1\}$ such that*

$$\widehat{x}_i = \frac{(\alpha_i^+)^2 \sum_{j=1}^{j_i-1} z_{i,j} + (\alpha_i^-)^2 \sum_{j=j_i}^{|U_i|} z_{i,j}}{(j_i - 1)(\alpha_i^+)^2 + (|U_i| - j_i + 1)(\alpha_i^-)^2} \tag{2.7}$$

(with the conventions $\sum_{j=1}^{0} \cdot = \sum_{j=|U_i|+1}^{|U_i|} \cdot = 0$).

Remark 2. When $\alpha_i^- = -\alpha_i^+$ with $i \in \{1,\ldots,n\}$ and $\alpha_i^+ \in]0,+\infty[$, then the expression of $\widehat{x}_i$ given in the above proposition reduces to the standard averaged value of $(y_j + w_{i,j})_{j \in U_i}$.

An image processing illustration of Proposition 2 is provided in the first two columns of Fig. 2. The original 8-bit images are noisy versions of a synthetic one of size 100×104 called Pyramid and of a natural scene of size 512×512 called Goldhill, displayed in 1. The graph consists here of a regular 4-connected grid. As expected, nonlinear filters ranging from a dilation (on the left) to an erosion (on the right) are generated, with the local averaging (third image) as a special case.

Fig. 1. Pyramid and Goldhill images

Another simple choice of functions $(\varphi_i)_{1 \leqslant i \leqslant n}$ is addressed next:

Proposition 3. *Suppose that Assumption 1 holds. Let $i \in \{1,\ldots,n\}$ and let $\widehat{x}_i$ be the i-th component of an optimal solution to Problem (2.1) where $\varphi_i \colon [0,+\infty[\mapsto \mathbb{R} \colon \xi \mapsto \xi$. Let us denote by $(z_{i,j})_{1 \leqslant j \leqslant |U_i|}$ the coefficients $(y_j + w_{i,j})_{j \in U_i}$ which have been re-indexed in increasing order. Assume that $(\alpha_i^-, \alpha_i^+) \in \mathbb{R}^2$ and set $j_i = \lfloor \frac{|U_i|}{1-\alpha_i^+/\alpha_i^-} \rfloor + 1$, where $\lfloor \cdot \rfloor$ denotes the lower rounding operation.*

Then, if $\frac{|U_i|}{1-\alpha_i^+/\alpha_i^-} \notin \mathbb{N}$, $\widehat{x}_i = z_{i,j_i}$.

If $\frac{|U_i|}{1-\alpha_i^+/\alpha_i^-} \in \mathbb{N}$, then $\widehat{x}_i$ can be chosen equal to any value in $[z_{i,j_i-1}, z_{i,j_i}]$.

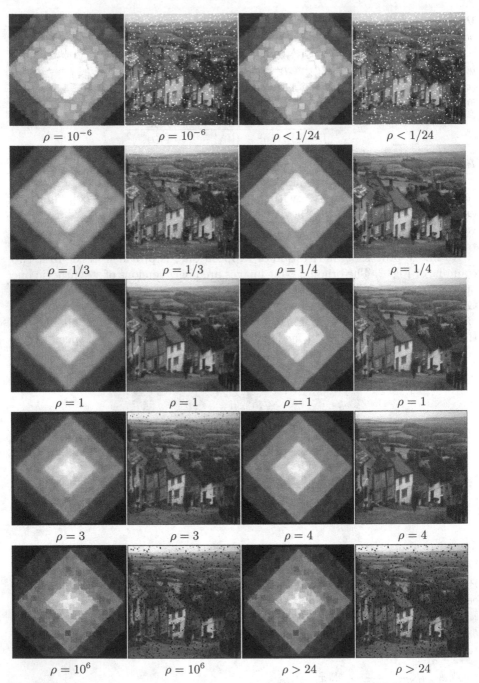

Fig. 2. First two columns: solutions to Problem (2.1) for Pyramid and Goldhill images when, for every pixel i, $\varphi_i \colon [0, +\infty[\mapsto \mathbb{R} \colon \xi \mapsto \xi^2$.

Last two columns: solutions to Problem (2.1) for Pyramid and Goldhill images when, for every pixel i, $\varphi_i \colon [0, +\infty[\mapsto \mathbb{R} \colon \xi \mapsto \xi$.

In both cases, $-\alpha_i^+/\alpha_i^- = \rho$, $(w_{i,j})_{j \in U_i} = 0$, and U_i is a 5×5 spatial neighborhood.

An illustration of Proposition 3 is provided in the last two columns of Fig. 2. The first (resp. last) row exacly corresponds to a dilation (resp. an erosion), as explained in the remark below. Note that we do not observe on the intermediate images the blur effect which is noticeable on those displayed in the first two columns of Fig. 2.

Remark 3. Under the assumptions of the above proposition, the following specific solutions can be obtained:

– median filtering
  If $\alpha_i^- = -\alpha_i^+$, $\alpha_i^+ \in]0,+\infty[$ and $U_i = N_i$ with $i \in \{1,\ldots,n\}$, then the i-th component $\widehat{x}_i$ of an optimal solution to Problem (2.1) is

$$\widehat{x}_i = \text{median}\{y_j + w_{i,j} \mid j \in N_i\}. \qquad (2.8)$$

– erosion
  If $-\alpha_i^+/\alpha_i^- > |N_i| - 1$ and $U_i = N_i$, then $\widehat{x}_i$ is given by (2.3).
– dilation
  If $-\alpha_i^+/\alpha_i^- (|N_i| - 1) < 1$ and $U_i = N_i$, then $\widehat{x}_i$ is given by (2.4).

More generally, by varying $-\alpha_i^+/\alpha_i^-$, any rank-order filtering operator between a dilation and an erosion can be generated, so putting more or less emphasis on large/small intensity values for vertices in the neighborhood N_i.

3   Towards a Variational Formulation of Opening and Closing

Let us now consider the more sophisticated convex minimization problem:

$$\underset{\substack{x=(x_i)_{1 \leqslant i \leqslant n} \in \mathbb{R}^n \\ t=(t_i)_{1 \leqslant i \leqslant n} \in \mathbb{R}^n}}{\text{minimize}} \sum_{i=1}^{n} \left(\sum_{j \in N_i} \varphi_i\big(\sigma_{\Omega_i}(t_i - y_j - w_{i,j})\big) + \sum_{j \in \check{N}_i} \psi_i\big(\sigma_{\Lambda_i}(x_i - t_j - \nu_{i,j})\big) \right)$$

$$(3.1)$$

where, for every $i \in \{1,\ldots,n\}$, $(w_{i,j})_{j \in N_i} \in \mathbb{R}^{|N_i|}$, $(\nu_{i,j})_{j \in N_i} \in \mathbb{R}^{|\check{N}_i|}$, $\varphi_i \colon [0,+\infty[\to \mathbb{R}$ and $\psi_i \colon [0,+\infty[\to \mathbb{R}$ are (strictly) increasing lower-semicontinuous convex functions, Ω_i and Λ_i are closed real intervals with $\alpha_i^- = \inf \Omega_i$, $\alpha_i^+ = \sup \Omega_i$, $\beta_i^- = \inf \Lambda_i$, and $\beta_i^+ = \sup \Lambda_i$.

Subsequently, we will suppose that:

Assumption 2. *For every* $i \in \{1,\ldots,n\}$, $-\infty \leqslant \alpha_i^- < 0 < \alpha_i^+ \leqslant +\infty$ *with* $(\alpha_i^-, \alpha_i^+) \neq (-\infty,+\infty)$, *and* $-\infty \leqslant \beta_i^- < 0 < \beta_i^+ \leqslant +\infty$ *with* $(\beta_i^-, \beta_i^+) \neq (-\infty,+\infty)$.

A first result on the general solutions to Problem (3.1) is as follows:

Proposition 4. *Suppose that Assumption 2 holds. Problem (3.1) admits a solution. Let* $(\widehat{x},\widehat{t})$ *be such a solution. Then,* $\widehat{t}$ *is a solution to the convex optimization problem:*

$$\underset{t=(t_i)_{1\leqslant i\leqslant n}\in\mathbb{R}^n}{\text{minimize}} \sum_{i=1}^{n}\Big(\sum_{j\in N_i}\varphi_i\big(\sigma_{\Omega_i}(t_i-y_j-w_{i,j})\big)+\Psi_i(t)\Big) \qquad (3.2)$$

where, for every $i\in\{1,\dots,i\}$, Ψ_i is the finite convex function given by

$$\Psi_i\colon t\mapsto \inf_{x_i\in\mathbb{R}}\sum_{j\in\check{N}_i}\psi_i\big(\sigma_{\Lambda_i}(x_i-t_j-\nu_{i,j})\big). \qquad (3.3)$$

In addition, if $(\widehat{t}_j)_{1\leqslant j\leqslant n}$ are the components of $\widehat{t}$, then the components $(\widehat{x}_i)_{1\leqslant i\leqslant n}$ of $\widehat{x}$ are such that

$$(\forall i\in\{1,\dots,n\})\qquad \widehat{x}_i\in\underset{x_i\in\mathbb{R}}{\text{Argmin}}\sum_{j\in\check{N}_i}\psi_i\big(\sigma_{\Lambda_i}(x_i-\widehat{t}_j-\nu_{i,j})\big). \qquad (3.4)$$

Moreover, if, for every $i\in\{1,\dots,n\}$, φ_i and ψ_i are strictly convex, $(\widehat{x},\widehat{t})$ is uniquely defined.

Let us now focus on the case when, for every $i\in\{1,\dots,n\}$, $\psi_i\colon[0,+\infty[\to\mathbb{R}\colon\xi\mapsto\xi^{\tau_i}$, where $\tau_i\in[1,+\infty[$. The next result shows that, under some asymptotic conditions, Problem (3.1) can basically be decoupled into two simpler optimization tasks:

Proposition 5. *Suppose that Assumption 2 holds. Let $(\mu^{(k)})_{k\in\mathbb{N}}$ be a sequence of positive reals such that*

$$\lim_{k\to+\infty}\mu^{(k)}=0. \qquad (3.5)$$

For every $i\in\{1,\dots,n\}$, let $(\Lambda_i^{(k)})_{k\in\mathbb{N}}$ be sequences of closed intervals such that, for every $k\in\mathbb{N}$,

$$\inf\Lambda_i^{(k)}=\mu^{(k)}\beta_i^-,\qquad \sup\Lambda_i^{(k)}=\mu^{(k)}\beta_i^+. \qquad (3.6)$$

For every $k\in\mathbb{N}$, let

$$(\widehat{x}^{(k)},\widehat{t}^{(k)})\in\underset{\substack{x=(x_i)_{1\leqslant i\leqslant n}\in\mathbb{R}^n\\ t=(t_i)_{1\leqslant i\leqslant n}\in\mathbb{R}^n}}{\text{Argmin}}\sum_{i=1}^{n}\Big(\sum_{j\in N_i}\varphi_i\big(\sigma_{\Omega_i}(t_i-y_j-w_{i,j})\big)$$
$$+\sum_{j\in\check{N}_i}\big(\sigma_{\Lambda_i^{(k)}}(x_i-t_j-\nu_{i,j})\big)^{\tau_i}\Big). \qquad (3.7)$$

Then, $(\widehat{x}^{(k)},\widehat{t}^{(k)})_{k\in\mathbb{N}}$ is a bounded sequence. In addition, if $(\widehat{x},\widehat{t})$ is a cluster point of $(\widehat{x}^{(k)},\widehat{t}^{(k)})_{k\in\mathbb{N}}$, then $\widehat{x}=(\widehat{x}_i)_{1\leqslant i\leqslant n}$ and $\widehat{t}=(\widehat{t}_i)_{1\leqslant i\leqslant n}$ where

$$(\forall i\in\{1,\dots,n\})\qquad \widehat{t}_i\in\underset{t_i\in\mathbb{R}}{\text{Argmin}}\sum_{j\in N_i}\varphi_i\big(\sigma_{\Omega_i}(t_i-y_j-w_{i,j})\big) \qquad (3.8)$$

$$\widehat{x}_i\in\underset{x_i\in\mathbb{R}}{\text{Argmin}}\sum_{j\in\check{N}_i}\big(\sigma_{\Lambda_i}(x_i-\widehat{t}_j-\nu_{i,j})\big)^{\tau_i}. \qquad (3.9)$$

Remark 4. Under the assumptions of Proposition 5, for every $i \in \{1, \ldots, N\}$, if $t_i \mapsto \sum_{j \in N_i} \varphi_i(\sigma_{\Omega_i}(t_i - y_j - w_{i,j})$ has a unique minimizer $\widehat{t}_i$, then the i-th component $(\widehat{t}_i^{(k)})_{k \in \mathbb{N}}$ of $(\widehat{t}^{(k)})_{k \in \mathbb{N}}$ converges to $\widehat{t}_i$. In addition, if $(\widehat{t}_j)_{j \in \check{N}_i}$ is uniquely defined and $x_i \mapsto \sum_{j \in \check{N}_i} \left(\sigma_{\Lambda_i}(x_i - \widehat{t}_j - \nu_{i,j})\right)^{\tau_i}$ has a unique minimizer $\widehat{x}_i$, then the i-th component $(\widehat{x}_i^{(k)})_{k \in \mathbb{N}}$ of $(\widehat{x}^{(k)})_{k \in \mathbb{N}}$ converges to $\widehat{x}_i$. The latter condition holds, in particular, when, for every $j \in \check{N}_i$, φ_j is a strictly convex function and $\tau_i > 1$.

Combining Propositions 1 and 5 with the above remark yields the following result:

Corollary 1. *Suppose that, for every $i \in \{1, \ldots, n\}$, $\alpha_i^- \in]-\infty, 0[$, $\alpha_i^+ = +\infty$, $\beta_i^- = -\infty$, and $\beta_i^+ \in]0, +\infty[$ (resp. $\alpha_i^- = -\infty$, $\alpha_i^+ \in]0, +\infty[$, $\beta_i^- \in]-\infty, 0[$, and $\beta_i^+ = +\infty$). Let $(\mu^{(k)})_{k \in \mathbb{N}}$ be a sequence of positive reals satisfying (3.5). For every $i \in \{1, \ldots, n\}$, let $(\Lambda_i^{(k)})_{k \in \mathbb{N}}$ be sequences of closed intervals such that, for every $k \in \mathbb{N}$, (3.6) holds. For every $k \in \mathbb{N}$, let $(\widehat{x}^{(k)}, \widehat{t}^{(k)})$ be given by (3.7). Then, $\lim_{k \to +\infty} \widehat{x}^{(k)} = (\widehat{x}_i)_{1 \leqslant i \leqslant n}$ and $\lim_{k \to +\infty} \widehat{t}^{(k)} = (\widehat{t}_i)_{1 \leqslant i \leqslant n}$, where*

$$(\forall i \in \{1, \ldots, n\}) \qquad \widehat{t}_i = \min \{y_j + w_{i,j} \mid j \in N_i\} \tag{3.10}$$

$$(resp. \ \widehat{t}_i = \max \{y_j + w_{i,j} \mid j \in N_i\})$$

$$\widehat{x}_i = \max \{\widehat{t}_j + \nu_{i,j} \mid j \in \check{N}_i\} \tag{3.11}$$

$$(resp. \ \widehat{x}_i = \min \{\widehat{t}_j + \nu_{i,j} \mid j \in \check{N}_i\}).$$

Remark 5.

(i) If

$$(\forall i \in \{1, \ldots, n\})(\forall j \in N_i) \qquad w_{i,j} = -\nu_{j,i} \tag{3.12}$$

then, under the assumptions of Corollary 1, we asymptotically obtain an opening (resp. a closing) operator with structuring element (2.5) (resp. (2.6)).

(ii) When, for every $i \in \{1, \ldots, n\}$, $\varphi_i \colon [0, +\infty[\mapsto \mathbb{R} \colon \xi \mapsto \xi$ and $\tau_i = 1$, it follows from Proposition 3 (see also Remark 3) that, if $(\forall i \in \{1, \ldots, n\})$ $(\alpha_i^-, \beta_i^-) \in]-\infty, 0[^2$ and $(\alpha_i^+, \beta_i^+) \in]0, +\infty[^2$ are such that

$$-\frac{\alpha_i^+}{\alpha_i^-} > |N_i| - 1 \quad \text{and} \quad -\frac{\beta_i^+}{\beta_i^-}(|\check{N}_i| - 1) < 1$$

$$(resp. \ -\frac{\alpha_i^+}{\alpha_i^-}(|N_i| - 1) < 1 \quad \text{and} \quad -\frac{\beta_i^+}{\beta_i^-} > |\check{N}_i| - 1), \tag{3.13}$$

then the conclusions of Corollary 1 also hold.

Illustrations of the effect of the proposed operators on the two image examples we already considered are shown in Fig. 3.

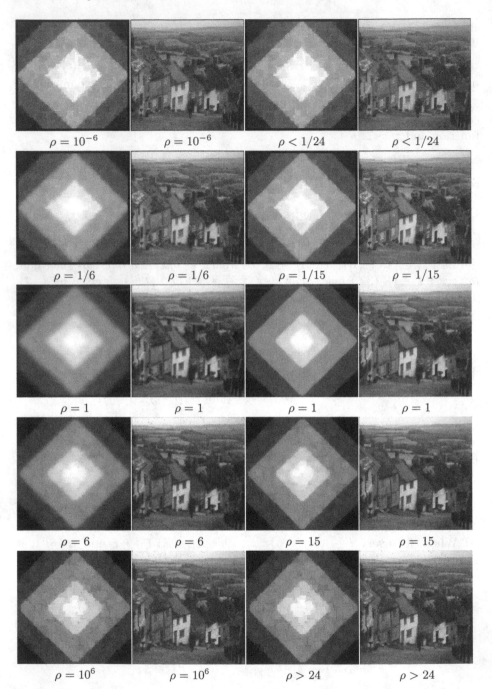

Fig. 3. Solutions to (3.8)-(3.9) for Pyramid and Goldhill images when $-\alpha_i^+/\alpha_i^- = -\beta_i^-/\beta_i^+ = \rho$, $(w_{i,j})_{j \in N_i} = (\nu_{i,j})_{j \in \check{N}_i} = 0$, and N_i (and $\check{N}_i$) is a 5×5 spatial neighborhood. For every pixel i, we have on the first two columns $\varphi_i \colon [0, +\infty[\mapsto \mathbb{R} \colon \xi \mapsto \xi^2$ and $\tau_i = 2$, whereas on the last two columns, $\varphi_i \colon [0, +\infty[\mapsto \mathbb{R} \colon \xi \mapsto \xi$ and $\tau_i = 1$.

4   Conclusion

In this paper, we have introduced what we may call *variational dilation, erosion, opening, and closing*. We have seen that these operators can be applied in graph processing and that they are defined thanks to support functions of closed real intervals and increasing convex functions. By varying these parameters, a wide class of operators can be defined, potentially leading to much flexibility in their use. In particular, we have proved that standard morphological dilation, erosion, opening, and closing are recovered as limit cases. For application purposes that we plan to investigate soon, this means that we are now able to use those operators for example as asymmetric regularizers, *e.g.* to favor bright contrast over dark contrast or conversely.

Building on the present paper, two main theoretical avenues can now be explored. The first one consists of more deeply analyzing the properties of these new variational operators and to see whether other classical morphological operators have their variational counterparts. The second direction is to look for extensions of the energy functions which have been set to define these operators in order to address more general problems of interest, *e.g.*, those involving local adaptive decision processes or some statistical knowledge on the target signal or the noise. Then, more attention should be paid to optimization algorithms allowing us to efficiently solve such problems.

Acknowledgments. This work received funding from the Agence Nationale de la Recherche, contract ANR-14-CE27-0001 GRAPHSIP) and through "Programme d'Investissements d'Avenir" (LabEx BEZOUT ANR-10-LABX-58).

References

1. Alvarez, L., Guichard, F., Lions, P.L., Morel, J.M.: Axiomatisation et nouveaux opérateurs de la morphologie mathématique. Compt. Rendus Acad. Sci. Math. 315(3), 265–268 (1992)
2. Alvarez, L., Guichard, F., Lions, P.L., Morel, J.M.: Axioms and fundamental equations of image processing. Arch. Ration. Mech. Anal. 123(3), 199–257 (1993)
3. Arehart, A.B., Vincent, L., Kimia, B.B.: Mathematical morphology: The Hamilton-Jacobi connection. In: Fourth IEEE Int. Conf. Comput. Vis., pp. 215–219 (1993)
4. Brockett, R.W., Maragos, P.: Evolution equations for continuous-scale morphological filtering. IEEE Trans. Signal Process. 42(12), 3377–3386 (1994)
5. Cartan, É.: Les systèmes différentiels extérieurs et leurs applications géométriques. Hermann, Paris (1945)
6. Couprie, C., Grady, L., Najman, L., Pesquet, J.C., Talbot, H.: Dual constrained TV-based regularization on graphs. SIAM J. Imaging Sci. 6(3), 1246–1273 (2013)
7. Cousty, J., Najman, L., Dias, F., Serra, J.: Morphological filtering on graphs. Comput. Vis. Image Understand. 117(4), 370–385 (2013)
8. Desbrun, M., Hirani, A.N., Leok, M., Marsden, J.E.: Discrete exterior calculus (2005), http://arxiv.org/abs/math/0508341
9. Desbrun, M., Hirani, A.N., Marsden, J.E.: Discrete exterior calculus for variational problems in computer vision and graphics. In: 42nd IEEE Conf. Decision Control, vol. 5, pp. 4902–4907 (2003)

10. Doyen, L., Najman, L., Mattioli, J.: Mutational equations of the morphological dilation tubes. J. Math. Imaging Vision 5, 219–230 (1995)
11. Elmoataz, A., Lezoray, O., Bougleux, S.: Nonlocal discrete regularization on weighted graphs: a framework for image and manifold processing. IEEE Trans. Image Process. 17(7), 1047–1060 (2008)
12. Grady, L.J., Polimeni, J.: Discrete calculus: Applied analysis on graphs for computational science. Springer (2010)
13. Guichard, F., Morel, J.M., Ryan, R.: Contrast invariant image analysis and PDEs. Tech. rep., Centre de Mathématiques et de Leurs Applications, Cachan, France (2004), http://dev.ipol.im/~morel/JMMBookOct04.pdf
14. Heijmans, H.J., Maragos, P.: Lattice calculus of the morphological slope transform. Signal Process 59(1), 17–42 (1997)
15. Heijmans, H., Nacken, P., Toet, A., Vincent, L.: Graph morphology. J. Vis. Comm. Image Repr. 3(1), 24–38 (1992)
16. Lézoray, O., Grady, L.: Image processing and analysis with graphs: theory and practice. CRC Press (2012)
17. Maragos, P.: Differential morphology and image processing. IEEE Trans. Image Process. 5(6), 922–937 (1996)
18. Maragos, P.: A variational formulation of PDE's for dilations and levelings. In: Mathematical Morphology: 40 Years On, pp. 321–332. Springer (2005)
19. Mattioli, J.: Relations différentielles d'opérations de la morphologie mathématique. Compt. Rendus Acad. Sci. Math. 316(9), 879–884 (1993)
20. Meyer, F., Angulo, J.: Micro-viscous morphological operators. In: Math. Morphology and Appl. Signal and Image Process, ISMM 2007, pp. 165–176 (2007)
21. Meyer, F., Maragos, P.: Nonlinear scale-space representation with morphological levelings. J. Vis. Comm. Image Repr. 11(2), 245–265 (2000)
22. Najman, L., Cousty, J.: A graph-based mathematical morphology reader. Pattern Recogn. Lett. 47, 3–17 (2014)
23. Najman, L., Talbot, H.: Mathematical Morphology. John Wiley & Sons (2013)
24. Paragios, N., Chen, Y., Faugeras, O.D.: Handbook of mathematical models in computer vision. Springer (2006)
25. Sapiro, G., Kimmel, R., Shaked, D., Kimia, B.B., Bruckstein, A.M.: Implementing continuous-scale morphology via curve evolution. Pattern Recogn. 26(9), 1363–1372 (1993)
26. Ta, V.T., Elmoataz, A., Lézoray, O.: Nonlocal PDEs-based morphology on weighted graphs for image and data processing. IEEE Trans. Image Process. 26(2), 1504–1516 (2011)
27. Van Den Boomgaard, R., Smeulders, A.: The morphological structure of images: The differential equations of morphological scale-space. IEEE Trans. Pattern Anal. Mach. Intell. 16(11), 1101–1113 (1994)
28. Vincent, L.: Graphs and mathematical morphology. Signal Process. 16(4), 365–388 (1989)

(max, min)-convolution
and Mathematical Morphology

Jesús Angulo$^{(\boxtimes)}$

MINES ParisTech, PSL-Research University,
CMM-Centre de Morphologie Mathématique, Paris, France
jesus.angulo@mines-paristech.fr

Abstract. A formal definition of morphological operators in (max, min)-algebra is introduced and their relevant properties from an algebraic viewpoint are stated. Some previous works in mathematical morphology have already encountered this type of operators but a systematic study of them has not yet been undertaken in the morphological literature. It is shown in particular that one of their fundamental property is the equivalence with level set processing using Minkowski addition and subtraction. Theory of viscosity solutions of the Hamilton-Jacobi equation with Hamiltonians containing u and Du is summarized, in particular, the corresponding Hopf-Lax-Oleinik formulas as (max, min)-operators. Links between (max, min)-convolutions and some previous approaches of unconventional morphology, in particular fuzzy morphology and viscous morphology, are reviewed.

Keywords: Minkowski addition · Adjunction · Hamilton–Jacobi PDE · Fuzzy morphology · Viscous morphology

1 Introduction

Let E be the Euclidean $\mathbb{R}^n$ or discrete space $\mathbb{Z}^n$ (support space) and let $\mathcal{T}$ be a set of grey-levels (space of values). For theoretical reasons it is typically assumed that $\mathcal{T} = \overline{\mathbb{R}} = \mathbb{R} \cup \{-\infty, +\infty\}$, but one often has $\mathcal{T} = [0, M]$. A grey-level image is represented by a function $f \colon E \to \mathcal{T}$, also noted as $f \in \mathcal{F}(E, \overline{\mathbb{R}})$, such that f maps each pixel $x \in E$ into a grey-level value in $\mathcal{T}$. Given a grey-level image, the two basic morphological mappings $\mathcal{F}(E, \mathcal{T}) \to \mathcal{F}(E, \mathcal{T})$ are the dilation and the erosion given respectively by

$$\begin{cases} (f \oplus b)(x) = \sup_{y \in E} \{f(y) + b(x - y)\}, \\ (f \ominus b)(x) = \inf_{y \in E} \{f(y) - b(y - x)\}, \end{cases} \tag{1}$$

where $b \in \mathcal{F}(E, \mathcal{T})$ is the structuring function which determines the effect of the operator. The other morphological operators, such as the opening and the closing, are obtained by composition of dilation/erosion [20,11]. The Euclidean framework has been recently generalized to images supported on Riemannian manifolds [2]. Operators (1) can be interpreted in nonlinear mathematics as the

© Springer International Publishing Switzerland 2015
J.A. Benediktsson et al. (Eds.): ISMM 2015, LNCS 9082, pp. 485–496, 2015.
DOI: 10.1007/978-3-319-18720-4_41

convolution in (max, +)-algebra (and in its dual algebra) [10]. This inherent connection of functional operators (1) with the supremal and infimal convolution of nonlinear mathematics and convex analysis has been extremely fruitful to the state-of-the-art on mathematical morphology (morphological PDE, slope transform, etc.). Nevertheless, the functional operators (1) do not extend all the fundamental properties of the dilation and erosion for sets, as formulated in Matheron's theory. Perhaps the most disturbing for us are, on the one hand, the lack of commutation with level set processing for nonflat structuring functions; on the other hand, the limitation of Matheron's axiomatic of granulometry to constant (i.e., flat) functions on a convex domain [12]. In addition, there are some unconventional morphological frameworks, such as the fuzzy morphology [8,14,7] or the viscous morphology [21,22,15] which do not fit in the classical (max, +)-algebra. Actually, the (max, +) is not the unique possible alternative to see morphological operators as convolutions. The idea in this paper is to consider the operation of convolution of two functions in the (max, min)-algebra. This is in fact our main motivation: to formally introduce the notion of (max, min)-mathematical morphology. As we show in the paper, this framework is not totally new in morphology since some fuzzy morphological operators are exactly the same convolutions that we introduce. But some of the key properties are ignored by in the fuzzy context, and the most important, they are not limited to fuzzy sets. By the way, even if much less considered than the supremal and infimal convolutions, convolutions in (max, min)-algebra have been the object of various studies in different branches of nonlinear applied mathematics, from quasi-convex analysis [24,19,25,9,13,17] to viscosity solutions of Hamilton-Jacobi equations [5,6,1,23]. Interested reader is also referred to the book [10] for a systematic comparative study of matrix algebra and calculus in the three algebras $(+, \times)$, (max, +) and (max, min).

The present work is exclusively a theoretical study and thus the practical interest of the operators is not illustrated here. Complete proofs and additional results can be found in [3].

We use the following representation of semicontinuous functions. Given an upper semicontinuous (USC) function $f \in \mathcal{F}(E, \overline{\mathbb{R}})$, it can be defined by means of its upper level sets $X_h^+(f)$ as follows $f(x) = \sup \left\{ h \in \overline{\mathbb{R}} : x \in X_h^+(f) \right\}$, or by its strict lower level sets $Y_h^-(f)$: $f(x) = \inf \left\{ h \in \overline{\mathbb{R}} : x \in Y_h^-(f) \right\}$, where

$$X_h^+(f) = \{x \in E : f(x) \geq h\}, \quad \text{and} \quad Y_h^+(f) = \{x \in E : f(x) > h\};$$
$$X_h^-(f) = \{x \in E : f(x) \leq h\}, \quad \text{and} \quad Y_h^-(f) = \{x \in E : f(x) < h\}.$$

A continuous function f can be decomposed/reconstructed using either its (strict) upper level sets or its (strict) lower level sets. One has $\left(X_h^+(f)\right)^c = Y_h^-(f)$.

2 (max, min)-convolutions: Definition and Properties

In this Section we define the alternative convolutions associated to a pair *(function f, structuring function b)* in the (max, min) mathematical framework. We also study their properties.

Definition 1. *Given a structuring function $b \in \mathcal{F}(\mathbb{R}^n, \overline{\mathbb{R}})$, for any function $f \in \mathcal{F}(\mathbb{R}^n, \overline{\mathbb{R}})$ we define the supmin convolution $f \triangledown b$ and the infmax convolution $f \triangle b$ of f by b as*

$$(f \triangledown b)(x) = \sup_{y \in \mathbb{R}^n} \{f(y) \wedge b(x-y)\}, \tag{2}$$

$$(f \triangle b)(x) = \inf_{y \in \mathbb{R}^n} \{f(y) \vee b^c(y-x)\}. \tag{3}$$

We also define the adjoint infmax $f \triangle^* b$ and the adjoint supmin $f \triangledown^* b$ convolutions as

$$(f \triangle^* b)(x) = \inf_{y \in \mathbb{R}^n} \{f(y) \wedge^* b(y-x)\}, \tag{4}$$

$$(f \triangledown^* b)(x) = \sup_{y \in \mathbb{R}^n} \{f(y) \vee^* b^c(x-y)\}, \tag{5}$$

where $\wedge^*$ is the adjoint operator to the minimum $\wedge$ and is given by

$$f(y) \wedge^* b(y-x) = \begin{cases} f(y) & \text{if } b(y-x) > f(y) \\ \top & \text{if } b(y-x) \leq f(y) \end{cases} \tag{6}$$

and $\vee^*$ the adjoint to $\vee$:

$$f(y) \vee^* b^c(x-y) = \begin{cases} f(y) & \text{if } b^c(x-y) < f(y) \\ \bot & \text{if } b^c(x-y) \geq f(y) \end{cases} \tag{7}$$

and where, if we define $\max g = \sup_{x \in \mathbb{R}^n} g(x)$ and $\min g = \inf_{x \in \mathbb{R}^n} g(x)$, the top and bottom elements for pair of functions f and b correspond to

$$\top = (\max f) \vee (\max b) \quad and \quad \bot = (\min f) \wedge (\min b^c).$$

Definitions remain valid if we replace $\mathbb{R}^n$ by a subset E or any subset of discrete space $\mathbb{Z}^n$. Similarly, the extended real line $\overline{\mathbb{R}}$ can be replaced by a bounded, eventually discrete, set of intensities $[0, M]$. Figure 1 illustrates the four (max, min)-convolutions for a given example of one dimensional functions defined in a bounded interval, i.e., $f, b \in \mathcal{F}(\mathbb{R}, [0, M])$.

Duality by Complement vs. Duality by Adjunction. From a morphological viewpoint, their most salient properties are summarized in this proposition (proof in [3]).

Proposition 1. *The supmin convolution $\triangledown$ and infmax convolution $\triangle$ are dual with respect to the complement. Similarly, the adjoint infmax convolution $\triangle^*$ and the adjoint supmin $\triangledown^*$ convolution are dual with respect to the complement, i.e., for $f, b \in \mathcal{F}(\mathbb{R}^n, \overline{\mathbb{R}})$ one has*

$$f \triangle b = \left(f^c \triangledown \check{b}\right)^c \quad and \quad f \triangledown b = \left(f^c \triangle \check{b}\right)^c \tag{8}$$

$$f \triangle^* b = \left(f^c \triangledown^* \check{b}\right)^c \quad and \quad f \triangledown^* b = \left(f^c \triangle^* \check{b}\right)^c \tag{9}$$

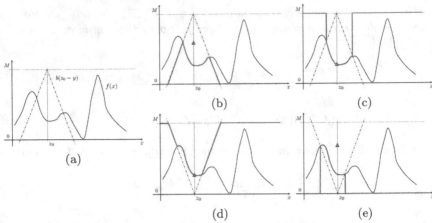

Fig. 1. Illustration of four (max, min)-convolutions for a given example of one dimensional functions defined in a bounded interval, i.e., $f, b \in \mathcal{F}(\mathbb{R}, [0, M])$: (a) original function $f(x)$ and translated structuring function b at point z_0; (b) in red, $f(y) \wedge b(z_0 - y)$ for all $y \in \mathbb{R}$, green triangle represents $(f \triangledown b)(x)$ the value of the supmin convolution at z_0; (c) in red, $f(y) \wedge^* b(y - z_0)$, green triangle, adjoint infmax at z_0: $(f \triangle^* b)(z_0)$; (d) in red, $f(y) \vee b^c(y - z_0)$, green triangle, infmax at z_0: $(f \triangle b)(z_0)$; (e) in red, $f(y) \vee^* b^c(z_0 - y)$, green triangle, adjoint supmin at z_0: $(f \triangledown^* b)(z_0)$.

The pair $(\triangle^*, \triangledown)$ forms an adjunction. Similarly, the pair $(\triangle, \triangledown^*)$ is also an adjunction, i.e., for $f, g, b \in \mathcal{F}(\mathbb{R}^n, \overline{\mathbb{R}})$ one has

$$f \triangledown b \le g \Longleftrightarrow f \le g \triangle^* b \tag{10}$$

$$f \triangledown^* b \le g \Longleftrightarrow f \le g \triangle b \tag{11}$$

Commutation with Level Set Processing. We can introduce now the fundamental property of (max, min)-convolutions (proof in [3]).

Proposition 2. *Let f and b in $\mathcal{F}(\mathbb{R}^n, \overline{\mathbb{R}})$. Then the four (max, min)-convolutions of f by b obey the following commutation rules of level sets with respect to Minkowski sum and substraction: for all $h \in \overline{\mathbb{R}}$*

$$X_h^+(f \triangledown b) = X_h^+(f) \oplus X_h^+(b) \tag{12}$$

$$Y_h^-(f \triangle b) = Y_h^-(f) \oplus Y_h^-(\check{b}^c) \tag{13}$$

$$X_h^+(f \triangle^* b) = X_h^+(f) \ominus X_h^+(b) \tag{14}$$

$$Y_h^-(f \triangledown^* b) = Y_h^-(f) \ominus Y_h^-(\check{b}^c) \tag{15}$$

This expression on strict lower level sets Y_h^- for $(f \triangle b)$ is valid for lower level sets X_h^- if $(f \triangle b)$ is *exact*, in the sense that, for each $x \in \operatorname{dom}^-(f \triangle b)$, there exists $y \in \mathbb{R}^n$ such that $(f \triangle b)(x) = f(y) \vee b^c(y - x)$ (i.e., the minimum is attained for any x in the domain) [19,13]. In particular, if f and b^c are both LSC quasiconvex functions, $(f \triangle b)$ and $(f \triangledown^* b)$ are exact, which involves $X_h^-(f \triangle b) = X_h^-(f) \oplus X_h^-(\check{b}^c)$ and $X_h^-(f \triangledown^* b) = X_h^-(f) \ominus X_h^-(\check{b}^c)$.

We need for the sequel an alternative formulation of the infmax and adjoint supmin convolution in terms respectively of Minkowski subtraction $\ominus$ and addition $\oplus$ of level sets. It is simply based on rewriting the infmax convolution using upper level sets:

$$(f \triangle b)(x) = \inf \left\{ h \in \mathbb{R} : x \in Y_h^-(f \triangle b) \right\}$$
$$= \sup \left\{ h \in \mathbb{R} : x \in \left(X_h^+(f) \ominus Y_h^-(b^c) \right) \right\}. \tag{16}$$

Analogously, one obtains the following equivalence for the adjoint supmin convolution:

$$(f \triangledown^* b)(x) = \inf \left\{ h \in \mathbb{R} : x \in \left(Y_h^-(f) \ominus Y_h^-(\breve{b}^c) \right) \right\}$$
$$= \sup \left\{ h \in \mathbb{R} : x \in \left(X_h^+(f) \oplus Y_h^-(\breve{b}^c) \right) \right\}. \tag{17}$$

Therefore, we can write

$$X_h^+(f \triangle b) = X_h^+(f) \ominus Y_h^-(b^c), \tag{18}$$
$$X_h^+(f \triangledown^* b) = X_h^+(f) \oplus Y_h^-(\breve{b}^c). \tag{19}$$

Further Properties. Other useful properties of (max, min)-convolutions are proven in [3].

Canonic Structuring Function. The conic structuring function plays a role similar to the multiscale quadratic structuring function in (max, +)-algebra.

Definition 2. *The multiscale conic structuring function is defined as the canonic structuring function in (max, min)-convolution:*

$$c_\lambda(x) = -\frac{\|x\|}{\lambda}. \tag{20}$$

In order to justify this canonicity, let us consider the upper level sets of $c_\lambda(x)$. First, we remind that a ball of radius centered at point x is given by the set $B_r(x) = \{ y \in \mathbb{R}^n : \|x - y\| \le r \}$.

Proposition 3. *The canonic structuring function in (max, min)-convolution satisfies the semi-group*

$$(c_\lambda \triangledown c_\mu)(x) = c_{\lambda+\mu}(x). \tag{21}$$

In the case of the L^∞ metric, a dimension separability is obtained for $c_\lambda^\infty(x) = -\|x\|_\infty/\lambda$; i.e., let us denote the coordinates of point as $x = (x_1, x_2, \cdots, x_n)$ and by $c_{\lambda;\,i}(x) = -|x_i|/\lambda$ the one dimensional conic structuring function, we have

$$c_\lambda^\infty(x) = (c_{\lambda;\,1} \triangledown c_{\lambda;\,2} \cdots \triangledown c_{\lambda;\,n}). \tag{22}$$

It is easy to see this property. We first note that $X_{-h}^+(c_\lambda) = B_{\lambda h}$. Second, we remind the Minkowski addition of balls: $B_{r_1} \oplus B_{r_2} = B_{r_1+r_2}$. Therefore, one has

$$X_{-h}^+(c_\lambda \triangledown c_\mu) = X_{-h}^+(c_\lambda) \oplus X_{-h}^+(c_\mu) = B_{\lambda h} \oplus B_{\mu h} = B_{(\lambda+\mu)h}.$$

Dimension separability in L^∞ metric is also a consequence of the Minkowski addition of segments. As a consequence of the L^∞ dimension separability, the classical theory of Minkowski decomposition of structuring elements [20].

3   Openings, Closings Using (max, min)-convolutions and Granulometries

The adjointness of the pairs $(\triangle^*, \triangledown)$ and $(\triangle, \triangledown^*)$ involves that from an algebraic viewpoint both the supmin convolution $\triangledown$ and the adjoint supmin convolution $\triangledown^*$ are a dilation; both the infmax convolution $\triangle$ and the adjoint infmax convolution $\triangle^*$ are an erosion. Therefore, their composition naturally yields openings and closings. Let us be more precise.

Definition 3. *Given any USC function $f \in \mathcal{F}(\mathbb{R}^n, \overline{\mathbb{R}})$, the (max, min)-opening and (max, min)-closing of f by the continuous structuring function $b \in \mathcal{F}(\mathbb{R}^n, \mathbb{R})$ are respectively given by*

$$(f \Diamond b) = ((f \triangle^* b) \triangledown b), \tag{23}$$

and

$$(f \blacklozenge b) = ((f \triangledown^* b) \triangle b), \tag{24}$$

such that their corresponding level sets representations, based on expressions (12), (14), and (13), (15), are given by

$$X_h^+ (f \Diamond b) = X_h^+ (f \triangle^* b) \oplus X_h^+ (b) = \left[X_h^+ (f) \ominus X_h^+ (b) \right] \oplus X_h^+ (b)$$
$$= X_h^+ (f) \circ X_h^+ (b), \tag{25}$$
$$Y_h^- (f \blacklozenge b) = Y_h^- (f \triangledown^* b) \oplus Y_h^- (\check{b}^c) = \left[Y_h^- (f) \ominus Y_h^- (\check{b}^c) \right] \oplus Y_h^- (\check{b}^c)$$
$$= Y_h^- (f) \circ Y_h^- (\check{b}^c). \tag{26}$$

We note that (max, min)-opening is defined from adjunction $(\triangle^*, \triangledown)$ whereas (max, min)-closing from $(\triangle, \triangledown^*)$. We can also switch roles and to formulate the so-called second family of dual (max, min)-opening and closing as

$$(f \Diamond^* b) = ((f \triangle b) \triangledown^* b), \tag{27}$$
$$(f \blacklozenge^* b) = ((f \triangledown b) \triangle^* b), \tag{28}$$

which has the following equivalent interpretation in terms of level sets:

$$Y_h^- (f \Diamond^* b) = Y_h^- (f) \bullet Y_h^- (\check{b}^c), \tag{29}$$
$$X_h^+ (f \blacklozenge^* b) = X_h^+ (f) \bullet X_h^+ (b). \tag{30}$$

Besides the duality by complement, classical properties of opening and closing hold in the (max, min) framework as a consequence of the adjunction [11]. See details in [3].

The extension of the granulometric theory [16] to the framework of (max, +)-based morphology was deeply studied in [12]. In particular, it was proven that one can build grey-level Euclidean granulometries with a multiscale structuring function if and only if structuring function has a convex compact domain and is constant there (i.e., flat function).

In the case of (max, min)-openings, we can naturally extend Matheron axiomatic of Euclidean granulometries without the flatness limitation (proof in in [3]).

Proposition 4. *Given a structuring function $b_1 \in \mathcal{F}(\mathbb{R}^n, \overline{\mathbb{R}})$ such that all its upper level sets $X_h^+(b_1)$ are convex sets, the family of multi-scale (max, min)-openings $\{f \Diamond b_\lambda\}_{\lambda \geq 1}$, where the structuring function at scale λ is given by*

$$b_\lambda(x) = b_1\left(\lambda^{-1}x\right),$$

forms an Euclidean granulometry on any image $f \in \mathcal{F}(\mathbb{R}^n, \overline{\mathbb{R}})$, i.e.,

$$(f \Diamond b_\lambda) = \lambda \star \left((\lambda^{-1} \star f) \Diamond b_1\right), \tag{31}$$

which involves compatibility with scaling in the spatial domain, in the sense of Matheron's axiomatic defined as follows

$$(\lambda \star f)(x) = f\left(\lambda^{-1}x\right), \quad \forall \lambda \geq 1.$$

In addition, we have the following semi-group properties, $\forall \lambda_1, \lambda_2 \geq 1$

$$b_{\lambda_1 + \lambda_2}(x) = (b_{\lambda_1} \triangledown b_{\lambda_2})(x), \tag{32}$$

$$((f \Diamond b_{\lambda_1}) \Diamond b_{\lambda_2})(x) = ((f \Diamond b_{\lambda_2}) \Diamond b_{\lambda_1})(x) = \left(f \Diamond b_{\sup(\lambda_1, \lambda_2)}\right)(x) \tag{33}$$

A good candidate of multi-scale isotropic structuring function leading to (max, min) granulometries is based on the canonic structuring function (21), as $b_\lambda(x) = c_\lambda(x) + \alpha$, which is equivalent to $b_\lambda(x) = \lambda^{-1}c_1(x) + \alpha$, $\lambda \geq 1$, $\alpha > 0$.

4   Hopf-Lax-Oleinik Formulas for Hamilton-Jacobi Equation $u_t \pm H(u, Du) = 0$

We study now the Hopf-Lax-Oleinik type formulas for Hamilton-Jacobi PDE of form $u_t \pm H(u, Du) = 0$ and its links to convolutions in (max, min)-algebra. The theory of this equation was developed by Barron, Jensen and Liu [5,6]. Other interesting results can be found in paper by Alvarez, Barron and Ishii [1] and the excellent survey paper by Van and Son [23]. The most relevant elements for us can be summarized in the following result.

Proposition 5. *Let us consider the two following Cauchy problems (first-order Hamilton-Jacobi PDEs):*

$$\begin{cases} u_t + H_1(u, Du) = 0, & in \ (x, t) \in \mathbb{R}^n \times (0, \infty), \\ u(x, 0) = f(x), & \forall x \in \mathbb{R}^n, \end{cases} \tag{34}$$

and

$$\begin{cases} u_t + H_2(u, Du) = 0, & in \ (x, t) \in \mathbb{R}^n \times (0, \infty), \\ u(x, 0) = g(x), & \forall x \in \mathbb{R}^n, \end{cases} \tag{35}$$

where the initial conditions are functions $f, g : \mathbb{R}^n \times \mathbb{R}$, such that f is a LSC proper function, bounded from below; and g an USC proper function, bounded from above. The Hamiltonians $H_1, H_2 : \mathbb{R} \times \mathbb{R}^n \to \mathbb{R}^n$ are assumed to satisfy the following conditions:

(A1) $H_1(\gamma, p)$ and $H_2(\gamma, p)$ are continuous;

(A2) $H_1(\gamma, p)$ and $H_2(\gamma, p)$ are nondecreasing in $\gamma \in \mathbb{R}$, $\forall p \in \mathbb{R}^n$;

(A3) $H_1(\gamma, p)$ is convex and $H_2(\gamma, p)$ is concave in $p \in \mathbb{R}^n$, $\forall \gamma \in \mathbb{R}$;

(A4) $H_1(\gamma, p)$ and $H_2(\gamma, p)$ are positively homogeneous of degree 1 in $p \in \mathbb{R}^n$,
i.e., $H_1(\gamma, \lambda p) = \lambda H_1(\gamma, p)$, $\forall \lambda \geq 0$.

The LSC viscosity solution of (34) is given by

$$u(x, y) = \inf_{y \in \mathbb{R}^n} \left[f(y) \vee H_1^\sharp \left(\frac{x - y}{t} \right) \right], \tag{36}$$

and the USC viscosity solution of (35) is

$$u(x, y) = \sup_{y \in \mathbb{R}^n} \left[f(y) \wedge H_{2\sharp} \left(\frac{x - y}{t} \right) \right], \tag{37}$$

where the conjugate operators $H^\sharp$ and $H_\sharp$ are defined as

$$H^\sharp(q) = \inf \left\{ \gamma \in \mathbb{R} : H(\gamma, p) \geq \langle p, q \rangle, \forall p \in \mathbb{R}^n \right\}, \tag{38}$$

$$H_\sharp(q) = \sup \left\{ \gamma \in \mathbb{R} : H(\gamma, p) \leq \langle p, q \rangle, \forall p \in \mathbb{R}^n \right\}. \tag{39}$$

The simplest case of admissible (A1)-(A4) convex Hamiltonian corresponds to $H(\gamma, p) = \gamma \|p\|$ such that, using Cauchy-Schwartz inequality, one gets

$$H^\sharp(q) = \inf \left\{ \gamma \in \mathbb{R} : \gamma \|p\| \geq \langle p, q \rangle \right\} = \|q\|.$$

The associated concave Hamiltonian is given by $H(\gamma, p) = -\gamma \|p\|$, whose conjugate is also $H_\sharp(q) = \|q\|$. Using this case as a starting point, a prototype of PDE in the framework of operators in (max, min)-algebra can be defined

Definition 4. Given any continuous and bounded function $f : E \to [a, b] \subset \mathbb{R}$, the canonic (Hamilton-Jacobi) PDE in (max, min)-morphology is defined as

$$\begin{cases} \frac{\partial u}{\partial t} = \pm u \|\nabla u\|, \ x \in E, \ t > 0 \\ u(x, 0) = f(x), \ x \in E \end{cases} \tag{40}$$

and its (unique weak) solutions at scale t are given by

$$u(x, t) = \sup_{y \in E} \left\{ f(y) \wedge \frac{\|x - y\|}{t} \right\} \quad (for + sign), \tag{41}$$

$$u(x, t) = \inf_{y \in E} \left\{ f(y) \vee \frac{\|x - y\|}{t} \right\} \quad (for - sign). \tag{42}$$

Therefore the viscosity solutions of Cauchy problem (40) are a supmin convolution and an infmax convolution using the conic structuring function $c_\lambda(x)$ given by (20), where the scale parameter is here the time; i.e., $\lambda = t$. More precisely, we note that these solutions

$$u(x, t) = (f \triangledown (-c_t))(x) \quad (for + sign),$$
$$u(x, t) = (f \triangle c_t)(x) \quad (for - sign),$$

are not adjoint in the sense of Section 2, consequently their composition does not lead to opening or closing.

The model (40) can be generalized to

$$\frac{\partial u}{\partial t} = \pm \alpha u \|\nabla u\|, \quad x \in E, \ t > 0$$

with initial condition $u(x, 0) = f(x)$ and $\alpha > 0$, such that we easily see that the corresponding solutions are

$$u(x, t) = (f \triangledown (-c_{\alpha t}))(x) \quad \text{(for + sign)},$$
$$u(x, t) = (f \triangle c_{\alpha t})(x) \quad \text{(for − sign)},$$

or in other words, multiplying u by α involves a scaling in time by α. This principle can be a clue to explore the notion of *spatially adaptive* (max, min)-operators based on using a scale depending on space x, i.e., a model of the form $u_t = \pm \alpha(x) u \|\nabla u\|$.

5 Ubiquity of (max, min)-convolutions in Mathematical Morphology

It is obvious the connection between (max, min)-convolutions and the distance function or the flat morphology. We discuss now links to fuzzy morphology and to viscous morphology. Relationships of (max, min)-convolutions with Boolean random function characterization and geodesic dilation/erosion are discussed in [3].

Links with Fuzzy Morphology. The state-of-the-art on morphological operators based on fuzzy logic is very extensive, see for instance [7]. Results on fuzzy morphology discussed here are mainly based on Deng and Heijmans [8], see also [14].

In fuzzy logic, the two basic (Boolean) logic operators, the conjunction $C(s, t) = s \wedge t$ and the implication $I(s, t) = s \Rightarrow t \ (= \neg s \vee t)$, are extended from the Boolean domain $\{0, 1\} \times \{0, 1\}$ to the rectangle $[0, 1] \times [0, 1]$. A fuzzy conjunction is a mapping from $[0, 1] \times [0, 1]$ into $[0, 1]$ which is increasing in both arguments and satisfies $C(0, 0) = C(1, 0) = C(0, 1) = 0$ and $C(1, 1) = 1$. A fuzzy implication is decreasing in the first argument, increasing in the second one and satisfies $I(0, 0) = I(0, 1) = I(1, 1) = 1$ and $I(1, 0) = 0$.

Given a fuzzy set μ, the dilation and erosion by a fuzzy structuring element ν are then defined as [8]:

$$\delta_{\nu, C}(\mu)(x) = \sup_y \left\{ C\left(\nu(x - y), \mu(y)\right) \right\}, \tag{43}$$

$$\varepsilon_{\nu, C}(\mu)(x) = \inf_y \left\{ I\left(\nu(y - x), \mu(y)\right) \right\}. \tag{44}$$

As shown in [8], (I, C) is an adjunction if and only if $(\varepsilon_{\nu, C}, \delta_{\nu, C})$ is an adjunction. Two particular cases of conjunction and adjoint implication widely used in fuzzy logic are the Gödel-Brower:

$$C_{GB}(a,t) = \min(a,t); \quad I_{GB}(a,t) = \begin{cases} s, \ s < a \\ 1, \ s \geq a \end{cases} \tag{45}$$

and the Kleen-Dienes:

$$C_{KD}(a,t) = \begin{cases} 0, \ t \leq 1-a \\ t, \ t > 1-a \end{cases}; \quad I_{KD}(a,s) = \max(1-a,s) \tag{46}$$

It is consequently straightforward to see that the four operators that we have defined in Section 2 are just fuzzy dilations and erosions when they are applied to fuzzy sets (i.e., functions valued in $[0,1]$):

$$\delta_{\nu,C_{GB}}(\mu)(x) = (\mu \bigtriangledown \nu)(x) \overset{adjoint}{\longleftrightarrow} \varepsilon_{\nu,C_{GB}}(\mu)(x) = (\mu \bigtriangleup^* \nu)(x),$$

$$\updownarrow \text{dual} \qquad\qquad \updownarrow \text{dual}$$

$$\varepsilon_{\nu,C_{KD}}(\mu)(x) = (\mu \bigtriangleup \nu)(x) \overset{adjoint}{\longleftrightarrow} \delta_{\nu,C_{KD}}(\mu)(x) = (\mu \bigtriangledown^* \nu)(x).$$

Links with Viscous Morphology. Theory and practice of morphological (flat) viscous operators was introduced by Vachier and Meyer [21,22]. The PDE formulation of these operators was done by Maragos and Vachier [15].

The idea of viscous operators is to apply a different scale (i.e., size) of structuring element at each upper level set. This principle can be seen now as an operator which locally adapts its activity with respect to the intensity. Let us formalize their definition according to [15]. For the sake of simplicity, let us consider a nonnegative bounded function $f : E \to [0,M]$. Viscous operators have been formulated as isotropic transforms, that is based on the use of balls B_λ as structuring elements.

Using intensity-adaptive operators and the two viscosity functions, two pairs of viscous dilation and erosion are defined for a given function f:

$$\delta_{\wedge}^{\text{visc}}(f) = \delta_{\lambda_\wedge(h)}(f) = \sup\left\{h \in [0,M] : x \in \left(X_h^+(f) \oplus B_{M-h}\right)\right\}, \tag{47}$$

$$\varepsilon_{\wedge}^{\text{visc}}(f) = \varepsilon_{\lambda_\wedge(h)}(f) = \sup\left\{h \in [0,M] : x \in \left(X_h^+(f) \ominus B_{M-h}\right)\right\}, \tag{48}$$

and

$$\delta_{\vee}^{\text{visc}}(f) = \delta_{\lambda_\vee(h)}(f) = \sup\left\{h \in [0,M] : x \in \left(X_h^+(f) \oplus B_h\right)\right\}, \tag{49}$$

$$\varepsilon_{\vee}^{\text{visc}}(f) = \varepsilon_{\lambda_\vee(h)}(f) = \sup\left\{h \in [0,M] : x \in \left(X_h^+(f) \ominus B_h\right)\right\}, \tag{50}$$

such that $\left(\varepsilon_{\wedge}^{\text{visc}}, \delta_{\wedge}^{\text{visc}}\right)$ and $\left(\varepsilon_{\vee}^{\text{visc}}, \delta_{\vee}^{\text{visc}}\right)$ form two adjunctions. The pairs $\left(\varepsilon_{\vee}^{\text{visc}}, \delta_{\wedge}^{\text{visc}}\right)$ and $\left(\varepsilon_{\wedge}^{\text{visc}}, \delta_{\vee}^{\text{visc}}\right)$ are dual by complement.

Let us introduce the following structuring function:

$$v(x) = \begin{cases} M - \|x\| & \text{if } \|x\| \leq M \\ 0 & \text{if } \|x\| > M \end{cases}$$

such that its complement structuring function is $v^c(x) = \|x\|$ if $\|x\| \leq M$ and M if $\|x\| > M$. We have $X_h^+(v) = B_{M-h}$ and $Y_h^-(v^c) = B_h$. Hence, viscous dilations and erosions (47)-(50) can be rewritten using the (max, min)-convolution (respectively expressions (12), (14), (18), (19)):

$$\delta_\wedge^{\mathrm{visc}}(f)(c) = (f \bigtriangledown v)(x) \overset{adjoint}{\longleftrightarrow} \varepsilon_\wedge^{\mathrm{visc}}(f)(x) = (f \bigtriangleup^* v)(x),$$

$$\updownarrow \text{dual} \qquad\qquad \updownarrow \text{dual}$$

$$\varepsilon_\vee^{\mathrm{visc}}(f)(x) = (f \bigtriangleup v)(x) \overset{adjoint}{\longleftrightarrow} \delta_\vee^{\mathrm{visc}}(f)(x) = (f \bigtriangledown^* v)(x).$$

In addition to the operator framework, a PDE formulation of viscous dilation and erosion was introduced in [15]. The proposed couple of PDEs are particular cases of the Hamilton-Jacobi models discussed above. More precisely, it corresponds to the case of the Hamiltonians given in expressions $H_1(\gamma, p) = (\alpha + \gamma)\|p\|$ and $H_2(\gamma, p) = -(\alpha + \gamma)\|p\|$, such that $H_1^\sharp(q) = H_{\sharp 2}(q) = \|q\| - \alpha$; or a pair $H_1(\gamma, p) = (\alpha - \gamma)\|p\|$ and $H_2(\gamma, p) = -(\alpha - \gamma)\|p\|$, with $H_1^\sharp(q) = H_{\sharp 2}(q) = \alpha - \|q\|$. Therefore solution $u(x, t)$ for $+$ sign of the PDE model is equivalent to viscous dilation $\delta_\wedge^{\mathrm{visc}}(f)$, but for $-$ sign it is not exactly equivalent to the viscous erosion $\varepsilon_\wedge^{\mathrm{visc}}(f)$. In our terminology, the latter is a case of adjoint infmax convolution while the solution for $-$ sign is an infmax convolution with the complemented structuring function.

6 Conclusion and Perspectives

Operators and filters underlying a formulation as (max, min)-convolutions are common in the state-of-the-art of mathematical morphology. However, their study *per se* has been neglected. From this epistemological viewpoint, we can conclude that the role of (max, min)-convolutions has been somewhat overshadowed by a multiplicity of viewpoints (fuzzy, viscous, "hitting of functions" in Choquet capacity, etc.) In order to address this theoretical lack, we have developed in our paper a rigorous formulation and characterization of the four convolution-like operators in (max, min)-algebra.

All the results on (max, min)-convolutions considered here are valid for functions supported in a general Banach space, consequently more general that the Euclidean space $\mathbb{R}^n$. In this generalization context, we plan to consider in particular the case of (max, min)-morphology for real-valued images on Riemannian manifolds.

References

1. Alvarez, O., Barron, E.N., Ishii, H.: Hopf-lax formulas for semicontinuous data. Indiana Univ. Math. J. 48, 993–1035 (1999)
2. Angulo, J., Velasco-Forero, S.: Riemannian Mathematical Morphology. Pattern Recognition Letters 47, 93–101 (2014)
3. Angulo, J.: Convolution in (max, min)-algebra and its role in mathematical morphology. HAL preprint, 59 p. (2014)
4. Bardi, M., Evans, L.C.: On Hopf's formulas for solutions of Hamilton- Jacobi equations. Nonlinear Analysis, Theory, Methods and Applications 8(11), 1373–1381 (1984)
5. Barron, E.N., Jensen, R., Liu, W.: Hopf-Lax formula for $u_t = H(u, Du) = 0$. J. Differential Equations 126, 48–61 (1996)

6. Barron, E.N., Jensen, R., Liu, W.: Hopf-Lax formula for $u_t = H(u, Du) = 0$, II. Comm. Partial Differential Equations 22, 1141–1160 (1997)

7. Bloch, I.: Duality vs. adjuntion for fuzzy mathematical morphology and general form of fuzzy erosions and dilations. Fuzzy Sets and Systems 160, 1858–1867 (2009)

8. Deng, T.-Q., Heijmans, H.J.A.M.: Grey-Scale Morphology Based on Fuzzy Logic. Journal of Mathematical Imaging and Vision 16(2), 155–171 (2002)

9. Gondran, M.: Analyse MINMAX. C.R. Acad. Sci. Paris, t. 323, Série I, 1249–1252 (1996)

10. Gondran, M., Minoux, M.: Graphs, Dioids and Semirings: New Models and Algorithms. Springer (2008)

11. Heijmans, H.J.A.M.: Morphological image operators. Academic Press, Boston (1994)

12. Kraus, E.J., Heijmans, H.J.A.M., Dougherty, E.R.: Gray-scale granulometries compatible with spatial scalings. Signal Processing 34(1), 1–17 (1993)

13. Luc, D.T., Volle, M.: Levels sets Infimal Convolution and Level Addition. Journal of Optimization Theory and Applications 94(3), 695–714 (1997)

14. Maragos, P.: Lattice image processing: a unification of morphological and fuzzy algebraic systems. Journal of Mathematical Imaging and Vision 22(2-3), 333–353 (2005)

15. Maragos, P., Vachier, C.: A PDE Formulation for Viscous Morphological Operators with Extensions to Intensity- Adaptive Operators. In: Proc. of 15th IEEE International Conference on Image Processing, ICIP 2008, pp. 2200–2203 (2008)

16. Matheron, G.: Random Sets and Integral Geometry. Wiley, New York (1975)

17. Penot, J.-P., Zălinescu, C.: Approximation of Functions and Sets. In: Lassonde, M. (ed.) Approximation, Optimization and Mathematical Economics, pp. 255–274. Physica-Verlag, Heidelberg (2001)

18. Rockafellar, R.T.: Convex analysis. Princeton University Press, Princeton (1970)

19. Seeger, A., Volle, M.: On a convolution operation obtained by adding level sets: classical and new results. Operations Research 29(2), 131–154 (1995)

20. Serra, J.: Image Analysis and Mathematical Morphology. Academic Press, London (1982)

21. Vachier, C., Meyer, F.: The Viscous Watershed Transform. Journal of Mathematical Imaging and Vision 22, 251–267 (2005)

22. Vachier, C., Meyer, F.: News from Viscous Land. In: Proc. of 8th Internation Symposium on Mathematical Morphology (ISMM 2007), Rio de Janeiro, Brazil, pp. 189–200 (2007)

23. Van, T.D., Son, N.D.T.: Hopf-Lax-Oleinik-Type Estimates for Viscosity Solutions to Hamilton-Jacobi Equations with Concave-Convex Data. Vietnam Journal of Mathematics 34(2), 209–239 (2006)

24. Volle, M.: The Use of Monotone Norms in Epigraphical Analysis. Journal of Convex Analysis 1(2), 203–224 (1994)

25. Volle, M.: Duality for the Level Sum of Quasiconvex Functions and Applications. ESAIM: Control, Optimisation and Calculus of Variations 3, 329–343 (1998)

Learning Watershed Cuts Energy Functions

Reid Porter[1]([✉]), Diane Oyen[1], and Beate G. Zimmer[2]

[1] Intelligence and Space Research Division, Los Alamos National Laboratory,
Los Alamos, New Mexico, 87545, USA
{rporter,doyen}@lanl.gov
[2] Department of Mathematics, Texas A&M University - Corpus Christi,
Corpus Christi, Texas, USA
Beate.Zimmer@tamucc.edu

Abstract. In recent work, several popular segmentation methods have been unified as energy minimization on a graph. In other work, supervised learning methods have been generalized from predicting labels to predicting structured, graph-like objects. A recent contribution to this second area showed how the Rand Index could be directly minimized when using Connected Components as a segmentation method. We build on this work and present an efficient mini-batch learning method for Connected Component segmentation and also show how it can be generalized to the Watershed Cuts segmentation method. We present initial results applying these new contributions to image segmentation problems in materials microscopy and discuss challenges and future directions.

Keywords: Segmentation · Watershed · Structured output prediction

1 Introduction

Image segmentation is a fundamental task in image and video processing that has wide ranging application. Segmentation has traditionally been an unsupervised problem, although incorporating user input into segmentation methods has also had a long history. In more recent years, a subfield of machine learning known as structured output prediction has made significant progress and there is now much interest in incorporating supervised learning into segmentation problems. A large number of these efforts focus on joint segmentation and labeling (also called semantic segmentation), where the training data (or labels) define object types or categories within the image [1, 2]. There have also been efforts to incorporate learning into the more traditional segmentation problem, where the training data defines a partition, or clustering, of the image. However, these efforts often treat learning and segmentation as independent steps within a multi-step process [3, 4].

In this paper, we describe recent work in learning to segment in the context of structured output prediction. This involves formulating segmentation as energy minimization on a graph in Section 2. We describe a learning method that was proposed for Connected Component segmentation in Section 3 and show how it can be extended to Watershed Cut segmentation in Section 4. In Section 5 we present our key

© Springer International Publishing Switzerland 2015
J.A. Benediktsson et al. (Eds.): ISMM 2015, LNCS 9082, pp. 497–508, 2015.
DOI: 10.1007/978-3-319-18720-4_42

contribution which is an efficient mini-batch learning algorithm for both types of segmentation. In Section 6 we report experimental results where we compare learning performance on segmentation tasks in materials microscopy.

2   Segmentation as Inference

A very useful way to describe a large number of segmentation methods is energy minimization on a graph $G = (\mathcal{V}, \mathcal{E})$. The vertices $\mathcal{V} = \{v_1, v_2, \dots, v_P\}$ are associated with spatial regions (or locations) and edges $\mathcal{E} = \{\dots, e_{ij}, e_{jk}, e_{kl}, e_{lm}, \dots\}$ associate an unordered pair of vertices. Often the vertices are associated with pixel locations and the structure of the graph is fixed in advance as a regular grid with edges connecting each vertex to its 4 (or 8) closest neighbors. However, the approach is equally applicable to super-pixels and irregular structures.

The image contains P pixels $X = (x_1, x_2, \dots, x_P)$, $x_i \in \mathbb{R}^D$, where each pixel is associated with a vertex (through its location) and has one (grayscale), three (color) or any number of dimensions (multi-, hyper-spectral). Also associated with the set of vertices is a set of discrete random variables (labels), $Y = (y_1, y_2, \dots, y_P)$, $y_i \in \{1, 2, \dots, K\}$, that identify the segment to which each location is assigned. The value of K determines the number of unique segments in the image, and typically depends on the image and segmentation method. When every pixel is assigned to its own segment $K = P$. Segmentation is defined as minimization of an energy function on the graph. This minimization is known as inference and is defined by Equation 1:

$$\hat{Y} = argmin_Y \, E(Y, X) \tag{1}$$

Several segmentation methods can be expressed as a pairwise energy function that sums terms associated with each edge in the graph:

$$E(Y, X) = \sum_{e_{ij} \in \mathcal{E}} g(y_i, y_j, X) \tag{2}$$

One choice of g, known as correlation clustering, has the following form [5]:

$$g(y_i, y_j, X) = I(y_i \neq y_j) A(X_{ij}) \tag{3}$$

Where I is the indicator function that returns 1 when its argument is true, and 0 otherwise. $A: X_{ij} \to \mathbb{R}$ is a real valued function (often called an affinity function), and X_{ij} is the subset of the image used by the affinity function to determine if pixels i and j should be in the same segment. In principle X_{ij} could be the whole image, but in practice it is typically a local neighborhood that includes i and j. $A(X_{ij}) > 0$ indicates pixels i and j should be in the same segment and $A(X_{ij}) < 0$ indicates they should be in different segments. Minimizing Equation 2 with Equation 3 is known to be NP hard.

In recent work [6], Couprie et. al. showed how a large number of segmentation methods can be obtained with particular choices for g. This included the watershed cut segmentation method [7], which can be defined for positive affinity functions, $A^+: X_{ij} \to \mathbb{R}^+$ as:

$$g(y_i, y_j, X) = I(y_i \neq y_j)\left(A^+(X_{ij})\right)^\infty \tag{4}$$

Couprie et. al. provide a computationally efficient algorithm for minimizing Equation 2 with Equation 4, called the Power Watershed. A case of particular interest in this paper, is the case when all affinities within the graph are distinct, or have unique values. In this case a labeling that minimizes Equation 2 with Equation 4 coincides with the labeling defined by the unique minimum spanning forest of the graph.

Another way to make inference tractable is to use binary edge weights:

$$g(y_i, y_j, X) = I(y_i \neq y_j)I(A(X_{ij}) > 0) \tag{5}$$

In this case a minimum labeling can be found by simply discarding the edges with non-positive affinity, and running connected components on the remaining graph. This solution is known as a *perfect clustering* in the correlation clustering literature, and in this paper, we refer to it as Connected Component segmentation.

3   Learning Connected Component Segmentation

A critical design choice in all of the segmentation methods defined in Section 2 is the affinity function. In traditional segmentation this function is fixed. For example, a popular choice is to estimate the magnitude of the gradient. However, this is an application specific choice, and this motivates supervised approaches which tailor the affinity function to the application through learning.

To apply learning to segmentation we must define a loss function to measure how well predicted segmentations match a ground-truth segmentation. We use a slight variation of the Rand Index, which we call the Rand Error (RE). It counts the number of pairwise differences between the predicted segmentation $\hat{Y} = (\hat{y}_1, \hat{y}_2, ..., \hat{y}_P)$ and the ground-truth segmentation $Y = (y_1, y_2, ..., y_P)$, both of size P pixels. This is defined in Equation 6.

$$RE(\hat{Y}, Y) = \binom{P}{2}^{-1} \sum_{\{i,j\} \in P \times P} I(\hat{y}_i \neq \hat{y}_j)I(y_i = y_j) + I(\hat{y}_i = \hat{y}_j)I(y_i \neq y_j) \tag{6}$$

One of the earliest uses of the Rand Index in learning was for clustering [8]. The correlation clustering energy function was used for inference (Equation 3) and the Structured Support Vector Machine framework was used for learning. We note that the only real difference between clustering and segmentation, in this context, is in the choice of the graph structure and in the choice of affinity function. In clustering the graph structure is typically fully connected, but in segmentation the graph is typically the 4 or 8-connected grid described previously. The correlation clustering energy function is intractable in both cases, and approximate inference methods are used within the learning algorithm.

In more recent work, Turaga et. al. [9] showed that for Connected Component segmentation, the gradient of the Rand Index could be calculated directly. One of the key ideas is to treat the binary edge weights in Equation 5 as a classifier. Turaga et. al. use a convolutional neural network for the classifier, but in this paper we use a simple linear model. Although the linear model is not as expressive as the convolutional network (and unlikely to do as well on real-world problems), it is easier to optimize which makes comparing learning objectives easier. Our linear edge classifier is defined in Equation 7 where ϕ calculates fixed features for each edge and $w \in \mathbb{R}^D$ is a vector of real-valued parameters.

$$\hat{y}_{ij} = I(A(X_{ij}) > 0) = I(\langle w, \phi(X_{ij}) \rangle > 0) \tag{7}$$

Note that the edge classifier, as defined, does not quite solve the segmentation problem defined in Section 2. The graph has been partitioned, but unique segment identifiers have not been assigned to the vertices. This is where connected components comes in.

To incorporate connected components within the edge classifier learning algorithm we need to relate the edge prediction $\hat{y}_{ij}$, made in Equation 7, to the vertex prediction $I(\hat{y}_i = \hat{y}_j)$, required in Equation 6. Turaga et. al.'s *maximin* procedure does exactly this. In words, if there exists a path (max) between vertices i and j that is completely connected (min) then connected components will assign the same label to i and j. More formally, if $\mathcal{P}_{ij}$ is the set of all paths in G between vertices i and j, then the connected components procedure guarantees:

$$I(\hat{y}_i = \hat{y}_j) = \max_{p \in \mathcal{P}_{ij}} \min_{e_{kl} \in p} \{\hat{y}_{kl}\} \tag{8}$$

A key property of the *maximin* procedure (Theorem 1 in [9]) is that it commutes with thresholding. This property enables the *maximin* procedure to be used within standard learning algorithms. Turaga et. al. propose an online stochastic gradient descent learning algorithm using a square-square loss function. At each iteration a random pair from Equation 6 is selected and the *maximin* procedure is used to estimate the gradient and update the classifier. In Section 5 we discuss learning algorithms in more detail and present a mini-batch version of stochastic gradient descent which makes more efficient use of the *maximin* procedure.

4   Watershed Cuts Segmentation

In recent years the traditional flood-filling metaphors and algorithms for watershed segmentation have been linked to the energy minimization frameworks described in Section 2, which has enabled a number of theorems and efficient algorithms to be developed [6]. In this context, segmentation by Equation 4 is known as Watershed Cuts [7]. When all the edge weights (affinities) within the graph are distinct, there is a unique minimum spanning forest (MSF). This MSF can be defined by a cut of the graph where each connected component produced by the cut corresponds to a tree within the MSF. The set of edges that define the cut have been called border edges, and can be identified by a local algorithm [10]. In this section we write this local algorithm in terms of a threshold function acting on the graph. This provides a direct link to the Connected Component segmentation method and suggests that we can use the same *maximin* procedure that was used to learn Connected Component segmentations to learn Watershed Cuts segmentations.

When all affinities are unique, the edges that are part of the watershed cut have an affinity that is larger than at least one neighbor on both sides with smaller affinity. We denote the neighbors of vertex v_i as N_i, and the set that excludes vertex v_j as $N_i \setminus \{j\}$. We also use the shorthand notation $A_{ij} = A(X_{ij})$ to represent the value of the affinity function at edge e_{ij}. The set of edges that belong to watershed basins (those not part of the watershed cut) can be identified with the indicator:

$$I(A_{ij}^* - A_{ij} > 0) \text{ where}$$

$$A_{ij}^* = max\left(min_{k \in N_i \setminus \{j\}} A_{ik}, \; min_{k \in N_j \setminus \{i\}} A_{kj}\right). \tag{9}$$

Given these definitions (and assumptions), we can implement the Watershed Cut segmentation as a two-step process, very similar to Connected Component segmentation:

1. Apply the threshold test (Equation 9) to each edge in the graph and throw away edges that are below threshold (in this case, below zero).

2. Run a connected component procedure on the remaining graph to obtain the vertex labels.

If $(A_{ij}^* - A_{ij})$ has a unique value for each edge in the graph, this two-step procedure will minimize Equation 2 using Equation 4. The only real difference between this two-step procedure and Connected Component segmentation is that the binary edge weight is a function of two edges instead of one. As we will see in the next section, this does not appear to introduce any significant new challenges for learning.

5   Algorithms for Maximin Learning

The *maximin* procedure is a function of the entire graph and is computationally expensive to execute for every pair in the Rand Error. Turaga et. al. show that the benefits of directly optimizing a segmentation loss can justify this expensive. One way of maximize the benefit / cost ratio of the *maximin* procedure is to use batch (or minibatch) stochastic gradient descent. To better understand how we can do this, we return to the Rand Error. If we write $y_{ij} = I(y_i = y_j)$ and $z_{ij} = I(\hat{y}_i = \hat{y}_j)$, then Equation 6 can be expressed as:

$$RE(Z, Y) = \binom{P}{2}^{-1} \sum_{\{i,j\} \in P \times P} I(z_{ij} \neq y_{ij}) \tag{10}$$

The Rand Error compares all possible pairs of pixels across both images. The set of variables that includes z_{ij} is therefore typically much larger than the set of variables that includes $\hat{y}_{kl}$. Furthermore, the maximin procedure guarantees that every term z_{ij} is mapped to the subset of pairs that belong to edges in the Minimum Spanning Tree (MST). This suggests a natural batch size for estimating the gradient of Equation 10 is the $(P - 1)$ pairs of the MST. The computation that counts the number of times a MST pair contributes to Equation 10 can be implemented efficiently with a simple modification of Kruskal's MST algorithm described in Section 5.1.

It is also useful to compare Equation 10 to the misclassification error of an edge classifier defined in Equation 11.

$$ME(\hat{Y}, Y) = \frac{1}{|\mathcal{E}|} \sum_{e_{ij} \in \mathcal{E}} I(\hat{y}_{ij} \neq y_{ij}) \tag{11}$$

Designing an edge classifier to minimize Equation 11 is straight forward since it assumes edges are Independent and Identically Distributed (IID) examples. However this does not include energy minimization within learning and the final classifier is not optimal for segmentation. In the particular case of Connected Component segmentation, minimizing Equation 11 spends unnecessary effort trying to correctly classify a subset of the positive edges. We compare performance of classifiers designed with Equations 10 and 11 in our experiments.

5.1   Playing with Kruskal

Kruskal's algorithm provides an efficient algorithm to count the number of times each MST edge prediction should be positive and how many times it should be negative with respect to the Rand Error. Inspired by the work in [11] we outline the approach in terms of disjoint set data-structures.

Initially, each edge is assigned to its own set. A running sum of the number of label pairs that are the same, and the number of label pairs that are different are calculated from the training data as the MST is formed. This requires a K-dimensional vector for each edge, where K is the number of distinct labels in the ground truth. The vector represents the number of times each label occurs within each set. We use Tarjan's Union-Find algorithm for FindCanonical and Union procedures.

Algorithm 1. Kruskal Algorithm for Rand Error counts

Input: An edge-weighted graph $G = (\mathcal{V}, \mathcal{E}, A)$ and labels Y
Output: A maximum spanning tree (MST)
Output: A collection of sets associated with connected components Q
Output: Count #same and #diff label pairs for each MST edge
// Collection Q is initialized to $\emptyset$
1. $e := 0$;
2. **for** all $x_i \in \mathcal{V}$ **do**
3.   MakeSet(i)
4.   labels[i] := InitLabelCountArray(y_i)
5. **for** all edges $\{u, v\} \in \mathcal{E}$ by decreasing weight A_{uv} **do**
6. c_u := Q.FindCanonical(u)
7. c_v := Q.FindCanonical(v)
8.   **if** $c_u \neq c_v$ **then**
9. Q.Union(c_u, c_v)
10. c_{new} := Q.FindCanonical(c_u)
11.   MST[e] := $\{u, v\}$
12.   #same[e] = labels[c_u]^T.labels[c_v]
13.   #diff[e] = $|c_u| * |c_v|$ − #same[e]
14.   labels[c_{new}] = labels[c_u] + labels[c_v]
15. e := $e + 1$;

Algorithm 1 produces a count of the positive and negative label pairs associated with each edge in the MST. We use these for learning in multiple different ways. First, we can calculate Equation 10 for Connected Component segmentation at any given threshold $\hat{y}_{ij} = I(A(X_{ij}) > T)$ with Equation 12.

$$RE(A,T,Y) = \sum_{e_{uv} \in MST} I(A_{uv} > T) \text{ \#diff}[e] + I(A_{uv} < T) \text{ \#same}[e] \qquad (12)$$

Second, we use it to calculate a ROC curve (Receiver-Operator Curve) for the Rand Error. A ROC curve traces out the fraction of positive pairs that are correctly predicted (the detection rate), and the fraction of negative pairs that are incorrectly predicted (the false alarm rate) for a number of different thresholds. Since the edges in the MST are sorted by Algorithm 1, we can maintain running sums for the two terms in Equation 12 and consider the set of thresholds that correspond to the MST affinities $T = \{..., A_{uv}, ...\}$. Note that in practice we typically use thresholds that lie half way between consecutive edge affinities since this avoids potential complications with equality. We show examples of these ROC curves in our experiments.

The third post-processing application for Algorithm 1 is to estimate gradients for learning algorithms. Typically, learning algorithms minimize a convex surrogate for misclassification error, known as a loss function, which would replace Equation 10. However the counts used in Equation 12 still play a role and we will provide a specific example of this for the Support Vector Machine loss function in the next section.

A final point worth mentioning is that Algorithm 1 can be applied hierarchically. This means learning does not have to start with pixels, but could start with any initial (over) segmentation of the image. Given an initial segmentation (also called a set of super-pixels) we initialize a K-dimensional vector for each segment based on the ground truth labels present in that segment. Algorithm 1 can then be applied to the super-pixel graph without modification. Note however there are offsets for the positive and negative counts that must be calculated. These correspond to the number of positive and negative pairs that are fixed by the initial segmentation. These terms are easily pre-calculated and included as constants in Equation 12 to obtain the pixel level Rand Error.

5.2   From Counts to (Sub-) Gradients

Many learning algorithms relax the hard thresholds associated with classification error with a convex loss function that is easier to minimize. In this paper we use the hinge loss used in support vector machines, but our approach is general. The Rand Error with hinge loss (which we call the Rand Loss) is defined in Equation 13.

$$RL(A,Y) = \binom{P}{2}^{-1} \sum_{\{i,j\} \in P \times P} max(0, 1 - y_{ij}A_{ij}) \qquad (13)$$

Note that from this point forward we assume the pairwise labels are expressed as $y_{ij} \in \{-1, +1\}$ instead of $\{0,1\}$. When the affinity function is a linear model $A_{ij} = \langle w, \phi(X_{ij}) \rangle$, the sub-gradient of Equation 13 with respect to w is defined by Equation 14.

$$\nabla(W,A,Y) = -\binom{P}{2}^{-1} \sum_{\{i,j\} \in P \times P} I(y_{ij}A_{ij} < 1)y_{ij}\phi(X_{ij}) \qquad (14)$$

For Connected Component segmentation Algorithm1 provides the mapping between every pair in Equation 14 to the MST. The #same and #diff variables can be used to determine how much each MST edge contributes to the gradient. We define a *meta-label* for each MST edge: $l_{uv} = sign(\text{\#same} - \text{\#diff})$. Equation 14 can then be calculated with Equation 15.

$$\nabla(W, A, Y) = - \binom{P}{2}^{-1} \sum_{e_{uv} \in MST} I(l_{uv}A_{uv} < 1)(\text{\#same} - \text{\#diff})\phi(X_{uv}) \qquad (15)$$

The algorithm and mappings described in the last few sections are equally applicable to Watershed Cuts segmentation. We first replace the affinities A used in Algorithm 1 with the difference in affinities as defined in Equation 9. We denote the difference in affinities as $A_{uv}^{\Delta} = A_{uv}^{*} - A_{uv}$, where $A_{uv}^{*} = \langle w, \phi(X_{uv}^{*}) \rangle$ and $\phi(X_{uv}^{*})$ is the feature vector associated with the A_{uv}^{*} edge. Given these definitions each term in Equation 15 can then be expressed with Equation 16.

$$\nabla_{uv} = I(l_{uv}A_{uv}^{\Delta} < 1)(\text{\#same} - \text{\#diff})\left(\phi(X_{uv}) - \phi(X_{uv}^{*})\right) \qquad (16)$$

5.3   Minimizing Loss with Stochastic Gradient Descent

We have now defined the sub-gradients for the Rand Loss in terms of the MST for both Connected Component and Watershed Cut segmentation methods. The next step is to use this gradient within an iterative sub-gradient descent algorithm. We use the mini-batch version of the Primal Estimated sub-GrAdient SOlver for SVM (Pegasos) [12] which provides good accuracy and runtime bounds. Note, this method also minimizes a regularization term and one of the few free parameters with the learning method is the amount of regularization. Because we are using a batch update procedure we use 1-dimensional line search to find the best threshold instead of finding it with the sub-gradient descent algorithm. Computationally, this is equivalent to tracing out the ROC curve, as described in Section 5.1.

6   Application to Microscopy

Material science has a large number of applications which would benefit from optimized segmentation methods. On the left in Figure 1 we show two different applications where ground truth was generated semi-manually by subject matter experts. It would be challenging to find a single segmentation method that would perform equally well on both applications. We have three images from each application. In our experiments we use one of the images for training and then apply the optimized segmentation to the remaining two images to estimate performance in that application. We repeat this 3 times, using a different image for training each time.

6.1   Learning Segmentation on a Super-Pixel Graph

To reduce the computation time of our experiments, we apply the learning algorithms to super-pixels instead of the original pixels. We use a standard Watershed algorithm,

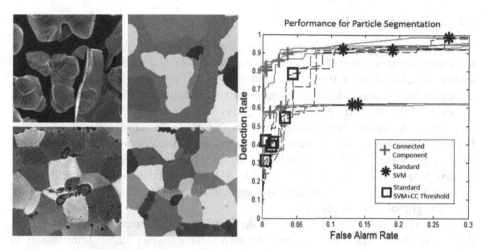

Fig. 1. Left) Image chips and ground truth segmentations for two different applications in materials microscopy referred to as Particle and Grain problems. Right) Performance comparison of *maximin* learning to IID learning.

applied to the magnitude of the image gradient, to produce the super-pixels. As described in Section 5, we precompute the count offsets for each image to ensure that our estimates for Rand Error are in terms of the pixel level segmentation problem.

We use a fixed set of features to represent each edge in the super-pixel graph. The features include the minimum, maximum, average and range of the gradient magnitude along the shared edge, the average and difference in pixel count between segments, as well as the average and difference in filter bank response between the segments. The filters themselves are learnt during training, by taking overlapping 5 by 5 image patches in the training image and using k-means to find 8 exemplars, or cluster centers. In test, we map each 5 by 5 patch of the test image into an 8 dimensional vector using triangle encoding to the exemplars (see [13] for details). The filter responses are pooled over each super-pixel and then normalized. In total we have 22 features (4 edge features, 2 size features and 16 texture features).

In our experiments we learn Connected Component and Watershed Cut segmentations on the super-pixel graph. It is interesting to note that these solutions correspond to two distinct methods for hierarchical segmentation that have reported in the literature. Using an initial Watershed followed by Connected Components is perhaps the most widely used approach and corresponds to making horizontal cuts of a merge tree based on attributes such as depth, area and volume [14]. Using an initial Watershed followed by a second Watershed has also been proposed and is known as the Waterfall algorithm [15]. In our learning experiments we focus exclusively on the second stage segmentation and the initial Watershed remains fixed. An interesting direction for future work would learn both stages of the segmentation hierarchy.

6.2　Comparison to IID Learning

In our first experiment we compare the performance of edge classifiers designed to minimize the Rand Error (Equation 10) compared to edge classifiers designed to minimize Misclassification Error (Equation 11). In the first case we use the Pegasos algorithm with the mini-batch gradient estimates in Equation 15. In the second case we use the LibLinear [16] package which provides traditional batch mode SVM learning. We set the amount of regularization in both cases to be very small and we ran the Pegasos algorithm for 10,000 iterations (Algorithm 1 is executed once per iteration). Our images were 512 by 512 pixels and produced super-pixel graphs with approximately 5000 vertices and 15000 edges. On a 2GHz workstation, our unoptimized C code took approximately 4 minutes to complete the training. The LibLinear training time is negligible for this problem size.

On the right in Figure 1 we compare the test image segmentation performance for the two methods. The solutions found by min-batch *maximin* learning are shown as red crosses and the SVM performance is shown as black stars. We also show the associated ROC curves for these solutions. Note, that in terms of the Rand Error, the black stars are orders of magnitude higher than the red crosses. Part of the problem is that the SVM threshold is not optimized for Connected Component segmentation and for reasons described in Section 5, this leads to higher false alarm rates. A potential solution to this problem is to use LibLinear to train the classifier and then use Algorithm 1 once, at the end, to fine tune the threshold for Connected Components. These solutions are shown as black boxes in Figure 1. This approach does improve the performance of the LibLinear solution significantly, but there is also still a significant gap to the *maximin* solutions found through min-batch sub-gradient descent.

6.3　Comparing Connected Components and Watershed Cuts

In the second set of experiments we compare the mini-batch *maximin* solutions for Equation 15 (Connected Component segmentation) and Equation 16 (Watershed Cuts segmentation). Each learning trial runs for 50,000 iterations at small levels of regularization and the results are shown in Figure 2.

Connected Components (CC) is generally able to find solutions with lower training error than Watershed Cuts (WC). In the Particle problem CC also had better performance on test data. Note that the CC thresholds (red crosses) were less reliable than for WC (blue circles). In fact, several of the crosses are not visible because the false alarm rate is much higher than 0.05. However by comparing the performance of CC in Figure 2 to Figure 1, we see this problem could be mitigated with the appropriate choice of the regularization parameter with a validation set. What Figure 2 does tell us is that the WC method appears to be less sensitive to the choice of regularization parameter.

In the Grains problem the WC method consistently outperformed the CC method. We observe that the segment boundaries are far less defined in the Grains problem than in the Particles problem. We suggest the flood-filling characteristics of Watershed Cuts may explain this performance improvement.

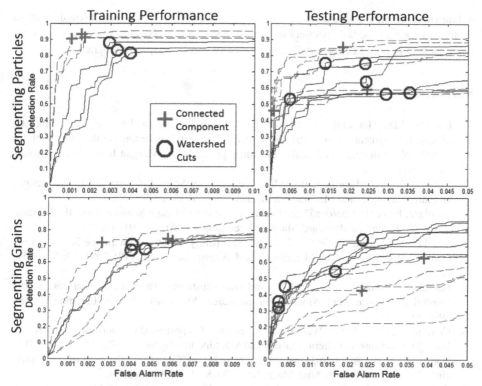

Fig. 2. Left) Training set performance and Right) test set performance on two different segmentation problems. Connected Components appears to be the preferred method for Particles and Watershed Cuts preferred for Grains.

7   Summary

This paper has expanded the *maximin* learning framework to provide mini-batch stochastic sub-gradient descent algorithms for Connected Component and Watershed Cuts segmentation methods. These algorithms are based a Kruskal-like procedure that can efficiently compute the gradients and sub-gradients as well as the Rand Error, and Rand Error ROC curves. The extension of the *maximin* learning to the Watershed Cuts method has been shown to be useful in practical applications and appears to the complement the strengths and weaknesses of Connected Component segmentation.

While the methods and tools for learning energy functions have progressed rapidly there is still much to be done to make learning to segment robust, fast and efficient. Specific areas of future research include: 1) Optimizing stochastic sub-gradient descent updates using as much information from the MST as possible. 2) More expressive features and classifiers, such as convolutional networks, that can incorporate more global information into the segmentation. 3) Extension, or generalization, of these methods to hierarchical energy functions would also be of much practical and theoretical interest.

Acknowledgement. We would like to thank Laurent Najman for very useful discussions about learning to segment and the reviewers for excellent feedback. This work was supported by the Department of Energy's Laboratory Directed Research and Development program.

References

1. Lafferty, J.D., McCallum, A., Pereira, F.C.N.: Conditional Random Fields: Probabilistic Models for Segmenting and Labeling Sequence Data. In: Proceedings of the Eighteenth International Conference on Machine Learning, pp. 282–289. Morgan Kaufmann Publishers Inc. (2001)
2. Farabet, C., et al.: Learning Hierarchical Features for Scene Labeling. IEEE Transactions on Pattern Analysis and Machine Intelligence 35(8), 1915–1929 (2013)
3. Arbelaez, P., et al.: Contour Detection and Hierarchical Image Segmentation. IEEE Transactions on Pattern Analysis and Machine Intelligence 33(5), 898–916 (2011)
4. Fowlkes, C., Martin, D., Malik, J.: Learning Affinity Functions for Image Segmentation: Combining Patch-based and Gradient-based Approaches. In: CVPR 2003, Madison, WI (2003)
5. Bansal, N., Blum, A., Chawla, S.: Correlation Clustering: Theoretical Advances in Data Clustering (Guest Editors: Nina Mishra and Rajeev Motwani). Machine Learning 56(1-3), 89-113 (2004)
6. Couprie, C., et al.: Power Watershed: A Unifying Graph-Based Optimization Framework. IEEE Transactions on Pattern Analysis and Machine Intelligence 33(7), 1384–1399 (2011)
7. Cousty, J., et al.: Watershed cuts: minimum spanning forests and the drop of water principle. IEEE Trans. Pattern Anal. Mach. Intell. 31(8), 1362–1374 (2009)
8. Finley, T., Joachims, T.: Supervised clustering with support vector machines. In: Proceedings of the 22nd International Conference on Machine Learning, pp. 217–224. ACM, Bonn (2005)
9. Turaga, S.C., et al.: Maximin affinity learning of image segmentation. In: NIPS (2009)
10. Cousty, J., et al.: Watershed Cuts: Thinnings, Shortest Path Forests, and Topological Watersheds. IEEE Transactions on Pattern Analysis and Machine Intelligence 32(5), 925–939 (2010)
11. Najman, L., Cousty, J., Perret, B.: Playing with Kruskal: Algorithms for Morphological Trees in Edge-Weighted Graphs. In: Hendriks, C.L.L., Borgefors, G., Strand, R. (eds.) ISMM 2013. LNCS, vol. 7883, pp. 135–146. Springer, Heidelberg (2013)
12. Shalev-Shwartz, S., Singer, Y., Srebro, N.: Pegasos: Primal Estimated sub-GrAdient SOlver for SVM. In: Proceedings of the Twenty-Fourth International Conference on Machine Learning, ICML (2007)
13. Coates, A., Ng, A.Y.: Learning Feature Representations with K-Means. In: Montavon, G., Orr, G.B., Müller, K.-R. (eds.) Neural Networks: Tricks of the Trade, 2nd edn. LNCS, vol. 7700, pp. 561–580. Springer, Heidelberg (2012)
14. Mangan, A.P., Whitaker, R.T.: Partitioning 3D surface meshes using watershed segmentation. IEEE Transactions on Visualization and Computer Graphics 5(4), 308–321 (1999)
15. Beucher, S.: Hierarchical Segmentation and Waterfall Algorithm. In: Serra, J., Soille, P. (eds.) Mathematical Morphology and Its Applications to Image Processing, pp. 69–76. Springer, Netherlands (1994)
16. Fan, R.-E., et al.: LIBLINEAR: A Library for Large Linear Classification. Journal of Machine Learning Research 9, 1871–1874 (2008)

Morphological PDE and Dilation/Erosion Semigroups on Length Spaces

Jesús Angulo[✉]

CMM-Centre de Morphologie Mathématique,
MINES ParisTech, PSL-Research University,
Paris, France
jesus.angulo@mines-paristech.fr

Abstract. This paper gives a survey of recent research on Hamilton–Jacobi partial differential equations (PDE) on length spaces. This theory provides the background to formulate morphological PDEs for processing data and images supported on a length space, without the need of a Riemmanian structure. We first introduce the most general pair of dilation/erosion semigroups on a length space, whose basic ingredients are the metric distance and a convex shape function. The second objective is to show under which conditions the solution of a morphological PDE in the length space framework is equal to the dilation/erosion semigroups.

Keywords: Hamilton–Jacobi PDE · Hamilton–Jacobi semigroup · Length space · Morphological PDE · Morphological semigroup

1 Introduction

Let us assume a Lipschitz continuous function $f : \mathbb{R}^n \to \mathbb{R}$. Consider now the following initial-value Hamilton–Jacobi first-order partial differential equation (PDE)

$$\begin{cases} u_t(x,t) \pm H\left(x, Du(x,t)\right) = 0, & \text{in } \mathbb{R}^n \times (0, +\infty), \\ u(x,0) = f(x), & \text{in } \mathbb{R}^n, \end{cases} \tag{1}$$

Such family of equations usually does not admit classic (i.e., everywhere differentiable) solutions but can be studied in the framework of the theory of viscosity solutions [14]. It is well known [7,19] that if the Hamiltonian has the properties: (i) $H(x,p) = H(p)$ is convex, (ii) superlinear growth in the sense of $\lim_{|p| \to +\infty} H(p)/|p| = +\infty$, and (iii) $H(0) = 0$, then the solution of Cauchy problem (1) is given for $+$ and $-$ respectively by the so-called Hopf–Lax–Oleinik formulas:

$$u(x,t) = \inf_{y \in \mathbb{R}^n} \left[f(y) + tL\left(\frac{x-y}{t}\right) \right], \quad u(x,t) = \sup_{y \in \mathbb{R}^n} \left[f(y) - tL\left(\frac{x-y}{t}\right) \right],$$

where the Lagrangian $L(q)$ is the one-dimensional Legendre–Fenchel transform of function $H(p)$, i.e.,

$$L(q) = H^*(q) = \sup_{p \in \mathbb{R}_+} \{p\,q - H(p)\}, \quad q \in \mathbb{R}_+. \tag{2}$$

© Springer International Publishing Switzerland 2015
J.A. Benediktsson et al. (Eds.): ISMM 2015, LNCS 9082, pp. 509–521, 2015.
DOI: 10.1007/978-3-319-18720-4_43

We note that, by standard results of the Legendre–Fenchel transform, L is increasing, convex, superlinear and satisfies $L(0) = 0$.

PDE (1) plays a central role in continuous mathematical morphology [1,5,12,26,10]. In particular, by taking $H(p) = 1/2\|p\|^2$, such that $L(q) = 1/2\|q\|^2$, a kind of canonic morphological PDE is formulated

$$\begin{cases} \frac{\partial u}{\partial t} = \pm\frac{1}{2}\|\nabla u\|^2, \ x \in \mathbb{R}^n, \ t > 0 \\ u(x,0) = f(x), \quad x \in \mathbb{R}^n \end{cases} \tag{3}$$

such that the corresponding viscosity solutions are given by

$$u(x,t) = \sup_{y \in \mathbb{R}^n} \left\{ f(y) - \frac{\|x-y\|^2}{2t} \right\} \quad \text{(for + sign)}, \tag{4}$$

$$u(x,t) = \inf_{y \in \mathbb{R}^n} \left\{ f(y) + \frac{\|x-y\|^2}{2t} \right\} \quad \text{(for − sign)}, \tag{5}$$

which just correspond to a dilation $(f \oplus b)$ and an erosion $(f \ominus b)$ of function $f(x)$ defined as

$$(f \oplus b)(x) = \sup_{y \in \mathbb{R}^n} \{ f(y) + b(y-x) \}, \tag{6}$$

$$(f \ominus b)(x) = \inf_{y \in \mathbb{R}^n} \{ f(y) - b(y+x) \}, \tag{7}$$

using as structuring function $b(x)$ the so-called multiscale quadratic (or parabolic) structuring function:

$$p_t(x) = -\frac{\|x\|^2}{2t}. \tag{8}$$

By the way, due to its properties of semigroup, dimension separability and invariance to transform domain [25,23,9], the structuring function $p_t(x)$ can be considered as the canonic one in morphology, playing a similar role to the Gaussian kernel in linear filtering. Other particularized forms of the Hamilton–Jacobi model (1) cover the flat morphology by disks [26]; i.e., $u_t = \pm\|\nabla u\|$, as well as operators with more general P-power concave structuring functions, i.e., $u_t = \pm\|\nabla u\|^P$. For the application of the latter model to adaptive morphology, see [15].

Morphological operators are classically defined for images supported on Euclidean spaces. We have recently introduced mathematical morphology for real valued images whose support space is a Riemannian manifold [2]. In fact, we have observed that the smoothness of the space (and its Riemannian structure) is not a fundamental requirement, since the counterpart of Euclidean quadratic operators (4) (5) are also sup/inf-convolutions where the Euclidean distance is replaced by the geodesic distance in the Riemannian manifold. Hence, dilation and erosion can be formulated for functions in a more general framework than the Euclidean or even the Riemannian case. We focus here on functions whose domain is a length (or geodesic) space and in particular we are interesting of relating the corresponding dilation/erosion with a Hamilton–Jacobi PDE formulation.

Morphological PDE on Graphs. The approximation of morphological operators using a PDE formulation has been already considered for the non-Euclidean case of weighted graphs [28,18]. The starting point is the definition of a gradient on the graph. Hence, the basic ingredient is an approximation of the first derivative in a vertex (or node) u in the direction to a vertex v as $\sqrt{w_{uv}}\,(f(v) - f(u))$, where w_{uv} is the weight in the edge linking u to v. Then, the gradient of a function f at a vertex u is defined as $\nabla f(u) = \sum_{v \in N(u)} \sqrt{w_{uv}}\,(f(v) - f(u))$, $N(u)$ being the set of vertices linked to u. Using this gradient, a counterpart of the classical morphological PDE is formulated. The weight function in [28] is generally a distance-based kernel used for adaptive/nonlocal filters. In general, this kind of weight does not involve a natural length structure on the graph and this can be a theoretical limitation in order to link such PDE with classical Hamilton–Jacobi PDE theory. In addition, existence of viscosity solutions, and their semigroups, for those morphological PDEs on graphs were not considered in [28,18].

Numerical Schemes for Hamilton–Jacobi Equations. There exists a large state-of-the-art on numerical schemes for Hamilton–Jacobi equations. The majority of numerical schemes which were proposed to solve Hamilton–Jacobi equations in Euclidean space are based on finite difference methods (upwind and centered discretizations, ENO or WENO schemes, etc.) The formulation of numerical discretization of Hamilton–Jacobi equations on general length spaces is out of the scope of the paper. We can nevertheless cite recent efforts on approximation schemes of Hamilton–Jacobi PDE on networks [13,22].

Hamilton–Jacobi Semigroups on Metric, Length and Geodesic Spaces. During the recent years, a series of works have considered the generalization of the Hopf–Lax–Oleinik formula to a class of Hamilton–Jacobi PDEs on a length space framework. The need of these technical results was motivated by the study of geometric inequalities related to concentration measure. More precisely, connections between logarithmic Sobolev type inequalities and optimal transport-entropy inequalities. See the book by Villani [29] for detailed overview on application of Hopf–Lax–Oleinik semigroup to optimal transport or papers by Ambrosio and co-workers [3,4] for the use of these semigroups in metric space calculus (heat flow, total variation, Ricci curvature bounds, etc.) on metric measure spaces. Nevertheless, up to the best of our knowledge, this theory has not been applied to practical problems in applied mathematics which use Hamilton–Jacobi PDEs, such as optimal control or mathematical morphology.

This series of works were inspired by the seminal contribution by Bobkov et al. [8] establishing the equivalence between logarithmic Sobolev inequality and hypercontractivity properties of classical Hamilton–Jacobi (semigroup) solutions. In our terminology, the semigroup used in [8] corresponds to the Euclidean erosion using a quadratic structuring function. The paper by Lott and Villani [24] is the pioneer work formulating Hamilton–Jacobi PDE acting on continuous functions on a compact measured length space and for a quadratic Hamiltonian. The approach in the same framework was extended to general convex Hamiltonians by Balogh et al. [6]. A different kind of generalization, studied more recently by Gozlan et al. [21] and Ambrosio et al. [3], involves the general case of a length space without the need of a measure structure. The particular case of the Hamilton–Jacobi semigroup on Riemannian manifolds is considered in [29]. We can mention also generalizations of Hamilton–Jacobi semigroups to specific differential geometry structures such as Heisenberg group [16]. Finally, reader is refereed to [20] for a depth insight to recent progresses on sub-solutions of Hamilton–Jacobi equations on Riemannian structures based on KAM theory.

Aim of the Paper. In this context, the goal of the present paper is to give a survey on this recent theory of Hamilton–Jacobi PDEs and associated semigroups on length spaces. Therefore, we do not provide new results, except from the adjunction viewpoint, since most of the proofs can be found in the above mentioned measure theory literature. Nevertheless, in our opinion, the paper has a relevant pedagogical interest in the mathematical morphology context since this theory is useful for the generalization of morphological PDEs for images and data supported on non-Euclidean spaces, such as surfaces, graphs, point clouds, and other length spaces which can be obtained by different image embeddings [2].

2   Preliminaries

Metric, Length and Geodesic Space [17]. A metric space is a set of points X endowed with a distance function $d : X \times X \to [0, \infty)$. In the paper is assumed that (X, d) is a complete separable metric space, locally compact (every closed ball or subset of X is compact).

A length space is a metric space (X, d) such that for any pair of points $x, y \in X$, we have $d(x, y) = \inf\{\text{Length}(\sigma)\}$, where the infimum is taken over all rectifiable curves $\sigma : [0, 1] \to X$ connecting x with y, i.e., $\sigma(0) = x$ and $\sigma(1) = y$.

A curve σ is called a geodesic if σ has constant speed and if $\text{Length}(\sigma|_{[t,t']}) = d(\sigma(t), \sigma(t'))$, $\forall t, t' \in [0, 1]$, $t \leq t'$. A curve σ is a geodesic if for every two points $x, y \in X$, with $\sigma(0) = x$ and $\sigma(1) = y$, one has $d(\sigma(t), \sigma(t')) = |t - t'| d(x, y)$, $\forall t, t' \in [0, 1]$. (X, d) is a geodesic space if for every pair of points $x, y \in X$ there exists a geodesic $\sigma : [0, 1] \to X$ joining x to y.

Note that every geodesic space is a length space. For the converse, we have the Hopf–Rinow Theorem: Let X be a length space, complete and locally compact, then X is a geodesic space.

Doubling Measure Space [6]. A Borel measure μ is doubling, if the measure of any open ball is positive and finite, and if there exists a constant $c_d \geq 1$ such that

$$\mu(B(x, 2r)) \leq c_d \mu(B(x, r))$$

for all $x \in X$ and $r > 0$. Here $B(x, r)$ denotes an open ball of radius r centered in x. A metric measure space (X, d, μ) satisfies a doubling condition if μ is a Borel doubling measure.

Metric Gradient and Subgradient [24,21,3]. We said that $f : X \to \mathbb{R}$ is d-Lipschitz if there exists $C \geq 0$ satisfying $|f(x) - f(y)| \leq Cd(x, y)$, $\forall x, y \in X$. The least constant C with this property will be denoted by $\text{Lip}(f)$. $\text{Lip}(X)$ denotes the set of real-valued Lipschitz functions on X.

Given $f : X \to \mathbb{R}$, we define the metric gradient of f at a point $x \in X$ by

$$|\nabla f|(x) = \limsup_{y \to x} \frac{|f(y) - f(x)|}{d(x, y)}. \tag{9}$$

If f is Lipschitz continuous then $|\nabla f| \in L^\infty(X)$.

We further introduce the metric subgradients of f at x defined as

$$|\nabla^- f|(x) = \limsup_{y \to x} \frac{[f(y) - f(x)]_-}{d(x, y)} = \limsup_{y \to x} \frac{[f(x) - f(y)]_+}{d(x, y)}, \tag{10}$$

and

$$|\nabla^+ f|(x) = \limsup_{y \to x} \frac{[f(y) - f(x)]_+}{d(x, y)} = \limsup_{y \to x} \frac{[f(x) - f(y)]_-}{d(x, y)}, \tag{11}$$

where $a_+ = \max(a, 0)$ and $a_- = \max(-a, 0)$. $|\nabla^- f|(x)$ is called descending slope and $|\nabla^+ f|(x)$ ascending slope. Notice that $|\nabla^- f|(x) = |\nabla^+(-f)|(x)$ and $|\nabla f|(x) = \max\{|\nabla^- f|(x), |\nabla^+ f|(x)\}$. We can therefore work exclusively with $|\nabla^- f|$. Finally, we observe that if d is finite, and (X, d, μ) is doubling, for any $f \in \mathrm{Lip}(X)$ then $|\nabla^- f|(x) = |\nabla^+ f|(x)$ μ-almost everywhere in X [3](Proposition 2.6). Clearly, we notice that $|\nabla^- f|(x) \leq |\nabla f|(x)$, thus the metric subgradient is a finer notion than the gradient norm and $|\nabla^- f|(x)$ vanishes if f has a local minimum at x. In a sense, $|\nabla^- f|(x)$ measures the downward pointing component of f near x: local variation of f taking into account only values less than $f(x)$.

If f is Lipschitz continuous then $|\nabla^\pm f|(x) \leq \mathrm{Lip}(f)$, $\forall x \in X$. Finally, when X is a Riemannian manifold and f is differentiable at x, metric subgradients $|\nabla^\pm f|(x)$ are equal to the norm of the vector $\nabla f(x) \in T_x X$ (the tangent space at x) [29].

3   Dilation and Erosion on Metric Spaces

Let us consider a metric space (X, d) and a given bounded function $f : X \mapsto \mathbb{R}$. We assume that f is Lipschitz continuous. Let us consider a one-dimensional (shape) function $L : \mathbb{R}_+ \to \mathbb{R}_+$, being increasing, superlinear, convex of class C^1 such that $L(0) = 0$. For all scales $t > 0$, we define the dilation $D_{L;t}f$ and the erosion $E_{L;t}f$ operators of f on (X, d) according to L as follows

$$D_{L;t}f(x) = \sup_{y \in X} \left\{ f(y) - tL\left(\frac{d(x, y)}{t}\right) \right\}, \quad \forall x \in X, \tag{12}$$

$$E_{L;t}f(x) = \inf_{y \in X} \left\{ f(y) + tL\left(\frac{d(x, y)}{t}\right) \right\}, \quad \forall x \in X. \tag{13}$$

We adopt the convention $D_{L;0}f = E_{L;0}f = f$. In the context of classical mathematical morphology operators (6)- (7), correspond respectively to the multi-scale dilation $(f \oplus b_t)(x) = D_{L;t}f(x)$ and erosion $(f \ominus b_t)(x) = E_{L;t}f(x)$ of function f by structuring function

$$b_t(x - y) = -tL\left(\frac{d(x, y)}{t}\right).$$

By the way, we note that by symmetry, one has $b_t(x - y) = b_t(y - x)$. A typical example of a shape function is $L(q) = q^P/P$, $P > 1$, such that

$$b_t(x - y) = -\frac{d(x, y)^P}{Pt^{P-1}}.$$

The canonic shape function corresponds to the case $P = 2$: $b_t(x - y) = -\frac{d(x,y)^2}{2t}$.

Properties. The following properties hold for any metric space (X, d).

1. (Adjunction) For any two real-valued functions f and g on (X, d), the pair $(E_{L;t}, D_{L;t})$ forms an adjunction, i.e.,

$$D_{L;t}f(x) \le g(x) \Leftrightarrow f(x) \le E_{L;t}g(x), \quad \forall x \in X.$$

2. (Duality by involution) For any function f and $\forall x \in X$, one has

$$D_{L;t}f(x) = -E_{L;t}(-f)(x); \quad \text{and } E_{L;t}f(x) = -D_{L;t}(-f)(x), \quad \forall t > 0.$$

3. (Increaseness) If $f(x) \le g(x)$, $\forall x \in X$, then

$$D_{L;t}f(x) \le D_{L;t}g(x); \quad \text{and } E_{L;t}f(x) \le E_{L;t}g(x), \quad \forall x \in X, \forall t > 0.$$

4. (Extensivity and anti-extensivity)

$$D_{L;t}f(x) \ge f(x); \quad \text{and } E_{L;t}f(x) \le f(x), \quad \forall x \in X, \forall t > 0.$$

5. (Ordering property) If $0 < s < t$ then $\forall x \in X$

$$\inf_X f \le E_{L;t}f(x) \le E_{L;s}f(x) \le f(x) \le D_{L;s}f(x) \le D_{L;t}f(x) \le \sup_X f.$$

6. (Convergence) For any function f and $\forall x \in X$, $D_{L;t}f(x)$ and $E_{L;t}f(x)$ converge monotonically to $f(x)$ as $t \to 0$. In particular $\lim_{t \to 0} D_{L;t}f = f$ and $\lim_{t \to 0} E_{L;t}f = f$.

7. (Lipschitz) The maps $(x, t) \mapsto D_{L;t}f(x)$ and $(x, t) \mapsto E_{L;t}f(x)$ are in $\text{Lip}(X \times \mathbb{R}_+)$.

8. (Semigroup) For any function f and $\forall x \in X$, and for all pair of scales $s, t > 0$,
   - If X is metric space:

$$D_{L;t}D_{L;s}f \le D_{L;t+s}f; \quad \text{and } E_{L;t}E_{L;s}f \ge E_{L;t+s}f.$$

   - If X is a length space:

$$D_{L;t}D_{L;s}f = D_{L;t+s}f; \quad \text{and } E_{L;t}E_{L;s}f = E_{L;t+s}f.$$

Proof. For property 1, on adjunction, we have that the inequality $D_{L;t}f(x) \le g(x)$ means that

$$\sup_{y \in X} \left\{ f(y) - tL\left(\frac{d(x, y)}{t}\right) \right\} \le g(x), \quad \forall x \in X,$$

It involves that $f(y) - tL\left(d(x,y)/t\right) \le g(x)$ for every $x, y \in X$. This is equivalent to rewrite $f(y) \le g(x) + tL\left(d(x,y)/t\right)$. Therefore, after substitution of $z = x$, we finally have

$$f(y) \le \inf_{z \in X}\left\{g(y) + tL\left(\frac{d(z,y)}{t}\right)\right\} = E_{L;\,t}g(y).$$

For the duality of 2, we have that $D_{L;\,t}(-f)(x)$ is equal to

$$\sup_{y \in X}\left\{-f(y) - tL\left(\frac{d(x,y)}{t}\right)\right\} = -\inf_{y \in X}\left\{f(y) + tL\left(\frac{d(x,y)}{t}\right)\right\} = -E_{L;\,t}(f)(x).$$

The properties 3 and 4 of increaseness and extensivity/anti-extensivity are obvious from the properties of supremum/infimum.

The proof of ordering property 5 is based on the following semigroup property [6](Theorem 2.5.(ii)): For $0 \le s < t$

$$E_{L;\,t}f(x) = \min_{y \in X}\left[E_{L;\,s}f(y) + (t-s)L\left(\frac{d(x,y)}{t-s}\right)\right].$$

Now for a fixed $z \in X$, we have

$$E_{L;\,t}f(z) = \min_{\zeta \in X}\left[E_{L;\,s}f(\zeta) + (t-s)L\left(\frac{d(\zeta,z)}{t-s}\right)\right] \le (t-s)L(0) + E_{L;\,s}f(\zeta).$$

By choosing $z = \zeta$ and using $L(0) = 0$, we have $E_{L;\,t}f(z) \le E_{L;\,s}f(z)$.

In order to prove the semigroup property 8, following [24], we consider for the sake of simplicity the case of the canonic shape function $L(q) = q^2/2$. Now, triangle inequality implies that for all $x, y \in X$ and $s, t > 0$,

$$\frac{d(x,y)^2}{2(t+s)} \le \inf_{z \in X}\left[\frac{d(x,z)^2}{2t} + \frac{d(z,y)^2}{2s}\right]. \tag{14}$$

The equality in (14) in length spaces comes from choosing a minimal geodesic between x and y, and a point z on this geodesic with $d(x,z) = \frac{t}{s+t}d(x,y)$. Finally, from (14), we obtain

$$E_{L;\,t+s}f(x) = \inf_{y \in X}\left[f(y) + \frac{d(x,y)^2}{2(t+s)}\right] = \inf_{y \in X}\inf_{z \in X}\left[f(y) + \frac{d(x,z)^2}{2t} + \frac{d(z,y)^2}{2s}\right]$$
$$= E_{L;\,t}E_{L;\,s}f(x).$$

For the a general function L, see [6].

The proof of properties 6 and 7 on convergence and Lipschitz are not included by the limited length of the paper, see [24,6].

Bibliographic Remark. Following [8] and [24], our metric erosion (13) corresponds to the semigroup $Q_t f$, which is the basic ingredient in the theory of geometric inequalities related to concentration measure. The dual and adjoint semigroup (our dilation (12)) is only considered in [21] and is denoted by $P_t f$. $Q_t f$ is named as Hamilton-Jacobi semigroup on length spaces in [24] whereas other works [6,21,3] use the most classical terminology from max-plus mathematics on Hilbert spaces: $Q_t f$ is the Hopf-Lax-Oleinik semigroup on length (or geodesic) spaces.

4   Morphological PDE on Metric Spaces

We introduce the morphological PDE on a metric space (X, d) as the the following initial-value Hamilton–Jacobi first-order equation:

$$\begin{cases} \frac{\partial}{\partial t} u(x,t) \pm H\left(|\nabla^- u(x,t)|\right) = 0, & \text{in } X \times (0, +\infty), \\ u(x,0) = f(x), & \text{in } X, \end{cases} \tag{15}$$

where the initial condition $f : X \to \mathbb{R}$ is a continuous bounded function and $H : \mathbb{R}_+ \to \mathbb{R}_+$ is the Legendre transform of function $L(q)$:

$$H(p) = \max_{q \in \mathbb{R}_+} \{pq - L(q)\}, \quad p \in \mathbb{R}_+.$$

Our objective now is to show under which conditions the solution of a Hamilton–Jacobi PDE in the metric space framework is equal to the dilation and erosion semigroups. We first consider the results from [24] and [6].

Theorem 1 (Lott and Villani, 2007; Balogh et al.,2012). *The solutions of PDE problem* (15) *are the dilation* (12) *and erosion* (13) *semigroups:*

$$u(x,t) = D_{L;\,t} f(x) \quad (for - sign), \tag{16}$$
$$u(x,t) = E_{L;\,t} f(x) \quad (for + sign), \tag{17}$$

in the following cases.

1. *If (X,d) is a length space: solutions hold for all $x \in X$ and for almost everywhere $t > 0$.*
2. *If (X, d, μ) satisfies a doubling condition and supports a local Poincaré inequality: solutions hold for μ-almost everywhere $x \in X$ and for all $t > 0$.*

Proof. For the sake of pedagogy, let us recall the proof of the solution as an erosion $u(x,t) = E_{L;\,t} f(x)$ in the case 1. The corresponding one for case 2 can be found in [24] and [6], where the role of doubling measure and Poincaré inequality are explained.

We first show that the inequality

$$\frac{\partial}{\partial t} u(x,t) + H(|\nabla^- u|(x,t)) \le 0 \tag{18}$$

holds for every $x \in X$ and a.e. $t \in \mathbb{R}_+$ for $u(x,t) = E_{L;\,t} f(x)$.

Fix $x \in X$ and let $t \in \mathbb{R}_+$ be a point of differentiability of $u(x, \cdot)$. If $|\nabla^- u|(x, t) = 0$, (18) reduces to $u_t(x, t) \leq 0$ since $H(0) = 0$. This clearly holds since $u(x, \cdot)$ is non-increasing. We can thus assume that $|\nabla^- u|(x, t) > 0$, and there exists a sequence $x_n \to x$ for which $u(x_n, t) < u(x, t)$ and $|\nabla^- u|(x, t) = \lim_{n \to \infty} \frac{u(x,t) - u(x_n, t)}{d(x_n, x)}$. For the moment, consider any positive sequence (h_n) with $h_n \to 0$. By the semigroup property [6](Theorem 2.5.(ii)): For $0 \leq s < t$

$$E_{L; t} f(x) = \min_{y \in X} \left[E_{L; s} f(y) + (t - s) L \left(\frac{d(x, y)}{t - s} \right) \right]. \tag{19}$$

we get

$$u(x, t + h_n) = \min_{y \in X} \left\{ h_n L \left(\frac{d(x, y)}{h_n} \right) + u(y, t) \right\} \leq h_n L \left(\frac{d(x, x_n)}{h_n} \right) + u(x_n, t),$$

which implies that

$$\frac{u(x, t + h_n) - u(x, t)}{h_n} \leq - \left[\frac{u(x, t) - u(x_n, t)}{h_n} - L \left(\frac{d(x, x_n)}{h_n} \right) \right]. \tag{20}$$

Since $H(p) = \max_{q \in \mathbb{R}_+} \{pq - L(q)\}$, $\forall p \in \mathbb{R}_+$, for each n it is possible to choose $h_n > 0$ such that

$$H \left(\frac{u(x, t) - u(x_n, t)}{d(x_n, x)} \right) = \frac{u(x, t) - u(x_n, t)}{h_n} - L \left(\frac{d(x, x_n)}{h_n} \right) \tag{21}$$

holds. Furthermore, it is easy to see directly from (21) that $x_n \to x$ implies $h_n \to 0$. Finally, combining (20) and (21) we obtain

$$\frac{u(x, t + h_n) - u(x, t)}{h_n} + H \left(\frac{u(x, t) - u(x_n, t)}{d(x_n, x)} \right) \leq 0.$$

As $x_n \to x$ and $h_n \to 0$, letting $n \to \infty$ gives us (18).

The converse inequality to (18) can be written as

$$\liminf_{s \to 0^+} \frac{E_{L; t+s} f(x) - E_{L; t} f(x)}{s} \geq -H \left(|\nabla^- E_{L; t} f|(x) \right). \tag{22}$$

Let us fix $x \in X$ and $t \in \mathbb{R}_+$. Since $(x, t) \mapsto E_{L; t} f(x)$ is a Lipschitz function, the limit inferior in (22) is finite and we can choose a positive sequence (h_n) such that $h_n \to 0$ and

$$\liminf_{s \to 0^+} \frac{E_{L; t+s} f(x) - E_{L; t} f(x)}{s} = \lim_{n \to \infty} \frac{E_{L; t+h_n} f(x) - E_{L; t} f(x)}{h_n}. \tag{23}$$

Next, applying again the semigroup property (19) we can write

$$E_{L; t+h_n} f(x) = \min_{y \in X} \left\{ h_n L \left(\frac{d(x, y)}{h_n} \right) + E_{L; t} f(y) \right\}. \tag{24}$$

For each n we choose a point $y_n \in X$ for which the minimum is attained. The superlinearity of L implies that $y_n \to x$. As $E_{L;\,t}f(x)$ is decreasing in t, we have $E_{L;\,t+h_n}f(x) \leq E_{L;\,t}f(x)$, and hence

$$E_{L;\,t}f(y_n) \leq h_n L\left(\frac{d(x,y)}{h_n}\right) + E_{L;\,t}f(y_n) \leq E_{L;\,t}f(x). \tag{25}$$

Since $H(p) = \max_{q \in \mathbb{R}_+}\{pq - L(q)\}$, we have $H(p) + L(q) \geq pq$, $\forall p, q \in \mathbb{R}_+$. Together with (25) this implies that

$$H\left(\frac{E_{L;\,t}f(x) - E_{L;\,t}f(y_n)}{d(x,y_n)}\right) + L\left(\frac{d(x,y_n)}{h_n}\right) \geq \frac{E_{L;\,t}f(x) - E_{L;\,t}f(y_n)}{h_n},$$

and we have

$$L\left(\frac{d(x,y_n)}{h_n}\right) + \frac{E_{L;\,t}f(y_n) - E_{L;\,t}f(x)}{h_n} \geq -H\left(\frac{[E_{L;\,t}f(x) - E_{L;\,t}f(y_n)]_+}{d(x,y_n)}\right).$$

Together with (24) this implies

$$\frac{E_{L;\,t+h_n}f(x) - E_{L;\,t}f(x)}{h_n} = \frac{1}{h_n}\left(h_n L\left(\frac{d(x,y_n)}{h_n}\right) + E_{L;\,t}f(y_n) - E_{L;\,t}f(x)\right)$$
$$\geq -H\left(\frac{[E_{L;\,t}f(x) - E_{L;\,t}f(y_n)]_+}{d(x,y_n)}\right).$$

Letting now $n \to \infty$ and using (23) we obtain

$$\liminf_{s \to 0^+} \frac{E_{L;\,t+s}f(x) - E_{L;\,t}f(x)}{s} \geq \limsup_{n \to \infty}\left(-H\left(\frac{[E_{L;\,t}f(x) - E_{L;\,t}f(y_n)]_+}{d(x,y_n)}\right)\right)$$
$$\geq -H\big(|\nabla^- E_{L;\,t}f|(x)\big).$$

Notice that, if $u(x,t) = E_{L;\,t}f(x)$, and t is a point of differentiability of $t \to u(x,t)$ for a fixed x, then it follows that $\frac{\partial}{\partial t}u(x,t) + H(|\nabla^- u|(x,t)) \geq 0$. Since u is Lipschitz–continuous, the above inequality holds for all $x \in X$ and a.e. $t \in \mathbb{R}_+$.

Combining inequalities (18) and (22), we obtain the equality.

Theorem 1 tell us that the solutions of the morphological PDE are the dilation and erosion for all $x \in X$, X being a length space, and for all t outside a set N_t of measure 0. In fact, it has been proven more recently [21] that the result holds without the need of measure theory.

Theorem 2 (Gozlan et al.,2014). *In a geodesic space (X,d), the solutions (16)-(17) hold for all $x \in X$ and for all $t > 0$.*

Finally, in analogy to the Euclidean case, the canonic morphological PDE in a length space (X,d) is given by

$$\begin{cases} \frac{\partial}{\partial t}u(x,t) = \pm\frac{1}{2}|\nabla^- u(x,t)|^2, & x \in X, \ t > 0 \\ u(x,0) = f(x), & x \in X, \end{cases} \tag{26}$$

such that the corresponding semigroup solutions are given by

$$u(x,t) = \sup_{y \in X} \left\{ f(y) - \frac{d(x,y)^2}{2t} \right\} \quad \text{(for + sign)}, \tag{27}$$

$$u(x,t) = \inf_{y \in X} \left\{ f(y) + \frac{d(x,y)^2}{2t} \right\} \quad \text{(for − sign)}. \tag{28}$$

Bibliographic Remark. We note that the case of real-valued extended functions $f : X \to \overline{\mathbb{R}}$, $\overline{\mathbb{R}} =$, with $\mathbb{R} \cup \{+\infty, -\infty\}$, requires a more technical treatment, see Section 3 in Ambrosio et al. [3].

5   Conclusions and Perspectives

We have introduced the most general pair of dilation/erosion operators on a metric space, whose basic ingredients are the metric distance and a convex shape function. We have stated that the families of scale-space dilations $\{D_{L;t}\}_{t>0}$ and erosions $\{E_{L;t}\}_{t>0}$ are semigroups acting on bounded functions only for length spaces. We have introduced the morphological PDE on length spaces and reviewed the theoretical results which provide us a complete transposition from the Euclidean to the geodesic counterpart, linking the morphological PDE to its viscous solutions as dilation/erosion semigroups.

The theory of this paper can be used in many practical situations under the assumption of working on a geodesic space, but without the need of any smoothness of the space or curvature constraints. Discretization and numerical schemas for the morphological PDE on useful cases such as graphs and meshes will be considered in future work. The starting point can be the recent approximation schemes of Hamilton–Jacobi PDE on networks [13,22].

From a theoretical viewpoint, we plan in our perspectives to explore three different lines. First, Eikonal equation is another Hamilton–Jacobi PDE which is the basic ingredient for morphological segmentation (computation of a weighted distance function and watershed segmentation [27]), the corresponding PDE on length spaces is therefore important for us. Second, we will focuss on the particular case of metric spaces of non-positive curvature and $CAT(0)$ spaces [11]. That includes combinatorial spaces such as trees and simplicial complexes. Third, we will study the counterpart of the theory for bounded functions on ultrametric spaces, which are also relevant to mathematical morphology (dendograms and hierarchies).

References

1. Alvarez, L., Guichard, F., Lions, P.-L., Morel, J.-M.: Axioms and fundamental equations of image processing. Arch. for Rational Mechanics 123(3), 199–257 (1993)
2. Angulo, J., Velasco-Forero, S.: Riemannian Mathematical Morphology. Pattern Recognition Letters 47, 93–101 (2014)

3. Ambrosio, L., Gigli, N., Savaré, G.: Calculus and heat flow on metric measure spaces and applications to spaces with Ricci curvature bounded below. Inventiones Mathematicæ 195(2), 289–391 (2014)
4. Ambrosio, L., Di Marino, S.: Equivalent definitions of BV space and total variation in metric measure spaces. Journal of Functional Analysis 266(7), 4150–4188 (2014)
5. Arehart, A.B., Vincent, L., Kimia, B.B.: Mathematical morphology: The Hamilton-Jacobi connection. In: Proc. of IEEE 4th Inter. Conf. on Computer Vision (ICCV 1993), pp. 215–219 (1993)
6. Balogh, Z.M., Engulatov, A., Hunziker, L., Maasalo, O.E.: Functional Inequalities and Hamilton–Jacobi Equations in Geodesic Spaces. Potential Analysis 36(2), 317–337 (2012)
7. Bardi, M., Evans, L.C.: On Hopf's formulas for solutions of Hamilton- Jacobi equations. Nonlinear Analysis, Theory, Methods and Applications 8(11), 1373–1381 (1984)
8. Bobkov, S.G., Gentil, I., Ledoux, M.: Hypercontractivity of Hamilton–Jacobi equations. J. Math. Pures Appl. 80(7), 669–696 (2001)
9. van den Boomgaard, R., Dorst, L.: The morphological equivalent of Gaussian scale-space. In: Proc. of Gaussian Scale-Space Theory, pp. 203–220. Kluwer (1997)
10. Breuß, M., Weickert, J.: Highly accurate PDE-based morphology for general structuring elements. In: Tai, X.-C., Mken, K., Lysaker, M., Lie, K.-A. (eds.) SSVM 2009. LNCS, vol. 5567, pp. 758–769. Springer, Heidelberg (2009)
11. Bridson, M.R., Haefliger, A.: Metric spaces of non-positive curvature. Grundlehren der mathematischen Wissenschaften, Vol. 319, Springer-Verlag (1999)
12. Brockett, R.W., Maragos, P.: Evolution equations for continuous-scale morphology. IEEE Trans. on Signal Processing 42(12), 3377–3386 (1994)
13. Camillia, F., Festab, A., Schiebornc, D.: An approximation scheme for a Hamilton–Jacobi equation defined on a network. Applied Numerical Mathematics 73, 33–47 (2013)
14. Crandall, M.G., Ishii, H., Lions, P.-L.: User's guide to viscosity solutions of second order partial differential equations. Bulletin of the American Mathematical Society 27(1), 1–67 (1992)
15. Diop, E.H.S., Angulo, J.: Multiscale Image Analysis Based on Robust and Adaptive Morphological Scale-Spaces. HAL preprint, hal-00975728 (2014)
16. Dragoni, F.: Metric Hopf–Lax formula with semicontinuous data. Discrete Contin. Dyn. Syst. 17(4), 713–729 (2007)
17. Burago, D., Burago, Y., Ivanov, S.: A course in metric geometry, Graduate Studies in Mathematics 33. AMS, Providence (2001)
18. Elmoataz, A., Desquesnes, X., Lézoray, O.: Non-Local Morphological PDEs and Laplacian Equation on Graphs With Applications in Image Processing and Machine Learning. IEEE Journal of Selected Topics in Signal Processing 6(7), 764–779 (2012)
19. Evans, L.C.: Partial differential equations. Graduate Studies in Mathematics, vol. 19. American Mathematical Society, Providence (1998)
20. Fathi, A.: Weak KAM Theorem in Lagrangian Dynamics. Cambridge Studies in Advanced Mathematics. Cambridge University Press (2014)
21. Gozlan, N., Roberto, C., Samson, P.-M.: Hamilton-Jacobi equations on metric spaces and transport-entropy inequalities. Revista Matematica Iberoamericana 30(1), 133–163 (2014)
22. Herty, M., Ziegler, U., Göttlich, S.: Numerical discretization of Hamilton-Jacobi equations on networks. Networks and Heterogeneous Media 8(3), 685–705 (2013)

23. Jackway, P.T., Deriche, M.: Scale-Space Properties of the Multiscale Morphological Dilation-Erosion. IEEE Trans. Pattern Anal. Mach. Intell. 18(1), 38–51 (1996)
24. Lott, J., Villani, C.: Hamilton–Jacobi semigroup on length spaces and applications. J. Math. Pures Appl. 88(3), 219–229 (2007)
25. Maragos, P.: Slope Transforms: Theory and Application to Nonlinear Signal Processing. IEEE Trans. on Signal Processing 43(4), 864–877 (1995)
26. Maragos, P.: Differential morphology and image processing. IEEE Trans. on Image Processing 5(1), 922–937 (1996)
27. Meyer, F., Maragos, P.: Multiscale Morphological Segmentations Based on Watershed, Flooding, and Eikonal PDE. In: Nielsen, M., Johansen, P., Fogh Olsen, O., Weickert, J. (eds.) Scale-Space 1999. LNCS, vol. 1682, pp. 351–362. Springer, Heidelberg (1999)
28. Ta, V.-T., Elmoataz, A., Lezoray, O.: Nonlocal PDEs-Based Morphology on Weighted Graphs for Image and Data Processing. IEEE Trans. on Image Processing 20(6), 1504–1516 (2011)
29. Villani, C.: Optimal transport. Old and new. Grundlehren der Mathematischen Wissenschaften [Fundamental Principles of Mathematical Sciences], vol. 338. Springer, Berlin (2009)

Hausdorff Distances Between Distributions Using Optimal Transport and Mathematical Morphology

Isabelle Bloch[1](✉) and Jamal Atif[2]

[1] CNRS LTCI, Institut Mines-Telecom, Telecom ParisTech, Paris, France
isabelle.bloch@telecom-paristech.fr
[2] LAMSADE, UMR 7243, PSL, Université Paris-Dauphine, Paris, France
atif@lamsade.dauphine.fr

Abstract. In this paper we address the question of defining and computing Hausdorff distances between distributions in a general sense. We exhibit some links between Prokhorov-Lévy distances and dilation-based distances. In particular, mathematical morphology provides an elegant way to deal with periodic distributions. The case of possibility distributions is addressed using fuzzy mathematical morphology. As an illustration, the proposed approaches are applied to the comparison of spatial relations between objects in an image or a video sequence, when these relations are represented as distributions.

Keywords: Comparison of distributions · Optimal transport · Mathematical morphology · Fuzzy mathematical morphology · Hausdorff · Prokhorov · Lévy distances · Spatial relations

1 Introduction

Comparing distributions is important in image processing and understanding. Typical applications concern the comparison of histograms of gray levels or colors, or of key points [12,21]. At a more structural level, spatial relations between objects, or between instances of objects at different times, are important to assess the spatial arrangement of objects on a scene and its evolution, thus requiring also comparison between representations, e.g. as distributions, of such spatial relations [4].

In this paper we consider the general framework of comparison of distributions in a general sense (related to image information or not), that can have a probabilistic or a possibilistic and fuzzy meaning. We focus on links between dilation-based distances and optimal transport ones.

The Hausdorff distance is a good choice for comparing sets or functions, since it has all the properties of a metric on compact sets. In this paper, we study this distance between distributions, from a mathematical morphology perspective. In particular we highlight links between existing metrics such as Prokhorov and Lévy, and existing or newly proposed expressions of the Hausdorff distance

© Springer International Publishing Switzerland 2015
J.A. Benediktsson et al. (Eds.): ISMM 2015, LNCS 9082, pp. 522–534, 2015.
DOI: 10.1007/978-3-319-18720-4_44

derived from morphological dilations. We consider distributions on the real line, as well as periodic distributions, which are important for comparing histograms of colors in some specific color spaces, or directional spatial relations. This problem has been addressed using the Wasserstein distance in [16], but not using the Hausdorff distance.

The Hausdorff distance has been defined between functions in [17], and by considering 1D functions as subsets of $\mathbb{R}^2$ in [18]. We will also investigate a similar approach in this work. This idea was then further studied in [7] by considering truncated umbras and dilations by a half ball, and in [13], where the case of discontinuous functions was also addressed.

When functions are membership functions of fuzzy sets or possibility distributions, different approaches for defining the Hausdorff distance have been proposed. Some of them define the distance as a number, by combining the values of the Hausdorff distances computed between α-cuts (thresholds of the functions, hence sets), either as a weighted sum, or using the extension principle [5,6,15,22]. Several generalizations of the Hausdorff distance have also been proposed under the form of fuzzy numbers [2,8]. Extensions of the Hausdorff distance based on fuzzy mathematical morphology have been developed, either as a number in [10] from the distance from a point to a fuzzy set [3], or as a fuzzy number [3]. This last approach will be exploited in the present work too.

Some preliminaries on periodic and non periodic distributions are first given in Section 2. Several types of dilations are then proposed in Section 3. Then we propose Hausdorff distances on distributions based on optimal transport and morphological methods in Section 4. The links between these two types of approaches allow us to address the case of non periodic distributions in Section 5. This case is illustrated in Section 6 for comparing directional relations between objects and their change in a synthetic video sequence.

2   Preliminaries

Distributions and cumulative distributions. Let f and g denote the distributions (in a broad sense) to be compared, via the computation of a distance between them. We denote by M the definition domain of these distributions. In this paper, we consider only one-dimensional domains, and M can be $\mathbb{R}$ or $\mathbb{R}^+$ for non-periodic distributions, and $[0, \rho]$ for periodic distributions of period ρ (for instance $[0, 2\pi]$ for the example of relative direction in Section 6). We denote points of M by $x, y...$, or $\theta, \alpha...$ when they are angles.

Normalized distributions are assumed in this paper. Two types of normalization are considered: by the sup or max, or by the sum. The first case goes with a fuzzy or possibilistic interpretation, while the second one corresponds to a probabilistic interpretation.

The cumulative distributions of f and g are denoted by F and G. Note that defining a distance between f and g from a distance between F and G actually provides a distance between distributions. For some definitions, we will consider F and G as sets in a 2D space, denoted by SF and SG. Cumulative distributions are right continuous. Jumps correspond to discontinuities in the underlying distributions. In such cases, SF and SG are completed by vertical segments corresponding to these jumps. In the sequel, we always assume that SF and SG are completed graphs. Therefore, the subset SF associated with a cumulative distribution F is the set defined as:

$$SF = \{(x, F(x)) \mid x \in M\} \cup \{(x, y) \mid x \in J(F) \text{ and } \lim_{x' \to x^-} F(x') \leq y \leq F(x)\}$$

where $J(F)$ denotes the set of points at which jumps occur (i.e. where the left limit of F at x is not equal to $F(x)$).

Ground distance. Existing methods for comparing histograms or probability distributions [9] are usually categorized into two classes: (i) bin-to-bin distances, and (ii) cross-bin distances, involving the distance on the support M (or ground-distance) [9,16,21]. In this paper, we only consider distances of the second class, keeping in mind the application to spatial relations. For instance, if two distributions are identical up to a translation and with disjoint supports, the distances of the first class will always provide the same value, while the second ones will differentiate situations with different translations.

Let us denote by d the ground distance on M. Its definition depends on M. If M is equal to $\mathbb{R}$ or $\mathbb{R}^+$, then d is defined from an L_p norm, for instance $d(x, y) = |x - y|$ in 1D. For periodic distributions (or defined on a circle), the geodesic distance is used. If the period is ρ, we will use $d(x, y) = \min(|x-y|, \rho-|x-y|) = \frac{\rho}{2} - ||x - y| - \frac{\rho}{2}|$. In case of distributions on the circle, with $\rho = 2\pi$, this ground distance is expressed as $d(\theta, \theta') = \min(|\theta-\theta'|, 2\pi-|\theta-\theta'|) = \pi-||\theta-\theta'|-\pi|$. This formulation allows us to consider that values close to 0 and 2π, respectively, are at a short distance from each other. The distance values can also be normalized, using for instance $\frac{d(\theta,\theta')}{\pi}$ or $\sin \frac{|\theta-\theta'|}{2}$.

3   Definition of Some Dilations of Distributions

3.1   Morphological Dilation of a Normalized Distribution

We assume in this section that the distributions are normalized by the sup (and we restrict this work to distributions with bounded sup), or at least that they all have the same maximum value. To simplify the presentation, we consider binary structuring elements, defined as subsets of M.

If the distributions are defined on the real line ($M = \mathbb{R}$ or $M = \mathbb{R}^+$), classical mathematical morphology applies and the dilation of f by a structuring element B is expressed by $\forall x \in M, \delta_B(f)(x) = \sup_{y \in B_x} f(y)$, where B_x denotes as usual the translation of B at x ($B_x = x + B$).

If the distributions are periodic, this periodicity should be taken into account in the dilation and the structuring element.

Definition 1. *Let f be a distribution on the unit circle. Its dilation is defined by:*

$$\forall \theta \in M = [0, 2\pi], \delta_{B^\alpha}(f)(\theta) = \sup_{\theta' \in B_\theta^\alpha} f(\theta') \tag{1}$$

where B^α is a structuring element of aperture α, defined as:

− *if* $\alpha \leq \pi$: $B_\theta^\alpha = [\theta - \alpha, \theta + \alpha]$ *if* $\theta - \alpha \geq 0$ *and* $\theta + \alpha \leq 2\pi$,
  $B_\theta^\alpha = [0, \theta + \alpha] \cup [\theta - \alpha + 2\pi, 2\pi]$ *if* $\theta - \alpha \leq 0$ *and* $\theta + \alpha \leq 2\pi$,
  $B_\theta^\alpha = [\theta - \alpha, 2\pi] \cup [0, \theta + \alpha - 2\pi]$ *if* $\theta - \alpha \geq 0$ *and* $\theta + \alpha \geq 2\pi$,
− *if* $\alpha \geq \pi$: $B_\theta^\alpha = [0, 2\pi]$. *(The case* $\theta - \alpha \leq 0$ *and* $\theta + \alpha \geq 2\pi$ *implies* $\alpha \geq \pi$.)

Note that Definition 1 extends directly to any periodic function.

The normalization ensures that the core of the distribution (set of points with maximum value) is extended according to the size of the structuring element. In particular, it is always possible to find a size of dilation such that a given point of the support of the distribution belongs to the core of the dilated distribution. This property will be used for Hausdorff distances defined from such dilations. The following proposition is easy to show (proofs are omitted due to lack of space):

Proposition 1. *For all* α, B^α *is a ball or radius* α *of the ground distance* d, *and for all* f *and* α, *we have* $\forall \theta, \delta_{B^\alpha}(f)(\theta) = \sup\{f(\theta') \mid d(\theta, \theta') \leq \alpha\}$.

3.2 Dilations of Cumulative Distributions

In this section we consider a cumulative distribution either as a function F from M into $[0, 1]$, or as a subset SF of $M \times [0, 1]$.

Let us consider as a structuring element a segment of length 2ε, with $\varepsilon \geq 0$. We denote by $B_x^\varepsilon = [x - \varepsilon, x + \varepsilon] \cap M$ the translation of this structuring element at x, restricted to the support.

Proposition 2. *The dilation of* F *by* B^ε *is expressed as:*

$$\forall x \in M, \delta_{B^\varepsilon}(F)(x) = \sup_{y \in B_x^\varepsilon} F(y) = \begin{cases} F(x + \varepsilon) & \text{if } x + \varepsilon \in M \\ 1 & \text{otherwise} \end{cases}$$

Let us now consider the dilation of SF, using different structuring elements, that will prove useful in the following. Let us first consider a ball of radius ε of the L^∞ distance, with a positive proportionality factor λ on M to account for the different scales of the two dimensions (i.e. the structuring element is a rectangle). It is expressed, when translated at (x, y), as:

$$(B_1^{\varepsilon,\lambda})_{(x,y)} = (\check{B}_1^{\varepsilon,\lambda})_{(x,y)} = [x - \lambda\varepsilon, x + \lambda\varepsilon] \times [y - \varepsilon, y + \varepsilon].$$

Proposition 3. *The dilation of any* SF *by* $B_1^{\varepsilon,\lambda}$ *is expressed as:*

$$\delta_1^{\varepsilon,\lambda}(SF) = \{(x, y) \in M \times [0, 1] \mid \exists x' \in M, \max(\frac{|x - x'|}{\lambda}, |y - F(x')|) \leq \varepsilon\}. \quad (2)$$

This dilation is illustrated in Figure 1, for $\lambda = 1$.

Let us now consider an asymmetric dilation, with the following structuring element centered at (x, y) and of size ε (still with the factor λ on M):
$(B_2^{\varepsilon,\lambda})_{(x,y)} = [x - \lambda\varepsilon, x + \lambda\varepsilon] \times [y - \varepsilon, 1]$. Its symmetrical with respect to (x, y) is then: $(\check{B}_2^{\varepsilon,\lambda})_{(x,y)} = [x - \lambda\varepsilon, x + \lambda\varepsilon] \times [0, y + \varepsilon]$.

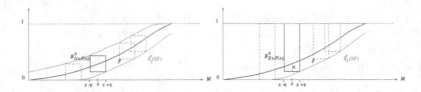

Fig. 1. Dilation with a symmetrical structuring element (left) and with a non-symmetrical one (right)

Proposition 4. *The asymmetric dilation of SF by $B_2^{\varepsilon,\lambda}$ is expressed as:*

$$\delta_2^{\varepsilon,\lambda}(SF) = \{(x,y) \in M \times [0,1] \mid \exists x' \in M, \max(\frac{|x-x'|}{\lambda}, F(x')-y) \leq \varepsilon\}.$$

It is illustrated in Figure 1.

3.3 Dilations of Cumulative Distributions in the Periodic Case

All the definitions introduced above apply also to the periodic case, using the following embedding of F into $\mathbb{R}$:

$$\forall x \in \mathbb{R}, F(x + \rho) = F(x) + 1 \tag{3}$$

and then normalizing the space. For instance if $\rho = 2\pi$, it is sufficient to consider an embedding in $]-\pi, 3\pi[\times[-1,2]$ since for $\lambda\varepsilon \geq \pi$, the dilation would provide the whole space $M \times [0,1]$. The extension of SF then writes:

$$SF^E = SF \cup \{(\theta, F(\theta+2\pi)-1), \theta \in]-\pi, 0]\} \cup \{(\theta, F(\theta-2\pi)+1), \theta \in [2\pi, 3\pi[\}. \tag{4}$$

Dilations can be expressed directly from this set, and we have the following simple form.

Proposition 5. *The dilation of SF with a symmetrical structuring element and $\lambda\varepsilon \leq \pi$ is expressed as:*

$$\delta_{c1}^{\varepsilon,\lambda}(SF) = \{(\theta,y) \in [0,2\pi]\times[0,1] \mid \exists\theta' \in [0,2\pi], |\theta-\theta'| \leq \lambda\varepsilon \text{ and } |F(\theta')-y| \leq \varepsilon\}. \tag{5}$$

For $\lambda\varepsilon > \pi$, then $\delta_{c1}^{\varepsilon,\lambda}(SF) = [0,2\pi] \times [0,1]$.

Note that the simple expression obtained in Proposition 5 corresponds to a geodesic way to process the boundaries of the domain, by truncating the translated structuring element to limit it to the part included in $[0,2\pi] \times [0,1]$. This dilation is illustrated in Figure 2.

Considering now the structuring element $B_2^{\varepsilon,\lambda}$ to dilate only the subgraph (and saturating its complement to 1) leads also to a simple expression:

Proposition 6. *The dilation of SF with an asymmetrical structuring element and $\lambda\varepsilon \leq \pi$ is expressed as:*

$$\delta_{c2}^{\varepsilon,\lambda}(SF) = \{(\theta,y) \in [0,2\pi]\times[0,1] \mid \exists\theta' \in [0,2\pi], |\theta-\theta'| \leq \lambda\varepsilon \text{ and } F(\theta')-y \leq \varepsilon\}. \tag{6}$$

For $\lambda\varepsilon > \pi$, we have $\delta_{c2}^{\varepsilon,\lambda}(SF) = [0,2\pi] \times [0,1]$.

Fig. 2. Dilation in the periodic case, for a symmetrical structuring element. The central circle corresponds to 0 and the larger one to 1. The dashed area is an example of structuring element centered at $(\theta, F(\theta))$. The dilation of SF includes SG.

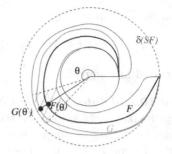

4   Distances Between Distributions on the Real Line

4.1   Morphological Approach

Haudorff distance from dilations of cumulative distributions. Let us first consider $\delta_1^{\varepsilon,\lambda}$ introduced in Section 3.2, and let us derive a Hausdorff distance from it (see Figure 3, for $\lambda = 1$).

Proposition 7. *The Hausdorff distance associated with δ_1 is:*

$$d_{H1}(F, G) = \max(\sup_{x \in M} \inf_{y \in M} \max(\frac{|x - y|}{\lambda}, |G(x) - F(y)|),$$

$$\sup_{y \in M} \inf_{x \in M} \max(\frac{|x - y|}{\lambda}, |F(y) - G(x)|)). \quad (7)$$

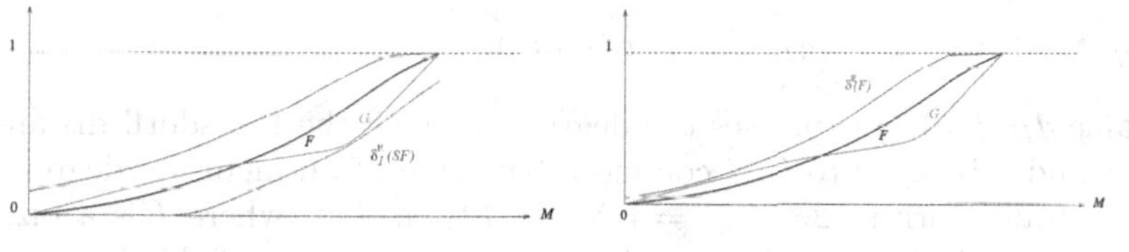

Fig. 3. Left: Minimal size of the dilation of SF such that it contains SG. Right: Computation of the Hausdorff distance by dilating the cumulative distributions considered as functions.

Let us now consider the asymmetric dilation δ_2.

Proposition 8. *The Hausdorff distance derived from δ_2 is:*

$$d_{H2}(F, G) = \max(\sup_{x \in M} \inf_{y \in M} \max(\frac{|x - y|}{\lambda}, G(y) - F(x)),$$

$$\sup_{y \in M} \inf_{x \in M} \max(\frac{|x - y|}{\lambda}, F(x) - G(y))). \quad (8)$$

Finally, let us derive the Hausdorff distance from cumulative distributions considered as functions.

Proposition 9. *We have:*

$$d_H(F,G) = \inf\{\varepsilon > 0 \mid \forall x \in M, G(x) \leq F(x+\varepsilon) \text{ and } F(x) \leq G(x+\varepsilon)\}. \quad (9)$$

This is illustrated in Figure 3.

Proposition 10. *All distances defined in this section are metrics (i.e. positive, separable, symmetrical and satisfy the triangular inequality). If the distributions are Dirac functions (with a unique non zero value at f_0 and g_0), the proposed distances are all equal to $d(f_0, g_0)$, where d is the ground distance.*

Hausdorff distance from dilations of distributions. The idea here is to exploit the link between morphological dilation and some distances, such as minimum and Hausdorff distances, in the case of sets [3,19]. Indeed, the Hausdorff distance between two sets is equal to the minimal size of the ball of the ground distance such that the dilation of each set by this ball contains the other set. We propose to use the same principle on distributions.

Definition 2. *[3] The fuzzy Hausdorff distance is defined from the dilation of the distributions, considered as fuzzy sets, and from an inclusion operator $\Delta_\subseteq(f,g)$, expressing the degree to which f is included in g:*

$$\forall \ell \in \mathbb{R}^{+*}, d_H(f,g)(\ell) = t(d'_H(f,g)(\ell), d'_H(g,f)(\ell)) \quad (10)$$

with

$$d'_H(f,g)(\ell) = t(\Delta_\subseteq(f, \delta_{B^\ell}(g)), \inf_{0 \leq \ell' < \ell} c(\Delta_\subseteq(f, \delta_{B^{\ell'}}(g)))),$$

and $d'_H(f,g)(0) = \Delta_\subseteq(f,g)$, with t a t-norm.

The value $d_H(f,g)(\ell)$ expresses the degree to which the Hausdorff distance between f and g is equal to ℓ. A common definition of an inclusion degree in the fuzzy set framework is $\Delta_\subseteq(f,g) = \inf_{x \in M} I(f(x), g(x))$ where I is a fuzzy implication. If a crisp number is needed, the center of gravity of this fuzzy number can be used: $\frac{\int_0^\infty d_H(f,g)(\ell)\ell d\ell}{\int_0^\infty d_H(f,g)(\ell)d\ell}$, or the following definition:

$$d_H(f,g) = \inf\{\ell \in \mathbb{R}^+ \mid \forall x \in M, \delta_{B^\ell}(f)(x) \geq g(x) \text{ and } \delta_{B^\ell}(g)(x) \geq f(x)\}, \quad (11)$$

which corresponds to a crisp version of the inclusion. This simplified expression corresponds to the definitions in [7,13] for flat structuring elements.

Proposition 11. *[3] The fuzzy distances introduced in Equations 10 and 11 are positive and symmetrical. The morphological Hausdorff distance between the distributions and computed with a crisp version of the inclusion degree (Equation 11) is separable and satisfies the triangular inequality, while the fuzzy version of the inclusion degree yields a distance (Equation 10) which is a fuzzy number, and separable for Lukasiewicz implication ($I(a,b) = \min(1, 1-a+b)$), but does not satisfy the triangular inequality.*

4.2   Lévy and Prokhorov Distances

An interesting distance between probability distributions, related to optimal transport problems [20] and which involves dilations, is the Prokhorov-Lévy metric $d_{Pr} : \mathcal{P}(M)^2 \to [0, +\infty[$ [14], defined for two distributions f and g as:

$$d_{Pr}(f,g) = \inf\{\varepsilon > 0 \mid \forall Z \in \mathcal{B}(M), f(Z) \le g(\delta^{\lambda\varepsilon}(Z)) + \varepsilon \text{ and } g(Z) \le f(\delta^{\lambda\varepsilon}(Z)) + \varepsilon\} \tag{12}$$

where $\delta^{\lambda\varepsilon}(Z)$ is the dilation of size $\lambda\varepsilon$ of Z (see Section 3.1, restricting functions to sets), and $\mathcal{B}(M)$ denotes the set of all Borel sets on M. The definition has been adapted here to introduce λ and thus to account for the potential different scales of M and $[0, 1]$, as in [17].

This distance generalizes the Lévy distance (also a metric), defined in 1D between two cumulative distributions F and G as:

$$d_L(F,G) = \inf\{\varepsilon > 0 \mid \forall x \in \mathbb{R}, G(x - \lambda\varepsilon) - \varepsilon \le F(x) \le G(x + \lambda\varepsilon) + \varepsilon\}. \tag{13}$$

By restricting the Borel sets of $\mathbb{R}$ to the intervals of the form $Z =]-\infty, x[$ (or equivalently $Z =]x, +\infty[$), which generate $\mathcal{B}(M)$, d_{Pr} is indeed equivalent to d_L in 1D. Note that if all Borel sets are considered, then we only have $d_L \le d_{Pr}$.

Hausdorffian expression of d_L. The Lévy distance can be expressed in a similar way as the Hausdorff distance [17] and we have:

$$d_L(F,G) = \max(\sup_{x \in M} \inf_{y \in M} \max(\frac{|x-y|}{\lambda}, G(y) - F(x)),$$

$$\sup_{y \in M} \inf_{x \in M} \max(\frac{|x-y|}{\lambda}, F(x) - G(y))). \tag{14}$$

Note that this expression involves explicitly the ground distance on M.

We now exhibit links with Hausdorff distances derived from the dilations proposed in Section 3.2. Note that d_{Pr} already involves a dilation and that the links between d_{Pr}, d_L and its Hausdorff-like expression already suggest that all these notions are closely related.

Proposition 12. *Let F and G be any two cumulative distributions. We have the following equivalences between their distances:*

- *the Lévy distance can be formulated as a Hausdorff-like expression (Equation 14);*
- *Equation 7 is similar to Equation 14, but with absolute values on $G(x) - F(y)$, providing one of the definitions in [17];*
- *Equation 8 is equivalent to Equation 14;*
- *Equation 9 is equivalent to Equation 13;*
- *Equation 11 is similar to d_{Pr} expressed on points.*

All these links make it easier to extend the definitions to the periodic case (next section).

Proposition 13. d_L *is a probability metric [17]. Similarly, the Hausdorff distances defined in Equations 7 and 9 are probability metrics.*

5   Distances Between Periodic Distributions

In this section we now assume periodic distributions. To fix the ideas, we set, without loss of generality, $\rho = 2\pi$.

5.1   Lévy and Prokhorov Distances

Let us start again from d_{Pr}. We propose to express this distance from a circular dilation and by restricting the Borelian sets to $Z = [0, \theta]$ (which are generating all Borelian sets on $[0, 2\pi]$), taking 0 as origin, arbitrarily[1]. Let us define a dilation of size ε, in the positive direction, as: $\delta^\varepsilon(Z) = [0, \theta + \varepsilon]$ if $\theta + \varepsilon \leq 2\pi$ and $[0, 2\pi]$ otherwise. This morphological expression allows us to derive easily the following result.

Proposition 14. *The Lévy distance, derived from the Prokhorov distance in 1D in the periodic case, is expressed as:*

$$d_L^c(F, G) = \inf\{\varepsilon > 0 \mid \forall \theta \in [0, 2\pi], F(\theta) \leq G(\theta + \lambda\varepsilon) + \varepsilon \text{ and } G(\theta) \leq F(\theta + \lambda\varepsilon) + \varepsilon\}.$$
(15)

by setting $G(\theta + \lambda\varepsilon) = F(\theta + \lambda\varepsilon) = 1$ *if* $\theta + \lambda\varepsilon \geq 2\pi$.

5.2   Morphological Approach

Haudorff distance from dilations of cumulative distributions. Let us consider symmetrical dilations.

Proposition 15. *The Hausdoff distance derived from* δ_{c1} *computed with a symmetrical structuring element is:*

$$d_{Hc1}(F, G) = \max(\sup_{\theta \in [0,2\pi]} \inf_{\theta' \in [0,2\pi]} \max(\frac{|\theta - \theta'|}{\lambda}, |F(\theta') - G(\theta)|),$$

$$\sup_{\theta \in [0,2\pi]} \inf_{\theta' \in [0,2\pi]} \max(\frac{|\theta - \theta'|}{\lambda}, |G(\theta') - F(\theta)|)).$$

The asymmetrical dilation δ_{c2} *leads to similar results, and the derived Hausdorff distance has a similar expression, without the absolute values:*

$$d_{Hc2}(F, G) = \max(\sup_{\theta \in [0,2\pi]} \inf_{\theta' \in [0,2\pi]} \max(\frac{|\theta - \theta'|}{\lambda}, G(\theta) - F(\theta')),$$

$$\sup_{\theta \in [0,2\pi]} \inf_{\theta' \in [0,2\pi]} \max(\frac{|\theta - \theta'|}{\lambda}, F(\theta) - G(\theta'))).$$

---

[1] If the origin is taken at θ_0, then the cumulative distribution is $\int_{\theta_0}^\theta f(t)dt = \int_0^\theta f(t)dt - \int_0^{\theta_0} f(t)dt = F(\theta) - F(\theta_0)$ if $\theta_0 \leq \theta \leq 2\pi$, and $\int_{\theta_0}^{2\pi} f(t)dt + \int_0^\theta f(t)dt = 1 - F_0(\theta_0) + F_0(\theta)$ if $0 \leq \theta \leq \theta_0$. If we want a distance which is independent of the choice of the origin, then $\inf_{\theta_0} d_L^c(F_{\theta_0}, G_{\theta_0})$ could be considered.

The computation of $d_{Hc1}(F, G)$ is illustrated in Figure 2, where the minimal size of dilation of SF such that it includes SG is shown.

Proposition 16. *As in the non-periodic case, the Hausdorff distance derived from asymmetrical dilation and the Lévy distance are equal:*

$$d_{Hc2}(F, G) = d_L^c(F, G). \tag{16}$$

Hausdorff distance from dilations of distributions. The definitions proposed in Equations 10 and 11 apply directly to periodic distributions, by considering appropriate dilations, taking the periodicity into account, as defined in Section 3.1.

An example of distribution on $[0, 2\pi]$ is given in Figure 4, with three translations. The Hausdorff distances values, computed using morphological dilations of the distributions (using Equation 11), between the first distribution of Figure 4 and the others, correspond to the distance between the cores of the distributions, as expected in this simple case.

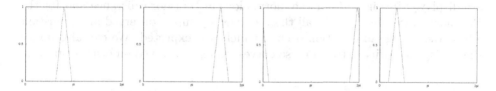

Fig. 4. Example of distribution on $[0, 2\pi]$ and three translations ($T = 2.45, T = 3.68, T = 4.9$). The distances values (in radians) are 0 for $T = 0$, 2.45 for $T = 2.45$, 2.60 for $T = 3.68$, and 1.37 for $T = 4.9$.

6   Comparison Between Spatial Relations

Observing the evolution of a pathology in medical images, or of soil occupation in remote sensing, detecting changes in video sequences, updating a spatial information system are examples that can all benefit from quantification and comparison of spatial relations between objects in the observed scenes. In this paper, to illustrate the proposed approaches, we consider spatial relations represented as distributions or fuzzy numbers, with the typical example of directional relations, represented as a periodic function on $[0, 2\pi]$ via the angle histogram [11]. The normalized angle histogram $ha_{A,B}$ between two 2D objects A and B is defined as: $\forall \theta \in [0, 2\pi], ha_{A,B}(\theta) = \frac{h'_{A,B}(\theta)}{\sup_{\theta' \in [0, 2\pi]} h'_{A,B}(\theta')}$, with $h'_{A,B}(\theta) = |\{(a, b), a \in A, b \in B \mid \angle(a, b) = \theta\}|$ and $\angle(a, b)$ the angle modulo 2π between the vector $\boldsymbol{ab}$ and the horizontal axis. This sum is further weighted by the membership values of a to A and of b to B if the objects are fuzzy.

Let us consider, as an example, the application of the proposed approach to quantify the evolution of directional relations between objects in a simulated

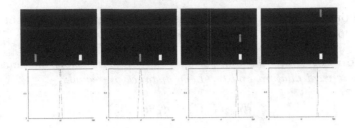

Fig. 5. Simulated video sequence (top, some frames) and angle histograms (bottom).

video sequence (Figure 5). The grey object gets close to the white one in a constant direction, and then changes direction and goes away. The angle histograms ha between these two objects are also illustrated in this figure.

These histograms have been compared using the different proposed measures, by computing the distance between the histogram at time t and the histogram in the first frame. The curves showing the evolution of this distance along time are displayed in Figure 6 for the morphological Hausdorff distance and for the Prokhorov-Lévy distance. In all these curves a jump is observed at the instant where the change in direction occurs, which was expected. We can also notice the strong similarity between these curves, as also observed on other examples.

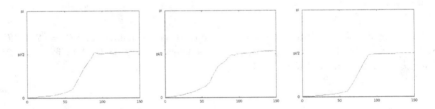

Fig. 6. Morphological Hausdorff distances between the histogram in each frame and the one in the first frame (left). Prokhorov-Lévy distance between the histogram in each frame and the one in the first frame, for histograms normalized by the sup (middle) and by the sum (right).

7   Conclusion

In this paper we have investigated several forms of Hausdorff distances for comparing distributions or cumulative distributions. Based on existing definitions and new ones proposed in this paper, we have exhibited interesting links between optimal transport metrics and morphological ones. In particular, these links have allowed adaptations and extensions to the case of periodic distributions. As an illustration, we have shown that the proposed distances allow comparing spatial relations between objects in images or videos, represented as distributions. This could lead to future applications for detection of ruptures in temporal sequences [1], for comparing different spatial configurations of objects, as a guide for structural recognition and scene understanding, and more generally for spatial reasoning. In our future work we will also go deeper in the formal properties of the proposed distances and their links.

Acknowledgement. This work has been supported by the ANR projects LOGIMA and DESCRIBE. The authors would like to thank Julie Delon for fruitful discussions, and Abdalbassir Abou-Elailah for the simulation of video sequences.

References

1. Abou-Elailah, A., Gouet-Brunet, V., Bloch, I.: Detection of ruptures in spatial relationships in video sequences. In: International Conference on Pattern Recognition Applications and Methods, ICPRAM, Lisbon, Portugal, pp. 110–120 (2015)
2. Aliev, R., Pedrycz, W., Fazlollahi, B., Huseynov, O.H., Alizadeh, A.V., Guirimov, B.G.: Fuzzy logic-based generalized decision theory with imperfect information. Information Sciences 189, 18–42 (2012)
3. Bloch, I.: On Fuzzy Distances and their Use in Image Processing under Imprecision. Pattern Recognition 32(11), 1873–1895 (1999)
4. Bloch, I., Atif, J.: Comparaison de relations spatiales floues - Approches par transport optimal et morphologie mathématique. In: Rencontres Francophones sur la Logique Floue et ses Applications, LFA, pp. 133–140. Cargèse, France (2014)
5. Boxer, L.: On Hausdorff-like Metrics for Fuzzy Sets. Pattern Recognition Letters 18, 115–118 (1997)
6. Chauduri, B.B., Rosenfeld, A.: On a Metric Distance between Fuzzy Sets. Pattern Recognition Letters 17, 1157–1160 (1996)
7. Dougherty, E.R.: Application of the Hausdorff metric in gray-scale mathematical morphology via truncated umbrae. Journal of Visual Communication and Image Representation 2(2), 177–187 (1991)
8. Dubois, D., Prade, H.: On Distance between Fuzzy Points and their Use for Plausible Reasoning. In: Int. Conf. Systems, Man, and Cybernetics, pp. 300–303 (1983)
9. Dudley, R.M.: Distances of probability measures and random variables. The Annals of Mathematical Statistics 39(5), 1563–1572 (1968)
10. Lindblad, J., Sladoje, N.: Linear time distances between fuzzy sets with applications to pattern matching and classification. IEEE Transactions on Image Processing 23(1), 126–136 (2014)
11. Miyajima, K., Ralescu, A.: Spatial Organization in 2D Images. In: Third IEEE Int. Conf. on Fuzzy Systems, FUZZ-IEEE 1994, Orlando, FL, pp. 100–105 (June 1994)
12. Pele, O., Werman, M.: A linear time histogram metric for improved SIFT matching. In: Forsyth, D., Torr, P., Zisserman, A. (eds.) ECCV 2008, Part III. LNCS, vol. 5304, pp. 495–508. Springer, Heidelberg (2008)
13. Popov, A.T.: Hausdorff distance and fractal dimension estimation by mathematical morphology revisited. In: NSIP, pp. 90–94 (1999)
14. Prokhorov, Y.: Convergence of random processes and limit theorems in probability theory. Theory of Probability & Its Applications 1(2), 157–214 (1956)
15. Puri, M.L., Ralescu, D.A.: Différentielle d'une fonction floue. C. R. Acad. Sc. Paris, Série I 293, 237–239 (1981)
16. Rabin, J., Delon, J., Gousseau, Y.: Transportation distances on the circle. Journal of Mathematical Imaging and Vision 41(1-2), 147–167 (2011)

17. Rachev, S.T.: Minimal metrics in the real random variables space. In: Stability Problems for Stochastic Models, pp. 172–190 (1983)
18. Sendov, B.: Hausdorff approximations, vol. 50. Springer (1990)
19. Serra, J.: Image Analysis and Mathematical Morphology. Academic Press, New York (1982)
20. Villani, C.: Optimal transport: old and new. Springer, Berlin (2003)
21. Werman, M., Peleg, S., Rosenfeld, A.: A distance metric for multidimensional histograms. Computer Vision, Graphics, and Image Processing 32(3), 328–336 (1985)
22. Zwick, R., Carlstein, E., Budescu, D.V.: Measures of Similarity Among Fuzzy Concepts: A Comparative Analysis. International Journal of Approximate Reasoning 1, 221–242 (1987)

The Power Laws of Geodesics in Some Random Sets with Dilute Concentration of Inclusions

François Willot[(✉)]

Center for Mathematical Morphology,
Mines ParisTech, PSL Research University,
35 rue St-Honoré, 77300 Fontainebleau, France
francois.willot@ensmp.fr
http://cmm.ensmp.fr/~willot

Abstract. A method for computing upper-bounds on the length of geodesics spanning random sets in 2D and 3D is proposed, with emphasis on Boolean models containing a vanishingly small surface or volume fraction of inclusions $f \ll 1$. The distance function is zero inside the grains and equal to the Euclidean distance outside of them, and the geodesics are shortest paths connecting two points far from each other. The asymptotic behavior of the upper-bounds is derived in the limit $f \to 0$. The scalings involve powerlaws with fractional exponents $\sim f^{2/3}$ for Boolean sets of disks or aligned squares and $\sim f^{1/2}$ for the Boolean set of spheres. These results are extended to models of hyperspheres in arbitrary dimension and, in 2D and 3D, to a more general problem where the distance function is non-zero in the inclusions. Finally, other fractional exponents are derived for the geodesics spanning multiscale Boolean sets, based on inhomogeneous Poisson point processes, in 2D and 3D.

Keywords: Geodesic · Shortest paths · Stochastic geometry · Boolean models · Multiscale random sets

1 Geodesics in Random Media

Among its many applications [1], geodesics have been linked to the transport properties of nonlinear random resistor networks. In the idealized problem considered by Roux and co-workers [2,3,4], each bond in the lattice is a conductor if the voltage drop across the bond is greater than a threshold $v(\boldsymbol{x})$. The values for $v(\boldsymbol{x})$ are uniformly distributed in $[0, 1]$. At the macroscopic scale, no current flows if the applied voltage is smaller than a macroscopic threshold V, determined by directed geodesics. More precisely:

$$V = \min \sum_i v(\boldsymbol{x}_i),$$

where the minimum is taken over all paths $(\boldsymbol{x}_i)_i$ spanning the lattice in the direction of the applied voltage. Similar random networks have been used to model the ductile fracture of porous materials with perfectly-plastic embedding

© Springer International Publishing Switzerland 2015
J.A. Benediktsson et al. (Eds.): ISMM 2015, LNCS 9082, pp. 535–546, 2015.
DOI: 10.1007/978-3-319-18720-4_45

medium [5]. In this problem, the effective plastic yield stress is determined by the length of minimal paths spanning the lattice. The length of the paths are weighted by the local plastic yield stress. Notably, several scaling laws are given in [5,6] for the first-order correction to the geodesics in the 2D square network with small concentration of "porous" bonds $f \ll 1$. The latter scale as $\sim f$ for minimal paths directed parallel to the bonds and $\sim f^{1/2}$ along their diagonals. A scaling law $\sim f^{2/3}$ has been derived in 2D for the geodesics, in the continuum [7], a result consistent with numerical computations for the plastic yield stress [8].

In this article, we extend the result in [7] to other inclusion shapes, distances and dimensions. We first consider a 2D Boolean set of disks (Sec. 2), as in [7]. We derive an upper-bound for its geodesics which is sharper than the one given in [7] but coincide with the latter in the dilute limit $f \to 0$. The rest of this work is concerned by other geometries. In Sec. (3), results for the Boolean model of disks are extended to other 2D models: aligned squares (3.1), disks where the distance function is non-zero (3.2) and multiscale 2D Boolean models (3.3). Sec. (4) is devoted to 3D Boolean models of spheres (4.1), to models of spheres where the distance function is non-zero (4.2) and to multiscale 3D models (4.3). The Boolean model of hyperspheres, in arbitrary dimension, is considered in Sec. (5). We conclude in Sec. (6).

2   Boolean Set of Disks

2.1   Distance Function

This section focuses on the geodesics, i.e. the minimal paths spanning a Boolean model [9] of disks in $\mathbb{R}^2$. As in [7], the distance between two points A and B is defined by:

$$d(A, B) = \inf_{p \in \mathcal{K}} \int_0^1 dt \, \chi(p(t)) \, ||\partial_t p(t)|| , \tag{1}$$

$$\chi(M) = \begin{cases} 0 & \text{if } M \text{ lies inside a disk,} \\ 1 & \text{otherwise.} \end{cases}$$

Therefore disks are crossed at no cost whereas the embedding medium is crossed at a unit cost. In the above, $|| \cdot ||$ is the Euclidean norm, $1 - \chi$ is the indicator function of the disks, and the paths are taken over the set of piecewise, continuously differentiable curves that connect A to B:

$$\mathcal{K} = \left\{ p \in \mathcal{C}^1 \left([0; 1], \mathbb{R}^2 \right), \quad p(0) = A, \quad p(1) = B \right\}. \tag{2}$$

We denote the disks surface fraction by f $(0 \leq f \leq 1)$ and their radius by $D > 0$. The "dilute limit" of inclusions is the limit $f \to 0$. The disks centers follow a homogeneous Poisson point process.

Any optimal path solution of (1) may be replaced by a union of line segments that join a set of disk centers C^i ($i = 1, ..., N$) of coordinates $(C_1^i; C_2^i)$. We denote by $(A; C^1; ...; C^N; B)$ ($N \geq 0$) such path. We are interested in the asymptotic limit of the normalized geodesic distance:

$$\xi = \frac{d(A, B)}{L}, \qquad L = |AB| \to \infty. \tag{3}$$

We assume hereafter that A is the center of a disk at the origin of a Cartesian coordinate system $(e_1; e_2)$ and that the line joining A and B is parallel to e_1. To obtain a upper-bound on ξ, we follow [7] and consider a set of disk centers defined by:

$$|C_1^{i+1} - C_1^i| = \inf \left\{ |C_1 - C_1^i|; \ C \text{ a disk center;} \right.$$

$$\left. C_1 > C_1^i, \ |C_2 - C_2^i| \leq \alpha \sqrt{D|C_1 - C_1^i|} \right\}. \tag{4}$$

where $\alpha > 0$ is a constant to be optimized on. This method amounts to choosing the next disk in a domain delimited by two curves of equation $x_2 - C_2^i = \pm \alpha \sqrt{D|x_1 - C_1^i|}$, with minimal coordinate x_1 along e_1 (see Fig. 1). The shape of this domain, elongated in the e_1 direction, is the result of a trade-off between following the direction e_1 from A to B and maximizing the chance to find a disk at a short distance. The curves that delimit the domain, with equation $|C_2 - C_2^i| \sim \sqrt{D|C_1 - C_1^i|}$, follow from geometrical considerations involving three discs [7].

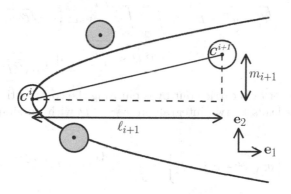

Fig. 1. Method for choosing the disk C^{i+1}, knowing C^i

2.2 Upper-Bound on the Length of Geodesics

We now compute the asymptotic length (3) of the path (4). The number N is chosen so that C^{N+1} is the first disk center with coordinate along e_1 larger than

B_1, i.e. $C_1^{N+1} > B_1$ and $C_1^N \leq B_1$. Define $\ell_i = C_1^i - C_1^{i-1}$ and $m_i = C_2^i - C_2^{i-1}$ ($i \geq 1$) (see Fig. 1). The path (4) provides the following upper bound on ξ:

$$\xi \leq \frac{\sum_{i=1}^N \max\left\{0; \sqrt{\ell_i^2 + m_i^2} - D\right\} + Z}{\sum_{i=1}^N \ell_i}, \tag{5}$$

where $Z = |C^N B|$ is the Euclidean distance from C^N to B. The m_i are uniform random variables in the interval $[-\alpha\sqrt{\ell_i D}; \alpha\sqrt{\ell_i D}]$. Using the Choquet capacity of a Poisson point process [10], the $\ell_i \in [0; \infty)$ follow the cumulative probability function:

$$P\{\ell_i \leq \ell\} = 1 - (1 - f)^{\frac{16\alpha}{3\pi}(\ell/D)^{3/2}}. \tag{6}$$

The above yields, for the denominator in (5):

$$\frac{1}{DN}\sum_{i=1}^N \ell_i \approx \int_{\ell \geq 0} \frac{\ell}{D} P\{\ell \leq \ell_i \leq \ell + d\ell\} = \frac{\Gamma\left(\frac{5}{3}\right)}{4}\left[\frac{3\pi}{-2\alpha\log(1-f)}\right]^{2/3} \tag{7}$$

where Γ is the Gamma (or extended factorial) function. Its asymptotic behavior in the dilute limit $f \to 0$ reads:

$$\frac{1}{DN}\sum_{i=1}^N \ell_i = \left(\frac{\pi}{4\alpha\sqrt{6}}\right)^{2/3}\frac{\Gamma\left(\frac{2}{3}\right)}{f^{2/3}} + O(f^{1/3}). \tag{8}$$

The numerator in (5) is computed as:

$$\frac{1}{N}\sum_{i=1}^N \max\left\{0; \sqrt{\ell_i^2 + m_i^2} - D\right\} \approx \int_{\ell=0}^\infty \int_{m=0}^{\alpha\sqrt{\ell D}} \max\left\{0; \sqrt{\ell^2 + m^2} - D\right\}$$
$$\times P\{\ell \leq \ell_i \leq \ell + d\ell\}\frac{dm}{\alpha\sqrt{\ell D}}. \tag{9}$$

The integration over m is carried out by separating the contribution from $\ell > D$ and $\ell \leq D$. This leads to two integrals in $x = \ell/D$ with no simple analytical solution:

$$\frac{1}{DN}\sum_{i=1}^N \max\left\{0; \sqrt{\ell_i^2 + m_i^2} - D\right\} \approx \int_{x \geq 1} \frac{2dx}{\pi}(1-f)^{\frac{16\alpha x^{3/2}}{3\pi}}\log(1-f)$$
$$\times \left[4\alpha\sqrt{x} - 2\alpha x\sqrt{\alpha^2 + x} - x^2\log\left(1 + \frac{2\alpha}{x}\left(\alpha + \sqrt{\alpha^2 + x}\right)\right)\right]$$
$$+ \int_{x=\sqrt{1+\frac{\alpha^4}{4}} - \frac{\alpha^2}{2}}^1 \frac{-4dx}{\pi}(1-f)^{\frac{16\alpha x^{3/2}}{3\pi}}\log(1-f)$$
$$\times \left[\alpha x\sqrt{\alpha^2 + x} + \sqrt{1 - x^2} - 2\alpha\sqrt{x} + x^2\log\left(\sqrt{x}\frac{\alpha + \sqrt{\alpha^2 + x}}{1 + \sqrt{1 - x^2}}\right)\right]. \tag{10}$$

The second integral in the above equation scales as $\sim f$ when $f \to 0$. The asymptotic behavior of the first one is computed by a Taylor expansion $x \to \infty$. We find at lowest order in f:

$$\frac{1}{DN} \sum_{i=1}^{N} \max\left\{0; \sqrt{\ell_i^2 + m_i^2} - D\right\} = \left(\frac{\pi}{4\alpha\sqrt{6}}\right)^{2/3} \frac{\Gamma\left(\frac{2}{3}\right)}{f^{2/3}} + \frac{\alpha^2}{6} - 1 + O(f^{1/3}).$$

(11)

Note that the term in Z in (5) becomes negligible when N is large. Indeed, $\ell_i \sim f^{-2/3}$ from (8) and so $m_i \sim \sqrt{\ell_i} \sim f^{-1/3}$ and Z scales as:

$$Z = \left|\sum_{i=1}^{N} m_i\right| \sim \sqrt{N} f^{-1/3} \sim \sqrt{L}.$$

Accordingly:

$$\frac{Z}{\sum_i \ell_i} \sim \frac{1}{\sqrt{L}} \to 0, \qquad L \to \infty.$$

Eqs. (8) and (11) then yield, for the normalized geodesic:

$$\xi \leq 1 - \frac{\alpha^{2/3}\left(6 - \alpha^2\right)}{\Gamma\left(\frac{2}{3}\right)} \left(\frac{2}{3\pi}\right)^{2/3} f^{2/3} + O(f^{4/3}).$$

(12)

The sharpest bound is obtained for $\alpha = \sqrt{3/2}$:

$$\xi \leq 1 - \frac{3}{\Gamma\left(\frac{2}{3}\right)} \left(\frac{3f}{2\pi}\right)^{2/3} + O(f^{4/3}) \approx 1 - 1.3534 f^{2/3},$$

(13)

a result identical to that derived in [7].

The upper-bound (5) is computed for the full range of porosity $0 \leq f \leq 1$ using (7) and (10). The two integrals in the right-hand side of (10) are solved numerically. Numerical experiments indicate that $\alpha = \sqrt{3/2}$ is optimal, i.e. produces the sharpest bounds, for all values of f. The bound is compared to numerical estimates of ξ in Fig. (2b). Boolean sets with increasing disks surface fractions $f = 0.06, 0.11, ..., 0.71$ are generated on images containing 4096^2 pixels. We fix the disk radius to 10 pixels and generate 10 realizations of the model for each value of f. We also generate 10 realizations of a model with 8192^2 pixels, disks of radius 4 voxels and surface fraction 0.007. The geodesic distance is computed in each pixel of all images using Matlab's graydist function [11,12]. The distance between the mid-points on two opposite faces are used to estimate geodesics.

Results are represented in Fig. (2), with error bars that indicate statistical fluctuations. As expected, the upper-bound (5) is significantly higher than the exact result when the surface fraction f of the disks is not small. For small values of f, however, the bound becomes a good estimate of the geodesics. When $f = 0.007$, the upper-bound provides $\xi \leq 0.9483$ and the numerical estimate $\xi \approx 0.9470$. As expected, the upper-bound percolates at $f = 1$, a value larger than the actual percolation threshold, which is about $f = f_p^{2D} \approx 0.68$ [13].

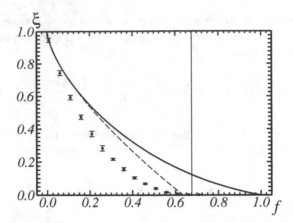

Fig. 2. Upper-bound (5) (solid line) vs. numerical estimates of ξ (circles with error bars) for increasing values of the disks surface fraction f. Dashed line: asymptotic expansion (13). Vertical solid line: percolation threshold $f_c \approx 0.68$.

3   Other 2D Boolean Sets

In this section, we extend result (13) obtained for a Boolean set of discs to some other random sets. The discs are replaced by aligned squares in Sec. (3.1). In Sec. (3.2) we let the distance function be non-zero inside the inclusions. Finally, we consider two-scales random media in Sec. (3.3).

3.1   Boolean Set of Aligned Squares

In this section, we suppose that the inclusions are aligned squares of side D and that points A and B are aligned with one of the direction of the squares. The path $(A; C^1; ...; C^N; B)$ is defined by:

$$|C_1^{i+1} - C_1^i| = \inf\{|C_1 - C_1^i|; \; C \text{ a square center};$$
$$C_1 > C_1^i + D, \; |C_2 - C_2^i| \leq \alpha\sqrt{D|C_1 - C_1^i|}\}. \quad (14)$$

Note that compared to Eq. (4), the C^i are now square centers and we have added the condition $C_1 > C_1^i + D$. Like in Sec. (2.2), we set $\ell_i = C_1^i - C_1^{i-1}$, $m_i = C_2^i - C_2^{i-1}$ with $\ell_i \geq D$, $|m_i| \leq \alpha\sqrt{\ell_i D}$. Again, α is a constant to be optimized on and we denote $f \ll 1$ the surface fraction of the squares. The distance function d_{sq} is defined as in (1) with χ replaced by the indicator function of the Boolean set of squares. Note that:

$$d_{\mathrm{sq}}(C^{i-1}, C^i) \leq \sqrt{(\ell_i - D)^2 + (m_i - D)^2},$$

which yields for the normalized geodesic distance:

$$\xi_{\mathrm{sq}} = \frac{d_{\mathrm{sq}}(A, B)}{L} \leq \frac{Z + \sum_{i=1}^{N} \sqrt{(\ell_i - D)^2 + (m_i - D)^2}}{\sum_{i=1}^{N} \ell_i}, \quad (15)$$

with $Z = |C^N B|$, $L = |AB|$. For squares, Eq. (6) now takes the form:

$$P\{\ell_i \leq \ell\} = 1 - (1-f)^{\frac{4\alpha}{3}[(\ell/D)^{3/2}-1]},$$

and the average of the ℓ_i reads:

$$\frac{1}{N} \sum_{i=1}^{N} \frac{\ell_i}{D} \approx 1 + \frac{2}{3}(1-f)^{-\frac{4\alpha}{3}} E_{1/3}\left(\frac{-4\alpha \log(1-f)}{3}\right) \qquad (16)$$

$$= \frac{\Gamma\left(\frac{2}{3}\right)}{(\sqrt{6}\alpha f)^{2/3}} + O(f^{1/3}), \qquad f \to 0. \qquad (17)$$

The mean of the term $\sqrt{(\ell_i - D)^2 + (m_i - D)^2}$ occurring in (15) is expressed as a double integral like in Sec. (2.2). We develop it for $\ell_i \gg D$ and integrate over ℓ_i and m_i:

$$\frac{1}{DN} \sum_{i=1}^{N} \sqrt{(\ell_i - D)^2 + (m_i - D)^2} = \frac{\Gamma\left(\frac{2}{3}\right)}{(\sqrt{6}\alpha f)^{2/3}} + \left(\frac{\alpha^2}{6} - 1\right) + O(f^{1/3}). \quad (18)$$

We neglect Z and choose $\alpha = \sqrt{3/2}$:

$$\xi_{sq} \leq 1 - \frac{3^{5/3}}{4\Gamma\left(\frac{2}{3}\right)} f^{2/3} + o(f^{2/3}) \approx 1 - 1.1521 f^{2/3}. \qquad (19)$$

This correction is smaller than that derived for the disks model. We emphasize that this model is anisotropic and that the geodesics are directed parallel to the squares's side. The correction is larger for geodesics oriented along the diagonal.

3.2   Boolean Set of Disks with Non-zero Distance Function Inside the Disks

In this section, we consider a Boolean model of disks with the following modified distance function:

$$d_p(A, B) = \inf_{p \in \mathcal{K}} \int_0^1 dt \, \chi_p(p(t)) \, \|\partial_t p(t)\|, \qquad (20)$$

$$\chi_p(M) = \begin{cases} p & \text{if } M \text{ lies inside a disk,} \\ 1 & \text{otherwise,} \end{cases}$$

where $0 \leq p < 1$ is the cost associated to the distance in the disks. The distance function d in (1) is recovered when $p = 0$. We consider a similar path $(A; C^1; ...; C^N; B)$ as in Sec. (2.2), defined by:

$$|C_1^{i+1} - C_1^i| = \inf\{|C_1 - C_1^i|; \; C \text{ a disk center;}$$

$$C_1 > C_1^i + D, \, |C_2 - C_2^i| \leq \alpha\sqrt{D|C_1 - C_1^i|}\}. \qquad (21)$$

with the extra condition $C_1 > C_1^i + D$. It provides a bound on $d_p(A, B)$:

$$\xi_p = \frac{d_p(A, B)}{L} \leq \frac{\sum_{i=1}^{N}\left[\sqrt{\ell_i^2 + m_i^2} - (1-p)D\right]}{\sum_{i=1}^{N}\ell_i}, \tag{22}$$

with $Z = |C^N B|$, $\ell_i = C_1^i - C_1^{i-1}$, $m_i = C_2^i - C_2^{i-1}$, $\ell_i \geq D$, $|m_i| \leq \alpha\sqrt{\ell_i D}$ and $\alpha > 0$. Note that the path $(A; C^1; ...; C^N; B)$ used to derive bound (22) consists in a set of segments joining the disks centers. Most geodesics will not pass through disk centers when $p > 0$. Nevertheless, in the dilute regime considered here we expect $|m_i| \ll \ell_i$ so that bound (22) should be a very good estimate of the length of the path $(A; C^1; ...; C^N; B)$. The asymptotic expansions for the means of the ℓ_i and of the quantity $\sqrt{\ell_i^2 + m_i^2} - D$ are the same as in (11) and (8). This yields:

$$\xi_p \leq 1 - \left(\frac{2\alpha}{3\pi}\right)^{2/3}\frac{6(1-p)-\alpha^2}{\Gamma\left(\frac{2}{3}\right)}f^{2/3} + o(f^{2/3}), \tag{23}$$

$$= 1 - \frac{3(1-p)^{4/3}}{\Gamma\left(\frac{2}{3}\right)}\left(\frac{3}{2\pi}\right)^{2/3}f^{2/3} + o(f^{2/3}) \approx 1.3534(1-p)^{4/3}f^{2/3}, \tag{24}$$

with $\alpha = \sqrt{3(1-p)/2}$. The upper-bound above is sharper than the trivial bound $\xi_p \leq 1 - (1-p)f$ in the domain $f \ll 1 - p$.

3.3   Multiscale Boolean set of Disks

Consider first two Boolean sets of disks with constant diameter, denoted $\mathcal{M}_1$ and $\mathcal{M}_2$. The disks of set $\mathcal{M}_1$ have constant diameter D_1 and that of set $\mathcal{M}_2$ have diameter $D_2 \ll D_1$. The centers of the disks in the Boolean sets $\mathcal{M}_1$ and $\mathcal{M}_2$ follow a homogeneous Poisson point process. We denote by f_1 and f_2 the surface fractions of sets $\mathcal{M}_1$ and $\mathcal{M}_2$ respectively and assume $f_1 \ll 1$, $f_2 \ll 1$. Hereafter we consider the intersection of the two sets $\mathcal{M} = \mathcal{M}_1 \cap \mathcal{M}_2$ which is a two-scales random set with surface fraction $f = f_1 f_2$. The model is assumed "symmetric" so that $f_1 = f_2 = \sqrt{f}$ and the distance function defined as in (1).

Accordingly to (13), the distance $d(A', B')$ between two points A' and B' that lie in a disk contained in $\mathcal{M}_1$ admits the following upper bound:

$$\frac{d(A', B')}{L'} \leq 1 - \frac{3}{\Gamma\left(\frac{2}{3}\right)}\left(\frac{3}{2\pi}\right)^{2/3}f_2^{2/3} + o(f_2^{2/3}),$$

when $L' = |A'B'| \gg D_2$. In the limit $D_2 \ll D_1$ the distance function in $\mathcal{M}$ is well approximated by that considered in Sec. (3.2) with $p = 1 - \frac{3}{\Gamma\left(\frac{2}{3}\right)}\left(\frac{3f_2}{2\pi}\right)^{2/3}$. Eq. (24) then provides the following bound, for two points A and B sufficiently far away from each other:

$$\xi_{ms} = \frac{d(A, B)}{L} \leq 1 - \frac{81}{(2\pi)^{14/9}\Gamma\left(\frac{2}{3}\right)^{7/3}}f^{7/9} + o(f^{7/9}) \approx 1 - 2.2892 f^{7/9}. \tag{25}$$

This correction is smaller than that derived in the one-scale model (13) and indicates that clustering tend to "constraint" the shortest paths and increase their lengths. Similarly, the 2D periodic model, which has a very homogeneous spatial distribution of voids, has an exponent $1/2$ and its geodesics are shorter than in the one-scale Boolean set of disks.

4   3D Boolean Sets

4.1   Boolean Set of Spheres

In this section, we study a Boolean set of spheres of volume fraction f. The distance function d^{3D} is defined in 3D as in (1):

$$d^{3D}(\boldsymbol{A}, \boldsymbol{B}) = \inf_{\boldsymbol{p} \in \mathcal{K}} \int_0^1 dt \, \chi^{3D}(\boldsymbol{p}(t)) \, \|\partial_t \boldsymbol{p}(t)\|, \tag{26}$$

with $1 - \chi^{3D}$ is the indicator of the Boolean set of spheres and $\mathcal{K}$ is given by (2). We define the path $(\boldsymbol{C}^0 = \boldsymbol{A}; \boldsymbol{C}^1; ...; \boldsymbol{C}^N; \boldsymbol{C}^{N+1} = \boldsymbol{B})$ by:

$$|C_1^{i+1} - C_1^i| = \inf \left\{ |C_1 - C_1^i|; \ \boldsymbol{C} \text{ a sphere center}; \right.$$
$$\left. C_1 > C_1^i + D, \ |\boldsymbol{C} - \boldsymbol{C}'| \le \alpha \sqrt{D|C_1 - C_1^i|} \right\}, \tag{27}$$

where $\boldsymbol{C}'$ is the orthogonal projection of $\boldsymbol{C}$ onto the line $(\boldsymbol{C}^i; \boldsymbol{C}^i + \mathbf{e}_1)$ and α is a constant to be optimized on. As in Sec. (2), the axis $\mathbf{e}_1$ is aligned with the line passing by $\boldsymbol{A}$ and $\boldsymbol{B}$. We set $\ell_i = C_1^i - C_1^{i-1}$ and $m_i = |\boldsymbol{C}^i \boldsymbol{C}^{i'}|$ so that $|\boldsymbol{C}^{i-1} \boldsymbol{C}^i|^2 = \ell_i^2 + m_i^2$. The path (27) leads to the bound:

$$\xi^{3D} = \frac{d^{3D}(\boldsymbol{A}, \boldsymbol{B})}{L} \le \frac{\sum_{i=1}^N \left(\sqrt{\ell_i^2 + m_i^2} - D \right) + Z}{\sum_{i=1}^N \ell_i}, \tag{28}$$

with $Z = |\boldsymbol{C}^N \boldsymbol{B}|$. Using the Choquet capacity [10], the variables $\ell_i \in [D; \infty)$ follow the cumulative probability function according t:

$$P\{\ell_i \le \ell\} = 1 - (1 - f)^{3\alpha^2 [(\ell/D)^2 - 1]}. \tag{29}$$

and so, for the mean of the ℓ_i:

$$\frac{1}{DN} \sum_{i=1}^N \ell_i \approx 1 + \frac{(1 - f)^{-3\alpha^2}}{2\alpha} \sqrt{\frac{\pi}{-3 \log(1 - f)}} \operatorname{erfc}\left(\alpha\sqrt{-3 \log(1 - f)}\right)$$

$$= \frac{1}{2\alpha} \sqrt{\frac{\pi}{3f}} + O(\sqrt{f}), \qquad f \to 0, \tag{30}$$

where $\mathrm{erfc}(z) = 2/\sqrt{\pi} \int_z^\infty dt\, e^{-t^2}$ is the complementary error function. The mean of the quantity $\sqrt{\ell_i^2 + m_i^2}$ reads:

$$\frac{1}{N} \sum_{i=1}^{N} \frac{\sqrt{\ell_i^2 + m_i^2}}{D} \approx \int_{\ell=D}^{\infty} \int_{m=0}^{\alpha\sqrt{\ell D}} \sqrt{\ell^2 + m^2}\, P\{\ell \le \ell_i \le \ell + d\ell\} \frac{2m\, dm}{\alpha^2 \ell D^2}$$

$$= -\int_{x=1}^{\infty} dx\, 4x^{3/2} \log(1-f)(1-f)^{3\alpha^2(x^2-1)} \left[(x+\alpha^2)^{3/2} - x^{3/2}\right]$$

$$= \frac{1}{2\alpha} \sqrt{\frac{\pi}{3f}} + \frac{\alpha^2}{4} + O(\sqrt{f}), \tag{31}$$

where the expression behind the integral has been expanded for $x \to \infty$. With $\alpha = 2/\sqrt{3}$ and $Z \ll 1$:

$$\xi^{3D} \le 1 - \frac{8}{3\sqrt{\pi}} \sqrt{f} + o(\sqrt{f}) \approx 1 - 1.5045\sqrt{f}, \qquad f \to 0. \tag{32}$$

The number of "possible choices" for picking C^{i+1} knowing C^i is greater in 3D than in 2D and results in a lower exponent in 3D.

4.2   Non-zero Distance Function in Spheres

The reasoning above extends to a distance function d_p^{3D} which is non-zero in the spheres. Define the distance d_p^{3D} as in (26) with χ^{3D} replaced by:

$$\chi_p^{3D}(M) = \begin{cases} p & \text{if } M \text{ lies inside a sphere,} \\ 1 & \text{otherwise,} \end{cases} \tag{33}$$

where $0 \le p < 1$ is a parameter. The following upper-bound on the normalized shortest paths is derived:

$$\xi_p^{3D} = \frac{d_p^{3D}(A, B)}{|AB|} \le 1 - \frac{8(1-p)^{3/2}}{3\sqrt{\pi}} \sqrt{f} + o(\sqrt{f}), \qquad f \to 0. \tag{34}$$

Again, this bound is non-trivial when $f \ll 1 - p$.

4.3   Multiscale Boolean Set of Spheres

The two-scales "symmetric" Boolean set of spheres is constructed similarly as in Sec. (3.3). The latter is the intersection of two Boolean models of spheres with scale separation. The indicator function χ_{ms}^{3D} of the two-scales Boolean model is used to define the distance d_{ms}^{3D} as in (26). The normalized shortest path is expanded in the dilute limit as:

$$\xi_{ms}^{3D} = \frac{d_{ms}^{3D}(A, B)}{|AB|} \le 1 - \left(\frac{8}{3\sqrt{\pi}}\right)^{5/2} f^{5/8} + o(f^{5/8}) \approx 1 - 2.7764 f^{5/8}, \qquad f \to 0. \tag{35}$$

Our conclusions are the same as that given in 2D. At fixed volume fraction of pores, geodesics are higher in the two-scales symmetric model than in the one-scale Boolean model. The shortest geodesics are found for the 3D periodic model, in which the spatial distribution of voids is very homogeneous and the voids well-separated.

5   Boolean Model of Hyperspheres

In this section we consider a Boolean model of hyperspheres in dimension $d \geq 2$. The geodesic distance d^{dD} and path $(A; C^1; ...; C^N; B)$ are defined as in (26) and (27) with spheres replaced by hyperspheres. A bound for the geodesic is given by the path from A to B. With $\ell_i = C_1^i - C_1^{i-1}$, $m_i = |C^i C^{i\prime}|$, $Z = |C^N B|$ and C' the orthogonal projection of C onto the line $(C^i; C^i + \mathbf{e}_1)$, we have:

$$\xi^{\mathrm{dD}} = \frac{d^{\mathrm{dD}}(A,B)}{L} \leq \frac{\sum_{i=1}^N \left(\sqrt{\ell_i^2 + m_i^2} - D\right) + Z}{\sum_{i=1}^N \ell_i} \approx 1 - \frac{-Z + \sum_{i=1}^N \left(D - \frac{m_i^2}{2\ell_i}\right)}{\sum_{i=1}^N \ell_i}. \tag{36}$$

Denote by $\pi_d(D)$ the volume of the hyperdimensional ball of diameter D and V_ℓ that of the domain:

$$\left\{ C; \quad D \leq C_1 \leq \ell, \quad 0 \leq |C'| \leq \alpha\sqrt{\ell D} \right\}.$$

The probability law for the $\ell_i \in [D, \infty($ depends on V_ℓ by:

$$P\{\ell_i \leq \ell\} = 1 - (1-f)^{\frac{V_\ell}{\pi_d(D)}} = 1 - (1-f)^{\frac{2^{d+1}\alpha^{d-1}\Gamma\left(1+\frac{d}{2}\right)}{(d+1)\sqrt{\pi}\Gamma\left(\frac{1+d}{2}\right)}\left[\left(\frac{\ell}{D}\right)^{\frac{d+1}{2}} - 1\right]}, \tag{37}$$

which provides the sum:

$$\frac{1}{DN} \sum_{i=1}^N \frac{m_i^2}{\ell_i} \approx \int_{\substack{\ell \geq D, \\ m \leq \alpha\sqrt{\ell D}}} P\{\ell \leq \ell_i \leq \ell + d\ell\} \frac{(d-1)m^d dm}{\alpha^{d-1}(D\ell)^{\frac{d+1}{2}}} = \alpha^2 \frac{d-1}{d+1}. \tag{38}$$

The mean of the ℓ_i, approximated by $\int_{\ell \geq D} \ell P\{\ell \leq \ell_i \leq \ell + d\ell\}$, is determined using the symbolic solver Mathematica [14]. The expression involves the function Γ as well as incomplete Γ functions (not shown). Carrying out a Taylor expansion of the latter and optimizing on α yield $\alpha = \sqrt{(1+d)/d}$. Finally:

$$\xi^{\mathrm{dD}} \leq 1 - \frac{(d+1)^{\frac{1+3d}{1+d}}}{\Gamma\left(\frac{2}{1+d}\right)} \left[\frac{\Gamma\left(1+\frac{d}{2}\right)}{2\sqrt{\pi}d^d\Gamma\left(\frac{3+d}{2}\right)}\right]^{\frac{2}{1+d}} f^{\frac{2}{1+d}} + o(f^{\frac{2}{1+d}}), \qquad f \to 0. \tag{39}$$

The above generalizes (13) and (32). Taking successively the limits $f \to 0$ and $d \to \infty$:

$$\xi^{\mathrm{dD}} \leq 1 - 2\left(1 - \frac{\log(2d\pi) - 2(1+\gamma)}{d}\right)f^{\frac{2}{1+d}} \sim 1 - \left(\frac{f}{\sqrt{f_p}}\right)^{2/d} \tag{40}$$

where $\gamma \approx 0.5772$ is Euler's constant and $f_p \sim 2^{-d}$ is the asymptotic percolation threshold in dimension $d \gg 1$ [15].

6 Conclusion

Powerlaws with fractional exponents $2/3$ and $1/2$ have been derived for the lowest-order corrections to the lengths of geodesics in 2D and 3D Boolean models of discs and spheres, respectively. The method is general and provides an upper-bound with lowest-order correction $\sim f^{\frac{2}{d-1}}$ in dimension $d \geq 2$.

The bounds obtained for multiscale models, which scale as $\sim f^{7/9}$ in 2D and $\sim f^{5/8}$ indicate lower variations of the geodesics near the point $f = 0$. These results underline that the the singularities for the geodesics are small for highly-heterogeneous dispersion of particles, and high when the dispersion is homogeneous.

Acknowledgement. This study was made with the support of A.N.R. (Agence Nationale de la Re- cherche) under grant 20284 (LIMA project).

References

1. Li, F., Klette, R.: Euclidean shortest paths – exact or approximate algorithms. Springer, London (2011)
2. Roux, S., Herrmann, H.J.: Disordered-induced nonlinear conductivity. Europhys. Lett. 4(11), 1227–1231 (1987)
3. Roux, S., Hansen, A., Guyon, É.: Criticality in non-linear transport properties of heteroegeneous materials. J. Physique 48(12), 2125–2130 (1987)
4. Roux, S., Herrmann, H., Hansen, A., Guyon, É.: Relation entre différents types de comportements non linéaires de réseaux désordonnés. C. R. Acad. Sci. Série II 305(11), 943–948 (1987)
5. Roux, S., François, D.: A simple model for ductile fracture of porous materials. Scripta Metall. Mat. 25(5), 1087–1092 (1991)
6. Derrida, B., Vannimenus, J.: Interface energy in random systems. Phys. Rev. B 27(7), 4401 (1983)
7. Willot, F.: The power law of geodesics in 2D random media with dilute concentration of disks. Submitted to Phys. Rev. E.
8. Willot, F.: Contribution à l'étude théorique de la localisation plastique dans les poreux. Diss., École Polytechnique (2007)
9. Matheron, G.: Random sets and integral geometry. Wiley, New-York (1975)
10. Matheron, G.: Random sets theory and its applications to stereology. J. Microscopy 95(1), 15–23 (1972)
11. MATLAB and Statistics Toolbox Release 2012b, The MathWorks, Inc., Natick, Massachusetts, United States
12. Soille, P.: Generalized geodesy vis geodesic time. Pat. Rec. Let. 15(12), 1235–1240 (1994)
13. Quintanilla, J., Torquato, S., Ziff, R.M.: Efficient measurement of the percolation threshold for fully penetrable discs. J. Phys. A: Math. Gen. 33(42), L399 (2000)
14. Wolfram Research, Inc.: Mathematica, Version 10.0. Champaign, IL (2014)
15. Torquato, S.: Effect of dimensionality on the continuum percolation of overlapping hyperspheres and hypercubes. J. of Chem. Phys. 136(5), 054106 (2012)

Topology and Discrete Geometry

A 3D Sequential Thinning Scheme Based on Critical Kernels

Michel Couprie[✉] and Gilles Bertrand

LIGM, Équipe A3SI, Université Paris-Est, ESIEE Paris, France
{michel.couprie,gilles.bertrand}@esiee.fr

Abstract. We propose a new generic sequential thinning scheme based on the critical kernels framework. From this scheme, we derive sequential algorithms for obtaining ultimate skeletons and curve skeletons. We prove some properties of these algorithms, and we provide the results of a quantitative evaluation that compares our algorithm for curve skeletons with both sequential and parallel ones.

Keywords: Topology preservation · 3D thinning algorithms · Asymmetric thinning · Curve and surface skeletons · critical kernels

1 Introduction

Topology-preserving transformations are used in many applications of 2D and 3D image processing. In discrete grids, they are used in particular to thin objects until obtaining curve or surface skeletons. The notion of simple point [1] allows for efficient implementations of topology-preserving transformations: intuitively, a point of an object X is simple if it may be removed from X without changing its topological characteristics. Thus, a transformation that iterates the detection and the deletion of a single simple point at each step, is topology-preserving. Simple points may be characterized locally in 2D, 3D and even in higher dimensions (see [2]).

In order to preserve the main geometrical features of the object, some simple points must be preserved from deletion, such points will be called *skeletal points* in the sequel. For example, curve extremities can be used as skeletal points if we want to obtain a curve skeleton. In this paper, we consider only algorithms that dynamically detect skeletal points during the thinning, as opposed to those in two passes, that first need to compute skeletal points (sometimes called anchor points) prior to the thinning process.

Furthermore, in order to obtain well-centered skeletons, a sequential thinning algorithm must consider simple points in a specific order. For example, a naive but natural idea consists of considering only points that are simple points for the original object in a first step. During this first step of thinning, new simple points may appear, they are considered in a second step, and so on.

This work has been partially supported by the "ANR-2010-BLAN-0205 KIDICO" project.

© Springer International Publishing Switzerland 2015
J.A. Benediktsson et al. (Eds.): ISMM 2015, LNCS 9082, pp. 549–560, 2015.
DOI: 10.1007/978-3-319-18720-4_46

Algorithm 1: SeqNaive(X)

Data: X, a set of voxels
Result: X
1 $K := \emptyset$;
2 **repeat**
3 | $K := K \cup \{x$ that is a skeletal voxel for $X\}$;
4 | $Y := \{x$ that is a simple voxel for $X\} \setminus K$;
5 | **foreach** $x \in Y$ **do**
6 | | **if** x *is simple for* X **then** $X := X \setminus \{x\}$;
7 **until** *stability* ;

Algorithm 2: SeqDir(X)

Data: X, a set of voxels
Result: X
1 $K := \emptyset; Y := X$;
2 **repeat**
3 | $K := K \cup \{x$ that is a skeletal voxel for $X\}$;
4 | **foreach** $\alpha \in DirSet$ **do**
5 | | **foreach** $x \in Y \setminus K$ *that is an α-point for* X *and that is simple for* Y **do**
6 | | | $Y := Y \setminus \{x\}$;
7 | $X := Y$;
8 **until** *stability* ;

The scheme SeqNaive uses this strategy in order to try to obtain centered skeletons. Variants of this scheme may be derived by using different criteria for defining skeletal points. For obtaining curve skeletons, one can for example use curve extremities as skeletal points (for an object X, a point of X is a curve extremity if it has exactly one neighbor in X).

Except for toy examples or small objects, this scheme yields noisy skeletons, see Fig. 1a. Furthermore, depending on the actual implementation, the centering may be quite bad. Consider for example an horizontal 2-pixel width ribbon in 2D: all its pixels are simple, and depending on the scanning order, the resulting skeleton can be just one pixel located in one of its extremities.

In order to make this order less arbitrary, one can use the so-called directional strategy that has been introduced by Rosenfeld in his seminal work on 2D parallel thinning [3]. Each iteration of the thinning algorithm is divided into several subiterations, and, in each subiteration, only points in a given direction are considered. For example in 2D, we may define a north (resp. south, east, west) point as an object point whose north (resp. south, east, west) neighbor belongs to the background. The scheme SeqDir is based on this strategy.

The directional scheme also leads to several variants, depending on the set of directions that are considered (*DirSet*), their order within the iteration, and the different criteria for defining skeletal points.

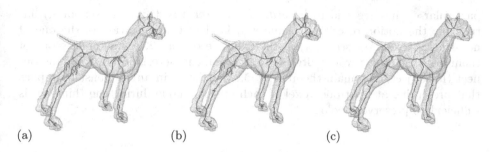

(a)   (b)   (c)

Fig. 1. (a): a skeleton produced by scheme `SeqNaive`. (b): a skeleton produced by scheme `SeqDir`. (c): a skeleton produced by the parallel thinning algorithm of [4]

It is important to note that the skeletons produced using either scheme `SeqNaive` or scheme `SeqDir` are thin, in the sense that they hold the following minimality property: the obtained skeleton contains no simple point, outside of those that have been detected as skeletal points.

While being better centered and a little less noisy than those produced using `SeqNaive`, the skeletons obtained using the directional strategy may also contain many spurious branches, see Fig. 1b.

This fact is a strong argument in favor of parallel thinning algorithms, which are known to produce more robust skeletons than sequential algorithms (see Fig. 1c). However, almost all of them fail to guarantee the minimality property, in other words, there may exist simple points that are not skeletal points in the obtained skeletons. Only some parallel thinning algorithms based on subgrids guarantee this property, but they are subject to some geometric artifacts, and they are not very good in terms of robustness to noise (see [5]).

The motivation of this work is to provide sequential thinning algorithms, that hold the minimality property, and that are as robust to noise as the best parallel algorithms. To achieve this goal, we use the framework of critical kernels.

Critical kernels constitute a framework that has been introduced by one of the authors [6] in order to study the topological properties of parallel thinning algorithms. It also allows one to design new algorithms, in which the guarantee of topology preservation is built in, and in which any kind of constraint may be imposed (see [7,8]). Recently, we showed in [5] that our parallel algorithm for computing thin 3D curve skeletons, based on critical kernels, ranks first in a quantitative evaluation of the robustness of the thinning algorithms of the same class proposed in the literature.

In the classical approach called digital topology [1], topological notions like connectivity are retrieved thanks to the use of two graphs or adjacency relations, one for the object and another one for the background. In our approach, instead of considering only individual points (or voxels) linked by an adjacency relation, we consider that a digital object is made of *cliques*, a clique being a set of mutually adjacent voxels. In Fig. 2, we show an object made of 14 voxels. Among the 48 different cliques that are subsets of this object, three are highlighted: the cliques C_1, C_2, and C_3 made of respectively 1, 2, and 3 voxels. These three particular cliques are said to be *critical* (this term is defined in section 4), in-tuitively the notion of critical clique is a kind of generalization of the one of non-simple voxel. We see that, in the example of Fig. 2, removing any one of them from the object would alter its topological characteristics: it would discon-nect this object. The main theorem of [6], that holds in any dimension, implies that preserving at least one voxel of each critical clique during the thinning, is sufficient to preserve topology.

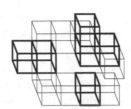

Fig. 2. An object made of 14 voxels, and in which one counts 48 different cliques, some overlapping some others. Three particular cliques, called critical cliques (see text), are highlighted.

The sequel of this paper is organized as follows. Sections 2, 3 and 4 provide all the necessary notions and results relative to, respectively, voxel complexes, sim-ple voxels and critical kernels. We introduce our new generic sequential thinning scheme in section 5, and we prove in section 6 that it guarantees both the preser-vation of topology and the minimality property. Finally, we describe in section 7 an experimental study that shows that our new 3D curve thinning algorithm outperforms the other sequential thinning methods and even the parallel ones in terms of robustness.

2   Voxel Complexes

In this section, we give some basic definitions for voxel complexes, see also [9,1]. Let $\mathbb{Z}$ be the set of integers. We consider the families of sets $\mathbb{F}_0^1$, $\mathbb{F}_1^1$, such that $\mathbb{F}_0^1 = \{\{a\} \mid a \in \mathbb{Z}\}$, $\mathbb{F}_1^1 = \{\{a, a+1\} \mid a \in \mathbb{Z}\}$. A subset f of $\mathbb{Z}^n$, $n \geq 2$, that is the Cartesian product of exactly d elements of $\mathbb{F}_1^1$ and $(n - d)$ elements of $\mathbb{F}_0^1$ is called a *face* or an *d-face* of $\mathbb{Z}^n$, d is the *dimension of f*.

A 3-face of $\mathbb{Z}^3$ is also called a *voxel*. A finite set that is composed solely of voxels is called a *(voxel) complex* (see Fig. 2 and Fig. 3). We denote by $\mathbb{V}^3$ the collection of all voxel complexes.

We say that two voxels x, y are *adjacent* if $x \cap y \neq \emptyset$. We write $\mathcal{N}(x)$ for the set of all voxels that are adjacent to a voxel x, $\mathcal{N}(x)$ is the *neighborhood of x*. Note that, for each voxel x, we have $x \in \mathcal{N}(x)$. We set $\mathcal{N}^*(x) = \mathcal{N}(x) \setminus \{x\}$.

Let $d \in \{0, 1, 2\}$. We say that two voxels x, y are *d-neighbors* if $x \cap y$ is a d-face. Thus, two distinct voxels x and y are adjacent if and only if they are d-neighbors for some $d \in \{0, 1, 2\}$.

Let $X \in \mathbb{V}^3$. We say that X is *connected* if, for any $x, y \in X$, there exists a sequence $\langle x_0, ..., x_k \rangle$ of voxels in X such that $x_0 = x$, $x_k = y$, and x_i is adjacent to x_{i-1}, $i = 1, ..., k$.

3 Simple Voxels

Intuitively a voxel x of a complex X is called a simple voxel if its removal from X "does not change the topology of X". This notion may be formalized with the help of the following recursive definition introduced in [8], see also [10,11] for other recursive approaches for simplicity.

Definition 1. Let $X \in \mathbb{V}^3$. We say that X is *reducible* if either:
i) X is composed of a single voxel; or
ii) there exists $x \in X$ such that $\mathcal{N}^*(x) \cap X$ is reducible and $X \setminus \{x\}$ is reducible.

Definition 2. Let $X \in \mathbb{V}^3$. A voxel $x \in X$ is *simple for* X if $\mathcal{N}^*(x) \cap X$ is reducible. If $x \in X$ is simple for X, we say that $X \setminus \{x\}$ is an *elementary thinning of* X.

Thus, a complex $X \in \mathbb{V}^3$ is reducible if and only if it is possible to reduce X to a single voxel by iteratively removing simple voxels. Observe that a reducible complex is necessarily non-empty and connected.

In Fig. 3 (left), the voxel a is simple for X ($\mathcal{N}^*(a) \cap X$ is made of a single voxel), the voxel d is not simple for X ($\mathcal{N}^*(d) \cap X$ is not connected), the voxel h is simple for X ($\mathcal{N}^*(h) \cap X$ is made of two voxels that are 2-neighbors and is reducible).

In [8], it was shown that the above definition of a simple voxel is equivalent to classical characterizations based on connectivity properties of the voxel's neighborhood [12,13,14,15,2]. An equivalence was also established with a definition based on the operation of collapse [16], this operation is a discrete analogue of a continuous deformation (a homotopy), see [10,6,2].

The notion of a simple voxel allows one to define thinnings of a complex, see an illustration Fig. 3 (right).

Let $X, Y \in \mathbb{V}^3$. We say that Y *is a thinning of* X or that X is *reducible to* Y, if there exists a sequence $\langle X_0, ..., X_k \rangle$ such that $X_0 = X$, $X_k = Y$, and X_i is an elementary thinning of X_{i-1}, $i = 1, ..., k$. Thus, a complex X is reducible if and only if it is reducible to a single voxel.

4 Critical Kernels

Let X be a complex in $\mathbb{V}^3$. It is well known that, if we remove simultaneously (in parallel) simple voxels from X, we may "change the topology" of the original object X. For example, the two voxels f and g are simple for the object X depicted Fig. 3 (left). Nevertheless $X \setminus \{f, g\}$ has two connected components whereas X is connected.

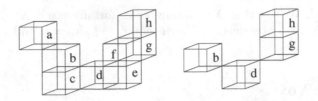

Fig. 3. Left: a complex X which is made of 8 voxels, Right: A complex $Y \subseteq X$, which is a thinning of X

In this section, we recall a framework for thinning in parallel discrete objects with the warranty that we do not alter the topology of these objects [6,7,8]. This method is valid for complexes of arbitrary dimension.

Let $d \in \{0, 1, 2, 3\}$ and let $C \in \mathbb{V}^3$. We say that C is a d-*clique* or a *clique* if $\cap\{x \in C\}$ is a d-face. If C is a d-clique, d is the *rank of C*.

If C is made of solely two distinct voxels x and y, we note that C is a d-clique if and only if x and y are d-neighbors, with $d \in \{0, 1, 2\}$.

Let $X \in \mathbb{V}^3$ and let $C \subseteq X$ be a clique. We say that C is *essential for X* if we have $C = D$ whenever D is a clique such that:

i) $C \subseteq D \subseteq X$; and

ii) $\cap\{x \in C\} = \cap\{x \in D\}$.

In other words, C is essential for X if it is maximal with respect to the inclusion, among all the cliques D in X such that ii) holds.

Observe that any complex C that is made of a single voxel is a clique (a 3-clique). Furthermore any voxel of a complex X constitutes a clique that is essential for X.

In Fig. 3 (left), $\{f, g\}$ is a 2-clique that is essential for X, $\{b, d\}$ is a 0-clique that is not essential for X, $\{b, c, d\}$ is a 0-clique essential for X, $\{e, f, g\}$ is a 1-clique essential for X.

Definition 3. Let $S \in \mathbb{V}^3$. The $\mathcal{K}$-*neighborhood of S*, written $\mathcal{K}(S)$, is the set made of all voxels that are adjacent to each voxel in S. We set $\mathcal{K}^*(S) = \mathcal{K}(S) \setminus S$.

We note that we have $\mathcal{K}(S) = \mathcal{N}(x)$ whenever S is made of a single voxel x. We also observe that we have $S \subseteq \mathcal{K}(S)$ whenever S is a clique.

Definition 4. Let $X \in \mathbb{V}^3$ and let C be a clique that is essential for X. We say that the clique C is *regular for X* if $\mathcal{K}^*(C) \cap X$ is reducible. We say that C is *critical for X* if C is not regular for X.

Thus, if C is a clique that is made of a single voxel x, then C is regular for X if and only if x is simple for X.

In Fig. 3 (left), the cliques $C_1 = \{b, c, d\}$, $C_2 = \{f, g\}$, and $C_3 = \{g, h\}$ are essential for X. We have $\mathcal{K}^*(C_1) \cap X = \emptyset$, $\mathcal{K}^*(C_2) \cap X = \{d, e, h\}$, and $\mathcal{K}^*(C_3) \cap X = \{f\}$. Thus, C_1 and C_2 are critical for X, while C_3 is regular for X.

The following result is a consequence of a general theorem that holds for complexes of arbitrary dimensions [6,8].

Theorem 5. *Let $X \in \mathbb{V}^3$ and let $Y \subseteq X$. The complex Y is a thinning of X if any clique that is critical for X contains at least one voxel of Y.*

See an illustration in Fig. 3 where the complexes X and Y satisfy the condition of theorem 5. For example, the voxel d is a non-simple voxel for X, thus $\{d\}$ is a critical 3-clique for X, and d belongs to Y. Also, Y contains voxels in the critical cliques $C_1 = \{b, c, d\}$, $C_2 = \{f, g\}$, and the other ones.

5   Generic Sequential Thinning Scheme

In this section, we introduce our new generic sequential thinning scheme, see algorithm 3. It is generic in the sense that any notion of skeletal point may be used, for obtaining, *e.g.*, ultimate, curve, or surface skeletons.

Our goal is to define a subset Y of a voxel complex X that is guaranteed to include at least one voxel of each clique that is critical for X. By theorem 5, this subset Y will be a thinning of X.

In order to compute curve or surface skeletons, we have to keep other voxels than the ones that are necessary for the preservation of the topology of the object X. In the scheme, the set K corresponds to a set of features that we want to be preserved by a thinning algorithm (thus, we have $K \subseteq X$). This set K, called *constraint set*, is updated dynamically at line 3. $Skel_X$ is a function from X on $\{True, False\}$ that allows us to detect some *skeletal voxels* of X, *e.g.*, some voxels belonging to parts of X that are surfaces or curves. For example, if we want to obtain curve skeletons, a frequently employed solution is to set $Skel_X(x) = True$ whenever x is a so-called *end voxel* of X: an end voxel is a voxel that has exactly one neighbor inside X. This is the criterion that we will use for this paper.

In the scheme, the set W stores the voxels that are selected to be preserved. At each iteration, W is constructed from scratch by gathering all elements that are selected in all critical cliques (note that a non-simple voxel form a critical 3-clique). Cliques are scanned in decreasing order of rank, and then, according to their orientation α. These orientations correspond to the three axes of the grid, see Fig. 4.

We illustrate our scheme in Fig. 5 on two different objects. For each one, we show an ultimate skeleton, obtained using a function $Skel_X$ that always returns the value *False*, and a curve skeleton, based on a function $Skel_X$ that detects end voxels.

6   Properties

Consider a single execution of the main loop of the algorithm (lines 3–10). It may be easily seen that, by construction, the set W at line 10 contains at least one voxel of all cliques that are critical for X and contained in $X \setminus K$. Furthermore, as all voxels of K are preserved during the execution of lines 3–9, the set W at line 10 contains at least one voxel of all cliques that are critical for X. Thus by theorem 5, we have the following property.

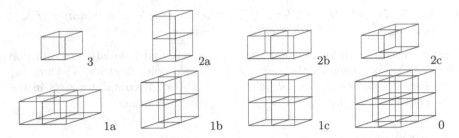

Fig. 4. Masks for cliques with rank $d \in \{3, 2, 1, 0\}$ and orientation $\alpha \in \{a, b, c\}$. An essential clique for a given voxel complex X is a subset of one of those masks. Note that the masks for 3-cliques and the 0-cliques have only one orientation, more precisely, they are invariant by $\pi/2$ rotations.

Algorithm 3: SeqThinningScheme(X, $Skel_X$)

Data: $X \in \mathbb{V}^3$, $Skel_X$ is a function from X on $\{True, False\}$
Result: X
1 $K := \emptyset$;
2 **repeat**
3      $K := K \cup \{x \in X \setminus K$ such that $Skel_X(x) = True\}$;
4      $W := \{x \in X \mid x$ is not simple for $X\} \cup K$;
5   **for** $d \leftarrow 2, 1, 0$ **do**
6   **for** $\alpha \leftarrow a, b, c$ **do**
7   **foreach** d-clique $C \subseteq X \setminus K$ critical for X with orientation α **do**
8   **if** $C \cap W = \emptyset$ **then**
9   Choose x in C; $W := W \cup \{x\}$;

10     $X := W$;
11 **until** stability ;

Proposition 6. Let $X \in \mathbb{V}^3$, let $Skel_X$ be a function from X on $\{True, False\}$, let $Z = SeqThinningScheme(X, Skel_X)$. Then, the complex Z is a thinning of X.

Next, we prove that the produced skeletons hold the minimality property, which is a direct consequence of the following proposition.

Proposition 7. Consider a single execution of the main loop of the algorithm SeqThinningScheme (lines 3–10). If $X \setminus K$ contains at least one simple voxel, then the steps 3 to 10 of the algorithm remove at least one voxel from X.

Proof: Since there is at least one simple voxel in $X \setminus K$, we know that W is different from X at the beginning of line 5. Suppose that, at the beginning of line 10, we have $W = X$. Let x be the last voxel that has been added to W. This voxel must have been added in line 9, thus the condition line 8: $C \cap W = \emptyset$ was fulfilled. But C is made of at least two voxels (it is not a 3-clique), thus there is another voxel $y \neq x$ in C that is not in W at this moment. This voxel y also

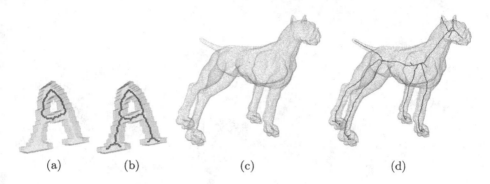

(a)　　　　　(b)　　　　　　(c)　　　　　　　(d)

Fig. 5. Skeletons obtained by using `SeqThinningScheme`. (a,c): ultimate skeletons. We set $Skel_X(x) = False$ for all x. Note that the ultimate skeleton of (c) is a single voxel, as the original object is connected and has no holes or cavities. (b,d): curve skeletons. We set $Skel_X(x) = False$ whenever x is a curve extremity.

has to be added to W since in the end, we have $W = X$. This contradicts the hypothesis that x is the last voxel added to W. Thus, we have $W \neq X$ at the beginning of line 10, and at least one voxel is removed from X. $\square$

Corollary 8. *At the end of* `SeqThinningScheme`, *the resulting complex X does not contain any simple voxel outside of the constraint set K.*

7　Experiments and Results

In these experiments, we used a database of 30 three-dimensional voxel objects. These objects were obtained by converting into voxel sets some 3D models freely available on the internet (mainly from the NTU 3D database, see http://3d.csie.ntu.edu.tw/~dynamic/benchmark). Our test set can be downloaded at http://www.esiee.fr/~info/ck/3DSkAsymTestSet.tgz. We chose these objects because they all may be well described by a curve skeleton, the branches of which can be intuitively related to object parts (for example, the skeleton of a human body in coarse resolution has typically 5 branches, one for the head and one for each limb). For each object, we manually indicated an "ideal" number of branches. Unnecessary branches are essentially due to noise. Thus, a simple and effective criterion for assessing the robustness of a skeletonization method is to

Table 1. Results of our experiments

Sequential			Parallel		
Method M	$E(M)$	$P(M)$	Method M	$E(M)$	$P(M)$
SeqNaive	29.5	0	Palágyi & Kuba 99 [4]	8.97	0.23
SeqDir	24.0	0	Lohou & Bertrand 05 [17]	11.3	0.003
SeqThinningScheme	6.53	0	Couprie & Bertrand [5]	6.73	0

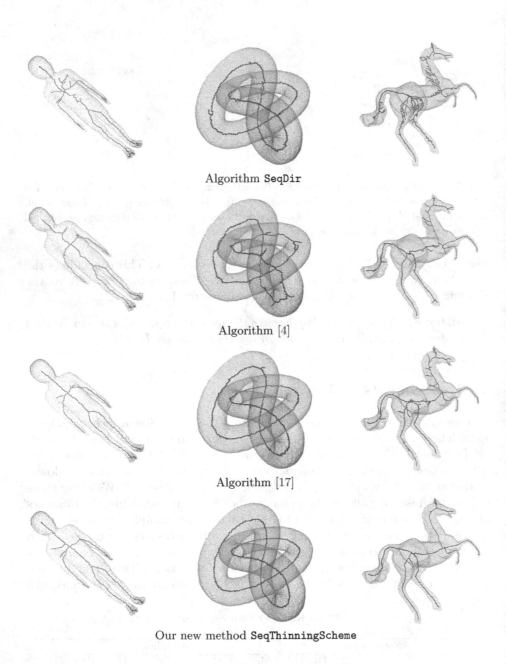

Algorithm SeqDir

Algorithm [4]

Algorithm [17]

Our new method SeqThinningScheme

Fig. 6. Some results with some selected images.

count the number of extra branches, or equivalently in our case, the number of extra curve extremities.

In order to compare methods, we mainly use the indicator $E(X, M) = |c(X, M) - c_i(X)|$, where $c(X, M)$ stands for the number of curve extremities for the result obtained from X after application of method M, and $c_i(X)$ stands for the ideal number of curve extremities to expect with the object X. Note that, for all objects in our database and all tested methods, the difference $c(X, M) - c_i(X)$ was positive, in other words the methods produced more skeleton branches than expected, or just the right number. We define $E(M)$ as the average, for all objects of the database, of $E(X, M)$. The lower the value of $E(M)$, the better the method M with respect to robustness.

The goal of sequential thinning is to provide "thin" skeletons. This means in particular that the resulting skeletons should contain no simple voxel, apart from the curve extremities. However, most parallel thinning algorithms may leave some extra simple voxels. We define our second indicator $P(X, M)$ as the percentage of voxels in the result obtained from X after application of method M that are simple voxels but not curve extremities. We define $P(M)$ as the average, for all objects of the database, of $P(X, M)$. The lower the value of $P(M)$, the better the method M with respect to thinness.

Table 1 gathers the results of our quantitative comparison. For a comparison with parallel methods, we chose the ones that produce thin skeletons (asymmetric methods) and that provide the best results, see [5] for a comparative study of all parallel algorithms of this kind. Note that the criterion for defining skeletal voxels is the same in all the tested method, that is, a skeletal voxel is a curve extremity.

In addition to those quantitative results, it is interesting to look at the results of some different methods for a same object (see Fig. 6). The example of the second column of Fig. 6 illustrates very well the sensitivity to contour noise of the methods. The original object is a solid cylinder bent in order to form a knot. Thus, its curve skeleton should ideally be a simple closed curve. Any extra branch of the skeleton must undoubtedly be considered as spurious. As can be seen in the figure, only our method (last row) among these four ones produces a skeleton of this object that is totally free of spurious branches.

8   Conclusion

We have presented a new generic sequential 3D thinning scheme. From this scheme, it is possible to derive algorithms for obtaining ultimate, curve or surface skeletons, that hold the minimality property. Any criterion for defining skeletal voxels can be employed to this aim. Here, we used curve extremities as skeletal voxels in order to obtain curve skeletons. In an other work [5], we showed that a criterion based on 1D and 2D isthmuses allows one to obtain robust curve and/or surface skeletons.

We showed experimentally that our new curve thinning algorithm has an excellent robustness to noise. Furthermore, our approach allows us do define a notion of "thinning iteration", as in parallel algorithms, that corresponds intuitively to the removal of one layer of voxels from the object. This feature, together with the use of isthmuses to characterize skeletal points, makes it possible to apply in the same approach, a strategy based on isthmus persistence (see [8,5]) in order to filter skeletons based on a single parameter.

References

1. Kong, T.Y., Rosenfeld, A.: Digital topology: introduction and survey. Comp. Vision, Graphics and Image Proc. 48, 357–393 (1989)
2. Couprie, M., Bertrand, G.: New characterizations of simple points in 2D, 3D and 4D discrete spaces. IEEE Transactions on Pattern Analysis and Machine Intelligence 31(4), 637–648 (2009)
3. Rosenfeld, A.: A characterization of parallel thinning algorithms. Information and Control 29(3), 286–291 (1975)
4. Palágyi, K., Kuba, A.: Directional 3D thinning using 8 subiterations. In: Bertrand, G., Couprie, M., Perroton, L. (eds.) DGCI 1999. LNCS, vol. 1568, pp. 325–336. Springer, Heidelberg (1999)
5. Couprie, M., Bertrand, G.: Asymmetric parallel 3d thinning scheme and algorithms based on isthmuses. Technical report hal-01104691 (2014) (preprint, to appear in Pattern Recognition Letters)
6. Bertrand, G.: On critical kernels. Comptes Rendus de l'Académie des Sciences, Série Math. I(345), 363–367 (2007)
7. Bertrand, G., Couprie, M.: Two-dimensional thinning algorithms based on critical kernels. Journal of Mathematical Imaging and Vision 31(1), 35–56 (2008)
8. Bertrand, G., Couprie, M.: Powerful Parallel and Symmetric 3D Thinning Schemes Based on Critical Kernels. Journal of Mathematical Imaging and Vision 48(1), 134–148 (2014)
9. Kovalevsky, V.: Finite topology as applied to image analysis. Computer Vision, Graphics and Image Processing 46, 141–161 (1989)
10. Kong, T.Y.: Topology-preserving deletion of 1's from 2-, 3- and 4-dimensional binary images. In: Ahronovitz, E. (ed.) DGCI 1997. LNCS, vol. 1347, pp. 1–18. Springer, Heidelberg (1997)
11. Bertrand, G.: New notions for discrete topology. In: Bertrand, G., Couprie, M., Perroton, L. (eds.) DGCI 1999. LNCS, vol. 1568, pp. 218–228. Springer, Heidelberg (1999)
12. Bertrand, G., Malandain, G.: A new characterization of three-dimensional simple points. Pattern Recognition Letters 15(2), 169–175 (1994)
13. Bertrand, G.: Simple points, topological numbers and geodesic neighborhoods in cubic grids. Pattern Recognition Letters 15, 1003–1011 (1994)
14. Saha, P., Chaudhuri, B., Chanda, B., Dutta Majumder, D.: Topology preservation in 3D digital space. Pattern Recognition 27, 295–300 (1994)
15. Kong, T.Y.: On topology preservation in 2-D and 3-D thinning. International Journal on Pattern Recognition and Artificial Intelligence 9, 813–844 (1995)
16. Whitehead, J.: Simplicial spaces, nuclei and m-groups. Proceedings of the London Mathematical Society 45(2), 243–327 (1939)
17. Lohou, C., Bertrand, G.: A 3D 6-subiteration curve thinning algorithm based on P-simple points. Discrete Applied Mathematics 151, 198–228 (2005)

How to Make nD Functions Digitally Well-Composed in a Self-dual Way

Nicolas Boutry[1,2(✉)], Thierry Géraud[1], and Laurent Najman[2]

[1] EPITA Research and Development Laboratory (LRDE), Paris, France
[2] LIGM, Équipe A3SI, ESIEE, Université Paris-Est, Paris, France
{firstname,lastname}@lrde.epita.fr, l.najman@esiee.fr

Abstract. Latecki *et al.* introduced the notion of 2D and 3D well-composed images, *i.e.*, a class of images free from the "connectivities paradox" of digital topology. Unfortunately natural and synthetic images are not *a priori* well-composed. In this paper we extend the notion of "digital well-composedness" to nD sets, integer-valued functions (gray-level images), and interval-valued maps. We also prove that the digital well-composedness implies the equivalence of connectivities of the level set components in nD. Contrasting with a previous result stating that it is not possible to obtain a discrete nD self-dual digitally well-composed function with a local interpolation, we then propose and prove a self-dual discrete (non-local) interpolation method whose result is always a digitally well-composed function. This method is based on a sub-part of a quasi-linear algorithm that computes the morphological tree of shapes.

Keywords: Well-composed functions · Equivalence of connectivities · Cubical grid · Digital topology · Interpolation · Self-duality

1 Introduction

Connectivities paradox is a well-documented issue in digital topology [7]: a connected component has to have a different connectivity whether it belongs to the background or to the foreground. Well-composed images have been introduced in 2D [9] and 3D [8] to solve that issue, as one of their main properties is the equivalence of connectivities. Intuitively, a major interest of well-composed image is to have functions (values) defined independently from the underlying space structure (graph). Indeed, life is easier if the connectivity of upper and lower level sets are the same. This is especially true when considering self-duality (recall that a transform φ is self-dual iff $\varphi(-u) = -\varphi(u)$, where φ acts on the space of functions): peaks and valleys are not processed with the same connectivity, and as a consequence the self-duality property is not "perfectly pure". The companion paper [4] discusses at length these questions, which have been largely ignored in the literature on self-duality.

Given that sequences of 3D images become more and more frequent, notably in the medical imaging field and in material sciences, it is important to extend well-composedness in 4D, and more generally in nD. However, extending the notion to higher dimension is not straightforward. The main objective of this paper is twofold.

© Springer International Publishing Switzerland 2015
J.A. Benediktsson et al. (Eds.): ISMM 2015, LNCS 9082, pp. 561–572, 2015.
DOI: 10.1007/978-3-319-18720-4_47

- First, we review and study some possible extensions of the well-composedness concept: based on the equivalence of connectivity, based on the continuous framework, based on the combinatorial definition of n-surfaces, and, most importantly in this paper, the digital well-composedness, based on some critical configurations. We prove in particular that digital well-composedness implies the equivalence of connectivities.
- Second, we propose a non-local interpolation producing, from a gray-level image defined on the nD cubical grid, an interpolated digital well-composed nD image, with $n \geq 2$. Recall that, in the same setting, we have proved [1] that, under some usual constraints, *no* local self-dual interpolation method can succeed in making nD digital well-composed functions for $n \geq 3$. Last, let us mention that another approach, based on changing the image values, has been investigated in [2]

The outline of this paper is the following. Section 2 presents several notions of well-composed sets and functions; as a side-effect, it also provides a disambiguation of the multiple definitions of "well-composedness". In Section 3, we propose an extension to nD of the notion of digital well-composedness, precisely on nD digitally well-composed sets, functions, and interval-valued maps. Section 4 studies a front propagation algorithm $\mathfrak{FP}$, and states that, when an nD interval-valued map U is digitally well-composed, then $\mathfrak{FP}(U)$ is digitally well-composed. In Section 5 we explain how practically we can turn an nD integer-valued function u into an nD well-composed function. Last we conclude in Section 6 and gives some perspectives of our work.

Note that, due to limited space, the proofs of theorems and propositions will be given in an extended version of this paper, available on the Internet from http://hal.archives-ouvertes.fr.

2   About Well-Composedness

2.1   Well-Composed Sets

An important result of the paper is the clarification of the terminology and of the various approaches dealing with well-composedness on cubical grids. There exist four different definitions:

- the *well-composedness based on the "equivalence of connectivities"* (or EoC well-composedness), EWC for short, which is the seminal definition of "well-composed 2D sets";
- the *digital well-composedness*, DWC for short, which relies on the definition of critical configurations (explained later in this paper);
- the *continuous well-composedness*, CWC for short, which relies on the continuous framework;
- the *Alexandrov well-composedness*, AWC for short, which relies on the combinatorial definition by Evako *et al.* [3] of n-surface.

As said in the introduction, digital topology is well known to force the practitioners to use a pair of connectivities [7], *e.g.* in 2D both c_4 and c_8. To avoid

Table 1. Different "flavors" of well-composedness: their definition is emphasized (underlined), and the relations between them are depicted. The bottom line is dedicated to their nD generalizations where $n \geq 2$. Additionally, note that the relation "EWC $\Leftarrow$ DWC" in nD comes from this present paper.

2D case:	**EWC** [9] $\Longleftrightarrow$	DWC	$\Longleftrightarrow$	AWC	$\Longleftrightarrow$	CWC
3D case:	EWC $\Longleftarrow$	DWC	$\Longleftrightarrow$	AWC	$\Longleftrightarrow$	**CWC** [8]
nD case:	$\boxed{\text{EWC}}$ $\Longleftarrow$	**DWC** (this paper)	$\Leftarrow?\Rightarrow$	**AWC** [12]	$\Leftarrow?\Rightarrow$	**CWC** [10]

this issue, Latecki, Eckhardt and Rosenfeld have introduced the notion of *well-composed* sets and gray-level images in [9] for the 2D case, and shortly afterwards, in [8] for the 3D case. Let us recall these seminal definitions.

Definition 1 (2D well-composed sets, 2D EWC). *A 2D digital set $X \in \mathbb{Z}^2$ is weakly well-composed if any 8-component of X is a 4-component. X is well-composed if both X and its complement $\mathbb{Z}^2 \setminus X$ are weakly well-composed.*

Starting from this definition, denoted by 2D EWC in Table 1, the seminal paper [9] shows that it is equivalent to the digital well-composedness, and to the continuous well-composedness. Though, a definition of the 3D well-composedness based on the notion of the "equivalence of connectivities" (3D EWC) does not lead to any interesting topological properties. So, in [8], Latecki introduced the "continuous well-composedness" (CWC) setting:

Definition 2 (3D well-composed sets, 3D CWC). *A 3D digital set $X \in \mathbb{Z}^3$ is well-composed if the boundary of its 3D continuous analog is a 2D manifold.*

As depicted in Table 1, both in 2D and in 3D, the digital well-composedness, the Alexandrov well-composedness, and the continuous well-composedness are equivalent (this property is used in [6]). A major difference between 2D and 3D comes from the fact that, in 3D and in any greater dimension, we lose the property "EWC $\Rightarrow$ CWC". Indeed, the 3D set $\left(\begin{smallmatrix} 1 & 0 \\ 0 & 1 \end{smallmatrix} \middle| \begin{smallmatrix} 1 & 1 \\ 1 & 1 \end{smallmatrix} \right)$ satisfies the "equivalence of connectivities" property—this set is actually both a 26-component and a 6-component—but it is not digitally well-composed.

The continuous well-composedness has been extended in nD by Latecki (Definition 4 in [10]). Yet, to the best of our knowledge, neither Latecki nor any other author have expanded on this definition. One reason might be that it is difficult to handle from a computational point of view. In contrast, the combinatorial notion of n-surface [3] seems highly adapted to model boundaries of subsets of $\mathbb{Z}^n$ (or of any subdivision of this space). Following these ideas, a definition of the Alexandrov well-composedness (AWC) has been proposed in [12].

A third approach, based on some forbidden critical configurations, is called the digital well-composedness (DWC), and is the focus of this paper. For practical purposes such as self-duality, the "well-composedness based on the equivalence of connectivities" (EWC) is the property we are looking for. It is a global property, while DWC is a local one. In this paper, we define the generalization to nD of the digital well-composedness (Section 3) and we show (Section 3.2) that "DWC $\Rightarrow$ EWC in nD". Hence, DWC is a practical way to check/enforce EWC.

Studying whether or not the continuous well-composedness, the digital well-composedness and the Alexandrov well-composedness are equivalent in nD for $n > 3$ is beyond the scope of the paper (in nD, we thus have "AWC $\Leftarrow?\Rightarrow$ CWC" and "DWC $\Leftarrow?\Rightarrow$ AWC" in Table 1).

2.2 2D Well-Composed Functions

Let us denote by $\mathbb{V}$ a finite ordered set of values; $\mathbb{V}$ can be a finite subset of $\mathbb{Z}$ or of $\mathbb{Z}/2$.

Given a 2D image $u : \mathcal{D} \subseteq \mathbb{Z}^2 \to \mathbb{V}$ and $\lambda \in \mathbb{V}$, the *upper and lower threshold sets* of u are respectively defined by $[u \geq \lambda] = \{\, x \in \mathcal{D} \mid u(x) \geq \lambda \,\}$ and $[u \leq \lambda] = \{\, x \in \mathcal{D} \mid u(x) \leq \lambda \,\}$. The *strict* threshold sets are obtained when replacing $\geq$ by $>$ and $\leq$ by $<$.

Definition 3 (2D well-composed functions). *A function u is* well-composed *iff all its upper threshold sets are well-composed.*

Note that relying on *lower* threshold sets (instead of upper ones) leads to an equivalent definition. Note that using strict threshold sets (instead of large ones) also leads to an equivalent definition; indeed, in digital topology with a finite set of values, we have: $\exists \epsilon \in \mathbb{R}$ such that $\forall \lambda$, $[u \geq \lambda] = [u > \lambda - \epsilon]$ and $[u \leq \lambda] = [u < \lambda + \epsilon]$.

A characterization of well-composed 2D images is that, any sub-part of 2×2 pixels of u valued as $\begin{pmatrix} a & b \\ c & d \end{pmatrix}$ shall satisfy $\mathrm{intvl}(a, d) \cap \mathrm{intvl}(b, c) \neq \emptyset$, where $\mathrm{intvl}(v_1, v_2) = [\![\min(v_1, v_2), \max(v_1, v_2)]\!]$.

3 Digital Well-Composedness in nD

This section presents a generalization to nD of the notions of digitally well-composed sets and functions, and an extension of digital well-composedness to interval-valued maps.

3.1 Notations

In the following, we consider sets and functions (typically gray-level images) defined on $(\mathbb{Z}/s)^n$, where $s \in \{1, 2\}$. When $s = 1$ we have the original space $\mathbb{Z}^n$, whereas with $s = 2$ we have a single subdivision of the original space (every coordinates of $z \in \left(\frac{\mathbb{Z}}{2}\right)^n$ are multiple of $\frac{1}{2}$). Practically, the definition domain will always be limited to an hyperrectangle $\mathcal{D} \subset (\mathbb{Z}/s)^n$. Let us denote by $\mathbb{B}_s$ the canonical basis of $(\mathbb{Z}/s)^n$. Given a point $z \in (\mathbb{Z}/s)^n$ and a subset $\mathcal{F} = \{f_1, \ldots, f_k\} \subseteq \mathbb{B}_s$ with $2 \leq k \leq n$, a *block* associated with z and $\mathcal{F}$ is defined as:

$$S(z, \mathcal{F}) = \{\, z + \sum_{i=1}^{k} \lambda_i f_i \mid \lambda_i \in \{0, \tfrac{1}{s}\}, \forall i \in [\![1, k]\!] \,\}.$$

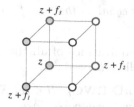

Fig. 1. A block of dimension 3 associated with z contains 8 points of $(\mathbb{Z}/s)^n$. Two blocks of dimension 2 associated with z are also depicted, each containing 4 points; their points are respectively filled in yellow and contoured in red. $(z + f_2, z + f_3)$ and $(z + f_1, z + f_2 + f_3)$ are two pairs of antagonist points, respectively for the red 2D block and for the 3D block.

Remark that a block $S(z, \mathcal{F}) \subset (\mathbb{Z}/s)^n$ actually belongs to a subspace of dimension k. In the following, we will thus say *a block S of dimension k*, meaning that we consider a block $S(z, \mathcal{F})$ such as $\mathrm{card}(\mathcal{F}) = k$ and whatever z. Figure 1 depicts some blocks.

Given a block $S \subset (\mathbb{Z}/s)^n$ of dimension k, and $p, p' \in S$, we say that p and p' are *antagonist in S* iff they maximize the distance L^1 between two points in S. Obviously an antagonist to a given point $p \in S$ exists and is unique; it is denoted by $\mathrm{antag}_S(p)$. We are now able to generalize the definition of critical configurations to any dimension $n \geq 2$.

3.2   Digitally Well-Composed nD Sets

A *primary critical configuration* of dimension k in $(\mathbb{Z}/s)^n$, with $2 \leq k \leq n$, is any set $\{p, \mathrm{antag}_S(p)\}$ with S being a block of dimension k. A *secondary critical configuration* of dimension k in $(\mathbb{Z}/s)^n$ is any set $S \setminus \{p, \mathrm{antag}_S(p)\}$ with S being a block of dimension k.

Definition 4 (nD DWC sets). *A set $X \subseteq (\mathbb{Z}/s)^n$ is digitally well-composed iff, for any $k \in [\![2, n]\!]$, and for any block S of dimension k, $X \cap S$ is neither a primary nor a secondary critical configuration.*

Notice that the definition of digital well-composedness is *self-dual*: any set $X \subseteq (\mathbb{Z}/s)^n$ is digitally well-composed iff $(\mathbb{Z}/s)^n \setminus X$ is digitally well-composed. An image is *a priori* not digitally well-composed; for instance, the classical gray-level "Lena" image contains 38039 critical configurations.

Let us now present a major result about digital well-composedness for nD sets.

Theorem 1 (Existence of a $2n$-path between antagonist points). *A set $X \subseteq (\mathbb{Z}/s)^n$ is digitally well-composed iff, for any block S and for any couple of points $(p, \mathrm{antag}_S(p))$ of $X \cap S$, resp. $S \setminus X$, there exists a $2n$-path between them in $X \cap S$, resp. $S \setminus X$.*

Definition 5 (nD EWC sets). *A set $X \subseteq (\mathbb{Z}/s)^n$ is well-composed based on the equivalence of connectivities (EWC) iff the set of $(3^n - 1)$-components*

of X, resp. of $(\mathbb{Z}/s)^n \setminus X$, is equal to the set of $2n$-components of X, resp. of
$(\mathbb{Z}/s)^n \setminus X$.

As corollary, we have the equivalence of all the classical connectivities on a
cubical grid for a digitally well-composed set:

Corollary 1 (nD DWC $\Rightarrow$ nD EWC). *If a set $X \subseteq (\mathbb{Z}/s)^n$ is digitally well-composed (DWC), then X is well-composed based on the equivalence of connectivities (EWC).*

3.3   Digitally Well-Composed nD Functions

The generalization of digital well-composedness from nD sets to nD functions is
the same as before.

Definition 6 (nD DWC functions). *Given $u : \mathcal{D} \subseteq (\mathbb{Z}/s)^n \to \mathbb{V}$ (a gray-level image), u is digitally well-composed iff all its upper threshold sets are digitally well-composed.*

Following the characterizations of 2D and 3D well-composed gray-level images
given respectively by Latecki in [9] and by the authors in [1], we can now express
a characterization of nD digitally well-composed images. Let us recall that the
span operator is defined on $V \subset \mathbb{V}$ by $\mathrm{span}(V) = [\![\min(V), \max(V)]\!] \in \mathbb{I}_{\mathbb{V}}$,
where $\mathbb{I}_{\mathbb{V}}$ denotes the set of intervals on $\mathbb{V}$.

Property 1 (Characterization of nD DWC functions). *Given a gray-level image $u : \mathcal{D} \subseteq (\mathbb{Z}/s)^n \to \mathbb{V}$, u is digitally well-composed iff, for any block S of dimension k and for any couple of points (p, p') with $p' = \mathrm{antag}_S(p)$, we have:*

$$\mathrm{intvl}(u(p), u(p')) \cap \mathrm{span}\{ u(p'') \mid p'' \in S \setminus \{p, p'\} \} \neq \emptyset.$$

3.4   Digitally Well-Composed nD Interval-Valued Maps

We call *interval-valued map* a map defined on $\mathcal{D} \subseteq (\mathbb{Z}/s)^n \to \mathbb{I}_{\mathbb{V}}$. Given an
interval-valued map U, we define two functions on $\mathcal{D} \to \mathbb{V}$, its lower bound $\lfloor U \rfloor$
and its upper bound $\lceil U \rceil$, such as $\forall z \in \mathcal{D}$, $U(z) = [\![\lfloor U \rfloor(z), \lceil U \rceil(z)]\!]$.

Remark that interval-valued maps are just a particular case of set-valued maps
defined on $\mathcal{D} \to 2^{\mathbb{V}}$ (also denoted by $\mathcal{D} \rightsquigarrow \mathbb{V}$). The threshold sets of set-valued
maps have been defined in [12,5]; let us recall their definitions, and derive a
simple characterization.

$$[U \unrhd \lambda] = \{ z \in \mathcal{D} \mid \exists v \in U(z), v \geq \lambda \}, \text{ and } [U \lhd \lambda] = \mathcal{D} \setminus [U \unrhd \lambda],$$

$$[U \unlhd \lambda] = \{ z \in \mathcal{D} \mid \exists v \in U(z), v \leq \lambda \}, \text{ and } [U \rhd \lambda] = \mathcal{D} \setminus [U \unlhd \lambda], \forall \lambda \in \mathbb{V}.$$

Definition 7 (nD DWC interval-valued maps). *An interval-valued map U is digitally well-composed iff all its threshold sets are digitally well-composed.*

Property 2 (Characterization of nD DWC interval-valued maps). *An nD interval-valued map $U : \mathcal{D} \subseteq (\mathbb{Z}/s)^n \to \mathbb{I}_{\mathbb{V}}$ is digitally well-composed iff both $\lceil U \rceil$ and $\lfloor U \rfloor$ are nD digitally well-composed functions (defined on $\mathcal{D} \to \mathbb{V}$).*

4   A Study of a Front Propagation Algorithm

In this section, we study a front propagation algorithm that takes a major role in transforming any nD function into a digitally well-composed function.

4.1   Origin of the Front Propagation Algorithm

The front propagation algorithm studied in this section is related to the algorithm, proposed in [5], which computes in quasi-linear time the morphological tree of shapes of a nD image. Schematically, the tree of shapes computation algorithm is composed of 4 steps:

$$u \xrightarrow{\;immersion\;} U \xrightarrow{\;sort\;} (u^\flat, \mathcal{R}) \xrightarrow{\;union-find\;} \mathcal{T}(u^\flat) \xrightarrow{\;emersion\;} \mathcal{T}(u).$$

The input is an integer-valued image u, defined on the nD cubical grid. First an immersion step creates an interval-valued map U, defined on a larger space $\mathcal{K}$. A front propagation step, based on a hierarchical queue, takes U and produces two outputs: an image $u^\flat$ and an array $\mathcal{R}$ containing the elements of $\mathcal{K}$. In this array, the elements are sorted so that the next step, an union-find-based tree computation, produces $\mathcal{T}(u^\flat)$ the tree of shapes of $u^\flat$. Actually $u^\flat|_{\mathbb{Z}^n} = u$ and $\mathcal{T}(u^\flat)|_{\mathbb{Z}^n} = \mathcal{T}(u)$. The last step, the emersion, removes from $\mathcal{T}(u^\flat)$ all the elements of $\mathcal{K} \setminus \mathbb{Z}^n$, and also performs a canonicalization of the tree. So $\mathcal{T}(u)$, the tree of shapes of u, is obtained [5].

The front propagation step (highlighted in red in the schematic description) acts as a *flattening* of an interval-valued map U into a function $u^\flat$, because we have $\forall z,\; u^\flat(z) \in U(z)$ [5]. In the following, we will denote by $\mathfrak{FP}$ both the front propagation algorithm (the part highlighted in red above) and the mathematical operator $\mathfrak{FP} : U \mapsto u^\flat$.

Last, let us give two important remarks. **1.** We are going to reuse the front propagation algorithm $\mathfrak{FP}$, yet in a *very different* way than it is used in the tree of shapes computation algorithm (see later in Section 5). Indeed, its input U will be different (both the structure and the values of U will be different), and its purpose also will be different (flattening versus sorting). **2.** Actually, the front propagation algorithm is *just a part* of the solution that we present to make nD functions digitally well-composed.

4.2   Brief Explanation of the Front Propagation Algorithm

Let us now explain shortly the $\mathfrak{FP}$ algorithm, which is recalled in Algorithm 1 (see [5] for the original version). This algorithm uses a classical front propagation on the definition domain of U. This propagation is based on a hierarchical queue, denoted by Q, the current level being denoted by ℓ. There are two notable differences with the well-known hierarchical-queue-based propagation. First the values of U are interval-valued so we have to decide at which (single-valued) level to enqueue the domain points. The solution is to enqueue a point h at the value of the interval $U(h)$ that is the closest to ℓ (see the procedure PRIORITY_PUSH). The image $u^\flat$ actually stores the enqueuing level of the points. Second, when

```
PRIORITY_PUSH(Q, h, U, ℓ)                    𝔉𝔓(U) : Image
/* modifies Q */                             /* computes u♭ */
begin                                        begin
  │ [lower, upper] ← U(h)                      │ for all h do
  │ if lower > ℓ then                          │ │ deja_vu(h) ← false
  │ │ ℓ' ← lower                               │ end
  │ end                                        │ PUSH(Q[ℓ∞], p∞)
  │ else if upper < ℓ then                     │ deja_vu(p∞) ← true
  │ │ ℓ' ← upper                               │ ℓ ← ℓ∞  /* start from root level */
  │ end                                        │ while Q is not empty do
  │ else                                       │ │ h ← PRIORITY_POP(Q, ℓ)
  │ │ ℓ' ← ℓ                                   │ │ u♭(h) ← ℓ
  │ end                                        │ │ for all
  │ PUSH(Q[ℓ'], h)                             │ │ n ∈ 𝒩(h) such as deja_vu(n) =
end                                            │ │ false do
                                               │ │ │ PRIORITY_PUSH(Q, n, U, ℓ)
                                               │ │ │ deja_vu(n) ← true
PRIORITY_POP(Q,  ℓ) : H                        │ │ end
/* modifies Q, and sometimes ℓ */              │ end
begin                                          return u♭
  │ if Q[ℓ] is empty then                    end
  │ │ │ ℓ' ← level next to ℓ such as Q[ℓ']
  │ │ │ is not empty
  │ │ │ ℓ ← ℓ'
  │ end
  │ return POP(Q[ℓ])
end
```

Algorithm 1: Computation of the function $u^♭$ from an interval-valued map U. Left: the routines PRIORITY_PUSH and PRIORITY_POP handle a hierarchical queue. Right: the queue-based front propagation algorithm $\mathfrak{FP}$.

the queue at the current level, $Q[\ell]$, is empty (and when the hierarchical queue Q is not yet empty), we shall decide what is the next current level. We have the choice of taking the next level, either less or greater than ℓ, such that the queue at that level is not empty (see the procedure PRIORITY_POP). Practically, choosing going up or down the levels does not change the resulting image $u^♭$. The neighborhood $\mathcal{N}$ used by the propagation corresponds to the $2n$-connectivity.

Such as in [5], the initialization of the front propagation relies on the definition of a point, p_∞ (first point enqueued), and of a value $\ell_\infty \in U(p_\infty)$, which is the initial value of the current level ℓ. Similarly to the case of the tree of shapes computation, p_∞ is taken in the outer boundary of the definition domain of U. The initial level ℓ_∞ is set to the median value of the points belonging to the inner boundary of the definition domain of U; more precisely, when the interval-valued U is constructed from an integer-valued function u, ℓ_∞ is computed from the values of the inner boundary of u. Using the median operator ensures that ℓ_∞ is set in a self-dual way: schematically $\ell_\infty(-u) = -\ell_\infty(u)$. An example is given later in Section 5.1.

Last, let us mention that a run of the $\mathfrak{FP}$ algorithm on a simple interval-valued map U is illustrated in the extended version of this paper, available on http://hal.archives-ouvertes.fr.

4.3 Properties of the Front Propagation Algorithm

The front propagation algorithm has three main properties: it is deterministic, it is self-dual, and its output is an nD digitally well-composed function if its input is an nD digitally well-composed interval-valued map.

Proposition 1 ($\mathfrak{FP}$ is deterministic). *Once given p_∞ and ℓ_∞, the front propagation algorithm $\mathfrak{FP}$ (Algorithm 1) is deterministic with respect to its input, the nD interval-valued map U.*

Proposition 2 ($\mathfrak{FP}$ is self-dual). *For any nD interval-valued map U, and whatever p_∞ and $\ell_\infty \in U(p_\infty)$ now considered as parameters, we have:*
$$\mathfrak{FP}_{(p_\infty, \ell_\infty)}(U) = -\mathfrak{FP}_{(p_\infty, -\ell_\infty)}(-U), \text{ so } \mathfrak{FP} \text{ is self-dual.}$$

Actually, the front propagation algorithm features some continuity properties due to the fact that the front propagation is spatially coherent, and due to the way the hierarchical queue is handled [5]. Consequently, we have for $\mathfrak{FP}$ the following strong result.

Theorem 2 ($\mathfrak{FP}(U)$ is DWC if U is DWC). *If the nD interval-valued map $U : \mathcal{D} \subset \left(\frac{\mathbb{Z}}{2}\right)^n \to \mathbb{I}_{\mathbb{Z}}$ is digitally well-composed, the resulting nD function $\mathfrak{FP}(U)$ is digitally well-composed.*

5 Making an nD Function Digitally Well-Composed

In this section we present a method to make any nD integer-valued function u (typically a gray-level image) digitally well-composed. This method is composed of two steps; the first one is an interpolation of u that gives an interval-valued map U_{DWC}, and the second one is the flattening $\mathfrak{FP}$ that gives the resulting single-valued function u_{DWC}:

$$u : \mathcal{D} \subset \mathbb{Z}^n \to \mathbb{Z} \xrightarrow{\text{interpolation}} U_{\text{DWC}} : \mathcal{D}_2 \subset \left(\frac{\mathbb{Z}}{2}\right)^n \to \mathbb{I}_{\mathbb{Z}} \xrightarrow{\text{flattening}} u_{\text{DWC}} : \mathcal{D}_2 \to \frac{\mathbb{Z}}{2}.$$

We are looking for an interpolation method that turns any function u into a digitally well-composed map U_{DWC}, so that eventually $u_{\text{DWC}} = \mathfrak{FP}(U_{\text{DWC}})$ is a digitally well-composed function (thanks to Theorem 2).

5.1 Interpolation

Let us consider a function (gray-level image) $u : \mathcal{D} \subset \mathbb{Z}^n \to \mathbb{V}$. We subdivide the space $\mathbb{Z}^n$ into $(\mathbb{Z}/2)^n$, and define a new map on $(\mathbb{Z}/2)^n$. The new definition domain is $\mathcal{D}_2 \subset (\mathbb{Z}/2)^n$ where $\mathcal{D}_2$ is the smallest hyperrectangle such as $\mathcal{D} \subset \mathcal{D}_2$. A sensible property of this new map is to be equal to u on $\mathcal{D}$. The values of this new map over $\mathcal{D}_2 \setminus \mathcal{D}$ are obtained by *locally* interpolating the values of u. With

{8}	{8}	{8}	{8}	{8}	{8}	{8}
{8}	{9}	$[\![9,11]\!]$	{11}	$[\![11,15]\!]$	{15}	{8}
{8}	$[\![7,9]\!]$	$[\![1,11]\!]$	$[\![1,11]\!]$	$[\![1,15]\!]$	$[\![13,15]\!]$	{8}
{8}	{7}	$[\![1,7]\!]$	{1}	$[\![1,13]\!]$	{13}	{8}
{8}	$[\![3,7]\!]$	$[\![1,7]\!]$	$[\![1,5]\!]$	$[\![3,13]\!]$	$[\![3,13]\!]$	{8}
{8}	{3}	$[\![3,5]\!]$	{5}	$[\![3,5]\!]$	{3}	{8}
{8}	{8}	{8}	{8}	{8}	{8}	{8}

9	11	15
7	1	13
3	5	3

(a) u **(b)** $\mathcal{I}_{\mathrm{span}}(u)$

Fig. 2. (a): A simple 2D integer-valued function. (b): Its span-based interpolation (the central part of 5×5 points of $(\mathbb{Z}/2)^n$ with values in $\mathbb{I}_{\mathbb{Z}}$); the external border (in gray) is required when passing $\mathcal{I}_{\mathrm{span}}(u)$ to the $\mathfrak{FP}$ algorithm (see text).

$B = \{-\frac{1}{2}, 0, \frac{1}{2}\}^n$, B_z the translation of B by z, and "op" an operator on subsets of $\mathbb{V}$, we can define the interpolating map:

$$\forall z \in \mathcal{D}_2, \quad (\mathcal{I}_{\mathrm{op}}(u))(z) = \begin{cases} \mathrm{op}\{u(z)\} & \text{if } z \in \mathcal{D}, \\ \mathrm{op}\{u(z'), \ z' \in B_z \cap \mathcal{D}\} & \text{otherwise.} \end{cases}$$

The following proposition, which could also be derived from [11], follows easily.

Proposition 3 ($\mathcal{I}_{\min}$ and $\mathcal{I}_{\max}$ give dual DWC functions). *For any $u : \mathcal{D} \subset \mathbb{Z}^n \to \mathbb{Z}$, the nD integer-valued functions $\mathcal{I}_{\min}(u)$ and $\mathcal{I}_{\max}(u)$ are digitally well-composed, and the interpolation operators $\mathcal{I}_{\min}$ and $\mathcal{I}_{\max}$ are dual (they verify $\forall u$, $\mathcal{I}_{\min}(u) = -\mathcal{I}_{\max}(-u)$).*

Since we have $\forall V \subset \mathbb{V}$, $\mathrm{span}(V) = [\![\min(V), \max(V)]\!]$, whatever an nD function u, the interval-valued map $\mathcal{I}_{\mathrm{span}}(u)$ is such as $\lfloor \mathcal{I}_{\mathrm{span}}(u) \rfloor = \mathcal{I}_{\min}(u)$ and $\lceil \mathcal{I}_{\mathrm{span}}(u) \rceil = \mathcal{I}_{\max}(u)$. Since these two functions are digitally well-composed, the interval-valued map $\mathcal{I}_{\mathrm{span}}(u)$ is digitally well-composed (thanks to Property 2). So we have:

Proposition 4 ($\mathcal{I}_{\mathrm{span}}$ is self-dual and gives DWC maps). *For any $u : \mathcal{D} \subset \mathbb{Z}^n \to \mathbb{Z}$, the nD interval-valued function $\mathcal{I}_{\mathrm{span}}(u) : \mathcal{D}_2 \subset (\mathbb{Z}/2)^n \to \mathbb{I}_{\mathbb{Z}}$ is digitally well-composed, and the interpolation operator $\mathcal{I}_{\mathrm{span}}$ is self-dual (it verifies $\forall u$, $\mathcal{I}_{\mathrm{span}}(u) = -\mathcal{I}_{\mathrm{span}}(-u)$).*

An example of the span-based interpolation is depicted in Figure 2. The outer/external boundary of the definition domain $\mathcal{D}_2$ of $\mathcal{I}_{\mathrm{span}}(u)$ is displayed in gray. This boundary is filled with a single value $\ell_\infty(u)$, which is actually the median value of the set of values of the inner/internal boundary of the definition domain of u (see Section 4.2). We have: $\ell_\infty(u) = \mathrm{med}\{3, 3, 5, 7, 9, 11, 13, 15\} = 8$. When we take $U_{\mathrm{DWC}} = \mathcal{I}_{\mathrm{span}}(u)$ as input to the $\mathfrak{FP}$ algorithm, p_∞ can be any point of the outer boundary. This way, which is similar to [5], we ensure that the propagation starts with the external boundary of U_{DWC}, and that all the points of the internal boundary are enqueued. Having $\ell_\infty(-u) = -\ell_\infty(u)$ guarantees that "$\mathcal{I}_{\mathrm{span}}$ with an outer boundary added" remains self-dual w.r.t. u.

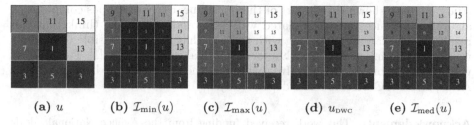

(a) u **(b)** $\mathcal{I}_{\min}(u)$ **(c)** $\mathcal{I}_{\max}(u)$ **(d)** u_{DWC} **(e)** $\mathcal{I}_{\mathrm{med}}(u)$

Fig. 3. Given an integer-valued function u, depicted in (a), we have the dual interpolations depicted in (b) and (c), and the self-dual digitally well-composed interpolation $u_{\mathrm{DWC}} = (\mathfrak{FP} \circ \mathcal{I}_{\mathrm{span}})(u)$ depicted in (d). Remark that $\mathcal{I}_{\mathrm{span}}(u)$ is depicted in Figure 2b. The rightmost sub-figure (e) depicts a *local* self-dual interpolation based on the median operator, $\mathcal{I}_{\mathrm{med}}(u)$; this interpolation is digitally well-composed in 2D but not in nD with $n > 2$ [1]. That contrasts with $\mathfrak{FP} \circ \mathcal{I}_{\mathrm{span}}$, which is a *non-local* interpolation being digitally well-composed in any dimension.

5.2   Considering $\mathfrak{FP} \circ \mathcal{I}_{\mathrm{span}}$ as a Solution

Now, let us consider $u_{\mathrm{DWC}} = (\mathfrak{FP} \circ \mathcal{I}_{\mathrm{span}})(u)$.

Proposition 5 ($\mathfrak{FP} \circ \mathcal{I}_{\mathrm{span}}$ is an nD self-dual DWC interpolation). *Given any nD integer-valued function (gray-level image) $u : \mathcal{D} \subset \mathbb{Z}^n \longrightarrow \mathbb{Z}$, the nD function $(\mathfrak{FP} \circ \mathcal{I}_{\mathrm{span}})(u) : \mathcal{D}_2 \subset (\mathbb{Z}/2)^n \longrightarrow \mathbb{Z}/2$ is a self-dual interpolation of u which is digitally well-composed.*

u_{DWC} **is digitally well-composed.** We know that $\mathcal{I}_{\mathrm{span}}(u)$ is a digitally well-composed map (see Proposition 4), and that $\mathfrak{FP}$ transforms such a map into a digitally well-composed function (see Theorem 2). Thus u_{DWC} is a digitally well-composed function.

u_{DWC} **is an interpolation of u.** Since the interpolation $\mathcal{I}_{\mathrm{span}}(u)$ satisfies $\forall z \in \mathcal{D}, \mathcal{I}_{\mathrm{span}}(z) = \{u(z)\}$, and since $\mathfrak{FP} : U \to u^b$ is such that $\forall z \in \mathcal{D}_2, u^b(z) \in U(z)$, we can deduce that $u_{\mathrm{DWC}|_{\mathcal{D}}} = u$. In addition, the values of u_{DWC} in $\mathcal{D}_2 \setminus \mathcal{D}$ are set "in-between" the ones of u because they belong to their span. Thus u_{DWC} is effectively an interpolation of u.

$-u_{\mathrm{DWC}}$ **is obtained from** $-u$. Both $\mathcal{I}_{\mathrm{span}}$ and $\mathfrak{FP}$ are self-dual (see respectively Propositions 4 and 2), so we have $(\mathfrak{FP} \circ \mathcal{I}_{\mathrm{span}})(-u) = -(\mathfrak{FP} \circ \mathcal{I}_{\mathrm{span}})(u)$. The transform $\mathfrak{FP} \circ \mathcal{I}_{\mathrm{span}}$ is therefore self-dual.

As a conclusion, $\mathfrak{FP} \circ \mathcal{I}_{\mathrm{span}}$ is an interpolation method that turns any nD integer-valued function into a DWC function in a self-dual way. Remark that it is a *non-local* interpolation method because the interpolated values are set by the propagation of a front in $\mathfrak{FP}$. An illustration is given by Figure 3.

6   Conclusion

In this paper, we studied several possible extension of the well-composedness concept. In particular, in the framework of digital topology, we prove that digital well-composedness implies the equivalence of connectivities. Based on this study, we have shown that a part of the quasi-linear tree-of-shapes computation algorithm produces an interpolated well-composed image. We think that this result is remarkable, contrasting with the fact that it is not possible to obtain a well-composed image in nD with a local self-dual interpolation [1]. Future work will build on this framework. In particular, the relationships between DWC, AWC, and CWC deserve an in-depth study.

Acknowledgments. This work received funding from the Agence Nationale de la Recherche, contract ANR-2010-BLAN-0205-03 and through "Programme d'Investissements d'Avenir" (LabEx BEZOUT n°ANR-10-LABX-58). We would also like to warmly thank Michel Couprie for enlightening discussions and the anonymous reviewers for their very insightful comments which have helped us to improve the paper.

References

1. Boutry, N., Géraud, T., Najman, L.: On making nD images well-composed by a self-dual local interpolation. In: Barcucci, E., Frosini, A., Rinaldi, S. (eds.) DGCI 2014. LNCS, vol. 8668, pp. 320–331. Springer, Heidelberg (2014)
2. Boutry, N., Géraud, T., Najman, L.: How to make nD images well-composed without interpolation (March 2015), http://hal.archives-ouvertes.fr/hal-01134166
3. Evako, A.V., Kopperman, R., Mukhin, Y.V.: Dimensional properties of graphs and digital spaces. Journal of Mathematical Imaging and Vision 6(2-3), 109–119 (1996)
4. Géraud, T., Carlinet, E., Crozet, S.: Self-duality and digital topology: Links between the morphological tree of shapes and well-composed gray-level images. In: Benediktsson, J.A., Chanussot, J., Najman, L., Talbot, H. (eds.) ISMM 2015. LNCS, vol. 9082, Springer, Heidelberg (2015)
5. Géraud, T., Carlinet, E., Crozet, S., Najman, L.: A quasi-linear algorithm to compute the tree of shapes of n-D images. In: Hendriks, C.L.L., Borgefors, G., Strand, R. (eds.) ISMM 2013. LNCS, vol. 7883, pp. 98–110. Springer, Heidelberg (2013)
6. Gonzalez-Diaz, R., Jimenez, M.-J., Medrano, B.: 3D well-composed polyhedral complexes. Discrete Applied Mathematics 183, 59–77 (2015)
7. Kong, T.Y., Rosenfeld, A.: Digital topology: Introduction and survey. Computer Vision, Graphics, and Image Processing 48(3), 357–393 (1989)
8. Latecki, L.: 3D well-composed pictures. Graphical Models and Image Processing 59(3), 164–172 (1997)
9. Latecki, L., Eckhardt, U., Rosenfeld, A.: Well-composed sets. Computer Vision and Image Understanding 61(1), 70–83 (1995)
10. Latecki, L.J.: Well-Composed Sets. In: Advances in Imaging and Electron Physics, vol. 112, pp. 95–163. Academic Press (2000)
11. Mazo, L., Passat, N., Couprie, M., Ronse, C.: Digital imaging: A unified topological framework. Journal of Mathematical Imaging and Vision 44(1), 19–37 (2012)
12. Najman, L., Géraud, T.: Discrete set-valued continuity and interpolation. In: Hendriks, C.L.L., Borgefors, G., Strand, R. (eds.) ISMM 2013. LNCS, vol. 7883, pp. 37–48. Springer, Heidelberg (2013)

Self-duality and Digital Topology: Links Between the Morphological Tree of Shapes and Well-Composed Gray-Level Images

Thierry Géraud[1(✉)], Edwin Carlinet[1,2], and Sébastien Crozet[1]

[1] EPITA Research and Development Laboratory (LRDE), Paris, France
{thierry.géraud,edwin.carlinet,sébastien.crozet}@lrde.epita.fr
[2] LIGM, Équipe A3SI, ESIEE, Université Paris-Est, Paris, France

Abstract. In digital topology, the use of a pair of connectivities is required to avoid topological paradoxes. In mathematical morphology, self-dual operators and methods also rely on such a pair of connectivities. There are several major issues: self-duality is impure, the image graph structure depends on the image values, it impacts the way small objects and texture are processed, and so on. A sub-class of images defined on the cubical grid, *well-composed* images, has been proposed, where all connectivities are equivalent, thus avoiding many topological problems. In this paper we unveil the link existing between the notion of well-composed images and the morphological tree of shapes. We prove that a well-composed image has a well-defined tree of shapes. We also prove that the only self-dual well-composed interpolation of a 2D image is obtained by the median operator. What follows from our results is that we can have a purely self-dual representation of images, and consequently, purely self-dual operators.

Keywords: Self-dual operators · Tree of shapes · Vertex-valued graph · Well-composed gray-level images · Digital topology

1 Introduction

Having a contrast invariant representation for images is of prime importance in computer vision. Indeed we often have to deal with illumination changes or parts of images being very poorly contrasted [6,5]. Some morphological contrast invariant trees have been successfully used for computer vision tasks; see *e.g.* [23].

In this paper we consider the settings of "digital topology" a-la Rosenfeld and Kong [12]. An image is considered as a vertex-valued graph: the underlying structure is a graph, and an image is a function associating values to vertices [18]. In the following, we only deal with images defined on the nD cubical grid, so on the square grid for the 2D case. The objective of this paper is to get a self-dual representation of gray-level images that is free of topological issues. Precisely, we want to guarantee some strong topological properties, to ensure a "pure" self-duality, and to be able consequently to process gray-level images easily and without trouble. Our work has two main motivations. First, we expect to obtain

© Springer International Publishing Switzerland 2015
J.A. Benediktsson et al. (Eds.): ISMM 2015, LNCS 9082, pp. 573–584, 2015.
DOI: 10.1007/978-3-319-18720-4_48

Fig. 1. In the image (a), the woman can be considered both as foreground (b) or as background (c)

a definition of a discrete tree of shapes that is theoretically sound. Indeed, since this morphological tree-based image representation is self-dual, it is subject to some topological problems. Second, considering a couple of connectivities has a deep impact when dealing with small objects and textures. The cornerstone of our proposal is to handle only one connectedness relationship *i.e.*, a unique topological structure. Put differently, we will consider that an image is *one* graph structure whose vertices are valued.

The contributions presented in this paper are the following. We show that the tree of shapes is "not purely" self-dual. We prove that, if a gray-level image is well-composed, then its tree of shapes is well defined. We also prove that, under some very reasonable assumptions, the only self-dual well-composed subdivision of a 2D image is obtained by the median operator. Last we propose a purely self-dual tree of shapes for gray-level images.

First Section 2 recalls the theoretical background of our work. Section 3 explains what we are looking for, and Section 4 gives the solution we propose. Section 5 is dedicated to related works. Last we conclude and give some perspectives in Section 6.

2   Theoretical Background

2.1   Digital Topology and Self-duality

A self-dual operator processes the same way the image contents whatever the contrast (*i.e.*, bright objects over dark background versus dark objects over bright background). That is often desirable when we *cannot* make an assumption about contrast, and/or when we *do not want* to make such an assumption because the notion of "object v. background" is not the appropriate one. Actually we often prefer the notion of "subject" (and its related context); this can be explained because the notions of foreground and background are highly contextual.

An illustration of this statement is depicted in Figure 1, where the colors green and red designate respectively the object (foreground) and the background. In the gray-level image, Fig. 1(a), if we take the woman as subject, then we obtain the representation of Fig. 1(b), and the image outer part is the background. Yet, if we take the baby as subject, then we obtain a different interpretation of the image; in Fig. 1(c), the woman is now the background.

Actually, if we take for granted that every part of the image can be a "subject", then we want a unique representation of the image: we do not want a different behavior based on the subject contrast (bright over dark, or the contrary). We thus want to process images in a self-dual way:

$$u \xrightarrow{\text{processing}} \varphi(u)$$

complementation $\downarrow$ $\qquad\qquad$ $\downarrow$ complementation

$$\complement u \xrightarrow{\text{processing}} \varphi(\complement u) \; = \; \complement \varphi(u)$$

In digital topology in the case of images defined on a regular cubical grid, a "Jordan pair" of connectivities (c_α, c_β) are required [12]: one for the object (foreground), and the other one for the background. Practically, in the 2D case for instance, the choice is (c_4, c_8) or (c_8, c_4); in nD, it can be (c_{2n}, c_{3^n-1}) or (c_{3^n-1}, c_{2n}). The use of the complementation in a self-dual operator thus forces to switch from one connectivity to the other one. That contrasts with the assertion of "processing the same way the image contents whatever the contrast..."; actually we should add "...except for their connectivity".

2.2 The Morphological Tree of Shapes

The tree of shapes has been defined in [17], even if its origin comes from [13] in the 2D case. We just briefly recall here its definition. Given an nD image $u : \mathbb{Z}^n \to \mathbb{Z}$, the lower level sets are defined as $[u < \lambda] = \{x \in X \mid u(x) < \lambda\}$, and the upper level sets as $[u \geq \lambda] = \{x \in X \mid u(x) \geq \lambda\}$. Considering the connected components of these sets[1], one can define a couple of dual trees, namely the min-tree $\mathcal{T}_<(u) = \{\Gamma \in \mathcal{CC}([u < \lambda])\}_\lambda$, and the max-tree $\mathcal{T}_\geq(u) = \{\Gamma \in \mathcal{CC}([u \geq \lambda])\}_\lambda$. The min-tree and the max-tree verify $\mathcal{T}_\geq(\complement u) = \mathcal{T}_<(u)$. They are said to be "dual" trees since the operators defined from them are dual.

With the cavity-fill-in operator, denoted by Sat, we have two new sets of components; they are the lower shapes $\mathcal{S}_<(u) = \{\mathrm{Sat}(\Gamma); \; \Gamma \in \mathcal{T}_<(u)\}$, and the upper shapes $\mathcal{S}_\geq(u) = \{\mathrm{Sat}(\Gamma); \; \Gamma \in \mathcal{T}_\geq(u)\}$. The tree of shapes is then defined as:

$$\mathfrak{S}(u) \; = \; \mathcal{S}_<(u) \cup \mathcal{S}_\geq(u)$$

and it (almost[2]) features:

$$\mathfrak{S}(\complement u) \; = \; \mathfrak{S}(u). \tag{1}$$

Such a tree is called *self-dual* since many self-dual operators can be derived from this tree [7,22].

Last let us recall that a quasi-linear algorithm exists to compute the tree of shapes, that has the property of also working in the nD case [10]. A parallel version of this algorithm is available [9].

---

[1] $\mathcal{CC}$ is an operator that takes a set and gives its set of connected components.

[2] We will see in Section 3 that the property $\mathfrak{S}(\complement u) = \mathfrak{S}(u)$ is not strictly correct.

2.3   Well-Composed Sets and Images

Let us start by recalling some seminal definitions from the paper [15] by Latecki, Eckhardt, and Rosenfeld.

A 2D set is *weakly well-composed* if any 8-component of this set is a 4-component. A set is *well-composed* if both this set and its complement are weakly well-composed. A very easy characterization of well-composed sets is based on the notion of "critical configurations"; a set is well-composed if the configurations ▧ and ▨ do not appear.

The notion of well-composedness has been extended from sets to functions, *i.e.*, gray-level images. A gray-level image u is well-composed if any set $[u \geq \lambda]$ is well-composed. A straightforward characterization of well-composed gray-level images is that every block $\begin{array}{|c|c|} \hline a & d \\ \hline c & b \\ \hline \end{array}$ should verify: $\text{intvl}(a,b) \cap \text{intvl}(c,d) \neq \emptyset$, where $\text{intvl}(v,w) = [\![\min(v,w), \max(v,w)]\!]$.

An image is not *a priori* well-composed; for instance, the classical "lena" image contains 38039 critical configurations. Two approaches exist to get a well-composed image from a primary image: changing its pixel values (yet it can alter the topology of its contents), or getting a well-composed interpolation. Below we give an example of an image (left), which is not well-composed, but whose interpolation (right) is well-composed:

3	2
1	8

3	2	2
2	2	5
1	4	8

The notion of well-composedness has also been defined for 3D sets and images [14], and it has recently been extended to nD [2,4]. Let us now recap those last results.

Given a point $z \in \mathbb{Z}^n$ and a subset $\mathcal{F} = \{f_1, \ldots, f_k\}$ of the canonical basis of $\mathbb{Z}^n$, with $k \in [\![2, n]\!]$, a block $S(z, \mathcal{F})$ associated with z and $\mathcal{F}$ is defined as: $S(z, \mathcal{F}) = \{z + \sum_{i=1}^{k} \lambda_i f_i \mid \lambda_i \in \{0,1\}, \forall i \in [\![1, k]\!]\}$. Just remark that a block $S(z, \mathcal{F}) \subset \mathbb{Z}^n$ actually belongs to a subspace of dimension k. In the following, we will thus say *a block S of dimension k*, meaning that we consider a block $S(z, \mathcal{F})$ such as $\text{card}(\mathcal{F}) = k$ and whatever z; see Figure 2 for some illustrations.

Given a block $S \subseteq \mathbb{Z}^n$ of dimension k, and $p, p' \in S$, we say that p and p' are *antagonist in S* iff they maximize the distance L^1 between two points in S. Obviously an antagonist to a given point $p \in S$ exists and is unique; it is denoted by $\text{antag}_S(p)$. A primary critical configuration of dimension k in $\mathbb{Z}^n$ is any set $\{p, \text{antag}_S(p)\}$ with S being a block of dimension k. A secondary critical configuration of dimension k in $\mathbb{Z}^n$ is any set $S \setminus \{p, \text{antag}_S(p)\}$ with S being a block of dimension k.

A set $X \subseteq \mathbb{Z}^n$ is *well-composed* iff, for any block S of dimension k, $X \cap S$ is neither a primary nor a secundary critical configuration. Just notice that the definition of well-composedness is *self-dual*: any set $X \subseteq \mathbb{Z}^n$ is well-composed iff $\mathbb{Z}^n \setminus X$ is well-composed.

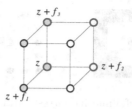

Fig. 2. A block of dimension 3 associated with z contains 8 points of $\mathbb{Z}^n$. Two blocks of dimension 2 associated with z are also depicted, each containing 4 points; their points are respectively filled in yellow and contoured in red.

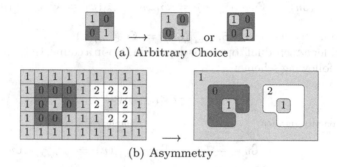

Fig. 3. Some abnormalities for a "self-dual" representation

3   A Quest for Self-duality

3.1   About the Impure Self-duality

Let us consider a connectivity c (in 2D it can be either c_4 or c_8). Let us denote by $-c$ its "dual" connectivity (the dual of c_4 is c_8, and conversely). Let us denote by $\mathcal{R}$ a relation such as $<$, $\leq$, $>$, or $\geq$. $\mathcal{R}^{-1}$ denotes the inverse relation (for instance $<^{-1}$ is $>$). $\neg\mathcal{R}$ denotes the relation corresponding to its negation (for instance $\neg <$ is $\geq$ since $\neg(a < b) \Leftrightarrow (a \geq b)$). From a set of components:

$$\mathcal{T}_{(\mathcal{R},c)}(u) = \{\, \Gamma \in \mathcal{CC}_c([u \,\mathcal{R}\, \lambda]) \,\}_\lambda,$$

we can get a set of shapes:

$$\mathcal{S}_{(\mathcal{R},c)}(u) = \{\, \mathrm{Sat}_{-c}(\Gamma); \ \Gamma \in \mathcal{T}_{(\mathcal{R},c)}(u) \,\}.$$

Note that, from a component Γ obtained with the connectivity c, the cavity-fill operator relies on the connectivity $-c$ to be topologically sound. Then just remark that the cavities of the components of $\mathcal{T}_{(\mathcal{R},c)}(u)$ are some shapes belonging to $\mathcal{S}_{(\neg\mathcal{R},-c)}(u)$. So we can define the tree of shapes by:

$$\mathfrak{S}_{(\mathcal{R},c)}(u) = \mathcal{S}_{(\mathcal{R},c)}(u) \,\cup\, \mathcal{S}_{(\neg\mathcal{R},-c)}(u).$$

Note that, in the discrete case, there is no difference between the sets $\mathcal{T}_{(\mathcal{R},c)}(u)$ and $\mathcal{T}_{(\neg\mathcal{R}^{-1},c)}(u)$ (*e.g.*, the same threshold sets are obtained using $<$ or $\leq$). We

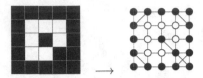

Fig. 4. A set (left) and its corresponding graph (right)

can finally observe that the tree of shapes is **not purely** self-dual:

$$\mathfrak{S}_{(\mathcal{R},\,c)}(\complement u) = \mathfrak{S}_{(\mathcal{R}^{-1},\,c)}(u) = \mathfrak{S}_{(\neg\mathcal{R}^{-1},\,-c)}(u) = \mathfrak{S}_{(\mathcal{R},\,-c)}(u).$$

For instance in 2D, we have: $\mathfrak{S}_{(<,\,c_4)}(\complement u) = \mathfrak{S}_{(<,\,c_8)}(u)$.

Now consider a "self-dual" operator depending upon a connectivity pair (c_α, c_β); we have the following scheme:

$$
\begin{array}{ccc}
u & \xrightarrow{\ \varphi\ } & \varphi_{(c_\alpha, c_\beta)}(u) \\[2pt]
\text{complementation}\ \Big\downarrow & & \Big\downarrow\ \text{complementation} \\[6pt]
\complement u & \xrightarrow{\ \psi\ } & \complement\varphi_{(c_\alpha, c_\beta)}(u) = \varphi_{(c_\beta, c_\alpha)}(\complement u)
\end{array}
$$

so precisely $\varphi_{(c_\alpha, c_\beta)}$ and $\varphi_{(c_\beta, c_\alpha)}$ are *dual*. We cannot formally say that φ is self-dual because we do not have $\complement\varphi_{(c_\alpha, c_\beta)} = \varphi_{(c_\alpha, c_\beta)}\complement$.

In both cases of the tree of shapes and of a self-dual operator, the underlying representation of an image depends upon the arbitrary choice[3] of taking either (c_α, c_β) or (c_β, c_α); such a choice is illustrated by Fig. 3(a). Consequently it yields an asymmetry in dealing with components, which is illustrated by Fig. 3(b). In those cases, we will say that self-duality is not pure.

3.2   Images as Vertex-Valued Graphs

If we consider an image as a vertex-valued graph, the use of a connectivity pair is *a priori* mandatory to avoid the connectivity paradoxes of digital topology. Figure 4 depicts a binary image (a set), and its corresponding graph, where the connectivities (c_4, c_8) have been chosen to represent respectively the foreground (white) and the background (black).

On the other hand, we want a purely self-dual tree, *i.e.*, a tree that strictly verifies $\mathfrak{S}(\complement u) = \mathfrak{S}(u)$. That starts with a first requirement: we shall have the same connectivity relation, say c, for both lower and upper shapes. Remind now that the lower (resp. upper) shapes come from the cavity-fill operator applied on some components of the upper (resp. lower) threshold sets. It leads to the conclusion that a unique connectivity shall be used everywhere (precisely to define the components of all threshold sets, and to define the cavity-fill operator

---

[3] Note that the classical workaround to ensure pure self-duality and avoid topological problems using 6-connectivity in 2D is out of scope, since it required a choice between 4 possible transforms (shifting either rows or columns, and either odd or even ones).

for both lower and upper components). The definitions of the previous section shall be rewritten as follows. From components we get shapes:

$$T_{(\mathcal{R},c)}(u) = \{\, \Gamma \in \mathcal{CC}_c([u \, \mathcal{R} \, \lambda]) \,\}_\lambda \; \longrightarrow \; \mathcal{S}_{(\mathcal{R},c)}(u) = \{\, \mathrm{Sat}_c(\Gamma); \; \Gamma \in T_{(\mathcal{R},c)}(u) \,\}$$

and we take the union of lower and upper shapes:

$$\mathfrak{S}_{(\mathcal{R},\,c)}(u) = \mathcal{S}_{(\mathcal{R},c)}(u) \, \cup \, \mathcal{S}_{(\mathcal{R},c)}(\complement u). \tag{2}$$

Finally, the components of $\mathfrak{S}_{(\mathcal{R},\,c)}(u)$, endowed with inclusion, do **not** form a tree, but a poset.

As a conclusion, if we consider that an image is a function, we cannot separate the structure of the definition domain from the values of the image. Put differently, it means that *the graph structure depends on the image values.* That is a major issue, since it is a very strong restriction.

3.3   Rationale

The rationale behind our proposal is the following. Given any gray-level nD image u, $(\mathfrak{S}_{(\mathcal{R},\,c)}(u), \subset)$ forms a poset. Taking $c = c_\alpha$ or $c = c_\beta$ is equivalent when an image is well-composed. We can define a self-dual interpolation $\mathfrak{I}(u)$ of u that is a well-composed image. So we can expect $\mathfrak{S}_{(\mathcal{R},\,c)}(\mathfrak{I}(u))$ to be a purely self-dual tree of shapes. To remain consistent with both digital topology and mathematical morphology, we will impose some reasonable constraints on $\mathfrak{I}$.

4   Getting a Purely Self-dual Tree of Shapes

4.1   Well-Composed $\Rightarrow$ Tree of Shapes

Actually there is a link between the notion of well-composed images and the definition of the tree of shapes:

Theorem. If a gray-level nD image u is well-composed, then the components of $\mathfrak{S}_{(\mathcal{R},\,c_{2n})}(u)$ form a (purely) self-dual tree of shapes.

Let us first remark that it is an implication, not an equivalence (being well-composed is a sufficient condition to get a tree of shapes; it is not a necessary one). Indeed, with:

$$u \;=\; \begin{array}{|c|c|c|c|}\hline 1 & 1 & 1 & 1 \\\hline 1 & 0 & 2 & 1 \\\hline 1 & 2 & 0 & 1 \\\hline 1 & 1 & 1 & 1 \\\hline\end{array}$$

$\mathfrak{S}_{(<,\,c_{2n})}(u)$ is a tree, while u is not well-composed.

Proof. A strong recent result, presented in [4], is the following: "if a set $X \subseteq \mathbb{Z}^n$ is well-composed, then its $2n$-components are identical to its $(3^n - 1)$-components". That implies that:

$$\mathfrak{S}_{(\mathcal{R},\,c_{2n})}(\complement u) = \mathfrak{S}_{(\mathcal{R},\,c_{3^n-1})}(u) = \mathfrak{S}_{(\mathcal{R},\,c_{2n})}(u),$$

which means that u and $\complement u$ have the same set of shapes. Consequently this set is (said) *self-dual*, since purely self-dual operators can be defined from it.

Let us show now that the elements of $\mathfrak{S}_{(\mathcal{R},\,c_{2n})}(u)$ form a tree. The proof that the shapes obtained with $(c_{2n},\,c_{3^n-1})$ form a tree can be found in [1,8]. Since the connectivities c_{2n} and c_{3^n-1} are equivalent for a well-composed image, this proof applies to our case.

$\square$

4.2   A Self-Dual 2D Morphological Interpolation

We are now going to study how to interpolate an image u defined on $\mathbb{Z}^n$ into an image $\mathfrak{I}(u)$ defined on $(\mathbb{Z}/2)^n$. The requirements over the interpolation $\mathfrak{I}$ are the following ones. **1.** It shall commute with the classical geometric reflections and rotations, *i.e.*, $\mathfrak{I} \circ T = T \circ \mathfrak{I}$, with T such a transform. **2.** $\mathfrak{I}(u)$ shall be considered as a rasterization equivalent to u (cf. [16,21]); in particular, it shall not create some new extremum. **3.** It shall be self-dual, *i.e.*, $\mathfrak{I}(\complement u) = \complement \mathfrak{I}(u)$. **4.** We shall ensure that the shapes are invariant to contrast changes (a classical morphological axiom): $\mathfrak{S}(g(\mathfrak{I}(u))) = \mathfrak{S}(\mathfrak{I}(g(u)))$, where g is a strictly increasing function. **5.** Last, the interpolation function, used to set values between the original pixels, shall actually be an operator. Note that we will also use the notation $\mathfrak{I}$ for this operator.

Consider a 3×3 pixels piece of $\mathfrak{I}(u)$ and a threshold set X. We will depict respectively by ◦ an element of X, by • an element of $\complement X$, and by ○ an element for which we do not know if it is in X or not. It yields 4 cases (modulo symmetries, rotations, and complementation). Using only the "no new extremum" constraint, we have:

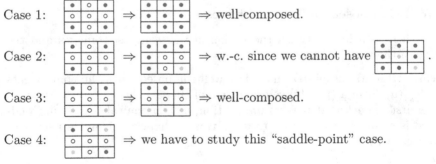

Case 1: $\Rightarrow$ well-composed.

Case 2: $\Rightarrow$ w.-c. since we cannot have .

Case 3: $\Rightarrow$ well-composed.

Case 4: $\Rightarrow$ we have to study this "saddle-point" case.

Let us assume that the pixel values are $a < b < c < d$, and that the interpolation of this piece of image is:

a	ad	d
ac	$abcd$	bd
c	bc	b

where we shorten a notation such as $\mathfrak{I}\{v, w\}$ into vw, and $\mathfrak{I}\{a, b, c, d\}$ into $abcd$. Now, just remark that the "no new extremum" and "self-dual" constraints imply that $a < d \Rightarrow a < ad < d$, and we have the similar ordering using $\{a, c\}$, $\{c, b\}$, and $\{b, d\}$. Last, this same constraint also implies that $ac <$

$abcd < bd$ and $ad < abcd < bc$. We can then deduce the Hasse diagram of the set of values, depicted on the left below:

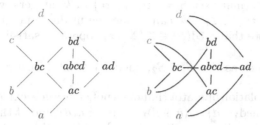

When only drawing between these values the 4-adjacencies of their pixels, we obtain the diagram depicted above on the right side. For instance, the point whose value is b is 4-adjacent to the points whose values are bc and bd. Remark that we have maintain the locations of values of the Hasse diagram, which allows us for keeping reasoning on value ordering.

Assume that the point of value ac is in X; so it is depicted in green below. Since we have $ac < abcd < bd$ the points with values $abcd$ and bd are in X, so they are depicted in green. We end up with the image piece, depicted below, which is well-composed (whatever its unknown part):

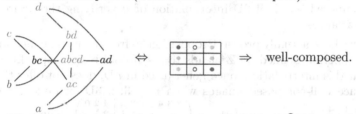

The assumption that the point of value bd is in $\complement X$ is the dual assumption of the previous one (with finally $abcd$ and ac in $\complement X$); it also leads to a well-composed image piece.

So there is only one remaining case, $bd \in X$ and $ac \in \complement X$, depicted below. For this image piece to be well-composed, we can see that the points depicted in blue ($\bullet$), corresponding to the values $abcd$ and bc, have to be in the same set (either X or $\complement X$). Finally we shall have $abcd = bc$:

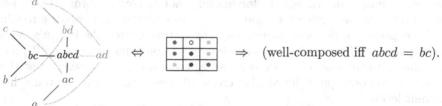

Being in the morphological framework, we want an operator so the result can be rewritten as $\mathrm{op}(\{a,b,c,d\}) = \mathrm{op}(\{b,c\})$. It should be true whatever the values, so the operator is a median.

Let us recall that we have assumed that, in the piece of image considered, the pixel values are all different ($a < b < c < d$). If we have now $a \leq b \leq c \leq d$, we actually get the same conclusions as before. Indeed, just remark that having some values equal means that the Hasse diagram is simplified, yet not really modified.

Let us consider a multi-set $S = \{z_1, \ldots, z_k\}$ of k integers, with k even, such as $\forall i \in [\![1, k-1]\!]$, $z_i \leq z_{i+1}$. We can define the median operator by $\mathrm{med}(S) = (z_{\frac{k}{2}} + z_{\frac{k}{2}+1})/2$; note that $\mathrm{med}(S) \in \mathbb{Z}/2$. This operator satisfies the property:

$$\forall\, S_1 \text{ and } S_2 \text{ such as } S = S_1 \cup S_2, \ \mathrm{med}(S) \in \mathrm{intvl}(\mathrm{med}(S_1), \mathrm{med}(S_2)),$$

so med is an interpolation operator that does not create new extremum. Indeed, we have for instance $\mathrm{med}\{a, d\} = (a+d)/2 \in \mathrm{intvl}(a, d)$, and at the center of an image piece: $\mathrm{med}\{a, b, c, d\} \in \mathrm{intvl}(\mathrm{med}\{a, c\}, \mathrm{med}\{b, d\})$, and $\mathrm{med}\{a, b, c, d\} \in \mathrm{intvl}(\mathrm{med}\{a, d\}, \mathrm{med}\{b, c\})$. In addition, the operator med is self-dual: $\mathrm{med}(S) = -\mathrm{med}\{-z_k, \ldots, -z_1\}$.

In 2D, we thus have a med-based interpolation operator, $\mathfrak{I}_{\mathrm{med}}$, that transforms an image $u : \mathbb{Z}^2 \to \mathbb{Z}$ into $\mathfrak{I}_{\mathrm{med}}(u) : (\mathbb{Z}/2)^2 \to \mathbb{Z}/2$. Formally, with $B = \{-\frac{1}{2}, 0, \frac{1}{2}\}^2$, and with B_z the translation of B by z, we have:

$$\forall\, z \in (\mathbb{Z}/2)^2, \ [\mathfrak{I}_{\mathrm{med}}(u)](z) = \begin{cases} u(z) & \text{if } z \in \mathbb{Z}^2, \\ \mathrm{med}\{\, u(z'), \ z' \in B_z \cap \mathbb{Z}^2\,\} & \text{otherwise,} \end{cases}$$

which is a well-composed self-dual 2D interpolation of u verifying the desired properties and invariances.

In this section, we have actually proven Proposition 25 from [19]: "The median interpolation of a function defined on $\mathbb{Z}^2$ leads to a self-dual plain map." Last, let us mention that this interpolation, once generalized in nD, does not offer the guaranty to produce well-composed images when $n \geq 3$; a 3D example given in [3] is: $u = \begin{pmatrix} 2\ 4 & 4\ 0 \\ 4\ 0 & 0\ 2 \end{pmatrix} \longrightarrow \mathfrak{I}_{\mathrm{med}}(u) = \begin{pmatrix} 2\ 3\ 4 & 3\ 3\ 2 & 4\ 2\ 0 \\ 3\ 3\ 2 & 3\ 2\ 1 & 2\ 1\ 1 \\ 4\ 2\ 0 & 2\ 1\ 1 & 0\ 1\ 2 \end{pmatrix}$, where a 2D critical configuration is depicted in italics .

5   Related Works

In [20], Ray and Acton give a proof that an inclusion tree exists when considering the connectivity pair (c_8, c_8). Unfortunately, not having a Jordan pair yields some results that are difficult to understand, since they do not conform to what can be expected in the continuous case. For instance, with the image given at the beginning of Section 4.1, the two central components (containing 2 points each, and with the respective levels 0 and 2) cross using c_8; it means that their respective contours (level lines) also cross, although these two contours have different levels.

In [1,8], the proof that the shapes obtained with (c_{2n}, c_{3^n-1}) form a tree relies on changing a discrete function $u : \mathbb{Z}^n \to \mathbb{R}$ into an upper semi-continuous function $u_{\mathrm{max}} : \mathbb{R}^n \to \mathbb{R}$. Actually, this latter function can be related to the interpolation $\mathfrak{I}_{\mathrm{max}}(u)$, obtained such as $\mathfrak{I}_{\mathrm{med}}(u)$ while replacing med by max. It is easy to see that $\mathfrak{I}_{\mathrm{max}}(u)$ is a well-composed discrete interpolation of u, which is not self-dual. Its dual interpolation, also well-composed, is $\mathfrak{I}_{\mathrm{min}}(u)$.

In [19], Najman and Géraud have proposed an interval-valued interpolation of nD images, based on the Khalimsky grid. They have given an alternative definition of well-composed nD images: the components of the boundaries of all threshold sets are discrete $(n - 1)$-manifolds. Yet, this definition is different from the one we use in the present paper (see Section 2.3) based on the notion of critical configurations [2]. To our knowledge, the equivalence (or non-equivalence) between both definitions has not been proven yet.

Several authors have proposed some methods to "repair" a 3D set so that the result is well-composed; a bibliography can be found in [11]. In these methods the notions of critical configurations are involved but, unlike us, they do not consider the simple setting of "digital topology".

In [4], Boutry et al. have presented a self-dual nD interpolation that is well-composed, whatever the dimension n. The major difference between their result and what we have proposed in Section 4.2 is that they do not impose the interpolation function to be local.

6 Conclusion

In this paper, we have shown that we need well-composed images to get (a really pure) self-duality, and we have proven that a median-based interpolation is the solution to get 2D well-composed images. A major consequence is that an nD image defined on the cubical grid can really be seen as a function valuing the vertices of a graph (otherwise the graph depends upon the image values). In particular, that means that we can have two different functions (set of values) defined on the same graph, since the domain structure can be truly uncorrelated from the valuation. The drawbacks of our proposal is that we need to subdivide the domain (yet it does not change any operator / method complexity because it is just a multiplicative factor, and RAM is cheap). The perspectives of this work are rather straightforward: finding if we can generalize the properties of well-composed sets and images to any graph (not only cubical grids), and evaluate the advantages of our 2D proposal versus an interpolation dedicated to 6-connectivity, which is intrinsically topology paradox free.

Acknowledgment. The authors want to thank Laurent W. Najman for his constant support and good advice.

References

1. Ballester, C., Caselles, V., Monasse, P.: The tree of shapes of an image. ESAIM: Control, Optimisation and Calculus of Variations 9, 1–18 (2003)
2. Boutry, N., Géraud, T., Najman, L.: A generalization of well-composedness to dimension n. Communication at Journée du Groupe de Travail de Géometrie Discrète (GT GeoDis, Reims Image 2014) (November 2014) (in French)
3. Boutry, N., Géraud, T., Najman, L.: On making nD images well-composed by a self-dual local interpolation. In: Barcucci, E., Frosini, A., Rinaldi, S. (eds.) DGCI 2014. LNCS, vol. 8668, pp. 320–331. Springer, Heidelberg (2014)

4. Boutry, N., Géraud, T., Najman, L.: How to make nD images well-composed in a self-dual way. In: Benediktsson, J.A., Chanussot, J., Najman, L., Talbot, H. (eds.) ISMM 2015. LNCS, vol. 9082, Springer, Heidelberg (2015)
5. Cao, F., Lisani, J.-L., Morel, J.-M., Musé, P., Sur, F.: A Theory of Shape Identification. Lecture Notes in Mathematics, vol. 1948. Springer (2008)
6. Caselles, V., Coll, B., Morel, J.M.: Topographic maps and local contrast changes in natural images. International Journal of Computer Vision 33(1), 5–27 (1999)
7. Caselles, V., Monasse, P.: Grain filters. Journal of Mathematical Imaging and Vision 17(3), 249–270 (2002)
8. Caselles, V., Monasse, P.: Geometric Description of Images as Topographic Maps. Lecture Notes in Mathematics, vol. 1984. Springer (2009)
9. Crozet, S., Géraud, T.: A first parallel algorithm to compute the morphological tree of shapes of nD images. In: Proceedings of the 21st International Conference on Image Processing (ICIP), Paris, France, pp. 2933–2937 (2014)
10. Géraud, T., Carlinet, E., Crozet, S., Najman, L.: A quasi-linear algorithm to compute the tree of shapes of n-D images. In: Hendriks, C.L.L., Borgefors, G., Strand, R. (eds.) ISMM 2013. LNCS, vol. 7883, pp. 98–110. Springer, Heidelberg (2013)
11. Gonzalez-Diaz, R., Jimenez, M.J., Medrano, B.: 3D well-composed polyhedral complexes. Discrete Applied Mathematics 183, 59–77 (March 2015), special Issue on Discrete Geometry for Computer Imagery
12. Kong, T.Y., Rosenfeld, A.: Digital topology: Introduction and survey. Computer Vision, Graphics, and Image Processing 48(3), 357–393 (1989)
13. Kronrod, A.: On functions of two variables. Uspehi Mathematical Sciences 5, 24–134 (1950) (in Russian)
14. Latecki, L.: 3D well-composed pictures. Graphical Models and Image Processing 59(3), 164–172 (1997)
15. Latecki, L., Eckhardt, U., Rosenfeld, A.: Well-composed sets. Computer Vision and Image Understanding 61(1), 70–83 (1995)
16. Latecki, L.J., Conrad, C., Gross, A.: Preserving topology by a digitization process. Journal of Mathematical Imaging and Vision 8(2), 131–159 (1998)
17. Monasse, P., Guichard, F.: Fast computation of a contrast-invariant image representation. IEEE Transactions on Image Processing 9(5), 860–872 (2000)
18. Najman, L., Cousty, J.: A graph-based mathematical morphology reader. Pattern Recognition Letters 47, 3–17 (2014)
19. Najman, L., Géraud, T.: Discrete set-valued continuity and interpolation. In: Hendriks, C.L.L., Borgefors, G., Strand, R. (eds.) ISMM 2013. LNCS, vol. 7883, pp. 37–48. Springer, Heidelberg (2013)
20. Ray, N., Acton, S.T.: Inclusion filters: A class of self-dual connected operators. IEEE Transactions on Image Processing 14(11), 1736–1746 (2005)
21. Tustison, N., Avants, B., Siqueira, M., Gee, J.: Topological well-composedness and glamorous glue: A digital gluing algorithm for topologically constrained front propagation. IEEE Transactions on Image Processing 20(6), 1756–1761 (2011)
22. Xu, Y., Géraud, T., Najman, L.: Morphological filtering in shape spaces: Applications using tree-based image representations. In: Proceedings of the International Conference on Pattern Recognition (ICPR), pp. 485–488. IAPR, Tsukuba Science City (2012)
23. Xu, Y., Monasse, P., Géraud, T., Najman, L.: Tree-based morse regions: A topological approach to local feature detection. IEEE Transactions on Image Processing 23(12), 5612–5625 (2014)

A Combinatorial 4-Coordinate System
for the Diamond Grid

Lidija Čomić[1]($\boxtimes$) and Benedek Nagy[2,3]

[1] Department of Fundamental Sciences, Faculty of Technical Sciences,
University of Novi Sad, Novi Sad, Serbia
`comic@uns.ac.rs`
[2] Department of Mathematics, Faculty of Arts and Sciences,
Eastern Mediterranean University, Mersin-10, Famagusta, North Cyprus, Turkey
[3] Department of Computer Science, Faculty of Informatics, University of Debrecen,
Debrecen, Hungary
`nbenedek.inf@gmail.com`

Abstract. A new combinatorial coordinate system for cells in the diamond grid is presented, and some of its properties are detailed. Four dependent coordinates are used to address the voxels (triakis truncated tetrahedra), their faces (hexagons and triangles), their edges and the points at their corners. The incidence (boundary and co-boundary) and adjacency relations of the cells can easily be captured by these coordinate values. Therefore, the new coordinate system can effectively by applied in morphological and topological operations.

Keywords: Diamond grid · Combinatorial coordinate system · Topological operations · Abstract cell complexes · Non-traditional grids

1 Introduction

Discrete spaces are usually represented by a discrete set of grid points. The induced tessellation of the underlying continuous space into an abstract cell complex is defined by the corresponding Voronoi cells associated with the grid points. Adjacency relation between the maximal grid cells in the tessellation reflects the neighborhood relation between grid points.

There are three regular tessellations of $\mathbb{R}^2$ and only one of $\mathbb{R}^3$. However, there are other tessellations known from crystallography, each with its advantages and drawbacks depending on the application. The most often used grids in image analysis and digital geometry in 2D and 3D are the square and cubic grid, respectively. The cells of all dimensions in the induced tessellation are easily represented through the Cartesian coordinate system [8].

Alternative regular grids in $\mathbb{R}^2$ are the (dual) hexagonal and triangular grids. A variety of different 2D (such as offset, trapezoidal, spiraling) and 3D coordinate schemes have been proposed for such grids [10], that address only the grid points, i.e., only the 2-cells in the induced tessellation of $\mathbb{R}^2$.

Alternative grids to the cubic grid in $\mathbb{R}^3$ are face centered and body centered cubic grids (that generalize hexagonal grid), and diamond grid (that generalizes

© Springer International Publishing Switzerland 2015
J.A. Benediktsson et al. (Eds.): ISMM 2015, LNCS 9082, pp. 585–596, 2015.
DOI: 10.1007/978-3-319-18720-4_49

triangular grid) [14,16]. Both 3- [18] and 4- [4,11] coordinate systems have been proposed for points in these grids, i.e., for 3-cells in the induced tessellations.

Many notions from image processing and discrete geometry, such as neighborhood sequences and distance transform, have been defined and investigated not only for square and cubic, but also for non-traditional 2D and 3D grids [19]. The implementation and ease of use of these notions depends on the chosen coordinate system.

For other applications, such as boundary tracking [5,7], watershed transform [2], thinning based on collapse [6], or morphological filters [9], the ability to access lower-dimensional cells is crucial. The pairing of these cells plays a critical role in Forman theory [3], which receives increasing interest as a versatile topological analysis tool. The ability to directly access all cells of a complex allows to avoid some topological paradoxes induced by discretization [7].

Combinatorial coordinate systems [8] have been developed to address the cells of all dimensions in a complex. Such systems for triangular and hexagonal plane tessellations have been proposed in [12,13]. We believe that combinatorial coordinate systems would be beneficial also for non-traditional tessellations in 3D. Here, as a first step in this direction, we present a combinatorial 4-coordinate system for the diamond grid.

The structure of the paper is as follows. In the next section we recall the symmetric 4-coordinate system [15,18,17] for the diamond grid, that defines the coordinate values of the 3-cells (voxels) of the grid. Then, in the following sections we extend this system to address the lower-dimensional cells of the grid, in decreasing order of dimension. We also give some possible applications of the defined coordinate system. Finally, we summarize our results and give some concluding remarks.

2   The Diamond Grid

The diamond grid is the grid of carbon atoms in the diamond crystals. Every carbon atom is connected to four neighbor atoms by covalent bonds, and these neighbors form the vertices of a tetrahedron, the simplest regular three-dimensional object. This structure makes the diamond the hardest material in the world.

The voxel (3-cell) of the tessellation induced by the diamond grid is a triakis truncated tetrahedron. Such a voxel has four regular hexagonal faces and twelve obtuse-angled isosceles triangular faces, and a number of edges and vertices (see Figure 1). The set of tetrahedra in the diamond grid can be partitioned into two subsets of different orientation, that we call up- and down-tetrahedra. Tetrahedra in each subset can be transformed into each other by using translation only, while the function that maps up-tetrahedra to down-tetrahedra is a mirroring.

There are only four closest neighbors through a hexagonal face (in comparison with six in the cubic grid), which can be beneficial in situations with a limited number of moving directions. Other two naturally defined neighbors, through a triangular face or a vertex, can be defined and used depending on application.

The diamond grid has the structure of an abstract cell complex, with naturally defined boundary and dimension. Abstract cell complexes have been used in [8,7] as a basis for topological analysis of digital images. An abstract cell complex C is defined as a set E of elements, called cells, together with a binary bounding relation B, and dimension function dim. Relation B is a partial order. Function dim assigns to each element in E a nonnegative integer, called dimension, such that $dim(e') < dim(e'')$ for all pairs $(e', e'') \in B$. Abstract cell complexes allow for the definition of classical notions from topology, such as open and closed sets, connectedness and neighborhood, and they enable the definition of incidence and adjacency relations, which are the basis for topological analysis of images and complexes.

Many data structures for representing cell complexes have been proposed in the literature. An overview can be found in [1].

3   Preliminaries: A 4-Coordinate Description of the Diamond Grid

Each tetrahedron (3-cell) in the diamond grid can be determined by four dependent coordinates (p, q, r, s) (in the direction of covalent bonds between carbon atoms), where $p, q, r, s \in \mathbb{Z}$, and $p + q + r + s$ is equal to 0 (for up tetrahedra) or 1 (for down tetrahedra) [15,14,16]. These are called even and odd points of the grid in [18,17]. For up-tetrahedra, "spikes" (truncated parts) are in the direction of negative axes. For down tetrahedra, spikes are in the direction of positive axes.

3.1   Tetrahedra Adjacencies: 1- and 2-Neighborhood of Voxels

Two tetrahedra are hex-adjacent (they are 1-neighbors) if they share a common hexagonal face. Each up-tetrahedron (p, q, r, s) is hex-adjacent to four down tetrahedra $(p+1, q, r, s)$, $(p, q+1, r, s)$, $(p, q, r+1, s)$ and $(p, q, r, s+1)$. Dually, each down-tetrahedron (p, q, r, s) is hex-adjacent to four up-tetrahedra $(p-1, q, r, s)$, $(p, q-1, r, s)$, $(p, q, r-1, s)$ and $(p, q, r, s-1)$.

Two tetrahedra are tri-adjacent (they are 2-neighbors) if they share a triangular face. Two tri-adjacent tetrahedra are of the same type: they are both up- or both down-tetrahedra.

Each up-tetrahedron (p, q, r, s) is tri-adjacent to 12 tetrahedra $(p-1, q+1, r, s)$ $(p-1, q, r+1, s)$, $(p-1, q, r, s+1)$ (in the direction of negative p axis), $(p+1, q-1, r, s)$, $(p, q-1, r+1, s)$, $(p, q-1, r, s+1)$ (negative q axis), $(p+1, q, r-1, s)$, $(p, q+1, r-1, s)$, $(p, q, r-1, s+1)$ (negative r axis), $(p+1, q, r, s-1)$, $(p, q+1, r, s-1)$, $(p, q, r+1, s-1)$ (negative s axis).

Each down-tetrahedron (p, q, r, s) is tri-adjacent to 12 tetrahedra $(p+1, q-1, r, s)$ $(p+1, q, r-1, s)$, $(p+1, q, r, s-1)$ (in the direction of positive p axis), $(p-1, q+1, r, s)$, $(p, q+1, r-1, s)$, $(p, q+1, r, s-1)$ (positive q axis), $(p-1, q, r+1, s)$, $(p, q-1, r+1, s)$, $(p, q, r+1, s-1)$ (positive r axis), $(p-1, q, r, s+1)$, $(p, q-1, r, s+1)$, $(p, q, r-1, s+1)$ (positive s axis).

Two hex-adjacent tetrahedra share three coordinate values, and they differ by 1 or -1 in one coordinate. Two tri-adjacent tetrahedra share two coordinate values and they differ by 1 and -1 in other two coordinates. The terms 1- and 2-neighbors reflect this fact [15].

4   Combinatorial Coordinates for Voxels

For the sake of convenience, to have only integer coordinate values for lower-dimensional cells, we make a rescaling. Thus, in a new coordinate system, an up-tetrahedron has coordinates (p, q, r, s), $p, q, r, s \in 6\mathbb{Z}$, $p + q + r + s = 0$. Hex-adjacent down-tetrahedra are obtained from (p, q, r, s) by increasing one of the coordinates by 6. Tri-adjacent up-tetrahedra are obtained by keeping two coordinates, and increasing one and decreasing the other of the remaining coordinates by 6.

For a down tetrahedron, $p + q + r + s = 6$. Adjacent tetrahedra can be obtained in a similar way.

In the next sections we show how this coordinate system can be extended to address lower dimensional cells.

5   Combinatorial Coordinates for Faces

Each tetrahedron is incident to four hexagonal faces (hex-faces) and to 12 triangular faces (tri-faces).

5.1   Hex-Faces

A hex-face is shared by an up-tetrahedron (p_1, q_1, r_1, s_1), $p_1 + q_1 + r_1 + s_1 = 0$, and a hex-adjacent down-tetrahedron (p_2, q_2, r_2, s_2), $p_2 + q_2 + r_2 + s_2 = 6$. We assign to such hex-face the average of the coordinates of the incident tetrahedra: $(\frac{p_1+p_2}{2}, \frac{q_1+q_2}{2}, \frac{r_1+r_2}{2}, \frac{s_1+s_2}{2})$. The sum of coordinates of a hex-face is 3. Three of its coordinates are $\equiv 0 \pmod 6$, and one is $\equiv 3 \pmod 6$. Figure 1 illustrates (in red) the coordinates of hex-faces incident to up-tetrahedron $(0, 0, 0, 0)$. For example, the common hex-face for up-tetrahedron $(0, 0, 0, 0)$ and hex-adjacent down-tetrahedron $(6, 0, 0, 0)$ is $(3, 0, 0, 0)$.

Tetrahedron - Hex-Face Incidences. The two tetrahedra incident to a hex-face (p, q, r, s), $p + q + r + s = 3$, are obtained by increasing and decreasing by 3 the coordinate of the hex-face that is $\equiv 3 \pmod 6$. Four hex-faces incident to an up-tetrahedron (p, q, r, s), $p + q + r + s = 0$, are $(p + 3, q, r, s)$, $(p, q + 3, r, s)$, $(p, q, r + 3, s)$ and $(p, q, r, s + 3)$. Four hex-faces incident to a down-tetrahedron (p, q, r, s), $p + q + r + s = 6$, are $(p - 3, q, r, s)$, $(p, q - 3, r, s)$, $(p, q, r - 3, s)$ and $(p, q, r, s - 3)$.

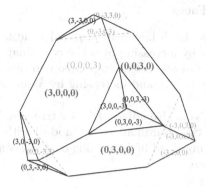

Fig. 1. Combinatorial coordinates for faces incident to up-tetrahedron $(0,0,0,0)$: hex-faces in red and tri-faces in black

5.2   Tri-faces

A tri-face is shared by two tri-adjacent up- or down-tetrahedra. We call such faces up-tri-faces and down-tri-faces, respectively. Let $(p_1, q_1, r_1, s_1), p_1+q_1+r_1+s_1 = 0$, and $(p_2, q_2, r_2, s_2), p_2 + q_2 + r_2 + s_2 = 0$, be two tri-adjacent up-tetrahedra. We assign to their common up-tri-face average coordinates $(\frac{p_1+p_2}{2}, \frac{q_1+q_2}{2}, \frac{r_1+r_2}{2}, \frac{s_1+s_2}{2})$. The sum of coordinates of an up-tri-face is 0. Figure 1 illustrates (in black) the coordinates of tri-faces incident to up-tetrahedron $(0,0,0,0)$.

If a tri-face is shared by two down-tetrahedra, we again take the arithmetic mean of their coordinates and assign them to the down-tri-face. The sum of coordinates of a down-tri-face is 6.

Two of the coordinates of a tri-face are $\equiv 0$, and two are $\equiv 3 \pmod 6$.

Tetrahedron - Tri-face Incidences. The two tetrahedra incident to a tri-face (p, q, r, s), $p + q + r + s = 0$ or 6, are obtained by increasing by 3 one of the coordinates that is $\equiv 3 \pmod 6$, and decreasing the other by 3.

The twelve up-tri-faces incident to an up-tetrahedron (p, q, r, s) are $(p-3, q+3, r, s)$ $(p-3, q, r+3, s)$, $(p-3, q, r, s+3)$ (these three tri-faces are opposite to hex-face $(p+3, q, r, s)$, i.e., in the direction of negative p axis), $(p+3, q-3, r, s)$, $(p, q-3, r+3, s)$, $(p, q-3, r, s+3)$ (in the direction of negative q axis), $(p+3, q, r-3, s)$, $(p, q+3, r-3, s)$, $(p, q, r-3, s+3)$ (direction of negative r axis), $(p+3, q, r, s-3)$, $(p, q+3, r, s-3)$, $(p, q, r+3, s-3)$ (direction of negative s axis).

The twelve down-tri-faces incident to a down-tetrahedron (p, q, r, s) are $(p+3, q-3, r, s)$ $(p+3, q, r-3, r)$, $(p+3, q, r, s-3)$ (in the direction of positive p axis, opposite to hex-face $(p-3, q, r, s)$), $(p-3, q+3, r, s)$, $(p, q+3, r-3, s)$, $(p, q+3, r, s-3)$ (in the direction of positive q axis), $(p-3, q, r+3, s)$, $(p, q-3, r+3, s)$, $(p, q, r+3, s-3)$ (direction of positive r axis), $(p-3, q, r, s+3)$, $(p, q-3, r, s+3)$, $(p, q, r-3, s+3)$ (direction of positive s axis).

5.3 Adjacencies of Faces

Hex-face Adjacencies. Each hex-face (p, q, r, s) is adjacent to six other hex-faces. They are obtained by increasing by 3 the coordinate that is $\equiv 3 \pmod 6$, and decreasing by 3 one of the remaining three coordinates, or by decreasing by 3 the coordinate that is $\equiv 3$, and increasing by 3 one of the remaining three coordinates.

Each hex-face (p, q, r, s) is also adjacent to six tri-faces. They share one coordinate $\equiv 3 \pmod 6$ and two coordinates $\equiv 0 \pmod 6$ with the hex-face, and the remaining coordinate is obtained by increasing or decreasing by 3 the remaining coordinate of the hex-face.

Tri-Face Adjacencies. Each tri-face is adjacent to two hex-faces, and to four tri-faces. The two hex-faces adjacent to an up-tri-face (p, q, r, s), $p+q+r+s = 0$ are obtained by increasing by 3 one of the coordinates that is $\equiv 3 \pmod 6$ and leaving the other three coordinates unchanged. The two hex-faces adjacent to a down-tri-face (p, q, r, s), $p + q + r + s = 6$ are obtained by decreasing by 3 one of the coordinates that is $\equiv 3 \pmod 6$ and leaving the other three coordinates unchanged.

The four tri-faces adjacent to up-tri-face (p, q, r, s) are found in the following way: Up-tri-face (p, q, r, s), $p + q + r + s = 0$, is the common face of two tri-adjacent up-tetrahedra $T_1 = (p_1, q_1, r_1, s_1)$ and $T_2(p_2, q_2, r_2, s_2)$, such that $p_1 + q_1 + r_1 + s_1 = p_2 + q_2 + r_2 + s_2 = 0$, the tetrahedra have coordinates that are $\equiv 0 \pmod 6$ and they differ by 6 in one coordinate and by -6 in another coordinate, while the other two coordinates are shared by the two tetrahedra. Let i be the coordinate in which T_1 differs by 6 from T_2, and let j be the coordinate in which T_1 differs from T_2 by -6. This means that tri-face (p, q, r, s) is on T_2 in the direction of negative i axis, and it is on T_1 in the direction of negative j axis.

We label by T_3 and T_4 the two up-tetrahedra tri-adjacent to both T_1 and T_2, that are tri-adjacent to T_1 in the direction of negative j axis and are tri-adjacent to T_2 in the direction of negative i axis. Their i coordinate is equal to the i coordinate of T_2, and their j coordinate equal to the j coordinate of T_1. Tri-faces between T_1 and T_3, T_1 and T_4, T_2 and T_3 and T_2 and T_4 are the four tri-faces adjacent to tri-face (p, q, r, s) (which is between T_1 and T_2).

For example, up-tri-face $(3, 0, -3, 0)$ is shared by two up-tetrahedra $T_1 = (0, 0, 0, 0)$ and $T_2 = (6, 0, -6, 0)$. It is in the negative direction of r axis for T_1, and in the negative direction of p axis for T_2. In this case, T_3 and T_4 are $(0, 6, -6, 0)$ and $(0, 0, -6, 6)$. The four up-tri-faces adjacent to up-tri-face $(3, 0, -3, 0)$ are $(0, 3, -3, 0)$, $(0, 0, -3, 3)$, $(3, 3, -6, 0)$, $(3, 0, -6, 3)$.

The argument for down-tri-faces is similar.

6 Combinatorial Coordinates for Edges

We distinguish between two types of edges: hex-edges are incident to two hex-faces and one tri-face and tri-edges are incident to three tri-faces.

6.1   Hex-Edges

A hex-edge is incident to two hex-faces, which both belong to one tetrahedron $T = (p, q, r, s)$, and each of them belongs to a hex-adjacent tetrahedron. If T is an up-tetrahedron, then T_1 and T_2 are down-tetrahedra. We will call such hex-edge an up-hex-edge. Tetrahedra T_1 and T_2 both share three coordinates with T, and one of the coordinates is greater by 6 than the corresponding coordinate of T. The two hex-faces are obtained from T by increasing one of the coordinates by 3, and leaving the other three coordinates unchanged. The (down-)tri-face, which is incident to the hex-edge, and which is adjacent to the two hex-faces, is the down-tri-face shared by T_1 and T_2; it has two coordinates $\equiv 0$ (mod 6), which are the same as the common coordinates $\equiv 0$ of the two hex-faces (and of T, T_1 and T_2) and the other two coordinates are the coordinates of T increased by 3. We take for coordinates of the up-hex-edge the average of the coordinates of the two incident hex-faces and the down-tri-face.

This is equivalent to taking the average of the three tetrahedra T, T_1 and T_2 that define the hex-edge. An up-hex-edge has two coordinates $\equiv 0$ (mod 6), and other two are $\equiv 2$ (mod 6). The sum of coordinates of an up-hex-edge is 4. Figure 2 illustrates (in blue and purle) the coordinates of hex-edges incident to up-tetrahedron $(0, 0, 0, 0)$.

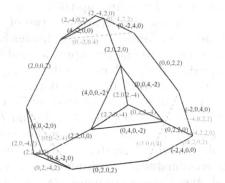

Fig. 2.   Combinatorial coordinates for edges incident to up-tetrahedron $(0, 0, 0, 0)$: hex-edges in blue/purple and tri-edges in green

For example, if $T_1 = (p + 6, q, r, s)$ and $T_2 = (p, q + 6, r, s)$, the two hex-faces have coordinates $(p + 3, q, r, s)$ and $(p, q + 3, r, s)$. The down-tri-face, which is shared by mutually tri-adjacent tetrahedra $(p + 6, q, r, s)$ and $(p, q + 6, r, s)$ has coordinates $(p + 3, q + 3, r, s)$. We assign coordinates $(p + 2, q + 2, r, s)$ to the common up-hex-edge, which is equal to the arithmetic mean of the coordinates of the incident faces, i.e., of the incident tetrahedra.

If a hex-edge is incident to two hex-faces, which both belong to a down-tetrahedron $T = (p, q, r, s)$, and each of them belongs to a hex-incident up-tetrahedron T_1 or T_2, we assign to such down-hex-edge the average of coordinates of T, T_1 and T_2. A down-hex-edge has two coordinates $\equiv 0$, and other two are $\equiv 4 \pmod 6$. The sum of coordinates of a down-hex-edge is 2.

For example, the six hex-edges incident to a hex-face shared by up-tetrahedron (p, q, r, s), $p + q + r + s = 0$, and an incident down tetrahedron e.g. $(p + 6, q, r, s)$ are $(p + 2, q \pm 2, r, s)$, $(p + 2, q, r \pm 2, s)$, $(p + 2, q, r, s \pm 2)$.

6.2   Tri-edges

A tri-edge is incident to and determined by three tri-faces, i.e., by three tetrahedra that are pairwise tri-adjacent. Such tetrahedra are all up- or all down-tetrahedra, and we call the tri-edges up- and down-tri-edges, respectively.

Let $T_1 = (p_1, q_1, r_1, s_1)$, $T_2 = (p_2, q_2, r_2, s_2)$ and $T_3 = (p_3, q_3, r_3, s_3)$ be three up-tetrahedra, each two of which are tri-adjacent. Thus, the sum of coordinates of all three tetrahedra is 0, all coordinates are $\equiv 0 \pmod 6$, and for each two tetrahedra the difference in coordinates forms the multi-set $\{0, 0, -6, 6\}$. This implies that there is one coordinate that is equal for all three tetrahedra.

We assign the average coordinates $\left(\frac{p_1+p_2+p_3}{3}, \frac{q_1+q_2+q_3}{3}, \frac{r_1+r_2+r_3}{3}, \frac{s_1+s_2+s_3}{3}\right)$ to the up-tri-edge shared by T_1, T_2 and T_3. The sum of the coordinates is equal to 0, one coordinate is $\equiv 0$, other three are $\equiv 2 \pmod 6$. Figure 2 illustrates (in green) the coordinates of tri-edges incident to up-tetrahedron $(0, 0, 0, 0)$.

Alternatively, the same coordinate values for an up-tri-edge can be obtained by taking the arithmetic mean of the three up-tri-faces incident to it. The three up-tri-faces have one common coordinate and any two of the up-tri-faces have another common coordinate. The sum of the coordinates for each up-tri-face is equal to 0, two coordinates are $\equiv 0$ and two are $\equiv 3 \pmod 6$.

For example, the up-tri-edge shared by tetrahedra $T_1 = (0, 0, 0, 0)$, $T_2 = (6, 0, -6, 0)$ and $T_3 = (0, 0, -6, 6)$ is $(2, 0, -4, 2)$. It can also be obtained as the average of three up-tri faces $(3, 0, -3, 0)$ (shared by T_1 and T_2), $(0, 0, -3, 3)$ (shared by T_1 and T_3) and $(3, 0, -6, 3)$ (shared by T_2 and T_3).

Coordinates for a down-tri-edge are defined analogously, as the arithmetic mean of the three incident down-tetrahedra, or of the three incident down-tri-faces. The sum of the coordinates of a down-tri-edge is equal to 6, one of its coordinates is $\equiv 0$, and other three coordinates are $\equiv 4 \pmod 6$.

For example, three pairwise tri-adjacent down-tetrahedra $T_1 = (6, 0, 0, 0)$, $T_2 = (12, -6, 0, 0)$ and $T_3 = (12, 0, -6, 0)$ define three down-tri-faces $(9, -3, 0, 0)$, $(9, 0, -3, 0)$ and $(12, -3, -3, 0)$, and the down-tri-edge $(10, -2, -2, 0)$.

6.3   Tri-face – Edges Incidence

An up-tri-face has two coordinates $\equiv 0$, two are $\equiv 3 \pmod 6$, and the sum of coordinates is 0. It is incident to three edges: two up-tri-edges and one hex-edge. The two incident up-tri-edges are obtained by keeping one of the coordinates that is $\equiv 0$, adding 2 to the remaining coordinate that is $\equiv 0$, and subtracting 1 from the remaining two coordinates. The hex-edge is obtained by keeping the coordinates of the up-tri-face that are $\equiv 0$ and adding 1 to the remaining coordinates.

6.4   Tri-edge – Tri-face and Tri-edge – Tetrahedron Incidence

An up-tri-edge has one coordinate $\equiv 0$, others are $\equiv 2$ (mod 6), and sum of coordinates is 0. It is incident to three tri-faces, that are obtained by keeping the coordinate that is $\equiv 0$, adding 1 to two of the coordinates that are $\equiv 2$ and subtracting 2 from the remaining coordinate.

It is also incident to three tetrahedra, that are obtained by keeping the coordinate that is $\equiv 0$, adding 4 to one of the coordinates that are $\equiv 2$ and subtracting 2 from the remaining two coordinates.

The co-boundary of a down-tri-edge is obtained in a similar manner.

7   Combinatorial Coordinates for Vertices

We distinguish between two types of vertices: hex-vertices are shared by six hex-edges and two tri-edges, tri-vertices are shared by four tri-edges.

7.1   Hex-vertices

Of the two tri-edges incident to the hex-vertex, one is up-tri-edge, and other is down-tri-edge. The two edges both have one (same) coordinate $\equiv 0$. Other coordinates of the up-tri-edge are $\equiv 2$, and of the down-tri-edge are $\equiv 4$ (mod 6). The coordinates of the down-tri-edge that are $\equiv 4$ are greater by 2 from the corresponding coordinates of the up-tri-edge. We assign the average value of the two tri-edge coordinates to the hex-vertex. One coordinate is $\equiv 0$, other coordinates are $\equiv 3$ (mod 6), and their sum is 3. The same coordinates for a hex-vertex can be obtained by taking the average of its six incident tetrahedra (three are up- and three are down-tetrahedra). Figure 3 illustrates (in green) the coordinates of hex-vertices incident to up-tetrahedron $(0,0,0,0)$.

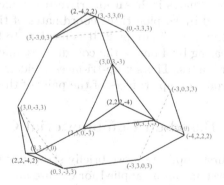

Fig. 3. Combinatorial coordinates for vertices incident to up-tetrahedron $(0,0,0,0)$: hex-vertices in green and tri-vertices in red

Hex-vertex – Co-boundary. A hex vertex is incident to two tri-edges, one up- and one down-tri-edge. The incident up-tri-edge is obtained by keeping the coordinate $\equiv 0$ of the hex-vertex, and decreasing by 1 the three other coordinates. For the incident down-tri-edge, the other three coordinates are increased by 1. The incident faces and tetrahedra can be obtained by using tri-edge co-boundary (tri-edge - tri-face and tri-edge - tetrahedron incidence).

7.2   Tri-vertices

An up-tri-vertex is defined by four up-tri edges that have it as end-point. Equivalently, it is defined by four up-tetrahedra, or by six up-tri-faces. The four up-tri-edges incident to the vertex have one coordinate $\equiv 0$, and other three are $\equiv 2 \pmod 6$. The sum of the coordinates is 0, and each three edges share three coordinates, and differ by one coordinate. For each vertex coordinate we choose the coordinate shared by three incident edges. The sum of the coordinates is 2, and each coordinate is $\equiv 2 \pmod 6$. Figure 3 illustrates (in red) the coordinates of hex-vertices incident to up-tetrahedron $(0,0,0,0)$.

We can get the same up-tri-vertex coordinates by considering tetrahedra: Any three of the four tetrahedra incident to the vertex share three coordinates. We add 2 to the shared coordinate and assign it to the vertex. For example, up-tetrahedra $(0,0,0,0)$, $(6,0,0,-6)$, $(0,6,0,-6)$ and $(0,0,6,-6)$ define the up-tri vertex $(2,2,2,-4)$.

A down-tri-vertex is obtained in a similar fashion. It has coordinates that are $\equiv 4 \pmod 6$, and the sum of coordinates is 4. For example, down-tetrahedra $(6,0,0,0)$, $(0,6,0,0)$, $(6,6,-6,0)$ and $(6,6,0,-6)$ define down-tri-edges $(4,4,0,-2)$, $(4,6,-2,-2)$, $(6,4,-2,-2)$ and $(4,4,-2,0)$ and down-tri-vertex $(4,4,-2,-2)$.

Tri-vertex Co-boundary. The coordinates of an up-tri-vertex are $\equiv 2 \pmod 6$ and the sum of coordinates is 2. An up-tri-vertex is incident to four up-tri-edges, that are obtained by keeping three coordinates of the up-tri-vertex, and decreasing the remaining coordinate by 2. It is incident to four tetrahedra, that are obtained by increasing by 4 one of the coordinates, and decreasing by 2 the remaining three coordinates. The six up-tri-faces incident to the up-tri-vertex can be found as the common up-tri-faces of the pairs of the incident tetrahedra.

8   Morphology Based on Combinatorial Coordinates

In this section, as a first application, we briefly show how the presented combinatorial coordinate system can be applied for closure and interior operations.

Closure of a cell complex Γ representing an object in the diamond grid can be obtained by applying the following operations in the given order:

- 2-closure: if a voxel (3-cell or tetrahedron) is included in Γ, then every face incident to it is also included (added to Γ).

- 1-closure: if a face (2-cell) is included in Γ, then each of its sides (edges or 1-cells) is added to Γ.
- 0-closure: if an edge (1-cell) is included in Γ, then both of its end-vertices (0-cells) is added to Γ.

Interior is similar to closure, but the roles of object and background cells are interchanged. In this way, 2-interior, 1-interior, 0-interior and interior can be defined as usual for three-dimensional cell complexes.

We present an algorithm that computes the 2-closure of a given cell complex Γ in the diamond grid.

ALGORITHM 2-CLOSURE (Γ):

1. **for** every voxel (p, q, r, s) in Γ **do**
2.   add the tri-faces $(p + 3, q - 3, r, s), (p - 3, q + 3, r, s), (p + 3, q, r - 3, s),$ $(p - 3, q + 3, r, s), (p + 3, q, r, s - 3), (p - 3, q, r, s + 3), (p, q + 3, r - 3, s), (p, q - 3, r + 3, s), (p, q + 3, r, s - 3), (p, q - 3, r, s + 3), (p, q, r + 3, s - 3), (p, q, r - 3, s + 3)$ to Γ
3.   **if** $p + q + r + s = 0$ **then** add the hex-faces $(p + 3, q, r, s), (p, q + 3, r, s), (p, q, r + 3, s), (p, q, r, s + 3)$ to Γ
4.   **else** add the hex-faces $(p - 3, q, r, s), (p, q - 3, r, s), (p, q, r - 3, s), (p, q, r, s - 3)$ to Γ
5. **end for**

Note that somewhat similar operations are the (ultimate) collapses for cell complexes [2,13].

9   Conclusions

We give a brief summary of our combinatorial coordinate system in Table 1.

We believe that this new coordinate system helps to implement various topological, morphological and other image processing/computer graphics algorithms on the diamond grid in a simple, elegant and efficient way.

Table 1. The numbers and possible values (mod 6) of the assigned coordinate values with their sum

cell	number	value	sum	cell	number	value	sum
voxel	4	$\equiv 0$	0 or 6	tri-edge	1	$\equiv 0$	
hex-face	3	$\equiv 0$		- up	3	$\equiv 2$	0
	1	$\equiv 3$	3	- down	3	$\equiv 4$	6
tri-face	2	$\equiv 0$		hex-vertex	1	$\equiv 0$	
	2	$\equiv 3$			3	$\equiv 3$	3
-up			0	tri-vertex			
-down			6	- up	4	$\equiv 2$	2
hex-edge	2	$\equiv 0$		-down	4	$\equiv 4$	4
- up	2	$\equiv 2$	4				
- down	2	$\equiv 4$	2				

References

1. Čomić, L., De Floriani, L.: Modeling and Manipulating Cell Complexes in Two, Three and Higher dimensions. In: Brimkov, V.E., Barneva, R.P. (eds.) Digital Geometry Algorithms: Theoretical Foundations and Applications to Computational Imaging, pp. 109–144. Springer (2012)
2. Cousty, J., Bertrand, G., Couprie, M., Najman, L.: Collapses and watersheds in pseudomanifolds of arbitrary dimension. Journal of Mathematical Imaging and Vision 50(3), 261–285 (2014)
3. Forman, R.: Morse Theory for Cell Complexes. Advances in Mathematics 134, 90–145 (1998)
4. Her, I.: Geometric transformations on the hexagonal grid. IEEE Transactions on Image Processing 4(9), 1213–1222 (1995)
5. Herman, G.T.: Geometry of Digital Spaces. Birkhauser, Boston (1998)
6. Kardos, P., Palágyi, K.: Topology-preserving hexagonal thinning. Int. J. Comput. Math. 90(8), 1607–1617 (2013)
7. Klette, R., Rosenfeld, A.: Digital geometry. Geometric methods for digital picture analysis. Morgan Kaufmann Publishers, San Francisco (2004)
8. Kovalevsky, V.A.: Geometry of Locally Finite Spaces (Computer Agreeable Topology and Algorithms for Computer Imagery). Editing House Dr. Bärbel Kovalevski, Berlin (2008)
9. Meyer, F., Angulo, J.: Micro-viscous morphological operators. In: Mathematical Morphology 8th International Symposium (ISMM), pp. 165–176 (2007)
10. Middleton, L., Sivaswamy, J.: Hexagonal Image Processing: A Practical Approach. Advances in Pattern Recognition. Springer (2005)
11. Nagy, B.: Generalized triangular grids in digital geometry. Acta Mathematica Academiae Paedagogicae Nyíregyháziensis 20, 63–78 (2004)
12. Nagy, B.: Cellular topology on the triangular grid. In: Barneva, R.P., Brimkov, V.E., Aggarwal, J.K. (eds.) IWCIA 2012. LNCS, vol. 7655, pp. 143–153. Springer, Heidelberg (2012)
13. Nagy, B.: Cellular topology and topological coordinate systems on the hexagonal and on the triangular grids. Annals of Mathematics and Artificial Intelligence (2014), http://dx.doi.org/10.1007/s10472-014-9404-z
14. Nagy, B., Strand, R.: A connection between Z^n and generalized triangular grids. In: Bebis, G., et al. (eds.) ISVC 2008, Part II. LNCS, vol. 5359, pp. 1157–1166. Springer, Heidelberg (2008)
15. Nagy, B., Strand, R.: Neighborhood sequences in the diamond grid. In: Proceedings of IWCIA 2008 Special Track on Applications Image Analysis - From Theory to Applications, pp. 187–195 (2008)
16. Nagy, B., Strand, R.: Non-traditional grids embedded in Z^n. International Journal of Shape Modeling 14(2), 209–228 (2008)
17. Nagy, B., Strand, R.: Neighborhood sequences in the diamond grid - algorithms with four neighbors. In: Wiederhold, P., Barneva, R.P. (eds.) IWCIA 2009. LNCS, vol. 5852, pp. 109–121. Springer, Heidelberg (2009)
18. Nagy, B., Strand, R.: Neighborhood sequences in the diamond grid: Algorithms with two and three neighbors. Int. J. Imaging Systems and Technology 19, 146–157 (2009)
19. Strand, R., Nagy, B., Borgefors, G.: Digital distance functions on three-dimensional grids. Theor. Comput. Sci. 412, 1350–1363 (2011)

Wiener Index on Lines of Unit Cells of the Body-Centered Cubic Grid

Hamzeh Mujahed[1,2] and Benedek Nagy[2,3(✉)]

[1]Al-Quds Open University, Hebron, Palestine
hmujahed@qou.edu
[2]Department of Mathematics, Eastern Mediterranean University
Mersin 10, Famagusta, North Cyprus, Turkey
[3]Department of Computer Science, Faculty of Informatics,
University of Debrecen, Debrecen, Hungary
nbenedek.inf@gmail.com

Abstract. The Wiener Index of a graph, known as the "sum of distances" of a connected graph, is the first topological index used in chemistry to sum the distances between all unordered pairs of vertices of a graph. In this paper, the lines of unit cells of the body-centered cubic grid are used. These graphs contain center points of the unit cells and other vertices, called border vertices. Closed formulae are obtained to calculate the sum of shortest distances between pairs of border vertices, between border vertices and centers and between pairs of centers. Based on these formulae, their sum, the Wiener Index of body-centered cubic grid with unit cells connected in a row graph is computed. Some relationships between formulae and integer sequences are also presented.

Keywords: Wiener Index · Body-centered cubic grid · Shortest paths · Non-traditional grids · Combinatorics on grids

1 Introduction

Graph theory is used in almost every field of science and it is also heavily used in practice for simulations and engineering solutions. Digital geometry deals with regular tessellations, i.e., graphs with regular, periodic structures. Digital geometry is close connection to image processing and computer graphics [8]. One of the main directions of research of digital geometry deals with descriptions and applications of non-traditional grids [13,17,18]. Non-traditional 3D grids, for instance, body-centered cubic (bcc), face-centered cubic and diamond cubic grids play an important role in physics and chemistry, as well, since various materials have these crystal structures, and the properties of the materials are closely related to their structures.

Wiener Index (*WI*) was introduced by Wiener in 1947 as a measure for various graphs [19]. This measure belongs to the set of molecular structure descriptors, called topological indices, that are used for the design of molecules with desired properties, therefore it is widely studied by chemists [7,11]. There are plenty of researches on Wiener indices, specially, about benzenoid hydrocarbons, hexagonal graphs [5,9].

© Springer International Publishing Switzerland 2015
J.A. Benediktsson et al. (Eds.): ISMM 2015, LNCS 9082, pp. 597–606, 2015.
DOI: 10.1007/978-3-319-18720-4_50

All the three regular tessellations of the plane were studied in [10]. Recently, the task to compute Wiener indices of various 3D structures is also of high importance [1,3].

In this paper, we consider special graphs that are built up from bcc unit cells in a row. The Wiener indices of these graphs are computed, and some interesting relations to integer sequences are also mentioned.

2   Basic Notions and Definitions

In this paper, all graphs are finite, simple, undirected and connected without loops or multiple edges. For a graph G, we denote by $V(G)$ and $E(G)$ its sets of vertices and edges, respectively. All paths are simple, i.e., they contain no repeated vertices. The length of a path P, denoted $|P|$, is the number of its edges [4]. Let $G=(V,E)$ be a simple connected graph. The distance $d_G(u,v)$ between two vertices u and v is defined as the number of edges on a shortest path connecting u and v. A molecular graph is a set of vertices representing the atoms in a molecule and a set of edges representing the covalent bonds between the atoms. To identify molecular structures of chemical compound, the molecular graph invariants, called topological indices could be used too. Topological indices are designed basically by transforming a molecular graph into a number. By these numbers some of the measured properties of the molecules can be predicted [19]. Not only molecules can be represented by graphs: there are some elements that form atomic grid, e.g. carbon and silicon. In a similar manner, crystals formed by ions can be modelled and measured by graphs underlining their structure, see, e.g., [1]. Moreover, in other crystals, such as in metals, the atoms (cations) are placed according to a well-defined arrangement. The most usual arrangements for metals are the body-centered cubic grid and the face-centered cubic grid. In body-centered cubic grid (bcc grid, in short) the atoms are located in a cubic structure and, additionally, there is an atom in the center of each unit cube.

2.1   Wiener Index

Wiener Index (*WI*) is a graph invariant that belongs to the molecular structure descriptors, called topological indices. These indices are widely used by chemists to design molecules with desired properties. In the initial applications, the *WI* is employed to predict physical parameters such as boiling points of the paraffins [19]. Other measurable physical quantities, e.g., heats of vaporization, molar volumes and molar refractions of various molecules can be characterized in a similar manner. The first mathematical definition of *WI* is based on the concept of graph theoretical distance. Topological indices are designed and used to assign a number to each (given type) molecular graph by some measure [19]. *WI* is used to study the relation between molecular structure and physical and chemical properties of certain hydrocarbon compounds. It is, generally, defined as the sum of the shortest distances between every pair of vertices of G. For molecules, in general, *WI* measures how compact a molecule is for its given weight. The molecule is more compact if its *WI* value is less. Wiener, originally, introduced the notion of path number of a graph as the sum of distances between any two carbon atoms in the molecules, in terms of carbon-carbon bonds [19]. However, the index named after him, the *WI* is defined as

$$WI(G) = \frac{1}{2} \sum_{u,v \in V(G)} d_G(u,v)$$ (1)

i.e., the sum of shortest distances for each pair of vertices of the graph G: the sum runs over all ordered pairs of vertices, and $d_G(u,v)$ denote the length of a shortest path in G between vertices u and v [12].

2.2   Body-Centered Cubic (bcc) Grid

By adding a grid point in the center of each cube with vertices on grid points in a cubic grid, a body-centered cubic (bcc) is obtained. The bcc unit cell is a cube (all sides are of the same length and all faces sharing a corner are perpendicular to each other) with an atom at each corner of the unit cell and an atom in the center of the unit cell [6,16,17]. Each of the corner atoms is the corner of another cube, thus the corner atoms are shared among eight unit cells. It is said that bcc has a coordination number of 8 and a bcc unit cell consists of a net total of two atoms; one in the center and eight eighths from corner atoms. Some of the materials that have a bcc structure include lithium, sodium, potassium, chromium, barium, vanadium, alpha-iron and tungsten. Metals which have a bcc structure are usually harder and less malleable than close-packed (e.g., face-centered cubic structured) metals such as gold. When the metal is deformed, the planes of atoms must slip over each other, and this is more difficult in the bcc structure [6]. In Fig. 1 a bcc unit cell is shown, moreover, the closest atoms are connected to each other. Cesium chloride and some other salts use also the same structure in their crystals having one type of atoms in the corners of a unit cell and the other type in the center. Thus, the neighbor relation in these salts contains only atoms (i.e., ions) of different kinds: an anion (e.g., Cl^-) and a cation (e.g., Cs^+). In salts, actually, the ionic bonds can be represented by connecting the neighbor ions. Connecting the closest atoms in a bcc grid, its usual graph-representation is obtained.

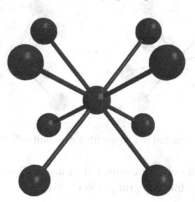

Fig. 1. A unit cell of body-centered cubic (bcc) grid showing the neighbor relation of the atoms

In this paper, we are using graphs that represent a row of unit cells of the bcc grid (i.e., the dimension of our space is $n \times 1 \times 1$ unit cells). We use the terms center points and border points/vertices for the points located on a center of a unit cell and on the corner of a cell, respectively.

3   Wiener Index for a Row of bcc Unit Cells

In this section we present our results. In the next subsections some subsums are computed that are needed later on. We start from a straightforward result:

Lemma 1. Let n be the number of bcc unit cells connected in a row, the number of vertices V in this graph is given as follows (the first term gives the number of border vertices, the second term is the number of center vertices):

$$|V| = (4n+4) + n. \tag{2}$$

3.1   Sum of Distances Between Center Points

Lemma 2. Let k bcc unit cells be connected in a row. And, a new unit cell is connected to the end of the row to form a graph that represents $k+1$ unit cells in a row. Then the sum of all distances between the new center and all old centers is

$$k(k+1). \tag{3}$$

Proof: The distances of the new center c_N to the old centers are: $d_G(c_N, c_1) = 2, d_G(c_N, c_2) = 4, \ldots, d_G(c_N, c_k) = 2k$, therefore the sum of the even numbers from 2 to $2k$ is needed, and it gives the result shown in (3). (See also Figure 2.)   □

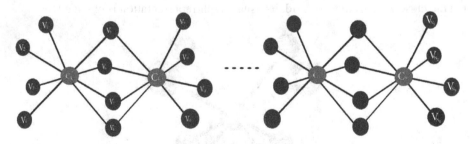

Fig. 2. k bcc unit cells connected in a row with a new unit cell attached the end of the row

Lemma 3. Let n bcc unit cells be connected in a row. Then the sum of all distances between center vertices in this bcc grid graph is given by

$$\frac{n^3 - n}{3}. \tag{4}$$

Proof: The proof goes by induction on the number of unit cells.

The base of the induction is the case $n = 1$. In this case, there is only 1 center, and thus there is no distance to sum up, consequently the sum has value 0, and the formula holds.

Now, let us assume that the formula is satisfied if $n = k$.

Let us prove that it also holds for the value $n = k + 1$. By Lemma 2, we know the sum of the distances obtained by the new center and old centers. Applying this, with the induction hypothesis we get

$$\frac{k^3 - k}{3} + k(k+1) = \frac{k^3 - k + 3k^2 + 3k}{3} = \frac{(k+1)^3 - (k+1)}{3}.$$

The proof of the induction is complete. By the induction, it follows that formula (4) is true for all (non-negative integer value of) n.   □

3.2   Sum of Distances Between Centers and Border Vertices

Lemma 4. Let k bcc unit cells be connected in a row and let a new bcc cell be connected to the end of this row. Then the sum of the distances between old centers and new border vertices plus the sum of the distances between the new center and old border vertices is

$$8(k+1)^2. \tag{5}$$

Proof: Observe that the 4 new border vertices (see also Figure 2, they are on the right) are connected to the new center and some of the old border vertices are also connected to the new center. We need to count the sum of the distances between the 4 new border vertices and the new and old centers, and between the old border vertices and the new center. The sum of these distances can be written in the form

$$\underbrace{4 \cdot 1 + 4 \cdot 3 + 4 \cdot 5 + 4 \cdot 7 + \ldots + 4(2k+1)}_{(new\ vertices\ to\ all\ centers)} + \underbrace{4 \cdot 1 + 4 \cdot 3 + 4 \cdot 5 + 4 \cdot 7 + \ldots + 4(2k+1)}_{(old\ vertices\ to\ new\ center)} =$$

$$= 2(4 \cdot 1 + 4 \cdot 3 + 4 \cdot 5 + 4 \cdot 7 + \ldots + 4(2k+1)) = 8(1 + 3 + 5 + 7 + \ldots + (2k+1)) = 8(k+1)^2. \quad □$$

Lemma 5. Let n bcc unit cells be connected in a row. Then the sum of all distances between center vertices and border vertices in this bcc grid graph is given by

$$2\binom{2n+2}{3}. \tag{6}$$

Proof: The proof goes by induction on n.

The base of the induction is the case $n = 1$. In this case, there is only 1 center, and it is connected to every of the 8 corners (border points) of the unit cell having unit distances. The sum is 8, and also, formula (6) gives this value.

Let us assume that the formula satisfies if $n = k$. Let us prove that it also satisfies if $n = k + 1$. By Lemma 4, we know the sum of the new distances obtained between old centers and new border vertices (of the $(k + 1)$st unit cell), between the $(k + 1)$st center and border vertices (of the previous k unit cells), and between the new $(k + 1)$st center and the new border vertices (of the $(k + 1)$st unit cell), see Fig. 2. Applying this, with the induction hypothesis gives the following statement that is needed to be proven:

$$2\binom{2k+2}{3} + 8(k+1)^2 = 2\binom{2(k+1)+2}{3} = 2\binom{2k+4}{3}.$$

By using the definition of the Binomial coefficients and applying mathematical simplifications, we get

$$\frac{(2(k+1))!}{3(2(k+1)-3)!} + 8(k+1)^2 = \frac{(2(k+2))!}{3(2(k+2)-3)!}.$$

Further, multiplying both sides by 3,

$$\frac{(2k+2)!}{(2k-1)!} + 24(k+1)^2 = \frac{(2k+4)!}{(2k+1)!}.$$

Now, our aim is to prove that the left hand side equals to the right hand side

$$\frac{(2k+2)(2k+1)(2k)(2k-1)!}{(2k-1)!} + 24(k+1)^2 = \frac{(2k+4)(2k+3)(2k+2)(2k+1)!}{(2k+1)!}.$$

Then, we have

$$8k^3 + 36k^2 + 52k + 24 = 8k^3 + 36k^2 + 52k + 24. \qquad \square$$

Observe that equation (6) can also be written in the form $\frac{8}{3}n^3 + 4n^2 + \frac{4}{3}n$.

3.3   Sum of Distances of Border Vertices

Lemma 6. Let k bcc unit cells be connected in a row. If a new bcc unit cell is connected to the previous k cells forming a row with $k + 1$ cells, then the sum of all distances between new and old border vertices is

$$16(k+1)(k+2)+12. \qquad (7)$$

Proof: Observe that the sum of distances between all pairs of the 4 new vertices is 12. (See Figure 2, for instance, for the distance between v_{N_1} and v_{N_2}: that is 2, i.e.,

$d_G(v_{N_1}, v_{N_2}) = 2$. Moreover there are $\binom{4}{2} = 6$ pairs).

Now, let us compute the distance between one of the new border vertices (e.g., v_{N_1}) and all old border vertices:

$$2 \cdot 4 + 4 \cdot 4 + 6 \cdot 4 + \ldots + (2k+1) \cdot 4 = 4(k+1)(k+2).$$

This result is multiplied by 4 since we have 4 new vertices $(v_{N_1}, v_{N_2}, v_{N_3}, v_{N_4})$. Thus we have:

$$16(k+1)(k+2).$$

Finally, the total sum of distances between all new and old border vertices is given by the sum of the previous two values:

$$16(k+1)(k+2)+12.$$

Thus, formula (7) is obtained. $\qquad\qquad\qquad\qquad\qquad\qquad\qquad\qquad\qquad\qquad$ □

Lemma 7. Let n bcc unit cells be connected in a row. Then the sum of all distances between pairs of border vertices is given by

$$\frac{16(n+1)^3 + 20(n+1)}{3}. \qquad\qquad\qquad\qquad (8)$$

Proof: The proof goes by induction on n.

The base of the induction is the case $n = 1$. In this case, there are 8 corners (border points) of the unit cell. Each pair of them has a distance 2 (by connecting them through the center), therefore the sum of distances between all pairs of border vertices is 56, and also, formula (8) gives this value.

Now, let us assume that the formula satisfies if $n = k$. Let us prove that it also satisfies if $n = k + 1$. By Lemma 6, we know the sum of the distances obtained by the old and new border vertices. Applying this, with the induction hypothesis gives the statement that is needed to be proven

$$\frac{16(k+1)^3 + 20(k+1)}{3} + [16(k+1)(k+2)+12] = \frac{16(k+2)^3 + 20(k+2)}{3}.$$

The result can be proven by the following mathematical simplifications/modifications starting from the left hand side. It equals to

$$\frac{16(k+1)^3 + 20(k+1)+48(k+1)(k+2)+36}{3} = \frac{16(k+2-1)^3 + (k+1)[20+48(k+2)]+36}{3} =$$

$$= \frac{16((k+2)^3 - 3(k+2)^2 + 3(k+2) - 1) + (k+1)[20+48(k+2)]+36}{3} =$$

$$= \frac{16(k+2)^3 - 48(k+2)^2 + 48(k+2) - 16 + 20(k+1) + 48(k+1)(k+2) + 36}{3} =$$

$$= \frac{16(k+2)^3 + (k+2)[-48(k+2)+48+48k+48]+20(k+1)+20}{3} =$$

$$= \frac{16(k+2)^3 + 20[(k+1)+1]}{3} = \frac{16(k+2)^3 + 20(k+2)}{3}.$$

This is exactly the formula on the right hand side.   □

3.4   Sum of All Distances: The Main Formula

Based on the results proven in the previous three subsections, we are able to state our main result.

Theorem 1. Let n be the number of bcc unit cells that are connected in a row. Then the formula to find *WI* for this graph is:

$$WI(n) = \frac{25}{3}n^3 + 20n^2 + \frac{71}{3}n + 12 . \tag{9}$$

Proof: The formula is the sum of equations (4), (6) and (8). All possible distances are considered in exactly one of the Lemmas 3, 5 and 7, and then, by simple calculation the sum of those formulae,

$$\frac{16(n+1)^3 + 20(n+1)}{3} + 2\binom{2n+2}{3} + \frac{n^3 - n}{3},$$

can be written in the form of equation (9).   □

Using formula (9) one can calculate *WI* for graph of bcc unit cells connected in a row, as we will present some examples in the next section.

4   Connection to Integer Sequences

In order to have the ability to compute *WI* for bcc graph, three different subsums are used. In this section we show some interesting connections between the subsums presented in equations (4), (6) and (8) and well-known integer sequences given in [15].

Equation (8), the subsum for border vertices, $\frac{16(n+1)^3 + 20(n+1)}{3}$, is identified in [15], as A001386. In [14] this sequence is described as a coordination sequence (giving the number of vertices that are located from a given distance from a chosen vertex of a lattice/grid) for 4-dimensional I-centered tetragonal orthogonal lattice (to obtain our sequence the first two elements of A001386 should be deleted).

The sequence defined by equation (4), $\frac{n^3 - n}{3}$, can also be found in Sloane's. It is A007290 and the values are, actually, the doubles of values of the binomials $\binom{n}{3}$. This sequence appear in various places in physics, mathematics, and specially, in graph theory, as well. Moreover, this sequence also gives the reverse Wiener index of the path graph with n vertices [2].

The integer sequences defined by equation (6) and by equation (9) are not found in [15].

Table 1 shows some of the first elements of the sequences we are working with, i.e., the values computed by equations (4), (6), (8) and (9) for some small values of n. The *WI* values are shown in the last row of the table.

Table 1. Some values of the subsums and *WI* for few bcc cells in a row

Number of bcc unit cells (n)	1	2	3	4	5	6	7
Equation (4)	0	2	8	20	40	70	112
Equation (6)	8	40	112	240	440	728	1120
Equation (8)	56	164	368	700	1192	1876	2784
Wiener Index *WI*	64	206	488	960	1672	2674	4016

5   Conclusions

Grids, and specially, non-traditional three-dimensional grids are lying in the intersection of digital and discrete geometry, graph and lattice theory, crystallography and other applied fields in physics and chemistry. One of the first and most important topological/geometrical indices of graph structure is the Wiener index. In this paper, the body-centered cubic grid is investigated in which a finite number of unit cells are placed next to each other at a line. We have formulated and proved the computation of Wiener index for these graphs. There are several ways to continue the line of the research that we have just started here:

— One can compute other topological indices, e.g., Szeged index for these graphs. The Szeged index of a graph G is computed as follows: for each edge $e_{u,v}$ of the graph G, let $n_u(e)$ be the number of vertices w of G that has smaller distance $d_G(u,w)$ from vertex u than from vertex v. Then the sum $\sum_{e_{u,v} \in E(G)} n_u(e) n_v(e)$ gives the Szeged index of G. See, for instance, [10], for some calculations of Szeged index of some two-dimensional regular grid graphs.
— One can extend the results to two and three dimensional rectangles and blocks of bcc unit cells.
— Moreover, other non-traditional grids, e.g., face-centered cubic grid can also be involved to similar studies.

References

1. Al-Kandari, A., Manuel, P., Rajasingh, I.: Wiener Index of Sodium Chloride and Benzenoid Structures. The Journal of Combinatorial Mathematics and Combinatorial Computing 79, 33–42 (2011)

2. Balaban, A.T., Mills, D., Ivanciuc, O., Basak, S.C.: Reverse Wiener indices. Croatica Chemica Acta 73, 923–941 (2000)
3. Bogdanov, B., Nikolić, S., Trinajstić, N.: On the three-dimensional Wiener number. Journal of Mathematical Chemistry 3, 299–309 (1989)
4. Bollobás, B.: Graph Theory: An Introductory Course. Springer, New York (1979)
5. Dobrynin, A.A., Gutman, I., Klavžar, S., Žigert, P.: Wiener index of hexagonal systems. Acta Appl. Math. 72, 247–294 (2002)
6. Kittel, C.: Introduction to Solid State Physics. Wiley, New York (2004)
7. Klavzar, S., Gutman, I.: Wiener number of vertex-weighted graphs and a chemical application. Discrete Applied Mathematics 80, 73–81 (1997)
8. Klette, R., Rosenfeld, A.: Digital geometry – geometric methods for digital picture analysis. Morgan Kaufmann (2004)
9. Knor, M., Skrekovski, R.: Wiener Index of generalized 4-stars and of their quadratic line graphs. Australasian Journal of Combinatorics 58, 119–126 (2014)
10. Manuel, P., Rajasingh, I., Arockiaraj, M.: Wiener and Szeged indices of Regular Tessellations. In: International Conference on Information and Network Technology (ICINT 2012). IPCSIT 37, pp. 210–214. IACSIT Press, Singapore (2012)
11. Mihalic, Z., Veljan, D., Amic, D., Nilkolic, S., Plavsic, D., Trianjstic, N.: The distance matrix in chemistry. J. Math. Chem. 11, 223–258 (1992)
12. Mohar, B., Pisanski, T.: How to compute the Wiener index of a graph. J. Math. Chem. 2, 267–277 (1988)
13. Nagy, B., Strand, R.: Non-Traditional Grids Embedded in Z^n. International Journal of Shape Modeling 14, 209–228 (2008)
14. O'Keeffe, M.: Coordination sequences for lattices. Zeit. f. Krist. 210, 905–908 (1995)
15. Sloane, N.: On-Line Encyclopedia of Integer Sequences (OEIS), http://oeis.org/
16. Strand, R., Nagy, B.: Distances based on neighbourhood sequences in non-standard three-dimensional grids. Discrete Applied Mathematics 155, 548–557 (2007)
17. Strand, R., Nagy, B.: Path-based distance functions in n-dimensional generalizations of the face-and body-centered cubic grids. Discrete Applied Mathematics 157, 3386–3400 (2009)
18. Strand, R., Nagy, B., Borgefors, G.: Digital distance functions on three-dimensional grids. Theoretical Computer Science 412, 1350–1363 (2011)
19. Wiener, H.: Structural determination of paraffin boiling points. Journal of American Chemical Society 69, 17–20 (1947)

Part-Based Segmentation by Skeleton Cut Space Analysis

Cong Feng[(✉)1], Andrei C. Jalba[2], and Alexandru C. Telea[1]

[1] Institute Johann Bernoulli, University of Groningen, Groningen, The Netherlands
c.feng@rug.nl, a.c.telea@rug.nl
[2] Department of Mathematics and Computer Science, TU Eindhoven,
Eindhoven, The Netherlands
a.c.jalba@tue.nl

Abstract. We present a new method for part-based segmentation of voxel shapes that uses medial surfaces to define a segmenting cut at each medial voxel. The cut has several desirable properties – smoothness, tightness, and orientation with respect to the shape's local symmetry axis, making it a good segmentation tool. We next analyze the space of all cuts created for a given shape and detect cuts which are good segment borders. Our method is robust to noise, pose invariant, independent on the shape geometry and genus, and is simple to implement. We demonstrate our method on a wide selection of 3D shapes.

Keywords: Part-based segmentation · Medial surfaces · Skeletonization

1 Introduction

Shape segmentation aims to decompose a 3D shape into a set of parts that obey certain application-related properties, and is used in many contexts such as image analysis, registration, and 3D modeling [27]. *Patch-based* segmentation detects quasi-flat segments whose borders follow local curvature maxima on the shape surface, and is most used for faceted shapes [24]. *Part-based* segmentation follows a semantics-oriented approach, aiming to find shape parts that one would intuitively perceive as being logically distinct, and is used for natural shapes [22].

For a shape $\Omega \subset \mathbb{R}^3$, part-based segmentations (PBS) using *partitioning cuts* create a set of cuts $c \subset \partial\Omega$ that divide the shape boundary $\partial\Omega$ into disjoint parts. Desirable PBS properties, *e.g.* smoothness, orientation, tightness, and position of cuts that create segments, can be stated in terms of the cut-set $\mathcal{B} = \{c\}$. Finding a good segmentation is thus mapped to finding a cut-set $\mathcal{B}$ having such properties, a hard problem due to the high dimensionality of the cut space.

We present a new way to produce PBS of 3D voxel shapes by *skeleton cuts*. First, we construct, at any shape point, a cut that is locally and globally smooth, tightly wraps around the surface, is self-intersection free, and is locally orthogonal to the shape's local symmetry axis. For this, we use the shape's medial surface. Next, we construct the cut-space $\mathcal{S} \subset \partial\Omega$ that contains all such cuts for

© Springer International Publishing Switzerland 2015
J.A. Benediktsson et al. (Eds.): ISMM 2015, LNCS 9082, pp. 607–618, 2015.
DOI: 10.1007/978-3-319-18720-4_51

a given shape. We extract the cut-set $\mathcal{B} \subset \mathcal{S}$ yielding our PBS by analyzing the global distribution of cut properties over $\mathcal{S}$. We demonstrate our method on a variety of 3D shapes and compare our results with eight existing PBS methods.

Section 2 reviews related work. Section 3 presents our method. Section 4 illustrates our method on a wide variety of 3D shapes and also compares it with related methods. Section 5 discusses our method. Section 6 concludes the paper.

2   Related Work

Two main shape segmentation types exist [1, 4, 28]: *Patch-based* methods segment a shape's surface into quasi-flat patches bounded by sharp surface creases, and are suitable for synthetic shapes. *Part-based* segmentation (PBS), our focus, cuts a shape's surface into its logical components, useful for shapes formed of articulated parts, *e.g.* human bodies. plants, and other natural structures.

Most PBS methods find segments along what a human would see as logical shape parts, in two steps: (a) find *where* to cut a shape to isolate a part; and (b) find *how* to build a cut, once its location is set. These steps are addressed in different ways. As the topology of the shape skeleton or medial axis matches the part-whole shape structure [31], many methods use medial axes to place cuts. Au *et al.* use curve skeletons [5], where each skeleton branch maps to a part. Cuts are built by optimizing for cut concavity and length via minimal cuts [12]. Golovinskiy *et al.* create a large randomized cut-set and find part borders as the cuts on which most surface edges lie [10]. Shapira *et al.* note that skeletonization and segmentation are related, and compute a scalar shape-diameter function (SDF) on the shape surface to segments as surface faces with similar SDF values [29]. Tierny *et al.* segment shapes hierarchically by topological and geometrical analysis of their Reeb graphs, which are similar to curve skeletons [33]. Chang *et al.* compute shape medial surfaces, separate their manifolds, and back project each manifold on the shape surface to find a segment [6]. Dey and Sun extract curve skeletons as the maxima of the medial geodesic function (MGF) which encodes the length of the shortest path between feature points of points in the shape [8], and segment tubular parts as those which minimize the eccentricity of such paths. Reniers *et al.* construct a part for each branch of a shape's curve skeleton [22]. Part borders correspond to curve-skeleton junction points, and are created by the shortest paths in [8]. However, curve skeletons can contain many spurious junctions which change widely when the shape is slightly perturbed. Reniers *et al.* alleviate this by heuristics that shift cut-points along the curve skeleton to optimize for cut stability and planarity [23]. Yet, this method cannot segment shapes of large geometric, but little topological, variability, like a pawn chess piece: Its curve skeleton has no junction points, so [23] cannot separate the pawn's head, body, and base, although these have different thicknesses.

Summarizing, the two elements of a good PBS (*where* to cut, and *how* to cut) are targeted in complement by different methods: Skeleton-based methods construct good partitioning cuts efficiently, *e.g.* by shortest-paths [8, 23]. Yet, curve skeletons do not encode enough of the shape geometry. Global search methods

that analyze a wide set of shape cuts offer good ways to select where to partition [10, 29]. Yet, they do not offer explicit constraints for the cut shapes, and
exhaustive cut-space search is expensive. Our method combines the advantages
of the two above classes of methods, while minimizing their limitations.

3 Method

Our method has a simple intuition: Say we want to cut the shape in Fig. 1 a close
to points $A \ldots E$. Which properties should these cuts have to yield a 'natural'
PBS? In other words: How would a human draw such cuts? Figure 1 a shows
five *undesirable* cuts: A is noisy, although it crosses a perfectly smooth surface
zone; B is self-intersecting; C and D are too loose (long); and E is unnaturally
slanted – a human asked to cut the shape at that point would arguably do it so
across the finger's symmetry axis. Figure 1 b shows five cuts for the same points,
computed with the method in this paper. We argue that these cuts are more
suitable for PBS than those in Fig. 1 a, as they are (1) tight, (2) locally smooth,
(3) self-intersection free, (4) and locally orthogonal to the shape's symmetry
axis. An additional property that cuts should satisfy is (5) being closed curves,
so that they divide the shape's surface into different parts. We construct such
cuts as follows: First, we compute a simplified medial surface of the input shape
(Sec. 3.1). For each medial point, we next construct a cut having the above
properties (Sec. 3.2). This answers the question "how to cut". By analyzing the
resulting cut-space, we next select a small cut-set that gives us the borders of
salient shape-parts (Sec. 3.3). This answers the question "where to cut".

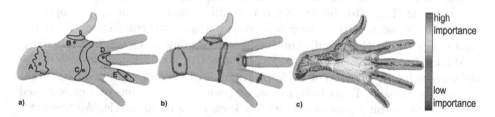

Fig. 1. Possible cuts for part-based segmentation. Suboptimal cuts (a). Cuts created
by our method (b). Medial surface colored by its importance metric (c).

3.1 Skeletonization

The Euclidean distance transform $DT_{\partial\Omega} : \Omega \to \mathbb{R}_+$ of a shape $\Omega \subset \mathbb{Z}^3$ with
boundary $\partial\Omega$ is

$$DT_{\partial\Omega}(\mathbf{x} \in \Omega) = \min_{\mathbf{y} \in \partial\Omega} \|\mathbf{x} - \mathbf{y}\|. \tag{1}$$

The medial surface, or surface skeleton, of $\partial\Omega$ is defined as

$$S_{\partial\Omega} = \{\mathbf{x} \in \Omega | \exists \{\mathbf{f}_1, \mathbf{f}_2\} \subset \partial\Omega, \mathbf{f}_1 \neq \mathbf{f}_2, \|\mathbf{x} - \mathbf{f}_1\| = \|\mathbf{x} - \mathbf{f}_2\| = DT_{\partial\Omega}(\mathbf{x})\} \tag{2}$$

where $\mathbf{f}_1$ and $\mathbf{f}_2$ are the contact (or feature) points with $\partial\Omega$ of the maximally inscribed ball in Ω centered at $\mathbf{x}$ [9, 26]. These define the feature transform $FT_{\partial\Omega} : \Omega \to \mathcal{P}(\partial\Omega)$

$$FT_{\partial\Omega}(\mathbf{x} \in \Omega) = \underset{\mathbf{y}\in\partial\Omega}{\operatorname{argmin}} \|\mathbf{x} - \mathbf{y}\|. \tag{3}$$

Medial surfaces are sensitive to small-scale noise on Ω, especially when using voxel-based models. To alleviate this, they can be *regularized* by a computing a metric $\rho : S_{\partial\Omega} \to \mathbb{R}_+$ such as the medial geodesic function (MGF) which sets $\rho(\mathbf{x})$ to the length of the shortest path on $\partial\Omega$ between the two feature points of $\mathbf{x}$ [8]. As the MGF monotonically increases from the medial surface boundary to its center, upper thresholding it yields connected and noise-free simplified medial surfaces (though tunnel preservation requires additional work) [25]. Figure 1 c shows a regularized medial surface using the MGF method in [25].

3.2 Cut Model

The first step of our PBS is to compute a rich set of cuts, or cut space $\mathcal{S}$, which all satisfy properties (1-5) listed in Sec. 3. To build a cut $c \in \mathcal{S}$, consider a point $\mathbf{x} \in S_{\partial\Omega}$. By definition, $\mathbf{x}$ has at least two feature points $\mathbf{f}_1$ and $\mathbf{f}_2$ on $\partial\Omega$ (Eqn. 2). Consider, for now, that there are precisely two such points. We first trace the shortest path $\gamma_1 \subset \partial\Omega$ between $\mathbf{f}_1$ and $\mathbf{f}_2$ (Fig. 2 a), whose length is the MGF value for $\mathbf{x}$ (Sec. 3.1). Next, we find the midpoint $\mathbf{m}$ of γ_1, *i.e.* the voxel of γ_1 furthest in arc-length distance from both $\mathbf{f}_1$ and $\mathbf{f}_2$. We then trace a ray through $\mathbf{x}$ and oriented in the direction $\mathbf{x} - \mathbf{m}$, and find the point $\mathbf{o}$ where this ray 'exits' Ω (Fig. 2 b). Intuitively, $\mathbf{o}$ is on the 'other side' of $S_{\partial\Omega}$ as opposed to $\mathbf{m}$. Finally, we construct the two shortest paths on $\partial\Omega$ connecting $(\mathbf{f}_1, \mathbf{o})$ and $(\mathbf{f}_2, \mathbf{o})$ respectively (Fig. 2 c,d). Our final cut c for point $\mathbf{x}$ is given by $\gamma_1 \cup \gamma_2 \cup \gamma_3$.

While c is piecewise geodesic (so locally smooth), it can be non-smooth at the three endpoints $\mathbf{f}_1, \mathbf{f}_2$ and $\mathbf{o}$ of γ_i. Also, our construction does not make c as tight as possible globally. To fix both issues, we perform 5 iterations of a constrained Laplacian smoothing pass over c, with a kernel size of 10 voxels. We prevent c leaving the surface, by reprojecting its voxels to their closest points on $\partial\Omega$ after each iteration. This smooths out possible 'kinks' at $\mathbf{f}_1, \mathbf{f}_2$ and $\mathbf{o}$, thus making c globally smooth and tight. If such kinks are very small or inexistent, smoothing has no effect, as c is globally geodesic. In that case, Laplacian smoothing shifts c's points along the surface normal, since c's acceleration c'' is normal to the surface, so reprojection moves the smoothed points back to their original location.

Cut properties: Our cuts meet the desired properties we require for PBS:

1. *tight:* Cut parts γ_i are piecewise-geodesic, thus shortest curves on $\partial\Omega$. Also, the constrained Laplacian smoothing shortens potential kinks present at the geodesic endpoints, thus making the entire c wrap tightly around the shape;
2. *smooth:* Guaranteed by the same properties as for tightness – piecewise geodesicness and constrained Laplacian smoothing;

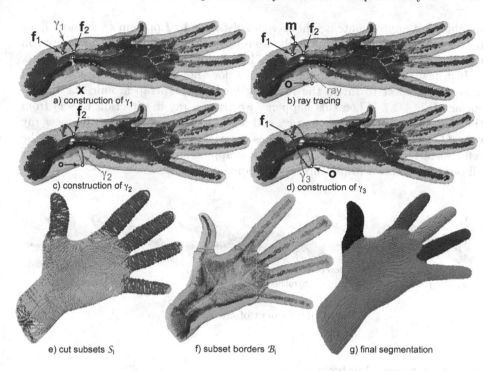

a) construction of γ_1

b) ray tracing

c) construction of γ_2

d) construction of γ_3

e) cut subsets S_i

f) subset borders $\mathcal{B}_i$

g) final segmentation

Fig. 2. Cut construction (a-d) and cut-space analysis (e-g) for part-based segmentation

3. *self-intersection free:* c is a geodesic triangle (three geodesics linking three *different* points on $\partial\Omega$) whose edges do not intersect except at endpoints;
4. *locally orthogonal* to the symmetry axis: The cut $c(\mathbf{x})$ surrounds the medial surface $S_{\partial\Omega}$ around point $\mathbf{x}$, by construction. Hence, it also surrounds the so-called *curve skeleton* of $\partial\Omega$, which is a 1D structure locally centered within $S_{\partial\Omega}$ with respect to its boundary $\partial S_{\partial\Omega}$. While we do not have a formal proof of local orthogonality, we observed in practice that our construction always creates cuts that are visually orthogonal to the curve skeleton;
5. *closed:* The cut c is a closed (Jordan) curve by construction.

Implementation: To build γ_1, we need two feature points $\mathbf{f}_1$ and $\mathbf{f}_2$. Two issues exist here: (1) Computing the feature transform $FT(\mathbf{x})$ on digital shapes cannot be done via Eqn. 3, given the finite voxel grid resolution [21, 25]. To fix this, we compute the so-called extended feature transform $EFT(\mathbf{x})$ which finds all closest-points on $\partial\Omega$ to all 26 neighbors of $\mathbf{x}$, and which is a superset of $FT(\mathbf{x})$ [25]. From this superset, we select exactly two feature points that best represent the symmetric embedding of $S_{\partial\Omega}$ in Ω. For this, we select the two feature points $\{\mathbf{f}_1, \mathbf{f}_2\} \subset EFT(\mathbf{x})$ that maximize the angle $\widehat{\mathbf{f}_1\mathbf{x}\mathbf{f}_2}$. We trace the ray used to find $\mathbf{o}$ by Bresenham's 3D line-tracing algorithm on the voxel shape. We compute geodesics by Dijkstra's shortest-path algorithm on the connectivity graph of voxels of $\partial\Omega$, using A^* heuristics to speed the search, and using edge weights that approximate neighbor-voxel distances by Eppstein's scheme [13] for better path-length accuracy. Finally, we reproject Laplacian-smoothed points on the shape surface by using the fast ANN library for finding nearest-neighbors [19].

In a few cases, point **o** found as above does not lie on the opposite side of $S_{\partial\Omega}$ with respect to **m**, so the resulting cut will not wrap around the medial surface (Fig. 3 a). When this happens, we trace a ray in direction $\mathbf{f_1} - \mathbf{f_2}$ from the midpoint **v** of the current ray, and set **o** to the voxel where this new ray exits Ω (Fig. 3 b). If the new **o** still does not yield a wrapping cut, we repeat the refinement (Fig. 3 c). This produces cuts wrapping around the medial surface for all our test shapes within 3 up to 4 refinement steps.

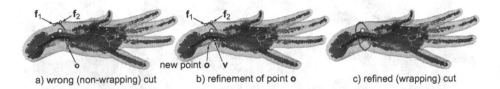

a) wrong (non-wrapping) cut   b) refinement of point **o**   c) refined (wrapping) cut

Fig. 3. Refinement of cut construction

3.3 Cut Space Analysis

We can create a cut $c(\mathbf{x})$ for any voxel **x** of a shape's medial surface $S_{\partial\Omega}$, which has good properties for PBS. Intuitively, $c(\mathbf{x})$ is a good way to cut the shape at point **x**, *if* we want a cut there. We now must decide *where* we want to cut to get a PBS with desired global properties. Let $\mathcal{S} = \{c(\mathbf{x}) | \mathbf{x} \in S_{\partial\Omega}\}$ be the space of all cuts created from $S_{\partial\Omega}$. Given our cut properties, cuts on the same shape-part share similar properties *e.g.* orientation and length. Cuts for different parts have different properties. Consider our hand model: Finger cuts are short; wrist cuts have average length; and palm cuts are longest. For a shape consisting of a rump and protruding parts, cuts for parts are shorter than cuts for the rump.

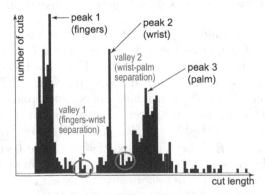

Fig. 4. Cut-length histogram for hand model (Sec. 3.3)

We use these insights to partition $\mathcal{S}$ in subsets $\mathcal{S}_i$ so that $\cup_i \mathcal{S}_i = \mathcal{S}$ and $\mathcal{S}_i \cup \mathcal{S}_{j \neq i} = \varnothing$ by using the histogram of cut lengths over $\mathcal{S}$. Histogram peaks show large similar-length cuts, so partitioning it by thresholds in the valleys between peaks gives our desired subsets $\mathcal{S}_i$. To find such thresholds robustly, we filter the histogram by mean shift [7] to 'sharpen' the cut-distribution and separate peaks from valleys more clearly. We define a peak as a histogram value exceeding λ times the cut count $\|\mathcal{S}\|$, and a valley as a value less than $\mu = \lambda/3$. Setting $\lambda = 0.01$ gave good results for all shapes in this paper. Figure 4 shows the cut-length histogram for the hand model. Its three main peaks describe cuts on the fingers, wrist, and palm; the two valleys give the two thresholds needed to separate fingers from the palm and the palm from the wrist.

Subsets $\mathcal{S}_i$ do not (yet) coincide with our desired segments. Indeed, an $\mathcal{S}_i$ can contain logically disjoint cuts of similar lengths – e.g. all cuts on the fingers (blue in Fig. 2 c) are in the same subset. Also, $\mathcal{S}$ does not fully cover $\partial\Omega$, since we compute it from the simplified medial surface. This is shown by the gaps between cuts in Fig. 2 d. To fix this, we first define a cut $c(\mathbf{x})$ as being a border $\mathcal{B}_i$ of subset $\mathcal{S}_i$ if $c(\mathbf{x})$ belongs to a different subset than any of the cuts $c(\mathbf{y})$, where $\mathbf{y}$ are the 26-neighbors of $\mathbf{x}$ on $S_{\partial\Omega}$. Using this definition, we find the set of cuts $\{\mathcal{B}_i\}$ that represent the borders of our final segments (Fig. 2 f). Note that, if a cut is marked as border, at least one of its neighbor cuts will be in a different cut subset, by definition. Hence, that neighbor cut will also be a border, so more than one border will be produced from a 3^2 voxel neighborhood. To remove such duplicates, we keep, for each such neighborhood, the shortest border. We compute our final segments by finding the connected components of $\partial\Omega$ separated by borders, via a simple flood-fill algorithm on $\partial\Omega$ (Fig. 2 g).

4   Results and Comparison

We have tested our method on several shapes provided as 3D polygon meshes, voxelized by *binvox* [20] at resolutions up to 400^3 voxels. Figure 5 compares our results with [23], the best medial-descriptor voxel PBS method we know. We get very similar results, but find more fine-grained segments than [23] – see finger and ear details of the animal models, pig tail, dragon spikes, and microscope lens. Segment borders are smooth and locally orthogonal to the shape's symmetry axis, *i.e.*, similar to how a human would cut the shape at the respective places. Our method finds segments of various sizes, ranging from details (dragon's tail, hound's ears), to large parts (limbs of various models). Figure 6 a-k compares our method with eight PBS methods on two shapes [3, 14–17, 22, 23, 33]. Here, Reniers *et al.* (1) denotes [22], and Reniers *et al.* (2) denotes [23]. These methods span from voxel-based to mesh-based, and use various segmentation heuristics (skeleton, curvature, salience, and topology-based). We argue that our method creates equally or, in some cases, more plausible PBSs. Since both our method and [23] use medial descriptors, computed by the same underlying method [25], a relevant question is how the two methods differ. We use (a) medial *surfaces*, while [23] uses *curve* skeletons; and (b) we find segment borders by analyzing *all*

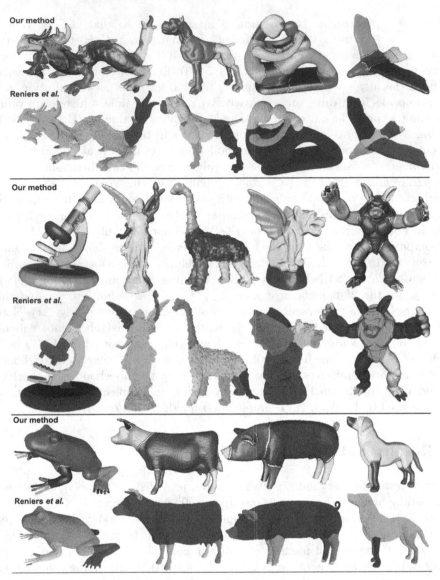

Fig. 5. Part-based segmentations of our method *vs* Reniers *et al.* [23] (Sec. 4)

possible cuts, while [23] places such borders around the curve-skeleton branch junctions. Fig. 6 l-p shows five examples where the public implementation of [23] fails to segment at all. We find two causes for this: The shape parts in Fig. 6 l cannot be well described by curve-skeleton branches, as they are nearly rotationally symmetric. As few (if any) such junctions exist, [23] fails. The shape in Fig. 6 n is described by a mix of medial surfaces (base plate) and curve skeletons (tubular parts). As [23] only uses curve skeletons, data on the base plate is incomplete or missing. For the shapes in Fig. 6 m-p, the many heuristics in [23]

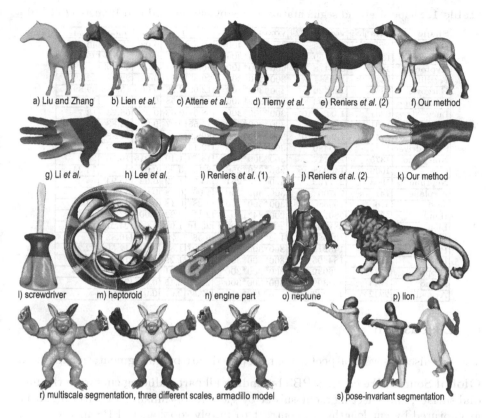

a) Liu and Zhang   b) Lien *et al.*   c) Attene *et al.*   d) Tierny *et al.*   e) Reniers *et al.* (2)   f) Our method

g) Li *et al.*   h) Lee *et al.*   i) Reniers *et al.* (1)   j) Reniers *et al.* (2)   k) Our method

l) screwdriver   m) heptoroid   n) engine part   o) neptune   p) lion

r) multiscale segmentation, three different scales, armadillo model   s) pose-invariant segmentation

Fig. 6. Comparison of our method with eight PBS methods (a-k). Our results for shapes where Reniers *et al.* fails (l-p). Multiscale (r) and pose-invariant (s) segmentations.

to select cuts centered on the curve-skeleton fail, as they imply that such cuts should be nearly planar. This does not happen for the above shapes.

We can also produce a *multiscale* PBS: For this, we simply change the values of λ and μ used to partition the cut space via its length histogram (Sec. 3.3). High λ values and low μ values yield fewer and more differentiated segments (in terms of local thickness); closer values of *lambda* and μ yield a finer-grained segmentation. Figure 6 r shows three such scales for the armadillo shape.

Our method is *pose invariant*, as illustrated in Fig. 6 s. Indeed, our cut space histogram captures local shape thickness, which does not depend on pose.

Table 1 shows the time for creating cuts (t_{cuts}), medial surfaces (t_{skel}), cut-space analysis (t_{space}), our total time (t_{total}), and total time for [23] ($t_{Reniers}$), for our method coded in C++ on an 8-core 3.5 GHz PC. Empty cells ($t_{Reniers}$) show shapes where [23] failed. As cuts are computed independently, we parallelized our method by *pthreads*, getting a speed boost factor of 7, close to the optimal value of 8 for our hardware. Compared to [23], on the same hardware, we are slightly faster in most cases, and can successfully segment all tested shapes.

Table 1. Shape sizes and segmentation times by our method and Reniers *et al.* [23]

Shapes	cuts $\|S\|$	voxels $\|\Omega\|$	voxel volume	t_{cuts}	t_{skel}	t_{part}	t_{total}	$t_{Reniers}$
Dragon	2789	283238	400*400*400	50.8	1.90	0.03	52.73	40.26
Hound	1530	245759	300*300*300	23.24	1.51	0.01	24.76	25.1
Hyptoroid	4873	651478	400*400*400	400.5	3.36	0.04	403.90	-
Fertility	1354	199581	300*300*300	20.85	2.02	0.01	22.88	22.89
Gargoyle	488	129420	300*300*300	12.62	3.26	0.005	15.885	69.89
Microscope	1397	307863	300*300*300	44.14	1.58	0.01	45.73	198.02
Lucy	6201	1.04×10^6	300*300*300	68.01	0.63	0.09	68.73	12.7
Engine part	1501	135416	300*300*300	15.55	0.27	0.01	15.83	-
Frog	41450	1.20×10^7	300*300*300	808.2	2.48	2.16	812.8	36.93
Screwdriver	1372	306480	300*300*300	13.14	0.60	0.01	13.75	-
Noisydino	1375	194117	300*300*300	14.79	1.19	0.015	16.00	20.2
Cow	1009	143938	256*256*256	8.15	0.96	0.01	9.12	14.34
Neptune	1908	211723	420*185*251	34.7	1.22	0.02	35.94	-
Airplane	741	76700	300*300*300	6.00	0.28	0.08	6.37	-
Bird	476	45638	300*300*300	2.28	0.18	0.003	2.47	7.98
Hand	584	58071	200*84*140	2.15	0.22	0.004	2.37	-
Lion	2181	381968	300*300*300	23.16	1.08	0.02	24.27	-
Horse	884	109555	142*300*251	9.58	1.24	0.008	10.83	-
Pig	959	145215	300*300*300	10.97	1.51	0.01	12.50	22.26
Dog	1241	184805	300*300*300	15.65	1.29	0.02	16.97	18.87
Hippo	838	166932	300*300*300	12.13	2.41	0.01	14.55	25.18
Rhino	1746	403399	300*300*300	25.20	2.15	0.03	27.39	-
Armadillo	2242	436933	300*229*252	47.55	2.67	0.03	50.26	-

5   Discussion

We next discuss several aspects of our proposed part-based segmentation method.

Global Search: We create a PBS by finding all part-inducing cuts from the medial surface, and selecting a cut-subset by globally optimizing for part-similarity as captured by cut lengths. In contrast to purely topological PBS methods [22, 23], we search a much wider space of possible partitionings; yet, our search space is much smaller than that of other methods which look for cuts of any possible orientation [10], thereby achieving a good flexibility-performance balance.

Simplicity: In our approach, we can use *any* medial surface skeletonization method, *e.g.* [2, 26, 30], as long as it outputs regularized skeletons. This makes our method applicable to mesh-based shapes (and their medial surfaces) [11].

Regularization: We use regularized medial surfaces (Sec. 3.1) having voxels with large MGF values, which have far-apart feature points f_1 and f_2. This ensures that the ray casting used to compute cuts robustly finds cuts that wrap around the medial surface (Sec. 3.2).

Multiscale: Multiscale PBS occurs at two levels: (1) Simplified medial surfaces yield cuts only for important shape parts; (2) The cut histogram analysis parameters λ and μ select the level-of-detail where we search for cut-length differences.

Pose invariance: Our method is pose-invariant [23, 32], as shown by the model in Fig. 6 s (which is also used in [32] to show pose invariance).

Robustness: We robustly segment noisy or detail-rich surfaces, *e.g. dragon* and *dino* (Fig. 5) or *lion* (Fig. 6). Segment borders are smooth by construction (Sec. 3.2). Since our segmentation uses a subset of these cuts, and only considers *integral* cut properties (length) rather than differential ones (*e.g.* curvature), noise and/or small-scale details are robustly handled.

Limitations: Our method's cost is $O(\|S_{\partial\Omega}\|\|\partial\Omega\|\log\|\partial\Omega\|)$. As our method parallelizes easily (Sec. 4), its practical cost is similar to other skeleton-based PBS method [22, 23] or cut-based methods [10]. For space constraints, we compare with only eight related methods. More PBS methods exist, and quantitative metrics can be further used to measure segmentation quality [18]. Yet, even without such extra insights, we argue that our goal of showing that *surface* skeletons have added value for PBS as opposed to *curve* skeletons is well defended.

6   Conclusions

We have presented a new method for part-based segmentation of 3D voxel shapes by analyzing the entire space of potential partitioning cuts constructed by using the shape's medial surface. To our knowledge, our approach is the first which uses medial surfaces for part-based segmentation, and thereby shows the added-value of medial surfaces for segmentation, as opposed to the well-known use of curve skeletons for the same task. We demonstrate our method on a wide variety of 3D shapes, and compare it with eight related segmentation methods.

Next, different ways to partition the cut space can be easily tried, *e.g.* bottom-up hierarchical clustering or cut similarities based on *e.g.* curvature, eccentricity, and orientation. This would lead to an entire family of PBS methods in a single simple implementation. Our cut-length histogram could be an effective shape descriptor for retrieval and matching. Finally, implementing our method for mesh-based shapes on the GPU should lead to massive scalability increases.

References

[1] Agates, A., Pratikakis, I., Perantonis, S., Sapidis, N., Azariadis, P.: 3D mesh segmentation methodologies for CAD applications. Computer-Aided Design & Applications 4(6), 827–841 (2007)

[2] Arcelli, C., Sanniti di Baja, G., Serino, L.: Distance-driven skeletonization in voxel images. IEEE TPAMI 33(4), 709–720 (2011)

[3] Attene, M., Falcidieno, B., Spagnuolo, M.: Hierarchical mesh segmentation based on fitting primitives. Visual Comput. 22(3), 181–193 (2006)

[4] Attene, M., Katz, S., Mortara, M., Patane, G., Spagnuolo, M., Tal, A.: Mesh segmentation - a comparative study. In: Proc. SMI, pp. 134–141 (2006)

[5] Au, O., Tai, C., Chu, H., Cohen-Or, D., Lee, T.: Skeleton extraction by mesh contraction. ACM TOG (Proc. ACM SIGGRAPH) 27(3), 441–449 (2008)

[6] Chang, M., Leymarie, F., Kimia, B.: Surface reconstruction from point clouds by transforming the medial scaffold. CVIU 113(11), 1130–1146 (2009)

[7] Comaniciu, D., Meer, P.: Mean shift: A robust approach toward feature space analysis. IEEE TPAMI 24(5), 603–619 (2002)

[8] Dey, T., Sun, J.: Defining and computing curve-skeletons with the medial geodesic function. In: Proc. SGP, pp. 143–152 (2006)

[9] Giblin, P., Kimia, B.: A formal classification of 3D medial axis points and their local geometry. IEEE TPAMI 26(2), 238–251 (2004)

[10] Golovinskiy, A., Funkhouser, T.: Randomized cuts for 3D mesh analysis. ACM TOG 27(5), 454–463 (2008)

[11] Jalba, A., Kustra, J., Telea, A.: Surface and curve skeletonization of large 3D models on the GPU. IEEE TPAMI 35(6), 1495–1508 (2013)

[12] Katz, S., Leifman, G., Tal, A.: Mesh segmentation using feature point and core extraction. Visual Comput. 21(8), 649–658 (2005)

[13] Kiryati, N., Szekely, G.: Estimating shortest paths and minimal distances on digitized three-dimensional surfaces. Pattern Recognition 26, 1623–1637 (1993)

[14] Lee, Y., Lee, S., Shamir, A., Cohen-Or, D., Seidel, H.P.: Mesh scissoring with minima rule and part salience. CAGD 22, 444–465 (2005)

[15] Li, X., Woon, T., Tan, T., Huang, Z.: Decomposing polygon meshes for interactive applications. In: Proc. I3D, pp. 35–42 (2001)

[16] Lien, J., Keyser, J., Amato, N.: Simultaneous shape decomposition and skeletonization. In: Proc. ACM SPM, pp. 219–228 (2005)

[17] Liu, R., Zhang, H.: Segmentation of 3D meshes through spectral clustering. In: Proc. Pacific Graphics, pp. 298–305 (2004)

[18] Liu, Z., Tang, S., Bu, S., Zhang, H.: New evaluation metrics for mesh segmentation. Computers & Graphics 37(6), 553–564 (2013)

[19] Mount, D., Arya, S.: Approximate nearest-neighbor search (2015), http://www.cs.umd.edu/~mount/ANN

[20] Nooruddin, F., Turk, G.: Simplification and repair of polygonal models using volumetric techniques. IEEE TVCG 9(2) (2003)

[21] Reniers, D., Telea, A.: Tolerance-based feature transforms. In: Braz, J., Ranchordas, A., Araújo, H., Jorge, J. (eds.) VISAPP and GRAPP 2007. CCIS, vol. 4, pp. 187–200. Springer, Heidelberg (2007)

[22] Reniers, D., Telea, A.: Hierarchical part-type segmentation using voxel-based curve skeletons. Visual Comput. 24(6), 383–395 (2008)

[23] Reniers, D., Telea, A.: Part-type segmentation of articulated voxel-shapes using the junction rule. CGF 27(7), 1845–1852 (2008)

[24] Reniers, D., Telea, A.: Patch-type segmentation of voxel shapes using simplified surface skeletons. CGF 27(7), 1837–1844 (2008)

[25] Reniers, D., van Wijk, J.J., Telea, A.: Computing multiscale skeletons of genus 0 objects using a global importance measure. IEEE TVCG 14(2), 355–368 (2008)

[26] Roerdink, J., Hesselink, W.: Euclidean skeletons of digital image and volume data in linear time by the integer medial axis transform. IEEE TPAMI 30(12), 2204–2217 (2008)

[27] Shamir, A.: A formulation of boundary mesh segmentation. In: Proc. 3DPVT (2004)

[28] Shamir, A.: A survey on mesh segmentation techniques. CGF 27(8), 1539–1556 (2008)

[29] Shapira, L., Shamir, A., Cohen-Or, D.: Consistent mesh partitioning and skeletonisation using the shape diameter function. Visual Comput. 24(4), 249–259 (2008)

[30] Siddiqi, K., Bouix, S., Tannenbaum, A., Zucker, S.: Hamilton-Jacobi skeletons. IJCV 48(3), 215–231 (2002)

[31] Siddiqi, K., Pizer, S.: Medial representations: mathematics, algorithms and applications. Springer (2008)

[32] Siddiqi, K., Zhang, J., Macrini, D., Shoukofandeh, A., Dickinson, S.: Retrieving articulated 3D models using medial surfaces. Mach. Vis. Appl. 19, 261–275 (2008)

[33] Tierny, J., Vandeborre, J., Daoudi, M.: Topology driven 3D mesh hierarchical segmentation. In: Proc. SMI, pp. 215–220 (2007)

Algorithms and Implementation

Fast Computation of Greyscale Path Openings

Jasper J. van de Gronde$^{(\boxtimes)}$, Herman R. Schubert*, and Jos B.T.M. Roerdink

Johann Bernoulli Institute for Mathematics and Computer Science,
University of Groningen, P.O. Box 407, 9700 AK Groningen, The Netherlands
{j.j.van.de.gronde,j.b.t.m.roerdink}@rug.nl, h.robert.schubert@gmail.com

Abstract. Path openings are morphological operators that are used to preserve long, thin, and curved structures in images. They have the ability to adapt to local image structures, which allows them to detect lines that are not perfectly straight. They are applicable in extracting cracks, roads, and more. Although path openings are very efficient to implement for binary images, the greyscale case is more problematic. This study provides an analysis of the main existing greyscale algorithm, and shows that although its time complexity can be quadratic in the number of pixels, this is optimal in terms of the output (if the full opening transform is created). Also, it is shown that under many circumstances the worst-case running time is much less than quadratic. Finally, a new algorithm is provided, which has the same time complexity, but is simpler, faster in practice and more amenable to parallelization.

Keywords: Path openings · Algebraic morphological operators · Attributes · Stack opening · Time complexity

1 Introduction

It is often useful to be able to extract long, but not necessarily thick, structures, for example: guide-wires in X-ray fluoroscopy [2], roads in remote sensing images [12], and cracks for non-destructive testing [7]. A possible way to do this is to apply openings using fixed line structural elements [8]. However, these openings can be inadequate if the features of interest are not perfectly straight, and they can be fairly expensive if needed for multiple directions. Path openings [4] solve these issues by looking for paths in a small number of purpose-made directed acyclic graphs (DAGs) (see Fig. 1).

Path openings were originally introduced by Heijmans et al. [4], along with an algorithm for the binary case that is roughly linear in the number of pixels. Unfortunately, the proposed algorithm does not transfer immediately to the greyscale case. To compute the opening in the greyscale case, we would have to perform the binary opening for every unique grey level, which becomes highly inefficient for images with a large number of grey levels. Talbot and Appleton [11] improved on this by only looking at the differences between adjacent

---

Jasper J. van de Gronde, H.R. Schubert—These authors contributed equally.

J.J. van de Gronde—This research is partially funded by The Netherlands Organisation for Scientific Research (NWO), project no. 612.001.001.

J.A. Benediktsson et al. (Eds.): ISMM 2015, LNCS 9082, pp. 621–632, 2015.
DOI: 10.1007/978-3-319-18720-4_52

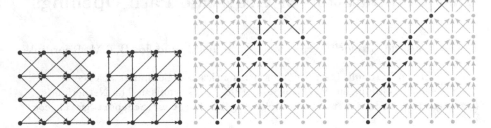

Fig. 1. Illustration of the DAGs used for path openings (our implementations only use graphs for the horizontal and vertical directions)

Fig. 2. A set $X \subseteq \Omega$ (black points on the left) and its path opening with $k = 7$ (black points on the right). Points only contained in paths of length 6 or less have been discarded (also see Fig. 3). The underlying adjacency graph is light grey, with black arrows highlighting the edges that are part of paths.

threshold levels, which can greatly reduce the amount of work needed. Unfortunately, despite several (additional) optimizations, the Talbot algorithm can take minutes when processing large images on modern desktop machines, raising the question whether it is possible to do better. Morard et al. [10] propose using 1D "path" openings on a carefully selected subset of possible paths to speed up path openings, but this only gives an approximation of the full path opening.

Despite some educated guesses [1, 3, 11] (of the *expected* time complexity), the time complexity of the algorithm developed by Talbot and Appleton (Talbot's algorithm for short) was never rigorously analysed. Here we show that the opening transform (the most general output of the algorithm) has a *space* complexity in $O(\min(d, |L|) |\Omega|)$, and that the *time* complexity of Talbot's algorithm is restricted purely by the size of the opening transform. Here d is the depth of the graph (typically the width/height of the image), L the set of grey levels in the image, and Ω the image domain. This implies the time complexity could be quadratic in the worst case, but only in the unlikely event that both the depth d *and* the number of grey levels $|L|$ can be considered of the same order as the number of pixels $|\Omega|$. We also introduce a new algorithm with the same time complexity as Talbot's algorithm, but which is simpler, faster in practice, and more amenable to parallelization. We demonstrate a considerable (additional) speedup when using a parallel implementation. Finally, it should be noted that our results are not limited to 2D images: in all our derivations and algorithms we simply assume the input is a weighted (sparse) directed acyclic graph.

2 Definitions

Path openings are constructed on directed acyclic graphs (DAGs). DAGs can be defined using a binary relation '$\mapsto$'. When $x \mapsto y$ (x adjacent to y) it means that there is an edge from x to y. We can also define the set of successors and predecessors for each pixel from the adjacency relationship.

Definition 1. *Let* $(\Omega, \mapsto)$ *form a DAG. We define the set of successors of a set* $X \in \mathcal{P}(\Omega)$ *as*

$$\delta(X) = \{y \in \Omega \mid x \mapsto y \text{ for some } x \in X\}. \tag{1}$$

The set of predecessors can similarly be defined as

$$\hat{\delta}(X) = \{y \in \Omega \mid y \mapsto x \text{ for some } x \in X\}. \tag{2}$$

Definition 2. *Let* $\mathbf{a} = (a_1, a_2, ..., a_k)$ *be a k-tuple of pixels, then* $\mathbf{a}$ *is called a path of length k iff* $a_i \mapsto a_{i+1}$ *for all* $i \in [1, k-1]$.

The set of elements in a path $\mathbf{a}$ is denoted by $\sigma(\mathbf{a})$. We can now define the concept of a path opening.

Definition 3. *Let* Π_k *be the set of all paths of length k, and let X be the foreground image. Then the path opening is defined by*

$$\alpha_k(X) = \bigcup \{\sigma(\mathbf{a}) \mid \mathbf{a} \in \Pi_k \text{ and } \sigma(\mathbf{a}) \subseteq X\}. \tag{3}$$

The path opening α_k gives the union of all sets of elements in k-tuple paths contained in X (see Fig. 2). It can be established that it is indeed an opening, i.e., it is increasing, anti-extensive and idempotent [4]. It is important to note that the final result is typically the union of the path openings for a set of different DAGs. However, this is immaterial to our discussion, so we will not stress this point further (a typical set of DAGs used is illustrated in Fig. 1).

The path opening can also be defined in an alternative manner. To this end, define the *opening transform* $\lambda_X : \Omega \to \mathbb{N}$ as the mapping that gives the maximum length of all paths restricted to X that visit a given pixel in Ω (so $\lambda_X(x) = 0$ for any $x \notin X$). Here $\mathbb{N}$ denotes the set of all non-negative integers. Using λ_X we then create the following definition of a path opening:

$$\alpha_k(X) = \{x \in X \mid \lambda_X(x) \geq k\}. \tag{4}$$

This simply preserves those pixels satisfying the path length criterion. Since only path *lengths* are important in the path opening, we can efficiently compute a path opening without keeping track of the paths themselves, as shown below.

In the greyscale case, we conceptually apply the binary algorithm to every upper level set. This can be expressed using the greyscale opening transform $\lambda_f : \Omega \times \mathbb{R} \to \mathbb{N}$, which returns the maximum path length for a certain position and threshold in a greyscale image $f : \Omega \to \mathbb{R}$.

3   Sizing Up the Opening Transform

The time complexity of an algorithm is always bounded from below by the space complexity of its output. After all, it has to have the time to construct this output. The algorithms for path openings that we discuss here (and that have been discussed, to our knowledge, in the literature) all output the full opening transform, or are all capable of outputting the opening transform (without this affecting their time complexity). In fact, we will see that their time complexities can be given solely in terms of the size of the opening transform. Theorem 1 and Corollary 1 give bounds for the size of the opening transform, and by extension for the time complexities of the presented algorithms.

Suppose we have a greyscale image $f : \Omega \to \mathbb{R}$ and the associated opening transform $\lambda_f : \Omega \times \mathbb{R} \to \mathbb{N}$. How much data is necessary to represent this opening transform? If we just look at a certain position x, then the mapping $\lambda_f(x) : \mathbb{R} \to \mathbb{N}$ given by $l \mapsto \lambda_f(x,l)$ can be seen to be weakly decreasing. That is, as the threshold level goes up, the maximum path length must go down (because the upper level sets become smaller). This means $\lambda_f(x)$ can be *represented* using any set $\Lambda_f(x) \in \mathcal{P}(\mathbb{R} \times \mathbb{N})$ of pairs of grey levels and their associated maximum path lengths, such that

$$\lambda_f(x,l) = \sup\{\lambda \mid (l',\lambda) \in \Lambda_f(x) \text{ and } l' \geq l\}. \tag{5}$$

When the set over which a supremum is computed is empty, the result is taken to be zero (there is no path at this position and threshold level). It should be clear that if $\Lambda_f(x)$ and $\Lambda'_f(x)$ are two sets satisfying Eq. (5) (so both give rise to the correct $\lambda_f(x)$), then $\Lambda_f(x) \cap \Lambda'_f(x)$ must also satisfy Eq. (5). In fact, it can be shown that this is true for the intersection of any set of sets that satisfy Eq. (5). We can thus speak of the smallest set of pairs of grey levels and maximum path lengths that satisfies Eq. (5), and in the remainder we will assume that $\Lambda_f(x)$ is in fact this smallest set of pairs. This means that it cannot contain two pairs with the same grey level or maximum path length. With a finite number of pixels $|\Omega|$, we can now bound the space needed to represent the opening transform.

Theorem 1. *If L is the set of grey levels in $f : \Omega \to \mathbb{R}$ and d is the maximum path length in the DAG given by '$\mapsto$' on Ω, then the total number of pairs in $\Lambda_f : \Omega \to \mathcal{P}(\mathbb{R} \times \mathbb{N})$ is bounded from below by $|\Omega|$ and from above by the class $O(\min(d, |L|) |\Omega|)$. Both bounds can be reached.*

Proof. The lower bound follows from the fact that for each position x the path length at threshold $f(x)$ is greater than zero, implying that $\Lambda_f(x)$ always contains at least one pair. For the upper bound we simply prove that the number of pairs in $\Lambda_f(x)$ is less than or equal to $\min(d, |L|)$ for all $x \in \Omega$, the statement then follows by multiplying this bound by $|\Omega|$. That $|\Lambda_f(x)|$ is less than or equal to the number of grey levels follows from the fact that we cannot have two pairs in $\Lambda_f(x)$ with the same grey level. Similarly, we cannot have more than d pairs in $\Lambda_f(x)$, since we cannot have more than d distinct positive integer path lengths less than or equal to d (zero path lengths would not occur explicitly in $\Lambda_f(x)$).

To see that the lower bound can be reached, just consider a constant image. There is then exactly one pair in $\Lambda_f(x)$ for all $x \in \Omega$. To prove that the upper bound can be reached, consider an image that consists of a sequence of rows with strictly increasing grey levels (but constant within each row), with edges (only) between adjacent rows (see Fig. 1). In this case the pixels on the first row have one pair in $\Lambda_f(x)$, the pixels on the second row two pairs, and so on, for a total numbers of pairs in $\Theta(d\,|\Omega|)$. If the grey levels stay constant after $|L| \le d$ rows, the total number of pairs is in $\Theta(\min(d,|L|)\,|\Omega|)$.   □

The above result shows that even on images with high bit depths, the size of the path opening transform will typically not be quadratic in the number of pixels, as d tends be $O(\sqrt{|\Omega|})$. On the other hand, for low bit depths the size will be linear in the number of pixels (albeit with a potentially high constant). Still, the resulting space complexity *can* be worse than linear. One optimization that has been applied in the literature is to consider all path lengths above a certain threshold to be equivalent. If we know the path length threshold we will be interested in, or at least some upper bound t, then we can define $\lambda_f^t(x,l) = \min(\lambda_f(x,l),t)$ and the associated Λ^t. Crucially, Λ^t would contain at most one pair with a path length greater than or equal to t (as it is known that the path length will be even greater for lower grey levels).

Corollary 1. *If L is the set of grey levels in f, d is the maximum path length in the DAG given by '$\hookrightarrow$' on Ω, and $t > 0$ is the maximum path length threshold, then the total number of pairs in $\Lambda_f^t : \Omega \to \mathcal{P}(\mathbb{R} \times \mathbb{N})$ is in $O(\min(t,d,|L|)\,|\Omega|)$.*

Proof. This follows from Theorem 1, except that we can now also constrain the number of positive path lengths by t: at most $t - 1$ positive path lengths less than t and at most one path length greater than or equal to t.   □

If we were to allow non-integer path lengths we get a bound in $O(|L|\,|\Omega|)$, but a lot of work on path openings does use integer path lengths. In either case, the reader might wonder whether it is possible to give better bounds. As shown above, it is possible to provide examples for which the path opening actually does require the amount of space suggested by the bounds derived above. On the other hand, we would expect a typical image to require much less storage, so there is definitely some room for making the above bounds more precise. Also, although it seems doubtful that much could be improved in terms of representing the opening transform at a specific position, it is definitely not beyond the realm of possibility that there exists a more efficient representation of the opening transform as a whole. We have not found one, however.

4   Algorithms

In this section we present three algorithms: the traditional binary algorithm, Talbot's algorithm, and our new stack-based algorithm.

4.1   Binary Images

An algorithm which computes the binary path opening efficiently is given by Heijmans et al. [4]. The idea is to do two sequential scans on the binary image, computing the largest path length up to each pixel in opposite directions, and then combining these results to compute λ. In the first scan we traverse all the rows (or columns, or ...) of the image from top to bottom. Let $\lambda_+ : \Omega \to \mathbb{R}$ be the map which gives the maximum path length for each pixel $x \in X$ based only on its *predecessors*. To compute such a map we use the following relations [4] (only for $x \in X$, for other pixels the λ's are set to zero):

$$\lambda_+(x) = \max_{y \in X \mid y \mapsto x} \lambda_+(y) + 1, \qquad \lambda_-(x) = \max_{y \in X \mid x \mapsto y} \lambda_-(y) + 1. \qquad (6)$$

It was shown by Heijmans et al. [4] that by combining $\lambda_+(x)$ and $\lambda_-(x)$, we can recover the maximum path length using the following relation (for $x \in X$):

$$\lambda(x) = \lambda_+(x) + \lambda_-(x) - 1. \qquad (7)$$

This notion is intuitive, as $\lambda_+(x)$ holds the maximum path length of all paths ending in x, and $\lambda_-(x)$ holds the maximum path length of all paths starting in x, so by combining them we recover the maximum path length through x. We subtract one from the sum of the two partial path lengths to avoid counting pixel x twice. Figure 3 shows an example of the computation of λ.

The above can be easily turned into an actual algorithm by topologically sorting [5] the DAG and then applying Eq. (6) in (reverse) topological order. This way the λ_+ and λ_- only need to be set once for each pixel.

4.2 Talbot's Algorithm

If we apply the algorithm described in the previous section to all upper level sets of an image $f : \Omega \to \mathbb{R}$, then we can find the path opening (transform) by combining all those results. However, this could be quite expensive. Luckily, as described by Talbot and Appleton [11], the algorithm described above can be modified to only update λ_+ and λ_- based on the changes resulting from going to one grey level to the next. This works by first sorting all grey levels and initializing all $\lambda_\pm$ to the largest possible path length, and then applying Algorithm 1 for each grey level (and an analogous algorithm for λ_-), in increasing order of grey level. For each grey level, after λ_+ and λ_- are updated, the algorithm goes through all the affected pixels and records any changes in per-pixel lists to

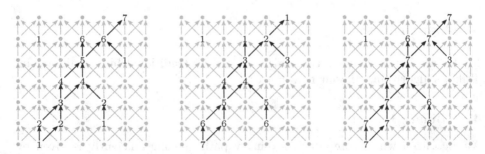

Fig. 3. Computation of λ in the example shown in Fig. 2. From left to right: the forward scan pass λ^+, the backward scan pass λ^-, the calculated length per pixel λ

Algorithm 1. Update of λ_+.

Input : λ_+ for the current grey level.
Output: λ_+ for the next grey level.

1 Initialize the priority queue Q with all pixels at the current grey level.
    `/* Priorities compatible with the topological order of the DAG. */`
2 **while** Q *not empty* **do**
3   remove smallest pixel x from Q
4      $\lambda \leftarrow 0$
5   **if** $f(x)$ *above current grey level* **then**
6          $\lambda \leftarrow \max_{y \in X | y \mapsto x} \lambda_+(y) + 1$
7   **if** $\lambda < \lambda_+(x)$ **then**
8          $\lambda_+(x) \leftarrow \lambda$
9   push successors of y onto Q

---

build up the opening transform. Although presented somewhat differently from the original, we will call the resulting algorithm "Talbot's algorithm". Instead of computing the entire opening transform, it can also directly compute the opening (but this does not affect the time complexity).

Theorem 2. *Assume each position has $O(1)$ predecessors/successors (the DAG induced by '$\mapsto$' is sparse), and that the priority queue in Algorithm 1 allows insertion and removal in $O(\log(|Q|))$. Talbot's algorithm then has a time complexity of $O(|\Lambda_f| \log(|\Lambda_f|))$, where $|\Lambda_f|$ is taken to mean the total number of pairs needed to represent Λ_f and $f : \Omega \to \mathbb{R}$ is the input image.*

Proof. Disregarding the time needed to sort the grey levels, it can be seen that asymptotically the running time is determined by the work done in Algorithm 1 (the amount of work done by the rest of the algorithm is dominated by the amount of work done in the update steps).

The crucial observation is now that $\Lambda_f(x)$ has a pair with the current grey level if and only if x is processed when updating λ_+ and/or λ_- *and* while doing so, the condition on Line 7 is true (or the analogous condition for λ_-). Each time this happens $O(1)$ elements are pushed onto Q (due to the sparse graph assumption), so if we look at all applications of Algorithm 1, the total number of queue pushes and (thus) executions of the while loop must be in $\Theta(|\Omega| + |\Lambda_f|) = \Theta(|\Lambda_f|)$. Assuming a priority queue with $O(\log(|Q|))$ insertion and removal, the total amount of work done is then in $O(|\Lambda_f| \log(|\Lambda_f|))$. Finally, we conclude that sorting all grey levels requires at most $O(|\Omega| \log(|\Omega|))$ time, so it does not alter the time complexity. $\qquad\qquad\Box$

The assumption that the DAG is sparse is fairly benign, as all existing use cases (to our knowledge) satisfy this assumption. That the priority queue allows $O(\log(|Q|))$ insertion and removal is also fairly standard. In some cases (like the typical DAGs used on images), it is even possible to get constant-time insertion and removal, by grouping pixels into "rows" based on their depth in the DAG. We conjecture that in general it may be possible to get (amortized) constant-time insertion and removal by using specialized data structures. In combination with linear time sorting of all (integer) grey levels, this would put the time complexity of Talbot's algorithm in $\Theta(|\Lambda_f|)$.

It should be noted that some tweaks to Talbot's algorithm can further reduce the time complexity by essentially restricting the opening transform as in Corollary 1, as well as ignoring pixels whose path length dropped *below* a certain threshold. It is currently not entirely clear how this affects the time complexity of the algorithm. Also, although Luengo Hendriks [6] presents a modification of Talbot's algorithm that is dimensionality independent, it does not necessarily process the pixels in optimal order, leading us to start from Talbot's algorithm instead (note that our presentation of Talbot's algorithm is also dimensionality independent). Similarly, if we look at Talbot's code, it loops over *all* rows/columns in each update step, while our code only visits those rows/columns where it has to do something. We do not expect these implementation differences to make a huge difference in performance, but it should be understood that due to these differences actual implementations might have slightly different time complexities from the one presented here.

4.3   Stack-Based Path Openings

Talbot's algorithm is essentially optimal in terms of its time complexity, but a truly optimal implementation (without the logarithmic factor) can be quite complex, and the algorithm has a fairly random memory access pattern, which is undesirable in most modern computer architectures. In this section we present an algorithm that suffers from none of these problems, based on the 1D algorithm presented by Morard et al. [9].

The basic idea is to use the traditional binary algorithm, but instead of having the scalar $\lambda_+(x)$ and $\lambda_-(x)$, we use sets $\Lambda_+(x)$ and $\Lambda_-(x)$, represented by non-redundant ordered lists of pairs of grey levels and path lengths. The 1D algorithm only needs to push and pop elements, so can use a traditional stack. Our algorithm does the same, but also needs to merge lists. We will still refer to the lists as stacks though. Algorithm 2 details the algorithm needed to compute Λ_+. Λ_- is computed in much the same way in a second pass over the data, and since all the lists are already sorted, merging Λ_+ and Λ_- to get the final opening transform can be done easily and efficiently. Note that the merge procedures should leave their output sorted and without any redundant pairs. This can be accomplished using a technique similar to the one used in merge sort.

Theorem 3. *Assume each position has $O(1)$ predecessors/successors (the DAG induced by '$\mapsto$' is sparse), the stack-based path opening then has a time complexity in $\Theta(|\Lambda_f|)$, where $f : \Omega \to \mathbb{R}$ is the input image.*

Proof. First of all topological sorting can be done in $O(|\Omega|)$ [5], so will not be a bottleneck. Similarly, merging the two partial path opening transforms into the final answer just requires iterating over the partial transforms (once), so also does not add anything to the time complexity. It remains to examine the time spent in Algorithm 2.

Algorithm 2. Computation of Λ_+.

Input   : The input image f.
Output: The partial opening transform Λ_+.

1 **for** x in Ω in topological order **do**
2    $\Lambda_+^{temp} \leftarrow$ merge($\{\Lambda_+(y) \mid y \mapsto x\}$)
3    $\lambda_+^{temp} \leftarrow \max(\{0\} \cup \{\lambda \mid (l', \lambda) \in \Lambda_+^{temp}$ and $l' \geq f(x)\}) + 1$
4    $\Lambda_+(x) \leftarrow \{(l, \lambda + 1) \mid (l, \lambda) \in \Lambda_+^{temp}$ and $l < f(x)\} \cup \{(f(x), \lambda_+^{temp})\}$

Algorithm 2 visits each pixel exactly once. For each pixel it first merges the partial transforms (stacks) from its predecessors (successors if computing Λ_-), then it computes $\lambda_+^{temp} = \lambda_+(x, f(x))$, and finally it makes sure $f(x)$ is the highest grey level in the new stack, while updating path lengths for grey levels below $f(x)$. Since the number of stacks being merged in Line 2 is in $O(1)$, we can assume the merge procedure to take time linear in its input. Since each stack is involved in $O(1)$ merges, the *total* time taken up by all merges (in both passes) is in $O(|\Lambda_+| + |\Lambda_-|)$. Lines 3 and 4 can be implemented together with an overall time complexity in $\Theta(|\Lambda_\pm(x)|)$. Summarizing, the total amount of work done by Algorithm 2 (in both passes, one for Λ_+ and one for Λ_-) is in $\Theta(|\Lambda_+| + |\Lambda_-|)$.

We now note that $|\Lambda_f| \leq |\Lambda_+| + |\Lambda_-| \leq 2|\Lambda_f|$. The lower bound follows from the fact that (by definition) every pair in Λ_f must correspond to a pair in Λ_+ or Λ_-. The upper bound can be shown similarly, by considering that every pair in Λ_+ and Λ_- must correspond to a pair in Λ_f. In particular, because of the monotonicity of the (partial) path opening transforms we cannot have a pair in Λ_+ "cancel out" a pair at the same grey level in Λ_-. We can now conclude that the time complexity of the stack-based path opening is indeed in $\Theta(|\Lambda_f|)$. ☐

The optimization referred to in Corollary 1 can be applied to the stack-based algorithm very easily, preserving the output sensitivity of the algorithm. The other optimization applied by Talbot and Appleton [11] seems to be harder to apply to the stack-based algorithm though. The problem is that this optimization discards any points whose *total* path length drops below the desired threshold, and in the stack-based algorithm we only have access to the total path length (for any grey level) after all computations have been done. For now it is not clear how this effects the time complexity of the algorithm. In terms of the space complexity, Talbot's algorithm is the clear winner though, as the stack-based algorithm always has to build at least part of the opening transform.

The stack-based algorithm will process an image row by row (instead of "row" one can also read column, or diagonal), and within each row the results only depend on the previous row. This allows us to compute the values within a single row in parallel. This kind of parallelization is somewhat limited by needing a synchronization point after each row, but as the next section demonstrates, it still allows for a very decent speedup using a small number of cores. Applying this technique to Talbot's algorithm would be possible, but would be complicated by not knowing beforehand what pixels need to be processed in each row. Also, this would involve (even) more synchronization, as Talbot's algorithm uses multiple (simpler) passes rather than one pass (per direction). It would of course also be possible to compute Λ_+ and Λ_- independently, as well as to process each of the directions independently, but these measures would apply equally well to either algorithm and are not evaluated here.

5   Results

To assess the performance of the path openings, we created an application in C++ which implements all of the previously discussed algorithms. In particular, we implemented the Talbot opening as discussed in Section 4.2, and the newly introduced stack opening as discussed in Section 4.3. Although we use our own implementation of Talbot's algorithm, it was verified that our implementation has similar or better performance. Although all the algorithms are applicable to arbitrary DAGs, we decided to implement the algorithms using the graphs illustrated in Fig. 1. Note that in the interest of simplicity we only used the horizontal and vertical ones, while Talbot originally included the diagonal ones as well. For the purposes of seeing which implementation was faster the diagonal ones were disabled in Talbot's implementation. This decision does not affect the above analysis in any way (since it holds for arbitrary sparse DAGs), and we also do not expect any major influence on the results in this section either (our parallelization strategy could be slightly less efficient on these other DAGs).

The algorithms were tested for their performance on an Intel® Xeon® E5-2630 with 32 logical cores and 512 GB of memory. The tests measure the performance of the complete opening transform. The differently sized images were generated by downscaling the test images using bilinear interpolation. Both 8-bit images and 32-bit images were assessed, where the 32-bit image is created from a gradient image where all the pixel values are strictly increasing from top to bottom and left to right. This allows us to see the worst-case behaviour of the algorithms and confirm our bound. The running times of the stack-based algorithm and Talbot's algorithm are shown in Fig. 4.

On sparse graphs, both the stack opening and the Talbot opening have a time complexity of $O(\min(d, |L|) |\Omega|)$, where $|L| \leq 256$, so we expect (roughly) linear behaviour. This is confirmed by Fig. 4a. Figure 4b, however, shows superlinear behaviour. This is expected, as the number of grey values is no longer the limiting factor. Since d is is equal to the width or the height of the image, the time complexity should be in $O(n\sqrt{n})$ (with n the number of pixels $|\Omega|$). The function $f(n) = c\,n\sqrt{n}$ was fitted to the 32-bit results using least squares, where c is the fitted parameter. We indeed see roughly $n\sqrt{n}$ behaviour in the results.

Both in the single-threaded case and in the multi-threaded case, the stack-based path opening outperforms Talbot's algorithm, by a factor of roughly 4 (or 25 using twelve threads). This is likely because of the data locality of the algorithm, as the algorithm walks more or less sequentially through the image, rather than having to reprocess certain sections repeatedly.

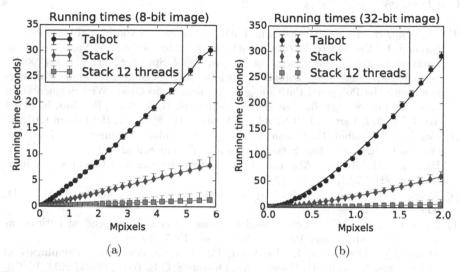

Fig. 4. Running times of the Talbot algorithm, and the stack-based opening (both generating the full opening transform). (a) On an 8-bit image, and (b) on an artificial 32-bit image representing a gradient. Code available at http://bit.ly/1BTC2Je.

6   Conclusion

We have shown that the space complexity of the path opening transform is in $O(\min(d, |L|) |\Omega|)$, with d the depth of the graph being processed, L the set of grey levels in the image, and Ω the image domain. We also showed that although there might be room for refinement, it is possible to construct graphs that reach this bound. Next we analysed the time complexity of a paraphrased version of the algorithm proposed by Talbot and Appleton [11], and found that it is (depending on the exact implementation) optimally output-sensitive, assuming the full opening *transform* is output. Finally, we presented a new algorithm that is easier to implement (at least in optimal fashion), is still optimally output-sensitive, allows for easier parallelization, and is significantly faster in practice. We presented results demonstrating our new algorithm outperforming an implementation of Talbot's algorithm by a factor of roughly 4 (and 25 when using parallelization on 12 cores). We also experimentally demonstrated that for high bit-depth images the performance of the algorithms can indeed scale superlinearly.

In future work it would be interesting to take a further look at various optimizations that one can apply if only part of the opening transform (or indeed just the actual opening) is needed. It would also be interesting to see whether the opening transform can be stored more efficiently than we propose here, and if so, whether this can actually lead to faster algorithms. Additionally, it would be interesting to try adapting the stack-based algorithm to robust [3] and incomplete [4] path openings, or other schemes for making path openings more robust to noise. Different schemes for parallelization could also be explored (for example by dividing the grey levels, rather than pixels, among different processors).

References

[1] Appleton, B., Talbot, H.: Efficient Path Openings and Closings. In: Ronse, C., Najman, L., Decencière, E. (eds.) Mathematical Morphology: 40 Years On, Computational Imaging and Vision, vol. 30, pp. 33–42. Springer Netherlands (2005)

[2] Bismuth, V., Vaillant, R., Talbot, H., Najman, L.: Curvilinear Structure Enhancement with the Polygonal Path Image - Application to Guide-Wire Segmentation in X-Ray Fluoroscopy. In: Ayache, N., Delingette, H., Golland, P., Mori, K. (eds.) MICCAI 2012, Part II. LNCS, vol. 7511, pp. 9–16. Springer, Heidelberg (2012)

[3] Cokelaer, F., Talbot, H., Chanussot, J.: Efficient Robust d-Dimensional Path Operators. IEEE J. Sel. Top. Signal. Process. 6(7), 830–839 (2012)

[4] Heijmans, H., Buckley, M., Talbot, H.: Path Openings and Closings. J. Math. Imaging Vis. 22(2), 107–119 (2005)

[5] Kahn, A.B.: Topological Sorting of Large Networks. Commun. ACM 5(11), 558–562 (1962)

[6] Luengo Hendriks, C.L.: Constrained and dimensionality-independent path openings. IEEE Trans. Image Process. 19(6), 1587–1595 (2010)

[7] Morard, V., Decencière, E., Dokladal, P.: Geodesic Attributes Thinnings and Thickenings. In: Soille, P., Pesaresi, M., Ouzounis, G.K. (eds.) ISMM 2011. LNCS, vol. 6671, pp. 200–211. Springer, Heidelberg (2011)

[8] Morard, V., Dokládal, P., Decencière, E.: Linear openings in arbitrary orientation in O(1) per pixel. In: IEEE International Conference on Acoustics, Speech and Signal Processing (ICASSP), pp. 1457–1460. IEEE (2011b)

[9] Morard, V., Dokládal, P., Decencière, E.: One-Dimensional Openings, Granulometries and Component Trees in O(1) Per Pixel. IEEE J. Sel. Top. Signal. Process. 6(7), 840–848 (2012)

[10] Morard, V., Dokládal, P., Decencière, E.: Parsimonious Path Openings and Closings. IEEE Trans. Image Process. 23(4), 1543–1555 (2014)

[11] Talbot, H., Appleton, B.: Efficient complete and incomplete path openings and closings. Image Vis. Comput. 25(4), 416–425 (2007)

[12] Valero, S., Chanussot, J., Benediktsson, J.A., Talbot, H., Waske, B.: Advanced directional mathematical morphology for the detection of the road network in very high resolution remote sensing images. Pattern Recognit. Lett. 31(10), 1120–1127 (2010)

Ranking Orientation Responses of Path Operators: Motivations, Choices and Algorithmics

Odyssée Merveille[1,2(✉)], Hugues Talbot[1], Laurent Najman[1], and Nicolas Passat[2]

[1] ESIEE-Paris, LIGM, CNRS, Université Paris-Est, Paris, France
odyssee.merveille@esiee.fr
[2] CReSTIC, Université de Reims Champagne-Ardenne, Reims, France

Abstract. A new morphological operator, namely RORPO (Ranking Orientation Responses of Path Operators), was recently introduced as a semi-global, morphological alternative to the local, Hessian-based operators for thin structure filtering in 3D images. In this context, a previous study has already provided experimental proof of its relevance by comparison to such differential operators. In this article, we present a methodological study of RORPO, which completes the presentation of this new morphological filter. In particular, we expose the motivations of RORPO with respect to previous morphological strategies; we present algorithmic developments of this filter and the underlying robust path operator; and we discuss computational issues related to parametricity and time efficiency. We conclude this study by a discussion on the methodological and applicative potentiality of RORPO in various fields of image processing and analysis.

Keywords: Antiextensive filtering · Robust path openings · Thin structure detection · 3D grey-level imaging · Vesselness

1 Introduction

Thin structures in nD images are characterized by a lower size in (at least) one of their n dimensions, compared to the others. Among them, linear structures, i.e., the patterns that present a low size in $n-1$ dimensions, are the most difficult to handle. The difficulty of coping with such patterns increases with the dimension of the surrounding image, as the ratio of the structures volume versus the image volume decreases – resulting in a high sensitivity to noise – while the degrees of freedom of their geometry increases. Dealing with these issues of sparcity and geometry complexity is still tractable in 2D but becomes difficult in 3D.

This has motivated the development of various filtering and segmentation methods for thin 1D structures in the fields of imaging that deal with 3D data, e.g., medical imaging or material sciences [1, Chapters 1, 2]. In this context, we recently introduced a new filter called RORPO (Ranking Orientation Responses of Path Operators) [2], as a semi-global, discrete, graph-based alternative to the local, continuous, second-order derivative Hessian filter proposed by Frangi et al. [3], and its numerous variants.

This research was funded by the French *Agence Nationale de la Recherche* (Grant Agreements ANR-10-BLAN-0205 and ANR-12-MONU-0010) and through *Programme d'Investissements d'Avenir* (LabEx Bézout, ANR-10-LABX-58).

© Springer International Publishing Switzerland 2015
J.A. Benediktsson et al. (Eds.): ISMM 2015, LNCS 9082, pp. 633–644, 2015.
DOI: 10.1007/978-3-319-18720-4_53

In [2], RORPO was described mainly from a behavioural point of view, motivated against continuous and local approaches, and experimentally validated in the field of 3D medical imaging, where Hessian-based filters are considered the gold standard.

In the present article we now complete our description of RORPO by presenting this operator in the context of graph-based and morphological filters (Section 2), in order to motivate its definition and justify its new contributions with respect to the morphological state of the art (Section 3). Then, we describe algorithmic key-points, hard and soft parametricity and the computational cost of RORPO (Section 4). An illustration of RORPO to materials image denoising and medical image segmentation is proposed (Section 5), and a short discussion of its further uses for segmentation guidance (Section 6) conclude this article, which can be viewed as a companion to [2].

2 Morphological Filtering of 1D Structures in Grey-Level Images

In grey-level images, thin 1D structures can present a complex geometry (e.g., curvature, tortuosity) and a complex topology (e.g., branches, cycles). However, we may assume that they are composed of *connected*, *locally straight* segments of *locally extremal intensity*. These hypotheses induced three main families of approaches.

2.1 Local (SE-based) Approaches

Following the local straightness and extremal intensity hypotheses, a basic idea is to develop filters from small kernels, fitting these specific properties. In the framework of mathematical morphology, structuring elements (SEs) play the role of such kernels.

Two dual approaches have been considered. The first models a 1D segment by a small straight SE [4,5], to carry out opening or closing operations by line segments of arbitrary orientation [6,7]. The second models the background in the orthogonal hyperplane of the segment, to carry out grey-level hit-or-miss transforms [8,9].

Even if the shape of a SE is unyielding, some degrees of freedom remain, for instance by varying its orientation. To represent the morphological analogue of a filter bank, several strategies have been experimented with: rotating structuring elements [10], knowledge-based parameterization [8], or content-based spatially-variant mathematical morphology [11]. Blurred operators [9] have been proposed to increase the robustness of the filtering.

Nevertheless, the inflexible geometry of such SEs remains a limitation to the accuracy of these approaches, that progressively led to consider more adaptive SEs [12].

2.2 Global (Connectivity-based) Approaches

Following the global connectedness and locally extremal intensity hypotheses, an alternative approach is to consider connected filters. In the framework of mathematical morphology, the notion of component-tree [13] has been specifically investigated.

In this context, attribute-based methods have been developed for antiextensive filtering. The attributes are mostly scalar [14,15], allowing threshold-based interaction. Recent effort have been directed towards the development of geodesic attributes [16], particularly designed for thin structures. Vectorial attributes are less frequently used [17], due to a more complex handling, often requiring a learning step.

The main drawback of connected filters is their unability to split connected components within an image, that may result in erroneous / undesired connections between 1D structures and artifacts, or between 1D branching segments. Various attempts at minimizing these drawbacks have been proposed, either via tilling approaches [17], or with undirected variants of component-trees [18].

It is worth mentionning that hybrid SE / connected strategies have been pursued, e.g., in [19] for reconnection purposes. Links between connectivity-based and path-based approaches (described below) were also investigated in [20].

2.3 Semi-global (path-based) Approaches

Following the global and local concepts of connectedness, straightness and extremal intensity hypotheses, approaches based on optimal paths lie at the convergence of the above two families. Global optimal paths consist of finding a series of successive vertices in a graph that minimizes a cost between two points, following standard strategies (e.g., Dijkstra algorithm; see [21] for a recent formal discussion).

In [22], a notion of local optimal path was pioneered. The purpose was to restrict path search to a given distance, and in a given cone of orientations, in order to find the best paths starting from a given point. This paradigm led to the development of a notion of Path Operators (PO) [23], which is an SE-based approach, where the considered SEs are a family of paths i.e., thin elongated connected sets, instead of a fixed shape, thus enabling a higher flexibility in geometry and size, while preserving a 1D semantics.

Algorithmic efforts were conducted to render such an approach dimension-independent [24] and robust to noise [25,26], leading to a notion of Robust Path Operator (RPO). Sparse paradigms [27] were also proposed to avoid redundant computation.

The use of PO and RPO, and a few other path-based paradigms (e.g., polygonal path image [28]), have been proposed in medical imaging and remote sensing [29,30,31], but for 2D applications only. In the sequel, we recall the basics of PO, and we discuss the challenges related to the 2D–3D transition, which constitute the genesis of RORPO.

3 Path Operators and the 2D–3D Transition Problem

3.1 Path Operators: Basic Notions

Path operators can be path openings or closings. Without loss of generality, we will discuss the opening case when convenient to avoid unnecessary repetitions.

Binary Path Operators. A binary image X is simply defined as a finite set of points. In the case of path operators, we are interested in the neighbourhood relations between the points of X. More precisely, we are interested in the notion of paths, defined as sets of connected points (i.e., successive neighbours). Given an adjacency (non necessarily symmetric) relation $\rightarrow$ on X, we can define the set $\Pi_L(X)$ of all the paths π of the graph

$(X, \rightarrow)$ of length equal to L. The binary path opening $\alpha_L(X) \subseteq X$ is defined as the union of all these paths:

$$\alpha_L(X) = \bigcup \{\sigma(\pi) \mid \pi \in \Pi_L(X)\} \tag{1}$$

where $\sigma(\pi) = \{x_1, x_2, \ldots, x_L\}$ is the set of points of X forming the path π.

A binary path opening preserves each point of X belonging to at least one path. From the above definitions, it is plain that α_L is increasing, antiextensive, and idempotent; from an algebraic point of view, it is a morphological opening. The α_L operator can also be viewed as an SE-based opening. Indeed, following a spatially-variant paradigm, it is an opening with respect to the family of all the SEs that can be defined from the graph $(X, \rightarrow)$. Practically, an elementary graph pattern periodically reproduced over a whole regular set $\Omega \supseteq X$ is used, resulting in a set of SEs that is invariant by translation. For instance, if we set $\Omega \subset \mathbb{Z}^2$, and $\rightarrow$ as an arc between (x, y) and $(x, y + 1)$, then α_L defines the standard opening by a linear vertical SE of L points.

Grey-Level Path Operators. A grey-level image I associates a value $I(x) \in \mathbb{R}$ to each point x of a finite set Ω. Practically, such an image is then a function $I : \Omega \rightarrow \mathbb{R}$. The notion of path opening can be easily extended from binary to grey-level images, following a flat morphology paradigm. By applying the binary path opening α_L on the binary level sets $I_\lambda = \{x \in \Omega \mid I(x) \geq \lambda\}$ of the image I, obtained by thresholding at value λ, we define the grey-level path opening operator A_L:

$$A_L(I) : x \mapsto \bigvee \{\lambda \mid x \in \alpha_L(I_\lambda)\} \tag{2}$$

3.2 The 2D–3D Transition Problem

To filter thin structures, PO are parameterized by a given adjacency $\rightarrow$ defined on Ω. As stated above, this adjacency is generally defined from an elementary pattern, reproduced periodically over Ω, leading to a regular graph structure. This pattern allows us to tune the general orientation, but also the tortuosity of the authorised paths by defining locally the angular area where each new point of the path can be defined with respect to its predecessor. For instance, the pattern $(x, y) \rightarrow (x, y + 1)$ in $\mathbb{Z}^2$, previously evoked, only authorises vertical paths without any tortuosity; by contrast, the pattern $(x, y) \rightarrow \{(x - 1, y + 1), (x, y + 1), (x + 1, y + 1)\}$, authorises vertical paths within an angular cone of $[\pi/4, 3\pi/4]$, around the vertical direction, and a maximal tortuosity of $\pi/2$ between three successive points.

Filtering approaches based on PO have been recently and successfully applied, e.g., in materials science, remote sensing or medical imaging. These applications [29,24,30,31] have been mostly limited to 2D images.

Indeed, in $\mathbb{Z}^2$, PO preserves the zones of the image in which the paths of authorised orientation, tortuosity and length, can lie; this results in preserving 1D linear structures, plus large 2D flat zones, the second being easily discriminated from the first by standard opening. In $\mathbb{Z}^3$, two kinds of thin structures coexist: 1D line and 2D plane structures; both are easily discriminated from 3D flat zones. However, PO cannot directly distinguish 1D lines from 2D planes. Indeed, a path lying into a 1D line also lies into a 2D plane in the same orientation.

Based on these facts, we now describe the process that led us to define RORPO as the natural way to use PO for 1D thin structure filtering from 3D images.

3.3 A First Approach: Top-Hat Path-Based Filtering

The main limitation of PO is its unability to discriminate 1D lines from 2D planes. A simple idea consists of applying a post-processing operator for performing such a discrimination. Let us consider an object X of $\mathbb{Z}^3$, and let us note L, L_2 and L_3 its size in its three principal directions, with $L \geq L_2 \geq L_3$. A 1D line structure will satisfy $L \gg L_2 \simeq L_3$; a 2D plane structure $L \simeq L_2 \gg L_3$, and a 3D structure $L \simeq L_2 \simeq L_3$.

Only 1D objects exhibit a significant difference between L and L_2. In other words, there exists a value $k > 0$ such that $L > kL_2$. Let d be the orientation of the 1D structure; a radial opening [7] with a segment SE of size $\frac{L}{k}$ in the plane P orthogonal to d, will preserve any structure in P that is longer than $\frac{L}{k}$ (Figure 1). Based on these considerations, applying a top-hat transformation involving such a radial opening on $\alpha_L(X)$ should be sufficient to remove the 2D planes while preserving the 1D structures.

This approach, although simple, presents several drawbacks. The first is quantitative: an object is considered 1D if it is at least k times longer than larger, but this parametric value k is not necessarily easy to tune. The second is geometric: the directions of the path operators are defined by angular cones. Nevertheless, the (radial) opening is not as flexible as PO: it is performed only in one plane P, namely the one orthogonal to the central axis of the cone. This may induce approximate or erroneous results. The third drawback is computational: calculating the radial openings is significantly costlier than the PO itself.

3.4 RORPO Filtering

We now seek a strategy that allows us not only to detect thin structures, but also to discriminate their 1D nature in a robust, non-parametered way, and without requiring costly post-processing. A key-idea that underlies the RORPO filter is *to use PO both for detecting thin structures and characterizing them as 1D*.

We now consider a grey-level image $I : \Omega \to \mathbb{R}$ and a point x of Ω. The grey-level path operator $A_L^d(I)$ for the direction d should associate a value $A_L^d(I)(x) \simeq I(x)$ (resp. a value $A_L^d(I)(x) \ll I(x)$) if the direction d lies inside (resp. outside) of a thin structure.

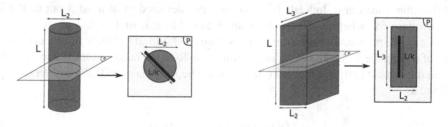

Fig. 1. A segment SE (in black) of length L/k cannot fit a 1D object (left) but can fit a 2D one (right), in the plane P orthogonal to the orientation of their largest dimension (see text).

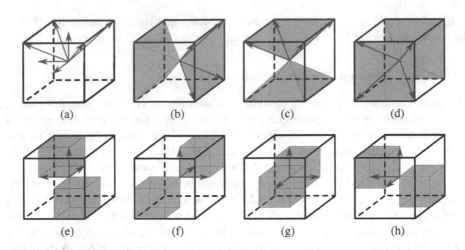

Fig. 2. (a) The 7 orientation vectors of $\mathcal{D}$ (principal orientations in red, principal diagonals in blue). The cones of orientations associated to the red vectors (b–d) and to the blue vectors (e–h).

The rationale of RORPO filter, noted γ_L, consists of determining the status of x by only analysing the set of PO responses $\{A_L^d(I)(x)\}_{d \in \mathcal{D}}$ for the set of orientations $\mathcal{D}$.

In theory, the analysis of these responses requires on the one hand a set $\mathcal{D}$ providing a dense, isotropic sampling of the 3D orientations; and on the other hand a qualitative analysis of the $A_L^d(I)(x)$ responses. In practice, when working on $\mathbb{Z}^3$, it is not tractable to define a large number of orientations. In the proposed version of RORPO [2], it was chosen to define 7 cones of orientations, corresponding to the 3 principal directions of $\mathbb{Z}^3$, plus the 4 principal diagonals (Figure 2). This choice is discussed in Section 4.2.

Our working hypothesis is that a *quantitative* analysis of $\{A_L^d(I)(x)\}_{d \in \mathcal{D}}$ should then be sufficient to decide wether x belongs to a thin 1D structure in I. Indeed, a 2D structure, being elongated in one more direction than a 1D structure, should be detected in more orientations $d \in \mathcal{D}$ than a 1D one. In order to experimentally confirm this intuition, we computed the number of high responses, $\{A_L^d(I)(x)\}_{d \in \mathcal{D}}$, for 100 synthetic binary straight tubes, planes and curved planes, respectively. The dimensions of the synthetic binary structures was set according to the fixed L chosen as the point of this experiment is to determine the influence of the direction. Results (Figure 3) tend to validate this conjecture. Indeed, 1D structures are detected in at most 3 out of the 7 orientations of $\mathcal{D}$ whereas 2D structures are detected in at least 4.

As a consequence, the discrimination between 1D and 2D objects in practice consists of sorting the 7 $\{A_L^d(I)(x)\}_{d \in \mathcal{D}}$, and defining the RORPO output as the difference between the highest (1st) and median (4th) values. The formula of RORPO is then simply defined as

$$\gamma_L(I)(x) = \gamma_L^1(I)(x) - \gamma_L^4(I)(x) \tag{3}$$

where $\gamma_L^i(I)(x)$ ($i \in [1, 7]$) is the ith greatest value within $\{A_L^d(I)(x)\}_{d \in \mathcal{D}}$.

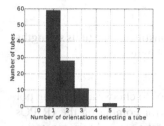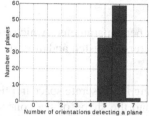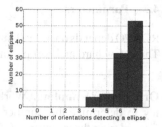

Fig. 3. Ratio of path opening high responses, computed from 100 binary synthetic examples: (left) a tube, (center) a plane, and (right) a curved plane, within the set $\mathcal{D}$ of 7 orientations illustrated in Figure 2.

4 Algorithmics, Parameters and Computational Cost

4.1 Robust Path-Opening

The core of RORPO is the computation of several path openings. RORPO then inherits the main weakness of PO, that is its non-robustness to disconnections induced by noise, a fortiori for high values of L. This fact motivated the development of algorithms for robust path-opening (RPO). Pioneering works were conducted in [23,25], and were finalised by the proposal of a RPO method by Cokelaer et al. in [26]. The principle is to tolerate K disconnected ("noisy") points between two successive points of the path.

From an applicative point of view, Cokelaer's RPO provides good results. Nevertheless, noisy points are not taken into account for determining the overall path length. The detected paths then have lengths between L (noise-free) and $L + K(L-1)$ (the most noisy). In this context, the parameter L becomes meaningless. In addition, the algorithmic layer devoted to handle the robustness is time and memory consuming.

Two alternatives of RPO were investigated. The first consists of explicitly modeling the jump between two non-noisy points via the adjacency relation, by considering the K-th power of $\rightarrow$, namely $\rightarrow^K$, instead of $\rightarrow$. This preserves L, but leads to complex and still time-consuming adaptations of PO processing. The second approach relies on a mask-based second-connectivity strategy [32], in order to "reconnect" the noisy parts of the graphs. Such approach was already proposed for 3D vessel segmentation in the framework of spatially-variant mathematical morphology [19]. The idea is here more basic and non-costly – but still effective – as a dilation by a cubical SE of size $K + 1$ is performed on the initial image I. In order to preserve the anti-extensivity of PO, an infimum operator is finally applied:

$$A_L^K(I) = \bigwedge \{I, A_L(\delta_{K+1}(I))\} \tag{4}$$

This approach, although not equivalent to [26] from a theoretical point of view, is quite similar in terms of results. As a benefit, it preserves the correct path length. It presents a computational cost that is equivalent to that of PO (only additional linear time operations are required), and is much lower than [26].

4.2   Parameters

RORPO depends on a small number of parameters, which constitutes one of its strengths. These few hard and soft parameters are listed and discussed hereafter.

Orientation Sampling. Working in a discrete space implies to choose a sampling policy to determine the number and the shape of the cones for computing the (R)POs. Considering the structure of $\mathbb{Z}^3$, isotropy requirements, and the algorithmic constraints of PO, three main families of sampling policies may be considered: along the 3 principal orientations; along the 3 + 4 principal orientations and principal diagonals; and along the 3 + 4 + 6 principal orientations and principal / secondary diagonals. It is plain that 3 orientations are not enough to accurately capture 1D structures. As discussed in Section 3.4 and [2], 7 orientations were experimentally proved to be sufficient to obtain satisfactory results. Experiments were also carried out to assess the relevance of considering 13 orientations. Three main reasons finally led us to reject this possibility. First, increasing the number of orientations complexifies a quantitative analysis necessary to define a simple formula similar to Equation (3). Second, the multiplication of angular cones also multiplies the handling of "limit cases" (see below). Third, the number of PO computations is nearly doubled, for a limited benefit.

Angular Cones. The basic patterns for each of the 7 orientations have to fully cover the direct neighbourhood of any point (x, y, z) of $\mathbb{Z}^3$, namely, the 26 points forming a $3 \times 3 \times 3$ cube around (x, y, z). Two policies may be considered: choosing patterns that induce either a partition or a cover of these 26 points. The main advantage of a partition is to avoid some overlapping effects between the angular cones, that may induce undesired side effects. Unfortunately, such a policy is not acceptable as it omits some paths whose global orientation lies in between the bounds of two successive angular cones. (For instance, in $\mathbb{Z}^2$, two successive patterns $(x, y) \rightarrow \{(x + 1, y), (x + 1, y + 1)\}$ and $(x, y) \rightarrow \{(x, y + 1), (x - 1, y + 1)\}$ generate graphs where some paths of orientation within $]\pi/4, \pi/2[$ will not be considered). It was then chosen to adopt a covering policy. In order to respect isotropy requirements and minimal overlapping, this cover was defined as illustrated in Figure 2(b–h). The counterpart of covering is the existence of limit cases, that correspond to the paths of length L that lie exactly at the frontier separating several cones. In such limit cases, Equation (3) is not valid. To correct this problem, a low cost but specific processing was devised (see [2] for a discussion on this topic).

Robustness Parameter. In our robust version of PO, the tolerance to disconnections, i.e., the maximal number of noisy points between two successive vertices of the path, is directly linked to the size of the cubical SE used for the dilation of the filtered image. We chose to fix this size to 3, i.e., $K = 2$, in order to allow a path propagation through a maximum of 2 successive noisy pixels. Setting this size to a larger value would increase the number of false propagations and the background noise reduction would be less effective.

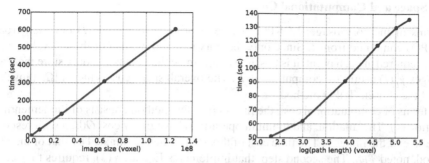

Fig. 4. Computational cost of RORPO, with respect to $|\Omega|$ (left) and L (right, log scale).

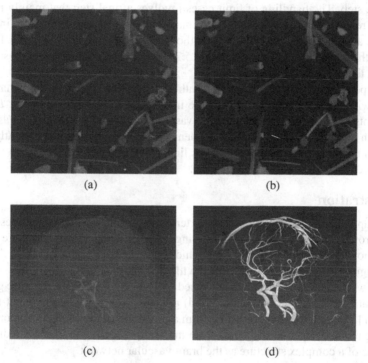

Fig. 5. A high resolution fibre image before (a) and after (b) RORPO filtering (maximum intensity projection). Brain magnetic resonance angiography (c) and component-tree segmentation with automatic markers selection using RORPO results (d) (volume rendering). (a) Courtesy Sébastien Moulinet, LPS, UMR 8550. (c) Courtesy In-Vivo Imaging Platform, Strasbourg University.

Path Length. The path length L is the only real (i.e., tunable) parameter of RORPO. Since it corresponds to the *minimal length* of the paths that *lie in a given angular cone*, it carries both a size and a curvature information. More precisely, when increasing L, RORPO will discard both small 1D structures, but also tortuous 1D structures whose part lying in a same angular cone does not exceed a length L. The choice of L should then derive from a trade-off between these two criteria.

4.3 Space and Computational Cost

In terms of memory usage, RORPO requires storing 7 images $A_L^d(I)$ obtained from the RPOs computed from I. Since only the maximal and median of these 7 values are finally required at each point, it is possible to optimize the process, to only store the first 4 images $\gamma_L^i(I)$ useful for computing $\gamma_L(I)$. The overall space cost is then $4.|\Omega|$, plus the cost of the chosen (R)PO method.

In terms of computational cost, the first step of the method consists of implementing Equation (4). The dilation, and infimum operations have linear cost $O(|\Omega|)$ with respect to the size of the image I. The time cost of this step is then dominated by that of the PO method, noted T_{PO}. The second step, that implements Equation (3), requires to access the sorted values of the 7 filtered images. This sorting has a linear cost $O(|\Omega|)$, and so has the step. The handling of limit cases involves a final step that mainly relies on greyscale reconstructions [33] on filtered images $\gamma_L(I)$, and also leads to a linear cost $O(|\Omega|)$. Finally, the total time cost of the method[1] is of the same order as $7.T_{PO}$. Between the algorithms proposed in the literature [25,24], we considered the latter, designed by Hendriks, that provides better genericity.

The experimental computation times gathered in Figure 4 assess the behaviour of RORPO with respect to the size $|\Omega|$ of the image and the length parameter L. These experiments were performed on synthetic vascular networks generated by Vascusynth [34] which allows us to tune the network density and image size independently. These results confirm that the time complexity is linear with respect to $|\Omega|$ (Figure 4(a)) and logarithmic with respect to L (Figure 4(b)).

5 Illustration

Medical applications, both on coronary arteries and brain vessels, were presented in [2]. We provide here an illustration of filtering by RORPO on a micro-CT fibre volume. Results show that small fibres which were hidden in the background noise in the initial image (Figure 5(a)) appear distinctly in the filtered image (Figure 5(b)).

As a low-level filter, information enhanced by RORPO may be used for the guidance of segmentation frameworks. Figure 5(c–d) presents results of a component-tree segmentation [35] using markers resulting from a threshold of the RORPO response. This automatic marker selection dispenses from manually selecting markers, a tedious task in the case of a complex structure as the brain vascular network.

6 Conclusion

This article, together with [2], provides a comprehensive description of RORPO as an efficient 1D structure detection filter. From a filtering point of view, the main perspectives are related to the handling of branching points (not dealt with here, as RORPO is a pure tube detector), and to the potential extension of RORPO in scale-space paradigms.

From a segmentation point of view, RORPO may be used as a guiding term in various energy-based, statistical or learning approaches for 3D image processing. In particular, research efforts have been initiated in the fields of Markovian and variational segmentation of 3D angiographic images.

---

[1] The C++ code of RORPO is available at: http://path-openings.github.io/RORPO.

From an applicative point of view, some extensions of RORPO may also be considered for explicitly detecting 2D thin structures. In this context, it may be compared to other recently proposed approaches [36] toward this goal in mathematical morphology.

References

1. Talbot, H.: Étude des directions en analyse d'image. Habilitation Thesis, Université Paris-Est (2013) (In English)
2. Merveille, O., Talbot, H., Najman, L., Passat, N.: Tubular structure filtering by ranking orientation responses of path operators. In: Fleet, D., Pajdla, T., Schiele, B., Tuytelaars, T. (eds.) ECCV 2014, Part II. LNCS, vol. 8690, pp. 203–218. Springer, Heidelberg (2014)
3. Frangi, A.F., Niessen, W.J., Vincken, K.L., Viergever, M.A.: Multiscale vessel enhancement filtering. In: Wells, W.M., Colchester, A.C.F., Delp, S.L. (eds.) MICCAI 1998. LNCS, vol. 1496, pp. 130–137. Springer, Heidelberg (1998)
4. Zana, F., Klein, J.C.: Segmentation of vessel-like patterns using mathematical morphology and curvature evaluation. IEEE Transactions on Image Processing 10(7), 1010–1019 (2001)
5. Tankyevych, O., Talbot, H., Dokládal, P.: Curvilinear morpho-Hessian filter. In: Proc. ISBI, pp. 1011–1014 (2008)
6. Soille, P., Breen, E., Jones, R.: Recursive implementation of erosions and dilations along discrete lines at arbitrary angles. IEEE Transactions on Pattern Analysis and Machine Intelligence 18(5), 562–567 (1996)
7. Soille, P., Talbot, H.: Directional morphological filtering. IEEE Transactions on Pattern Analysis and Machine Intelligence 23(11), 1313–1329 (2001)
8. Naegel, B., Passat, N., Ronse, C.: Grey-level hit-or-miss transforms—Part II: Application to angiographic image processing. Pattern Recognition 40(2), 648–658 (2007)
9. Bouraoui, B., Ronse, C., Baruthio, J., Passat, N., Germain, P.: 3D segmentation of coronary arteries based on advanced mathematical morphology techniques. Computerized Medical Imaging and Graphics 34(5), 377–387 (2010)
10. Thackray, B.D., Nelson, A.C.: Semi-automatic segmentation of vascular network images using a rotating structuring element (ROSE) with mathematical morphology and dual feature thresholding. IEEE Transactions on Medical Imaging 12(3), 385–392 (1993)
11. Tankyevych, O., Talbot, H., Dokladál, P., Passat, N.: Spatially variant morpho-Hessian filter: Efficient implementation and application. In: Wilkinson, M.H.F., Roerdink, J.B.T.M. (eds.) ISMM 2009. LNCS, vol. 5720, pp. 137–148. Springer, Heidelberg (2009)
12. Buckley, M., Talbot, H.: Flexible linear openings and closings. In: Proc. ISMM. CIV, vol. 18, pp. 109–118. Springer (2000)
13. Salembier, P., Oliveras, A., Garrido, L.: Antiextensive connected operators for image and sequence processing. IEEE Transactions on Image Processing 7(4), 555–570 (1998)
14. Wilkinson, M.H.F., Westenberg, M.A.: Shape preserving filament enhancement filtering. In: Niessen, W.J., Viergever, M.A. (eds.) MICCAI 2001. LNCS, vol. 2208, pp. 770–777. Springer, Heidelberg (2001)
15. Xu, Y., Géraud, T., Najman, L.: Two applications of shape-based morphology: Blood vessels segmentation and a generalization of constrained connectivity. In: Hendriks, C.L.L., Borgefors, G., Strand, R. (eds.) ISMM 2013. LNCS, vol. 7883, pp. 390–401. Springer, Heidelberg (2013)
16. Morard, V., Decencière, E., Dokládal, P.: Efficient geodesic attribute thinnings based on the barycentric diameter. Journal of Mathematical Imaging and Vision 46(1), 128–142 (2013)
17. Caldairou, B., Naegel, B., Passat, N.: Segmentation of complex images based on component-trees: Methodological tools. In: Wilkinson, M.H.F., Roerdink, J.B.T.M. (eds.) ISMM 2009. LNCS, vol. 5720, pp. 171–180. Springer, Heidelberg (2009)

18. Perret, B., Cousty, J., Tankyevych, O., Talbot, H., Passat, N.: Directed connected operators: Asymmetric hierarchies for image filtering and segmentation. IEEE Transactions on Pattern Analysis and Machine Intelligence, doi:10.1109/TPAMI.2014.2366145
19. Dufour, A., Tankyevych, O., Naegel, B., Talbot, H., Ronse, C., Baruthio, J., Dokládal, P., Passat, N.: Filtering and segmentation of 3D angiographic data: Advances based on mathematical morphology. Medical Image Analysis 17(2), 147–164 (2013)
20. Wilkinson, M.H.F.: Hyperconnectivity, attribute-space connectivity and path openings: Theoretical relationships. In: Wilkinson, M.H.F., Roerdink, J.B.T.M. (eds.) ISMM 2009. LNCS, vol. 5720, pp. 47–58. Springer, Heidelberg (2009)
21. Stawiaski, J.: Optimal path: Theory and models for vessel segmentation. In: Soille, P., Pesaresi, M., Ouzounis, G.K. (eds.) ISMM 2011. LNCS, vol. 6671, pp. 417–428. Springer, Heidelberg (2011)
22. Vincent, L.: Minimal path algorithms for the robust detection of linear features in gray images. In: Proc. ISMM. CIV, vol. 12, pp. 331–338. Springer (1998)
23. Heijmans, H.J.A.M., Buckley, M., Talbot, H.: Path openings and closings. Journal of Mathematical Imaging and Vision 22(2-3), 107–119 (2005)
24. Luengo Hendriks, C.L.: Constrained and dimensionality-independent path openings. IEEE Transactions on Image Processing 19(6), 1587–1595 (2010)
25. Talbot, H., Appleton, B.: Efficient complete and incomplete path openings and closings. Image and Vision Computing 25(4), 416–425 (2007)
26. Cokelaer, F., Talbot, H., Chanussot, J.: Efficient robust d-dimensional path operators. IEEE Journal of Selected Topics in Signal Processing 6(7), 830–839 (2012)
27. Morard, V., Dokládal, P., Decencière, E.: Parsimonious path openings and closings. IEEE Transactions on Image Processing 23(4), 1543–1555 (2014)
28. Bismuth, V., Vaillant, R., Talbot, H., Najman, L.: Curvilinear structure enhancement with the polygonal path image – Application to guide-wire segmentation in X-Ray fluoroscopy. In: Ayache, N., Delingette, H., Golland, P., Mori, K. (eds.) MICCAI 2012, Part II. LNCS, vol. 7511, pp. 9–16. Springer, Heidelberg (2012)
29. Valero, S., Chanussot, J., Benediktsson, J.A., Talbot, H., Waske, B.: Advanced directional mathematical morphology for the detection of the road network in very high resolution remote sensing images. Pattern Recognition Letters 31(10), 1120–1127 (2010)
30. Rossant, F., Badellino, M., Chavillon, A., Bloch, I., Paques, M.: A morphological approach for vessel segmentation in eye fundus images, with quantitative evaluation. Journal of Medical Imaging and Health Informatics 1(1), 42–48 (2011)
31. Sigurðsson, E.M., Valero, S., Benediktsson, J.A., Chanussot, J., Talbot, H., Stefánsson, E.: Automatic retinal vessel extraction based on directional mathematical morphology and fuzzy classification. Pattern Recognition Letters 47, 164–171 (2014)
32. Ouzounis, G.K., Wilkinson, M.H.F.: Mask-based second-generation connectivity and attribute filters. IEEE Transactions on Pattern Analysis and Machine Intelligence 29(6), 990–1004 (2007)
33. Vincent, L.: Morphological grayscale reconstruction in image analysis: Applications and efficient algorithms. IEEE Transactions on Image Processing 2(2), 176–201 (1993)
34. Hamarneh, G., Jassi, P.: Vascusynth: Simulating vascular trees for generating volumetric image data with ground truth segmentation and tree analysis. Computerized Medical Imaging and Graphics 34(8), 605–616 (2010)
35. Passat, N., Naegel, B., Rousseau, F., Koob, M., Dietemann, J.L.: Interactive segmentation based on component-trees. Pattern Recognition 44(10-11), 2539–2554 (2011)
36. Urbach, E.R., Pervukhina, M., Bischof, L.: Segmentation of cracks in shale rock. In: Soille, P., Pesaresi, M., Ouzounis, G.K. (eds.) ISMM 2011. LNCS, vol. 6671, pp. 451–460. Springer, Heidelberg (2011)

Exact Linear Time Euclidean Distance Transforms of Grid Line Sampled Shapes

Joakim Lindblad[1](✉) and Nataša Sladoje[2,3]

[1] University of Novi Sad, Faculty of Technical Sciences, Novi Sad, Serbia
joakim@cb.uu.se
[2] Centre for Image Analysis, Uppsala University, Sweden
natasa.sladoje@it.uu.se
[3] Mathematical Institute, Serbian Academy of Sciences and Arts, Belgrade, Serbia

Abstract. We propose a method for computing, in linear time, the exact Euclidean distance transform of sets of points s.t. one coordinate of a point can be assigned any real value, whereas other coordinates are restricted to discrete sets of values. The proposed distance transform is applicable to objects represented by grid line sampling, and readily provides sub-pixel precise distance values. The algorithm is easy to implement; we present complete pseudo code. The method is easy to parallelize and extend to higher dimensional data. We present two ways of obtaining approximate grid line sampled representations, and evaluate the proposed EDT on synthetic examples. The method is competitive w.r.t. state-of-the-art methods for sub-pixel precise distance evaluation.

Keywords: Exact Euclidean distance transform · Grid line sampling · Shape representation · Sub-pixel precision

1 Introduction

Distance transform (DT) assigns to each image point its smallest distance to a selected subset of image points. This operator serves as a basis for a variety of methods and tools in image processing and analysis. A survey of algorithms can be found in, e.g., [2] for the 2D case, and in [8] for the 3D case. DT can be defined for any distance function, but the Euclidean metric is most often considered.

In this paper we explore representations of real continuous objects by point sets $\{x_i\}$ where one coordinate of each point x_i may have a real value, whereas remaining coordinates are restricted to discrete sets of values. Such representations allow preservation of more information about the original shape than if relying on the standard approach where all coordinates are restricted to discrete sets of values. We present a method for computation of the exact EDT for such sets of points. The method fully captures the increased precision of the representation without sacrificing processing speed.

The input to the distance transform may either be directly acquired in the desired format, e.g. from a range sensor, where the data naturally has one coordinate which is not restricted to a particular grid, or can be estimated from

© Springer International Publishing Switzerland 2015
J.A. Benediktsson et al. (Eds.): ISMM 2015, LNCS 9082, pp. 645–656, 2015.
DOI: 10.1007/978-3-319-18720-4_54

image intensity values as to provide sub-pixel location of object boundaries. We present two methods for sub-pixel boundary estimation; one based on linear interpolation to find iso-level transitions (i.e. sub pixel precise thresholding), and one based on precise grid intersections computed from coverage representations. We evaluate the proposed method for EDT computation on synthetic data and confirm its competitive performance in comparisons with other methods.

2   Sampling Schemes and Shape Representation

Digitization of continuous shapes can be decomposed into sampling and reconstruction. When representing crisp objects by bi-level images, the reconstruction step is most often a nearest neighbour interpolation, effectively achieved by replacing each sampling point with its Voronoi neighbourhood, i.e., a finite sized pixel. For the sampling step, several options are at hand; common sampling schemes include Gauss (a.k.a. subset), (outer) Jordan (a.k.a. supercover), and grid intersection sampling. Gauss centre point sampling contains all grid points lying inside the object; this scheme is most common in practice. The supercover sampling contains all points with corresponding pixels having a non-empty intersection with the object. The grid intersection sampling, which is best suited for representing thin sets, such as lines and curves, dates back to Freeman [5]: "Thus, one may lay a rectangular grid over the curve and describe the curve by identifying the grid points which lie closest to it."

Let us denote the union of horizontal and vertical grid lines of $\mathbb{Z}^2$, with $L(\mathbb{Z}^2) = (\mathbb{R} \times \mathbb{Z}) \cup (\mathbb{Z} \times \mathbb{R})$ and refer to them as *integer grid lines*. A more formal definition of the grid intersection sampling scheme can then be expressed:

Definition 1. *The* grid intersection sampling *of a curve C in $\mathbb{R}^2$ is the set of points* $S = \left\{ (i,j) \mid (i,j) = (\lfloor x + \frac{1}{2} \rfloor, \lfloor y + \frac{1}{2} \rfloor), (x,y) \in C \cap L(\mathbb{Z}^2) \right\}$.

Instead of rounding to an end point of the intersected grid line segment, the mid-point of that segment, the *midcrack point* [21], can be considered as the sampling point; the *crack* being the line segment that borders two pixels. Alternatively, *endcrack points* that lie on the pixel corners can be used [21]. In some cases, both midcrack and endcrack points are considered (e.g., as in [1]).

In this paper we propose to improve the precision of the representation by using a sampling scheme that we refer to as *grid line sampling*. We suggest to describe a curve (e.g., an object boundary) by identifying the exact positions of points in $\mathbb{R}^2$ where the curve intersects each grid line. Such a curve representation is defined by points for which one coordinate assumes a real value, whereas the other is restricted to integer values.

Definition 2. *The* grid line sampling *of a curve C in $\mathbb{R}^2$ is the set S of (non-rounded) grid line intersection points* $S = C \cap L(\mathbb{Z}^2)$.

An alternative approach towards a more information rich object representation is by using the *Coverage model*, proposed in [19] and further elaborated in [20]. Coverage digitization can be seen as a fuzzy version of Gauss centre

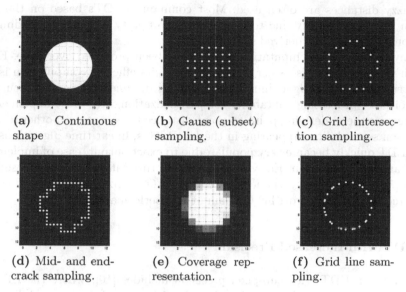

(a) Continuous shape   **(b)** Gauss (subset) sampling.   **(c)** Grid intersection sampling.

(d) Mid- and end-crack sampling.   **(e)** Coverage representation.   **(f)** Grid line sampling.

Fig. 1. Different object (and object boundary) representations, resulting from different sampling schemes

point digitization, where a pixel is allowed to partly belong to the object and its membership is given by the area of overlap between pixel and object.

Definition 3. [20] *For a given continuous object $S \subset \mathbb{R}^2$, inscribed into an integer grid with pixels $p_{(i,j)}$, the coverage digitization of S is*

$$\mathcal{D}(S) = \left\{ \left((i,j), \frac{A(p_{(i,j)} \cap S)}{A(p_{(i,j)})} \right) \middle| (i,j) \in \mathbb{Z}^2 \right\},$$

where $A(X)$ denotes the area of a set X.

Examples of different object representations are illustrated in Fig. 1, where a continuous disk is represented by: (b) Gauss (subset) sampling points; (c) Grid intersection sampling of the disk boundary; (d) Midcrack and endcrack points corresponding to the scheme suggested in [1]; (e) Area coverage values (illustrated with shades of grey); (f) Grid line sampling of the object boundary.

2.1 Euclidean Distance Transforms

The distance transform of a set S is an image where each pixel p is assigned a value corresponding to the distance from p (according to a distance d) to S:

$$DT_S(p) = \min\{d(p,q) \mid q \in S\}. \tag{1}$$

The underlying pointwise distance d can be of different types. Manhattan, Euclidean, Chessboard, and different weighted distances, but also grey-weighted and fuzzy distances are often used. Most common are DTs based on the Euclidean metric, primarily due to its appealing property of isotropy, leading to rotational invariance of derived shape descriptors.

Many methods for computation of DT have been proposed over time. Even though the idea of DT is rather straightforward, its efficient computation is far from trivial. Existing algorithms differ in terms of time complexity, accuracy, order of pixel processing, suitability for parallelization, etc. Most implementations require the underlying point-to-point distance to be a metric; other rely on dimension separability. Appearing in the late 1990's, linear time algorithms for exact EDT quickly became very popular, due to exact output, ease of implementation and parallelization. Early examples from this category are proposed by Hirata [7], Meijster et al. [17] and Maurer et al. [16]. According to the evaluation presented in [2] these algorithms remain at top performance position.

2.2 DT with Increased Precision

Early work on EDT with sub-pixel precision includes [10], where the EDT is estimated by a fast-marching method (an algorithm based on an equal distance contour evolution process, and level-set representation). Sub-pixel precision is achieved by interpolation of distance values from neighbouring grid points. Extension to 3D is studied in [11], where a sub-voxel positioning of a wavefront that corresponds to the evolving level set is observed, and an algorithm of linear complexity is given.

The anti-aliased Euclidean distance transform (AAEDT) is introduced in [6]. Utilizing additional information provided by coverage representation of an object, the positions of the (straight) edges of the object are estimated within the boundary pixels. The AAEDT method refines a vector-propagation based EDT by considering these approximate edge positions and computing the contribution of the boundary pixels to the propagated distance with subpixel precision. The AAEDT approach is further analysed in [13], where a graph-based implementation is proposed, and in [12] where the method is extended to 3D, as well as to other types of grids, such as Body-Centered and Face-Centered Cubic grids.

2.3 Signed Distance Transform

The *signed distance transform* of a set S is usually defined as $\text{SDT}_S(x) = (-1)^{S(x)}\text{DT}_{\partial S}(x)$, where ∂S denotes the boundary of the set S, and $S(x)$ is equal to 0 outside and 1 inside the set. For digital sets, the pixel boundary ∂S in general does not coincide with the grid. This leads to complications and imprecisions when SDT is computed on representations which are restricted to grid points. Often, asymmetric boundaries are used, leading to jumps or plateaus in the SDT at the zero-level, see Fig. 2b. To avoid this problem, it is proposed in [1] to increase the grid resolution by a factor of two and compute the exact SDT for binary digital shapes from a combination of mid- and endcrack points, Fig. 2d.

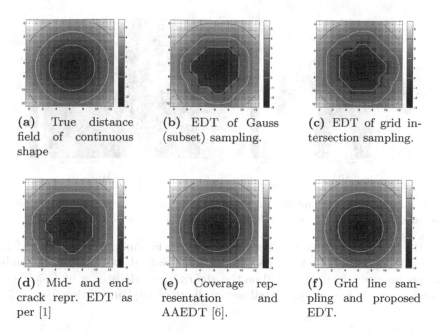

(a) True distance field of continuous shape

(b) EDT of Gauss (subset) sampling.

(c) EDT of grid intersection sampling.

(d) Mid- and end-crack repr. EDT as per [1]

(e) Coverage representation and AAEDT [6].

(f) Grid line sampling and proposed EDT.

Fig. 2. Signed Distance transforms computed from different object representations. A few iso-distance curves are overlayed. Representations used in (c) and (f) provide results closest to the true SDT, shown in (a).

3   Method

All above mentioned methods for computing DT with increased precision are approximative. The contribution of this paper is a method for computing the exact EDT of a grid line sampling of the shape boundary. Our idea is based on the observation that many of the popular methods for computing the exact EDT [16,17,4] can, with a minor modification, handle the case where the binary image is replaced by a sampled real valued function f; the resulting exact DT can still be computed in linear time.

The DT of a sampled function is in [4] (first published as [3]) defined as follows: Let $\mathcal{G}$ be a regular grid and $f : \mathcal{G} \to \mathbb{R}$, then the DT of f is

$$DT_f(p) = \min_{q \in \mathcal{G}}(d(p, q) + f(q)). \tag{2}$$

Letting f represent the (squared) distance to a shape in the $(n+1)$-th dimension, as captured by a range image, this approach allows us to compute an nD slice of a $(n+1)$D distance transform (without having to generate any $(n+1)$ dimensional object representation). An example of such is shown in Fig. 3, for $n=2$. Note that there is no requirement that the range values are integer valued. This can be further extended to handle the case of grid line sampled data.

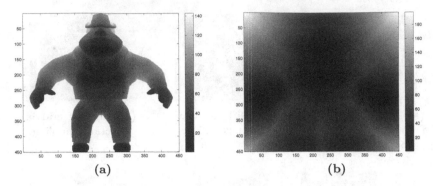

Fig. 3. (a) Range image and (b) Euclidean 3D distance transform computed for the plane at depth $z = 0$. (3D image volume courtesy of E. Remy.)

In the following we present a detailed algorithm for computing the exact EDT for sets of points such that one coordinate is real valued, whereas the remaining ones are integer valued. We build on the algorithm of Meijster et al. [17], due to its high speed and stability, as well as simplicity of implementation.

3.1 Algorithm

Assume a given set of points $S = \{(x, y) \mid x \in \mathbb{Z}, y \in \mathbb{R}\}$. We here present an algorithm that computes the exact EDT w.r.t the set S, assigning distance values to the points of the integer grid, in linear time (w.r.t. the number of grid points where the distance is computed). The algorithm is easy to parallelize and can also be easily adjusted to compute the Manhattan (MDT), and the chessboard distance (CDT) transforms following [17].

The algorithm consists of a minor modification of the method presented in [17]. We have adjusted the algorithm by Meijsters so that the procedure, during the "first phase" (following the naming in [17]), computes the real valued distance to the closest point in S in each column, generalizing by that the original version, where only integer valued distance to a binary digital set B is computed.

In a first phase each column C_x of the image is separately scanned. For each point (x, y) in C_x, the distance $G(x, y)$ of (x, y) to the nearest points of $C_x \cap S$ is determined. In a second phase each row R_y is separately scanned, and for each point (x, y) on R_y the minimum of $(x - x')^2 + G(x', y)^2$ is determined where (x', y) ranges over the row R_y. This "second phase" is identical to Meijsters original algorithm. The algorithm is easily extended to n-dimensional distance transforms by repeated applications of the second phase along each of the additional dimensions. The overall process expressed in pseudo code is presented in Algorithm 1 (again, we follow notation of [17]). For the Euclidean distance we define $\text{Sep}(i, u) = (u^2 - i^2 + g(u)^2 - g(i)^2) \textbf{ div } (2(u - i))$, where **div** denotes integer division with rounding off towards zero.

Changing the order of coordinates, the same algorithm is applicable for point sets where the x-coordinate is non-integer instead. For grid line sampling, any one of the coordinates may be non-integer. This case is solved by repeated applications of Algorithm 1, followed by a step in which for each pixel the minimal distance of the computed distance maps is taken.

Algorithm 1.

Input: A sorted 2D array-list structure, $s[x, i]$, containing, for $x, i \in \mathbb{Z}$, the real valued y-coordinate of the i-th point in $S \cap C_x$, i.e., with integer coordinate x. A 1D array, $e[x]$, indicating the number of points in $S \cap C_x$ for each x.
Output: A 2D array, dt, containing the exact EDT of S evaluated at points $\{0, 1, \ldots, m - 1\} \times \{0, 1, \ldots, n - 1\}$.

```
forall x ∈ {0,...,m − 1} do
  (* scan 1 *)
  i := 0; p := −∞;
  for y:=1 to n-1 do
    while i < e[x] ∧ s[x,i] ≤ y do
      p := s[x,i]; i := i + 1
    endwhile
    g[x,y] := y − p
  end for

  (* scan 2 *)
  i := e[x]; p := ∞;
  for y:=n-1 downto 0 do
    while i >= 0 ∧ s[x,i] ≥ y do
      p := s[x,i]; i := i − 1
    endwhile
    g[x,y] := min(g[x,y],p − y)
  end for
end forall
```

```
forall y ∈ {0,...,n − 1} do
  q:=0; s[0]:=0; t[0]:=0
  for u:=1 to m-1 do (* scan 3 *)
    while q ≥ 0 ∧ f(t[q], s[q]) > f(t[q], u) do
      q:=q-1
    if q < 0 then
      q:=0; s[0]:=u
    else
      w:=1+Sep(s[q],u)
      if w < m then
        q:=q+1; s[q]:=u; t[q]:=w
      end if
    end if
  end for
  for u:=m-1 downto 0 do (* scan 4 *)
    dt[u,y]:=f(u,s[q])
    if u = t[q] then q:=q-1
  end for
end forall
```

(a) Phase one   (b) Phase two

Straightforwardly derived properties of the proposed EDT include:

- The computed distances to the point set S are exact, up to precision of the used numerical (floating-point) representation.
- The algorithm is of complexity $\mathcal{O}(N)$, where N is the number of points where the distance is evaluated. (This follows directly from [17].)
- For an arbitrary shape $O \in \mathbb{R}^2$ with grid line sampling S, the proposed method never underestimates the distance, since every point in S is in O.
- For straight boundaries, it is easy to conclude that the worst case error appears for diagonal edges. This maximal error is $1 - \frac{\sqrt{2}}{2}$.

4   Estimation of Grid Line Sampling

If knowing the true continuous object boundary, then grid line sampling is easily facilitated. However, usually an imaging sensor provides an array of pixel values, which represents our object. One path forward is the plethora of contour-

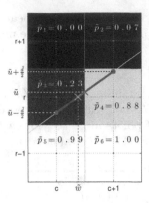

Fig. 4. A 3×2 region from a pixel coverage image, with coverage values $\tilde{p}_1 \ldots \tilde{p}_6$, where the location of a straight edge segment (blue) is estimated. The intersections with vertical grid lines $x = c$ and $x = c + 1$ (red dots) and the intersection with horizontal grid line $y = r$ (green cross) are included in the estimated grid line sampling.

or surface-based segmentation methods, e.g. Active contour approaches [9]. An alternative is to directly aim for sub-pixel estimation of edge location based on gradient estimation, e.g. as in Canny's algorithm, see [18].

A common situation is when the object boundary corresponds to a level set of the image function (i.e. an isocurve or an isosurface). In that case objects can be extracted by thresholding, or by α-cutting of a fuzzy set. For this case, a good approximation of the grid line sampling is readily achieved by linear interpolation along the grid lines (a.k.a. adaptive sampling in [21]) to reach a sub-pixel precise level set at a given threshold value. This is similar to the interpolation performed in the Marching cubes algorithm [14] or the graph based segmentation in [15].

For a coverage representation, linear interpolation is not optimal. For such a representation we instead suggest to use the end points of local edge segments, as computed in the perimeter estimation method presented in [19]. Due to lack of space, we cannot provide all details, but refer the reader to [19]. The method uses regions of 3×2 pixels, as the one shown in Fig. 4. W.l.o.g. we assume that an edge intersecting the region has a slope $k \in [0, 1]$; other cases are solved by symmetry of the square grid. The height $\tilde{u}$ of intersection between the object boundary and the middle vertical edge, indicated with a red cross in Fig. 4, is computed as: $\tilde{u} = \frac{1}{2} \sum_{i=1}^{6} \tilde{p}_i + (r - \frac{3}{2})$, where $\tilde{p}_i$ are pixel values for the 3×2 block and with pixel $\tilde{p}_3$ located at coordinates (c, r). If $\tilde{u}$ is in $(r - \frac{1}{2}, r + \frac{1}{2}]$ then the intersection is valid for the given block. Representing the object boundary by points $(c + \frac{1}{2}, \tilde{u})$, computed for each such block, provides a grid line sampling in the half-shifted grid. To remain in the original grid, we instead use the end points of the estimated edge segments, i.e. the red points in Fig. 4, with coordinates $(c, \tilde{u} - \frac{\tilde{d}}{2})$ and $(c + 1, \tilde{u} + \frac{\tilde{d}}{2})$, where $\tilde{d} = \sum_{i \in \{2,4,6\}} \tilde{p}_i - \sum_{i \in \{1,3,5\}} \tilde{p}_i$. To increase the sampling density, we also add a point where an edge segment intersects a horizontal grid line, i.e., we include the green cross in Fig. 4, located at $(\tilde{w}, r)$.

5   Evaluation

We evaluate the proposed method by computing distance maps for a straight edge and a disk. We compare results with exact analytically derived distance values of the respective continuous shape. We also observe different methods for estimation of grid line sampling. For the proposed method we present signed EDTs (SEDT) computed for: i) exact grid line sampling of a known continuous shape, ii) grid line sampling estimated by linear interpolation to find the grid line intersections of the zero level isocurve of the exact distance field of the continuous object, iii) local edge line intersections derived from a coverage representation of the shape. In addition we compare with: iv) a standard SEDT computed from a grid intersection digitization, v) SEDT of the interpixel boundaries, as proposed in [1], and vi) the signed AAEDT of the best performing (3rd) method in [6].

We first evaluate the method on a segment of a half-plane. Results are summarized in Figs. 5-8, where we present: representation, distance field, signed error, and histogram of errors, for the different methods. Note that two different intensity scales are used when visualizing the error (3rd column of images) due to large difference in performance between the methods. For straight boundaries

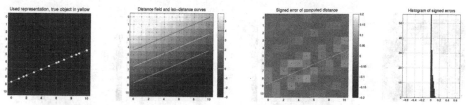

Fig. 5. Approaches (i-iii): Proposed sub-pixel SEDT of grid line sampling

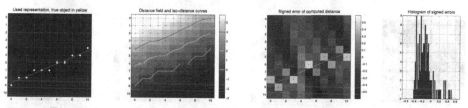

Fig. 6. Approach (iv): Binary SEDT of grid intersection digitization

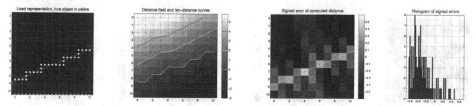

Fig. 7. Approach (v): SEDT of crack points at twice original resolution [1]

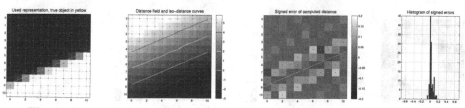

Fig. 8. Approach (vi): Signed AAEDT [6] of exact coverage digitization [20]

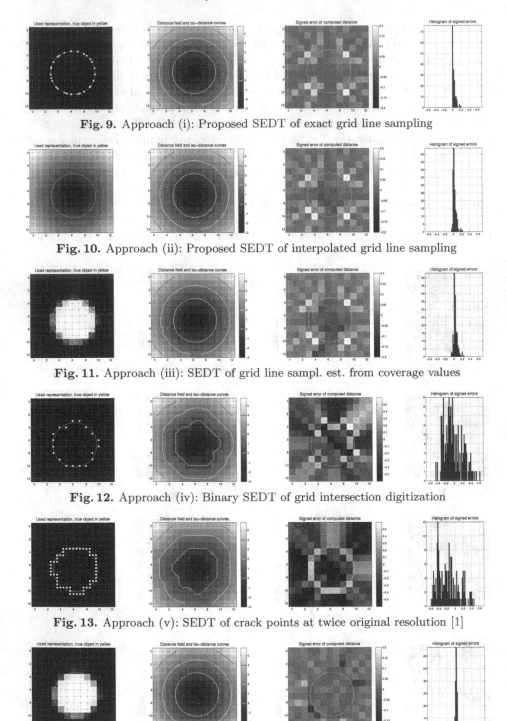

Fig. 9. Approach (i): Proposed SEDT of exact grid line sampling

Fig. 10. Approach (ii): Proposed SEDT of interpolated grid line sampling

Fig. 11. Approach (iii): SEDT of grid line sampl. est. from coverage values

Fig. 12. Approach (iv): Binary SEDT of grid intersection digitization

Fig. 13. Approach (v): SEDT of crack points at twice original resolution [1]

Fig. 14. Approach (vi): Signed AAEDT [6] of exact coverage digitization [20]

both the linear interpolation and the coverage based approach provide the exact grid line sampling, so there is no difference in result between methods (i-iii). We conclude that the proposed approach has the overall best performance on this object, with the AAEDT method not being far behind. The proposed EDT never underestimates the distance, as can be verified in the histogram in Fig 5.

Our second object of evaluation is a disk of radius 3.43 pixels. Results are summarized in Figs. 9-14. For curved edges, neither linear interpolation nor the estimation derived from coverage data, provide exact grid line sampling. For this reason underestimated distances do appear in Fig 10 and 11. For this object, the AAEDT method performs slightly better, with a mean absolute error (MAE) of 0.026 pixels, vs. 0.029 for the proposed method.

6   Conclusions

We propose to consider grid line sampled representations of continuous shapes, and suggest methods for approximating such sampling: by linear interpolation of a level-set function (sub-pixel precise thresholding), or by estimation of local edge segments from a coverage representation.

We present a novel approach for computing the exact Euclidean Distance Transform (EDT) of a grid line sampling. The method extends the algorithm of Meijster to handle real valued point positions in one dimension. The algorithm computes the EDT in linear time. It is easy to parallelize and extend to higher dimensional data. The resulting DT of the set of sample points is exact. The combination of grid line sampling and EDT computation therefore never underestimates the distance to a real continuous object. For straight boundaries the maximal overestimate is $1 - \frac{\sqrt{2}}{2} \approx 0.29$ pixels.

The conducted evaluation confirms the excellent performance of the EDT method, for true grid line sampled shapes as well as in combination with proposed methods for estimating grid line sampling. The Anti-aliased Euclidean distance transform [6] performs roughly equally well as the proposed method. We conclude that neither is a clear winner, but that they both have good properties. In favour of the proposed method are easier extension to higher dimensions as well as better knowledge of errors. By use of an *irregular matrix*, as proposed in [22], the method can be extended to arbitrary point sets in $\mathbb{R}^n$ (at the cost of higher computational complexity). The proposed EDT method can be appreciated for its generality and wide applicability to different types of object representations. Methods based on classic binary object representations have consistently much higher errors, and should clearly be avoided whenever precision is of interest.

Acknowledgements. The Authors are supported by the Ministry of Science of the Republic of Serbia through Projects ON174008 and III44006 of the Mathematical Institute of the Serbian Academy of Sciences and Arts. N. Sladoje is supported by the Swedish Governmental Agency for Innovation Systems (VINNOVA).

References

1. Ciesielski, K.C., Chen, X., Udupa, J.K., Grevera, G.J.: Linear time algorithms for exact distance transform. Journal Math. Imaging and Vision 39(3), 193–209 (2011)
2. Fabbri, R., Costa, L.D.F., Torelli, J.C., Bruno, O.M.: 2D Euclidean distance transform algorithms: A comparative survey. ACM Computing Surveys 40(1), 2 (2008)
3. Felzenszwalb, P., Huttenlocher, D.: Distance transforms of sampled functions. Tech. rep., Cornell University (2004)
4. Felzenszwalb, P.F., Huttenlocher, D.P.: Distance transforms of sampled functions. Theory of Computing 8(1), 415–428 (2012)
5. Freeman, H.: On the encoding of arbitrary geometric configurations. Electronic Computers, IRE Trans. (2), 260–268 (1961)
6. Gustavson, S., Strand, R.: Anti-aliased Euclidean distance transform. Pattern Recognition Letters 32(2), 252–257 (2011)
7. Hirata, T.: A unified linear-time algorithm for computing distance maps. Information Processing Letters 58(3), 129–133 (1996)
8. Jones, M., Baerentzen, J., Sramek, M.: 3D distance fields: A survey of techniques and applications. IEEE Trans. Visualization Comp. Graphics 12(4), 581–599 (2006)
9. Kass, M., Witkin, A., Terzopoulos, D.: Snakes: Active contour models. International Journal of Computer Vision 1(4), 321–331 (1988)
10. Kimmel, R., Kiryati, N., Bruckstein, A.: Sub-pixel distance maps and weighted distance transforms. Journal Math. Imaging and Vision 6(2-3), 223–233 (1996)
11. Krissian, K., Westin, C.F.: Fast sub-voxel re-initialization of the distance map for level set methods. Pattern Recognition Letters 26(10), 1532–1542 (2005)
12. Linnér, E., Strand, R.: Anti-aliased Euclidean distance transform on 3D sampling lattices. In: Proc. Discrete Geom. for Comp. Imagery., Siena, Italy, pp. 88–98 (2014)
13. Linnér, E., Strand, R.: A graph-based implementation of the anti-aliased Euclidean distance transform. In: Proc. Int. Conf. Pattern Recognition, Stockholm, Sweden, pp. 1025–1030 (2014)
14. Lorensen, W.E., Cline, H.E.: Marching cubes: A high resolution 3D surface construction algorithm. SIGGRAPH Comput. Graph. 21(4), 163–169 (1987)
15. Malmberg, F., Lindblad, J., Sladoje, N., Nyström, I.: A graph-based framework for sub-pixel image segmentation. Theoretical Comp. Sci. 412(15), 1338–1349 (2011)
16. Maurer Jr., C.R., Qi, R., Raghavan, V.: A linear time algorithm for computing exact Euclidean distance transforms of binary images in arbitrary dimensions. IEEE Trans. Pattern Anal. Mach. Intell. 25(2), 265–270 (2003)
17. Meijster, A., Roerdink, J.B., Hesselink, W.H.: A general algorithm for computing distance transforms in linear time. In: Proc. Math. Morph. and its Appl. to Image Signal Process. Comput. Imaging and Vision, vol. 18, pp. 331–340. Springer (2000)
18. Rockett, P.: The accuracy of sub-pixel localisation in the canny edge detector. In: Proc. British Machine Vision Conf., pp. 1–10 (1999)
19. Sladoje, N., Lindblad, J.: High-precision boundary length estimation by utilizing gray-level information. IEEE Trans. Patt. Anal. Mach. Intell. 31(2), 357–363 (2009)
20. Sladoje, N., Lindblad, J.: The coverage model and its use in image processing. In: Zbornik Radova, vol. (23), pp. 39–117. Matematički institut SANU (2012)
21. Stelldinger, P.: Image digitization and its influence on shape properties in finite dimensions, vol. 312. Ph.D. thesis. IOS Press (2008)
22. Vacavant, A., Coeurjolly, D., Tougne, L.: Distance transformation on two-dimensional irregular isothetic grids. In: Coeurjolly, D., Sivignon, I., Tougne, L., Dupont, F. (eds.) DGCI 2008. LNCS, vol. 4992, pp. 238–249. Springer, Heidelberg (2008)

Fast Estimation of Intrinsic Volumes in 3D Gray Value Images

Michael Godehardt[✉], Andreas Jablonski, Oliver Wirjadi, and Katja Schladitz

Fraunhofer-Institut für Techno- und Wirtschaftsmathematik,
D-67663 Fraunhofer-Platz 1, Kaiserslautern, Germany
michael.godehardt@itwm.fraunhofer.de
www.itwm.fraunhofer.de/en/departments/image-processing/

Abstract. The intrinsic volumes or their densities are versatile structural characteristics that can be estimated efficiently from digital image data, given a segmentation yielding the structural component of interest as foreground. In this contribution, Ohser's algorithm is generalized to operate on integer gray value images. The new algorithm derives the intrinsic volumes for each possible global gray value threshold in the image. It is highly efficient since it collects all neccesary structural information in a single pass through the image.

The novel algorithm is well suited for computing the Minkowski functions of the parallel body if combined with the Euclidean distance transformation. This application scenario is demonstrated by means of computed tomography image data of polar ice samples. Moreover, the algorithm is applied to the problem of threshold selection in computed tomography images of material microstructures.

Keywords: Intrinsic volumes · Gray value images · Minkowsi functions · Connectivity analysis · Segmentation · Microstructure analysis

1 Introduction

The possibilites and the demand to spatially image materials microstructures has grown tremendiously during the last decade. Image sizes, complexity of the imaged structures, and detail of the analysis tasks grow at even higher speed, increasing the demand for time and memory efficient algorithms yielding quantitative structural information.

One very general image analysis tool are the intrinsic volumes. The intrinsic volumes, also known as Minkowski functionals or quermass integrals [1], are in some sense a basic set of geometric structural characteristics [2]. In 3D, they yield information about the volume, surface, mean and Gaussian curvatures of analyzed structures. Various other characteristics describing e.g. shape [3] or structure specific features e. g. for open cell foams [4] can be derived. The densities of the intrinsic volumes, combined with erosions and dilations of the structure under consideration have been studied, e. g. to quantify connectivity [5]. Mecke [6] called them Minkowski functions.

© Springer International Publishing Switzerland 2015
J.A. Benediktsson et al. (Eds.): ISMM 2015, LNCS 9082, pp. 657–668, 2015.
DOI: 10.1007/978-3-319-18720-4_55

For a given segmentation of the gray value image into the component of interest (foreground) and its complement (background), the intrinsic volumes can be efficiently derived from the resulting binary image by Ohser's algorithm [7,8]. In this paper, we generalize Ohser's approach to gray value images. That is, we introduce an algorithm for fast simultaneous calculation of the intrinsic volumes for all possible threshold values in an integer gray value image.

The usefulness of the new algorithm is proved by applying it to X-ray computed tomography images of materials microstructures: In the first scenario, we apply our new algorithm to the Euclidean distance transformed image. This yields immediately the Minkowski functions widely used in physical applications [6]. For polar ice, they reveal structural differences of the multiply connected pore system.

In the second scenario, we exploit our new algorithm to find the optimal global gray value threshold: Segmentation remains a notorius problem, even in the simplest case of porous materials consisting of one homogeneous solid component with a clear contrast to the air filled pore space. Identification of the solid phase in the image data is an essential prerequisite for the majority of geometric analyses as well as for numerical simulation of macroscopic materials properties using the segmented image data as computational domain. In engineering, fast, objective segmentation methods are sought after. Global gray value thresholding is the easiest choice and competes well as long as global gray value fluctuations have been avoided or removed. It remains to devise a strategy to determine the threshold. There is a variety of threshold selection schemes, most prominent Otsu's [9] and the isodata method [10].

Here, we pursue the obvious idea of choosing the threshold such that previously known characteristics of the imaged structure are met. Examples for such characteristics range from simple prior knowledge such as the solid volume fraction of the component of interest to more complex properties of microstructures such as surface density or mean thickness. Of course, this approach is endangered by noise, porosity on a scale finer than the image resolution and imaging artefacts. Discretization effects further complicate matters, in the case of X-ray computed tomography in particular the partial volume effect. Nevertheless, one could hope for a range of possible thresholds, where essential geometric characteristics do not change drastically. This could be identified by monitoring the dependence of these characteristics on the gray value threshold. This idea is pursued for a glass fiber reinforced composite.

2   Estimation of Intrinsic Volumes in 3D Binary Images

The intrinsic volumes (or their densities) are a system of basic geometric characteristics for microstructures. In 3D, there are four intrinsic volumes – the volume V, the surface area S, the integral of mean curvature M and the Euler number χ. For a convex object, M is up to a constant the mean width. The Euler number is a topological characteristic alternately counting the connected components, the tunnels, and the holes. For a convex body, we have $\chi = 1$, for a torus $\chi = 1 - 1 = 0$, and for a sphere $\chi = 1 + 1 = 2$. For macroscopically homogeneous structures, the densities of the intrinsic volumes are considered instead – the volume fraction V_V, the specific surface area S_V, the density of the integral of mean curvature M_V, and the density of the Euler number χ_V. That is, the respective quantities are divided by the total sample volume measured. These characteristics can be estimated based on observations restricted to a compact window.

An efficient algorithm for the simultaneous measurement of all intrinsic volumes from 3D image data is based on weighted local $2 \times 2 \times 2$ pixel configurations in a binary 3D image [8, Chapter 5]. The restriction to these small configurations allows to code them in an 8bit gray value image of the same size using the convolution with the mask shown in Figure 1. All further steps of the algorithm are based solely on the gray value histogram h of this image whose size does not depend on image size or content. Thus it is simple and fast to compute the intrinsic volumes. The algorithm is deduced from a discrete version of the

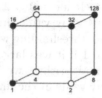

Fig. 1. Mask used for coding the $2 \times 2 \times 2$ pixel configurations. Here, the black colored pixels are set resulting in configuration code $c(p) = 182$

integral geometric Crofton formulae [8,11] boiling down computing the intrinsic volumes to computing Euler numbers in lower dimensional intersections. The Euler numbers in turn can be estimated efficiently in the discrete setting by the Euler-Poincaré formula as the alternating sum of numbers of cells, faces, edges, and vertices. We shortly summarize the algorithm:

1. Given the 3D image of a microstructure, binarize it, e. g. using a global gray value threshold.
2. Convolve the binary image with the following $2 \times 2 \times 2$ mask:

$$\left(\begin{pmatrix} 1 & 2 \\ 4 & 8 \end{pmatrix}, \begin{pmatrix} 16 & 32 \\ 64 & 128 \end{pmatrix} \right). \tag{1}$$

3. Compute the gray value histogram h of the convolution result.
4. Calculate the intrinsic volumes as scalar products of h with tabulated weight vectors $v^{(k)}$, $k = 0, \ldots, 3$ derived from discretizing the Crofton formulae, [8, Section 5.3.5].

Note that the specific choice of a weight vector v in step 4 depends on the discrete connectivities assumed for foreground and background of the image.

3   Estimation of Intrinsic Volumes as Function of Gray Value Threshold

In this section, we generalize the algorithm sketched in the previous section to gray value images. As before, the actual calculation of the intrinsic volumes consists of scalar multiplication of suitable weight vectors with the vector h of frequencies of $2 \times 2 \times 2$ black-or-white pixel configurations. The generalization is achieved by an algorithm for efficiently collecting and coding the local black-or-white pixel configurations induced by thresholding the local pixel configurations in gray value images.

The naive generalization of the algorithm from Section 2 would loop through the image's gray value range $R = \{0, \ldots, 255\}$ for 8bit gray values or $R = \{0, \ldots, 2^{16} - 1\}$ for 16bit gray values, use each value $t \in R$ as global gray value threshold, and apply the algorithm sketched above to each of the resulting binary images. Clearly, this is not practical. Instead, the matrix of the frequency vectors $h(t), t \in R$, is created and exploited. The main observation yielding an efficient algorithm is the following: For each local $2 \times 2 \times 2$ pixel configuration $p = (p_1, \ldots, p_8)$, only those eight pixels' gray values are threshold values, for which the contribution of p changes. Here, thresholding with $t \geq 0$ means the following: The gray value of pixel p_i is set to 1 if it was larger or equal to t and to 0 else.

Before describing details of the algorithm, we introduce some notation: Let $M = \max\{R\}$ be the maximal gray value in the image. Write $c(p, t)$ for the code of pixel configuration p after thresholding with threshold t. Without loss of generality, we assume the structures of interest to be represented by bright values, i.e., with larger gray values. Then, in particular, $c(p, 0) = 255$ and $c(p, M+1) = 0$ for all p as thresholding by 0 creates the configuration completely contained in the foreground and thresholding by more than the maximal possible gray value creates the configuration completely contained in the background. For all remaining threshold values, $c(p, t)$ represents the result of the convolution with the mask from (1) if the image has been thresholded at t.

1. All changes are stored in the difference matrix $\Delta(c, t)$, $c = 0, \ldots, 255$, $t = 0, \ldots, M + 1$ initialized with 0.
2. For local pixel configuration p do
   (a) Sort the current eight gray values $t_1, \ldots, t_8$ and let $0 = t_0 \leq t_1 \leq \ldots \leq t_8 \leq M$.
   (b) For $i = 0, \ldots, 7$ do
       – Increase $\Delta(c(p, t_i), t_i + 1)$ by one.
       – Decrease $\Delta(c(p, t_i), t_{i+1} + 1)$ by one.
   (c) Increase $\Delta(0, t_8 + 1)$ by one.
3. Initialize the final matrix h by $h(c, t) = 0$ for all $c \in \{0, \ldots, 255\}$, $t \in \{0, \ldots, M + 1\}$.
4. $h(c, t) = \sum_{\tau=0}^{t} \Delta(c, \tau)$.

This results in $h(\cdot, t)$ being the configuration frequency vector for the image thresholded at t. Note that in general, for d-dimensional images, $\Delta(c, t)$ has

$2^{2^d} \times M + 1$ elements, independent of the actual size of the image. E.g., for the case of 3D 8bit gray value images, this amounts to a memory requirement of $256 \cdot 256 = 65536$ integers to process any such image, regardless of its size.

Example 1. For the sake of clarity and brevity of the presentation, we just consider a 2D 8bit gray value image, that is, the 2×2 configurations are coded by convolution with

$$\begin{pmatrix} 1 & 2 \\ 4 & 8 \end{pmatrix}$$

and the difference matrix $\Delta(c,t)$ has dimensions 16×256. In the 2D case, $c(p,0) = 1 + 2 + 4 + 8 = 15$ and $c(p,256) = 0$ for all p. Let the image consist of just one 2×2 pixel configuration

$$\begin{pmatrix} 26 & 4 \\ 128 & 17 \end{pmatrix}.$$

Sorting yields $t_1 = 4$, $t_2 = 17$, $t_3 = 26$, $t_4 = 128$. Now the algorithm proceeds as follows

1. Increase $\Delta(15,0)$ by one. Decrease $\Delta(15,5)$ by one.
2. Binarization with threshold $t_1 = 4$ yields the configuration

$$\begin{pmatrix} 1 & 0 \\ 1 & 1 \end{pmatrix}$$

   whose code is $1 + 4 + 8 = 13$. Thus increase $\Delta(13,5)$ by one. Decrease $\Delta(13,18)$ by one.
3. Binarization with threshold $t_2 = 17$ yields the configuration

$$\begin{pmatrix} 1 & 0 \\ 1 & 0 \end{pmatrix}$$

   whose code is $1 + 4 = 5$. Thus increase $\Delta(5,18)$ by one. Decrease $\Delta(5,27)$ by one.
4. Binarization with threshold $t_3 = 26$ yields the configuration

$$\begin{pmatrix} 0 & 0 \\ 1 & 0 \end{pmatrix}$$

   whose code is 4. Thus increase $\Delta(4,27)$ by one. Decrease $\Delta(4,129)$ by one.
5. Binarization with threshold $t_4 = 128$ yields the configuration

$$\begin{pmatrix} 0 & 0 \\ 0 & 0 \end{pmatrix}$$

   whose code is 0. Thus increase $\Delta(0,129)$ by one.

This results in

$$\Delta(c,t) = \begin{array}{c} \\ 0 \\ \vdots \\ 4 \\ 5 \\ \vdots \\ 13 \\ \vdots \\ 15 \end{array} \begin{array}{ccccccccccccc} 0 & \cdots & 5 & \cdots & 18 & \cdots & 27 & \cdots & 129 & \cdots & 256 \\ \left(0 \right. & \cdots & 0 & \cdots & 0 & \cdots & 0 & \cdots & 1 & \cdots & 0 \\ & & & & & & & & & & \\ 0 & \cdots & 0 & \cdots & 0 & \cdots & 1 & \cdots & -1 & \cdots & 0 \\ 0 & \cdots & 0 & \cdots & 1 & \cdots & -1 & \cdots & 0 & \cdots & 0 \\ & & & & & & & & & & \\ 0 & \cdots & 1 & \cdots & -1 & \cdots & 0 & \cdots & 0 & \cdots & 0 \\ & & & & & & & & & & \\ \left. 1 \right. & \cdots & -1 & \cdots & 0 & \cdots & 0 & \cdots & 0 & \cdots & 0 \end{array} \right)$$

Summing up yields $h(\cdot,t) = (0,\ldots,0,1)$ for $t = 0,\ldots,4$. Then, the configuration changes and thus $h(\cdot,t) = (\underbrace{0,\ldots,0}_{13\text{times}},1,0,0)$ for $t = 5,\ldots,17$. At threshold 18 the next change happens, resulting in $h(\cdot,t) = (0,0,0,0,0,1,0,\ldots,0)$ for $t = 18,\ldots,26$. The configuration changes next at threshold 27 and yields $h(\cdot,t) = (0,0,0,0,1,0,\ldots,0)$ for $t = 27,\ldots,128$. Finally $h(\cdot,t) = (1,0,\ldots,0)$ for $t = 129,\ldots,256$.

4 Application Examples

In the following, the just derived algorithm is applied to two real 3D images of microstructures. For the pore system in polar ice samples from two different depths, the densities of the intrinsic volumes with respect to dilations and erosions are derived, revealing structural differences. For a glass fiber reinforced composite, a suitable global gray value threshold is found using prior knowledge on the imaged structure.

4.1 Polar Firn: Erosion-Dilation Analysis

Polar ice is a climate information archive and therefore of considerable interest for climate research. During the last decades a couple of deep polar ice cores were drilled through the Antarctic and Greenlandic ice sheets. The upper 50 to 100m of these ice sheets are formed by so-called firn – sintered ice grains with an inter-connected air-filled pore system. Here, we consider samples of firn core B26 drilled during the the North-Greenland-Traverse in 1993-1995 from depths between 56m and 72m, first studied in [12]. Volume renderings are shown in Figure 2.

The Minkowski functions as used e.g. in [6] are the intrinsic volumes of parallel body $X \oplus B_r$ resp. $X \ominus B_r$ of the structure X over the radius r, that is, the intrinsic volumes of stepwise erosions/dilations of X. From these functions, 3d structural information can be directly derived. In particular the specific Euler number as a function of successive erosions was called connectivity function

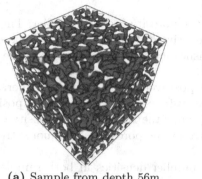
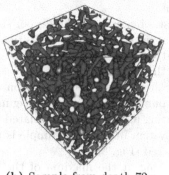

(a) Sample from depth 56m.   (b) Sample from depth 72m.

Fig. 2. Volume rendering of the pore system of firn core B26 from North Greenland. Pixel edge length 40μm, visualized are $400 \times 400 \times 400$ pixels corresponding to 16mm $\times$ 16mm $\times$ 16mm

and used to evaluate bond sizes [5]. In [13], the Euler number combined with successive erosions was used to study the connectivity of firn from the B35 core, see also [8].

Here, we exploit the algorithm derived in 3 and the fact that erosions and dilations with a ball can be efficently computed via thresholding the result of the Euclidean distance transformation (EDT). That is, the Minkowski functions are computed using the novel algorithm on the EDT image. Edge effects are avoided by minus-sampling the distance images. Figure 3 shows the results for

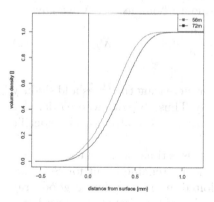
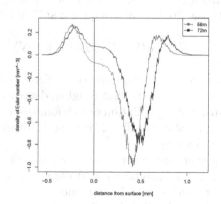

(a) Volume density function   (b) Euler number density function

Fig. 3. Minkowski functions (a) $V_V(X \ominus B_r)$ and $V_V(X \oplus B_r)$, and (b) $\chi_V(X \ominus B_r)$ and $\chi_V(X \oplus B_r)$, where X denotes the pore system of the firn sample. That means, negative distances correspond to erosions of the pore system or dilations of the ice, while positive distances correspond to dilations of the pore system or erosions of the ice.

the volume fraction and the Euler number density, both applied to the pore system.

First, obviously, in the deeper sample the porosity is sligtly smaller. This can be seen from the volume densities for negative distances in Fig. 3a, where the porosity in the deeper sample (72m) vanishes in earlier erosion steps of the pore space than in case of 56m.

Recall that isolated pores contribute positively to the Euler number. The Euler number density of the original pore system in the 72m sample is positive, indicating a prevelance of isolated pores, while the Euler number density of the pore system in the 56m sample is negative as the pore system is more strongly connected (Fig. 3b).

Looking at the maxima of the Euler number densities in both samples for negative distances, we see that connections between pores are broken by smaller erosions of the pore space at 72m than at 56m. In other words, the inter-pore connections appear to be thinner in the lower sample.

The minimum of the Euler number density is reached later for the 72m sample, showing that the ice grains are isolated in a later dilation step. This indicates locally thicker ice grains in the deeper sample, ie., larger pore-to-pore distances.

These findings are consistent with the findings in earlier studies of such systems [12,14,15].

4.2   Glass Fiber Reinforced Composite: Surface Density-Based Threshold Selection

Here, we briefly summarize an algorithm for threshold selection which incorporates prior knowledge exploiting the efficiency of the fast estimation of intrinsic volumes for gray valued data that was proposed in this paper. First, observe that due to the Steiner formula, for a sufficiently smooth bounded set X we have

$$\lim_{r \to 0}(V_V(X \oplus B_r) - V_V(X))/r = \frac{dV_V}{dr}(X) = S_V(X). \tag{2}$$

For a rigourous derivation see [2, Section 14.4] or [3].

Moreover, observe that roughly, decreasing or increasing the threshold slightly is equivalent to a dilation or erosion, respectively. Thus, dV_V/dt, where t denotes the gray value threshold, can be interpreted as an estimator for the specific surface area S_V.

However, our experiments have shown that this estimator is far too sensitive to noise and discretization effects in order to yield a reliable result for S_V. This relation can nevertheless be successfully exploited in order to find a good gray value threshold. To this end, our surface density-based threshold selection scheme exploits that $S_V(X(t))/(dV_V(X(t))/dt) = 1$, where $X(t)$ denotes the foreground set induced through applying threshold t to the input image, when we assume that Δt gets infinitesimally small. Thus, we may choose t such that this condition will be fulfilled. The rationale behind it is that if the foreground structures depicted in some gray value image fulfill the prerequisites of the Steiner formula, the ratio should approach one for the correct threshold. Wrong thresholds, on the other hand, are expected to destroy the structure (e.g. due to noise), thus invalidating that ratio.

Given a 3D image, we implement that idea in the following algorithm.

1. Compute the intrinsic volumes of the image for all thresholds t using the algorithm from above.
2. Apply a backward difference filter on this result to obtain $dV_V(X(t))/dt$.
3. Choose the mode of the ratio of $S_V(X(t))$ and $dV_V(X(t))/dt$ to obtain the surface density-based threshold.

The third and last step of this algorithm is particularly important. First note, that we cannot get an infinitesimally small ball B_r, but we have a difference quotient in the second step. Therefore, there are remaining terms in the Steiner formulae, so the ratio may not reach one. So we choose instead the mode of the ratio. This is necessary since for real data, discretization effects have a major impact on the results. E.g., increasing the threshold by one can result in quite strong erosions, thus violating equation 2, where the dilation should be with an infinitesimally small structuring element.

Nevertheless, we observe that in practice, there will be a pronounced peak in $S_V(X(t))/(dV_V(X(t))/dt)$. This phenomenon is caused by the contrast of the images, which should always be highest around the structures under investigation. Thus, $dV_V(X(t))/dt$ can be expected to be low at the threshold value of interest, while $S_V(X(t))$ should approach the true value of $S_V(X)$ at that point.

We finally consider a glass fiber reinforced polymer sample, more precisely a polypropylene matrix reinforced by long glass fibers (average length of 7 mm, diameter of about 16 μm). The sample of size $6 \times 4 \times 6$ mm^3 was imaged by X-ray computed tomography with a pixel edge length of 4 μm. For details on sample production parameters and μCT imaging see [16], where the local fiber orientations were analyzed in order to identify the misoriented region in the central part.

(a) Volume rendering

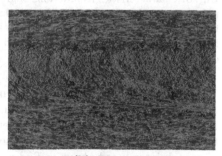

(b) Slice view

Fig. 4. Volume rendering of and section through the glass fiber reinforced polymer sample. Sample preparation and μCT imaging IVW Kaiserslautern. Pixel edge length 4 μm. Visualized are $1\,100 \times 1\,500 \times 1\,500$ pixels corresponding to 4mm $\times$ 6mm $\times$ 6mm.

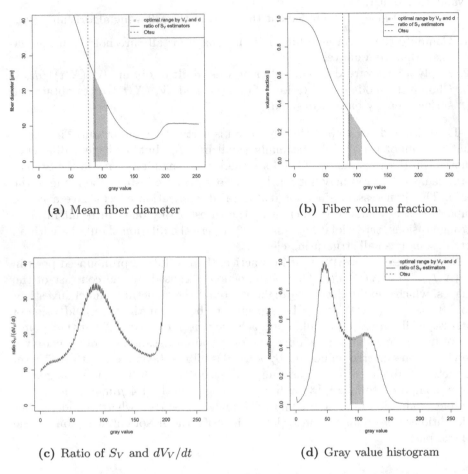

(a) Mean fiber diameter

(b) Fiber volume fraction

(c) Ratio of S_V and dV_V/dt

(d) Gray value histogram

Fig. 5. Characteristics as functions of gray value threshold for the glass fiber reinforced polymer sample. The known fiber thickness $d = 16\mu m$ and the known fiber volume fraction $V_V = 35\%$ yield an interval of reasonable thresholds. Otsu's threshold (79) lies left of this interval, while the threshold defined by the mode of the ratio of S_V and dV_V/dt (89) lies well within.

The fiber weight content of the material is 60 %, resulting in an expected fiber volume fraction of 35 %. The image quality is very good, nevertheless it is impossible to segment the fiber system such that both the known fiber volume fraction and the known fiber thickness are perfectly fit. We therefore use the approach introduced above to find a good trade off between these two targets.

Note, that the noise in the plots is caused by the preprocessing of spreading the original 16bit image data to 8bit data.

Our numerical experiments deviated from the predictions derived from the Steiner formula, namely that the ratio between specific surface and first deriva-

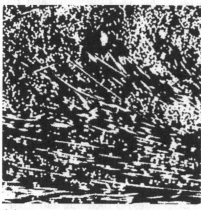

(a) Region of slice through image binarized using Otsu's threshold

(b) Region of slice through image binarized using derived threshold

Fig. 6. Section through the thresholded images of the glass fiber reinforced polymer sample (detail of 4b). The resulting images are similar, note, that the fibers in some bundles are slightly better separated in 6b.

tive of the volume density should be one. We attribute this to discretization effects. Nevertheless, the surface density based threshold selection rule proposed in the present paper yields very promising results and we expect this method to be useful for segmentation of a wide range of microstructures.

5   Discussion

In this paper, we described an efficient algorithm for calculating the intrinsic volumes for each possible gray value threshold in an integer valued image and demonstrated two of its possible applications. Namely, the method can be used to derive efficiently the Minkowski functions of parallel body of a structure when applied to the Euclidean distance transformed image. Also, it can be used to find the global gray value threshold yielding the best agreement of the segmentation result with pre-known geometric characteristics of the structure.

Our algorithm is of linear complexity in the number of pixels. Even more important are the two other features: It works purely locally on the image and requires fixed memory space, independent of the size of the original image. This is extremely valuable, as with growing flat-panel detectors, alternative scanning geometries like helical (or spiral) computed tomography 3D image data reach sizes larger than 100GB.

References

1. Stoyan, D., Kendall, W.S., Mecke, J.: Stochastic Geometry and its Applications, 2nd edn. Wiley, Chichester (1995)

2. Schneider, R., Weil, W.: Stochastic and Integral Geometry. Probability and Its Applications. Springer, Heidelberg (2008)
3. Ohser, J., Redenbach, C., Schladitz, K.: Mesh free estimation of the structure model index. Image Analysis and Stereology 28(3), 179–186 (2009)
4. Schladitz, K., Redenbach, C., Sych, T., Godehardt, M.: Model based estimation of geometric characteristics of open foams. Methodology and Computing in Applied Probability, pp. 1011–1032 (2012)
5. Vogel, H.J.: Morphological determination of pore connectivity as a function of pore size using serial sections. European Journal of Soil Science 48(3), 365–377 (1997)
6. Mecke, K.: Additivity, convexity, and beyond: Application of minkowski functionals in statistical physics. In: Mecke, K.R., Stoyan, D. (eds.) Statistical Physics and Spatial Statistics. LNP, vol. 554, pp. 111–184. Springer, Heidelberg (2000)
7. Ohser, J., Nagel, W., Schladitz, K.: Miles formulae for Boolean models observed on lattices. Image Anal. Stereol. 28(2), 77–92 (2009)
8. Ohser, J., Schladitz, K.: 3d Images of Materials Structures – Processing and Analysis. Wiley VCH, Weinheim (2009)
9. Otsu, N.: A threshold selection method from gray level histograms. IEEE Trans. Systems, Man and Cybernetics 9, 62–66 (1979)
10. Ridler, T., Calvard, S.: Picture thresholding using an iterative selection method. IEEE Transactions on Systems, Man and Cybernetics 8(8), 630–632 (1978)
11. Schladitz, K., Ohser, J., Nagel, W.: Measurement of intrinsic volumes of sets observed on lattices. In: Kuba, A., Nyúl, L.G., Palágyi, K. (eds.) DGCI 2006. LNCS, vol. 4245, pp. 247–258. Springer, Heidelberg (2006)
12. Freitag, J., Wilhelms, F., Kipfstuhl, S.: Microstructure-dependent densification of polar firn derived from x-ray microtomography. Journal of Glaciology 50(169), 243–250 (2004)
13. Freitag, J., Kipfstuhl, S., Faria, S.H.: The connectivity of crystallite agglomerates in low-density firn at Kohnen station, Dronning Maud land, Antarctica. Ann. Glaciol. 49, 114–120 (2008)
14. Redenbach, C., Särkkä, A., Freitag, J., Schladitz, K.: Anisotropy analysis of pressed point processes. Advances in Statistical Analysis 93(3), 237–261 (2009)
15. Kronenberger, M., Wirjadi, O., Freitag, J., Hagen, H.: Gaussian curvature using fundamental forms for binary voxel data. Graphical Models (accepted, 2015)
16. Wirjadi, O., Godehardt, M., Schladitz, K., Wagner, B., Rack, A., Gurka, M., Nissle, S., Noll, A.: Characterization of multilayer structures in fiber reinforced polymer employing synchrotron and laboratory X-ray CT. International Journal of Materials Research 105(7), 645–654 (2014)

Viscous-Hyperconnected Attribute Filters: A First Algorithm

Ugo Moschini[✉] and Michael H.F. Wilkinson

Institute for Mathematics and Computing Science,
University of Groningen, P.O. Box 407, 9700 Groningen, AK, The Netherlands
{u.moschini,m.h.f.wilkinson}@rug.nl

Abstract. In this paper a hyperconnectivity class that tries to address the leakage problem typical of connected filters is used. It shows similarities with the theory of viscous lattices. A novel algorithm to perform attribute filtering of viscous-hyperconnected components is proposed. First, the max-tree of the image eroded by a structuring element is built: it represents the hierarchy of the cores of the hyperconnected components. Then, a processing phase takes place and the node attributes are updated consistently with the pixels of the actual hyperconnected components. Any state-of-the-art algorithm can be used to build the max-tree of the component cores. An issue arises: edges of components are not always correctly preserved. Implementation and performance are presented. A possible solution is put forward and it will be treated in future work.

Keywords: Hyperconnected components · Max-tree · Attribute filtering

1 Introduction

In image analysis the notion of connectivity [12] defines how pixels are grouped together into connected components. In classical connectivity, those are regions of the image made of pixels that share the same intensity and are path-wise connected. Such regions are also called flat-zones. Attribute filters are operators that filter an image at the level of connected components: flat-zones are removed or preserved through filtering operations [3, 9, 10]. Issues might arise with connected filters based on the definition of connectivity mentioned above. For example, a structure that a human observer would consider as a single connected component might be broken into multiple components by noise or other artefacts. Another example is the leakage problem. It occurs when two different objects are connected by some bridging elements, such as noise or irrelevant image structures. Several solutions have been proposed, some based on connectivity [2, 15, 17], and some which are related but not strictly connected [16]. Connectivity-based solutions first modify the image with some anti-extensive operator and compute connected components based on the modified image [4]. A drawback of this approach is that it boils down to performing an attribute filter based on standard connectivity on the modified image [20] in all practical cases. Leakage is related with

---

This work was funded by the Netherlands Organisation for Scientific Research (NWO) under project number 612.001.110.

© Springer International Publishing Switzerland 2015
J.A. Benediktsson et al. (Eds.): ISMM 2015, LNCS 9082, pp. 669–680, 2015.
DOI: 10.1007/978-3-319-18720-4_56

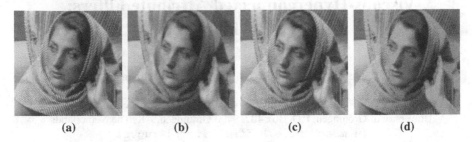

(a)	(b)	(c)	(d)

Fig. 1. (a) original image; images after deletion of viscous-hyperconnected components (b) and standard connected components (c) and (d) with area smaller than 10, 10 and 1000 pixels, respectively. Even for large values of area threshold, the thin white stripes of the veil are preserved in (c) and (d) due to the leakage problem. The cores of hyperconnected components were obtained eroding the image with a 5x5 disk.

a more general problem present in standard connectivity: spatial relationships among components as well as their grey level connection are not taken into account. That is, overlapping among flat-zones at the same intensity is not possible because the zones would be considered as one component. Hyperconnectivity [8, 14, 21, 23] and attribute-space connectivity [20] were created to overcome these limitations. In this paper, we will use a hyperconnectivity class as a possible solution to the problem of leakage. It shows similarities with viscous lattices [13] and it is referred to as viscous hyperconnection. Different viscosity indexes are given by different sizes of structuring element used in the extensive dilation operation that defines a viscous lattice. Fig. 1 shows the different output of connected and hyperconnected filtering, which illustrates that leakage is stopped by using viscous-hyperconnected components. In this work, we propose an algorithm based on max-trees [9] to perform attribute filtering of viscous-hyperconnected components. The implementation is analysed and performance and drawbacks are discussed.

2   Viscous-Hyperconnectivity Class

Two axiomatics for hyperconnectivity were defined in the recent years by Wilkinson in [21, 23] and by Perret et al. in [8]. It was noted in [8] that the definition of an overlap criterion as in [21] is not either required or needed to derive any new property for hyperconnections. In this paper we restrict the definition of a hyperconnectivity class to the following three axioms:

Definition 1. *A hyperconnectivity class* $\mathcal{H} \subseteq \mathcal{L}$, *with* $\mathcal{L}$ *a complete lattice is a class with the following properties:*

1. $0 \in \mathcal{H}$
2. $\mathcal{H}$ *is sup-generating*
3. $\mathcal{H}$ *is chain-sup complete,*

Informally, the second property states that every possible subset of the image space can be constructed as supremum of elements in $\mathcal{H}$. The third property states that the supremum of any totally ordered subset $\mathcal{H}$ also is in $\mathcal{H}$. We refer to the works in [14,23] for a more thorough explanation. Hyperconnectivity can generalise a large number of operators from edge-preserving connected filters to structural filters [5, 21, 22] into a single framework. This vast generalization leads to the problem of defining meaningful hyperconnections useful for image processing purposes since a very broad range of possibilities is open up. A recent theory [6] has been developed that unifies connectivity and hyperconnectivity defining axiomatics for both.

In this section a hyperconnectivity class is illustrated as a possible solution to the problem of leakage typical of connected filters. This class and its connection to viscous lattices [13] have been already put forward in [21]. It is inspired by the process of constrained reconstruction from markers in [19], a solution to prevent objects linked by narrow structures from being erroneously reconstructed. To define this hyperconnectivity class, let us start from defining a structuring element B as a ball centred on the origin, and C some connectivity class on $\mathcal{P}(\mathcal{L})$, the power set of a lattice $\mathcal{L}$. The element B will be considered a flat and connected structuring element in the rest of the paper. Consider the following set:

$$\mathcal{H}_B = \{\emptyset\} \cup S \cup \{\mathcal{H} \in \mathcal{P}(\mathcal{L}) \mid \exists C \in \mathcal{C} : \mathcal{H} = \delta_B C\} \tag{1}$$

It represents the set of all dilates by B of all connected sets, augmented with the empty set $\emptyset$ and all the singletons S, in the same way as in the theory of viscous lattices. This set is the viscous-hyperconnectivity class. This is equivalent to stating that the intersection of elements belonging to $\mathcal{P}(\mathcal{L})$ contains at least one translate of B. In viscous-hyperconnectivity, any image is constructed from a series of hyperconnected components which all lie within the opening $\gamma_B f$ and a series of singletons which lie in $f - \gamma_B f$. Fig. 2b shows a connected component in standard connectivity that it is split in the two hyperconnected components in Fig. 2d: in fact, the structuring element in Fig. 2c does not fit at the locations where the two square structures intersect, thus forming two components. Although viscous reconstruction (reconstruction with criteria) from markers has already been studied and linked to viscous hyperconnected

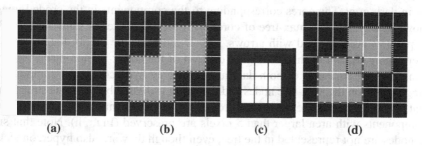

(a)　　　　　(b)　　　　　(c)　　　　　(d)

Fig. 2. (b) the dashed line highlights the boundary of the connected component in (a); (c) 3x3 square structuring element; dashed and dotted lines highlight the two viscous-hyperconnected components in (a) according to the structuring element in (c)

filtering in [11], there is currently no method to perform *hyperconnected attribute filtering* [23] based on viscous-hyperconnectivity. A solution is proposed in the next section.

3   Viscous-Hyperconnected Filtering

The max-tree [9] structure is commonly used to perform efficient attribute computation and filtering on the image connected components. Each of the nodes of the tree represents a peak component of a grey level image f. Peak components at a grey level h are connected components of threshold sets at level h for the image f. The image is a set of nested peak components: every node has a single parent pointer to a node at a grey level below its own. Nodes contain usually auxiliary data used to calculate measures of the component, e.g. area or shape features. These measures are referred to as attributes. The root of the tree represents the image background, representing the component with lowest intensity in the image. The leaves of the tree represent the highest intensities. The viscous hyperconnected components of an image are also ordered in the same way as their counterparts in standard connectivity. Indeed, the ordering is still given by the flat-zone intensity. The fact that components at the same intensity can overlap, without being merged together, does not interfere with the ordering. The cores of viscous-hyperconnected components can readily be represented in a hierarchical structure like the max-tree. According to the hyperconnectivity class in Section 2, a viscous-hyperconnected set is either a singleton or a dilate of a connected set. A viscous-hyperconnected set can be seen as made of a connected *core* and a *shell*, that is the difference between a core and its dilate. Cores of components are created by eroding the original image with some structuring element that drives the viscosity. The computation of the tree of viscous-hyperconnected components is summarised in the following. First, a max-tree of the eroded image containing the cores is built and attributes are computed. Then, for every pixel p, nodes n are determined so that p is in the shell of the core components corresponding to the nodes n. The attributes of such nodes are updated consistently with those pixels belonging to the shell. Fig. 3a shows a grayscale image with the connected component in light grey highlighted by a red dashed line. With B the structuring element as in Fig. 3b, the light grey component corresponds to two hyperconnected components, shown in Fig. 3c. The same happens for the component in dark grey. The cores corresponding to the components in the eroded image are shown in Fig. 3d. The max-tree of cores with area attributes is shown in Fig. 3e. Parent pointers are indicated with arrows. Fig. 3f indicates the pixels that are part of the overlapping section of the shells after dilation of the cores: their contribution will be counted in the attributes of both dark and light grey hyperconnected components. Filtering is done on the max-tree nodes and the result is dilated by the same structuring element. In our example, the tree of hyperconnected components in Fig. 3g is filtered and components with area larger than 5 pixels are preserved (Fig. 3h). Note that singleton nodes are not represented in the tree, even though they are also hyperconnected components. If they satisfy the filter criterion, they are added directly to the filtered image. Fig 3i shows that the singleton node of unit area corresponding to the white pixel at the centre of the image would have two parent nodes. That is, in general, the actual structure would not be a tree but a directed acyclic graph, if singletons are considered.

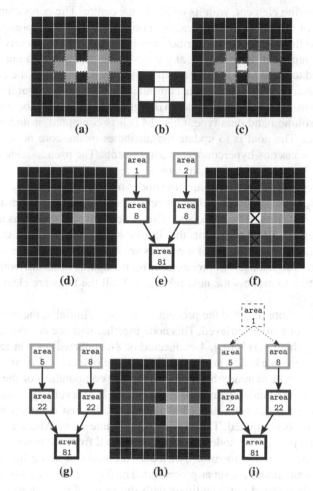

Fig. 3. (a) grayscale image with highlighted one connected component with area=12; (b) cross structuring element; (c) two viscous-hyperconnected components given (b); (d) cores of (a); (e) max-tree of the cores of hyperconnected components in (d); (f) the black crosses indicate the pixels shared by the hyperconnected components; (g) max-tree of hyperconnected components with updated attributes; (h) filtered image: hyperconnected components with area ≥ 5 pixels are preserved; (i) directed acyclic graph if the singleton node is added.

A more rigorous definition of the graph structure that originates from hyperconnected components will be given in future work. For the purpose of this work, we analyse the tree of viscous-hyperconnected components that are not singletons.

3.1   The Algorithm

The implementation of the tree of cores of hyperconnected components of an image f starts with building a max-tree $MT\epsilon_B$ of the eroded image $\epsilon_B f$, where B is a flat and connected structuring element, with its origin in the centre. The cores contain a subset of all the pixels of the whole hyperconnected components. $MT\epsilon_B$ represents the nodes corresponding to the cores: just the attributes are updated with the pixels belonging to the shell of a component. In principle, any state-of-the-art algorithm that builds max-trees can be used to build $MT\epsilon_B$ and store the auxiliary data (area of the component in the rest of the paper) needed for attribute computation. Sequential or parallel algorithms that follow bottom-up flooding or top-down merging approaches can be used according to the image resolution and data type. Once $MT\epsilon_B$ is computed, it undergoes a post-processing phase. The goal is to update the attributes of the core nodes consistently with the complete viscous-hyperconnected components. The pseudo-code is detailed in Alg. 1 and Alg. 2 and it has three main steps.

At first, for every pixel p in the image, the nodes of $MT\epsilon_B$ that intersect B_p are determined, where B_p is the structuring element centred on p. Only the nodes at level larger than or equal to $\epsilon_B f(p)$ will be processed. This procedure can be seen as a scanning of the opened image $\delta_B \epsilon_B f$: it represents the set of the dilates by B of the connected sets (cores) in $\epsilon_B f$. Later, the attributes of the nodes are updated. Every node contains a flag isProcessed to signal itself as processed after the update of its attributes. In the last step before starting to process the next pixel $p + 1$, all the flags are cleared and set to false.

Let us examine more in detail the procedure in Alg. 2. Initially, the node n_p that the current pixel p belongs to is retrieved. This node together with the nodes that correspond to the direct neighbour pixels (e.g. 4-connected or 8-connected) of p at level less than or equal to $\epsilon f(p)$ are marked as processed. Since these pixels are directly connected to p, their contribution has already been counted in the computation of the attributes of n_p in $MT\epsilon_B$. For the remaining nodes n lying within B at level larger than or equal to $\epsilon f(p)$, if a node n has not been flagged as processed, the first common ancestor n_{anc} between n and n_p is computed. The while loop at line 14 of Alg. 2 implements the attribute updating phase. The node structure is traversed from the node n down to (and not including) the common ancestor. Attributes are updated along the way with the contribution of p and nodes are set as processed. The flag isProcessed prevents that a node attribute is updated multiple times with the value of pixel p, when descending from the other nodes lying within B.

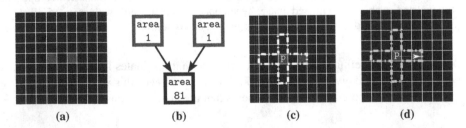

(a)   (b)   (c)   (d)

Fig. 4. (a) erosion by a 5x5 cross B; (b) max-tree of (a); (c) B overlapped on the location of pixel p; (d) the node at the dark grey intensity not containing p in (b) must be updated with the contribution of p, in spite of being at the same intensity level

Algorithm 1. Process the max-tree of cores of viscous-hyperconnecteced components $MT\epsilon_B$.

1: **procedure** PROCESSTREE(Image $\epsilon_B f$, Nodelist $nlist$)
2:     **for all** pixels $p \in \epsilon_B f$ **do**
3:         PutNodesWithinSEIntoNodeList(p, $nlist$, $\epsilon_B f$);
4:         ComputeAttributes(p, $nlist$, $\epsilon_B f$);
5:         ClearIsProcessedFlag($nlist$);
6:     **end for**
7: **end procedure**

The descent could stop at the level of pixel p in $\epsilon_B f$ only for the nodes that have n_p in their root path. In fact, a node m different from n_p may exists at the same level of p. Fig. 4 illustrates that. In this case, in $MT\epsilon_B$ the contribution of pixel p was not counted in the attributes of m and of all the other nodes down to the common ancestor. Once the attributes are computed, the updated max-tree structure is filtered using the attributes of the hyperconnected components. The result is then dilated by B, to have a correct image restitution of the filtered image. Finally, if the singletons meet the filtering criterion, they are merged with the result in a manner consistent with the filtering rule used.

4   Implementation Notes

4.1   Eroding the Input Image

The first step of the algorithm is the creation of $\epsilon_B f$, for some structuring element B. The algorithm in [18] was used to compute efficiently erosion operations with arbitrary flat structuring elements: it performs independently of the image content and the number of intensities. Shapes tested in our algorithm are crosses, rectangles and ellipses of different sizes. The same algorithm was used to perform the final dilation to generate the output image.

4.2   Computing the Nodes Within B_p and the Common Ancestor

An important issue is to update the list of nodes that intersect B_p in an efficient way. Assume G is the number of levels in the image. We use an array *nodeList* of length G, where each entry contains a linked list of pointers to nodes in the max-tree. The linked lists store the nodes currently within B_p. The max-tree node structure must have two fields: an integer numPxInNode that indicates how many pixels currently within B_p belong to the given node and a pointer listpos that points to its entry in *nodeList*. Initially, all lists in *nodeList* are empty, all numPxInNode fields are set to 0 and all listpos pointers are set to null. Starting from the first pixel p at the first location, when a node within B_p must be added to *nodeList*, it is checked if its numPxInNode field is set to 0. If so, a new list entry is created in *nodeList* at the proper level, and the pointer listpos is updated to point to this entry node. If numPxInNode is not zero, the node is already in *nodeList* and the counter numPxInNode is simply incremented.

Algorithm 2. Compute the attributes of the viscous-hyperconnected components.

 1: **procedure** COMPUTEATTRIBUTES(Pixel p, NodeList $nlist$, Image $\epsilon_B f$)
 2:     $n_p \leftarrow$ GetTreeNode(p);
 3:     n_p.isProcessed $\leftarrow$ `true`;
 4:     **for all** 4-/8-connected neighbour pixels $pneigh$ of pixel p **do**
 5:         **if** $\epsilon_B f(pneigh) \leq \epsilon_B f(p)$ **then**
 6:             $n_{pn} \leftarrow$ GetTreeNode($pneigh$);
 7:             n_{pn}.isProcessed $\leftarrow$ `true`;
 8:         **end if**
 9:     **end for**
10:     **for all** nodes $n \in nlist$ **do**
11:         **if** $n \ != $ ROOT $\wedge$ n.isProcessed $==$ `false` $\wedge$ n.Level $\geq \epsilon_B f(p)$ **then**
12:             $n\_anc \leftarrow$ GetFirstCommonAncestor(n, n_p);
13:             $n\_curr \leftarrow$ n;
14:             **while** $n_curr \ != $ ROOT $\wedge$ n_curr.isProcessed $==$ `false` **do**
15:                 **if** n_curr.Level $> n\_anc$.Level **then**
16:                     UpdateAttributes(n_curr, p);
17:                     n_curr.isProcessed $\leftarrow$ `true`;
18:                     $n\_curr \leftarrow n\_curr$.Parent;
19:                 **end if**
20:             **end while**
21:         **end if**
22:     **end for**
23: **end procedure**

A similar situation occurs when a node must be removed. The counter `numPxInNode` is decremented. If `numPxInNode` is zero, through the pointer `listpos` the entry in *nodeList* is accessed and removed, while `listpos` is set to `null`. Adding and removing entries is in constant time. Summarising, at each first pixel p for every image scan line, all the nodes within B_p are inserted into *nodeList*. Each time the processing move on to the next pixel, the nodes at the left-hand edge of B_p are removed and those ones at the right edge of B_{p+1} are added. Function `getFirstCommonAncestor()` finds the common ancestor of two nodes examining all the parent pointers starting from them until a common node is reached. Retrieving the ancestor node ought to be implemented more efficiently in constant time after a linear pre-processing of the tree as in [1]. The algorithm has roughly a complexity equal to $O(N_f \cdot G \cdot N_B)$, with N_f and N_B the number of pixels in the image f and in B, respectively.

5 Filtering Results and Performance

Fig. 1 shows an example of hyperconnected filtering. The structuring element used is a disk of 5 pixel diameter. The filter applied is an area opening, preserving the components with area larger than some threshold value. In Fig. 1b and Fig. 1c, the difference between connected and hyperconnected filtering is evident. Both images were filtered with area threshold larger than 10 pixels. The thin white stripes on the veil of the woman are preserved in Fig. 1c where connected filtering was applied. Even for

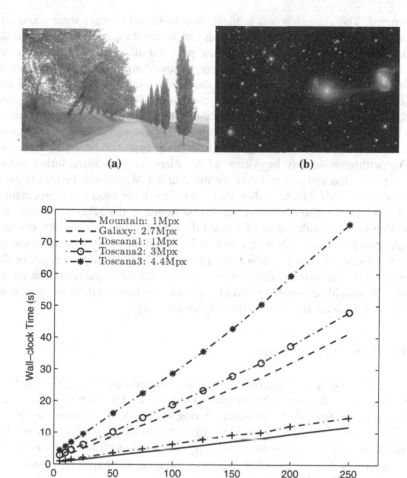

Fig. 5. (a) and (b) show the images referred to as Toscana and Galaxy in the plot in (c). The Toscana image was down-sampled at three different resolutions. The plot shows the time spent to update the max-tree of cores.

larger area threshold values (1000 pixels in the case of Fig. 1d), the leakage problem prevents the stripes from being removed. Leakage occurs linking the stripes together to other bright structures in the face or in the background, thus making a single component with large area. With viscous-hyperconnected filtering, an area threshold equal to 10 pixels achieves the result in Fig. 1b. Even smaller area thresholds could have been used because the thin structures have a width of a few pixels and they would be anyway split into singletons by the structuring element used. Leakage through the thin white stripes in the veil is therefore stopped. The plot in Fig. 5 shows the wall-clock times needed to perform the hyperconnected filtering for a few images. Area attribute was computed. The algorithm was implemented in C and timings were taken on an Intel Core i7-2670QM@2.20Ghz laptop with second level cache of 6MB. The code is sequential and it runs on a single core. In the plot, the diameter of a disk structuring element varies from 5 to 250 pixels. The images tested are 8-bit images portraying a mountain landscape, a merger galaxy from the Sloan catalogue Data Release 7 and a Tuscan countryside road at three different resolutions. The time to create the eroded image, the max-tree of the cores, and the final dilation is a fraction of the total time, on average five per cent of the total execution time. Erosions and dilations are computed with the algorithm in [18] and $MT\epsilon_B$ with a flooding approach based on hierarchical queue. Yet with considerably large sizes of the diameter of the structuring element, about 100 pixels, the updated max-tree for the 2 and 3 Megapixels images is created in less than 20 seconds. Much smaller diameters are usually used when processing an image. A test was run also on the original 16-bit image of the galaxy. The time spent in processing the tree with a disk of 5 pixel diameter is long, about 300 seconds. It goes slightly up to 350 seconds with a disk of 250 pixel diameter. Accessing the array *nodeList* dominates the computation time. In case of high bit-depth integer or floating point images, the *nodeList* data structure should not be implemented as an array of length G. It should be possible to avoid any data structure at all, by flagging conveniently the nodes within B_p. This is currently under study.

6   Edge Preservation Issues

A drawback of the proposed algorithm is that the filtered output image is a subset of the opened $\gamma_B f$ image, except when singletons are included in the result. This reduces the edge-preserving feature of connected filtering approaches. The effect is visible in Fig. 6d. The way attributes are computed has to be modified. The attribute accumulation phase should be changed by looking in $\bar{\delta}_f^1 \gamma_B f$ for pixels within a translate of $\delta^1 B$, with δ^1 the unitary dilation and $\bar{\delta}^1$ the geodesic dilation. Furthermore, we need to do both a dilation by B and a geodesic dilation in the last step of the algorithm, before dealing with the singletons. This was proposed in the case of reconstruction in [16, 19]. A complication arises since $\bar{\delta}_f^1 \gamma_B f$ may contain grey levels not present in $\epsilon_B f$,

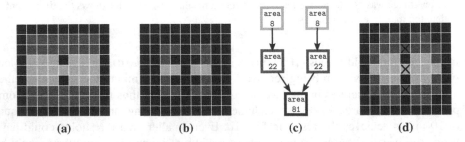

Fig. 6. (b) cores of (a) after erosion with a 3x3 cross structuring element; (c) max-tree of hyperconnected components; (d) image representation of the tree in (c): edges differs from the ones in (a).

and therefore absent in the max-tree. Nodes ought to be added, and their attributes consistently updated. There is currently ongoing work that will address this issue. A possible solution could be also found in the frame of *direct* connected operators [7], in which the definition of strong connectivity resembles the idea of core of a component presented in Section 3.

7 Conclusions

In this paper we used a hyperconnectivity class that addresses the leakage problem, as an alternative of working on a viscous lattice. A novel algorithm to perform hyperconnected attribute filtering on the viscous-hyperconnected components belonging to this class is proposed. First, the max-tree of the image eroded by a structuring element B is built. It contains the cores of the viscous-hyperconnected components. Their attributes are updated at a later stage with the pixels that lie between the core components and their dilate, through accessing the nodes within B_p for every image pixel p. Any existing algorithm can be used to build the max-tree of the eroded image. The proposed filtering strategy can also be used with high bit-depth integer or floating point images. Performance tests showed the importance of efficiently retrieving the nodes within B, at every location in the image. The filtered image is dilated by B and singletons are merged, according to the filtering rule used. A rigorous definition of directed acyclic graph that originates when singletons are added must be given. A drawback of this approach is that the filters do not preserve the edges. A possible solution has been found and it will be subject for a further work in the near future.

References

1. Bender, M.A., Farach-Colton, M.: The lca problem revisited. In: Gonnet, G.H., Viola, A. (eds.) LATIN 2000. LNCS, vol. 1776, pp. 88–94. Springer, Heidelberg (2000)
2. Braga-Neto, U., Goutsias, J.: A theoretical tour of connectivity in image processing and analysis. J. Math. Imaging Vis. 19(1), 5–31 (2003), http://dx.doi.org/10.1023/A:1024476403183
3. Breen, E.J., Jones, R.: Attribute openings, thinnings and granulometries. Comp. Vis. Image Understand. 64(3), 377–389 (1996)
4. Ouzounis, G.K., Wilkinson, M.H.F.: Mask-based second generation connectivity and attribute filters. IEEE Transactions on Pattern Analysis and Machine Intelligence 29(6), 990–1004 (2007)
5. Ouzounis, G.K., Wilkinson, M.H.F.: Hyperconnected attribute filters based on k-flat zones. IEEE Transactions on Pattern Analysis and Machine Intelligence 33(2), 224–239 (2011)
6. Perret, B.: Inf-structuring functions: A unifying theory of connections and connected operators. J. Math. Imaging Vis. 51(1), 171–194 (2015)
7. Perret, B., Cousty, J., Tankyevych, O., Talbot, H., Passat, N.: Directed connected operators: Asymmetric hierarchies for image filtering and segmentation. IEEE Transactions on Pattern Analysis and Machine Intelligence (2014)
8. Perret, B., Lefevre, S., Collet, C., Slezak, E.: Hyperconnections and hierarchical representations for grayscale and multiband image processing. IEEE Transactions on Image Processing 21(1), 14–27 (2012)
9. Salembier, P., Oliveras, A., Garrido, L.: Anti-extensive connected operators for image and sequence processing. IEEE Transactions on Image Processing 7, 555–570 (1998)

10. Salembier, P., Serra, J.: Flat zones filtering, connected operators, and filters by reconstruction. IEEE Transactions on Image Processing 4, 1153–1160 (1995)
11. Santillán, I., Herrera-Navarro, A.M., Mendiola-Santibáñez, J.D., Terol-Villalobos, I.R.: Morphological connected filtering on viscous lattices. J. Math. Imaging Vis. 36(3), 254–269 (2010), http://dx.doi.org/10.1007/s10851-009-0184-8
12. Serra, J.: Image Analysis and Mathematical Morphology. II: Theoretical Advances. Academic Press, London (1988)
13. Serra, J.: Viscous lattices. In: Proc. Int. Symp. Math. Morphology (ISMM 2002), pp. 79–90 (2002)
14. Serra, J.: Connectivity on complete lattices. J. Math. Imag. Vis. 9(3), 231–251 (1998), http://dx.doi.org/10.1023/A:1008324520475
15. Sofou, A., Tzafestas, C., Maragos, P.: Segmentation of soilsection images using connected operators. In: Int. Conf. Image Proc. 2001, pp. 1087–1090 (2001)
16. Terol-Villalobos, I.R., Vargas-Vzquez, D.: Openings and closings with reconstruction criteria: A study of a class of lower and upper levelings. J. Electronic Imaging 14(1), 013006 (2005), http://dblp.uni-trier.de/db/journals/jei/jei14.html#Terol-VillalobosV05
17. Tzafestas, C.S., Maragos, P.: Shape connectivity: Multiscale analysis and application to generalized granulometries. J. Math. Imag. Vis. 17, 109–129 (2002)
18. Urbach, E.R., Wilkinson, M.H.F.: Efficient 2-D gray-scale morphological transformations with arbitrary flat structuring elements. IEEE Transactions on Image Processing 17, 1–8 (2008)
19. Wilkinson, M.H.F.: Connected filtering by reconstruction: Basis and new advances. In: 15th IEEE International Conference on Image Processing, ICIP 2008, pp. 2180–2183 (October 2008)
20. Wilkinson, M.H.F.: Attribute-space connectivity and connected filters. Image Vision Comput. 25(4), 426–435 (2007), http://dx.doi.org/10.1016/j.imavis.2006.04.015
21. Wilkinson, M.H.F.: An axiomatic approach to hyperconnectivity. In: Wilkinson, M.H.F., Roerdink, J.B.T.M. (eds.) ISMM 2009. LNCS, vol. 5720, pp. 35–46. Springer, Heidelberg (2009), http://dx.doi.org/10.1007/978-3-642-03613-2_4
22. Wilkinson, M.H.F.: Hyperconnectivity, attribute-space connectivity and path openings: Theoretical relationships. In: Wilkinson, M.H.F., Roerdink, J.B.T.M. (eds.) ISMM 2009. LNCS, vol. 5720, pp. 47–58. Springer, Heidelberg (2009)
23. Wilkinson, M.H.F.: Hyperconnections and openings on complete lattices. In: Soille, P., Pesaresi, M., Ouzounis, G.K. (eds.) ISMM 2011. LNCS, vol. 6671, pp. 73–84. Springer, Heidelberg (2011), http://dx.doi.org/10.1007/978-3-642-21569-8_7

Incremental and Efficient Computation of Families of Component Trees

Alexandre Morimitsu[1], Wonder A.L. Alves[1,2], and Ronaldo F. Hashimoto[1(✉)]

[1] Department of Computer Science, University of São Paulo,
Institute of Mathematics and Statistics, São Paulo, Brazil
[2] Department of Informatics, Nove de Julho University, São Paulo, Brazil
{alem,wonder,ronaldo}@ime.usp.br

Abstract. Component tree allows an image to be represented as a hierarchy of connected components. These components are directly related to the neighborhood chosen to obtain them and, in particular, a family of component trees built with increasing neighborhoods allows the linking of nodes of different trees according to their inclusion relation, adding a sense of scale as we travel along them. In this paper, we present a class of neighborhoods obtained from second-generation connectivities and show that this class is suited to the construction of a family of trees. Then, we provide an algorithm that benefits from the properties of this class, which reuses computation done in previously built tree in order to construct the entire family of component trees efficiently.

Keywords: Connected component · Connected operator · Component trees · Second-generation connectivity

1 Introduction

The set of connected components (CCs) of the level sets of an image can be represented in a hierarchical way based on the inclusion relations of these CCs [1]. This hierarchy can be represented as a tree structure known as component tree. The component tree can be computed efficiently and attributes from each CC can be extracted quickly, making it a popular choice for applications like image filtering [2], text localization [3, 4], among others.

In the last decade, the use of component trees under second-generation connectivities [5] has been explored [6, 7]. The use of these connectivities allow clustering originally disjoint CCs into a single component and, when two increasing neighborhoods are considered, nodes of the corresponding two component trees can be linked according to their inclusion relation. This idea can be extended to a sequence of increasing neighborhoods, allowing the construction of a sequence of trees with inclusion of nodes between consecutive trees (see Fig. 1). This approach was first used by Passat and Naegel [7] in a segmentation method and is useful because it allows the extraction of measures of how the attributes of the merged nodes varies.

When building independently many component trees, however, the use of the usual algorithms [8, 9] might incur in high processing time, since most of

© Springer International Publishing Switzerland 2015
J.A. Benediktsson et al. (Eds.): ISMM 2015, LNCS 9082, pp. 681–692, 2015.
DOI: 10.1007/978-3-319-18720-4_57

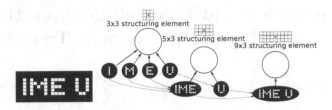

Fig. 1. Component trees built built with different neighborhoods. Black arrows indicate inclusion relation within the tree while blue arrows indicate inclusion of CCs of different trees.

them has complexity that depends on the size of the considered neighborhood. So in this paper we will focus on the development of an efficient algorithm to build these trees associated with increasing neighborhood. Our approach differs from [7] because the authors were more focused on attaining a fast algorithm for segmentation instead of tree construction. In this sense, this paper share similarities with the one presented by Ouzounis and Wilkinson [6], that is also focused on efficient construction of component trees using second-generation connectivity but it is not focused on a family of trees. In addition, our approach is also different because we are focusing on clustering based connectivity [5, 10] instead of a mask-based one.

After this brief introduction, this paper is divided in the following way: in Section 2, we present the theoretical background. Properties about the family of neighborhoods chosen in order to allow fast computation of the family is presented in Section 3. In Section 4 is shown our proposed way of implementing the construction algorithm. We then show some preliminary results in Section 5 and conclude the paper in Section 6.

2   Theoretical Background

In this study, we consider a gray-scale image f as a mapping from a subset of a rectangular grid $\mathcal{D} \subset \mathbb{Z}^2$ to a set of gray levels $\mathbb{K} = \{0, 1, \ldots, K-1\}$, i.e., $f : \mathcal{D} \to \mathbb{K}$. An element $p \in \mathcal{D}$ is called pixel and the gray level associated with the p in image f is denoted by $f(p)$. In addition, a pixel $p \in \mathcal{D}$ can be a neighbor of others pixels of $\mathcal{D}$ in their surroundings. This notion of neighborhood is defined by a structuring element (SE) $\mathcal{A} \subset \mathbb{Z}^2$ as follows: a pixel $p \in \mathcal{D}$ is neighbor of a pixel $q \in \mathcal{D}$ under $\mathcal{A}$ if and only if $q \in (\{p\} \oplus \mathcal{A})$, where $\oplus$ denotes the Minkowski addition and it is defined as $\mathcal{P} \oplus \mathcal{A} = \{p + a \in \mathcal{D} : p \in \mathcal{P} \text{ and } a \in \mathcal{A}\}$. For simplicity, we denote the neighbor pixels of p under $\mathcal{A}$ by $p \oplus \mathcal{A}$. Now, we can define notion of connectivity between pixels: let $\mathcal{A}$ be a SE and let $\mathcal{X} \subseteq \mathcal{D}$ be a set. We say that two pixels p and q that belong to $\mathcal{X}$ are connected under $\mathcal{A}$ if and only if there exists a sequence of pixels $(p = p_1, p_2, \ldots, p_\ell = q)$ within $\mathcal{X}$ satisfying $p_{k+1} \in (p_k \oplus \mathcal{A})$ for $1 \leq k < \ell$. A maximal subset C of $\mathcal{X} \subseteq \mathcal{D}$ such that any two pixels $p, q \in C$ are connected under $\mathcal{A}$ is called *connected component* (CC). In this way, let $CC(\mathcal{X}, \mathcal{A})$ denote the set of all CCs of $\mathcal{X}$ under $\mathcal{A}$.

From an image f we define, for any $\lambda \in \mathbb{K}$, the sets $\mathcal{X}_\downarrow^\lambda(f) = \{p \in \mathcal{D} : f(p) < \lambda\}$ and $\mathcal{X}_\lambda^\uparrow(f) = \{p \in \mathcal{D} : f(p) \geq \lambda\}$ as the *lower* and *upper level sets* at value λ of the image f, respectively. These level sets are nested, i.e., $\mathcal{X}_\downarrow^0(f) \subseteq \mathcal{X}_\downarrow^1(f) \subseteq \dots \subseteq \mathcal{X}_\downarrow^{K-1}(f)$ and $\mathcal{X}_{K-1}^\uparrow(f) \subseteq \mathcal{X}_{K-2}^\uparrow(f) \subseteq \dots \subseteq \mathcal{X}_0^\uparrow(f)$.

From the lower and upper level sets of an image f, one can define two other sets $\mathcal{L}_\mathcal{A}(f)$ and $\mathcal{U}_\mathcal{A}(f)$ composed by the CCs of the lower and upper level sets of f under $\mathcal{A}$, i.e., $\mathcal{L}_\mathcal{A}(f) = \{C \in \mathcal{CC}(\mathcal{X}_\downarrow^\lambda(f), \mathcal{A}) : \lambda \in \mathbb{K}\}$ and $\mathcal{U}_\mathcal{A}(f) = \{C \in \mathcal{CC}(\mathcal{X}_\lambda^\uparrow(f), \mathcal{A}) : \lambda \in \mathbb{K}\}$. The ordered pairs consisting of the CCs under $\mathcal{A}$ of the lower and upper level sets and the usual inclusion set relation, i.e., $(\mathcal{L}_\mathcal{A}(f), \subseteq)$ and $(\mathcal{U}_\mathcal{A}(f), \subseteq)$, induce two dual trees [11, 12] called *component trees*. The trees $(\mathcal{L}_\mathcal{A}(f), \subseteq)$ and $(\mathcal{U}_\mathcal{A}(f), \subseteq)$ can be represented by non-redundant data structures known, respectively, as min-tree and max-tree [1].

In the context of max-trees, CCs may also be referred as *nodes* and the inclusion relation of these nodes may also be referred as *parent relation*. In these trees, each pixel $p \in \mathcal{D}$ is associated to the smallest CC of the tree containing it which, by the inclusion relationship, it is also associated to all the ancestors CCs. Since there is a duality regarding upper and level sets, all the theory applied to one set can be applied to the other in its dual form. For simplicity, from now on all definitions and propositions used will refer only to upper level sets. Then, a max-tree will be generically denoted by $\mathcal{T}_\mathcal{A}$ and $\mathcal{SC}(\mathcal{T}_\mathcal{A}, p)$ will denote the smallest CC that contains p in $\mathcal{T}_\mathcal{A}$.

3   Incremental Computation of Component Trees

As explained in Sect. 1, we want to efficiently build a family of component trees of an image f satisfying the following property: given a set $I = \{1, 2, \dots, N\}$ and two trees $\mathcal{T}_{\mathcal{A}_i}$ and $\mathcal{T}_{\mathcal{A}_j}$ such that $i < j$ and $i, j \in I$, $\mathcal{A}_i \subset \mathcal{A}_j$. Let C_j be a CC in $\mathcal{CC}(\mathcal{X}_\lambda^\uparrow(f), \mathcal{A}_j) \subseteq \mathcal{T}_{\mathcal{A}_j}$ for a fixed λ. Then, C_j is the union of some CCs belonging to $\mathcal{CC}(\mathcal{X}_\lambda^\uparrow(f), \mathcal{A}_i) \subseteq \mathcal{T}_{\mathcal{A}_i}$.

In particular, given a family of structuring elements $\mathbb{A} = \{\mathcal{A}_i : i \in I\}$ indexed by the set I, we are interested in constructing incrementally a family of N component trees $\mathbb{T} = \{\mathcal{T}_{\mathcal{A}_i} : i \in I\}$ such that $\mathcal{T}_{\mathcal{A}_i} = (\mathcal{U}_{\mathcal{A}_i}(f), \subseteq)$.

For that, we need to guarantee that we can efficiently build the next tree using the previous one. In this paper, we will construct the family $\mathbb{A}$ by using an auxiliary set of structuring elements $\mathbb{B} = \{\mathcal{B}_1, \dots, \mathcal{B}_N\}$, with $\mathcal{B}_i$ satisfying $o \in \mathcal{B}_i$, for any $i \in I$. Then, the family $\mathbb{A}$ will be defined in the following way:

$$\begin{cases} \mathcal{A}_1 = (\mathcal{B}_1 \oplus \mathcal{B}_1^t), \\ \mathcal{A}_i = (\mathcal{A}_{i-1} \oplus \mathcal{B}_i \oplus \mathcal{B}_i^t), \text{ for } 2 \leq i \leq N, \end{cases} \tag{1}$$

where $\mathcal{B}^t = \{-b : b \in \mathcal{B}\}$ is the transposed of set $\mathcal{B}$.

Regarding these families, we call the family $\mathbb{A}$ as *generated family* and $\mathbb{B}$ as *generator family*. We note that, for the generated family $\mathbb{A}$, for all $1 \leq i < N$, the following properties hold: (a) $\mathcal{A}_i \subseteq \mathcal{A}_{i+1}$; (b) $\mathcal{A}_i$ is symmetric;

Although any $\mathcal{A}_i \in \mathbb{A}$ is symmetric, there are some symmetric elements that can not be built by using Eq. 1 (for example, 4-connected SEs). Fig. 2 depicts some examples of generated families, with the one in the right depicting the neighborhood used in Fig. 1.

Even though not any symmetric SE can be constructed by combining different SEs in $\mathbb{B}$ according to Eq. 1, we still can generate many different kinds of SEs for $\mathbb{A}$. For instance, we can obtain any SE that has a rectangular shape by combining SEs in $\mathbb{B}$ that expand neighborhood horizontally (the one shown in the middle of Fig. 2) followed by SEs that expands the neighborhood vertically.

Fig. 2. Examples of generated and generator families $\mathbb{A}$ and $\mathbb{B}$

For the generator family $\mathbb{B} = \{\mathcal{B}_i, i \in I\}$, it will be convenient to define a notation for a cumulative result of the Minkowski addition of consecutive SEs. Then, we denote $\mathcal{B}_i^\Sigma$ as $\mathcal{B}_i^\Sigma = (\mathcal{B}_1 \oplus \ldots \oplus \mathcal{B}_i)$.

Using this notation, one can easily prove the following proposition.

Proposition 1. *Let $\mathbb{B} = \{\mathcal{B}_i, i \in I\}$ be a generator family. If $\mathbb{A} = \{\mathcal{A}_i, i \in I\}$ is the corresponding generated family, then $\mathcal{A}_i = (\mathcal{B}_i^\Sigma \oplus (\mathcal{B}_i^\Sigma)^t)$, for any $i \in I$.*

The main reason why we are using this specific family is because of the following theorem:

Theorem 1. *Let $\mathbb{B} = \{\mathcal{B}_i \in I\}$ be a generator family and $\mathbb{A} = \{\mathcal{A}_i, i \in I\}$ be the corresponding generated family. Then, two pixels p and q are neighbor under $\mathcal{A}_i$ if and only if $(p \oplus \mathcal{B}_i^\Sigma) \cap (q \oplus \mathcal{B}_i^\Sigma) \neq \emptyset$, for any $i \in I$.*

Note that from definition of $\mathcal{B}_i^\Sigma$, we have $(p \oplus \mathcal{B}_i^\Sigma) = ((p \oplus \mathcal{B}_{i-1}^\Sigma) \oplus \mathcal{B}_i)$. Thus, by Theorem 1, it is possible to find neighbor pixels of any $p \in \mathcal{D}$ under $\mathcal{A}_i$ using the previous neighborhood $p \oplus \mathcal{B}_{i-1}^\Sigma$ and the current SE $\mathcal{B}_i$.

In order to use Theorem 1, we need an efficient way of finding intersection between two Minkowski additions. We can solve this problem by using a labeling algorithm: given a SE $\mathcal{B}$ and a pixel p of an image f, we label all elements in $p \oplus \mathcal{B}$ with the element p (we will refer to it as the *propagation* of the label p). and repeat this process to all other pixels in $\mathcal{D}$. We note that some pixels can have more than one label. For example, given two pixels p and q, a pixel r will have labels p and q if and only if $r \in (p \oplus \mathcal{B}) \cap (q \oplus \mathcal{B})$. Thus, we know that an intersection occurred between $p \oplus \mathcal{B}$ and $q \oplus \mathcal{B}$ if one of the elements in $q \oplus \mathcal{B}$ already has the label p and, according to Theorem 1, we know that p and q are neighbors under $\mathcal{A} = \mathcal{B} \oplus \mathcal{B}^t$.

Formally, given an image f and a SE $\mathcal{B}$, we can define a function $labels(f, \mathcal{B})$: $\mathcal{D} \to P(\mathcal{D})$ (where $P(\mathcal{D})$ is the power set of $\mathcal{D}$) as follows: for any $q \in \mathcal{D}$,

$$[labels(f, \mathcal{B})](q) = \{p \in \mathcal{D} : q \in (p \oplus \mathcal{B})\}. \tag{2}$$

It is easy to prove that $[labels(f, \mathcal{B})](q) = \{p \in \mathcal{D} : p \in (q \oplus \mathcal{B}^t)\}$, but we will prefer the definition given by Eq. 2 because it defines the process of the propagation of the label p to its neighbors in $p \oplus \mathcal{B}$.

We can obtain the connected components of any level set $\mathcal{X}_\lambda^\uparrow(f)$, $\lambda \in \mathbb{K}$ under a SE $\mathcal{A}$ by using the information contained in $labels(f, \mathcal{B})$ and the gray-level of the pixels in $X_\lambda^\uparrow(f)$. By definition, we know that two pixels r_1 and r_2 in $\mathcal{X}_\lambda^\uparrow(f)$ are connected under $\mathcal{A}$ if there is a sequence of pixels $(r_1 = p_1, p_2, ..., p_\ell = r_2)$ in $\mathcal{X}_\lambda^\uparrow(f)$ such that $p_{k+1} \in (p_k \oplus \mathcal{A})$ for $1 \le k < \ell$. Neighboring information can be retrieved from $labels(f, \mathcal{B})$ function: we know that p_k and p_{k+1} in $\mathcal{X}_\lambda^\uparrow(f)$ are neighbors under $\mathcal{A} = \mathcal{B} \oplus \mathcal{B}^t$ if there is a pixel q_k such that p_k and p_{k+1} are in $[labels(f, \mathcal{B})](q_k)$ and the foreground information can be obtained simply by comparing the gray-level of the elements with λ. Moreover, from Eq. 2, $labels(f, \mathcal{B}_{i+1})$ can be easily obtained from $labels(f, \mathcal{B}_i)$ by making use of the following property:

Proposition 2. *If $q \in \mathcal{D}$ and $\mathbb{B} = \{\mathcal{B}_i \in I\}$ is a generator family, then*
$$[labels(f, \mathcal{B}_{i+1}^\Sigma)](q) = \bigcup_{q' \in (q \oplus \mathcal{B}_{i+1}^t)} [labels(f, \mathcal{B}_i^\Sigma)](q')$$

That means we have a fast way of building $labels(f, \mathcal{B}_i^\Sigma)$ for any $i \in I$ and obtaining the CCs of all levels sets when we change our SE.

4   Fast Implementation

In this section we will discuss data structures and algorithms that will be used in order to build the family of trees associated with a generator family $\mathbb{B}$.

4.1   Data Structure

Thanks to the inclusion relation of the level sets, it is possible to efficiently store the whole max-tree $\mathcal{T}_{\mathcal{A}_i}$ under $\mathcal{A}_i$ in a union-find structure [13, 9, 14] for any $i \in I$. We can then use an array $parent[\ ]$ to represent the max-tree: in this structure, each CC is associated with one of its pixel, known as the canonical element [13]. Thus, if a pixel $p \in \mathcal{D}$ is the canonical element of $\mathcal{SC}(\mathcal{T}_\mathcal{A}, p)$ then $parent[p]$ points to the canonical element of the parent of $\mathcal{SC}(\mathcal{T}_\mathcal{A}, p)$. For all other pixels $q \in \mathcal{SC}(\mathcal{T}_\mathcal{A}, p)$ such that $p \ne q$ and $f(q) = f(p)$, we will have $parent[q] = p$. So we can find the canonical element of each pixel $p \in \mathcal{D}$ by calling the function $\texttt{levroot}(p)$ as described in [14].

Using this $parent[\ ]$ array, we can also cluster originally disjoint CCs efficiently: suppose that we have two pixels p and $q \in \mathcal{D}$ that had not been neighbors under $\mathcal{A}_i$ but became neighbors under $\mathcal{A}_{i+1}$. At all level sets that the CCs containing

p and q were disjoint, we need to cluster them and update this information in the tree structure. We can do so by using the `connect()` algorithm in [14]. That algorithm adapted to our problem is presented in Algorithm 1.

Algorithm 1. Merging disjoint nodes containing p and q

Data: Two canonical pixels p and $q \in \mathcal{D}$, an array $parent[\ ]$ and a list of pairs $merge$ that stores changes in the $parent[\ ]$ array.

Result: At the end of all recursive calls: $parent[\ ]$ updated with the clustering of all disjoint nodes containing p and q, $merge$ updated with pairs representing changes in the $parent[\ ]$ array.

```
 1  union (p, q, parent, merge) begin
 2  │    fP ← −1, fQ ← −1
 3  │    if p ≠ −1 then fP ← f(p)
 4  │    if q ≠ −1 then fQ ← f(q)
 5  │    if fP < fQ then
 6  │    └    union(q, p, parent, merge)
 7  │    else
 8  │    │    if p ≠ q then
 9  │    │    │    parentP ← −1, fParent ← −1
10  │    │    │    if parent[p] ≠ −1 then
11  │    │    │    │    parentP ← canonical element of parent[p]
12  │    │    │    └    fParent ← f(parentP)
13  │    │    │    if fParent ≥ fQ then
14  │    │    │    └    union(parentP, q, parent, merge)
15  │    │    │    else
16  │    │    │    │    parent[p] ← q
17  │    │    │    │    merge ← merge ∪ (p, q)
18  │    │    │    └    union(q, parentP, parent, merge)
```

In order to guaranteed that the algorithm stops we will initialize the $parent[\ ]$ array with all pixels pointing to an artificial node with id number -1. This is only needed at step $i = 1$, while we do not know which pixel will represent the root node, but we will keep it at other steps $i > 1$ for consistency.

By immediately calling this function **union** every time a pair of pixels p and q become neighbors, then it will suffice to store only the pixel $s \in [labels(f, \mathcal{A}_{i+1})](r)$ with the highest level to represent $[labels(f, \mathcal{A}_{i+1})](r)$, since all other elements of the set will belong to CCs that are ancestors of $\mathcal{SC}(\mathcal{T}_{\mathcal{A}_i}, s)$ in the tree and they can be obtained by traversing the tree starting from $\mathcal{SC}(\mathcal{T}_{\mathcal{A}_i}, s)$ towards the root node.

This means that the function $labels(f, \mathcal{A}_i)$ can be stored simply by using an array of pixels which we will call $lbl[\ ]$.

4.2 Further Optimizations

In the algorithm, since we only want to keep the label with the highest gray-level for each pixel, ordering the pixels according to their gray-levels before processing them is a good strategy. This way, once a label of a pixel changes at a step i, we know that it can not change again before the next step $i + 1$ starts.

Moreover, suppose family $\mathbb{B}$ has a cycle, i.e., there is a $c \in \mathbb{N}$ such that $\mathcal{B}_{i'} = \mathcal{B}_{i'+c}$ for any $1 \leq i' \leq N - c$. Then suppose we are processing $q \in \mathcal{D}$ and let p be its label. If after n steps the label does not change, then we do not need to keep propagating the label p to the neighbors of q, $q \oplus \mathcal{B}_n^{\Sigma}$, since this was already done in the previous cycle and we know that all of its neighbors under $\mathcal{B}_n^{\Sigma}$ already received it. This will reduce the number of pixels that needs to propagate their labels and therefore will reduce consumption time.

4.3   Main Algorithm

By combining all the ideas seen along this section, it is possible to develop an efficient algorithm to build the family of component trees associated to $\mathbb{A}$. The code is presented in Algorithm 2.

It is important to note that the resulting $parent[\,]$ array may not be canonized. If needed, this can be done by calling the function that finds the canonical element to all elements p that had their parents changed. These elements are stored in the $merge$ structure.

Algorithm 2. Building the trees associated to $\mathbb{A}$.

Data: The family of structuring element $\mathbb{B} = (\mathcal{B}_1, \ldots, \mathcal{B}_N)$, an integer $cyclesize$ that indicates the size of the cycle existing in $\mathbb{B}$ (if there is no cycle, it is set to $N = |\mathbb{B}|$) and $merge$, an array of N lists of pairs.

Result: The array of pairs $merge$ updated with $merge[i]$ containing all changes in parent relation during step i; $parent[\,]$ representing the max-tree under $\mathcal{A}_N$.

```
 1  buildTrees (B, cyclesize, merge) begin
 2      foreach p ∈ D do
 3          lbl[p] ← p, add p to a priority queue Q with priority f(p)
 4          parent[p] ← −1, iStop[p] ← 1 + cyclesize, fLbl[p] ← f(p)
 5          nextLbl[p] ← null, added ← ∅, addToQ[p] ← false
 6      foreach 1 ≤ i ≤ N do
 7          while Q is not empty do
 8              p ← remove element with the highest priority of Q
 9              foreach q ∈ p ⊕ Bᵢ do
10                  if nextLbl[q] = null then   nextLbl[q] ← lbl[p]
11                  else
12                      canonicalP ← canonical element of lbl[p]
13                      canonicalQ ← canonical element of nextLbl[q]
14                      union(canonicalP, canonicalQ, parent, merge[i])
15                  if f(lbl[p]) > fLbl[q] then
16                      fLbl[q] ← f(lbl[p]), iStop[q] ← i + 1 + cyclesize
17                      nextLbl[q] ← lbl[p]
18                      if addToQ[q] = false then
19                          addToQ[q] ← true, added ← added ∪ {q}
20                  if iStop[p] > i + 1 and addToQ[p] = false then
21                      addToQ[p] ← true, added ← added ∪ {p}
22              foreach p ∈ added do
23                  add p to Q with priority fLbl[p]
24                  added ← added \ {p}, addToQ[p] ← false
25              lbl ← copy of nextLbl
26      return merge
```

As seen in Algorithm 2, there is a lot of auxiliary data structures, but the algorithm can be explained briefly as follows: we dequeue a pixel p with the highest priority and propagate its label to its neighbors in $p \oplus \mathcal{B}_i$. These labels are stored in $nextLbl[\ ]$, while $lbl[\ ]$ stores $labels(f, \mathcal{B}_{i-1})$. For $i = 1$, we simply initialize $lbl[p] \leftarrow p$.

Intersections are found when $nextLbl[\ ] \neq null$ (Line 11). The union algorithm is called and is updates the tree accordingly. If a pixel had its label changed or still did not complete an entire cycle propagating its label, then it must be added to queue and will be processed again in step $i + 1$. This process is repeated until the step N.

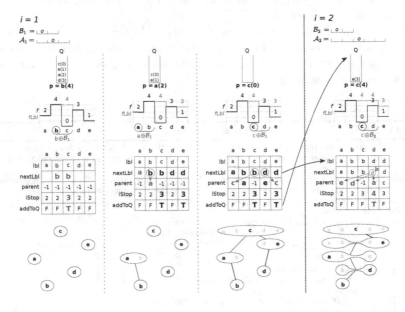

Fig. 3. Simulation of Algorithm 2 for the given families $\mathbb{A}$, $\mathbb{B}$ and image f

Fig. 3 depicts a simulation of Algorithm 2 for a 1-dimensional image f. The value of the step i and the $\mathcal{A}_i$ and $\mathcal{B}_i$ used are presented at the top, with $\mathcal{A}_i$ being the 1-dimensional SE with width $2i + 1$ (we are assuming $N = 2$ and that implies $cyclesize = 1$). The four columns below them represents 4 moments of the algorithm that we judged useful for better understanding of the algorithm. Q denotes the priority queue with its elements orderer by their priorities (between parenthesis). Below Q is shown a graphical representation of the image f (black line) and the array $fLbl$ (red line). The values of each structure and a graphic representation of the updated max-tree are shown in the bottom.

Starting from left to right, the first column shows an example of a pixel (b) that propagates its label but does not change the parent relation, it simply updates the other data structures (shown in red). When the propagation of b ends, d will be dequeued and the same process will be repeated. The changes made while $p = d$ are analog to the ones made by b so we do not show them explicitly in Fig. 3.

The second column shows an example of a change in the parent relation. When $p = a$ we will have an intersection between a and b (more precisely, an intersection of $lbl[a]$ with $nextLbl[b]$). In this case, we will need to update the parent relations between the CCs that contain a and b. This process is repeated until Q is empty, and the final state of each data structure is shown in the third column. The elements satisfying $addToQ[p] = true$ are re-added to Q and we start the step $i = 2$ with $lbl \leftarrow nextLbl$.

In the rightmost column is shown an example of change in parent relation that affects more than one node. We detect an intersection between $lbl[c] = b$ and $nextLbl[d] = d$ (it will be changed to b later, as shown in Fig 3). The union algorithm then perform changes in the parent relation in order to reflect the fact that b and d now are neighbors.

4.4 Linking Nodes of Different Trees

Using the *merge* structure, we can rebuild any max-tree $\mathcal{T}_{\mathcal{A}_i}$ starting with a *parent* array with all positions assigned to -1 and updating it with changes from $merge[1]$ to $merge[i]$. Moreover, if we build two max-trees $\mathcal{T}_{\mathcal{A}_i}$ and $\mathcal{T}_{\mathcal{A}_j}$, it is possible to link nodes of these trees based on inclusion relation between nodes of each tree: suppose, without loss of generality, that $\mathcal{A}_i \subset \mathcal{A}_j$. We will then refer to $\mathcal{T}_{\mathcal{A}_i}$ as the 'smaller' tree and $\mathcal{T}_{\mathcal{A}_j}$ as the 'bigger' tree. Let $parent_j[\;]$ be the array representing the max-tree $\mathcal{T}_{\mathcal{A}_j}$. Then, if p is a canonical element of a CC in the smaller tree, the canonical element of p in $parent_j[\;]$ will indicate the smallest CC that contains it in the bigger tree, making it possible to link all nodes in both max-trees.

5   Experimental Results

This section shows some preliminary results that were obtained using this family of component trees:

5.1   Comparison of Attributes Between Merged Nodes

Each node of a tree represents a CC of the image given an SE $\mathcal{A}_i$. For each CC, we can compute attributes like area, perimeter or width and, since each node consists of merges of nodes of previous trees, we can evaluate how the attributes of the merged nodes vary and extract some information comparing the variance of these attribute.

This allows us to detect some nodes that have attributes that differ too much from the other merged nodes. For example, if we have words with different sizes, we can segment them by calculating the variance of the horizontal spacing between a pair of characters (Fig. 4, image from the ICDAR database [15]).

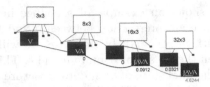

Fig. 4. Some nodes and the variance of the heights of the characters. Black lines are edges linking nodes of the tree, red and blue lines indicate links between nodes from different trees. The numbers indicate the variance of the spacing between the characters in the merged node. The red number indicates a big variance, indicating that it is likely that the word mixed with another with different size.

5.2 Complexity and Time Consumption

The cost of the algorithms used to build the trees are strongly related to how many pairs of pixels they need to examine in order to update the parent relation. It is difficult to give a precise complexity of the proposed algorithm because the cost of the function **union** and the amount of steps a same pixel will be added to the priority queue both vary according to the topology of the image and the generator family chosen. For each step $1 \leq i \leq N$, our algorithm has complexity depending on the number of pixels in the queue (n^2 in the worst case, where n^2 is the size of the image), the number of neighbors processed ($|\mathcal{B}_i|$), and the cost of the **union** algorithm (the cost consists of the size of the path linking p to q in the tree, bounded by $2K$). If the ordering algorithm has complexity $O(n^2 \log n)$, a worst case analysis will give us a complexity of either $O(Nn^2BK)$, where $B = \max_{i \in I} |\mathcal{B}_i|$ or $O(Nn^2 \log n)$, depending on how BK compares to $\log n$.

In the case we have a cycle in $\mathbb{B}$, however, the expected amount of pixels in the priority queue will be reduced as the step i increases. To give an estimation of the complexity of the algorithm in this case, we will fix a family $\mathbb{B}$ and calculate how many times a pixel is re-added to the queue to a randomized image (each pixel is assigned a gray-level from $\mathbb{K}$ with probability $\frac{1}{K}$). Assuming the best case when the cycle has size 1 ($\mathcal{B}_i = \mathcal{B}_1, \forall i \in I$), the expected amount of times a pixel is added to the queue is bounded by $c \log m$, where c is a positive constant and $m = |\mathcal{A}_N|$. This changes the complexity to the biggest factor between $O(Kn^2 \log m |\mathcal{B}_1|)$ and $O(n^2 \log n \log m)$.

Known union-find based algorithms used in simple connectivities have a complexity of $O(\alpha n^2 m)$ (α time if the inverse Ackermann function) if the ordering step can be performed in linear time. So the size of m directly impacts the complexity of these algorithms: if m increases linearly, so does the time consumption and if m increases quadratically, then complexity can become as high as $O(\alpha n^4)$. The dual-input algorithm [6] is expected to perform better in these conditions. However, this would require further investigation on how to create the mask images to generate the same trees in order to provide a fair comparison.

So to analyze the complexities obtained in an actual image, we fixed $\mathbb{B}$ as the family shown on the left of Fig 2, considered $K = 256$ and analyzed a 2-dimensional image shown in the left side of Fig. 5 (res. 1280x912, taken from the ICDAR[15] database). Using these parameters, the expected complexity for our algorithm is $O(n^2 \log n \log m)$ when considering that we know it contains a cycle and $O(Nn^2 \log n)$ otherwise, while the complexity of a union-find based algorithm is $O(\alpha n^2 m)$. Fig. 5 depicts the time consumption of both algorithms: the blue curve shows the total execution time for the given image for the proposed algorithm to build the max-tree restricted to $\mathcal{A}_j$ considering $cyclesize = 1$ and the gray curve shows the time consumption without the cyclic assumption (this was done by calling the algorithm with $cyclesize = N$). The red curve shows the time spent by the algorithm shown in [9]. It can be seen that the curves follow the complexities estimated. These results were obtained using a notebook with 2.4GHz processor and 4GB of RAM, coded in Java and without using parallelism.

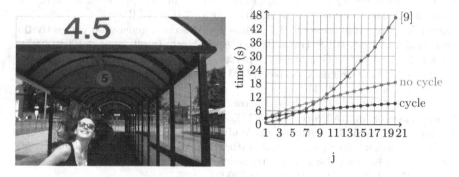

Fig. 5. Time spent to built the max-tree restricted to $\mathcal{A}_i$ (the square structuring element with side $2i + 1$) for the image in the left side

6   Conclusion

In this paper, we explored the idea of using a family of component trees generated from a family of structuring elements, which allowed us to establish relations between CCs of different trees. We then presented an algorithm to build this family of component trees and showed that part of the computation used to build a tree can be reused in the construction of the next ones, making the algorithm more efficient, specially under the condition of cycling structuring elements.

Acknowledgements. We would like to thank the financial support from CAPES, CNPq, and FAPESP (grant #2011/50761-2).

References

[1] Salembier, P., Oliveras, A., Garrido, L.: Anti-extensive connected operators for image and sequence processing. IEEE Transactions on Image Processing 7(4), 555–570 (1998)

[2] Xu, Y., Géraud, T., Najman, L.: Morphological filtering in shape spaces: Applications using tree-based image representations. In: 2012 21st International Conference on Pattern Recognition (ICPR), pp. 485–488. IEEE (2012)

[3] Sun, L., Huo, Q.: A component-tree based method for user-intention guided text extraction. In: 2012 21st International Conference on Pattern Recognition (ICPR), pp. 633–636. IEEE (2012)

[4] Alves, W., Hashimoto, R.: Text regions extracted from scene images by ultimate attribute opening and decision tree classification. In: Proceedings of the 23rd SIBGRAPI Conference on Graphics, Patterns and Images, pp. 360–367 (2010)

[5] Serra, J.: Connectivity on complete lattices 9(3), 231–251 (1998)

[6] Ouzounis, G.K., Wilkinson, M.H.F.: Mask-based second generation connectivity and attribute filters. IEEE Trans. Pattern Anal. Mach. Intell. 29, 990–1004 (2007)

[7] Passat, N., Naegel, B.: Component-hypertrees for image segmentation. In: Soille, P., Pesaresi, M., Ouzounis, G.K. (eds.) ISMM 2011. LNCS, vol. 6671, pp. 284–295. Springer, Heidelberg (2011)

[8] Najman, L., Couprie, M.: Building the component tree in quasi-linear time. IEEE Transactions on Image Processing 15(11), 3531–3539 (2006)

[9] Berger, C., Géraud, T., Levillain, R., Widynski, N., Baillard, A., Bertin, E.: Effective component tree computation with application to pattern recognition in astronomical imaging. In: IEEE International Conference on Image Processing, ICIP 2007, vol. 4, pp. IV – 41–IV – 44 (2007)

[10] Braga-Neto, U., Goutsias, J.: A theoretical tour of connectivity in image processing and analysis. Journal of Mathematical Imaging and Vision 19(1), 5–31 (2003)

[11] Alves, W.A.L., Morimitsu, A., Hashimoto, R.F.: Scale-space representation based on levelings through hierarchies of level sets. In: Benediktsson, J.A., Chanussot, J., Najman, L., Talbot, H. (eds.) ISMM 2015. LNCS, vol. 9082, pp. 265–276. Springer, Heidelberg (2015)

[12] Caselles, V., Meinhardt, E., Monasse, P.: Constructing the tree of shapes of an image by fusion of the trees of connected components of upper and lower level sets. Positivity 12(1), 55–73 (2008)

[13] Najman, L., Couprie, M.: Building the component tree in quasi-linear time. IEEE Transactions on Image Processing 15(11), 3531–3539 (2006)

[14] Wilkinson, M.H.F., Gao, H., Hesselink, W.H., Jonker, J.E., Meijster, A.: Concurrent computation of attribute filters using shared memory parallel machines. IEEE Trans. Pattern Anal. Mach. Intell. 30(10), 1800–1813 (2008)

[15] Lucas, S.M., Panaretos, A., Sosa, L., Tang, A., Wong, S., Young, R.: Icdar 2003 robust reading competitions. In: 2013 12th International Conference on Document Analysis and Recognition, vol. 2, pp. 682–682. IEEE Computer Society (2003)

Efficient Computation of Attributes and Saliency Maps on Tree-Based Image Representations

Yongchao Xu[1,2]([✉]), Edwin Carlinet[1,2], Thierry Géraud[1], and Laurent Najman[2]

[1] EPITA Research and Development Laboratory (LRDE), Lekremlin-Bicetne, France
{yongchao.xu,edwin.carlinet,thierry.geraud}@lrde.epita.fr
[2] Université Paris-Est, LIGM, Équipe A3SI, ESIEE Paris, France
l.najman@esiee.fr

Abstract. Tree-based image representations are popular tools for many applications in mathematical morphology and image processing. Classically, one computes an attribute on each node of a tree and decides whether to preserve or remove some nodes upon the attribute function. This attribute function plays a key role for the good performance of tree-based applications. In this paper, we propose several algorithms to compute efficiently some attribute information. The first one is incremental computation of information on region, contour, and context. Then we show how to compute efficiently extremal information along the contour (*e.g.*, minimal gradient's magnitude along the contour). Lastly, we depict computation of extinction-based saliency map using tree-based image representations. The computation complexity and the memory cost of these algorithms are analyzed. To the best of our knowledge, except information on region, none of the other algorithms is presented explicitly in any state-of-the-art paper.

Keywords: Min/Max-tree · Tree of shapesa · Algorithm · Attribute · Saliency map

1 Introduction

In a large number of applications, processing relies on objects or areas of interest. Therefore, region-based image representations have received much attention. In mathematical morphology, several region-based image representations have been popularized by attribute filters [2,17] or connected operators [13,14], which are filtering tools that act by merging flat zones. Such operators rely on transforming an image into an equivalent region-based representation, generally a tree of components (*e.g.*, the Min/Max-trees [13] or the tree of shapes [9]). Such trees are equivalent to the original image in the sense that the image can be reconstructed from the associated tree. Filtering then involves the design of an attribute function that weighs how important/meaningful a node of the tree is or how much a node of the tree fits a given shape. The filtering is achieved by preserving and removing some nodes of the tree according to the attribute function. This filtering process is either performed classically by thresholding the attribute function [14]

© Springer International Publishing Switzerland 2015
J.A. Benediktsson et al. (Eds.): ISMM 2015, LNCS 9082, pp. 693–704, 2015.
DOI: 10.1007/978-3-319-18720-4_58

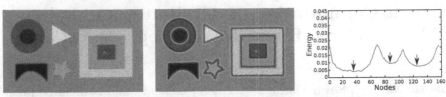

(a) Illustration on a synthetical image. Green: exterior region; Blue: interior region.

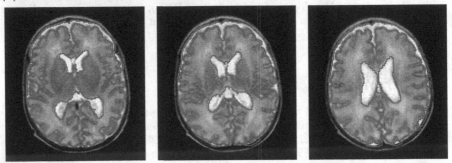

(b) Illustration of cerebrospinal fluid detection on MRI images of a newborn's brain.

Fig. 1. Examples of object detection using the context-based energy estimator [21] relying on contour and context information. An evolution of this attribute along a branch starting from the yellow point to the root is depicted on the right side of (a).

or by considering the tree-based image representations as graphs and applying some filters on this graph representation [23,20].

There exist many applications in image processing and computer vision that rely on tree-based image representations (see [20] for a short review). All these applications share a common scheme: one computes a tree representation and an attribute function upon which the tree analysis is performed. The choice of tree representation and the adequacy of attribute function mainly determine the success of the corresponding applications.

Many algorithms for computing different trees have been proposed (see Section 2.2 for a short review). In this paper, we focus on attribute computation, which is also an important step for the tree-based applications. To the best of our knowledge, only the algorithms for information computed on region have been presented [19] so far, none of the existing papers gives explicitly the algorithms computing the other attribute information employed in tree-based applications. In this paper, firstly, we detail explicitly how to incrementally compute some information on region, contour, and context. These informations form the basis for many classical attribute functions (*e.g.*, area, compactness, elongation). Let us remark that contextual information is very adequate for object detection, such as the context-based energy estimator [21] that relies on information computed on contour and context. Two examples of object detection using this attribute are shown in Fig. 1. Another type of interesting information is extremal information along the contour (*e.g.*, the minimal gradient's magnitude along the boundary). An example employing this information is the number of false alarms (NFA) for meaningful level lines extraction [7,3]. Here we propose

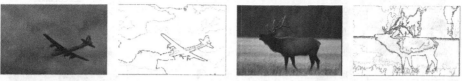

(a) Extinction-based saliency map using color tree of shapes [5]

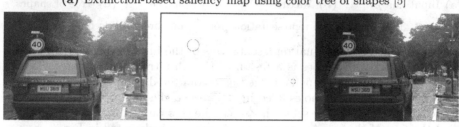

(b) Circular object oriented extinction-based saliency map

Fig. 2. Illustrations of extinction-based saliency maps from the tree of shapes

an efficient algorithm that does not require much memory to compute this kind of information. Lastly, we depict an algorithm computing the extinction-based saliency map [20] representing a hierarchical morphological segmentation using tree-based image representations (two examples are illustrated in Fig. 2). These algorithms form the main contribution of this paper.

The rest of the paper is organized as follows: A short review of some tree-based image representations and their computations using immersion algorithm are provided in Section 2. Our proposed algorithms to compute some attribute information and saliency maps are detailed in Section 3, and we analyze in Section 4 the complexity and the memory cost of the proposed algorithms. Finally, we conclude and give some perspectives in Section 5.

2 Review of Morphological Trees and Their Computations

Region-based image representations are composed of a set of regions of the original image. Those regions are either disjoint or nested, and they are organized into a tree structure thanks to the inclusion relationship. There are two types of such representations: fine to coarse hierarchical segmentations and threshold decomposition-based trees. In this paper, we only consider the threshold decomposition-based trees.

2.1 Tree-Based Image Representations

Let f be an image defined on domain Ω and with values on ordered set V (typically $\mathbb{R}$ or $\mathbb{Z}$). For any $\lambda \in V$, the upper level sets $\mathcal{X}_\lambda$ and lower level sets $\mathcal{X}^\lambda$ of an image f are respectively defined by $\mathcal{X}_\lambda(f) = \{p \in \Omega \mid f(p) \geq \lambda\}$ and $\mathcal{X}^\lambda(f) = \{p \in \Omega \mid f(p) \leq \lambda\}$. Both the upper and lower level sets have a natural inclusion structure: $\forall \lambda_1 \leq \lambda_2$, $\mathcal{X}_{\lambda_1} \supseteq \mathcal{X}_{\lambda_2}$ and $\mathcal{X}^{\lambda_1} \subseteq \mathcal{X}^{\lambda_2}$, which

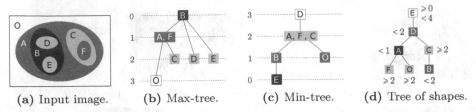

(a) Input image.   (b) Max-tree.   (c) Min-tree.   (d) Tree of shapes.

Fig. 3. Tree-based image representations relying on threshold decompositions

leads to two distinct and dual representations of the image: Max-tree and Min-tree [13]. The tree of shapes is a fusion of the Max-tree and Min-tree via the notion of *shapes* [9]. A shape is defined as a connected component of an upper or lower level set with its holes filled in. Thanks to the inclusion relationship of both kinds of level sets, the set of shapes can be structured into a tree structure, called the tree of shapes. An example of these trees is depicted in Fig. 3.

2.2   Tree Computation and Representation

There exist three types of algorithms to compute the Min/Max-tree (see [4] for a complete review): flooding algorithms [13,18,11], merge-based algorithms [19,12], and immersion algorithms [1,10]. In this paper, we employ the immersion algorithm to construct the Min/Max-tree. Concerning the tree of shapes [9], there are four different algorithms [9,15,6,8]. We use the one proposed by Géraud *et al.* [8]. It is similar to the immersion algorithms used for the Min/Max-tree computation. All these trees feature a common scheme of process: they start with considering each pixel as a singleton and sorting the pixels in decreasing tree order (*i.e.*, root to leaves order), followed by an union-find process (in reverse order) to merge disjoint sets to form a tree structure.

Let $\mathcal{R}$ be the vector of the N sorted pixels, and $\mathcal{N}(p)$ be neighbors (*e.g.*, 4- or 8-connectivity) of the pixel p. The union-find process is then depicted in Fig. 4 (a), where *parent* and *zpar* are respectively the parenthood image and the root path compression image. The whole process of tree computation is given in Fig. 4 (b), where SORT_PIXELS is a decreasing tree order sorting. The algorithms for computing the Min/Max-tree and the tree of shapes differ in this pixel sorting step. For the Min/Max-tree, they are either sorted in decreasing order (Min-tree) or increasing order (Max-tree). If the image f is low quantized, we can use the Bucket sort algorithm to sort the pixels. Concerning the tree of shapes, the sorting step is more complicated. It first interpolates the scalar image to an image of range using a simplicial version of the 2D discrete grid: the Khalimsky grid as shown in Fig. 5. We note $\mathcal{K}_\Omega$, the domain Ω immersed on this grid. In Fig. 5 (a), the original points of the image are the 2-faces, the boundaries are materialized with 0-faces and 1-faces. The algorithm in [8] ensures that shapes are open connected sets (*e.g.*, the purple shape in Fig. 5 (a)) and that shapes' borders are composed of 0-faces and 1-faces only (*e.g.*, the dark curve in Fig. 5 (a)). We refer the interested reader to the work of Géraud *et al.* [8] for more details on this pixel sorting step.

The tree structure is encoded through the image *parent* : $\Omega \to \Omega$ or $\mathcal{K}_\Omega \to \mathcal{K}_\Omega$ that states the parenthood relationship between nodes. In *parent*, a node is

```
 1  FIND_ROOT(zpar, x)
 2  if zpar(x) = x then return x
 3  else
 4  |   zpar(x) ←
    |       FIND_ROOT(zpar, zpar(x));
 5  |   return zpar(x)

 6  UNION_FIND(R)
 7  for all p do zpar(p) ← undef;
 8  for i ← N − 1 to 0 do
 9  |   p ← R[i], parent(p) ← p,
    |       zpar(p) ← p;
10  |   for all n ∈ N(p) if
    |       zpar(n) ≠ undef do
11  |   |   r ←
    |   |       FIND_ROOT(zpar, n);
12  |   |   if r ≠ p then
13  |   |   |   parent(r) ← p,
    |   |   |       zpar(r) ← p;
14  return parent
```

(a) Union-find process.

```
 1  CANONIZE_T(f, R, parent)
 2  for i ← 0 to N − 1 do
 3  |   p ← R[i];
 4  |   q ← parent(p);
 5  |   if f(parent(q)) = f(q)
    |       then
 6  |   |   parent(p) ← parent(q);
 7  return parent

 8  COMPUTE_TREE(f)
 9  R ← SORT_PIXELS(f);
10  parent ← UNION_FIND(R);
11  parent ←
        CANONIZE_T(f, R, parent);
12  return parent
```

(b) Complete tree construction.

Fig. 4. Tree construction relying on union-find process

represented by a single pixel (a 2-face of the Khalimsky grid in the case of the tree of shapes) called the canonical element, and each non-canonical element is attached to the canonical element representing the node it belongs to. In the following, we denote by getCanonical : $\Omega \to \Omega$ or $\mathcal{K}_\Omega \to \mathcal{K}_\Omega$, the routine that returns the canonical element of each point in the image.

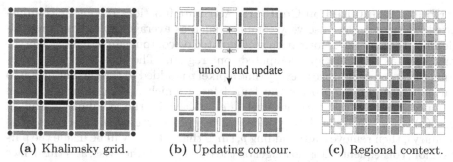

(a) Khalimsky grid.   (b) Updating contour.   (c) Regional context.

Fig. 5. (a): A point in a 2D image is materialized with 0-faces (blue disks), 1-faces (green strips), and 2-faces (red squares). (b): Updating contour information when an union between two components (yellow and blue) occurs thanks to a pixel (gray). (c): The approximated interior and exterior regional context of the red level line is respectively the dark gray region and the light gray region.

3   Proposed Algorithms

In this section, we detail several algorithms related to some applications using tree-based image representations, including computation of some classical information used in many attribute functions (accumulated information in Section 3.1, and extremal information along the contour in Section 3.2), and computation of extinction-based saliency maps [20] in Section 3.3. For the sake of simplicity, we consider the Min-tree or Max-tree representation. The algorithms for the tree of shapes construction share the same principle.

3.1   Incremental Computation of Some Accumulated Information

There are three main types of accumulated information: computed on region A (*e.g.*, area), on contour L (*e.g.*, length), and on context X (interior context X^i or exterior context X^e).

Attributes Computed on Regions. During the tree construction process, the algorithm starts with the pixels lying on the leaves, and the union-find acts as a region merging process. The connected components in the tree are built during this region growing process. We are able to handle information computed on region efficiently, such as its size, the sum of gray level or sum of square of gray level that can be used to compute the mean and the variance inside each region, the moments of each region based on which we can compute some shape attribute that measures how much a node fits a specific pattern. The algorithm for computing these information is depicted in Fig. 6 by adding some additional operations (red lines) to the union-find process during the tree construction, where i_A encodes information on pixels (*i.e.*, 2-faces). For example if A is the size or the sum of gray level, then i_A would be 1 (size of a pixel) or the pixel value. The operator $\widehat{+}$ is a binary commutative and associative operator having a neutral element $\widehat{0}$ [19]. For example, if A is the size, then the operator $\widehat{+}$ and $\widehat{0}$ would be the classical operator $+$ and 0 for the initialization.

Attributes Computed on Contours. Attribute functions relying on contour-related information are also very common, such as average of gradient's magnitude along the contour. Information accumulated on contour can be managed in the same way as information computed on region. The basic idea is that during the union-find process, every time a pixel p is added to the current region to form a parent region, process the four 1-faces which are the four neighbors (4-connectivity) of the current pixel (*i.e.*, 2-face in the Khalimsky grid in Fig. 5 (a)). If a 1-face e is already added to the current region (*i.e.*, belongs to its boundary), then remove e after adding p, since that 1-face e will be inside the parent region, consequently it is no longer on the boundary. Otherwise, add this 1-face e. This process is illustrated in Fig. 5 (b). It relies on an image *is_boundary* defined on the 1-faces that indicates if the 1-face belongs to the boundary of some region. Information on contour is computed by adding some supplementary process (green and gray lines in Fig. 6) to the union-find process, where i_L encodes information defined on 1-faces. For example if L is the contour length or the sum of gradient's

```
 1  UNION_FIND(R)
 2  for all p do
 3  │   zpar(p) ← undef;
 4  │   A(p) ← 0̂; //information computed on region (e.g., area, sum of gray level)
 5  │   L(p) ← 0̂; //information computed on contour (e.g., contour length)
 6  │   Xⁱ(p) ← 0̂, Xᵉ(p) ← 0̂; //information computed on context
 7  │   V_L(p) ← M̂; //extremal information along the contour
 8  for all e do is_boundary(e) ← false;
 9  for i ← N − 1 to 0 do
10  │   p ← R[i], parent(p) ← p, zpar(p) ← p;
11  │   A(p) ← A(p) +̂ i_A(p); //i_A: information on pixels (i.e., 2-faces)
12  │   for all n ∈ N(p) such as zpar(n) ≠ undef do
13  │   │   r ← FIND_ROOT(zpar, n);
14  │   │   if r ≠ p then
15  │   │   │   parent(r) ← p, zpar(r) ← p;
16  │   │   │   A(p) ← A(p) +̂ A(r);
17  │   │   │   L(p) ← L(p) +̂ L(r);
18  │   │   │   Xⁱ(p) ← Xⁱ(p) +̂ Xⁱ(r), Xᵉ(p) ← Xᵉ(p) +̂ Xᵉ(r);
19  │   for all e ∈ N₄(p) do
20  │   │   if not is_boundary(e) then
21  │   │   │   is_boundary(e) ← true;
22  │   │   │   L(p) ← L(p) +̂ i_L(e); //i_L: information on 1-faces
23  │   │   │   //i_X^{tr} and i_X^{dl}: top-right and down-left context of 1-faces
24  │   │   │   if e is above or on the right of p then
25  │   │   │   │   Xⁱ(p) ← Xⁱ(p) +̂ i_X^{dl}(e), Xᵉ(p) ← Xᵉ(p) +̂ i_X^{tr}(e);
26  │   │   │   else Xⁱ(p) ← Xⁱ(p) +̂ i_X^{tr}(e), Xᵉ(p) ← Xᵉ(p) +̂ i_X^{dl}(e);
27  │   │   │   appear(c) ← p;
28  │   │   else
29  │   │   │   is_boundary(e) ← false;
30  │   │   │   L(p) ← L(p) −̂ i_L(e);
31  │   │   │   if e is above or on the right of p then
32  │   │   │   │   Xⁱ(p) ← Xⁱ(p) −̂ i_X^{tr}(e), Xᵉ(p) ← Xᵉ(p) −̂ i_X^{dl}(e);
33  │   │   │   else Xⁱ(p) ← Xⁱ(p) −̂ i_X^{dl}(e), Xᵉ(p) ← Xᵉ(p) −̂ i_X^{tr}(e);
34  │   │   │   vanish(e) ← p;
35  for all e do
36  │   N_a ← appear(e), N_v ← vanish(e);
37  │   while N_a ≠ N_v do
38  │   │   V_L(N_a) ← update(V_L(N_a), i_L(e)); //update: either min or max
39  │   │   N_a ← parent(N_a);
40  return parent
```

Fig. 6. Incremental computation of information on region (in red), contour (in green), and context (in blue). The computation of extremal information is in magenta. The black lines represent the original union-find process, and the gray lines are used for the computation of contour, context, and extremal information.

magnitude, then i_L would be 1 (size of a 1-face) or the gradient's magnitude on the 1-faces. The operator $\hat{-}$ is the inverse of the operator $\hat{+}$.

Attributes Computed on Contexts. In [21], we have presented a context-based energy estimator that is adequate for object detection (see Fig. 1 for some examples). It relies on regional context information. The interior and exterior contextual region of a given region S (*e.g.*, a shape) is defined as the set of pixels respectively inside and outside the region with a distance to the boundary less than a given threshold ε. More formally, given a ball B_ε of radius ε, the exterior and interior of the shape S are defined as $Ext_B(S) = \delta_B(S) \setminus S$ and $Int_B(S) = S \setminus \epsilon_B(S)$ where δ and ϵ denote the dilation and erosion.

An approximated interior and exterior contextual region is illustrated in Fig. 5 (c) with $\varepsilon = 2$. As shown in this figure, we approximate the interior region and the exterior region of each level line by only taking into account the pixels which are aligned perpendicularly to each 1-face of the level line. Note that some pixels may be counted several times. Information on context can be computed in the same way as information on contour. But one has to attend closely to interior and exterior information while doing the update operation. The algorithm for computing interior (resp. exterior) contextual information X^i (resp. X^e) is shown in Fig. 6 by adding the gray and blue lines to the union-find process. This algorithm relies on two pre-computed images defined on 1-faces: i_X^{tr} and i_X^{dl} that encode information of ε pixels above (horizontal 1-face) or on the right side (vertical 1-face) of e, and respectively below (horizontal 1-face) or on the left side (vertical 1-face) of e.

Contextual information can be retrieved exactly at cost of a higher computation complexity. For every point p, we aim at finding all the shapes for which p is in the interior or the exterior. Given two points p and q such that $q \in B(p)$, we note S_p and S_q their respective shapes (nodes). We also note $Anc = LCA(S_p, S_q)$ where LCA stands for the least common ancestor of the two nodes and finally, let $[A \rightsquigarrow B) = \{S \mid A \subseteq S \subset B\}$ denotes the path from A to B in the tree. For all shapes $S \in [S_p \rightsquigarrow LCA(S_p, S_q))$, we have $p \in S$, but $q \notin S$, thus $p \in Int_B(S)$ and $q \in Ext_B(S)$ (see Fig. 7). The algorithms in Fig. 8 use the above-mentioned idea to compute contextual information, where i_X stands for information on pixels. A set of nodes $DjVu$ is used to track the shapes for which the current point has already been considered. If for neighbors q_1 and q_2, $[S_p \rightsquigarrow LCA(S_p, S_{q1}))$ and $[S_p \rightsquigarrow LCA(S_p, S_{q2}))$ have shapes in common, they will not be processed twice.

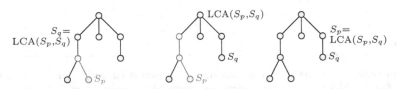

Fig. 7. Three cases for contextual computation. p and q are two neighbors (w.r.t. B). The red path denotes the nodes in $[S_p \rightsquigarrow LCA(S_p, S_q))$ for which p is in the interior and q in the exterior. Left: case $S_p \subset S_q$, middle: case S_p and S_q are in different paths, right: case $S_q \subset S_p$.

1 EXTERNAL_CONTEXT($parent$)
2 **foreach** $node$ x **do** $X^e(x) \leftarrow \widehat{0}$;
3 **foreach** point q in Ω **do**
4      $DjVu \leftarrow \emptyset$;
5   **foreach** point p in $B_\varepsilon(q)$ **do**
6          $N_p \leftarrow$ getCanonical(p);
7          $N_q \leftarrow$ getCanonical(q);
8          $Anc \leftarrow$ LCA(N_p, N_q);
9   **while** $N_p \neq Anc$ **do**
10   **if** $N_p \notin DjVu$ **then**
11               $X^e(N_p) \leftarrow$
                   $X^e(N_p) \widehat{+} i_X(q)$;
12             $DjVu \leftarrow DjVu \cup \{N_p\}$;
13             $N_p \leftarrow parent(N_p)$;
14 **return** X^e
	1 INTERNAL_CONTEXT($parent$)
	2 **foreach** $node$ x **do** $X^i(x) \leftarrow \widehat{0}$;
	3 **foreach** point p in Ω **do**
	4 $DjVu \leftarrow \emptyset$;
	5 **foreach** point q in $B_\varepsilon(p)$ **do**
	6 $N_p \leftarrow$ getCanonical(p);
	7 $N_q \leftarrow$ getCanonical(q);
	8 $Anc \leftarrow$ LCA(N_p, N_q);
	9 **while** $N_p \neq Anc$ **do**
	10 **if** $N_p \notin DjVu$ **then**
	11 $X^i(N_p) \leftarrow$
	                   $X^i(N_p) \widehat{+} i_X(p)$;
	12 $DjVu \leftarrow DjVu \cup \{N_p\}$;
	13 $N_p \leftarrow parent(N_p)$;
	14 **return** X^i

Fig. 8. Algorithms for exact computation of contextual information X^i and X^e

3.2 Computation of Extremal Information along the Contour

Apart from those attributes based on accumulated information, the number of false alarms (NFA) [7,3] (see [3] for several examples of meaningful level lines selection using NFA) requires to compute the minimal gradient's magnitude along the boundary of each region. Here we propose an efficient algorithm that requires low memory to handle this extremal information along the contour V_L. It relies on two images $appear$ and $vanish$ defined on the 1-faces. $appear(e)$ encodes the smallest region $\mathcal{N}_a$ in the tree for which the 1-face e lies on its boundary, while $appear(e)$ stands for the smallest region $\mathcal{N}_v$ for which e is inside it. Note that $\mathcal{N}_a$ and $\mathcal{N}_v$ might be equal, $e.g.$, in the case of 1-faces in the interior of a flat zone. The computation of extremal information along the contour V_L is depicted in Fig. 6 by adding the gray and magenta lines to the union-find process, where $\widehat{M}$ in the initialization step is the maximal (resp. minimal) value for minimal (resp. maximal) information computation, and the operator "update" is a "min" (resp. "max") operator for the minimal (resp. the maximal) information.

3.3 Computation of the Saliency Map

As shown in [22,20], the saliency map introduced in the framework of shape-based morphology relies on the extinction values $\mathcal{E}$ defined on the local minima [16]. Once the extinction values computed for all the minima (see [16] for details about the computation of the extinction values COMPUTE_EXTINCTION), we can weigh the extinction values on the region boundaries corresponding to the minima. Each 1-face takes the maximal extinction value of those minima for which this 1-face is on their boundaries. This can be achieved via two images $appear$ and $vanish$ that have been used in the computation of extremal information along the contour (as shown in Fig. 6). For each 0-face o, it takes the maximal value among the four 1-faces e_1, e_2, e_3, and e_4 that are neighbors (4-connectivity) of o in the Khalimsky grid. Finally, the extinction-based saliency map $\mathcal{M}_\mathcal{E}$ is obtained. The computation of the saliency map is given in Fig. 9.

```
 1  COMPUTE_SALIENCY_MAP(f)
 2  (T, A) ← COMPUTE_TREE(f);
 3  E ← COMPUTE_EXTINCTION(T, A);
 4  for all e do M_E(e) ← 0;
 5  for all e do
 6  |    N_a ← appear(e), N_v ← vanish(e);
 7  |    while N_a ≠ N_v do
 8  |    |    M_E(N_a) ← max(E(N_a), M_E(e)), N_a ← parent(N_a);
 9  for all 0-face o do M_E(o) ← max(M_E(e_1), M_E(e_2), M_E(e_3), M_E(e_4));
10  return M_E
```

Fig. 9. Computation of extinction-based saliency map $\mathcal{M_E}$

4   Complexity Analysis

We use the algorithms based on the Tarjan's Union-Find process to construct the Min-tree and Max-tree [10,1,4] and the tree of shapes [8]. These approaches would take $O(n \log(n))$ time, where n is the number of pixels of the image f. For low quantized images (typically 12-bit images or less), the complexity of the computation of these trees is $O(n\,\alpha(n))$, where α is a very slow-growing diagonal inverse of the Ackermann's function. In this section, we analyze the additional complexity and the memory usage of the algorithms proposed in Section 3.

4.1   Accumulated Information on Region, Contour, and Context

As described in Section 3.1 and shown in Fig. 6, information computed on regions, contours, and contexts (the approximated version) are computed incrementally during the union-find process. Consequently, they have the same complexity as the union-find which is $O(n\,\alpha(n))$. Besides, the pre-computed images ($e.g.$, i_L or i_X^{tr}) can be obtained in linear time, so the $O(n\,\alpha(n))$ complexity is maintained. To compute exactly contextual information as described in Fig. 8, for each pixel p, we have to compute the least common ancestor Anc of p and any $q \in B_\varepsilon(p)$ and propagate from N_p to Anc. The computation of the least common ancestor has a $O(h)$ complexity if a depth image is employed, where h is the height of the tree. Consequently, the total complexity is $O(n\varepsilon^2 h)$.

Apart from the necessary memory of the union-find process, the computation of information on regions does not require auxiliary memory. For information computed on contours and contexts (approximated), the auxiliary memory usage is $4n$ for the intermediate image $is\_boundary$ (defined on the Khalimsky grid). For the exact computation of contextual information, we need the depth image (n pixels) used by the least common ancestor algorithm and the intermediate set $DjVu$ ($O(h)$ elements). The total auxiliary memory cost is thus $n + h$.

4.2   Extremal Information along the Contour

The algorithm computing extremal information along the contour relies on two auxiliary images $appear$ and $vanish$. As described in Section 3.2 and shown in

Fig. 6, these two images are computed incrementally during the union-find process. The complexity of this step is $O(n\,\alpha(n))$. Then, to compute the final extremal information, for each 1-face e, we have to propagate the value to a set of node (from $appear(e)$ to $vanish(e)$). In the worst case, we have to traverse the whole branch of the tree. Consequently, the complexity would be $O(nh)$. In terms of auxiliary memory cost, it would take $4n$ for each intermediate image $appear$, $vanish$, and $is\_boundary$. So the total additional memory cost would be $12n$. Such extra cost is acceptable for 2D cases, but become prohibitive for very large or 3D images. Actually, we could avoid the extra-memory used for the storage of $appear$ and $vanish$ as the information they provide could be computed on the fly in each algorithm. Nevertheless, for the purpose of clarification, we have chosen to compute these information one for all to avoid code redoundancy in the algorithms we have proposed.

4.3 Saliency Map

The computation of extinction-based saliency map given in Section 3.3 and depicted in Fig. 9 also relies on the two temporary images $appear$ and $vanish$. Suppose that we have the extinction values $\mathcal{E}$ for all the local minima. In the same way as the computation of extremal information along the contour, for each 1-face e, we have to propagate from $appear(e)$ to $vanish(e)$. The worst time complexity would be $O(nh)$. The computation of extinction values $\mathcal{E}$ relies on a Max-tree computation process, which is quasi-linear. The auxiliary memory cost would be $12n$ ($4n$ for each temporary image $appear$, $vanish$, and $is\_boundary$). Yet, the remark about the memory usage given in Section 4.2 holds for this complexity analysis.

5 Conclusion

In this paper, we have pesented several algorithms related to some applications using tree-based image representations. First of all, we have shown how to incrementally compute information on region, contour, and context which forms the basis of many widely used attribute functions. Then we have proposed an algorithm in order to compute extremal information along the contour (required for some attribute functions, such as the number of false alarms (NFA)), which requires few extra memory. Finally, we have depicted how to compute extinction-based saliency maps from tree-based image representations. The time complexity and the memory cost of these algorithms are also analyzed. To the best of our knowledge, this is the first time that these algorithms (except for information computed on region) are explicitly depicted, which allows reproducible research and facilitates the development of some novel interesting attribute functions. In the future, extension of these algorithms to 3D images will be studied. And we would like to study some more attribute functions: learning attribute functions in particular would be one interesting future work.

References

1. Berger, C., Géraud, T., Levillain, R., Widynski, N., Baillard, A., Bertin, E.: Effective component tree computation with application to pattern recognition in astronomical imaging. In: Proc. of IEEE ICIP., vol. 4, pp. 41–44 (2007)

2. Breen, E., Jones, R.: Attribute openings, thinnings, and granulometries. CVIU 64(3), 377–389 (1996)
3. Cao, F., Musé, P., Sur, F.: Extracting meaningful curves from images. JMIV 22, 159–181 (2005)
4. Carlinet, E., Géraud, T.: A comparative review of component tree computation algorithms. IEEE Transactions on Image Processing 23(9), 3885–3895 (2014)
5. Carlinet, E., Géraud, T.: A color tree of shapes with illustrations on filtering, simplification, and segmentation (submitted for publication, 2015)
6. Caselles, V., Monasse, P.: Geometric Description of Images as Topographic Maps, 1st edn. Springer Publishing Company, Incorporated (2009)
7. Desolneux, A., Moisan, L., Morel, J.: Edge detection by helmholtz principle. JMIV 14(3), 271–284 (2001)
8. Géraud, T., Carlinet, E., Crozet, S., Najman, L.: A quasi-linear algorithm to compute the tree of shapes of nD images. In: Hendriks, C.L.L., Borgefors, G., Strand, R. (eds.) ISMM 2013. LNCS, vol. 7883, pp. 98–110. Springer, Heidelberg (2013)
9. Monasse, P., Guichard, F.: Fast computation of a contrast-invariant image representation. IEEE Trans. on Image Processing 9(5), 860–872 (2000)
10. Najman, L., Couprie, M.: Building the component tree in quasi-linear time. IEEE Trans. on Image Processing 15(11), 3531–3539 (2006)
11. Nistér, D., Stewénius, H.: Linear time maximally stable extremal regions. In: Forsyth, D., Torr, P., Zisserman, A. (eds.) ECCV 2008, Part II. LNCS, vol. 5303, pp. 183–196. Springer, Heidelberg (2008)
12. Ouzounis, G.K., Wilkinson, M.H.F.: A parallel implementation of the dual-input max-tree algorithm for attribute filtering. In: ISMM, pp. 449–460 (2007)
13. Salembier, P., Oliveras, A., Garrido, L.: Antiextensive connected operators for image and sequence processing. ITIP 7(4), 555–570 (1998)
14. Salembier, P., Wilkinson, M.H.F.: Connected operators. IEEE Signal Processing Mag. 26(6), 136–157 (2009)
15. Song, Y.: A topdown algorithm for computation of level line trees. IEEE Transactions on Image Processing 16(8), 2107–2116 (2007)
16. Vachier, C., Meyer, F.: Extinction values: A new measurement of persistence. In: IEEE Workshop on Non Linear Signal/Image Processing, pp. 254–257 (1995)
17. Westenberg, M.A., Roerdink, J.B.T.M., Wilkinson, M.H.F.: Volumetric attribute filtering and interactive visualization using the max-tree representation. ITIP 16(12), 2943–2952 (2007)
18. Wilkinson, M.H.F.: A fast component-tree algorithm for high dynamic-range images and second generation connectivity. In: Proc. of ICIP, pp. 1021–1024 (2011)
19. Wilkinson, M.H.F., Gao, H., Hesselink, W.H., Jonker, J.E., Meijster, A.: Concurrent computation of attribute filters on shared memory parallel machines. PAMI 30(10), 1800–1813 (2008)
20. Xu, Y.: Tree-based shape spaces: Definition and applications in image processing and computer vision. Ph.D. thesis, Université Paris Est, Marne-la-Vallée, France (December 2013)
21. Xu, Y., Géraud, T., Najman, L.: Context-based energy estimator: Application to object segmentation on the tree of shapes. In: ICIP, pp. 1577–1580. IEEE (2012)
22. Xu, Y., Géraud, T., Najman, L.: Two applications of shape-based morphology: Blood vessels segmentation and a generalization of constrained connectivity. In: Hendriks, C.L.L., Borgefors, G., Strand, R. (eds.) ISMM 2013. LNCS, vol. 7883, pp. 390–401. Springer, Heidelberg (2013)
23. Xu, Y., Géraud, T., Najman, L.: Morphological Filtering in Shape Spaces: Applications using Tree-Based Image Representations. In: ICPR, pp. 485–488 (2012)

Fast Evaluation
of the Robust Stochastic Watershed

Bettina Selig[✉], Filip Malmberg, and Cris L. Luengo Hendriks

Centre for Image Analysis,
Swedish University of Agricultural Sciences and Uppsala University, Uppsala, Sweden
{bettina,filip,cris}@cb.uu.se

Abstract. The stochastic watershed is a segmentation algorithm that estimates the importance of each boundary by repeatedly segmenting the image using a watershed with randomly placed seeds. Recently, this algorithm was further developed in two directions: (1) The exact evaluation algorithm efficiently produces the result of the stochastic watershed with an infinite number of repetitions. This algorithm computes the probability for each boundary to be found by a watershed with random seeds, making the result deterministic and much faster. (2) The robust stochastic watershed improves the usefulness of the segmentation result by avoiding false edges in large regions of uniform intensity. This algorithm simply adds noise to the input image for each repetition of the watershed with random seeds. In this paper, we combine these two algorithms into a method that produces a segmentation result comparable to the robust stochastic watershed, with a considerably reduced computation time. We propose to run the exact evaluation algorithm three times, with uniform noise added to the input image, to produce three different estimates of probabilities for the edges. We combine these three estimates with the geometric mean. In a relatively simple segmentation problem, F-measures averaged over the results on 46 images were identical to those of the robust stochastic watershed, but the computation times were an order of magnitude shorter.

Keywords: Stochastic watershed · Watershed cuts · Monte Carlo simulations

1 Introduction

The *watershed transform* [5, 15] is a powerful method for image segmentation. Intuitively, the watershed of a gray scale image (seen as a topographic landscape) is the set of locations from which a drop of water could flow toward different minima. The watershed thus partitions the image such that each region contains precisely one local minimum, and the region boundaries run along crest lines. The watershed transform is typically applied to the gradient magnitude of an image, yielding regions of homogeneous intensity. Normally, the number of local minima is larger than the number of desired objects, so that the watershed transform leads to an over-segmentation of the image.

© Springer International Publishing Switzerland 2015
J.A. Benediktsson et al. (Eds.): ISMM 2015, LNCS 9082, pp. 705–716, 2015.
DOI: 10.1007/978-3-319-18720-4_59

Seeded watersheds [11] solve the over-segmentation problem by requiring a set of markers, one for each desired region. The image is transformed so that it only has local minima at the markers, prior to applying the watershed transform. This does, however, not solve the segmentation problem; it merely transforms it to a problem of object detection.

The *stochastic watershed* (SW) [1] is a method for identifying salient contours in an image. The method computes a *probability density function* (PDF), assigning to each piece of contour in the image the probability to appear as a segmentation boundary in a seeded watershed segmentation with randomly placed seeds. Contours that appear with high probability are assumed to be more important. The resulting PDF can be segmented with a standard watershed transform, combined with and H-minima transform [13]. The H-minima transform reduces the number of local minima, so that only contours with a probability larger than h remain.

In the original publication by Angulo and Jeulin [1], the PDF was estimated by Monte Carlo simulation, i.e., repeatedly selecting random markers and performing seeded watershed segmentation. Meyer and Stawiaski [12] showed that the PDF can be calculated exactly, without performing any Monte Carlo simulations. Their work was later extend by Malmberg et al. [9, 10] who proposed an efficient (pseudo-linear) algorithm for computing the exact PDF. Here, we refer to this method as the *exact evaluation of the stochastic watershed* (ESW).

One problem with the SW is that it gives high probability to lines in large uniform areas, where no apparent boundary exists. This is caused by the way seeded watersheds choose boundaries: Starting with the seeds as initial regions, pixels adjacent to a region are iteratively added until each pixel has been assigned to a region. Pixels are added in order of increasing intensity, so that at each iteration the adjacent pixel with the lowest intensity is added to a region. The resulting regions will therefore meet at pixels with a high intensity, and ridges in the image corresponding to object boundaries are therefore consistently found. The boundaries returned by the seeded watershed algorithm are incredibly stable with respect to the position of the seeds. Audigier and Lotufo [3] studied the robustness of seeded watersheds with respect to seed position, and showed that typically each seed can be moved freely within a large region without affecting the segmentation result. While this stability is usually a desirable feature, it also means that the algorithm will consistently place segmentation lines in places where there are no actual ridges. Thus, within regions of uniform intensity, the SW may assign an unreasonably high probability to an arbitrarily selected false boundary. Bernander et al. [4] proposed to overcome this problem by adding a small amount of uniform noise to the image in each iteration of the Monte Carlo simulation. This changes the order in which the pixels are processed by the algorithm, and therefore each iteration finds different false boundaries. Here, we refer to this method as the *robust stochastic watershed* (RSW).

In this paper, we combine the RSW with the ESW. The resulting method produces segmentation results comparable to those of the RSW, but is substantially faster. We refer to this method as the *fast evaluation of the robust stochastic watershed* (FRSW).

2 Watershed Cuts and Minimum Spanning Forests

Many different formulations of seeded watersheds on digital images have been proposed, see e.g. Cousty et al. [7] for a review of different approaches to this problem. Here, we consider the *watershed cut* approach [7, 8], which formalizes watersheds on edge weighted graphs. Within this framework, an image is represented by a graph whose vertex set is the set of pixels, and whose edge set encodes an adjacency relation among the pixels. The edges of the graph are weighted to reflect changes in image intensity, color, or other features.

As established by Cousty et al. [7, 8], a watershed cut on such a graph is equivalent to the computation of a *minimum spanning forest* relative to the seeds. This forest consists of multiple connected components, each containing precisely one seed. The watershed cut is then defined as the set of edges whose endpoints are in different components. Additionally, we say that a pixel is a boundary pixel if at least one of its neighbors is in a different component for a given watershed cut. This allows us to define the stochastic watershed PDF as a function over the pixels of the image, representing the probability that a given pixel is a boundary pixel for a watershed cut relative to randomly selected seeds.

3 Method

As noted in the introduction, Bernander et al. [4] used Monte Carlo techniques to evaluate the stochastic watershed PDF, but also added a small amount of noise at each iteration. This corresponds to replacing the original fixed pixel values by stochastic variables, whose expected values correspond to the original pixel intensities. In contrast to the original stochastic watershed method [1], the RSW thus randomizes both the image data and the seed positions.

The seeded watershed with randomized image data, but with fixed seeds, was studied by Straehle et al. [14]. For a graph with randomized edge weights, they showed that no polynomial time algorithm can calculate the probability of a vertex belonging to a specified component of the MSF relative to the seeds. Thus, it seems plausible that no polynomial time algorithm exists for the even more challenging problem of finding the exact PDF for the RSW.

Instead of attempting to evaluate the PDF of the RSW exactly, the proposed FRSW method uses a hybrid approach. We generate N images by adding a small amount of uniform noise to the original image. For each of these images, we use the ESW to calculate the exact stochastic watershed PDF for the given image. From these N intermediate PDFs, a final PDF is generated. In other words, our method uses Monte Carlo sampling to handle the stochastic pixel values while the randomness of the seed positions is handled by the exact evaluation method. Compared to the full Monte Carlo sampling employed by Bernander et al. [4], our hybrid approach drastically reduces the number of samples needed to obtain a good solution. In the context of image segmentation, we will show that even three samples are enough to approximate the result of the RSW.

We explore several methods for merging the N intermediate PDFs into a final PDF. To match the method of Bernander et al. [4], the final PDF should be defined as the arithmetic mean of the intermediate PDFs. Our experiments indicate, however, that other merging functions can give better results. We seek a merging function that produces good results for small N.

As a final step, the combined PDF can be segmented with a standard watershed in combination with an H-minima transform, as in SW and RSW.

4   Experiments

4.1   Method to Merge Resulting PDFs

It is essential to combine the intermediate PDFs in a suitable way to be able to extract the important information. A simple way is to use the mean or the median. But since it is rare that exactly the same false boundaries are found repeatedly and since the true boundaries are present in almost all PDFs, a measure favoring small values is beneficial. Suitable for this is for example the power mean

$$M_p = \left(\frac{1}{N} \sum_{i=1}^{N} x_i{}^p \right)^{\frac{1}{p}}, \tag{1}$$

where parameter p determines if the result tends towards the maximum x_i (when $p > 1$) or the minimum x_i (when $p < 1$). We define the power mean for $p = 0$ as

$$M_0 = \lim_{p \to 0} M_p = \left(\prod_{i=1}^{N} x_i \right)^{\frac{1}{N}}. \tag{2}$$

The power mean with $p = 1$ corresponds to the arithmetic mean. In Fig. 1, we show how the power mean with $p \leq 1$ performs when merging three test images.

To find an optimal merging method that combines the N intermediate PDFs efficiently, we compare different approaches based on the power mean M_p, and the median.

For the experiments, we created three different synthetic images. These images of 460×460 pixels show a set of curves on a black background. The intensity of the foreground decreases linearly from the left to the right of the images. That means that the foreground has its maximal value 255 on the left most column and 0, the gray value of the background on the right most column. Normally distributed noise with a standard deviation of 32 was applied to the images. Since the result might be negative after this step, the images were normalized to values between 0 and 1. The resulting test images *Cells*, *Lines* and *Bubbles* and the ground truth segmentations are shown in Fig. 2.

We compared the following six merging methods: arithmetic mean M_1, geometric mean M_0, harmonic mean M_{-1}, power mean M_{-2}, minimum $M_{-\infty}$ and median.

For each method and each image, we first needed to determine the optimal values for the number of seeds n, the strength of noise s and threshold h for the H-minima transform, which is applied to receive the final segmentation. For computational costs, it is desirable to keep N, the number of realizations of the

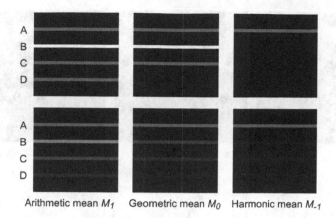

Arithmetic mean M_1 Geometric mean M_0 Harmonic mean M_{-1}

Fig. 1. Input images (top) and results of merging methods (bottom). The arithmetic mean M_1 yields a strong response for lines B and C, and line D is quite prominent. The geometric mean M_0 suppresses more strongly lines that are not present in all three images (B-D). This effect is more pronounced for the harmonic mean M_{-1}.

ESW, as small as possible. The smallest, reasonable value that worked for all methods was $N = 3$.

For each method, we segmented each of the three images ten times with each possible set of training parameters. The parameter n could take ten different values between 2 and 200, s seven different values between 0.005 and 0.5 and h seven different values between 0.0005 and 0.03.

Next, the mean F-measures were calculated as in Arbelaez et al. [2]. The F-measure states how well the resulting segmentations match the ground truth, and is represented by a value between 0 and 1. An F-measure close to 1 indicates a high agreement. It is calculated by the harmonic mean of precision (fraction of correctly assigned boundary pixels of segmentation) and recall (fraction of correctly assigned boundary pixels of ground truth). Here, we considered pixels located four or more pixels away from the corresponding boundary a mismatch.

The optimal set of parameters was determined as the set with the largest mean F-measure. To judge how well each method possibly can perform, we plotted these maximal F-measures in Fig. 3. The approach using the geometric mean yielded the highest F-measure on average.

4.2   Number of Realizations

As mentioned before, it is preferable to choose a small value for N for computational reasons. But a small N does not necessary lead to an optimal result. In this section, we examine the performance of the geometric mean, the method that performed best in the previous section, when modifying N.

We tested the segmentation results for $N = \{1, 2, 3, ..., 10\}$ for the three synthetic images using the geometric mean to merge the intermediate PDFs. In a separate experiment, we observed that the optimal parameters do not change

Fig. 2. Synthetic test images *Cells*, *Lines* and *Bubbles* (top) and their ground truths (bottom)

significantly when changing N (data not shown). Therefore, we used the optimal parameters as determined in Section 4.1.

We segmented each of the synthetic test images ten times with each value for N. Then, we determined the means of the resulting F-measures, as described before. The results are displayed in Fig. 4.

The F-measure improves slightly as N increases, but we feel it does not improve sufficiently after $N = 3$ to justify the additional computational cost.

4.3   Evaluation

In this section, we compare the FRSW, using the geometric mean and $N = 3$, to previously published stochastic watershed algorithms. For this, we used a data set of fluorescence microscope images of nuclei with hand-drawn ground-truth segmentation, see Fig. 5. Coelho et al. [6] provided two different collections of images, of which we chose the first one with 46 images. We determined the gradient magnitude (Gaussian gradient with $\sigma = 1$ px) for each image, normalized the intensities to values between 0 and 1, and finally subsampled each dimension by a factor 2.

We compared the FRSW using $N = 3$ and the geometric mean, the SW, the ESW, and the RSW. For both the SW and the RSW, we used 50 realizations of seeded watershed with random seeds.

First, the parameters n, s and h needed to be trained for all methods. We used the leave-one-out method to be able to do the testing on all images in the data set. During training, the parameter n could take eleven different values between 2 and 300, s five different values between 0.025 and 0.2, and h eleven different values between 0.0005 and 0.4. We segmented all images with the different

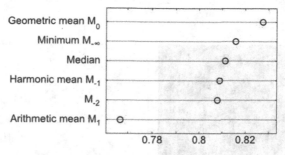

Fig. 3. Mean F-measure for tested merging methods using optimal parameters. (Mean values for all three synthetic images.)

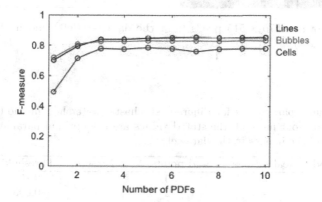

Fig. 4. Mean F-measures of segmentation results using $N = 1, 2, ..., 10$ PDFs merged with geometric mean

approaches and corresponding optimal parameters (Table 1) and calculated the F-measures, as described before. The results are shown in Fig. 6. Segmentations of the example image (Fig. 5) are shown in Fig. 7.

As expected, the SW and the ESW had trouble segmenting the images properly. Their median F-measures were 0.67 and 0.79, respectively. In contrast, the RSW and the FRSW scored high median F-measures of 0.91 and 0.92, respectively.

4.4   Convergence

In Section 3, we reasoned that the FRSW needs far fewer realizations than the SW and the RSW to produce a good result. In this section, we examine the convergence behavior of these algorithms. For this, we chose one representative image from the data set, see Fig. 5. During the evaluation in Section 4.3, it yielded average F-measures with all four methods, see Fig. 7.

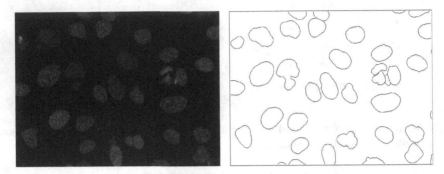

Fig. 5. Image of size 675 × 515 pixels from the data set (left) and its hand-drawn ground truth (right)

Table 1. Optimal parameters for different stochastic watershed methods. Since we used the leave-one-out method, the stated values are the optimal parameters for the great majority of the images in the data set.

Method	Realizations	Seed points N	Noise s	Threshold h
SW	50	50	-	0.15
ESW	-	2	-	0.005
RSW	50	250	0.1	0.2
FRSW	3	100	0.05	0.05

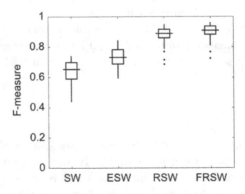

Fig. 6. F-measures for segmentation results using SW, ESW, RSW and FRSW with $N = 3$ and geometric mean

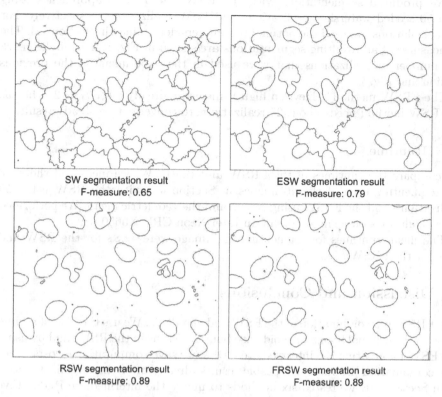

SW segmentation result
F-measure: 0.65

ESW segmentation result
F-measure: 0.79

RSW segmentation result
F-measure: 0.89

FRSW segmentation result
F-measure: 0.89

Fig. 7. Final segmentation of example image (Fig. 5)

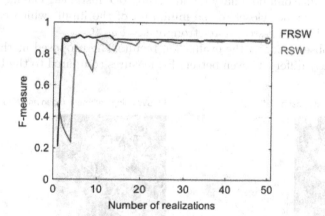

Fig. 8. Comparison of convergence for the methods RSW and FRSW for 1 to 50 realizations. The circles mark the segmentation used during the experiments.

We produced segmentations with the RSW and FRSW approaches using 1 to 50 seeded watershed realizations and ESW realizations, respectively. For the calculations, we used the optimal parameters determined in Section 4.3. The F-measures of all resulting segmentations are shown in Fig. 8. The circles mark the number or realizations that were used in the experiment in the previous section and Fig. 7.

The FRSW method achieves a high F-measure already after $N = 2$, whereas the RSW method needs about 25 realizations to converge to a steady result.

4.5   Runtime

To compare the runtimes of the FRSW and the RSW, we measured the time for segmenting each of the 46 images in Section 4.3 with the RSW using 50 realizations, and the FRSW using $N = 3$ and the geometric mean. We performed the calculations on a single core of an Intel Xeon CPU X5650.

The mean runtimes for segmenting one image were 42.8 s for the RSW and 6.4 s for the FRSW.

5   Discussion and Conclusions

The FRSW is a combination of the ESW and the RSW. We use the idea of adding noise to the original image to avoid false boundaries, as in the RSW, and by using the ESW for creating the intermediate PDFs, we save computational costs, since this combination produces acceptable results already after three realizations.

In Section 4.1, we tested six methods to merge the intermediate PDFs. Five of them yielded acceptable results, but we achieved the best results with the geometric mean M_0. The arithmetic mean M_1 performed worse than the other methods, because it required a very large number of intermediate PDFs to bring down the relevance of a boundary found in only one instance. The methods with $p < 1$ produced values closer to the minimum of the input, which causes false boundaries to be suppressed more strongly, see Fig. 9.

On closer observation of the evaluation results, the proposed method FRSW achieved slightly different (even better) F-measures compared to the RSW. This

FRSW using arithmetic mean and N=3:   FRSW using arithmetic mean and N=3:

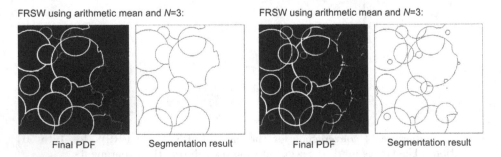

Final PDF   Segmentation result   Final PDF   Segmentation result

Fig. 9. Example of final PDFs and segmentation results of synthetic image *Bubbles*

is because one realization of the ESW (in the FRSW) contains the information of an infinite number of seeded watershed segmentations, whereas the RSW yields only 50 in total. The FRSW has only three samples of the noise rather than 50, this does not seem to affect its ability to suppress false boundaries.

We did not compare the various versions of the stochastic watershed with other segmentation methods, such as the classical watershed with an H-minima transform. The purpose here was to show that the proposed FRSW produces results comparable to the RSW. For the relatively easy segmentation problem that we used in the test, many segmentation algorithms work really well. The SW was originally developed to avoid the difficulties that arise with the classical H-minima/watershed combination, where it is often impossible to find a value for h that yields a correct segmentation: some superfluous local minima are deeper in this case than some other relevant local minima. The SW family of algorithms does not fully solve this problem, but brings us in the right direction.

The experiments shown in Figs. 4 and 8 show that only a few instances of the SW are needed in the FRSW to overcome a segmentation result with false boundaries. In both experiments, the F-measure did not significantly increase when merging more than three PDFs. A key insight for this result was the use of the power mean to merge the PDFs. For one particular non-edge pixel, possible PDF values form a highly skewed distribution, where most PDFs have a low value, but some have a high value (these are the false boundaries seen in the individual PDFs). By taking a power mean with $p < 1$, lower values are favoured, and a better estimate of the population mean is obtained than through the arithmetic mean (given only a few samples of the highly skewed population, where one of the samples is a high value). It is likely that, for a very large number of PDFs, a correct result can be obtained with the arithmetic mean.

When inspecting the development for the RSW for an increasing number of realizations, in Fig. 8, we observed that the F-measure jumped up every fifth realization and then slowly declined. This was caused by the chosen threshold for the H-minima, which was $h = 0.2$, optimized for 50 realizations. This means that, for the first four realizations, the number of false boundaries increased, as we added segmentations on top of each other, but no boundaries were removed because h was too small. After five realizations, many of the false boundaries had a value of $1/5$, equal to h, and were therefore removed. The F-measure sharply increased. For the tenth realization, the same happened, as two overlapping boundaries then yielded a value of 0.2, and were removed. This process repeats with a period of $1/h$. Thus, the sharp jumps in the F-measure are an artifact of training h for a large number of realizations.

Furthermore, we inspected the convergence of the proposed method in comparison to the RSW. We observed that when segmenting the fluorescence microscope images from the experiment in Section 4.3, the FRSW already reached a high F-measure with $N = 2$, whereas RSW needed more than 20 realizations to converge.

Each realization of the FRSW, which is the ESW, is computationally more expensive than one realization of the seeded watershed in the RSW. But because the FRSW needs far fewer realizations than the RSW to produce a similar result, it is considerably faster. With the method proposed here, we obtain similar, or slightly better, segmentations in a fraction of the time.

References

[1] Angulo, J., Jeulin, D.: Stochastic watershed segmentation. In: 8th International Symposium on Mathematical Morphology (ISMM), pp. 265–276 (2007)

[2] Arbelaez, P., Maire, M., Fowlkes, C., Malik, J.: Contour detection and hierarchical image segmentation. IEEE Transactions on Pattern Analysis and Machine Intelligence 33(5), 898–916 (2011)

[3] Audigier, R., Lotufo, R.: Seed-relative segmentation robustness of watershed and fuzzy connectedness approaches. In: 20th Brazilian Symposium on Computer Graphics and Image Processing 2007, pp. 61–70 (October 2007)

[4] Bernander, K.B., Gustavsson, K., Selig, B., Sintorn, I.M., Luengo Hendriks, C.L.: Improving the stochastic watershed. Pattern Recognition Letters 34(9), 993–1000 (2013)

[5] Beucher, S., Lantuejoul, C.: Use of Watersheds in Contour Detection. In: International Workshop on Image Processing: Real-time Edge and Motion Detection/Estimation, Rennes, France (1979)

[6] Coelho, L.P., Shariff, A., Murphy, R.F.: Nuclear segmentation in microscope cell images: a hand-segmented dataset and comparison of algorithms. In: IEEE International Symposium on Biomedical Imaging: From Nano to Macro, ISBI 2009, pp. 518–521. IEEE (2009)

[7] Cousty, J., Bertrand, G., Najman, L., Couprie, M.: Watershed cuts: Minimum spanning forests and the drop of water principle. IEEE Transactions on Pattern Analysis and Machine Intelligence 31(8), 1362–1374 (2009)

[8] Cousty, J., Bertrand, G., Najman, L., Couprie, M.: Watershed cuts: Thinnings, shortest path forests, and topological watersheds. IEEE Transactions on Pattern Analysis and Machine Intelligence 32(5), 925–939 (2010)

[9] Malmberg, F., Luengo Hendriks, C.L.: An efficient algorithm for exact evaluation of stochastic watersheds. Pattern Recognition Letters 47, 80–84 (2014), Advances in Mathematical Morphology

[10] Malmberg, F., Selig, B., Luengo Hendriks, C.L.: Exact evaluation of stochastic watersheds: From trees to general graphs. In: Barcucci, E., Frosini, A., Rinaldi, S. (eds.) DGCI 2014. LNCS, vol. 8668, pp. 309–319. Springer, Heidelberg (2014)

[11] Meyer, F., Beucher, S.: Morphological segmentation. Journal of Visual Communication and Image Representation 1(1), 21–46 (1990)

[12] Meyer, F., Stawiaski, J.: A stochastic evaluation of the contour strength. In: Goesele, M., Roth, S., Kuijper, A., Schiele, B., Schindler, K. (eds.) Pattern Recognition. LNCS, vol. 6376, pp. 513–522. Springer, Heidelberg (2010)

[13] Salembier, P., Serra, J.: Flat zones filtering, connected operators, and filters by reconstruction. Trans. Img. Proc. 4(8), 1153–1160 (1995)

[14] Straehle, C., Koethe, U., Knott, G., Briggman, K., Denk, W., Hamprecht, F.A.: Seeded watershed cut uncertainty estimators for guided interactive segmentation. In: 2012 IEEE Conference on Computer Vision and Pattern Recognition (CVPR), pp. 765–772. IEEE (2012)

[15] Vincent, L., Soille, P.: Watersheds in digital spaces: an efficient algorithm based on immersion simulations. IEEE Transactions on Pattern Analysis and Machine Intelligence 13(6), 583–598 (1991)

A Watershed Algorithm Progressively Unveiling Its Optimality

Fernand Meyer[✉]

CMM – Centre de Morphologie Mathématique,
MINES ParisTech – PSL Research University,
35 rue St Honoré, 77300 Fontainebleau, France
fernand.meyer@mines-paristech.fr

Abstract. In 1991 I described a particularly simple and elegant watershed algorithm, where the flooding a topographic surface was scheduled by a hierarchical queue. In 2004 the watershed line has been described as the skeleton by zone of influence for the topographic distance. The same algorithm still applies. In 2012 I defined a new distance based on a lexicographic ordering of the downstream paths leading each node to a regional minimum. Without changing a iota, the same algorithm does the job.

Keywords: Watershed · Flooding · Hierarchical queue · Topographic distance · Weighted graphs · Steepest lexicographic distance

1 Introduction

This paper is singular as it analyses a paper published in 1991 [5], known and used by many, describing an algorithm for constructing watershed partitions (for a literature review on the watershed, see [9]). In the sequel we refer to it as the "HQ algo."

Classically considering a grey tone function as a topographic surface, the algorithm constructs a watershed partition by flooding this surface, the darker levels being flooded before the brighter levels. At the same time it provides a correct treatment of the plateaus of uniform grey tone: the flooding starts at its downward boundary and progresses inwards with uniform speed. The flooding is controlled by a hierarchical queue, i.e. a series of hierarchical FIFO queues. Each queue is devoted to the flooding of a particular level. The FIFO structure cares for the flooding of the plateaus.

If one considers by which other pixel each pixel is flooded, one obtains a flooding trajectory. A pixel is assigned to the catchment basin of a particular minimum m if there exists a flooding trajectory between this pixel and m. A pixel may be linked to several minima by distinct flooding trajectories. The hierarchical queues selects a particular flooding trajectory for each pixel and assigns this pixel to the minimum at the beginning of the trajectory. The quality of the final watershed partition entirely relies on the choice of the best trajectories.

© Springer International Publishing Switzerland 2015
J.A. Benediktsson et al. (Eds.): ISMM 2015, LNCS 9082, pp. 717–728, 2015.
DOI: 10.1007/978-3-319-18720-4_60

The algorithm has been developed for a lexicographic order of flooding: between two pixels, the first to be flooded is the lowest ; and if they have the same gray tone, the first to be flooded is the nearest to the lower boundary of a plateau.

Later the topographic distance has been developed and a more selective order relation defined. If a node has lower neighbors, it may be flooded only by one of the lowest of them. The behavior on plateaus remains identical. The "HQ algo". remains valid for this new lexicographic distance derived from the topographic distance [8,6]. A good review on watershed algorithms may be found in [10].

Finally, in 2012, I introduced a new distance for which the geodesics are the steepest possible [7]. The paths linking a node with the regional minima are compared using an infinite lexicographic order relation based on the gray tone values of all pixels along the path. It was a real surprise to discover that, again, without the slightest modification, the algorithm of 1991 does the job.

The paper explains why the HQ algorithm remains valid for these 3 types of very different distances. A slight modification of the HQ algorithm permits to prune the graph in order to keep only the infinitely steep flooding paths. We conclude by some applications using such steepest paths.

We conclude, recognizing that the HQ algorithm was perhaps not optimal, but did much more as initially expected.

2 The Early Days of the Watershed

2.1 The Geodesic Distance and the Skeleton by Zones of Influence

Christian Lantuejoul, in order to model a polycrystalline alloy, defined and studied the skeleton by zones of influence of a binary collection of grains in his thesis [3] ; he studied the geodesic metric used for constructing a SKIZ in [4]. The shortest distance $d_Z(x, y)$ between two pixels x and y in a domain Z is the length of the shortest path linking both pixels within the domain. If X is a subset of Z, we likewise define the geodesic distance between a point $x \in Z$ and the set X, as the shortest path between x and a pixel belonging to X.

Consider now a set $X \subset Z$, union of a family of connected components X_i. The skeleton by zone of influence of X, or skiz(X, Z) assigns to each connected component X_i the pixels which are closer to X_i than to any other set X_j for the geodesic distance d_Z.

The first implementation of the SKIZ was made on the TAS, a hardwired processor developed at the CMM and used homotopic thickenings for constructing the SKIZ. With the advent of cheaper memories and the personal computer, it was possible to represent images on random access memories (whereas the TAS permitted only raster scan access to the images). Luc Vincent proposed a very efficient implementation of a SKIZ [11]; the growing of the various germs being governed by a FIFO. The algorithm for constructing the SKIZ skiz(X, Z) of a family of particles $X = (X_i)$ within a domain Z is the algorithm 1.

Algorithme 1. A fifo based algorithm for constructing the geodesic SKIZ

Input : A family of seeds $(K_i)_{i \in I}$ within a domain Z

Result : The geodesic SKIZ of the of seeds $(K_i)_{i \in I}$ within Z

1 **Initialisation:** Create a FIFO Q
Introduce the inside boundary pixels of the seeds K_i in Q

2 **while** Q *not empty* **do**
3 | extract the node j with the highest priority from Q
   | **for** *each unlabeled neighboring node* $i \in Z$ *of* j **do**
4 | | $label(i) = label(j)$
   | | put i in the queue Q

2.2 The Level by Level Construction of the Watershed

The watershed partition π_f of a grey tone image represented by a function f, defined on a domain D and taking its value in the interval $[0, N]$ is then easily obtained with the algorithm 2, by iteratively constructing a series of SKIZ in increasing domains depending on the successive thresholds of f [1]. In algorithm 2, the notation $\{f = \lambda\}$ means the set of pixels of D for which f takes the value λ

Algorithme 2. A level by level watershed algorithm

Input : A function f

Result : the watershed partition π_f of f

1 **Initialisation:** Initialize the set π_f with the regional minima of f

2 **for** *each gray level* $\lambda = 1, N$ **do**
3 | $\pi_f = skiz(\pi_f, \pi_f \cup \{f = \lambda\})$

One gets the watershed algorithm by introducing the algorithm for the SKIZ into the algorithm of C. Lantuéjoul. L. Vincent and P.Soille completed the algorithm by adding a clever mechanism for finding the successive thresholds of a grey tone function [11].

2.3 A Hierarchical Queue Algorithm for the Watershed

The algorithm 2 repeats the SKIZ construction for all successive levels of the grey tone function sequentially. For each level it uses the algorithm 1. When a pixel is dequeued from the FIFO, its neighbors are labeled if they belong to the same level set, and discarded if not. They will be processed later, when the level set corresponding to their greytone value is processed. The principal innovation of the "HQ algo" [5] is to create all FIFOs at once, each being devoted to a distinct grey tone, with the advantage to put all pixels which are met in a waiting position before they flood neighboring pixels. When a node x is dequeued all its neighbors may be processed: the neighbors with the same grey tone as x are put in the same FIFO as x, the others are put in the FIFOs corresponding to their grey tone. The resulting algorithm 3 is particularly simple.

Algorithme 3. The hierarchical queue based watershed algorithm

---

Input : A gray tone function

Result : A partition of labeled catchment basins

1 **Initialisation**: Detect and label the regional minima
  Create a hierarchical queue HQ
  Introduce the inside boundary pixels of the minima in the HQ

2 **while** *HQ not empty* **do**
3   extract the node j with the highest priority from the HQ
      for *each unlabeled neighboring node i of j* **do**
4         $label(i) = label(j)$
          put i in the queue with priority f_i

Analysis of the Algorithm. The nodes are processed according to a dual order: nodes with a lower altitude are flooded before nodes with a higher altitude. And nodes within a plateau are processed in an order proportional to the distance to the lower border of the plateau.

3 The Watershed as the SKIZ for the Topographic Distance

Each image defined on a grid may be considered as a particular graph. The pixels are the nodes of the graph ; neighboring pixels are linked by a node. All algorithms defined so far immediately apply to node weighted graphs. We now give a few reminders on graphs

3.1 Reminders on Node Weighted Graphs

A *non oriented graph* $G = [N, E]$ contains a set N of vertices or nodes and a set E of edges ; an edge being a pair of vertices. The nodes are designated with small letters: $p, q, r...$The edge linking the nodes p and q is designated by e_{pq}.

A *path*, ϖ, is a sequence of vertices and edges, interweaved in the following way: ϖ starts with a vertex, say p, followed by an edge e_{pq}, incident to p, followed by the other endpoint q of e_{pq}, and so on.

Denote by $\mathcal{F}_n$ the sets of non negative weight functions on the nodes. The function $\nu \in \mathcal{F}_n$ takes the weight ν_p on the node p. The operator $\varepsilon_n : \mathcal{F}_n \longrightarrow \mathcal{F}_n$ assigning to each node the weight of its lowest neighboring node is an erosion.

A subgraph G' of a node weighted graph G is a *flat zone*, if any two nodes of G' are connected by a path along which all nodes have the same weight. If furthermore all neighboring nodes have a higher altitude, it is a *regional minimum*. A *flooding path* is a path along which the node weights is never increasing.

Definition 1. *The catchment zone of a minimum m is the set of nodes linked by a flooding path with a minimum.*

Obviously, a node may be linked with several regional minima through distinct flooding paths. The catchment zones may overlap and do not necessarily form a partition. They form a partition if each node is linked with one and only one basin through a flooding path. As there may be several paths towards distinct minima, the "HQ algo." selects for each node a particular flooding path linking this node with are regional minimum. The next section shows how to restrict even more the admissible flooding paths.

3.2   The Topographic Distance

Consider an arbitrary path $\varpi = (x_1, x_2, ..., x_p)$ of the node weighted graph G between two nodes x_1 and x_p. The weight ν_p at node x_p can be written:

$$\nu_p = \nu_p - \nu_{p-1} + \nu_{p-1} - \nu_{p-2} + \nu_{p-2} - \nu_{p-3} + ... + \nu_2 - \nu_1 + \nu_1$$

The node $k - 1$ is not necessarily the lowest node of node k, therefore $\nu_{k-1} \geq \varepsilon_n \nu_k$ and $\nu_k - \nu_{k-1} \leq \nu_k - \varepsilon_n \nu_k$.

Replacing each increment $\nu_k - \nu_{k-1}$ by $\nu_k - \varepsilon_n \nu_k$ will produce a sum $\nu_p - \varepsilon_n \nu_p + \nu_{p-1} - \varepsilon_n \nu_{p-1} + ... + \nu_2 - \varepsilon_n \nu_2 + \nu_1$ which is larger than ν_p. It is called the topographic length of the path $\varpi = (x_1, x_2, ..., x_p)$.

The path with the shortest topographic length between two nodes is called the topographic distance between these nodes. It will be equal to ν_p if and only if the path $(x_1, x_2, ..., x_p)$ precisely is a path of steepest descent, from each node to its lowest neighbor. The topographic distance has been introduced independently in [8,6].

Consider again the "HQ algo". The regional minima are labeled. During the execution of the algorithm, each node p, when dequeued, assigns its label to its neighbors without label. If a node q has no label when its neighboring node p is dequeued, it means that q has no other neighbor which has been dequeued before. If the nodes p and q have distinct weights, it means that p is one of its lowest neighbors: between nodes with distinct grey tones, it is the topographic distance which prevails. If the nodes p and q have the same weight, it means that p and q both belong to the same plateau and p is closer to the lower border of the plateau than q. We obtain like that a more restrictive lexicographic distance, where the first term is the topographic distance and the second term the distance to the lower boundaries of the plateaus.

3.3   Illustration of the HQ Algorithm Applied to Node Weighted Graphs

Fig.1 shows how the hierarchical queue algorithm constructs the watershed partition of a node weighted graphs.

A hierarchical queue is created. The regional minima, the nodes a, b and c are detected and labeled each with a distinct color ; having the grey tone 1, they are put into the FIFO of priority 1.

The first node to be extracted is a. Its neighboring node d, has no label ; d gets the label of a and is introduced in the FIFO of priority 2.

The node b is extracted. It has no neighbor. The node c propagates its label to e, which is introduced in the FIFO of priority 2.

The node d gives its labels to its unlabeled neighbors f, g and i. The nodes f and g are introduced in the FIFO of priority 3 and the node i in the FIFO of priority 4.

The node f has no unlabeled neighbor. The node g gives its label to the node j ; the node j is introduced in the FIFO with a priority 5 as shown in the last configuration of the HQ in fig.1. All nodes of the graph are now labeled and constitute the watershed partition of the graph. The last nodes are extracted from the HQ without introducing new nodes in the HQ. When the HQ is empty, the algorithm stops.

4   The Watershed as the SKIZ for Infinitely Steep Lexicographic Distance

The hierarchical queue algorithm presented so far makes 2 jobs. The first job is selecting a family of flooding paths which are all geodesics of the same distance function. If several geodesics link a node to several regional minima, it furthermore selects one of them, the first which is able to label a node without label.

We now establish, that in fact the HQ algorithm makes a much more severe selection between flooding paths as it appears so far. It does in fact implement a much more selective distance function, based on an infinite lexicographic order.

4.1   A Lexicographic Preorder Relation Between Steepest Path

Consider now a flooding path, i.e. a path along which the node weights are never increasing, ending at a node m belonging to a regional minimum. If m is an isolated regional minimum node, we artificially add a loop edge linking the node m with itself. Like that the flooding path, as it reaches the node m may be prolonged into a path of infinite length ; if m is an isolated regional minimum

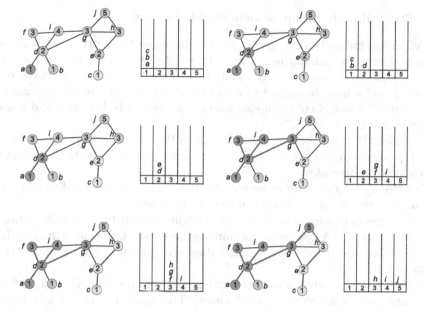

Fig. 1. The hierarchical queue algorithm constructing the watershed partition of a node weighted graph

node, the cycles around in the added loop edge ; if m has a neighboring node p belonging to the same regional minimum, the path oscillates between m and p.

A **lexicographic preorder relation** compares the infinite paths $\varpi = (p_1, p_2, ...p_k, ...)$ and $\chi = (q_1, q_2, ...q_k, ...)$:

* $\varpi \prec \chi$ if $\nu_{p_1} < \nu_{q_1}$ or there exists t such that $\begin{array}{l} \forall l < t : \nu_{p_l} = \nu_{q_l} \\ \nu_{p_t} < \nu_{q_t} \end{array}$

* $\varpi \preceq \chi$ if $\varpi \prec \chi$ or if $\forall l : \nu_{p_l} = \nu_{q_l}$.

The preorder relation $\preceq$ is total, as it permits to compare all paths with the same origin p. The lowest of them are called the *steepest* paths of origin p. It is obvious that if ϖ is a steepest path of origin p towards a regional minimum m, then any subpass obtained by suppressing a given number of nodes at the beginning of the path also is a steepest path.

A node may be linked by flooding paths with 2 or more distinct regional minima. In this case the watershed zones overlap. If one considers only the *steepest* paths, this will rarely happen, as the weights all along the paths should be absolutely identical. In particular, steepest paths reaching regional minima with different altitude necessarily have a distinct steepness. One obtains like that highly accurate watershed partitions, whereas the classical algorithms, being myopic as they use only the adjacent edges of each node, pick one solution out of many.

4.2   The Infinitely Steep Lexicographic Distance

We define the infinitely steep lexicographic distances (in short ∞sld) between nodes and regional minima as follows. The distance $\chi(p, m)$ between a node p and a regional minimum m is equal to ∞ if there exists no steepest path between p and a node belonging to m. Otherwise it is equal to the sequence of never increasing weights of the nodes along a steepest path linking p and a node belonging to m.

The infinitely steep skeleton by influence is the SKIZ associated to this distance [7]. Surprisingly enough, the HQ algo.precisely implements the infinitely steep lexicographic distance.

We write χ_p for the ∞sld of each node. It is an infinite series of weights, the first in the series being the weight of the node p itself.

For a regional minimum node m, this weight is equal to an infinite series of identical values ν_m. As the regional minima nodes are introduced in the FIFO with the same priority as their weights, they are also ranked according to their distances χ_m.

Suppose now that when a node p is extracted from the HQ, all nodes labeled so far have got their correct ∞sld distance. The ∞sld distance of p is χ_p and has been correctly estimated. If q is a neighboring node of p without label, this means that p is the neighbor of q with the lowest ∞sld distance. The ∞sld distance of q is then obtained by appending the weight ν_q of q at the beginning of the ∞sld distance of p : $\chi_p = \nu_q \triangleright \chi_p$. The node q gets the same label as the node p, and is introduced in the FIFO with the priority equal to its weight ν_q. It takes a place in this FIFO after all nodes with the same weight but with lower ∞sld distances and before all nodes with the same weight but with higher ∞sld distances.

The order in which each node floods its neighbors is much more subtle as imagined in 1991: the nodes flood their neighbors according to their ∞sld distances. The order in which the plateaus are flooded is only a particular case. When a plateau is flooded, its lower boundary pixels are not equivalent, but enter into play according to their own ∞sld distances.

4.3   The Watershed for Edge Weighted Graphs

There are situations where one desires obtaining the watershed partition associated to an edge weighted graph. For instance, the relations between the catchment basins of a topographic surface also may be modelled as a graph. The catchment basins are the nodes of the graph : neighboring basins are linked by an edge, which is weighted by the altitude of the pass point leading from one basin to the other.

The definition of regional minima, flooding paths and catchment basins are similar to those defined for node weighted graphs.

Denote by $\mathcal{F}_e$ the sets of non negative weight functions on the edges. The function $\eta \in \mathcal{F}_e$ takes the value η_{pq} on the edge e_{pq}, and the graph holding only edge weights is designated by $G(\eta, nil)$.

A subgraph G' of an edge weighted graph G is a *flat zone*, if any two nodes of G' are connected by a path with uniform edge weights. If in addition, all edges in its cocycle have a higher altitude, it is a *regional minimum*. A path $\varpi = \{v_1.e_{12}.v_2.e_{23}.v_3 \cdots\}$ is a *flooding path*, if each edge $e_k = (v_k, v_{k+1})$ is **one of the lowest edges of its extremity** v_k, and if along the path the edge weights are never increasing.

Definition 2. *The catchment zone of a regional minimum m is the set of nodes linked by a flooding path having its end-node within m.*

Jean Cousty et al. studied the watershed on edge weighted graphs, calling it watershed cuts, as each catchment basins is represented by a connected subgraph or a tree [2]. Cousty found that in some respects, the watershed on edge weighted graphs is superior that the watershed on node weighted graphs.

In fact, there is no such superiority and I established in [7] the perfect equivalence of node or edge weighted graphs with respect to the watershed. Figure 2 shows how to transform an edge weighted graph into a node weighted graph having the same regional minima, flooding paths and catchment zones as the initial edge weighted graph. Figure 2A presents an edge weighted graph. The edges which are not the lowest of one of their extremities do not belong to any flooding path. In figure 2B, each edge which is not the lowest edge of one of its extremities has been suppressed. Each node is assigned a weight equal to the weight of its lowest adjacent edge. Finally in figure 2C, the edge weights have been dropped and the regional minima detected and labeled. The HQ algorithm for node weighted graphs may be applied on this graph, yielding the correct watershed partition for the initial edge weighted graph.

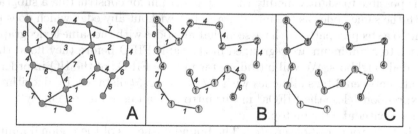

Fig. 2. A: an edge weighted graph
B: each edge which is not the lowest edge of one of its extremities has been suppressed. Each node is assigned a weight equal to the weight of its lowest adjacent edge.
C: The edge weights have been dropped and the regional minima detected and labeled.

Algorithme 4. Construction of the steepest partial subgraph of G

Input : $G = [N, E]$, a node weighted graph

Result : $\widetilde{G} = [N, \widetilde{E}]$, the steepest partial graph of G

1 **Initialisation:** $HQ =$ hierarchical queue
   Create a set S containing the regional minima of G
   Put the inside boundaries of the regional minima in HQ
   Create $\widetilde{G} = [N, \varnothing]$
   Close the FIFOs

2 **while** HQ *not empty* **do**
3   extract all nodes up to the first tag into a set P
     Suppress this tag
     for *each* $q \in \partial^+ P \cap \overline{S}$ **do**
4   **for** *each* $p \in P$, *such that* (p, q) *neighbors* **do**
5           $\widetilde{E} = \widetilde{E} \cup e_{qp}$
6         $S = S \cup \{q\}$
       put q in the queue with a priority equal to its weight
7   Close all FIFOs

4.4 Constructing the Steepest Subgraph of a Node or Edge Weighted Graph

We have established that the HQ algorithm chooses among all flooding paths the steepest. If there are 2 paths with identical weights leading to 2 distinct minima, it does an arbitrary choice between both for constructing a watershed partition.

It is possible to slightly modify the HQ algorithm for constructing a subgraph $\widetilde{G}$ with the same nodes as the initial graph but without any edge which does not belong to a steepest path. We have seen that all nodes with the same lexicographic distance to a minimum are regrouped in the same FIFO before they flood their unflooded neighbors. We will introduce tags in the FIFOs of the HQ in order to separate between 2 tags all nodes with the same ∞sld distance. In algorithm 4, the expression "close the FIFOs" means introduce a tag in all non empty FIFOs which are without a tag on top.

At initialization a HQ is created. The boundary nodes of the regional minima are introduced in a set S and in the FIFOs corresponding to their weight and all non empty FIFOs are closed. The set S will contain the union of all nodes which are or have been in the HQ. The algorithm stops when S contains all nodes of N.

As long as the hierarchical queue is not empty, all nodes with the same highest priority are extracted, including the tag itself. They are put into a set P. We call $\partial^+ P$ the set of nodes which are not in P but have a neighbor in P. The set $\partial^+ P \cap \overline{S}$ represents the nodes of $\partial^+ P$ which are not yet in the set S. For each couple of nodes (p, q), $p \in \partial^+ P \cap \overline{S}$ and $q \notin S$, the edge e_{pq} is added to the edges $\widetilde{E}$ of

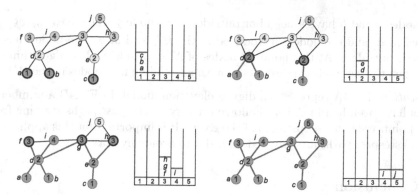

Fig. 3. A hierarchical queue algorithm for constructing the steepest paths (bold red edges) in a node weighted graph

the graph $\widetilde{G}$ (these edges are indicated in red in fig.3) . The complete algorithm is described in algorithm 4.

Illustration of the algorithm (fig.3):

A hierarchical queue is created. The boundary nodes of the regional minima, here the nodes a, b and c are put in the HQ. A tag (red line) is added to the FIFO.

The first set of most priority nodes (a, b, c) are extracted from the HQ including the tag in top of them. The node d is neighbor of a and of b. The edges e_{ad} and e_{bd} are created in $\widetilde{E}$. The node e is neighbor of the node c. An edge e_{ce} is created in $\widetilde{E}$. The nodes d and e are put in the FIFO of priority 2 and in the set S; this FIFO is then closed, by adding a tag (second figure in fig.3). The next group of nodes below a tag to be extracted are the nodes d and e. The node d has three neighbors (f, i, g). The node g is also neighbor of the node e. The node h is a second neighbor of e. The edges e_{df}, e_{di}, e_{dg} and e_{eg}, e_{eh} are created in $\widetilde{E}$, the nodes f, i, g, h are introduced in the HQ in their corresponding FIFOs and in the set S. The open FIFOs are then closed (third figure in fig.3). The next group of nodes are (f, g, h).

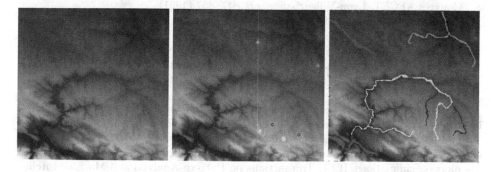

Fig. 4. Left : a digital elevation model
Center: Hand marked extremities of rivers
Right: The union of the steepest paths having these nodes as extremities

The nodes g and h have a neighbor outside S, the node j The edges e_{gj}, e_{hj} are created in $\widetilde{E}$, and the node j introduced in the HQ and in S. The node i has no neighbor outside S. At this point all nodes of N are inside S. The algorithm will stop when the last nodes are extracted without sending new nodes in the HQ.

Application: Fig.4A represents a digital elevation model. In Fig.4B a number of points have been hand marked. Following the $\infty - steepest$ paths starting from these points creates the upstream of the rivers. It is important for this application that all steepest paths are detected, in order to avoid any bias.

5   Conclusion

The HQ algorithm has been developed with a given definition of the watershed in mind and revealed to be still valid for two different mathematical formulations of the watershed. This paper illustrates the major and often ignored role of the data structure used for implementing a mathematical concept: among all possible solutions compatible with a particular definition, it does a hidden sampling.

References

1. Beucher, S., Lantuéjoul, C.: Use of watersheds in contour detection. In: Watersheds of Functions and Picture Segmentation, Paris, pp. 1928–1931 (May 1982)
2. Cousty, J., Bertrand, G., Najman, L., Couprie, M.: Watershed cuts: Minimum spanning forests and the drop of water principle. IEEE Transactions on Pattern Analysis and Machine Intelligence 31, 1362–1374 (2009)
3. Lantuéjoul, C.: La squelettisation et son application aux mesures topologiques de mosaäßques polycristallines. PhD thesis, École nationale supérieure des mines de Paris (1978)
4. Lantuéjoul, C., Beucher, S.: On the use of the geodesic metric in image analysis. J. Microsc. (1981)
5. Meyer, F.: Un algorithme optimal de ligne de partage des eaux. In: Proceedings $8^{\grave{e}me}$ Congrès AFCET, Lyon-Villeurbanne, pp. 847–857 (1991)
6. Meyer, F.: Topographic distance and watershed lines. Signal Processing, 113–125 (1994)
7. Meyer, F.: Watersheds on weighted graphs. Pattern Recognition Letters (2014)
8. Najman, L., Schmitt, M.: Watershed of a continuous function. Signal Processing 38(1), 99–112 (1994), Mathematical Morphology and its Applications to Signal Processing
9. Najman, L., Talbot, H. (eds.): Mathematical morphology. Wiley editor (2012)
10. Roerdink, J.B.T.M., Meijster, A.: The watershed transform: Definitions, algorithms and parallelization strategies. Fundamenta Informaticae 41, 187–228 (2001)
11. S.P., Vincent, L.: Watersheds in digital spaces: An efficient algorithm based on immersion simulations. IEEE Transactions on Pattern Analysis and Machine Intelligence 13(6), 583–598 (1991)

An Approach to Adaptive Quadratic Structuring Functions Based on the Local Structure Tensor

Anders Landström[✉]

Luleå University of Technology, Luleå, Sweden
anders.p.landstrom@ltu.se

Abstract. Classical morphological image processing, where the same structuring element is used to process the whole image, has its limitations. Consequently, adaptive mathematical morphology is attracting more and more attention. So far, however, the use of non-flat adaptive structuring functions is very limited. This work presents a method for defining quadratic structuring functions from the well known local structure tensor, building on previous work for flat adaptive morphology. The result is a novel approach to adaptive mathematical morphology, suitable for enhancement and linking of directional features in images. Moreover, the presented strategy can be quite efficiently implemented and is easy to use as it relies on just two user-set parameters which are directly related to image measures.

Keywords: Adaptive morphology · Quadratic structuring functions · Local structure tensor

1 Introduction

1.1 Background

The aim of this work is to present a method for adaptive non-flat morphological filtering, suitable for enhancing or linking directional features and patterns. Classical morphological operators are non-adaptive, i.e. the whole image is probed by the same structuring element without taking variations in structure into account. This is often not ideal, however. In particular; the resulting morphological operations risk stretching over edges, which may destroy important structure information. This has led to the development of adaptive mathematical morphology, where the structuring element (in the adaptive case often called structuring function) may change for each point in the image. For a review of the field of adaptive mathematical morphology, the interested reader is referred to [9] and [14].

Of particular interest to this work are adaptive methods that rely on image structure, i.e. methods where structuring elements are defined based on edges and contours rather than restricted by e.g. measures of similarity. These methods are suitable for prolonging shapes or bridging gaps, thereby emphasizing directional structures in the processed image. Some methods of this type work

© Springer International Publishing Switzerland 2015
J.A. Benediktsson et al. (Eds.): ISMM 2015, LNCS 9082, pp. 729–740, 2015.
DOI: 10.1007/978-3-319-18720-4_61

in multiple scales in order to adapt the size of the structuring element to the local scale of structures in the image [2,18]. Other methods address structure by considering local orientation only [17], or by combining local orientation with other factors such as distances to edges [21] or degree of anisotropy [12]. It should be noted, however, that one can always find an orientation even though it may very well be completely irrelevant (i.e. where there is no prevalent orientation). Imposing an orientation in such cases may introduce bias from the method itself. Consider, for instance, the simple case of unintentionally assigning a line of random orientation as structuring element. Hence the degree of anisotropy is an important aspect.

The Local Structure Tensor (LST) is a well known method for representing image structure, containing information about both local orientation and degree of anisotropy [8]. Some methods for adaptive morphology, such as the line-shaped or rectangular structuring elements presented by Verdú-Monedero et al. [21] or the continuous PDE-based morphology presented by Breuß et al. [6], use the LST components implicitly or explicitly, but without using the anisotropy information it contains. Only the Elliptical Adaptive Structuring Elements (EASE) method [11,12] takes advantage of this property of the LST.

The methods listed above are all presented for the so called *flat* case, where structuring elements are defined by sets of points rather than functions. While flat morphology may certainly be highly useful it does have limitations, and it is natural to consider an extension to the *non-flat* case, where structuring elements become structuring functions. Non-flat adaptive mathematical morphology is far from common in literature, but some work have been published. Bouaynaya and Schonfeld [4,5] have presented a base for adaptive structuring functions. The non-flat case has also been considered in non-local morphology by Salembier [16] and Velasco-Forero and Angulo [20]. Angulo and Velasco-Forero [3] have used structuring functions based on random walks, and Ćurić and Luengo Hendriks [10] have presented a method for salience-adaptive morphology based on paraboloidal structuring functions. Moreover, Quadratic Structuring Functions (QSFs) have been used by Angulo [1] as a base for flat adaptive morphology by thresholding them at a given level.

This work builds on the previously presented EASE method [11,12], presenting a strategy for non-flat mathematical morphology where QSFs are set from the Local Structure Tensor.

1.2   Contribution

It has been shown that Quadratic Structuring Functions (QSFs) play a similar role to morphology as Gaussian functions do to standard convolution filtering, i.e. QSFs are to the morphological (max,+) and (min,+) algebras what the Gaussian function is to the standard (+,×) algebra used in linear filtering [7,19]. QSFs are thereby or particular interest – and importance – to mathematical morphology.

As QSFs are paraboloids with elliptical level contours, the previously presented concept of Elliptical Adaptive Structuring Elements (EASE) [12] can be quite naturally extended into the non-flat case, following the general theoretical work presented by Bouaynaya and Schonfeld [4,5]. This article presents a method that sets QSFs based on the information contained in the LST, building on previous work [12] but enabling non-flat structuring functions. The presented strategy yields a novel straight-forward approach to non-flat structure-based adaptive morphology based on QSFs, which can be quite efficiently implemented. Moreover; only two user-set parameters are required, which makes the method easy to use.

2   Method

2.1   Adaptive Structuring Functions

Let f denote an image with values $f(\mathbf{x}) \in [0, 1]$ defined for points $\mathbf{x} \in \mathcal{D}(f)$ and, following the theoretical work by Bouaynaya et al. [4,5], let

$$s_{\mathbf{x}}(\mathbf{u}) = s[\mathbf{x}](\mathbf{u} - \mathbf{x}) \quad \forall \mathbf{u} \in \mathcal{D}(s_{\mathbf{x}}) \subseteq \mathcal{D}(f) \tag{1}$$

denote an adaptive structuring function for a point $\mathbf{x}$. Note that both $\mathbf{x}$ and $\mathbf{u}$ are given in global coordinates. The notation $s[\mathbf{x}]$ denotes that the function s itself, rather than just its values, varies with $\mathbf{x}$. The corresponding so called *reflected* (or *transposed*) structuring function, used for defining proper morphological operations, is defined as

$$s_{\mathbf{x}}^*(\mathbf{u}) = s_{\mathbf{u}}(\mathbf{x}) = s[\mathbf{u}](\mathbf{x} - \mathbf{u}). \tag{2}$$

The complete set of structuring functions $\{s_{\mathbf{x}} \mid \mathbf{x} \in \mathcal{D}(f)\}$ is known as a *Structuring Element Map* (SEM) for the image. This SEM also implicitly contains the reflected (or transposed) structuring functions.

   The morphological erosion and dilation for the adaptive case are then given by (see Ref. [4]):

$$\varepsilon_s(f) = \bigwedge\nolimits_{\mathbf{u} \in \mathcal{D}(s_{\mathbf{x}})} \{f(\mathbf{u}) - s_{\mathbf{x}}(\mathbf{u})\}, \tag{3}$$

$$\delta_s(f) = \bigvee\nolimits_{\mathbf{u} \in \mathcal{D}(s_{\mathbf{x}}^*)} \{f(\mathbf{u}) + s_{\mathbf{x}}^*(\mathbf{u})\}. \tag{4}$$

Given that the SEM remains constant over the operations, these definitions ensure that the erosion and dilation are adjunct – a property often overseen within adaptive morphology, as noted by Roerdink [15]. The opening and closing can now be properly defined as

$$\gamma_s(f) = (\delta_s \circ \varepsilon_s)(f) = \delta_s(\varepsilon_s(f)), \tag{5}$$

$$\varphi_s(f) = (\varepsilon_s \circ \delta_s)(f) = \varepsilon_s(\delta_s(f)). \tag{6}$$

In practice the dilation $\delta_s(f)$ can be calculated by the following algorithm, which goes through all points $\mathbf{x}$ and updates the value at point $\mathbf{u}$ in $\delta_s(f)$ when $f(\mathbf{x}) + s_{\mathbf{x}}(\mathbf{u})$ yields a higher value (similar to previous usage for the flat case, see e.g. Ref. [13]):

$$(\delta_s(f))(\mathbf{u}) \longleftarrow \bigvee \{(\delta_s(f))(\mathbf{u}), f(\mathbf{x}) + s_{\mathbf{x}}(\mathbf{u})\}, \ \forall \mathbf{u} \in s_{\mathbf{x}}, \forall \mathbf{x} \in \mathcal{D}(f). \tag{7}$$

We now simply need to define our SEM.

2.2   The Local Structure Tensor

As demonstrated in previous work [12], the Local Structure Tensor (LST) can be used to define elliptical adaptive structuring elements. The LST $\mathbf{T}(\mathbf{x})$ can be constructed from the image gradient by

$$\mathbf{T}(\mathbf{x}) = \begin{pmatrix} T_{11} & T_{12} \\ T_{12} & T_{22} \end{pmatrix} (\mathbf{x}) = G_\sigma * \left(\nabla f(\mathbf{x}) \nabla^\mathsf{T} f(\mathbf{x}) \right), \tag{8}$$

where $\nabla = \left(\frac{\partial}{\partial x_1} \; \frac{\partial}{\partial x_2} \right)^\mathsf{T}$ and G_σ is a Gaussian kernel with standard deviation σ [8]. The image gradient can be estimated by applying standard gradient filters on a slightly smoothed version of the input image. The parameter σ sets the scale for which the LST should be representative, and can be set implicitly by defining a radial bandwidth r_w for the filter (so that G_σ decreases to half of its maximum value at distance r_w from its center), i.e.

$$\sigma = \frac{r_w}{\sqrt{2 \ln 2}}. \tag{9}$$

The eigenvalues $\lambda_1(\mathbf{x})$ and $\lambda_2(\mathbf{x})$ ($\lambda_1(\mathbf{x}) \geq \lambda_2(\mathbf{x})$) and corresponding eigenvectors $\mathbf{e}_1(\mathbf{x})$ and $\mathbf{e}_2(\mathbf{x})$ of $\mathbf{T}(\mathbf{x})$ hold information about structures (edges) in the image. Eigenvalues can be interpreted based on Table 1, while $\mathbf{e}_2(\mathbf{x})$ represents the direction of the smallest variation [8].

Table 1. Interpretation of the eigenvalues λ_1 and λ_2 of the LST $\mathbf{T}$

$\lambda_1 \approx \lambda_2 \gg 0$	No dominant direction
	(edge crossing or point)
$\lambda_1 \gg \lambda_2 \approx 0$	Strong dominant direction (edge)
$\lambda_1 \approx \lambda_2 \approx 0$	No dominant direction (no edge)

2.3   Flat Structuring Elements

Previous work [12,11] has defined flat elliptical adaptive structuring elements. The axes $a(\mathbf{x})$ and $b(\mathbf{x})$ are set from the eigenvalues of $\mathbf{T}(\mathbf{x})$ by the expressions

$$a(\mathbf{x}) = \frac{\lambda_1(\mathbf{x})}{\lambda_1(\mathbf{x}) + \lambda_2(\mathbf{x})} \cdot M, \quad b(\mathbf{x}) = \frac{\lambda_2(\mathbf{x})}{\lambda_1(\mathbf{x}) + \lambda_2(\mathbf{x})} \cdot M, \tag{10}$$

where M denotes the maximum allowed semi-major axis. Numerical stability is addressed by adding a small positive number (i.e. machine epsilon) to the eigenvalues. The orientation $\theta(\mathbf{x})$ is retrieved from the corresponding eigenvectors by

$$\theta(\mathbf{x}) = \begin{cases} \arctan\left(\dfrac{e_{2,x_2}(\mathbf{x})}{e_{2,x_1}(\mathbf{x})} \right), & e_{2,x_1}(\mathbf{x}) \neq 0, \\[2ex] \pi/2, & e_{2,x_1}(\mathbf{x}) = 0, \end{cases} \tag{11}$$

where $e_{2,x_1}(\mathbf{x})$ and $e_{2,x_2}(\mathbf{x})$ denote the components of the eigenvector $\mathbf{e}_2(\mathbf{x})$. The resulting structuring elements range dynamically from lines of length M where $\lambda_1(\mathbf{x}) \gg \lambda_2(\mathbf{x}) \approx 0$, i.e. near strong dominant edges in the data, to disks with radius $\frac{M}{2}$ where $\lambda_1(\mathbf{x}) \approx \lambda_2(\mathbf{x})$, i.e. where no single direction represents the local image structure.

2.4 Quadratic Structuring Functions

The change from flat structuring elements to Quadratic Structuring Functions (QSFs) is quite straight-forward, yet yields a substantial change: the LST is still used to calculate a, b, and θ, but the parameters are now used to set QSFs rather than flat elliptical structuring elements. The result is a method for non-flat adaptive morphology based on a well known method for estimating structure in images.

Given a, b, and θ we define the quadratic structuring function (with elliptical level contours)

$$s_{\mathbf{x}}(\mathbf{u}) = -\frac{1}{2}\left(\left(\frac{(x_1 - u_1)\cos\theta + (x_2 - u_2)\sin\theta}{a}\right)^2\right.$$
$$\left. + \left(\frac{(x_1 - u_1)\sin\theta - (x_2 - u_2)\cos\theta}{b}\right)^2\right) \tag{12}$$

$$= -\frac{1}{2}(\mathbf{u} - \mathbf{x})^T \mathbf{R}\begin{pmatrix} a^{-2} & 0 \\ 0 & b^{-2} \end{pmatrix}\mathbf{R}^T(\mathbf{u} - \mathbf{x}) \tag{13}$$

$$= -\frac{1}{2M^2}(\mathbf{u} - \mathbf{x})^T \mathbf{R}\begin{pmatrix} \alpha^{-2} & 0 \\ 0 & \beta^{-2} \end{pmatrix}\mathbf{R}^T(\mathbf{u} - \mathbf{x}) \tag{14}$$

where

$$\mathbf{R} = \begin{pmatrix} \cos\theta & -\sin\theta \\ \sin\theta & \cos\theta \end{pmatrix}, \quad \alpha = \frac{\lambda_1}{\lambda_1 + \lambda_2}, \quad \beta = \frac{\lambda_2}{\lambda_1 + \lambda_2}. \tag{15}$$

For numerical stability, Eqs. (12) or (13) should be used for setting the QSFs. To avoid division by zero in the case when $b = 0$, a and b (which are measured in pixel units) can be increased by $\frac{1}{2}$. If image function values within the range $[0, 1]$ can be assumed, a non-flat structuring function $s_{\mathbf{x}}$ will not cause any change where $s_{\mathbf{x}} < -1$. The values of the QSFs can therefore be precomputed on a fixed spatial support $\mathcal{D}(s_{\mathbf{x}})$ large enough to ensure $s_{\mathbf{x}}(\mathbf{u}) \leq -1 \, \forall \mathbf{u} \notin \mathcal{D}(s_{\mathbf{x}})$.

Equation (14) clearly shows how M becomes a scale factor for the structuring functions. It should be noted that the set $\{\mathbf{u} \,|\, s_{\mathbf{x}}(\mathbf{u}) \geq -\frac{1}{2}\}$ yields the flat elliptical structuring element in the previous section.

3 Results

3.1 Structuring Function Shapes

Figure 1 shows a close-up of an input image and a subset of its SEM for $M = r_w = 8$ (Fig. 1a). The red ellipses in Fig. 1a show the contours of $s_{\mathbf{x}} = -\frac{1}{2}$ for

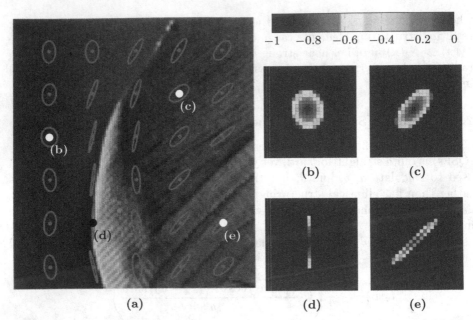

Fig. 1. Detailed examples of structuring functions. 1a: An original image with the contours $s_x = -\frac{1}{2}$ drawn in red for a subset of the SEM. 1b–1e: The QSFs for the corresponding points marked in 1a.

the selected subset of points, and are thereby equivalent to the flat ellipses in Sect. 2.3. Four detailed examples of the resulting QSFs, corresponding to the points marked with letters are presented in detail in Figs. 1b–1e.

The close connection to the flat elliptical structuring elements is obvious, and it is clear that the structuring functions follow the image structure well. One potential issue, which does not show up in this particular subset of structuring functions but may be easily anticipated, may appear as a result from going from a continuous definition to a discrete domain of pixels: in the case of a QSF with a domain in the shape of a line which is not aligned with the axes, the maximum of the expression may end up in-between pixels. Consider, for instance, a small shift of the orientation of the structuring function in Fig. 1d. The values in the outer pixels would soon end up in between the pixel positions, as the structuring function is very thin. This would make the implemented structuring function shorter than the theoretical.

3.2   Comparison to Flat EASE

Figure 2 compares the presented QSF method to flat EASE. Looking at the top and left parts of the result for flat EASE (Fig. 2b), sharp edges are introduced

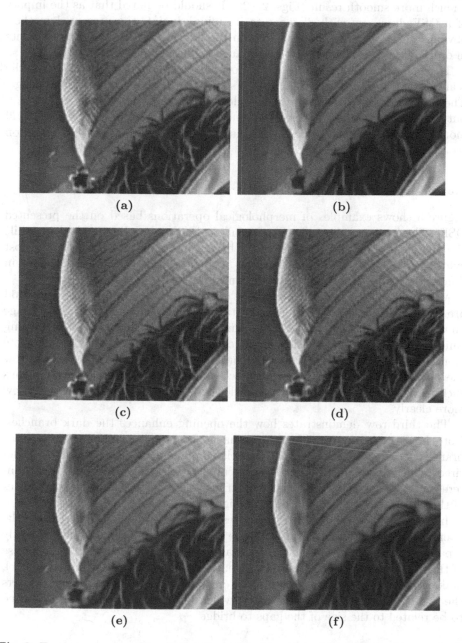

(a)   (b)

(c)   (d)

(e)   (f)

Fig. 2. Top row: close-up (size 175×145 pixels) of an original image 2a and its opening by flat EASE with $M = r_w = 4$ 2b. Second and third rows: Openings by the presented QSFs for $M = \{2, 4, 8, 16\}$, $r_w = 4$ (2c,2d,2e, and 2f).

by the flat structuring elements while the quadratic structuring functions leave a much more smooth result (Figs. 2c–2f). It should be noted that as the impact of a QSF decreases with distance to its middle point, the presented method obviously need a larger value of M than the flat elliptical structuring elements in order to achieve a similar level of change in the image.

The effect of different scales of the morphological operations are clear: as the scale for QSFs is increased (Figs. 2c–2f), so are the changes to the original image. The result for $M = 2$ is more or less identical to the original image (Fig. 2c), but as M is increased the result becomes more and more similar to (yet still more smooth than) the more severe changes resulting from the flat operation (compare Fig. 2f with Fig. 2b).

3.3   Morphological Operations

Figure 3 shows examples of morphological operations based on the presented QSFs. The erosion and opening successfully enhances dark lines and edges, while the dilation and closing do the same for bright directional features. This is most evident in the hat, hair, and feathers. Also note how the mouth changes in different ways in the erosion and dilation.

More examples of adaptive openings and closings by the presented method are presented in Fig. 4. The dark structures around the windows in the image on the first row (original in the left column) are clearly enhanced by the opening (middle row), while the closing (right column) instead links the bright spots along the left edge of the roof into a continuous bright line. In the second row the opening removes bright features such as the whiskers, while darks regions such as the mouth are enhanced. The closing instead makes the whiskers show more clearly.

The third row demonstrates how the opening enhances the dark branches. Note how the structuring functions change from anisotropic to isotropic as branches cross, which means that there is no obvious direction to follow. If desired (which may depend on the specific application), this effect could be countered by changing the size of the structuring functions based on the magnitudes of the eigenvalues (which will both be high at crossings).

In the fourth row, the linking properties of the presented method are clear: the opening enhances the dark rail while the closing connects the originally slightly separated parts of the bright train. Finally, the last row demonstrates the use of the method in a specific practical case: the opening enhances a crack in steel, making it easier to find and classify the crack. Note that the parameter r_w sets the scale for the structures to align to, which means that both M and r_w need to be related to the size of the gaps to bridge.

Fig. 3. An original image (size 512×512 pixels) and examples of the erosion 3b, dilation 3c, opening 3d, and closing 3e with $M = 16$, $r_w = 8$.

Fig. 4. Five original images of size 400×400 pixels (left column), and their openings (middle column) and closings (right column) using $M=16$, $r_w=8$.

4 Conclusion

In this work an approach for interpreting the LST into well defined QSFs in a straight-forward manner has been investigated. The results show that the presented method successfully enhances edges and lines, and adapts well to structure. It builds on the previously presented EASE method, with the important contribution of allowing non-flat adaptive morphology.

The strategy does not only demonstrate a mathematically solid method for non-flat morphological filtering, based on well known theory, but presents a method which can be quite easily implemented with a low number of user-set parameters (the scale for the LST and the scale for the QSF, and possibly the level of prefiltering) which are easily related to measures in the image. The implementation can be done quite efficiently by pre-calculating the weights of the quadratic structuring functions. There are some issues to consider when going from continuous theory to discrete practice, however, as discussed in Sect. 3.1.

The presented method demonstrates a solid interpretation of the LST which yields well defined structuring functions, but there are of course other ways in which the LST could be converted into QSFs. Future work should investigate this direction further. In particular, the relation to the covariance matrix (which under some assumptions is equivalent to the LST) and the Mahalanobis distance should be pursued. A quantitative comparison to other methods based on adaptive structuring functions should also be conducted.

Acknowledgment. The author would like to thank Dr. Jesús Angulo for highly appreciated input, which inspired this work.

References

1. Angulo, J.: Morphological bilateral filtering and spatially-variant adaptive structuring functions. In: Soille, P., Pesaresi, M., Ouzounis, G.K. (eds.) ISMM 2011. LNCS, vol. 6671, pp. 212–223. Springer, Heidelberg (2011)
2. Angulo, J., Velasco-Forero, S.: Structurally adaptive mathematical morphology based on nonlinear scale-space decompositions. Image Analysis & Stereology 30(2), 111–122 (2011)
3. Angulo, J., Velasco-Forero, S.: Stochastic morphological filtering and bellman-maslov chains. In: Hendriks, C.L.L., Borgefors, G., Strand, R. (eds.) ISMM 2013. LNCS, vol. 7883, pp. 171–182. Springer, Heidelberg (2013)
4. Bouaynaya, N., Schonfeld, D.: Spatially variant morphological image processing: theory and applications. In: Proceedings of SPIE, vol. 6077, pp. 673–684 (2006)
5. Bouaynaya, N., Schonfeld, D.: Theoretical foundations of spatially-variant mathematical morphology part II: Gray-level images. IEEE Transactions on Pattern Analysis and Machine Intelligence 30(5), 837–850 (2008)
6. Breuß, M., Burgeth, B., Weickert, J.: Anisotropic continuous-scale morphology. In: Martí, J., Benedí, J.M., Mendonça, A.M., Serrat, J. (eds.) IbPRIA 2007. LNCS, vol. 4478, pp. 515–522. Springer, Heidelberg (2007)
7. Burgeth, B., Weickert, J.: An explanation for the logarithmic connection between linear and morphological system theory. International Journal of Computer Vision 64(2-3), 157–169 (2005)

8. Cammoun, L., Castaño-Moraga, C.A., Muñoz-Moreno, E., Sosa-Cabrera, D., Acar, B., Rodriguez-Florido, M.A., Brun, A., Knutsson, H., Thiran, J.P.: A review of tensors and tensor signal processing. In: Tensors in Image Processing and Computer Vision, pp. 1–32. Springer (2009)
9. Ćurić, V., Landström, A., Thurley, M.J., Luengo Hendriks, C.L.: Adaptive mathematical morphology – a survey of the field. Pattern Recognition Letters 47(0), 18–28 (2014), Advances in Mathematical Morphology
10. Ćurić, V., Hendriks, C.L.L.: Salience-based parabolic structuring functions. In: Hendriks, C.L.L., Borgefors, G., Strand, R. (eds.) ISMM 2013. LNCS, vol. 7883, pp. 183–194. Springer, Heidelberg (2013)
11. Landström, A.: Elliptical Adaptive Structuring Elements for Mathematical Morphology. Doctoral thesis, Luleå University of Technology, Sweden (2014)
12. Landström, A., Thurley, M.J.: Adaptive morphology using tensor-based elliptical structuring elements. Pattern Recognition Letters 34(12), 1416–1422 (2013)
13. Lerallut, R., Decencière, É., Meyer, F.: Image filtering using morphological amoebas. In: Mathematical Morphology: 40 Years On, pp. 13–22 (2005)
14. Maragos, P., Vachier, C.: Overview of adaptive morphology: trends and perspectives. In: 16th IEEE International Conference on Image Processing (ICIP), pp. 2241–2244. IEEE (2009)
15. Roerdink, J.B.T.M.: Adaptivity and group invariance in mathematical morphology. In: 16th IEEE International Conference on Image Processing (ICIP), pp. 2253–2256. IEEE (2009)
16. Salembier, P.: Study on nonlocal morphological operators. In: 2009 16th IEEE International Conference on Image Processing (ICIP), pp. 2269–2272. IEEE (2009)
17. Shih, F.Y., Cheng, S.: Adaptive mathematical morphology for edge linking. Information Sciences 167(1), 9–21 (2004)
18. Tankyevych, O., Talbot, H., Dokládal, P., Passat, N.: Direction-adaptive grey-level morphology. application to 3d vascular brain imaging. In: 16th IEEE International Conference on Image Processing (ICIP), pp. 2261–2264. IEEE (2009)
19. Van Den Boomgaard, R., Dorst, L., Makram-Ebeid, S.: John Schavemaker. Quadratic structuring functions in mathematical morphology. In: Mathematical Morphology and its Applications to Image and Signal Processing, pp. 147–154. Springer (1996)
20. Velasco-Forero, S., Angulo, J.: On nonlocal mathematical morphology. In: Hendriks, C.L.L., Borgefors, G., Strand, R. (eds.) ISMM 2013. LNCS, vol. 7883, pp. 219–230. Springer, Heidelberg (2013)
21. Verdú-Monedero, R., Angulo, J., Serra, J.: Anisotropic morphological filters with spatially-variant structuring elements based on image-dependent gradient fields. IEEE Transactions on Image Processing 20(1), 200–212 (2011)

Adaptive Hit or Miss Transform

Vladimir Ćurić[1]([✉]), Sébastien Lefèvre[2], and Cris L. Luengo Hendriks[3]

[1] Department of Cell and Molecular Biology, Uppsala University, Uppsala, Sweden
vladimir.curic@icm.uu.se
[2] IRISA, University of Bretagne-Sud, Vannes, France
sebastien.lefevre@irisa.fr
[3] Centre for Image Analysis, Uppsala University, Uppsala, Sweden
cris@cb.uu.se

Abstract. The Hit or Miss Transform is a fundamental morphological operator, and can be used for template matching. In this paper, we present a framework for adaptive Hit or Miss Transform, where structuring elements are adaptive with respect to the input image itself. We illustrate the difference between the new adaptive Hit or Miss Transform and the classical Hit or Miss Transform. As an example of its usefulness, we show how the new adaptive Hit or Miss Transform can detect particles in single molecule imaging.

Keywords: Hit or Miss Transform · Adaptive morphologya · Adaptive structuring elements · Template matching

1 Introduction

At its early development, mathematical morphology was dedicated to binary images using the notion of structuring element [1–3]. A structuring element was a fixed shape used to probe every point in the image. Morphological operators defined with these fixed structuring elements are still very commonly used for many different problems in image analysis. Nevertheless, the basic morphological operators can be extended to adaptive morphological operators using adaptive structuring elements, which are structuring elements that change size and shape according to local characteristics of the image. The construction of different adaptive structuring elements and consequently adaptive morphological operators has attracted a lot of attention in the last decade [4–10]. Theoretical advances and limitations of adaptive mathematical morphology have been explored by several researchers [11–14]. Adaptivity can be included into morphological operators in different ways [15], and the recent overview paper on adaptive mathematical morphology presents a comparison of different methods for adaptive structuring elements [16].

Despite adaptive mathematical morphology has been developed extensively in the last decade, its development is mostly based on the construction of new types of adaptive structuring elements. These structuring elements are then applied to basic morphological operators (erosions, dilations, openings and closings) or

© Springer International Publishing Switzerland 2015
J.A. Benediktsson et al. (Eds.): ISMM 2015, LNCS 9082, pp. 741–752, 2015.
DOI: 10.1007/978-3-319-18720-4_62

noise filtering techniques. More complex operators or filters have not been extensively studied. In this paper, we focus on one such operator, the Hit or Miss Transform (HMT).

The HMT has most often been used for detecting a given pattern in a binary image (this problem is also called template matching) and has a number of different applications [17–19]. Two survey papers on HMT have been presented recently [20, 21]. HMT relies on a pair of disjoint structuring elements, where one fits the foreground (object) and the other fits the background. The common approach to ensure robustness to variations in the shape or size of the sought pattern is to consider a set of structuring element pairs; a match is allowed if one of the pairs match the considered pattern [17]. Instead of using a limited, pre-defined set of structuring element pairs, we propose to use a single adaptive structuring element pair in order to limit the computation burden.

In this paper, we propose a framework for adaptive Hit or Miss Transform, where the HMT is defined for gray level images using the notion of adaptive structuring elements. We present a way to define the adaptive HMT, and experimentally show its advantages over the classical HMT (i.e. with fixed structuring elements). In the rest of this paper, we will denote as *classical HMT* the HMT defined using fixed structuring elements, and as *adaptive HMT* the HMT defined using adaptive structuring elements. In addition, we illustrate the usefulness of adaptive HMT to a real life imaging problem, i.e. finding fluorescent dots in single molecule imaging.

2   Background

In this section, we recall the basics of HMT and adaptive mathematical morphology with adequate definitions and notations. First, we present the HMT for binary images as well as its extension to gray level images. Second, we recall the main concepts of adaptive mathematical morphology and briefly present the method for adaptive structuring elements that will be used in this paper.

2.1   Hit or Miss Transform

The Hit or Miss Transform aims to extract objects (i.e. parts of the image) that fit two distinct criteria. These criteria are defined as structuring elements that are respectively associated with foreground (object) and background. Several equivalent formulations have been given to define the HMT, and we use the following one to denote the HMT of an image I by the couple of structuring elements (SE_O, SE_B):

$$\text{HMT}_{SE_O, SE_B}(I) = \varepsilon_{SE_O}(I) \setminus \delta_{SE'_B}(I), \tag{1}$$

where $\varepsilon_{SE_O}(I)$ and $\delta_{SE_B}(I)$ represent erosion and dilation of I with structuring elements SE_O and SE_B respectively, and $A' : x \to -x$ denotes the reflexion of the structuring element A with respect to the its origin.

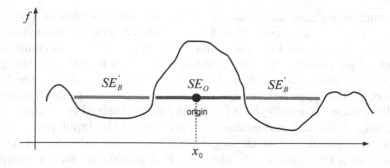

Fig. 1. Example of 1D function f where SE_O fits the region (foreground) around the point x_0 and SE'_B fits the neighbourhood (background) of the same point

Conversely to the binary case, various non equivalent definitions have been given for the gray scale HMT. The reader will find in [22] a common theoretical framework, and in [20] a review of these existing works. In this paper, and for the sake of illustration, we will consider the definition from Soille [23] assuming flat structuring elements (see Fig. 1), i.e.,

$$\text{HMT}_{SE_O, SE_B}(I)(x) = \max\left\{\varepsilon_{SE_O}(I)(x) - \delta_{SE'_B}(I)(x), 0\right\}, \quad x \in D \quad (2)$$

where D is the image domain.

Let us note, however, that nothing prevents the method proposed in this paper to be applied to other HMT definitions, including those that are dealing with structuring functions, since adaptive structuring functions have been explored before [24, 25].

2.2 Adaptive Structuring Elements

As mentioned in the introduction, adaptive mathematical morphology has attracted a lot of attention in recent years and is a topic of ongoing research. Adaptive structuring elements adapt to the image structures by taking into account different image attributes such as gray level values, geodesic distances between points in the image, the image gradient, the Hessian, etc. There exist many ways to include these information in the construction of adaptive structuring elements, and hence there exist many different methods to construct adaptive structuring elements [16].

Any of these existing methods could be used to define an adaptive HMT. However, for the sake of illustration, we will use adaptive structuring elements that have a fixed shape and varying size [26]. We have chosen these adaptive structuring elements since they can be computed in linear time with respect to the number of pixels in the image. The size of the structuring element is adjusted using the salience map SM.

The salience map SM is computed from the salience distance transform [27] of the edge image, where edges are weighted according to their importance, i.e., salience, containing information about the important structures in the image. To preserve most of the edges in the image, we use the gradient estimation and non-maximal suppression from the Canny edge detector [28]. To compute the non-maximal suppression, we use Gaussian derivatives to estimate the gradient in the input image, but exclude the hysteresis thresholding from the Canny edge detector. This approach preserves even the edges with a small response in the gradient image. Formally, NMS(f) is the image obtained by computing the gradient magnitude and non-maximal suppression of the input image f. The edge pixels are initialized with the negative values of their salience and the non-edge pixels are set to infinity [27]. The salience distance transform is computed with the classical two-pass chamfering algorithm [29]. After the salience distance transform is propagated from $-$ NMS(f), the distance image is offset to all positive values. Then, by inverting these values, we obtain the salience map SM(f), which can be formally written as

$$[\mathrm{SM}(f)](y) = \mathrm{Offset} + \bigvee_{x \in D} \Big(\mathrm{NMS}(f)(x) - d(x,y) \Big), \quad y \in D, \qquad (3)$$

where

$$\mathrm{Offset} = \bigwedge_{y \in D} \bigvee_{x \in D} \Big(\mathrm{NMS}(f)(x) - d(x,y) \Big), \qquad (4)$$

and $d(x,y)$ is a spatial distance.

This salience map SM contains the information about the spatial distance between points in the image and preserves the information about the salience of the edges in the image, where the largest values in the salience map SM correspond to the strongest edges and lower values to weaker ones, and its value decreases with the distance to the edges in the image [9].

In this paper, we consider adaptive structuring elements that preserve better strong edges than weak ones [26]. We find the value of the largest local maximum of SM, and the value of its largest local minimum, denoted here with M and m respectively. Then, the size (radius) of the adaptive structuring element is defined by

$$r(x) = \left| \mathrm{SM}(x) - \frac{m+M}{2} \right| \cdot c, \quad \text{for all } x \in D, \qquad (5)$$

where c is a constant that additionally adjusts the size of the structuring elements.

3   Method

The selection of appropriate structuring elements determines the performance of the HMT and its applications to template matching. Template matching using the HMT is often not trivial for the case of binary images, and it is even more complicated for gray level images [17].

A specific object in the image can be extracted using the HMT with two structuring elements: one defines the object and another defines what is not the

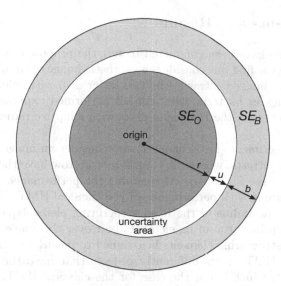

Fig. 2. Example of structuring elements SE_O, SE_B used for the extraction of objects

object (the object's surrounding, or often called background). Here, the following steps are necessary:

1. The structuring element representing the object is fully defined (SE_O), as well as the structuring element dedicated to background (SE_B).
2. HMT is defined using adaptive structuring elements that depend on the object and it properties. In order to be able to match the object even with discretization artefacts (e.g., stairing effect), it might be necessary to define an uncertainty area located between the two structuring elements [30] (see Fig. 2).

In this paper, we consider the adaptive structuring elements such that the size of adaptive structuring elements SE_O and SE_B is defined in the following way (see Fig. 2)

$$SE_O(x) = \{y \in D : |x - y| < r(x)\} \tag{6}$$
$$SE_B(x) = \{y \in D : r(x) + u \leq |x - y| < r(x) + u + b\} \tag{7}$$

where the adaptive radius $r(x)$, $x \in D$ is computed using the definition by Eq. (5), and u and b define the size of the uncertainty area and the size of the background area, respectively.

Note that, in this paper, we keep the uncertainty area u and the background b fixed for all points in the image. These two parameters that determine the position and size of the background structuring element SE_B can also have different values dependent on the image. We feel that selection of these parameters issue deserves further studies, and therefore is placed high on our list of future work.

4   Experiments and Results

Three experiments have been conducted to test the proposed method for adaptive HMT. While the first one considers a synthetic image of circular shapes with different sizes and image contrasts, the two latter deal with microscopy images of single molecules in bacteria cells. For all performed experiments, adaptive structuring elements are the Euclidean disks with adaptive radius computed by Eq. (5).

In the first experiment (Experiment 1), we consider an image of disks of different sizes and contrasts, where the shapes in each row have the same size and different contrast (see Fig. 3(a)). We examine the performance of the adaptive HMT and compare it with performance of the classical HMT. For the comparison, we defined the radius of the fixed structuring element pair as the mean radius of the adaptive structuring elements pair over the whole image, i.e., the size of the fixed structuring elements is assigned to mean$\{r(x) : x \in D\}$.

The adaptive HMT treats differently objects that have the same size but different contrast, which is not the case for the classical HMT, as depicted in Fig. 3. While the latter finds objects of the same size (Fig. 3(b) and (c)), the former detects object with respect to their contrast and not their size (Fig. 3(d) and (e)).

In the next two experiments (Experiment 2 and 3), we consider a realistic use of adaptive HMT dealing with microscopy images of the single molecules in living cells, in particular in bacteria E-coli (see Figs. 4 and 5). Single molecules appear visible as fluorescent dots and are brighter than the very noisy inhomogeneous background. For live cell imaging, for which fluorescence microscopy is often used, the signal-to-noise ratio (SNR) if often very low, making automated spot detection a very challenging task [31]. Also, since these are images of single molecules, fluorescent spots are only a few nanometers wide; however, due to diffraction limit, they appear as spots that cover a few pixels.

In Experiment 2, we examine how the size of adaptive structuring elements influences the performance of adaptive HMT. We use an adaptive structuring element defined by Eq. (5), and consider different values for the constant c that determines the size of the adaptive structuring element. For this experiment, we have no uncertainty region between two structuring elements SE_O and SE_B, i.e. $u = 0$. Also, we keep the size of the background structuring element SE_B fixed with $b = 3$. The results are shown in Fig. 4. Despite the fact that ground truth does not exist for these data, it is obvious that the two fluorescent dots can be found in the presented image (see Fig. 4(a)): one of them is close to the top edge of the image, the other is near the middle. These two fluorescent dots are only found when $c = 0.1$ (Fig. 4(b)), while for larger c the dimmer of the two fluorescent dots cannot be found (see Fig. 4(c) and (d)). In contrast, for the classical HMT there is no parameter setting that allows it to find both dots simultaneously.

In Experiment 3, we investigate how the selection of the background structuring element SE_B influences the adaptive HMT. We vary the size of the uncertainty region, which is the region between the foreground structuring element

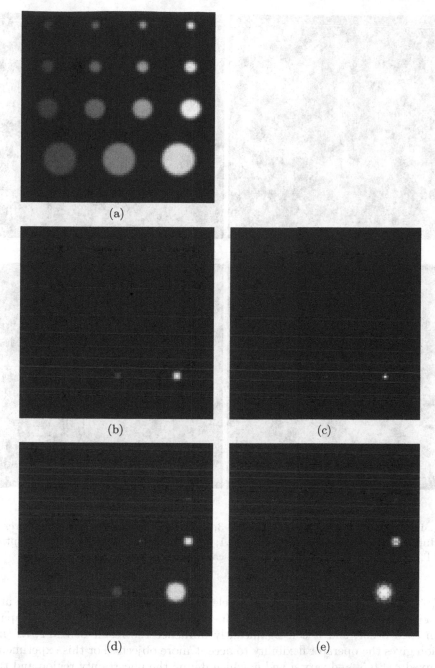

Fig. 3. Comparison between the classical HMT and adaptive HMT using different radii $r(x)$, and for $u = 2$ and $b = 1$. The radius of fixed structuring element is equal to $\text{mean}\{r(x) : x \in D\}$. (a) Input image; (b) Classical HMT that is compared to the adaptive HMT in (d); (c) Classical HMT that is compared to the adaptive HMT in (e); (d) Adaptive HMT with $c = 0.2$; (e) Adaptive HMT with $c = 0.4$

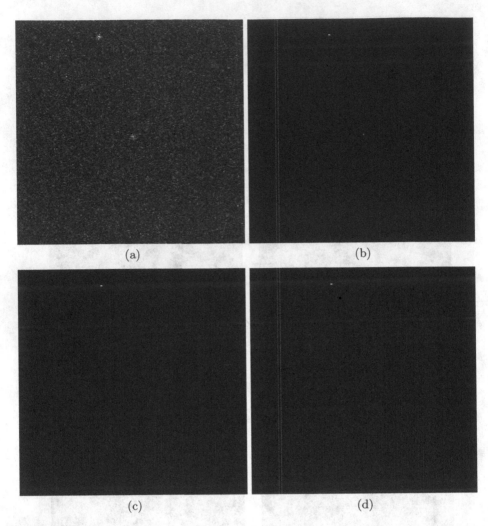

(a)

(b)

(c)

(d)

Fig. 4. Adaptive HMT for $u = 0$ and $b = 3$, with various values for size parameter c. (a) Input image; (b) Adaptive HMT, $c = 0.1$; (c) Adaptive HMT, $c = 0.5$; (d) Adaptive HMT, $c = 1$. Note that the classical HMT can find at most one dot at the time.

SE_O and the background structuring element SE_B (see Fig. 2). The size and shape of the uncertainty region determines how far SE_O and SE_B will be apart from each other, and it can significantly influence the result of HMT, as this region gives the operator flexibility to accept more objects. For this experiment, we fixed $c = 0.5$, and vary u and b, which define the uncertainty region and the background structuring element SE_B. Out of the four fluorescent dots in the image, three were detected in all three cases, whereas the fourth was detected only with a small uncertainty region.

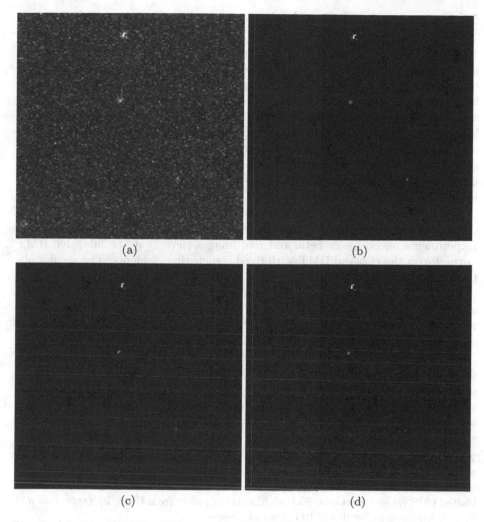

Fig. 5. Adaptive HMT for different sizes of the uncertainty region between structuring elements SE_O and SE_B, i.e., for different values of u. Here, $c = 0.1$ and $b = 2$; (a) Input image; (b) Adaptive HMT, $u = 1$; (c) Adaptive HMT, $u = 2$; (d) Adaptive HMT, $u = 3$. Note that the classical HMT can find at most one dot at the time.

We can conclude that besides the background and foreground size parameters c and b, the width of the uncertainty area u also impacts the properties of the pair of adaptive structure elements (foreground, background) and thus the result of the adaptive HMT. Similarly to any template matching solution, these three parameters are context-dependent and need to be set depending on the application considered.

5   Summary and Conclusions

We have proposed the Hit or Miss Transform (HMT) that is defined with an adaptive structuring element pair. We used adaptive structuring elements based on the salience information in the image [26]. Nevertheless, any other method for adaptive structuring elements can be used. Our approach uses an isotropic structuring element that scales with the distance to and strength of edges in the image. Closer to a strong edge, the structuring element is smaller. The result is an adaptive HMT that detects small, bright objects, and larger, dimmer objects at the same time. This operator, in contrast to the classical HMT, does not detect objects based on size and shape alone, but at the same time includes information of their contrast.

We have presented an application for the adaptive HMT operator in the context of template matching, illustrating the usefulness of such an operator. Experiments on both synthetic and real images show that the adaptive HMT outperforms the classical HMT with fixed structuring elements. The results obtained on microscopy images illustrated the potential of this new operator.

Our approach to adaptive HMT requires the definition of three constants: r scales the object structuring element, b scales the background structuring element, and u scales the uncertainty area, the space in between the two structuring elements. The best values for these three parameters are application dependent.

We have used Soille's definition for gray scale HMT [23]. There exist other definitions for gray scale HMT in the literature and we plan to explore these other definitions, as well as alternative methods for adapting the structuring elements. We are also thinking about how to apply the adaptive HMT to color and multivariate images.

Acknowledgements. The images used in Figs. 4 and 5 are courtesy of Prof. Johan Elf, Department of Cell and Molecular Biology, Uppsala University, Sweden.

This work has been initiated when Vladimir Ćurić and Cris L. Luengo Hendriks visited IRISA/Univ. Bretagne-Sud, with mobility grants from Univ. Bretagne-Sud and French Embassy in Sweden (FRÖ program) respectively.

References

1. Matheron, G.: Random sets and integral geometry. Willey, New York (1975)
2. Serra, J.: Image analysis and mathematical morphology. Academic Press, London (1982)
3. Serra, J.: Image analysis and mathematical morphology. Theoretical advances, vol. 2. Academic Press, New York (1988)
4. Lerallut, R., Decencière, E., Meyer, F.: Image filtering using morphological amoebas. Image and Vision Computing 25(4), 395–404 (2007)
5. Debayle, J., Pinoli, J.: General adaptive neighborhood image processing – part I: Introduction and theoretical aspects. Journal of Mathematical Imaging and Vision 25(2), 245–266 (2006)

6. Tankyevych, O., Talbot, H., Dokládal, P.: Curvilinear morpho-Hessian filter. In: Proc. of IEEE International Symposium on Biomedical Imaging: From Nano to Macro, pp. 1011–1014 (2008)

7. Angulo, J., Velasco-Forero, S.: Structurally adaptive mathematical morphology based on nonlinear scale-space decompositions. Image Analysis & Stereology 30(2), 111–122 (2011)

8. Verdú-Monedero, R., Angulo, J., Serra, J.: Anisotropic morphological filters with spatially-variant structuring elements based on image-dependent gradient fields. IEEE Transactions on Image Processing 20(1), 200–212 (2011)

9. Ćurić, V., Luengo Hendriks, C.L., Borgefors, G.: Salience adaptive structuring elements. IEEE Journal of Selected Topics in Signal Processing 6(7), 809–819 (2012)

10. Landström, A., Thurley, M.J.: Adaptive morphology using tensor-based elliptical structuring elements. Pattern Recognition Letters 34(12), 1416–1422 (2013)

11. Roerdink, J.B.T.M.: Adaptive and group invariance in mathematical morphology. In: Proc. of IEEE International Conference on Image Processing, pp. 2253–2256 (2009)

12. Bouaynaya, N., Charif-Chefchaouni, M., Schonfeld, D.: Theoretical foundations of spatially-variant mathematical morphology part I: Binary images. IEEE Transactions on Pattern Analysis and Machine Intelligence 30(5), 823–836 (2008)

13. Bouaynaya, N., Schonfeld, D.: Theoretical foundations of spatially-variant mathematical morphology part II: Gray-level images. IEEE Transactions on Pattern Analysis and Machine Intelligence 30(5), 837–850 (2008)

14. Velasco-Forero, S., Angulo, J.: On nonlocal mathematical morphology. In: Proceedings of the International Symposium on Mathematical Morphology, pp. 219–230. Springer, Heidelberg (2013)

15. Maragos, P., Vachier, C.: Overview of adaptive morphology: Trends and perspectives. In: Proceedings of the International Conference on Image Processing, pp. 2241–2244. IEEE (2009)

16. Ćurić, V., Landström, A., Thurley, M.J., Luengo Hendriks, C.L.: Adaptive mathematical morphology – A survey of the field. Pattern Recognition Letters 47, 18–28 (2014)

17. Perret, B., Lefèvre, S., Collet, C.: A robust hit-or-miss transform for template matching applied to very noisy astronomical images. Pattern Recognition 42(11), 2470–2480 (2009)

18. Weber, J., Tabbone, S.: Symbol spotting for technical documents: An efficient template-matching approach. In: Proceedings of the IEEE International Conference on Pattern Recognition, pp. 669–672 (2012)

19. Lefèvre, S., Weber, J.: Automatic building extraction in VIIR images using advanced morphological operators. In: Proceedings of the Urban Remote Sensing Joint Event, pp. 1–5. IEEE (2007)

20. Murray, P., Marshall, S.: A review of recent advances in the hit-or-miss transform. Advances in Imaging and Electron Physics 175, 221–282 (2012)

21. Lefèvre, S., Aptoula, E., Perret, B., Weber, J.: Morphological template matching in color images. In: Advances in Low-Level Color Image Processing, pp. 241–277. Springer Netherlands (2014)

22. Naegel, B., Passat, N., Ronse, C.: Grey-level hit-or-miss transforms – part I: Unified theory. Pattern Recognition 40(2), 635–647 (2007)

23. Soille, P.: Advances in the analysis of topographic features on discrete images. In: Braquelaire, A., Lachaud, J.-O., Vialard, A. (eds.) DGCI 2002. LNCS, vol. 2301, pp. 175–186. Springer, Heidelberg (2002)

24. Angulo, J.: Morphological bilateral filtering and spatially-variant adaptive structuring functions. In: Soille, P., Pesaresi, M., Ouzounis, G.K. (eds.) ISMM 2011. LNCS, vol. 6671, pp. 212–223. Springer, Heidelberg (2011)
25. Ćurić, V., Hendriks, C.L.L.: Salience-based parabolic structuring functions. In: Hendriks, C.L.L., Borgefors, G., Strand, R. (eds.) ISMM 2013. LNCS, vol. 7883, pp. 183–194. Springer, Heidelberg (2013)
26. Ćurić, V., Luengo Hendriks, C.L.: Adaptive structuring elements based on salience information. In: Bolc, L., Tadeusiewicz, R., Chmielewski, L.J., Wojciechowski, K. (eds.) ICCVG 2012. LNCS, vol. 7594, pp. 321–328. Springer, Heidelberg (2012)
27. Rosin, P., West, G.: Salience distance transforms. Graphical Models and Image Processing 57(6), 483–521 (1995)
28. Canny, J.: A computational approach to edge detection. IEEE Transactions on Pattern Analysis and Machine Intelligence (6), 679–698 (1986)
29. Borgefors, G.: Distance transformations in digital images. Computer Vision, Graphics, and Image Processing 34(3), 344–371 (1986)
30. Weber, J., Lefèvre, S.: Spatial and spectral morphological template matching. Image and Vision Computing 30(12), 934–945 (2012)
31. Smal, I., Loog, M., Niessen, W., Meijering, E.: Quantitative comparison of spot detection methods in fluorescence microscopy. IEEE Transactions on Medical Imaging 29(2), 282–301 (2010)

Author Index

Printed in the United States
By Bookmasters